Lecture Notes in Computer Science 12679

More information about this subseries at http://www.springer.com/series/7412

Abderrahim Elmoataz · Jalal Fadili ·
Yvain Quéau · Julien Rabin ·
Loïc Simon (Eds.)

Scale Space and Variational Methods in Computer Vision

8th International Conference, SSVM 2021
Virtual Event, May 16–20, 2021
Proceedings

Springer

Editors
Abderrahim Elmoataz
UNICAEN, GREYC –
Normandy University
Caen, France

Jalal Fadili
ENSICAEN, GREYC –
Normandy University
Caen, France

Yvain Quéau
CNRS, GREYC –
Normandy University
Caen, France

Julien Rabin
UNICAEN, GREYC –
Normandy University
Caen, France

Loïc Simon
ENSICAEN, GREYC –
Normandy University
Caen, France

ISSN 0302-9743   ISSN 1611-3349   (electronic)
Lecture Notes in Computer Science
ISBN 978-3-030-75548-5   ISBN 978-3-030-75549-2   (eBook)
https://doi.org/10.1007/978-3-030-75549-2

LNCS Sublibrary: SL6 – Image Processing, Computer Vision, Pattern Recognition, and Graphics

This Springer imprint is published by the registered company Springer Nature Switzerland AG
The registered company address is: Gewerbestrasse 11, 6330 Cham, Switzerland

Preface

The International Conference on Scale Space and Variational Methods in Computer Vision (SSVM) series was born in 2007 in Ischia, from the fusion of the biannual Conference on Scale Space and the workshop on Variational, Geometric and Level Set Methods (VLSM). Since then, it has become a major event in the mathematical imaging community. Following the tradition of biannual events at beautiful, remote places, the 8th edition of this conference (SSVM 2021, https://ssvm2021.sciencesconf.org/) was initially planned to take place in the romantic city of Cabourg, on the Normandy coast in France. However, the conference was eventually held online due to the COVID-19 pandemic situation.

The conference provides a platform for state-of-the-art scientific exchange on topics related to computer vision and image analysis, including diverse themes such as 3D-vision, convex and non-convex variational modeling, image analysis, inverse problems in imaging, optimization methods in imaging, machine learning in imaging, PDEs in image processing, registration, restoration and reconstruction, scale-space methods, and segmentation. The 45 contributions (15 selected for oral presentation and 30 for poster presentation) in this proceedings demonstrate a strong and lively community. The contributions underwent a rigorous peer-review process similar to that of a high-ranking journal in the field.

Following the tradition of previous SSVM conferences, four outstanding researchers were invited to give a keynote presentation: Daniel Cremers (TU Munich, Germany), Julie Delon (Université Paris Descartes, France), Carola-Bibiane Schönlieb (University of Cambridge, UK) and Jean-Luc Starck (CEA, France).

We would like to thank all those who contributed to the success of this conference. We are grateful to the authors for their contributions. We are indebted to the reviewers for their commitment, hard work and enthusiasm that made this event possible at the highest scientific standards. We thank the CNRS and NORMASTIC for financial support. Lastly, we thank the steering committee and the organizers of all the previous SSVM conferences.

April 2021

Abderrahim Elmoataz
Jalal Fadili
Yvain Quéau
Julien Rabin
Loïc Simon

Organization

Conference Chairs

Abderrahim Elmoataz	UNICAEN, GREYC – Normandy University, France
Jalal Fadili	ENSICAEN, GREYC – Normandy University, France
Yvain Quéau	CNRS, GREYC – Normandy University, France
Julien Rabin	UNICAEN, GREYC – Normandy University, France
Loïc Simon	ENSICAEN, GREYC – Normandy University, France

Program Committee

Luis Alvarez	Universidad de Las Palmas de Gran Canaria, Spain
Jean-François Aujol	Institut Universitaire de France, France
Coloma Ballester	Universitat Pompeu Fabra, Spain
Marcelo Bertalmìo	Universitat Pompeu Fabra, Spain
Kristian Bredies	University of Graz, Austria
Michael Breuß	BTU Cottbus, Germany
Andres Bruhn	University of Stuttgart, Germany
Aurélie Bugeau	LaBRI, France
Martin Burger	FAU Erlangen-Nürnberg, Germany
Stéphane Canu	INSA Rouen, France
Antonin Chambolle	CMAP, École Polytechnique, France
Raymond Honfu Chan	Chinese University of Hong Kong, China
Ke Chen	University of Liverpool, UK
Laurent Cohen	CNRS, Université Paris Dauphine, France
Anders Bjorholm Dahl	DTU Compute, Denmark
Vedrana Dahl	DTU Compute, Denmark
Sune Darkner	University of Copenhagen, Denmark
Agnès Desolneux	CMLA, ENS Paris-Saclay, France
Nicolas Dobigeon	IRIT, University of Toulouse, France
Yiqiu Dong	DTU Compute, Denmark
Remco Duits	Eindhoven University of Technology, Netherlands
Vincent Duval	Inria, France
Paul Escande	CNRS, Aix-Marseille University, France
Virginia Estellers	UCLA, USA
Maurizio Falcone	Sapienza Università di Roma, Italy
Michael Felsberg	Linköpings Universitet, Sweden
Luc Florack	Eindhoven University of Technology, Netherlands
Guy Gilboa	Technion, Israel
Yann Gousseau	Telecom ParisTech, France
Markus Grasmair	NTNU, Norway
Lewis Griffin	University College London, UK

Weihong Guo	Case Western Reserve University, USA
Atsushi Imiya	Chiba University, Japan
Sung Ha Kang	Georgia Tech, USA
Charles Kervrann	Inria, France
Arjan Kuijper	TU Darmstadt, Germany
Stefan Kunis	University of Osnabrueck, Germany
François Lauze	University of Copenhagen, Denmark
Carole Le Guyader	INSA Rouen, France
Jan Lellman	University of Luebeck, Germany
Tony Lindeberg	KTH Royal Institute of Technology, Sweden
Marco Loog	TU Delft, Netherlands
Dirk Lorenz	TU Braunschweig, Germany
Russel Luke	Universität Göttingen, Germany
Julien Mairal	Inria, France
Simon Masnou	University of Lyon, France
Jan Modersitzki	University of Luebeck, Germany
Michael Möller	Universität Siegen, Germany
Michael Ng	Hong Kong Baptist University, China
Peter Ochs	Saarland University, Germany
Nicolas Papadakis	CNRS, University of Bordeaux, France
Gabriel Peyré	CNRS, ENS Paris, France
Gerlind Plonka-Hoch	Universität Göttingen, Germany
Thomas Pock	TU Graz, Austria
Otmar Scherzer	RICAM, Austria
Christoph Schnörr	Universität Heidelberg, Germany
Fiorella Sgallari	University of Bologna, Italy
Stefan Horst Sommer	University of Copenhagen, Denmark
Emmanuel Soubies	CNRS, University of Toulouse, France
Jon Sporring	University of Copenhagen, Denmark
Gabriele Steidl	Universität Kaiserlautern, Germany
Xue-Cheng Tai	University of Bergen, Norway
Samuel Vaiter	CNRS, Université de Bourgogne, France
Joachim Weickert	Saarland University, Germany
Pierre Weiss	CNRS, University of Toulouse, France
Martin Welk	UMIT TIROL, Austria
Xiaoqun Zhang	Shanghai Jiao Tong University, China

Other Referees

Danielle Bednarski	University of Luebeck, Germany
Ghada Laribi	UMIT TIROL, Austria
Rodrigue Siry	SAFRAN, France
Ryan Webster	Normandy University, France

Financial Management Staff

Sophie Rastello   CNRS, GREYC – Normandy University, France
Virginie Desnos-Carreau   UNICAEN, GREYC – Normandy University, France

Invited Speakers

Daniel Cremers   Professor and chair of Computer Vision and Artificial
                              Intelligence, Technical University of Munich,
                              Germany
Julie Delon   Professor at MAP5, Université Paris Descartes, France
Carola-Bibiane Schönlieb   Professor and head of the Cambridge Image Analysis
                              group, University of Cambridge, UK
Jean-Luc Starck   Director of CosmoStat, CEA, France

Sponsoring Institutions

Centre National de la Recherche Scientifique (CNRS)
CNRS national network "Mathematics in Imaging sciences and Applications"
(GDR MIA)
Fédération Normande de Recherche en Sciences et Technologies de l'Information et de
la Communication (NORMASTIC)

Contents

Scale Space and Partial Differential Equations Methods

Scale-Covariant and Scale-Invariant Gaussian Derivative Networks   3
  Tony Lindeberg

Quantisation Scale-Spaces 15
  Pascal Peter

Equivariant Deep Learning via Morphological and Linear Scale Space
PDEs on the Space of Positions and Orientations 27
  Remco Duits, Bart Smets, Erik Bekkers, and Jim Portegies

Nonlinear Spectral Processing of Shapes via Zero-Homogeneous Flows 40
  Jonathan Brokman and Guy Gilboa

Total-Variation Mode Decomposition 52
  Ido Cohen, Tom Berkov, and Guy Gilboa

Fast Morphological Dilation and Erosion for Grey Scale Images Using
the Fourier Transform 65
  Marvin Kahra, Vivek Sridhar, and Michael Breuß

Diffusion, Pre-smoothing and Gradient Descent 78
  Martin Welk

Local Culprits of Shape Complexity 91
  Mazlum Ferhat Arslan and Sibel Tari

Extension of Mathematical Morphology in Riemannian Spaces 100
  El Hadji S. Diop, Alioune Mbengue, Bakary Manga, and Diaraf Seck

Flow, Motion and Registration

Multiscale Registration 115
  Noémie Debroux, Carole Le Guyader, and Luminita A. Vese

Challenges for Optical Flow Estimates in Elastography 128
  Ekaterina Sherina, Lisa Krainz, Simon Hubmer, Wolfgang Drexler,
  and Otmar Scherzer

An Anisotropic Selection Scheme for Variational Optical Flow Methods
with Order-Adaptive Regularisation 140
  Lukas Mehl, Cedric Beschle, Andrea Barth, and Andrés Bruhn

Low-Rank Registration of Images Captured Under Unknown,
Varying Lighting 153
  Matthieu Pizenberg, Yvain Quéau, and Abderrahim Elmoataz

Towards Efficient Time Stepping for Numerical Shape Correspondence. 165
  Alexander Köhler and Michael Breuß

First Order Locally Orderless Registration 177
  Sune Darkner, José D. T. Vidarte, and François Lauze

Optimization Theory and Methods in Imaging

First-Order Geometric Multilevel Optimization for Discrete Tomography 191
  Jan Plier, Fabrizio Savarino, Michal Kočvara, and Stefania Petra

Bregman Proximal Gradient Algorithms for Deep Matrix Factorization 204
  Mahesh Chandra Mukkamala, Felix Westerkamp, Emanuel Laude,
  Daniel Cremers, and Peter Ochs

Hessian Initialization Strategies for ℓ-BFGS Solving Non-linear
Inverse Problems. 216
  Hari Om Aggrawal and Jan Modersitzki

Inverse Scale Space Iterations for Non-convex Variational Problems Using
Functional Lifting 229
  Danielle Bednarski and Jan Lellmann

A Scaled and Adaptive FISTA Algorithm for Signal-Dependent Sparse
Image Super-Resolution Problems. 242
  Marta Lazzaretti, Simone Rebegoldi, Luca Calatroni,
  and Claudio Estatico

Convergence Properties of a Randomized Primal-Dual Algorithm
with Applications to Parallel MRI. 254
  Eric B. Gutiérrez, Claire Delplancke, and Matthias J. Ehrhardt

Machine Learning in Imaging

Wasserstein Generative Models for Patch-Based Texture Synthesis 269
  Antoine Houdard, Arthur Leclaire, Nicolas Papadakis, and Julien Rabin

Sketched Learning for Image Denoising. 281
  Hui Shi, Yann Traonmilin, and Jean-François Aujol

Translating Numerical Concepts for PDEs into Neural Architectures 294
  Tobias Alt, Pascal Peter, Joachim Weickert, and Karl Schrader

CLIP: Cheap Lipschitz Training of Neural Networks.   307
  Leon Bungert, René Raab, Tim Roith, Leo Schwinn,
  and Daniel Tenbrinck

Variational Models for Signal Processing with Graph Neural Networks   320
  Amitoz Azad, Julien Rabin, and Abderrahim Elmoataz

Synthetic Images as a Regularity Prior for Image Restoration
Neural Networks.   333
  Raphaël Achddou, Yann Gousseau, and Saïd Ladjal

Geometric Deformation on Objects: Unsupervised Image Manipulation
via Conjugation   346
  Changqing Fu and Laurent D. Cohen

Learning Local Regularization for Variational Image Restoration   358
  Jean Prost, Antoine Houdard, Andrés Almansa, and Nicolas Papadakis

Segmentation and Labelling

On the Correspondence Between Replicator Dynamics
and Assignment Flows.   373
  Bastian Boll, Jonathan Schwarz, and Christoph Schnörr

Learning Linear Assignment Flows for Image Labeling
via Exponential Integration.   385
  Alexander Zeilmann, Stefania Petra, and Christoph Schnörr

On the Geometric Mechanics of Assignment Flows for Metric
Data Labeling.   398
  Fabrizio Savarino, Peter Albers, and Christoph Schnörr

A Deep Image Prior Learning Algorithm for Joint Selective Segmentation
and Registration   411
  Liam Burrows, Ke Chen, and Francesco Torella

Restoration, Reconstruction and Interpolation

Inpainting-Based Video Compression in FullHD.   425
  Sarah Andris, Pascal Peter, Rahul Mohideen Kaja Mohideen,
  Joachim Weickert, and Sebastian Hoffmann

Sparsity-Aided Variational Mesh Restoration   437
  Martin Huska, Serena Morigi, and Giuseppe Antonio Recupero

Lossless PDE-based Compression of 3D Medical Images.   450
  Ikram Jumakulyyev and Thomas Schultz

Splines for Image Metamorphosis   463
   Jorge Justiniano, Marko Rajković, and Martin Rumpf

Residual Whiteness Principle for Automatic Parameter Selection in ℓ_2-ℓ_2
Image Super-Resolution Problems   476
   *Monica Pragliola, Luca Calatroni, Alessandro Lanza,
   and Fiorella Sgallari*

Inverse Problems in Imaging

Total Deep Variation for Noisy Exit Wave Reconstruction in Transmission
Electron Microscopy   491
   *Thomas Pinetz, Erich Kobler, Christian Doberstein, Benjamin Berkels,
   and Alexander Effland*

GMM Based Simultaneous Reconstruction and Segmentation
in X-Ray CT Application   503
   Shi Yan and Yiqiu Dong

Phase Retrieval via Polarization in Dynamical Sampling   516
   Robert Beinert and Marzieh Hasannasab

Invertible Neural Networks Versus MCMC for Posterior Reconstruction
in Grazing Incidence X-Ray Fluorescence   528
   *Anna Andrle, Nando Farchmin, Paul Hagemann,
   Sebastian Heidenreich, Victor Soltwisch, and Gabriele Steidl*

Adversarially Learned Iterative Reconstruction for Imaging
Inverse Problems   540
   Subhadip Mukherjee, Ozan Öktem, and Carola-Bibiane Schönlieb

Towards Off-the-grid Algorithms for Total Variation Regularized
Inverse Problems   553
   Yohann De Castro, Vincent Duval, and Romain Petit

Multi-frame Super-Resolution from Noisy Data   565
   Kireeti Bodduna, Joachim Weickert, and Marcelo Cárdenas

Author Index   579

Scale Space and Partial Differential Equations Methods

Scale-Covariant and Scale-Invariant Gaussian Derivative Networks

Tony Lindeberg$^{(\boxtimes)}$ (iD)

Computational Brain Science Lab, Division of Computational Science
and Technology, KTH Royal Institute of Technology, Stockholm, Sweden
`tony@kth.se`

Abstract. This paper presents a hybrid approach between scale-space theory and deep learning, where a deep learning architecture is constructed by coupling parameterized scale-space operations in cascade. By sharing the learnt parameters between multiple scale channels, and by using the transformation properties of the scale-space primitives under scaling transformations, the resulting network becomes provably scale covariant. By in addition performing max pooling over the multiple scale channels, a resulting network architecture for image classification also becomes provably scale invariant. We investigate the performance of such networks on the MNISTLargeScale dataset, which contains rescaled images from original MNIST over a factor of 4 concerning training data and over a factor of 16 concerning testing data. It is demonstrated that the resulting approach allows for scale generalization, enabling good performance for classifying patterns at scales not present in the training data.

1 Introduction

A problem with traditional deep networks is that they are not covariant with respect to scaling transformations in the image domain. In deep networks, non-linearities are performed relative to the current grid spacing, which implies that the deep network does not commute with scaling transformations.

One way of achieving scale covariance in a brute force manner is by applying the same deep net to multiple rescaled copies of the input image. Such an approach has been developed and investigated in [1]. When working with such a scale-channel network it may, however, be harder to combine information between different scale channels, unless the multi-resolution representations at different levels of resolution are also resampled to a common reference frame when information from different scale levels is to be combined.

Another approach to achieve scale covariance is by applying multiple rescaled non-linear filters to the same image. For such an architecture, it will specifically be easier to combine information from multiple scale levels, since the image data at all scales have the same resolution.

The support from the Swedish Research Council (contract 2018-03586) is gratefully acknowledged.

A. Elmoataz et al. (Eds.): SSVM 2021, LNCS 12679, pp. 3–14, 2021.
https://doi.org/10.1007/978-3-030-75549-2_1

For the primitive discrete filters in a regular deep network, it is, however, not obvious how to rescale the primitive components, in terms of *e.g.*, local 3×3 or 5×5 filters or max pooling over 2×2 neighbourhoods in a sufficiently accurate manner over continuous variations of spatial scaling factors. For this reason, it would be preferable to have a continuous model of the image filters, which are then combined together into suitable deep architectures, since the support regions of the continuous filters could then be rescaled in a continuous manner. Specifically, if these filters are chosen as scale-space filters, which are designed to handle scaling transformations in the image domain, there is a potential of constructing a rich family of hierarchical networks based on scale-space operations that are provably scale covariant, as shown in [2].

The subject of this article is to develop and experimentally investigate one such hybrid approach between scale-space theory and deep learning. The idea that we shall follow is to define the layers in a deep architecture from scale-space operations, and then use the closed-form transformation properties of the scale-space primitives under scaling transformations to achieve provable scale covariance and scale invariance of the resulting continuous deep network. Specifically, we will demonstrate that this will give the deep network the ability to generalize to previously unseen scales not present in the training data.

Technically, we will experimentally explore this idea for one specific type of architecture, where the layers are parameterized linear combinations of Gaussian derivatives up to order two. The overall principle for obtaining scale covariance and scale invariance is, however, much more general and applies to much wider classes of possible ways of defining layers from scale-space operations.

2   Relations to Previous Work

In classical computer vision, it has been demonstrated that scale-space theory constitutes a powerful paradigm for constructing scale-covariant and scale-invariant feature detectors and making visual operations robust to scaling transformations [3–8]. In the area of deep learning, a corresponding framework for handling general scaling transformations has so far not been as well established.

Concerning the relationship between deep networks and scale, several researchers have observed robustness problems of deep networks under scaling variations [9,10]. There have been some approaches developed to aim at scale invariant CNNs [11–15]. These approaches have, however, not been experimentally evaluated on the task of generalizing to scales not present in the training data [12,13,15], or only over a very narrow scale range [11,14].

Spatial transformer networks have been proposed as a general approach for handling image transformations in CNNs [16]. Plain transformation of feature maps by a spatial transformer will, however, in general, not deliver a correct transformation, and will therefore not support truly invariant recognition [17].

Concerning the notion of scale covariance and its relation to scale generalization, a general sufficiency result was presented in [2] that guarantees provable scale covariance for hierarchical networks that are constructed from continuous layers defined from partial derivatives or differential invariants expressed in terms of scale-normalized derivatives. This idea was developed in more detail for a hand-crafted quasi quadrature network, with the layers representing idealized models of complex cells, and experimentally applied to the task of texture classification. It was demonstrated that the resulting approach allowed for scale generalization on the KTH-TIPS2 dataset, enabling classification of texture samples at scales not present in the training data.

The idea of modelling layers in neural networks as continuous functions instead of discrete filters has also been advocated in [18] and [19].

Concerning scale generalization for CNNs, [1] presented a multi-scale-channel approach, where the same discrete CNN was applied to multiple scaled copies of each input image. It was demonstrated that the resulting scale-channel architectures had much better ability to handle scaling transformations compared to a regular vanilla CNN, and lead to good scale generalization, for classifying image patterns at scales not present in the training data.

The subject of this article is to complement these latter works, and specifically combine a specific instance of the general class of scale-covariant networks in [2] with deep learning, where we will choose the continuous layers as linear combinations of Gaussian derivatives and demonstrate how such an architecture allows for scale generalization.

3 Gaussian Derivative Networks

In a traditional deep network, the filter weights are usually free variables with few additional constraints. In scale-space theory, on the other hand, theoretical results have been presented showing that Gaussian kernels and their corresponding Gaussian derivatives constitute a canonical class of image operations [20–26]. In classical computer vision based on hand-crafted image features, it has been demonstrated that a large number of visual tasks can be successfully addressed by computing image features and image descriptors based on Gaussian derivatives, or approximations thereof, as the first layer of image features [3–8]. One could therefore raise the question if such Gaussian derivatives could also be used as computational primitives for constructing deep networks.

Motivated by the fact that a large number of visual tasks have been successfully addressed by first- and second-order Gaussian derivatives, which are the primitive filters in the Gaussian 2-jet, let us explore the consequences of using linear combinations of first- and second-order Gaussian derivatives as the class of possible filter weight primitives in a deep network. Thus, given an image f, which could either be the input image to the deep net, or some higher layer F_k in the deep network, we first compute its scale-space representation by smoothing with the Gaussian kernel

$$L(x, y;\ \sigma) = (g(\cdot, \cdot;\ \sigma) * f(\cdot, \cdot))(x, y), \tag{1}$$

where

$$g(x, y;\ \sigma) = \frac{1}{2\pi\sigma^2} e^{-(x^2+y^2)/2\sigma^2}. \tag{2}$$

Then, for simplicity with the notation for the scale parameter σ suppressed, we consider arbitrary linear combinations of first- and second-order Gaussian derivatives as the class of possible linear filtering operations on f (or correspondingly for F_k):

$$J_{2,\sigma}(f) = C_0 + C_x L_\xi + C_y L_\eta + \frac{1}{2}(C_{xx} L_{\xi\xi} + 2C_{xy} L_{\xi\eta} + C_{yy} L_{\eta\eta}), \quad (3)$$

where L_ξ, L_η, $L_{\xi\xi}$, $L_{\xi\eta}$ and $L_{\eta\eta}$ are scale-normalized derivatives [3] according to $L_\xi = \sigma L_x$, $L_\eta = \sigma L_y$, $L_{\xi\xi} = \sigma^2 L_{xx}$, $L_{\xi\eta} = \sigma^2 L_{xy}$ and $L_{\eta\eta} = \sigma^2 L_{yy}$, and C_0 is an offset term for later use in non-linearities between adjacent layers.

Since directional derivatives can be computed as linear combinations of partial derivatives, for first- and second-order derivatives we have

$$L_\varphi = \cos\varphi\, L_x + \sin\varphi\, L_y, \quad (4)$$

$$L_{\varphi\varphi} = \cos^2\varphi\, L_{xx} + 2\cos\varphi\,\sin\varphi\, L_{xy} + \sin^2\varphi\, L_{yy}, \quad (5)$$

it follows that the parameterized second-order kernels of the form (3) span all possible linear combinations of first- and second-order directional derivatives.

The corresponding affine extension of such receptive fields, by replacing the rotationally symmetric Gaussian kernel (2) for scale-space smoothing by a corresponding affine Gaussian kernel, does also constitute a good idealized model for the receptive fields of simple cells in the primary visual cortex [27].

In contrast to previous work in computer vision or functional modelling of biological visual receptive fields, where Gaussian derivatives are used as a first layer of linear receptive fields, we will, however, here investigate the consequences of coupling such receptive fields in cascade to form deep image representations.

When coupling several of such smoothing stages in cascade, for example in combination with non-linear ReLU stages in between, it is natural to let the scale parameter for layer k be proportional to an initial scale level σ_0, such that the scale parameter σ_k in layer k is $\sigma_k = \beta_k \sigma_0$ for some set of $\beta_k \geq \beta_{k-1} \geq 1$ and some minimum scale level $\sigma_0 > 0$. Specifically, it is natural to choose the relative scale parameter factors according to a geometric distribution

$$\sigma_k = r^{k-1}\sigma_0 \quad (6)$$

for some $r \geq 1$.

A similar idea of using Gaussian derivatives as structured receptive fields in convolutional networks has also been explored in [28], although not in the relation to scale covariance or using a self-similar sampling of the scale levels.

3.1 Provable Scale Covariance

To prove that a deep network constructed by coupling linear filtering operations of the form (3) with pointwise non-linearities in between is scale covariant, let us consider two images f and f' that are related by a scaling transformation $f'(x', y') = f(x, y)$ for $x' = S\,x$, $y' = S\,y$ and some spatial scaling factor $S > 0$.

If the scale parameters σ and σ' in the two domains are related according to $\sigma' = S\sigma$, then it follows from a general result in [3, Eq. (20)] that the scale-normalized derivatives will be equal (when using scale normalization power $\gamma = 1$ in the scale-normalized derivative concept):

$$L'_{\xi^\alpha \eta^\beta}(x', y'; \sigma') = L_{\xi^\alpha \eta^\beta}(x, y; \sigma). \tag{7}$$

Thus, given that the image positions and the scale levels are appropriately matched according to $x' = Sx$, $y' = Sy$ and $\sigma' = S\sigma$, it holds that the first layers F_1 and F'_1 will be equal up to a scaling transformation:

$$F'_1(x', y'; \sigma'_1) = J_{2,\sigma'_1}(f')(x', y'; \sigma'_1) = J_{2,\sigma_1}(f)(x, y; \sigma_1) = F_1(x, y; \sigma_1). \tag{8}$$

By continuing a corresponding construction of higher layers, by applying similar operations in cascade, $F_{k+1}(x, y; \sigma_{k+1}) = J_{2,\sigma_{k+1}}(\text{ReLU}(F_k(\cdot, \cdot; \sigma_k)))$ $(x, y; \sigma_{k+1})$ for $k \geq 1$, with the initial scale levels σ_0 and σ'_0 in (6) related according to $\sigma'_0 = S\sigma_0$, it follows that also the higher layers will be equal up to a scaling transformation:

$$F'_k(x', y'; \sigma'_k) = F_k(x, y; \sigma_k). \tag{9}$$

A pointwise non-linearity, such as a ReLU stage, trivially commutes with scaling transformations and does therefore not affect the scale covariance properties.

4   Experiments with a Single-Scale-Channel Network

To investigate the ability of such a deep hierarchical Gaussian derivative network to capture image structures with different shape, we first did initial experiments with the regular MNIST dataset [29]. We constructed a 6-layer network in PyTorch with 12-14-16-20-64 channels in the intermediate layers and 10 output channels, intended to learn each type of digit. We chose the initial scale level $\sigma_0 = 0.9$ pixels and the relative scale ratio $r = 1.25$ in (6), implying that the maximum value of σ is $0.9 \times 1.25^5 \approx 2.7$ pixels relative to the image size of 28×28 pixels. The individual receptive fields do then have larger spatial extent because of the spatial extent of the Gaussian kernels used for image smoothing and the larger positive and negative side lobes of the first- and second-order derivatives.

We used regular ReLU stages between the filtering steps, but no spatial max pooling or spatial stride, and no fully connected layer, since such operations would destroy the scale covariance. Instead, the receptive fields are solely determined from linear combinations of Gaussian derivatives, with successively larger receptive fields of size $\sigma_0 r^{k-1}$, which enable a gradual integration from local to regional image structures (Fig. 1). In the final layer, only the value at the central pixel is extracted, or here for an even image size, the average over the central 2×2 neighbourhood, which, however, destroys full scale covariance. To ensure full scale covariance, the input images should instead have odd image size.

The network was trained on 50 000 of the training images in the dataset, with the offset term C_0 and the Gaussian derivative weights C_x, C_y, C_{xx}, C_{xy} and

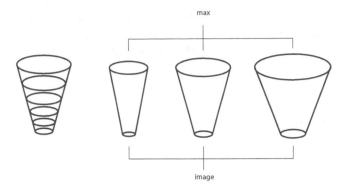

Fig. 1. (left) Schematic illustration of the architecture of the single-scale-channel network, with 6 layers of receptive fields at successively coarser levels of scale. (right) Schematic illustration of the architecture of a multi-scale-channel network, with multiple parallel scale channels over a self-similar distribution of the initial scale level.

C_{yy} in (3) initiated to random values and trained individually for each layer and feature channel by stochastic gradient descent over 40 epochs using the Adam optimizer [30] set to minimize the binary cross-entropy loss. A cosine learning curve was used with maximum learning rate of 0.01 and minimum learning rate 0.00005 and using batch normalization over batches with 50 images. The experiment lead to 99.93% training accuracy and 99.43% test accuracy on the test dataset containing 10 000 images.

Notably, the training accuracy does not reach 100.00%, probably because of the restricted shapes of the filter weights, as determined by the *a priori* shapes of the receptive fields in terms of linear combinations of first- and second-order Gaussian derivatives. Nevertheless, the test accuracy is quite good given the moderate number of parameters in the network ($6 \times (12 + 12 \times 14 + 14 \times 16 + 16 \times 20 + 20 \times 64 + 64 \times 10) = 15\ 864$).

4.1   Discrete Implementation

In the numerical implementation of scale-space smoothing, we used separable smoothing with the discrete analogue of the Gaussian kernel $T(n;\ s) = e^{-s}I_n(s)$ for $s = \sigma^2$ in terms of modified Bessel functions I_n of integer order [31]. The discrete derivative approximations were computed by central differences, $\delta_x = (-1/2, 0, 1/2)$, $\delta_{xx} = (1, -2, 1)$, $\delta_{xy} = \delta_x\delta_y$, etc., implying that the spatial smoothing operation can be shared between derivatives of different order, and implying that scale-space properties are preserved in the discrete implementation of the Gaussian derivatives [32]. This is a general methodology for computing Gaussian derivatives for a large number of visual tasks.

By computing the Gaussian derivative responses in this way, the scale-space smoothing is only performed once for each scale level, and there is no need for repeating the scale-space smoothing for each order of the Gaussian derivatives.

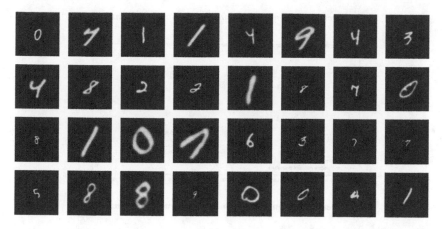

Fig. 2. Sample images from the MNISTLargeScale dataset [1,33]. This figure shows digits for sizes in the range [1,4], for which there are training data. In addition, the MNISTLargeScale dataset contains testing data over the wider size range [1/2,8].

5 Experiments with a Multi-Scale-Channel Network

To investigate the ability of a multi-scale-channel network to handle spatial scaling transformations, we made experiments on the MNISTLargeScale dataset [1,33]. This dataset contains rescaled digits from the original MNIST dataset [29] embedded in images of size 112×112, see Fig. 2 for an illustration. For training, we used either of the datasets containing 50 000 rescaled digits with relative scale factors 1, 2 or 4, respectively, henceforth referred to as sizes 1, 2 and 4. For testing, the dataset contains 10 000 rescaled digits with relative scale factors between 1/2 and 8, respectively, with a relative scale ratio of $\sqrt[4]{2}$ between adjacent sizes.

To investigate the properties of a multi-scale-channel architecture experimentally, we created a multi-scale-channel network with 8 scale channels with their initial scale values σ_0 between $1/\sqrt{2}$ and 8 and a scale ratio of $\sqrt{2}$ between adjacent scale channels. For each channel, we used a Gaussian derivative network of similar architecture as the single-scale-channel network, with 12-14-16-20-64 channels in the intermediate layers and 10 output channels, and with a relative scale ratio $r = 1.25$ (6) between adjacent layers, implying that the maximum value of σ in each channel is $\sigma_0 \times 1.25^5 \approx 3.1\ \sigma_0$ pixels.

Importantly, the parameters C_0, C_x, C_y, C_{xx}, C_{xy} and C_{yy} in (3) are shared between the scale channels, implying that the scale channels taken together are truly scale covariant, because of the parameterization of the receptive fields in terms of scale-normalized Gaussian derivatives. The batch normalization stage is also shared between the scale channels. The output from max pooling over the scale channels is furthermore truly scale invariant, if we assume an infinite number of scale channels so that scale boundary effects can be disregarded.

Scale generalization performance when training on size 1

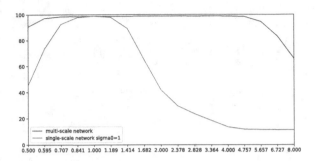

Scale generalization performance when training on size 2

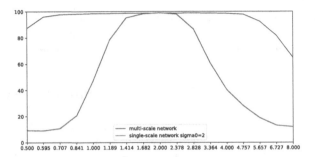

Scale generalization performance when training on size 4

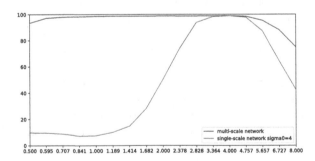

Fig. 3. Experiments showing the ability of a multi-scale-channel network to generalize to new scale levels not present in the training data. In the top row, all training data are for size 1, whereas we evaluate on multiple testing sets for each one of the sizes between $1/2$ and 8. The red curve shows the generalization performance for a single-scale-channel network for $\sigma_0 = 1$, whereas the blue curve shows the result for a multi-scale-channel network covering the range of σ_0-values between $1/\sqrt{2}$ and 8. As can be seen from the results, the generalization ability is much better for the multi-scale-channel network compared to the single-scale-channel network. In the middle row, a similar type of experiment is repeated for training size 2 and with $\sigma_0 = 2$ for the single-scale-channel network. In the bottom row, a similar experiment is performed for training size 4 and with $\sigma_0 = 4$ for the single-scale-channel network. (Horizontal axis: Scale of testing data.) (Colour figure online)

Scale generalization performance when training on different sizes: Default network

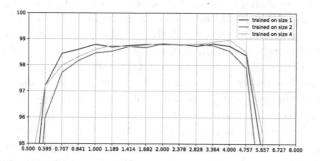

Scale generalization performance when training on different sizes: Larger network

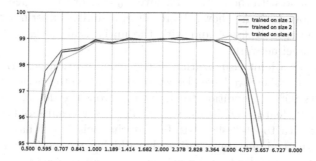

Fig. 4. Joint visualization of the generalization performance when training a multi-scale-channel Gaussian derivative network on training data with sizes 1, 2 and 4, respectively. As can be seen from the graphs, the performance is rather similar for all these networks over the size range between 1 and 4, a range for which the discretization errors in the discrete implementation can be expected to be low (a problem at too fine scales) and the influence of boundary effects causing a mismatch between what parts of the digits are visible in the testing data compared to a training data (a problem at too coarse scales). The top figure shows the results for the default network with 12-16-24-32-64-10 feature channels. The bottom figure shows the results for a larger network with 16-24-32-48-64-10 feature channels. (Horizontal axis: Scale of testing data.) (Colour figure online)

Figure 3 shows the result of an experiment to investigate the ability of such a multi-scale-channel network to generalize to testing at scales not present in the training data. For the experiment shown in the top figure, we have used 50 000 training images of size 1, and 10 000 testing images for each one of the 19 sizes between 1/2 and 8 with a relative size ratio of $\sqrt[4]{2}$ between adjacent sizes. The red curve shows the generalization performance for a single-scale-channel network with $\sigma_0 = 1$, whereas the blue curve shows the generalization performance for the multi-scale-channel network with 8 scale channels between $1/\sqrt{2}$ and 8.

As can be seen from the graphs, the generalization performance is very good for the multi-scale-channel network, for sizes in roughly in the range $1/\sqrt{2}$ and $4\sqrt{2}$. For smaller sizes near $1/2$, there are discretization problems due to sampling artefacts and too fine scale levels relative to the grid spacing in the image, implying that the transformation properties under scaling transformations of the underlying Gaussian derivatives are not well approximated in the discrete implementation. For larger sizes near 8, there are problems due to boundary effects and that the entire digit is not visible in the testing stage, implying a mismatch between the training data and the testing data. Otherwise, the generalization performance is very good over the size range between 1 and 4.

For the single-scale-channel network, the generalization performance to scales far from the scales in the training data is on the other hand very poor.

In the middle figure, we show the result of a similar experiment for training images of size 2 and with the initial scale level $\sigma_0 = 2$ for the single-scale-channel network. The bottom figure shows the result of a similar experiment performed with training images of size 4 and with the initial scale level $\sigma_0 = 4$ for the single-scale-channel network.

Figure 4 shows a joint visualization of the generalization performance for all these experiments, where we have also zoomed in on the top performance values in the range 98–99%. In addition to the results from the default network with 12-16-24-32-64-10 feature channels, we do also show results obtained from a larger network with 16-24-32-48-64-10 feature channels, which has more degrees of freedom in the training stage and leads to higher top performance and also somewhat better generalization performance. As can be seen from the graphs, the performance values for training sizes 1, 2 and 4, respectively, are quite similar for testing data with sizes in the range between 1 and 4, a size range for which the discretization errors in the discrete implementation can be expected to be low (a problem at too fine scales) and the influence of boundary effects causing a mismatch between what parts of the digits are visible in the testing data compared to the training data (a problem at too coarse scales).

To conclude, the experiment demonstrates that it is possible to use the combination of (i) scale-space features as computational primitives for a deep learning method with (ii) the closed-form transformation properties of the scale-space primitives under scaling transformations to (iii) make a deep network generalize to new scale levels not spanned by the training data.

5.1   Scale Selection Properties

Since the Gaussian derivative network is expressed in terms of scale-normalized derivatives over multiple scales, and the max-pooling operation over the scale channels implies detecting maxima over scale, the resulting approach shares similarities to classical methods for scale selection based on local extrema over scales of scale-normalized derivatives [3,4]. The approach is also closely related to the scale selection approach in [34,35] based on choosing the scales at which a supervised classifier delivers class labels with the highest posterior.

6   Summary and Discussion

We have presented a hybrid approach between scale-space theory and deep learning, where the layers in a hierarchical network architecture are modelled as continuous functions instead of discrete filters, and are specifically chosen as scale-space operations, here in terms of linear combinations of scale-normalized Gaussian derivatives. Experimentally, we have demonstrated that the resulting approach allows for scale generalization and enables good performance for classifying image patterns at scales not spanned by the training data.

The work is intended as a proof-of-concept of the idea of using scale-space features as computational primitives in a deep learning method, and of using their closed-form transformation properties under scaling transformations to perform extrapolation or generalization to new scales not present in the training data.

Concerning the choice of Gaussian derivatives as computational primitives in the method, it should, however, be emphasized that the necessity results that specify the uniqueness of these kernels only state that the first layer of receptive fields should be constructed in terms of Gaussian derivatives. Concerning higher layers, further studies should be performed concerning the possibilities of using other scale-space features in the higher layers, that may represent the variability of natural image structures more efficiently, within the generality of the sufficiency result for scale-covariant continuous hierarchical networks in [2].

References

1. Jansson, Y., Lindeberg, T.: Exploring the ability of CNNs to generalise to previously unseen scales over wide scale ranges. In: International Conference on Pattern Recognition (ICPR 2020), pp. 1181–1188 (2021)
2. Lindeberg, T.: Provably scale-covariant continuous hierarchical networks based on scale-normalized differential expressions coupled in cascade. J. Math. Imaging Vis. **62**, 120–148 (2020)
3. Lindeberg, T.: Feature detection with automatic scale selection. Int. J. Comput. Vis. **30**, 77–116 (1998)
4. Lindeberg, T.: Edge detection and ridge detection with automatic scale selection. Int. J. Comput. Vis. **30**, 117–154 (1998)
5. Mikolajczyk, K., Schmid, C.: Scale and affine invariant interest point detectors. Int. J. Comput. Vis. **60**, 63–86 (2004)
6. Lowe, D.G.: Distinctive image features from scale-invariant keypoints. Int. J. Comput. Vis. **60**, 91–110 (2004)
7. Bay, H., Ess, A., Tuytelaars, T., van Gool, L.: Speeded up robust features (SURF). CVIU **110**, 346–359 (2008)
8. Lindeberg, T.: Image matching using generalized scale-space interest points. J. Math. Imaging Vis. **52**, 3–36 (2015)
9. Fawzi, A., Frossard, P.: Manitest: are classifiers really invariant? In: British Machine Vision Conference (BMVC 2015) (2015)
10. Singh, B., Davis, L.S.: An analysis of scale invariance in object detection – SNIP. In: Proceedings Computer Vision and Pattern Recognition (CVPR 2018), pp. 3578–3587 (2018)

11. Xu, Y., Xiao, T., Zhang, J., Yang, K., Zhang, Z.: Scale-invariant convolutional neural networks. arXiv preprint arXiv:1411.6369 (2014)
12. Kanazawa, A., Sharma, A., Jacobs, D.W.: Locally scale-invariant convolutional neural networks. arXiv preprint arXiv:1412.5104 (2014)
13. Marcos, D., Kellenberger, B., Lobry, S., Tuia, D.: Scale equivariance in CNNs with vector fields. arXiv preprint arXiv:1807.11783 (2018)
14. Ghosh, R., Gupta, A.K.: Scale steerable filters for locally scale-invariant convolutional neural networks. arXiv preprint arXiv:1906.03861 (2019)
15. Worrall, D., Welling, M.: Deep scale-spaces: equivariance over scale. In: Advances in Neural Information Processing Systems, pp. 7366–7378 (2019)
16. Jaderberg, M., Simonyan, K., Zisserman, A., Kavukcuoglu, K.: Spatial transformer networks. In: Proceedings of Neural Information Processing Systems (NIPS 2015), pp. 2017–2025 (2015)
17. Finnveden, L., Jansson, Y., Lindeberg, T.: Understanding when spatial transformer networks do not support invariance, and what to do about it. In: International Conference on Pattern Recognition (ICPR 2020), pp. 3427–3434 (2021)
18. Roux, N.L., Bengio, Y.: Continuous neural networks. In: Artificial Intelligence and Statistics (AISTATS 2007), vol. 2, pp. 404–411 (2007)
19. Shocher, A., Feinstein, B., Haim, N., Irani, M.: From discrete to continuous convolution layers. arXiv preprint arXiv:2006.11120 (2020)
20. Iijima, T.: Basic theory on normalization of pattern (in case of typical one-dimensional pattern). Bull. Electrotech. Lab. **26**, 368–388 (1962)
21. Koenderink, J.J.: The structure of images. Biol. Cybern. **50**, 363–370 (1984)
22. Koenderink, J.J., van Doorn, A.J.: Generic neighborhood operators. IEEE-TPAMI **14**, 597–605 (1992)
23. Lindeberg, T.: Scale-Space Theory in Computer Vision. Springer, New York (1993). 10.1007/978-1-4757-6465-9
24. Florack, L.M.J.: Image Structure. Springer, Dordrecht (1997). 10.1007/978-94-015-8845-4
25. ter Haar Romeny, B.: Front-End Vision and Multi-Scale Image Analysis. Springer, Dordrecht (2003). 10.1007/978-1-4020-8840-7
26. Lindeberg, T.: Generalized Gaussian scale-space axiomatics comprising linear scale-space, affine scale-space and spatio-temporal scale-space. J. Math. Imaging Vis. **40**, 36–81 (2011)
27. Lindeberg, T.: A computational theory of visual receptive fields. Biol. Cybern. **107**, 589–635 (2013)
28. Jacobsen, J.J., van Gemert, J., Lou, Z., Smeulders, A.W.M.: Structured receptive fields in CNNs. In: Proceedings of Computer Vision and Pattern Recognition (CVPR 2016), pp. 2610–2619 (2016)
29. LeCun, Y., Bottou, L., Bengio, Y., Haffner, P.: Gradient-based learning applied to document recognition. Proc. IEEE **86**, 2278–2324 (1998)
30. Kingma, P.D., Ba, J.: Adam: a method for stochastic optimization. In: International Conference for Learning Representations (ICLR 2015) (2015)
31. Lindeberg, T.: Scale-space for discrete signals. IEEE-TPAMI **12**, 234–254 (1990)
32. Lindeberg, T.: Discrete derivative approximations with scale-space properties: a basis for low-level feature extraction. J. Math. Imaging Vis. **3**, 349–376 (1993)
33. Jansson, Y., Lindeberg, T.: MNISTLargeScaledataset. Zenodo (2020)
34. Loog, M., Li, Y., Tax, D.M.J.: Maximum membership scale selection. In: Benediktsson, J.A., Kittler, J., Roli, F. (eds.) MCS 2009. LNCS, vol. 5519, pp. 468–477. Springer, Heidelberg (2009). https://doi.org/10.1007/978-3-642-02326-2_47
35. Li, Y., Tax, D.M.J., Loog, M.: Scale selection for supervised image segmentation. Image Vis. Comput. **30**, 991–1003 (2012)

Quantisation Scale-Spaces

Pascal Peter[✉]

Mathematical Image Analysis Group, Faculty of Mathematics and Computer Science,
Campus E1.7, Saarland University, 66041 Saarbrücken, Germany
peter@mia.uni-saarland.de

Abstract. Recently, sparsification scale-spaces have been obtained as a sequence of inpainted images by gradually removing known image data. Thus, these scale-spaces rely on spatial sparsity. In the present paper, we show that sparsification of the co-domain, the set of admissible grey values, also constitutes scale-spaces with induced hierarchical quantisation techniques. These quantisation scale-spaces are closely tied to information theoretical measures for coding cost, and therefore particularly interesting for inpainting-based compression. Based on this observation, we propose a sparsification algorithm for the grey-value domain that outperforms uniform quantisation as well as classical clustering approaches.

Keywords: Quantisation · Scale-space · Inpainting · Compression

1 Introduction

Image inpainting [11] reconstructs an image from a mask that specifies known pixel data at a subset of all image coordinates. Cárdenas et al. [3] have shown that a wide variety of inpainting operators can create scale-spaces: Sequentially removing points from the mask yields inpainting results that form a scale-space with the mask density as a discrete scale dimension. The order in which the data is removed, the sparsification path, can impact the reconstruction quality significantly. Implicitly, many inpainting-based compression approaches (e.g. [6,13,16]) and associated mask optimisation strategies (e.g. [1,4,10]) rely on these sparsification scale-spaces: They choose sparse masks as a compact representation of the image and aim for an accurate reconstruction from this known data.

However, there is another aspect of compression, that has hitherto not been explored from a scale-space perspective: The known data has to be stored, which means the information content of the grey or colour values plays an important role. In particular, all contemporary compression codecs use some form of quantisation combined with variants of entropy coding, no matter if they rely on transforms [12,19], inpainting [6,13,16], or neural networks [14].

This work has received funding from the European Research Council (ERC) under the European Union's Horizon 2020 research and innovation programme (grant agreement no. 741215, ERC Advanced Grant INCOVID).

A. Elmoataz et al. (Eds.): SSVM 2021, LNCS 12679, pp. 15–26, 2021.
https://doi.org/10.1007/978-3-030-75549-2_2

Quantisation reduces the amount of different admissible values in the co-domain to lower the Shannon entropy [18], the measure for information content.

Sparsification scale-spaces have demonstrated that there are viable scale-space concepts beyond the classical ideas that rely on partial differential equations (PDEs) [2,8,15,21] or evolutions according to pseudodifferential operators [5,17]. In the present paper, we show that quantisation methods can also imply scale-spaces and that this leads to practical consequences for inpainting-based compression. In this application context, quantisation has so far only been the main focus in the work of Hoeltgen et al. [7] who compared multiple different strategies. However, they did not consider scale-space ideas and concluded that non-uniform quantisation does not offer advantages over uniform ones in inpainting-based compression.

Our contributions. Our first results are of theoretical nature: We propose quantisation scale-spaces based on the broad class of hierarchical quantisation methods inspired by Ward clustering [20]: The image is gradually simplified by merging level sets. Not only do these sequences of quantised images satisfy all essential scale space-properties, we also show that our Lyapunov criterion guarantees a decrease of the coding cost.

This observation motivates our practical contributions: We demonstrate that committed (i.e. structure-adaptive) quantisation scale-spaces are particularly useful. They allow the design of flexible, task-specific quantisation approaches. As a concrete application we use the natural ties of quantisation scale-spaces to information theory for improved inpainting-based compression.

Organisation. In Sect. 2, we propose quantisation scale-spaces and their properties, compare them to sparsification scale-spaces in Sect. 3, and discuss practical applications in Sect. 4. Section 5 concludes with a final discussion.

2   Quantisation Scale-Spaces

While there is a plethora of quantisation approaches designed for specific purposes, not all of them correspond to scale-spaces. Thus, we first define a suitable class of quantisation methods without imposing too many restrictions on their design. To this end, we use a very general hierarchical clustering idea inspired by the classical algorithm of Ward [20]. Intuitively, these approaches reduce the quantisation levels by mapping exactly two of the fine grey values to the same coarse quantisation value. In the following we consider only 1-D vectors, but these represent images of arbitrary dimension with N pixels (e.g. a row by row representation in 2-D).

Definition 1: Hierarchical Quantisation Function
Consider a quantised image $\boldsymbol{f} \in \{v_1, ..., v_Q\}^N$ with N pixels and Q quantisation levels $v_1 < v_2 < \cdots < v_Q$. A function $q : V := \{v_1, ..., v_Q\} \to \hat{V} := \{\hat{v}_1, ..., \hat{v}_{Q-1}\}$ mapping to coarse quantisation levels $\hat{V}$ with $v_1 \leq \hat{v}_1 < \hat{v}_2 < \cdots < \hat{v}_{Q-1} \leq v_Q$ is a *hierarchical quantisation* function if it fulfils

$$\exists s, t \in \{1, ..., Q\}, \, s \neq t : \forall i, j \in \{1, ..., Q\}, \, i \neq j :$$
$$q(v_i) = q(v_j) \iff i, j \in \{s, t\}. \tag{1}$$

Equation (1) guarantees that exactly v_s and v_t from the fine grey value range V are mapped to the same value $v_r := q(v_s)$ from the coarse range $\hat{V}$. Since $\hat{V}$ contains exactly one value less, this simultaneously implies a one-to-one mapping for all remaining $Q - 2$ values in $V \setminus \{v_s, v_t\}$ and $\hat{V} \setminus \{v_r\}$.

For our purposes we also consider quantisation as an operation on level sets. Let a level set L_k of the image $f = (f_i)_{i=1}^N$ be given by the locations with grey value v_k:

$$L_k = \{i \mid f_i = v_k\}, \quad k \in \{1, ..., Q\}. \tag{2}$$

With this notation, our definition of a hierarchical quantisation step is equivalent to merging exactly the two level sets L_s and L_t, while leaving the rest untouched.

Note that we do not specify which level sets need to be merged. Moreover, the only restriction that we impose to the target range $\hat{V}$ is that it does not exceed the original grey level range. As we demonstrate in Sect. 4, hierarchical quantisation covers a very broad range of quantisation methods that can be tailored to specific applications. With the formal notion of hierarchical quantisation functions we are now suitably equipped to define quantisation scale-spaces.

Definition 2: Quantisation Scale-Spaces
Consider the original image $f = (f_i)_{i=1}^N \in V^0 := \{v_1, ..., v_Q\}^N$, $v_1 < v_2 < \cdots < v_Q$ containing Q initial grey levels. Let $f^0 := f$ and consider hierarchical quantisation functions q^ℓ with $\ell \in \{1, ..., Q - 1\}$. Each function q^ℓ maps $V^{\ell-1}$ with $|V^{\ell-1}| = Q - \ell + 1$ to a quantised range V^ℓ with $Q - \ell$ grey values. The quantisation scale-space is then given by the family of images

$$f^0 = f, \tag{3}$$

$$f^\ell = \left(q^\ell\left(f_1^{\ell-1}\right), ..., q^\ell\left(f_N^{\ell-1}\right)\right), \quad \ell \in \{1, ..., Q - 1\}, \tag{4}$$

with discrete scale parameter $\ell \in \{0, ..., Q - 1\}$.

Since we use hierarchical quantisation functions in Definition 2, we make sure that the range of admissible grey values is reduced by one when we transition from scale ℓ to scale $\ell + 1$ with $q^{\ell+1}$. Note that we do not require that f actually contains all values of the range V^0. Thus, some of its level sets might be empty. For 8-bit images, we have $Q = 256$.

2.1   Scale-Space Properties

In the following we show that quantisation scale-spaces in the sense of Definition 2 fulfil six essential scale-space properties [2]. For some of these properties, it is helpful to argue on level sets. Therefore, we introduce the notation L_k^ℓ for the level set of the value v_k^ℓ (as in Eq. (2)) on scale ℓ.

Property 1: Original Image as Initial State
For $\ell = 0$, Definition 2 directly yields $f^0 = f$ which can be seen as quantisation with the identity function.

Property 2: Semigroup Property

Due to Definition 1, we can construct the scale-space in a cascadic manner by subsequent merging of level sets. In particular, the hierarchical quantisation criterion from Eq. (1) implies that $q^{\ell+1}$ merges exactly two level sets L_s^ℓ and L_t^ℓ into a new level set $L_r^{\ell+1}$, while the rest remains unchanged. Therefore, a quantised image $f^{\ell+n}$ can be obtained from f^0 in $\ell + n$ merging steps or, equivalently, from f^ℓ in n merging steps.

Property 3: Maximum–Minimum Principle

According to Definition 1, hierarchical quantisation functions do not expand the grey value range. Thus, all quantised values lie in the range $[v_1, v_Q]$ and the maximum–minimum principle is fulfilled.

Property 4: Lyapunov Sequences

Lyapunov sequences play a vital role in describing the sequential simplification of the image with increasing scale parameter [21]. Multiple different Lyapunov sequences can be specified that emphasise specific types of simplification. In analogy to sparsification scale spaces [3], for $J = \{1, ..., N\}$, we can use the total image contrast

$$V(f^\ell) := \max_{i \in J} f_i^\ell - \min_{i \in J} f_i^\ell \tag{5}$$

to define a Lyapunov sequence based on the maximum–minimum principle. Thus, the image is visually simplified by shrinking the contrast successively.

However, for progressive quantisation, it is more meaningful to consider a different Lyapunov sequence that directly reflects the intention behind quantisation in compression applications: the entropy as a measure of information content and storage cost. For the Shannon entropy [18], we need the probabilities of the grey values occurring in the image f^ℓ. We can define them via the level sets as

$$p_k^\ell := \frac{|L_k^\ell|}{N},$$

where $|\cdot|$ denotes the cardinality. Then, the binary entropy H, with $\log_2$ denoting the logarithm to the base 2, is given by

$$H(f^\ell) = - \sum_{\substack{k=1, \\ p_k \neq 0}}^{Q-\ell} p_k \log_2 p_k . \tag{6}$$

It constitutes the minimal average bit cost of storing a grey value of the image f^ℓ that can be achieved by any binary zeroth order entropy coder. Since Definition 2 allows empty level sets, these are excluded from the sum in Eq. (6).

Note that this entropy criterion for simplification differs significantly from the notion of entropy used in PDE-based scale-spaces [21]: There, the entropy is defined directly on the grey values instead of their histogram and thus does not reflect storage cost. In the following, we prove that the Shannon entropy constitutes a Lyapunov sequence.

Proposition 1: Shannon Entropy Constitutes a Lyapunov Sequence
For a quantisation scale-space according to Definition 2, the Shannon entropy decreases monotonically with the scale parameter, i.e. for all ℓ

$$H(\boldsymbol{f}^\ell) \geq H(\boldsymbol{f}^{\ell+1}). \tag{7}$$

Proof. Let L_i^ℓ, L_j^ℓ denote the two level sets that are merged to obtain $L_r^{\ell+1} = L_i^\ell \cup L_j^\ell$ according to Eq. (1). The probabilities p_i and p_j are additive under merging, i.e.

$$p_r^{\ell+1} = \frac{|L_r^{\ell+1}|}{N} = \frac{|L_i^\ell|}{N} + \frac{|L_j^\ell|}{N} = p_i^\ell + p_j^\ell. \tag{8}$$

This implies equality in Eq. (7) if L_i^ℓ or L_j^ℓ are empty. Assuming that both level sets are non-empty, the entropy is strictly decreasing for increasing scale parameter ℓ:

$$H(\boldsymbol{f}^{\ell+1}) = -p_r^{\ell+1} \log_2 p_r^{\ell+1} - \sum_{k=1, k \neq r}^{Q-\ell} p_k^{\ell+1} \log_2 p_k^{\ell+1} \tag{9}$$

$$= -(p_i^\ell + p_j^\ell) \log_2 (p_i^\ell + p_j^\ell) - \sum_{k=1, k \notin \{i,j\}}^{Q-\ell} p_k^\ell \log_2 p_k^\ell \tag{10}$$

$$< -p_i^\ell \log_2(p_i^\ell) - p_j^\ell \log_2(p_j^\ell) - \sum_{k=1, k \notin \{i,j\}}^{Q-\ell} p_k^\ell \log_2 p_k^\ell = H(\boldsymbol{f}^\ell) \tag{11}$$

Note that in the first step of the computation above we have used that there is a one-to-one mapping between all coarse and fine level sets not affected by the merge. The last step follows from the fact that $p_i^\ell > 0$, $p_j^\ell > 0$, and $\log_2$ is monotonically increasing, hence

$$(p_i^\ell + p_j^\ell) \log_2(p_i^\ell + p_j^\ell) > p_i^\ell \log_2(p_i^\ell) + p_j^\ell \log_2(p_j^\ell). \tag{12}$$

Thus, we have shown that each step in a quantisation scale-space does not increase information content and, if no level sets are empty, it even strictly decreases the entropy. □

Property 5 (Invariances)
A quantisation according to Definition 1 is a point operation, i.e. it acts independently of the spatial configuration of image pixels. Thus, it is invariant under any permutation of the image pixels. Moreover, it is invariant under any brightness rescaling operations that keep the histogram, and thus the level sets, intact.

Property 6 (Convergence to a Flat Steady-State)
According to Definition 2, the final image $\boldsymbol{f}^{Q-1}$ contains $Q - (Q - 1) = 1$ grey value and is thus flat.

Now that the important properties of quantisation scale-spaces have been verified, we briefly compare them to their spatial relatives, the sparsification scale-spaces [3]. Then, in Sect. 4, we also consider practical applications for this novel type of scale-space.

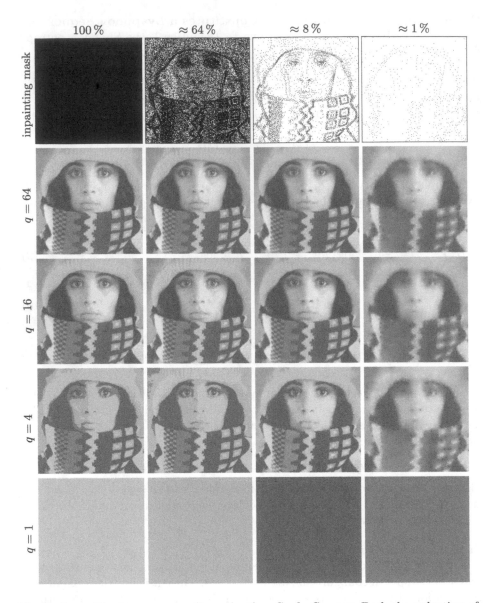

Fig. 1. Sparsification versus Quantisation Scale-Spaces: Both the reduction of known data (from left to right) and the reduction of grey values (from top to bottom) simplifies the test image *trui*. Here, we used probabilistic sparsification [3,10] for the spatial domain and our quantisation by sparsification for the grey level domain. Interestingly, for very low quantisation levels q, results might be visually more pleasant for lower mask densities, compare e.g. 8 % to 64 % at 4 grey levels. This impression results from the smooth interpolation by diffusion, that avoids unpleasant discontinuities and simultaneously expands the number of grey levels in the reconstructed image beyond those of the mask.

3   Relations to Sparsification Scale-Spaces

Sparsification scale-spaces [3] rely on a series of nested inpainting masks that specify known data for an image with domain Ω with $|\Omega| = N$. They start with a full mask $K^0 = \Omega$ and successively reduce K^ℓ to $K^{\ell+1}$ by removing exactly one pixel coordinate from this set. From each of these sets, an image u^ℓ can be inpainted by solving

$$C^\ell(u^\ell - f) - (I - C^\ell) A(u^\ell) u^\ell = 0 \qquad (\ell = 0, ..., N-1). \qquad (13)$$

Here, $C^\ell \in \mathbb{R}^{N \times N}$ is a diagonal matrix that corresponds to the known data set K^ℓ, and $A \in \mathbb{R}^{N \times N}$ implements an inpainting operator including boundary conditions. While there are many viable choices for A, we only consider homogeneous diffusion inpainting [8] here, where A corresponds to a finite difference discretisation of the Laplacian with reflecting boundary conditions. For more details on sparsification scale-spaces we refer to Cárdenas et al. [3].

Both sparsification and quantisation scale-spaces have a discrete scale parameter. The corresponding amount of steps is determined by the image resolution in the spatial setting and by the grey value range in the quantisation setting. Additionally, sparsification scale-spaces do not have an ill-posed direction: For a known order of pixel removals, one can easily go from coarse scales to fine scales. Due to the many-to-one mapping by merging level sets, this is not the case for quantisation scale-spaces. In compression applications, this is no issue, since the original image is available. However, one cannot go to finer quantisations, i.e. from an 8-bit to a high dynamic range image.

Interestingly, we can combine sparsification and quantisation scale-spaces. For an image $f \in V^N$ with $|V| = Q$, consider the sparsification path given by $(K^\ell)_{\ell=0}^{N-1}$. Now, we can define a quantisation scale-space on $K^\ell = \{i_1, ..., i_n\}$ with $n := N - \ell$ known pixels. Thereby, following Definition 2, we consider the known grey values $g^\ell := (f_{i_1}, ..., f_{i_n})$ instead of the full image f. Quantisation with scale parameter m yields n pixels from V^m with $q := |V^m| = Q - m$ quantisation values. The reconstruction $u^{\ell,m}$ now depends on the respective scale parameters ℓ of the sparsification and m of the quantisation scale-space.

This does not affect our theoretical results, since none of the properties relies on the spatial configuration of the image pixels. If an inpainting operator fulfils the maximum–minimum principle, this carries over to the inpainted image, and thereby also the total contrast of the known data. We can thus even extend Properties 3 and 4 to the reconstructions $u^{\ell,m}$.

An investigation of both scale dimensions in Fig. 1 yields surprising results: A lower amount of known data at the same quantisation scale can yield visually more pleasing results (e.g. $q = 4$ and 64% vs. 8% mask density). This results from the fact that smooth inpainting can fill in additional grey levels, hence the inpainted areas are less coarsely quantised than the known data.

Finally, both types of scale-spaces can be committed to the image, i.e. adapted to the image structure. In the following section, we show that this allows the design of highly task-specific quantisation approaches.

4   Applications to Quantisation and Compression

The results of practical experiments depend significantly on the order in which known data are removed and level sets are merged. For the spatial scale-space, we use an adaptive sparsification path obtained with the algorithm from [3]. First, we propose several committed and uncommitted quantisations. We show some exemplary results on the test image *trui* that has been also used in [3].

4.1   Uncommitted and Committed Quantisation

Uncommitted Uniform Quantisation: Hierarchical uniform quantisation can be seen as a pyramidal approach. On a level with $a := 2^k$ level sets, we have $b := 2^{k-1}$ pairs of neighbouring level sets. We progress to b grey values by merging L_{2i-1} and L_{2i} for $i = 1, ..., b$ in b steps. A bin containing the minimum grey value v_{min} and maximum value v_{max} is associated to the new value $v_{min} + \frac{1}{2}(v_{max} - v_{min})$. We round the intermediate results to the next integer.

Committed Ward Clustering: While the method of Ward [20] describes a general clustering approach, not a quantisation technique, it can be used easily as such by choosing the mean squared error (MSE) on the quantised data as an optimality criterion. It simply merges two level sets that minimise this criterion.

Committed Quantisation by Sparsification: Inspired by the spatial sparsification for adaptive scale-spaces [1,3], we successively remove the one grey level that has the lowest impact on the global inpainting error. This merging strategy can equivalently be interpreted as an inpainting-based merging criterion for Ward clustering that is designed for cases where parts of the image are unknown. If all pixels are known this corresponds to regular Ward clustering.

For both committed quantisations, we assign the value of the corresponding coarse level set with the largest histogram occurrence to the newly merged set.

Experiments on the test image *trui* with 8% known data in Fig. 3(a) verify that the entropy monotonically decreases with increasing scale for all three approaches. However, this is quite irregular for uniform quantisation, since it does not respect the actual distribution of grey levels in the image. The non-uniform methods do not yield any changes until they reach 186 grey levels, since this is the number actually occurring in the image. Sparsification leads to a slightly quicker descent in entropy, but for coarse scales, all methods yield similar results.

The quantisation error in Fig. 3(b) shows Ward clustering as a clear winner: Unsurprisingly, it yields much better results than uniform quantisation due to its adaptivity. Interestingly, sparsification exhibits the worst results for medium quantisation levels. The next section reveals that this is the intended result of its merging criterion and not a design flaw.

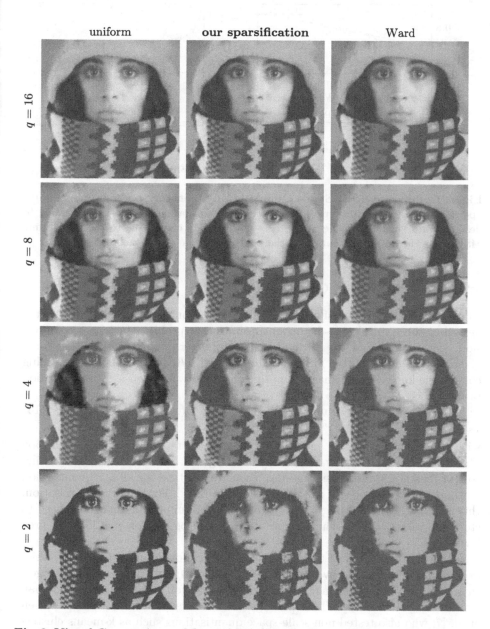

Fig. 2. Visual Comparison of Quantisation Scale-Spaces: The uncommitted uniform quantisation scale-space does not preserve details of *trui* as well as the committed sparsification and Ward approaches for low amounts of quantisation levels q. The quantisation by sparsification scale-space preserves structures in the facial area, and the hat even down to only two quantisation levels (mouth, nose). The Ward method preserves a few more patterns on the left side of the scarf, but looses detail in all other areas.

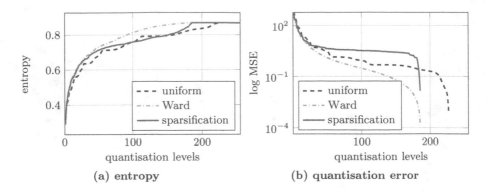

(a) entropy   (b) quantisation error

Fig. 3. Comparing Scale-Space Quantisation Operators on *trui.* **(a)** All three operators guarantee decreasing entropy for less quantisation levels. **(b)** According to design, Ward clustering achieves the best quantisation error. Due to the logarithmic MSE scale, the graphs are cut off at the point where the error reaches zero.

4.2   Inpainting-Based Compression

The combination of sparsification and quantisation scale-spaces from Sect. 3 describes the natural rate-distortion optimisation problem in image compression: We want to find the optimal scale parameters $\tilde{\ell}$ and $\tilde{m}$ such that for a given storage budget t, the MSE is minimised. If $B(K^{\ell}, g^{l,m})$ denotes the coding cost of the known data, image compression can now be seen as a minimisation problem

$$(\tilde{\ell}, \tilde{m}) = \underset{(\ell,m)}{\operatorname{argmin}} \|f - u^{\ell,m}\|^2, \quad \text{s.t. } B(K^{\ell}, g^{l,m}) < t \tag{14}$$

with Euclidean norm $\| \cdot \|$ and cost threshold t. Here, we consider the entropy $H(g^{l,m})$ of the known grey values for the budget B. We can neglect the coding cost of the known data positions since they are identical for all our quantisation methods. However, we still need to consider overhead. For uniform quantisation, this is only the number of grey levels (8 bit). Non-uniform methods also need to store the values in V^m, which comes down to $-q \log_2(q)$ for $q = 256 - m$ grey levels.

Figure 4(a) shows that the overhead of non-uniform quantisation can easily exceed uniform overhead by a factor 190. Consequentially, this is a game changer: In Fig. 4(b) Ward clustering is consistently worse for actual compression than a uniform quantisation. These findings are consistent with those of Hoeltgen et al. [7] who also tested non-scale-space quantisations such as k-means clustering [9]. We did not consider those here since they do not offer better results.

However, our quantisation by sparsification does not only beat Ward clustering by more than 20%, but also uniform quantisation by more than 10%. This is a direct result of the correct error measure for compression during the choice of the quantisation path. Minimising the MSE on the inpainted image is the true goal of this application, while Ward clustering only ensures accurate known values. Thus, committed quantisation scale-spaces can be useful to adapt to different applications.

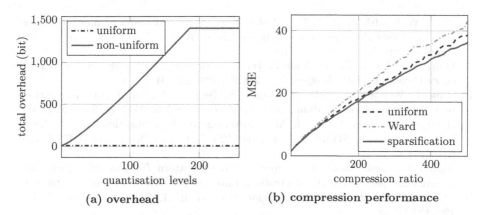

(a) overhead   (b) compression performance

Fig. 4. Compression with Quantisation Scale-Spaces on *trui*. (a) Non-uniform quantisation creates much higher overhead than uniform approaches. (b) Despite the overhead, quantisation by sparsification yields the best compression performance, while Ward clustering falls behind uniform quantisation.

5 Conclusions

We have established quantisation scale-spaces as the co-domain counterpart of spatial sparsification scale-spaces: They act on the grey value domain instead of the pixel domain. Both of these approaches successively sparsify their corresponding discrete domains and we can even combine them in a meaningful way for concrete applications in compression.

However, there are also notable differences: In contrast to spatial sparsification, the Lyapunov criterion for quantisation scale-spaces simplifies the known data set in terms of its Shannon entropy, thus yielding a guaranteed reduction of the coding cost. From the viewpoint of inpainting-based image compression, we can now interpret rate-distortion optimisation as the selection of suitable scale parameters in both scale-spaces.

In future research, we are going to consider quantisation scale-spaces on colour spaces and further investigate their impact on compression in a more practical setting, including a larger scale evaluation.

References

1. Adam, R.D., Peter, P., Weickert, J.: Denoising by inpainting. In: Lauze, F., Dong, Y., Dahl, A.B. (eds.) SSVM 2017. LNCS, vol. 10302, pp. 121–132. Springer, Cham (2017). https://doi.org/10.1007/978-3-319-58771-4_10
2. Alvarez, L., Guichard, F., Lions, P.L., Morel, J.M.: Axioms and fundamental equations in image processing. Arch. Ration. Mech. Anal. **123**, 199–257 (1993)
3. Cárdenas, M., Peter, P., Weickert, J.: Sparsification scale-spaces. In: Lellmann, J., Burger, M., Modersitzki, J. (eds.) SSVM 2019. LNCS, vol. 11603, pp. 303–314. Springer, Cham (2019). https://doi.org/10.1007/978-3-030-22368-7_24

4. Chen, Y., Ranftl, R., Pock, T.: A bi-level view of inpainting-based image compression. In: Proceedings of 19th Computer Vision Winter Workshop, Křtiny, Czech Republic, pp. 19–26, February 2014
5. Duits, R., Florack, L., de Graaf, J., ter Haar Romeny, B.: On the axioms of scale space theory. J. Math. Imaging Vis. **20**, 267–298 (2004)
6. Galić, I., Weickert, J., Welk, M., Bruhn, A., Belyaev, A., Seidel, H.P.: Image compression with anisotropic diffusion. J. Math. Imaging Vis. **31**(2–3), 255–269 (2008)
7. Hoeltgen, L., Peter, P., Breuß, M.: Clustering-based quantisation for PDE-based image compression. SIViP **12**(3), 411–419 (2017). https://doi.org/10.1007/s11760-017-1173-9
8. Iijima, T.: Basic theory on normalization of pattern (in case of typical one-dimensional pattern). Bull. Electrotech. Lab. **26**, 368–388 (1962). in Japanese
9. Lloyd, S.P.: Least squares quantization in PCM. IEEE Trans. Inf. Theory **28**(2), 129–137 (1982)
10. Mainberger, M., et al.: Optimising spatial and tonal data for homogeneous diffusion inpainting. In: Bruckstein, A.M., ter Haar Romeny, B.M., Bronstein, A.M., Bronstein, M.M. (eds.) SSVM 2011. LNCS, vol. 6667, pp. 26–37. Springer, Heidelberg (2012). https://doi.org/10.1007/978-3-642-24785-9_3
11. Masnou, S., Morel, J.M.: Level lines based disocclusion. In: Proceedings of 1998 IEEE International Conference on Image Processing, Chicago, IL, vol. 3, pp. 259–263, October 1998
12. Pennebaker, W.B., Mitchell, J.L.: JPEG: Still Image Data Compression Standard. Springer, New York (1992)
13. Peter, P.: Fast inpainting-based compression: combining Shepard interpolation with joint inpainting and prediction. In: Proceedings of 2019 IEEE International Conference on Image Processing, Taipei, Taiwan, pp. 3557–3561, September 2019
14. Rippel, O., Bourdev, L.: Real-time adaptive image compression. In: Proceedings of 34th International Conference on Machine Learnin, Sydney, Australia, pp. 2922–2930, August 2017
15. Scherzer, O., Weickert, J.: Relations between regularization and diffusion filtering. J. Math. Imaging Vis. **12**(1), 43–63 (2000)
16. Schmaltz, C., Peter, P., Mainberger, M., Ebel, F., Weickert, J., Bruhn, A.: Understanding, optimising, and extending data compression with anisotropic diffusion. Int. J. Comput. Vis. **108**(3), 222–240 (2014)
17. Schmidt, M., Weickert, J.: Morphological counterparts of linear shift-invariant scale-spaces. J. Math. Imaging Vis. **56**(2), 352–366 (2016)
18. Shannon, C.E.: A mathematical theory of communication - Part 1. Bell Syst. Tech. J. **27**(3), 379–423 (1948)
19. Taubman, D.S., Marcellin, M.W. (eds.): JPEG 2000: Image Compression Fundamentals Standards and Practice. Kluwer, Boston (2002)
20. Ward, J.H.: Hierarchical grouping to optimize an objective function. J. Am. Stat. Assoc. **58**(301), 236–244 (1963)
21. Weickert, J.: Anisotropic Diffusion in Image Processing. Teubner, Stuttgart (1998)

Equivariant Deep Learning via Morphological and Linear Scale Space PDEs on the Space of Positions and Orientations

Remco Duits[1], Bart Smets[1(✉)], Erik Bekkers[2], and Jim Portegies[1]

[1] Department Mathematics and Computer Science,
TU/e, Eindhoven, The Netherlands
`b.m.n.smets@tue.nl`
[2] Informatics Institute, UvA, Amsterdam, The Netherlands

Abstract. We present PDE-based Group Convolutional Neural Networks (PDE-G-CNNs) that generalize Group equivariant Convolutional Neural Networks (G-CNNs). In PDE-G-CNNs a network layer is a set of PDE-solvers where geometrically meaningful PDE-coefficients become trainable weights. The underlying PDEs are morphological and linear scale space PDEs on the homogeneous space $\mathbb{M}_d$ of positions and orientations. They provide an equivariant, geometrical PDE-design and model interpretability of the network.

The network is implemented by morphological convolutions with approximations to kernels solving morphological α-scale-space PDEs, and to linear convolutions solving linear α-scale-space PDEs. In the morphological setting, the parameter α regulates soft max-pooling over balls, whereas in the linear setting the cases $\alpha = 1/2$ and $\alpha = 1$ correspond to Poisson and Gaussian scale spaces respectively.

We show that our analytic approximation kernels are accurate and practical. We build on techniques introduced by Weickert and Burgeth who revealed a key isomorphism between linear and morphological scale spaces via the Fourier-Cramér transform. It maps linear α-stable Lévy processes to Bellman processes. We generalize this to $\mathbb{M}_d$ and exploit this relation between linear and morphological scale-space kernels.

We present blood vessel segmentation experiments that show the benefits of PDE-G-CNNs compared to state-of-the-art G-CNNs: increase of performance along with a huge reduction in network parameters.

Keywords: Convolutional neural networks · Scale space theory · Cramér transform · Geometric deep learning · Morphological convolutions and PDEs

1 Introduction

Current deep learning with convolutional neural networks (CNNs) [10,16,17] performs superbly on classification tasks like vessel segmentations, but requires

© Springer Nature Switzerland AG 2021
A. Elmoataz et al. (Eds.): SSVM 2021, LNCS 12679, pp. 27–39, 2021.
https://doi.org/10.1007/978-3-030-75549-2_3

massive annotated datasets for training because it optimizes inefficiently over huge parameter spaces. Additionally they lack geometric model interpretability because network weights in CNNs have a limited geometric meaning, and it is hard to control or understand the dynamics.

To improve the efficiency of the optimization, one can perform geometric reduction by hard-coding in the system roto-translation equivariance of the network (i.e. roto-translation of input should yield the same roto-translation of output) and roto-translation invariance of classification. To improve the interpretability of the network dynamics, one would like to avoid the use of ad-hoc (yet effective) non-linearities such as ReLU's and max-pooling, as they hamper (stochastic) analysis of the network dynamics via PDEs and central limit theorems.

Recently, inclusion of basic geometric Lie group designs into CNNs has had significant success in group equivariant CNNs (G-CNNs) [3,6,20,23], but lacks a geometric PDE description of crossing-preserving network dynamics on the homogeneous space $\mathbb{M}_d$ of positions and orientations.

We therefore propose a PDE-based equivariant convolutional neural network (PDE-G-CNN) with a linear and morphological scale space design that:

- allows for a serious geometric reduction of network parameters without loss of classification performance (e.g. pixel-wise blood vessel classification).
- allows for a clear geometric analysis of geometric scale space flows through the network, where we build upon our homogeneous space generalization of the core isomorphism [1,5,19] of scale space theory: the Cramér transform.

The overall idea is visualized in the context of vessel segmentation in Fig. 1. Here the image data is first lifted to the 3D space $\mathbb{M}_2 \equiv SE(2) := \mathbb{R}^2 \rtimes SO(2)$ of positions and orientations after which we apply PDE-layers strictly involving linear and morphological convolution layers with analytic kernels implementing linear and morphological PDEs on $\mathbb{M}_2$. We conclude with a projection layer that integrates over all angles before pixel-wise classification.

Our geometric PDEs relate to α-stable Lévy processes [7] and cost-processes akin to [1], but then on $\mathbb{M}_d$ rather than $\mathbb{R}^d$. This relates to probabilistic equivariant numerical neural networks [13] that use anisotropic convection-diffusions on $\mathbb{R}^d$. In contrast to these networks, the PDE-G-CNNs that we propose allow for *simultaneous* spatial and angular diffusion on $\mathbb{M}_d$. Furthermore we include Bellman processes [1] for max pooling over Riemannian balls.

The main contributions of this article are:

- We propose a new PDE-based design of equivariant convolutional neural networks: PDE-G-CNNs where standard nonlinearities are replaced by morphological scale space convolutions.
- We present an (approximate) Cramér transform that generalizes the fundamental isomorphism [1,5,19] between linear scale spaces and morphological scale spaces from $\mathbb{R}^d$ towards homogeneous spaces such as $\mathbb{M}_d$. We use it for new analytic approximation of morphological scale space kernels in Definition 3.

- We assess the quality of the approximative linear scale space kernels in Theorem 1 and the quality of the approximative morphological scale space kernels in Proposition 1 with 3 subsequent motivations. We use them in our PDE-G-CNNs.
- We show that PDE-G-CNNs outperform state-of-the-art (awarded) G-CNNs [3] on vessel segmentation while having 10 times less parameters.

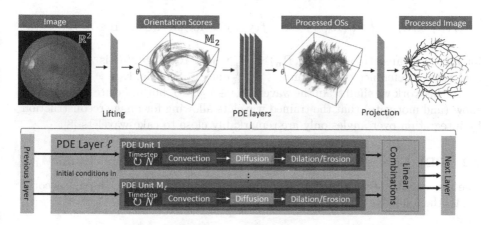

Fig. 1. Illustration of PDE-based CNNs for vessel segmentation. Here convection and diffusion flows (for transport and regularization) will be implemented by linear group convolution, dilation (for max-pooling) and erosion (for data-sharpening) by morphological group convolutions [21]

2 Design of PDE-Based Equivariant Neural Network

2.1 The Lifting Layer: Extending the Image Domain from $\mathbb{R}^d$ to $\mathbb{M}_d$

In order to disentangle all local orientations in an image we will lift the data from $\mathbb{R}^d$ to the homogeneous space $\mathbb{M}_d$ of positions and orientations given by

$$\mathbb{M}_d = \mathbb{R}^d \rtimes S^{d-1} := G/H, \qquad d \in \{2, 3\},$$

with the subgroup $H = \{g \in G \mid g \odot \mathbf{p}_0 = \mathbf{p}_0\}$ of the roto-translation group $G = SE(d) = \mathbb{R}^d \rtimes SO(d)$, with origin $\mathbf{p}_0 := (\mathbf{0}, \mathbf{a})$ where $\mathbf{a} \in S^{d-1}$ sets an a priori fixed reference axis. The Lie group G carries the following group product:

$$g_1 g_2 = (\mathbf{x}_1, \mathbf{R}_1)(\mathbf{x}_2, \mathbf{R}_2) = (\mathbf{x}_1 + \mathbf{R}_1 \mathbf{x}_2, \mathbf{R}_1 \mathbf{R}_2)$$

for all $g_1, g_2 \in G$. The Lie group G acts on $\mathbb{R}^d \rtimes S^{d-1}$ by

$$g \odot \mathbf{p} := (\mathbf{x} + \mathbf{R}\mathbf{x}', \mathbf{R}\mathbf{n}'), \quad \text{for } g = (\mathbf{x}, \mathbf{R}) \in SE(d), \ \mathbf{p} = (\mathbf{x}', \mathbf{n}') \in \mathbb{R}^d \rtimes S^{d-1}, \quad (1)$$

The case $d = 2$ is of primary interest. Then $H = \{e\}$, with unity element $e = (\mathbf{0}, I)$ so that $\mathbb{M}_2 \equiv SE(2)$. Then we set a priori reference orientation $\mathbf{a} = (1, 0)$ and identify counter-clockwise rotation $\mathbf{R}_\theta \in SO(2)$ with orientation $\mathbf{n}(\theta) = (\cos\theta, \sin\theta) \in S^1$, and angle $\theta \in \mathbb{R}/(2\pi\mathbb{Z})$.

We lift the image data to an orientation score $\mathcal{W}_\psi f : \mathbb{M}_d \to \mathbb{R}$ that reveals how an image is decomposed out of local orientations. It extends the image domain from $\mathbb{R}^d$ to $\mathbb{M}_d$ and is obtained by convolving the image $f \in \mathbb{L}_2(\mathbb{R}^d)$ by a family of (anisotropic) group-coherent wavelets:

$$\mathcal{W}_\psi f(\mathbf{x}, \mathbf{n}) = \int_{\mathbb{R}^d} \overline{\psi(\mathbf{R}_\mathbf{n}^{-1}(\mathbf{y} - \mathbf{x}))}\, f(\mathbf{y})\, \mathrm{d}\mathbf{y}, \quad \text{for all } \mathbf{x} \in \mathbb{R}^d, \mathbf{n} \in S^{d-1}, \qquad (2)$$

for all roto-translations $g = (\mathbf{x}, \mathbf{R}_\mathbf{n}) \in G$, where $\mathbf{R}_\mathbf{n} \in SO(d)$ is any rotation that rotates reference orientation $\mathbf{a} \in S^{d-1}$ onto $\mathbf{n}$. See Fig. 1.

In this work we shall train the wavelets $\psi \in \mathbb{L}_2(\mathbb{R}^d)$, but in future work we will show (and motivate) that the trained wavelets allowing for image reconstruction by integration over angles only are remarkably close to cake-wavelets [8].

2.2   PDE Layers by Linear and Morphological Scale Spaces on $\mathbb{M}_d$

The processing of feature maps on $\mathbb{M}_d$ in our proposed network happens according to the following PDE:

$$\overset{\textbf{convection}}{} \quad \overset{\text{frac. diffusion}}{} \quad \overset{\text{dilation/erosion}}{}$$

$$\frac{\partial W}{\partial t} = -\mathbf{c} \cdot \nabla_{\mathcal{G}_1} W + |\Delta_{\mathcal{G}_2}|^\alpha W \pm \|\nabla_{\mathcal{G}_3^\pm} W\|^{2\alpha}, \text{ on } \mathbb{M}_d \times \mathbb{R}^+, \qquad (3)$$

$$W|_{t=0} = \mathcal{W}_\psi f, \qquad\qquad\qquad\qquad\qquad \text{ on } \mathbb{M}_d,$$

where the orientation score (2) serves as the initial condition, and where the gradient $\nabla_{\mathcal{G}_1} W$ is relative to a left-invariant metric tensor $\mathcal{G}_1$ on $\mathbb{M}_d$, and where $\Delta_{\mathcal{G}_2}$ is a left-invariant Laplacian for diffusion indexed by a left-invariant metric tensor $\mathcal{G}_2$. For the dilation (+ case) and erosion (− case) part we use metric tensors $\mathcal{G}_3^\pm$, as we explain below. The convection takes care of equivariant transport, the fractional diffusion for equivariant regularization, the dilation for data-propagation, and the erosion for data-sharpening.

In principle any evolution type PDE can be chosen as the basis of a PDE layer, as long as the PDE is equivariant the resulting layer will also be equivariant. The PDE we propose here is the result of scale space axiomatics for linear [18] and non-linear scale spaces [19] combined with left invariance [7].

Geometric Parameter Reduction: The left-invariant metric must be well-defined on $\mathbb{M}_d$. This allows for geometric parameter reduction [21, Cor. 2.7]. E.g. if data-adapation is omitted one must use [21, Prop. 2.8] metric tensors:

$$\mathcal{G}_m\big|_\mathbf{p}(\dot{\mathbf{p}}, \dot{\mathbf{p}}) = \xi^2|\dot{\mathbf{x}} \cdot \mathbf{n}|^2 + \nu^2\|\dot{\mathbf{n}}\|^2 + \xi^2\epsilon_m^{-2}\|\dot{\mathbf{x}} \wedge \mathbf{n}\|^2, \qquad (4)$$

for all $\mathbf{p} = (\mathbf{x}, \mathbf{n}) \in \mathbb{M}_d$, and tangent vectors $\dot{\mathbf{p}} = (\dot{\mathbf{x}}, \dot{\mathbf{n}}) \in T_\mathbf{p}(\mathbb{M}_d)$, for $m \in \{1, 2, 3\}$. Parameter $\nu > 0$ regulates angular motion costs. By default we set $\nu = 1$ as the cases $\nu > 0$ follow by direct scaling arguments. For $\nu \to \infty$ and $d = 2$ we arrive at related work [13]. Regarding the anisotropies in $\mathcal{G}_m$:

- For $m = 1$ (convection) we set $\epsilon_1 = \xi = 1$ and only train multiple off-sets **c**.
- For $m = 2$ (diffusion) and $m = 3$ (dilation) we train respectively $\mathcal{G}_2$ and $\mathcal{G}_3^+$. One expects $0 < \epsilon_2, \epsilon_3 \ll 1$ in order to diffuse/dilate primarily along lines.
- For $m = 3$ (erosion), i.e. the - case in (3) we train $\mathcal{G}_3^-$. Here one expects $\xi \gg 1$ and $\xi/\epsilon_3 \ll 1$ so that one sharpens across lines (and not along them).

We will train the metric tensors so that the data determines optimal anisotropies.

Distances on $\mathbb{M}_d$: The symmetric left-invariant metric tensor fields $\mathcal{G}_m$ on $\mathbb{M}_d$ induces a symmetric left-invariant distance ('metric') by

$$d_{\mathcal{G}_m}(\mathbf{p}_0, \mathbf{p}_1) = \inf_{\substack{\gamma(0) = \mathbf{p}_0, \ \gamma(1) = \mathbf{p}_1, \\ \gamma \in \text{Lip}([0,1], \mathbb{M}_d)}} \int_0^1 \sqrt{\mathcal{G}_m|_{\gamma(t)} (\dot{\gamma}(t), \dot{\gamma}(t))} \, dt. \qquad (5)$$

For all $g \in G = SE(d)$, $\mathbf{p}_1, \mathbf{p}_2 \in \mathbb{M}_d$ one has $d_{\mathcal{G}_m}(\mathbf{p}_1, \mathbf{p}_2) = d_{\mathcal{G}_m}(g \odot \mathbf{p}_1, g \odot \mathbf{p}_2)$.

Logarithmic Approximations of Distances on $\mathbb{M}_d$: The metric tensor fields also induce a norm on the tangent space $T_e(G)$ of the unity element $e = (\mathbf{0}, I)$:

$$\|\mathbf{v}\|_{\mathcal{G}} := \left\| \frac{\partial}{\partial t} \exp_G (t\mathbf{v}) \odot p_0 \Big|_{t=0} \right\|_{\mathcal{G}|_{\mathbf{p}_0}} \qquad (6)$$

for all $\mathbf{v} \in T_e(G)$, where $\exp_G : T_e(G) \to G$ is the exp map of G. We define

$$\rho_{\mathcal{G}}(\mathbf{p}) := \inf_{g \in G_{\mathbf{p}}} \|\log_G g\|_{\mathcal{G}} \quad \text{with} \quad G_{\mathbf{p}} := \{g \in G \mid g \odot \mathbf{p}_0 = \mathbf{p}\}. \qquad (7)$$

For simple formulas that we used in our code for $d = 2$ see [21]. For formulas for $d = 3$ (identifying the minimum element of $G_{\mathbf{p}}$ above) see [7].

2.3 PDE-Based Deep Learning by G-CNNs on $\mathbb{M}_2$

Usually CNNs on $\mathbb{R}^d$ [16] iterate 1) possibly off-centered convolution kernels, 2) max-pooling, 3) regularization, 4) ReLU's. We see them as sampled operator splittings of a PDE evolution combining convection, *fractional* morphology and diffusion on $\mathbb{R}^d$, with multiple convection-vectors for training center off-sets.

We replace $\mathbb{R}^d$ by $\mathbb{M}_d$ and obtain equivariant PDE-based CNNs. This allows for the geometric reduction of parameters, and we obtain equivariance. Both regular CNNs on $\mathbb{R}^2$ and equivariant PDE-based CNNs on $\mathbb{M}_d$ are special instances of PDE-based CNNs on homogeneous spaces as shown in Fig. 2.

Recall Fig. 1 where our PDE-based CNNs on $\mathbb{M}_d$ are depicted. The depth of the network is created by $\underline{\Phi}_l$ which maps $U_l \in \mathbf{L}_2(\mathbb{M}_d)^{N_l}$ to $U_{l+1} \in (\mathbf{L}_2(\mathbb{M}_d))^{N_{l+1}}$ by

$$U_{l+1} = \underline{\Phi}_l(U_l), \quad \text{with} \quad U_{l+1}^{k'} := \sum_{k=1}^{N_l} \omega_{k'k}^l \cdot \Phi_t^k(U_l^k), \quad \text{and} \quad \Phi_t^k(U) := W(\cdot, T), \qquad (8)$$

where $\omega_{k',k}^l$ are trainable weights. At the first layer of the network we set the orientation score $U = \mathcal{W}_\psi f$ as initial condition. In subsequent layers we use the output of the previous layer as the initial conditions of the PDEs and take the solutions of the PDEs at a fixed time $t = T$. The choice $t = T$ is not crucial as the α-homogeneity of the generators in PDE (3) ensures time-scale invariance [18, 19].

3   Linear and Morphological Kernel Implementation

The operator splitting of the equivariant PDE evolution (3) as depicted in Fig. 2, boils down to iteratively activating one of the terms in their generator.

Firstly, the convection PDE part in Fig. 2 is solved by a transport along an exponential curve, for details see [21, Prop. 5.1].

Secondly, fractional diffusion in (3) is solved by a linear group convolution:

$$W(\mathbf{p}, t) = (K_t^\alpha * U)(\mathbf{p}) = \int_G K_t^\alpha(g^{-1} \odot \mathbf{p}) \, U(g \odot \mathbf{p}_0) \, dg \tag{9}$$

with α-scale-space kernel K_t^α, where $\mathbf{p}_0 = (\mathbf{0}, \mathbf{a})$ is the origin in the homogeneous space $\mathbb{M}_d$ using the action (1) of the Lie group $G = SE(d)$, and Haar measure dg. Thirdly, viscosity solutions [12] of dilations/erosions are

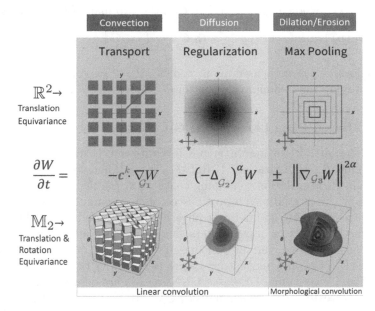

Fig. 2. Geometric design of neurons by PDEs in homogeneous space $\mathbb{R}^2$ (top, resulting in translation equivariance) and in homogeneous space $\mathbb{M}_2$ (bottom, resulting in roto-translation equivariance), the trainable parameters are highlighted in red. Transport and regularization are solved by linear convolutions (9), and dilation/erosion are solved by morphological convolutions (10). For dilation with $\alpha = \frac{1}{2}$ this boils down to max pooling over geometric balls that grow with $t > 0$, as depicted bottom right.

morphological convolutions:

$$W(\mathbf{p}, t) = (k_t^\alpha \,\square\, U)(\mathbf{p}) = \pm \inf_{g \in G} \left\{ k_t^\alpha(g^{-1} \odot \mathbf{p}) \pm U(g \odot \mathbf{p}_0) \right\}. \qquad (10)$$

Here the kernels k_t^α are positive and the $-$ cases above solve the dilation PDE, and the $+$ cases solves the erosion PDE, cf. (3). Tangible approximations for the kernels will be obtained further on by approximations of the Cramér transform.

The next subsections (3.1 and 3.2) are necessarily technical. The main message is that linear kernels K_t^α in (9) are well-approximated by $K_t^{\alpha,appr}$ (Definitions 1 and 2), whereas morphological kernels k_t^α in (10) are well-approximated by $k_t^{\alpha,appr}$ (Definition 3).

3.1 Analytic Approximations of α-Scale-Space Kernels on $\mathbb{M}_2$

For exact solutions of fractional diffusion kernels see [7]. For more tangible analytic approximations for $\alpha = 1$ see [8]. They are due to a general theory of estimating heat kernels [4, Thm. 3.1], [15, Thm. 7.6, p.53], [11] on Riemannian manifolds and allow us to provide more explicit bounds. First we summarize these existing mathematical results for $\alpha = 1$ on $(\mathbb{M}_2, \mathcal{G}_2)$ and then generalize, to the fractional diffusion case where $\alpha \in (0, 1]$.

Lemma 1. *Consider Riemannian manifold* $(\mathbb{M}_2, \mathcal{G} = \mathcal{G}_2)$ *with metric* $d_\mathcal{G}$. *Let* $M_{opt} := \mathbb{M}_2 \setminus cut(\mathbf{p}_0)$ *denote the complement of the closure of the cut-locus. Then there exists* $c_j \in C^\infty(M_{opt})$ *with bounded partial derivatives s.t.*

$$K_t^1(\mathbf{p}) = \frac{e^{-\frac{d_\mathcal{G}(\mathbf{p}, \mathbf{p}_0)^2}{4t}}}{(4\pi t)^{\frac{3}{2}}} \left(\sum_{j=0}^N c_j(\mathbf{p}) t^j + O(t^{N+1}) \right) \qquad (11)$$

for all $\mathbf{p} \in M_{opt}$, *where* $N \geq 0$, *and where we may approximate:*

$$d_\mathcal{G}(\mathbf{p}, \mathbf{p}_0) = \rho_\mathcal{G}(\mathbf{p}) + \epsilon(\mathbf{p}), \qquad (12)$$

with $\epsilon(\mathbf{p}) \to 0$ *quadratically and* $c_0(\mathbf{p}) \to 1$ *for* $\mathbf{p} \to \mathbf{p}_0$. *For* $N = 0$ *we have:*

$$K_t^1(\mathbf{p}) = \frac{e^{-\frac{(d_\mathcal{G}(\mathbf{p}, \mathbf{p}_0))^2}{4t}}}{(4\pi t)^{\frac{3}{2}}} \left(c_0(\mathbf{p}) + O\left(\frac{t}{(\rho(\mathbf{p}))^2} \right) \right) \text{ for } 0 < \frac{\sqrt{t}}{\rho(\mathbf{p})} \ll 1,$$

$$\exists_{C \geq 1, D \geq 1} \forall_{t > 0} : \; C^{-1} \frac{e^{-\frac{(\rho(\mathbf{p}))^2}{4D^{-1}t}}}{(4\pi D^{-1}t)^{\frac{3}{2}}} \leq K_t^1(\mathbf{p}) \leq C \frac{e^{-\frac{(\rho(\mathbf{p}))^2}{4Dt}}}{(4\pi Dt)^{\frac{3}{2}}},$$

$$\qquad (13)$$

for $\mathbf{p} \in C$, *with* $C \subset M_{opt}$ *compact around* $\mathbf{p}_0$, *where we set* $\rho(\mathbf{p}) = \rho_\mathcal{G}(\mathbf{p})$.

Remark 1 **(assumption).** We assume that $\mathcal{G} = \mathcal{G}_2$ is chosen such that the exact kernels are concentrated on $C \subset M_{opt} \subset \tilde{\mathbb{M}}_2 := \mathbb{R}^2 \times [-\pi, \pi]$. Then by (4) we have constant $\det(\mathcal{G}) = \xi^4 \nu^2 \epsilon_2^{-2}$ and

$$1 = \int_{\mathbb{M}_2} K_t^1(x, y, \theta) \sqrt{\det \mathcal{G}} \; dxdyd\theta \approx \int_{C \subset M_{opt}} K_t^1(x, y, \theta) \sqrt{\det \mathcal{G}} \; dxdyd\theta. \qquad (14)$$

This assumption makes estimates (13) meaningful and avoids periodicity hassles.

Lemma 1 motivates the following approximation kernel for $\alpha = 1$.

Definition 1. *Now we define the approximate Gaussian kernel*

$$K_t^{1,appr}(\mathbf{p}) := \frac{J(\mathbf{p})}{(4\pi t)^{3/2}} e^{-\frac{(\rho_\mathcal{G}(\mathbf{p}))^2}{4t}}, \mathbf{p} \in \mathbb{M}_2, t > 0. \tag{15}$$

with Jacobian $J(\mathbf{p}) := \frac{(\theta/2)^2}{\sin^2(\theta/2)}$ *for* $\mathbf{p} = (\mathbf{x}, \theta) \in \mathbb{M}_2$, *and* $J(\mathbf{x}, 0) = 1$.

Regarding the Jacobian we note that the logarithm on $G = SE(2)$ equals [21]:

$$Log_G(\mathbf{p}) = \left(\frac{\theta}{2} \left(y + x \cot \left(\frac{\theta}{2} \right) \right), \frac{\theta}{2} \left(-x + y \cot \left(\frac{\theta}{2} \right) \right), \theta \right), \theta \neq 0, \tag{16}$$

for all $\mathbf{p} = (x, y, \theta)$. Thereby $J(\mathbf{p}) = |\det D_\mathbf{p} Log_G(\mathbf{p})|$, and by assumption in Remark 1 kernel $K_t^{1,appr}$ is nearly $\mathbb{L}_1$-normalised w.r.t. Haar-measure $\sqrt{\det \mathcal{G}}\, dxdyd\theta$.

From semi-group theory [24, Ch. IX] it follows that semi-groups generated by taking fractional powers of the generator (in our case $\Delta_\mathcal{G} \mapsto -|\Delta_\mathcal{G}|^\alpha$) amount to the following **key relation** between the α-kernel and the diffusion kernel:

$$K_t^\alpha(\mathbf{p}) := \int_0^\infty q_{t,\alpha}(\tau) K_\tau^1(\mathbf{p}) \, d\tau, \text{ for } \alpha \in (0,1], \ t > 0. \tag{17}$$

where $q_{t,\alpha} \geq 0$ is the inverse Laplace transform of $r \mapsto e^{-tr^\alpha}$ [24, Ch. IX, p.260].

Definition 2. *Akin to (17) we set for* $0 < \alpha < 1, t > 0$ *the kernel*

$$K_t^{\alpha,appr}(\mathbf{p}) := \int_0^\infty q_{t,\alpha}(\tau) K_\tau^{1,appr}(\mathbf{p}) \, d\tau \geq 0, \qquad \mathbf{p} \in \mathbb{M}_2. \tag{18}$$

Their $\mathbb{L}_1$-norms satisfy $\|K_t^\alpha\| = \int_0^\infty \|K_\tau^1\| \, d\tau \approx \int_0^\infty q_{t,\alpha}(\tau) 1 d\tau = e^{-tr^\alpha}\big|_{r=0} = 1$.

Theorem 1. *The approximative Poisson kernel on* $\mathbb{M}_2$ *given by (18) equals:*

$$\boxed{K_t^{\frac{1}{2},appr}(\mathbf{p}) = \frac{1}{\pi^2} \frac{t \, J(\mathbf{p})}{(t^2 + \rho(\mathbf{p})^2)^2}} \tag{19}$$

for all $\mathbf{p} \in \mathbb{M}_2, t > 0$, *where* $\rho(\mathbf{p}) := \rho_\mathcal{G}(\mathbf{p})$. *For* $\mathbf{p}$ *in a compact set* $\mathcal{C} \subset \mathbb{M}_{opt}$:

$$\boxed{\begin{array}{l} \exists_{C \geq 1, D \geq 1} \forall_{t > 0} : \ C^{-1} K_{tD^{-1}}^{\alpha,appr}(\mathbf{p}) \leq K_t^\alpha(\mathbf{p}) \leq C \, K_{tD}^{\alpha,appr}(\mathbf{p}), \\ K_t^{\alpha,appr}(\mathbf{p}_0) = \frac{\Gamma\left(\frac{3}{2\alpha}\right)}{\alpha \Gamma\left(\frac{3}{2}\right) (4\pi t^{\frac{1}{\alpha}})^{\frac{3}{2}}}, \text{ for all } \alpha \in (0,1]. \end{array}} \tag{20}$$

For all $\alpha \in (0,1]$ *the approximation error can be estimated by*

$$\mathcal{E}^\alpha(\mathbf{p}, t) := |K_t^\alpha(\mathbf{p}) - K_t^{\alpha,appr}(\mathbf{p})| \leq C \frac{1}{4\pi(\rho(\mathbf{p}))^2} K_t^{\alpha,1D}(\rho(\mathbf{p})) \tag{21}$$

for $C > 0$ and t/ρ^2 small. Here $K_t^{\alpha,1D}$ is the α-stable Lévy process kernel on $\mathbb{R}$. For $0 < \alpha \leq \frac{1}{2}$ we have asymptotic expansion for $\frac{t^{\frac{1}{2\alpha}}}{\rho(\mathbf{p})} \ll 1$:

$$\mathcal{E}^\alpha(\mathbf{p}, t) \equiv \frac{-2\alpha}{4\pi^2(\rho(\mathbf{p}))^3} \sum_{m=0}^\infty \sin((m+1)\pi\alpha)\left(\frac{-t}{\rho(\mathbf{p})^{2\alpha}}\right)^{m+1} \frac{\Gamma((m+1)2\alpha)}{m!} \sim O(t). \tag{22}$$

For $\frac{1}{2} \leq \alpha \leq 1$ we also have $\mathcal{E}^\alpha(\mathbf{p}, t) \sim O(t)$,

Proof. First we derive (19). From Yosida's computations on (17), using contour integration in $\mathbb{C}$ for the inverse Laplace transform for $q_{t,\frac{1}{2}}$ one has on $\mathbb{R}^n$:

$$\int_0^\infty q_{t,\alpha=\frac{1}{2}}(\tau) \frac{e^{\frac{-\rho^2}{4\tau}}}{(4\pi t)^{\frac{n}{2}}} d\tau = \frac{\Gamma\left(\frac{n+1}{2}\right)}{\pi^{\frac{n+1}{2}}} \frac{t}{(t^2+r^2)^{\frac{n+1}{2}}}$$

with radius $r = \|\mathbf{x}\| > 0$, $\mathbf{x} \in \mathbb{R}^n$. Now by Definition 2 we must essentially re-do the same computations but by Definition 1 we must now substitute $r = \rho_{\mathcal{G}}(\mathbf{p})$ $n = \dim(\mathbb{M}_2) = 3$ and multiply by the Jacobian $J(\mathbf{p})$ above. This yields (19).

The top-formula in (20) follows by Lemma 1, $q_{t,\alpha} \geq 0$ and linearity in (17). The formula for $K_t^{\alpha,Appr}(\mathbf{p}_0)$ in (20) comes by inverse Fourier transform:

$$K_t^{\alpha,appr}(\mathbf{p}_0) = \left(\mathcal{F}_{\mathbb{R}^3}^{-1}(\omega \mapsto \frac{1}{(2\pi)^{3/2}} e^{-t\|\omega\|^{2\alpha}})\right)(0) = \frac{4\pi}{(2\pi)^3} \int_0^\infty e^{-\left(Rt^{\frac{1}{2\alpha}}\right)^{2\alpha}} R^2 dR$$

with $\|\omega\| = R$. The tangible relation (21) comes by (17) and (18) and Lemma 1:

$$|K_t^\alpha(\mathbf{p}) - K_t^{\alpha,Appr}(\mathbf{p})| = \int_0^\infty q_{t,\alpha}(\tau) |K_\tau^1(\mathbf{p}) - K_\tau^{1,Appr}(\mathbf{p})| \, d\tau$$

$$\overset{\text{Lemma 1}}{\leq} C \int_0^\infty q_{t,\alpha}(\tau) \frac{\tau}{(\rho(\mathbf{p}))^2} K_\tau^{1,appr}(\mathbf{p})) \, d\tau \text{ for } t/\rho^2 \text{ small}$$

$$= \frac{C}{4\pi(\rho(\mathbf{p}))^2} \int_0^\infty q_{t,\alpha}(\tau) \frac{(4\pi\tau)}{(4\pi\tau)^{\frac{3}{2}}} e^{-\frac{\rho^2(\mathbf{p})}{4\tau}} \, d\tau = C \frac{K_t^{\alpha,1D}(\rho(\mathbf{p}))}{4\pi(\rho(\mathbf{p}))^2}.$$

For (22) we rely on (21) and asymptotic formulas for the 1D α-stable Lévy process kernels in [14, Eq. 2.3]. The order term for the cases $1 \geq \alpha > \frac{1}{2}$ follows by similar asymptotic behavior [14, Eq. 2.3] for 1D α-kernels when $x = \rho/t^{\frac{1}{2\alpha}} \to \infty$ □

From [8, Ch. 5.4.1] it follows that one can set $(C, D) = (1, 2)$. If $\mathcal{G}_2$ is spatially isotropic ($\epsilon_2 = 1$), one has $(C, D) \approx (1, 1)$, for t small, recall Remark 1.

3.2   Analytic Approximations of α-Dilation/Erosion Kernels on $\mathbb{M}_2$

For functions on $\mathbb{M}_2$ that have the property that the Fourier transform $\mathcal{F}[\mathbf{v} \mapsto f((\exp_G \mathbf{v}) \odot \mathbf{p}_0)]$ is real-valued and non-negative for all $\mathbf{v} \in T_e G \equiv \mathbb{R}^3$, $G = SE(2)$, we define the approximate Cramér-Fourier transform as

$$\mathcal{C}^{appr} f(\mathbf{p}) := \mathcal{C}[\mathbf{v} \mapsto f((\exp_G \mathbf{v}) \odot \mathbf{p}_0)](\log_G g_{\mathbf{p}}), \tag{23}$$

where $\exp_G$ and $\log_G$, recall (16), denote the exponential and logarithm in $G = SE(2)$, where $\mathbf{p} = (\mathbf{x}, \theta) = g_\mathbf{p} \odot \mathbf{p}_0 \in \mathbb{M}_2$ with $g_\mathbf{p} = (\mathbf{x}, \mathbf{R}_\theta) \in SE(2)$. Here $\mathcal{C}$ is the Fourier-Cramér transform on $T_e(G) \equiv \mathbb{R}^3$ given in [19]. It concatenates Fourier transform $\mathcal{F}$, logarithm and Fenchel transform so that $\mathcal{C}(f * g) = \mathcal{C}f \,\square\, \mathcal{C}g$.

Definition 3. *Let $\rho_{\mathcal{G}_3}$ be the estimate (7) of the Riemannian distance (5) between $\mathbf{p}$ and $\mathbf{p}_0$ induced by metric tensor field $\mathcal{G}_3$ (4). Then we define*

$$k_t^{\alpha,appr}(\mathbf{p}) := c_\alpha \left(\frac{(\rho_{\mathcal{G}_3}(\mathbf{p}))^{2\alpha}}{t} \right)^{\frac{1}{2\alpha-1}}, \quad k_t^{1/2,appr}(\mathbf{p}) := \begin{cases} 0 & \text{if } \rho_{\mathcal{G}_3}(\mathbf{p}) \le t, \\ \infty & \text{if } \rho_{\mathcal{G}_3}(\mathbf{p}) > t, \end{cases} \quad (24)$$

with $c_\alpha = \frac{2\alpha-1}{(2\alpha)^{2\alpha/(2\alpha-1)}}$ for $\frac{1}{2} < \alpha \le 1$.

Proposition 1. *The approximate morphological convolution kernels relate to the linear convolution kernels as follows: $\mathcal{C}^{appr} K_t^{\alpha,appr} = k_t^{\alpha,appr}$.*

Proof. The right-hand side of (24) is the pullback of the morphological kernel on $\mathbb{R}^3$ via the isomorphism of Lie-algebra $T_e(G)$ and $\mathbb{R}^3$ given by $c^1 \, \partial_x|_e + c^2 \, \partial_y|_e + c^3 \, \partial_\theta|_e \leftrightarrow (c^1, c^2, c^3)$. Then by [19, Tab. 2] one has $\mathcal{C}^{appr} K_t^{\alpha,appr} = k_t^{\alpha,appr}$ $\square$

We have 3 main motivations for the kernel approximations in Definition 3:

1) For $\alpha = \frac{1}{2}$ it follows by Riemannian wavefront propagation on $\mathbb{M}_2$ [9] that

$$k_t^{1/2}(\mathbf{p}) = \begin{cases} 0 & \text{if } d_{\mathcal{G}_3}(\mathbf{p}, \mathbf{p}_0) \le t, \\ \infty & \text{if } d_{\mathcal{G}_3}(\mathbf{p}, \mathbf{p}_0) > t. \end{cases} \quad (25)$$

The logarithmic norm *locally* approximates the Riemannian distance, recall (12). Now compare (25) with (24), then for $t > 0$ small there exists $D > 1$ such that

$$k_{tD^{-1}}^{\alpha=1/2,appr} \le k_t^{\alpha=1/2} \le k_{tD}^{\alpha=1/2,appr}. \quad (26)$$

For $\alpha = \frac{1}{2}$ the morphological convolutions (10) are 'max- pooling' over balls in $\mathbb{M}_d$. This becomes a 'soft-max pooling' for $\alpha \in (\frac{1}{2}, 1]$, see [21, Figs. 7&8], where the approximation $k_t^{\alpha,appr}$ is still close enough in practice [21, Fig. 10].

2) $k_t^{\alpha,appr}$ is nearly equal to k_t^α if $\epsilon_3 = 1$, $\nu \to \infty$ or $\xi \to \infty$ (where the non-commutative group structure is irrelevant allowing us to rely on exact morphological kernels [19] on $\mathbb{R}^3$), if t is small such that fronts hardly reach $|\theta| = \pi$.

3) The function $\tilde{S}(\mathbf{c}, t) := k_t^{\alpha,appr}(\exp_G(\mathbf{c}))$ solves $\partial_t \tilde{S}(\mathbf{c}, t) = \|\nabla_\mathbf{c} \tilde{S}(\mathbf{c}, t)\|^{2\alpha}$, whereas $S(\cdot, t) = k_t^\alpha$ solves $\partial_t S(\mathbf{p}, t) = \|\nabla_{\mathcal{G}_3} S(\mathbf{p}, t)\|^{2\alpha}$ for $t > 0$ and $\mathbf{p} \in \mathbb{M}_2$. Extensions to $t \ge 0$ require limits of viscosity solutions [2] ('Lax-Oleinik' solutions [12]). By the formula of the logarithm (16) and the chain-law one can express $\nabla_{\mathcal{G}_3}$ into the gradient in logarithmic coordinates and after technical asymptotics one can show $\|\nabla_{\mathcal{G}_3} S\|^{2\alpha} = \|(\log_G)_* \nabla_\mathbf{c} S\|^{2\alpha} (1 + O(\alpha \nu^{-2}(1 - \epsilon_3)\xi^{-2} t^{\frac{1}{2\alpha}}))$. This reveals the nearly exact cases in 2) and supports our *conjecture* that (26) generalizes to $\alpha \in [\frac{1}{2}, 1]$, which may be shown via comparison principles [2].

Ad-hoc nonlinearities in CNNs are obsolete and excluded in our PDE-G-CNNs:

Theorem 2. *Let* $f : \mathbb{M}_d \to \mathbb{R}$ *be continuous and compactly supported.*
Let $\emptyset \neq S \subset \mathbb{M}_d$. *Let* $M = \max\limits_{\mathbf{y} \in \mathbb{M}_d} f(\mathbf{y})$. *Define* $k_S, k^M : \mathbb{M}_d \to \mathbb{R} \cup \{\infty\}$ *by:*

$$k_S(\mathbf{p}) := \begin{cases} 0 & \text{if } \mathbf{p} \in S, \\ \infty & \text{else,} \end{cases} \qquad k^M(\mathbf{p}) := \begin{cases} 0 & \text{if } \mathbf{p} = \mathbf{p}_0, \\ M & \text{else.} \end{cases} \tag{27}$$

Then max-pooling and ReLU's boil down to morphological convolutions:

$$\sup_{g \in G : g^{-1} \odot \mathbf{p} \in S} f(g \odot \mathbf{p}_0) = -(k_S \square - f)(\mathbf{p}), \quad \max\{0, f(\mathbf{p})\} = -(k^M \square - f)(\mathbf{p})$$

Proof. Follows by computation of (10); for details see [21, Prop. 5.16, 5.18]. □

4 Experimental Observations and Analysis

We test the viability of PDE-G-CNNs for automatic segmentation of vasculature, recall Fig. 1. To this end we consider a roto-translation equivariant CNN on $\mathbb{M}_2$ where convolution layers involve split-operator solvers (as depicted in the bottom part of Fig. 1) solving the following evolution equation:

$$\frac{\partial W}{\partial t} = -\mathbf{c} \cdot \nabla W + \|\nabla_{\mathcal{G}_3^+} W\|^{2\alpha} - \|\nabla_{\mathcal{G}_3^-} W\|^{2\alpha}, \tag{28}$$

wherein we set $\alpha = 0.65$ and train the convection vectors $\mathbf{c}$ and Riemannian metric tensor fields $\mathcal{G}_3^-$ and $\mathcal{G}_3^+$ (4). Since the PDE contains a convection, a dilation and an erosion term we refer to this layer as a *CDE* layer. The convection is implemented by linear interpolation, the dilation and erosion is implemented with morphological convolution (10) with kernels from Definition 3. A batch normalization and linear combination step complete each layer, see Fig. 1.

The PDE-based network operators, as described by (8),(9),(10) and (28), are implemented as an extension to the PyTorch deep learning framework. Training is performed by standard back-propagation using the ADAM stochastic gradient descent algorithm with a continuous DICE loss function. The same learning rate is used for all parameters and a small amount of L^2 regularization is added to the loss. The choice for $\alpha = 0.65$ has been made experimentally.

We consider three networks: a traditional CNN, a Group CNN and a PDE-G-CNN and apply these to the retinal vessel segmentation problem, specifically the publicly available DRIVE dataset [22], where we classify each pixel of the input image as either being a vessel or background. The accuracy of the model is measured by the average DICE coefficient of the test dataset.

The key properties of these networks, all having 6 layers, are compared in Table 1. Each model was trained 10 times, the distribution of DICE performance per realization is shown in Fig. 3. We observe better performance of our proposed PDE-G-CNNs over G-CNNs [3] even with 10 times less network parameters.

Table 1. Metrics for the 3 tested models.

Model	Spatial CNN	Group CNN	(CDE) PDE-G-CNN
Parameters	**47352**	**39258**	**4128**
Test time(s)	1.7	6.5	6.9
Avg. DICE	0.8058	0.8085	0.8115

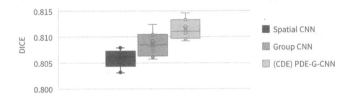

Fig. 3. Distribution of DICE coefficients of the test dataset for the three models.

5   Conclusion

We have presented a new PDE-framework for equivariant deep learning. Our PDE-G-CNNs only involve morphological and linear convolutions with scale space kernels. These kernels allow for close analytic approximations (Theorem 1, Definition 3) connected by the Cramér transform (23) in Proposition 1. Theorem 2 and α-stability on $\mathbb{M}_d$ [7] allow us to cut all ad-hoc nonlinearities from CNNs. Practical vessel segmentation experiments show that PDE-G-CNNs outperform state-of-the-art G-CNNs [3] even while reducing the network-complexity by a factor of 10 when using a depth of 6 layers. Ongoing tests by Smets even indicate that the reduction factor increases by a factor of 40 for 12-layer depths, but a full analysis of benefits of PDE-G-CNNs over (G-)CNNs is left for future work.

References

1. Akian, M., Quadrat, J., Viot, M.: Bellman processes. LNCIS **199**, 302–311 (1994)
2. Bardi, M., Capuzzo-Dolcetta, I.: Discontinuous viscosity solutions and applications. In: Optimal Control and Viscosity Solutions of Hamilton-Jacobi-Bellman Equations. Systems and Control: Foundations and Applications. Birkhäuser, Boston, MA (1997) https://doi.org/10.1007/978-0-8176-4755-1_5
3. Bekkers, E.J., Lafarge, M.W., Veta, M., Eppenhof, K.A.J., Pluim, J.P.W., Duits, R.: Roto-translation covariant convolutional networks for medical image analysis. In: Frangi, A.F., Schnabel, J.A., Davatzikos, C., Alberola-López, C., Fichtinger, G. (eds.) MICCAI 2018. LNCS, vol. 11070, pp. 440–448. Springer, Cham (2018). https://doi.org/10.1007/978-3-030-00928-1_50
4. Ben Arous, G.: Development asymptotique du noyau de la chaleur hypoelliptique sur la diagonale. In: Annales de l'institut Fourier, pp. 73–99 (1989)
5. Burgeth, B., Weickert, J.: An explanation for the logarithmic connection between linear and morphological systems. In: Proceedings 4th SSVM pp. 325–339 (2003)

6. Cohen, T., Welling, M.: Group equivariant convolutional networks. In: Proceedings of the 33rd International Conference on Machine Learning, pp. 2990–2999 (2016)
7. Duits, R., Bekkers, E.J., Mashtakov, A.: Fourier transform on $\mathbb{M}_3$ for exact solutions to linear PDEs. Entropy (SI: 250 year of Fourier) **21**(1), 1–38 (2019)
8. Duits, R., Franken, E.M.: Left invariant parabolic evolution equations on $SE(2)$ and contour enhancement via orientation scores. QAM-AMS **68**, 255–331 (2010)
9. Duits, R., Meesters, S., Mirebeau, J.M., Portegies, J.M.: Optimal paths of the reeds-shepp car with applications in image analysis. JMIV **60**(6), 816–848 (2018)
10. Elad, M.: Deep, Deep Trouble. SIAM-NEWS p. 12 (2017)
11. ter Elst, A.F.M., Robinson, D.W.: Weighted subcoercive operators on Lie groups. J. Funct. Anal. **157**, 88–163 (1998)
12. Evans, L.C.: Partial differential equations. AMS, Providence, R.I. (2010)
13. Finzi, M., Bondesan, R., Welling, M.: Probabilistic numeric convolutional neural networks. https://arxiv.org/pdf/2010.10876.pdf pp. 1–22 (2020)
14. Garoni, T., Frankel, N.: Lévy flights: exact results and asymptotics beyond all orders. J. Math. Phys. **43**(5), 2670–2689 (2002)
15. Grigorian, A.: Heat Kernel and Analysis on Manifolds. Math. Dep, Bielefeld (2009)
16. LeCun, Y., Bengio, Y., Hinton, G.: Deep learning. Nat. Res. **521**(7553), 436–444 (2015)
17. Litjens, G., Bejnodri, B., van Ginneken, B., Sánchez, C.: A survey on deep learning in medical image analysis. Med. Image Anal. **42**, 60–88 (2017)
18. Pauwels, E., van Gool, L., Fiddelaers, P., Moons, T.: An extended class of scale-invariant and recursive scale space filters. IEEE Trans. Pattern Anal Mach. Intell. **17**(7), 691–701 (1995)
19. Schmidt, M., Weickert, J.: Morphological counterparts of linear shift-invariant scale-spaces. J. Math. Imag. Vision **56**(2), 352–366 (2016)
20. Siffre, L.: Rigid-motion scattering for image classification. Ph.D. thesis, Ecole Polyechnique, Paris (2014)
21. Smets, B., Portegies, J., Bekkers, E., Duits, R.: PDE-based group equivariant convolutional neural networks. arXiv preprint arXiv:2001.09046 (2020)
22. Staal, J., Abramoff, M., Niemeijer, M., Viergever, M., van Ginneken, B.: Ridge-based vessel segmentation in images of the retina. IEEE TMI **23**(4), 501–509 (2004)
23. Weiler, M., Geiger, M., Welling, M., Boomsma, W., Cohen, T.: 3D steerable CNNs: learning equivariant features in volumetric data. In: NeurIPS, pp. 1–12 (2018)
24. Yosida, K.: Functional Analysis. CM, vol. 123. Springer, Heidelberg (1995). https://doi.org/10.1007/978-3-642-61859-8

Nonlinear Spectral Processing of Shapes via Zero-Homogeneous Flows

Jonathan Brokman[✉] and Guy Gilboa[✉]

Technion - Israel Institute of Technology, 3200003 Haifa, Israel
sjonatha@campus.technion.ac.il, guy.gilboa@ee.technion.ac.il

Abstract. In this work we extend the spectral total-variation framework, and use it to analyze and process 2D manifolds embedded in 3D. Analysis is performed in the embedding space - thus "spectral arithmetics" manipulate the shape directly. This makes our approach highly versatile and accurate for feature control. We propose three such methods, based on non-Euclidean zero-homogeneous p-Laplace operators. Each method satisfies distinct characteristics, demonstrated through smoothing, enhancing and exaggerating filters.

Keywords: 1-Laplacian · 3-Laplacian · Spectral TV · cMCF · Mesh processing · Nonlinear spectral processing

1 Introduction

In 1995 Taubin [22] proposed to utilize the shape-induced Laplacian eigenfunctions as a basis for shape filtering, in an analogue manner to classical signal processing techniques. A transform is computed by projecting the shape onto the basis, where filtering is obtained by weighted reconstruction via this basis. Many variations of this method were utilized for different tasks (e.g. [21]). Over time, the Laplace-Beltrami became the standard Laplacian of choice, for spectral applications, and in general [24][1].

Nonlinear spectral processing has been developed in recent years for image analysis and manipulation. Spectral total-variation (TV) was introduced in [13,14] facilitating nonlinear edge-preserving filtering. Essentially, the idea is to decompose a signal into nonlinear spectral elements related to eigenfunctions of the total-variation subgradient. The method is based on evolving gradient descent with respect to the TV functional (TV-flow [1]). The spectral elements decay linearly in this flow. Different decay rates correspond to different scales

---

[1] Such an adaptation of [22] can be found e.g. in [23], which proposed a computationally efficient shape filtering, and demonstrated some core filtering capabilities: Shape exaggeration, detail enhancement, shape smoothing and regularization.

We acknowledge support by grant agreement No. 777826 (NoMADS), by the Israel Science Foundation (Grant No. 534/19) and by the Ollendorff Minerva Center.

A. Elmoataz et al. (Eds.): SSVM 2021, LNCS 12679, pp. 40–51, 2021.
https://doi.org/10.1007/978-3-030-75549-2_4

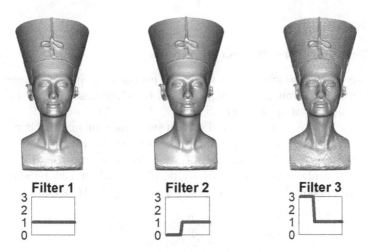

Fig. 1. Bust of Queen Nefertity reconstructed and filtered via Directional Shape TV (M3)

(in the case of a single eigenfunction the rate is exactly the eigenvalue). Theoretical underpinning was performed for the spatial discrete case [7] and continuous case [5]. For the one-dimensional TV setting, it was shown that the spectral elements are orthogonal to each other. Various applications were suggested for image enhancement, manipulation and fusion [2,15]. Only recently, in 2020, a spectral TV framework for shape processing was proposed, for the first time, by Fumero et al. [12]. They advocate applying spectral TV to the normals of the shape, thus gradients are calculated on the normals' domain (unit sphere), and spectral processing of the embedded shape is done implicitly.

Burger et al. [7] generalized the concepts of spectral TV to decompositions based on general convex absolutely one-homogeneous functionals. Decompositions based on minimizations with the Euclidean norm, as well as with inverse-scale-space flows [6] were also proposed. The space-continuous setting was recently analyzed in [5]. A common thread related to gradient flows of one-homogeneous functionals is that they are based on zero-homogeneous operators.

In this work we propose three new methods, extending nonlinear spectral processing of shapes using zero-homogeneous flows. We modify existing Laplacian-based flows of shapes such that they will comply with the prerequisites of nonlinear spectral processing. We demonstrate how each flow induces distinct spectral properties. On one hand, our work extends ideas from [12] to other zero-homogeneous flows. On the other hand - we examine a complementary approach, as our proposed spectral processing methods are performed explicitly on the shape in its embedding space.

2   Background

2.1   Differential Operators on Manifolds

Continuous Formulation. In this work, the processed shape is assumed to be a smooth 2D manifold embedded in 3D, denoted by M, represented as $S(\omega_1, \omega_2) = (x(\omega_1, \omega_2), y(\omega_1, \omega_2), z(\omega_1, \omega_2))$, i.e. $S : \Omega \subset \mathbb{R}^2 \to M$. Let $f : S \to \mathbb{R}$, and $\tilde{f} = f \circ S(\omega_1, \omega_2)$ i.e. $\tilde{f} : \Omega \to \mathbb{R}$. For instance, we can map each point $q = S(\omega_1, \omega_2) \in M$ to $x(\omega_1, \omega_2)$. This function is termed the *x-coordinate function*. If f is the x-coordinate function, then $\tilde{f}$ maps each point ω_1, ω_2 to $x(\omega_1, \omega_2)$. Let $T_q M$ be the plane tangent to M at point $q \in M$, it can be shown that $\frac{\partial S}{\partial \omega_1}, \frac{\partial S}{\partial \omega_2} \in \mathbb{R}^3$, denoted $S_{\omega_1}, S_{\omega_2}$, at point q span $T_q M$. M is equipped with a metric g,

$$g(\omega_1, \omega_2) = \begin{pmatrix} S_{\omega_1}^T S_{\omega_1} & S_{\omega_1}^T S_{\omega_2} \\ S_{\omega_2}^T S_{\omega_1} & S_{\omega_2}^T S_{\omega_2} \end{pmatrix}, \tag{1}$$

which, given that $S_{\omega_1}, S_{\omega_1}$ are linearly independent, can be shown to be positive semi-definite and invertible. g induces the inner product $\langle a, b \rangle_g = a^T g b = A^T B$, where $A, B \in T_q M$ are the mapping of $a, b \in \Omega$, considering mapped vectors to be velocities of mapped routes. The squared-root determinant satisfies the "area elements" $\sqrt{|g|} dudv$. The gradient operator $\nabla_g f(q) = g^{-1} \nabla_{\omega_1, \omega_2} \tilde{f}(\omega_1, \omega_2)$ satisfies $\langle \nabla_g f(q), w \rangle = \lim_{h \to 0} \frac{f(q+hw)-f(q)}{h}$, and the divergence $\nabla_g \cdot F = \frac{1}{\sqrt{|g|}} \nabla_{\omega_1, \omega_2} \cdot (\sqrt{|g|} \tilde{F})$ is obtained as an adjoint of $\nabla_g(f)$. Full details are available at [10]. Finally, the $\mathcal{P}$-Laplace-Beltrami is defined by,

$$\Delta_{g,\mathcal{P}} f := \nabla_g \cdot (|\nabla_g f|^{\mathcal{P}-2} \nabla_g f). \tag{2}$$

For $\mathcal{P} = 2$ the Laplace-Beltrami is obtained. Other special cases we will discuss are $\mathcal{P} = 3$, and $\mathcal{P} = 1$.

Discretization. We use the discretization framework of [16]. M is approximated as a triangulated mesh and f as a function on vertices. Discrete matrix operators are derived, including $[\sqrt{|g|} \cdot]$ as mass matrix denoted $[M]$, $[\nabla_g]$ denoted $[G]$ and $[\sqrt{|g|} \cdot \nabla_g \cdot]$ denoted $[D]$. Thus $\nabla_g \cdot$ is descretized as $[M^{-1}][D]$. We remark that both $[G]$ and $[M^{-1}][D]$ satisfy linearity, are Hermitian conjugate of one another, and $[M^{-1}][D][G]$ is a common discretization of the Laplace-Beltrami operator. Finally, a semi-discrete diffusion process of f on M can be formulated as,

$$\frac{\partial u(t)}{\partial t} = [M^{-1}][D][G] \cdot u(t), \qquad u(0) = f. \tag{3}$$

2.2   Vectorial Total Variation

Total variation (TV) has been used extensively for the past three decades in image processing and computer vision (see [8,9] for theory and applications). For smooth functions, the TV functional over the domain $\Omega \subset \mathbb{R}^n$ is $TV(u) = \int_\Omega |\nabla u(x)| dx$, where $x = \{x_1, \cdot, x_n\}$. When dealing with a vectorial function (of several channels) it is desired to take into account the inter-component correlations. A common definition for vectorial TV (see e.g. [3]) is: $VTV(u) = \int_\Omega \sqrt{\sum_c |\nabla u_c(x)|^2}\, dx$, where u_c is channel $c = 1, \cdot, C$ in $u = \{u_1, \cdot, u_C\}$. The gradient descent flow evolves each channel by the VTV flow: $\frac{\partial u_c}{\partial t} = \nabla \cdot (\frac{\nabla u_c}{\sqrt{\sum_{\tilde{c}} |\nabla u_{\tilde{c}}|^2}})$. Note, as in the scalar TV-flow, this flow is zero-homogeneous.

2.3   Laplacian-Based Flows

Mesh smoothing is often achieved by some form of a discrete diffusion process. A plethora of such algorithms were proposed, here we present the three most relevant to this paper.

Heat Flow. The simplest of the three, often termed *Heat Flow*, smooths each coordinate function via Eq. (3), where $[M], [D], [G]$ are induced by the initial shape's metric, denoted g_0 and are fixed throughout the process. The smoothed shape is given by the final three smoothed coordinate functions.

Mean Curvature Flow. The most notable shape smoothing process, *Mean Curvature Flow* (MCF) is derived as an area minimizing process. A common implementation of it resembles the heat flow method, however, at each time step the operator matrices $[M], [D], [G]$ are re-calculated according to the present metric of the smoothed shape, denoted g_t [19].

Conformalized Mean Curvature Flow. In [17] cMCF was introduced as a modified version of MCF. The metric is updated at each time step, but unlike MCF, the metric is conformalized to the initial shape's metric. This conformal-ized metric, $\tilde{g}_t = \sqrt{|g_0^{-1} g_t|} g_0$, induces the "conformalized Laplace-Beltrami". Proposed implementation is similar to the former whereas $[M]$ is updated w.r.t g_t while $[D], [G]$ are fixed w.r.t g_0. cMCF is significantly more immune to singularities than MCF, making it more suitable for editing surface extremities such as head and limbs of humanoid models. Adding rescaling for numerical stability, the shape converges to a sphere (not a point). In [17] they demonstrate that the resulting smoothed shape admits a conformal mapping of the initial shape.

2.4   TV Mesh Processing

A suitable TV flow for M is a nontrivial task, as its immediate representation $S(u, v) = x(u, v), y(u, v), z(u, v)$ is a vectorial function of correlated components. We briefly summarize two methods relevant to our work, related to graph smoothing and to spectral processing of normals.

$\mathcal{P}$-Laplace Flows on Weighted Graphs. Elmoataz et al. [11] proposed a gen-eralization of TV regularization of functions on weighted graphs. Their discrete gradient and divergence operators, induced by the graph topology, provide a non-Euclidean $\mathcal{P}$-Laplace. Treating mesh triangle sides as graph edges, $\mathcal{P}$-Laplace mesh smoothing was proposed. Shape's inter correlations were accounted via combined gradient magnitude as in VTV.

Normal TV: Flow and Shape Spectral Analysis. Recently, in [12] a spectral TV framework for shape analysis was suggested. The authors proposed spectral TV analysis of the shape's normals, followed by shape-from-normals reconstruction. This achieves rotation invariance while operating on a fixed metric (the unit-sphere). Another interesting property of this implicit approach is the convergence to a plain, which is a translation of normals converging to a point. They show the method is compatible with a range of applications.

3   Proposed Methods

We propose a general framework for nonlinear spectral filtering of shapes (meshes). First, we present our zero-homogeneous framework. Then we suggest three methods which utilize this framework to filter shapes. Each method is inspired by a different flow: Heat Flow, cMCF and MCF. The shape is represented in its embedding space, allowing excellent feature control as spectral manipulations directly manipulate embedded structures.

Spectral Processing via (Any) Zero-Homogeneous Flow. Let X be a space of functions on a general Euclidean space. Let $p : X \to X$ be a zero-homogeneous operator, i.e.,

$$p(\alpha f) = \text{sign}(\alpha)p(f), \quad \alpha \in \mathbb{R}, f \in X. \tag{4}$$

We examine the following zero-homogeneous flow,

$$u_t = -p(u(t)), \quad u(0) = f_0 \in X, \ t \geq 0, \tag{5}$$

where $u_t = \frac{\partial u}{\partial t}$. We assume the flow exists and that the solution is unique. We also assume the second time-derivative of u exists (in the distributional sense) a.e. and define $\phi : t \to X$ by,

$$\phi(t) = t \cdot u_{tt}. \tag{6}$$

Let f_0 be an eigenfunction with respect to p, with a positive eigenvalue, i.e. $\exists \lambda \in (0, \Lambda < \infty) : p(f_0) = \lambda \cdot f_0$, then

$$u(t) = (1 - \lambda t)^+ f_0, \tag{7}$$

satisfies Eq. (5), where $(q)^+ = \max(0, q)$. This can be verified by taking the time derivative on both sides and using the zero-homogeneity of p. We note that since we are examining smoothing processes, p in general is a positive semidefinite operator, $\langle f, p(f) \rangle \geq 0, \forall f \in X$. Thus the eigenvalues are positive. In the case of negative eigenvalues, the flow diverges (but for a finite stopping time can still have a solution). Thus, for eigenfunctions of positive eigenvalues we get $\phi(t) = \delta(t - \frac{1}{\lambda})f_0$, i.e. ϕ's energy is concentrated in a single scale ("frequency") which corresponds to the eigenvalue of f_0, $\lambda = \frac{1}{t}$. For a general $f_0 \in X$, this motivates the interpretation of ϕ as a spectral transform of f_0, where the spectral components are positive eigenfunctions of p, in a similar manner to [5,7,13].

We can compute the reconstruction formula, for a general stopping time T, using integration by parts (and assuming $u_t(0)$ is bounded), by

$$\int_0^T \phi(t)\, dt = tu_t|_0^T - \int_0^T u_t\, dt = Tu_t(T) - u(T) + f_0 = -Tp(u(T)) - u(T) + f_0.$$

Denoting the residual $R = Tp(u((T)) + u(T)$, the following reconstruction identity holds $f_0 = \int_0^T \phi(t)\, dt + R$. Similar to gradient flows of one-homogeneous functionals, which are a special case of zero-homogeneous flows, this spectral framework resembles in some sense classical Fourier analysis, e.g. - we can filter f_0 using a transfer function (window) $H : \mathbb{R}^+ \to \mathbb{R}$ as follows:

$$f_0^{filtered} = \int_0^T H(t)\phi(t)\, dt + R, \tag{8}$$

where for $H(t) \equiv 1$ ("All-pass filter") we obtain the reconstruction formula. This is the core capability we use for shape processing in all three methods proposed below. Note that contrary to previous studies, the above properties rely on very mild assumptions regarding the flow. We do not assume the flow minimizes a one-homogeneous functional (and for a finite stopping time, even strict convergence is not necessary, as all diverging components are in R and the reconstruction formula is valid). Our findings are straightforward extensions of observations done in [7], where further discussion takes place, including other key properties, such as orthogonality of the spectral components and decomposition into eigenfunctions by the flow in certain settings.

Modifying Flows for Nonlinear Spectral Processing. Our framework requires a zero-homogeneous flow evolving on a fixed metric (to induce spectral linear decay). Examining Heat Flow, MCF and cMCF we find that none of these flows is zero-homogeneous, and Heat Flow is the only one performed on a fixed metric. Hence, adaptations of these flows are required.

Technical Briefing. A 2D manifold M embedded in 3D is given by its 3 coordinate functions $S_0(\omega_1, \omega_2) = x_0(\omega_1, \omega_2), y_0(\omega_1, \omega_2), z_0(\omega_1, \omega_2)$ where $\omega_1, \omega_2 \in \Omega \subset \mathbb{R}^2$, inducing the intrinsic metric g_0. M is discretized as a triangular mesh, and S_0 as the vertex coordinates. Gradient and divergence operators $\nabla_{g_0}, \nabla_{g_0} \cdot$ on M are discretized as in [16]. In all modifications of Eq. (2), if $\mathcal{P} < 2$ we replace the magnitude by $\sqrt{|\nabla_g f|^2 + \epsilon^2}$ for stability. Our flows are implemented using semi-implicit time steps, and differ by their operator. The evolving shape at time t is denoted $S(t)$, and $c(t)$ denotes any of $S(t)$'s evolving coordinate functions $x(t), y(t), z(t)$.

3.1   Naive Method: Unpaired Coordinate Spectral TV

The naive approach utilizes a modification of Heat Flow for our framework. Heat Flow processes each coordinate function independently via Eq. (3), utilizing the Laplace-Beltrami on the fixed metric g_0 throughout the flow. Thus it satisfies a fixed metric, but it is not zero-homogeneous, and a modification is required. By replacing the Laplace-Beltrami with the 1-Laplace-Beltrami of Eq. (2) zero-homogeneity is achieved, which results in the operator $-p_{Naive}(c) := \Delta_{g_0,1}c$, and a per-coordinate flow is defined by setting $p(u(t)) = -p_{Naive}(c(t))$ in Eq. (5). Each channel evolves separately, hence the name "unpaired coordinates". We can now perform nonlinear spectral filtering as in Eq. (8), demonstrated in Fig. 2. While processing each of $c = x, y, z$ independently is sub-optimal, this approach has good feature control, as spectral decomposition is applied directly to the embedded shape.

3.2   Method 1 (M1): Shape Spectral TV

Here we take into account shape coordinate inter-correlations, i.e. we go from coordinate to shape processing. We apply a vectorial flow (as in VTV) on meshes, previously suggested by [11], which results in the operator,

$$ - p_{M1}(c) := \nabla_{g_0} \cdot \left(\frac{\nabla_{g_0} c}{\sqrt{\sum_{\tilde{c}=x,y,z} |\nabla_{g_0} \tilde{c}|^2}} \right). \tag{9} $$

Note that the metric is fixed as g_0. We can also verify that the operator is zero-homogeneous. The flow is followed by per-coordinate spectral processing as in Eq. (6), (8) - thus x, y, z inter-correlation is preserved. We obtain good explicit feature control (see Fig. 2).

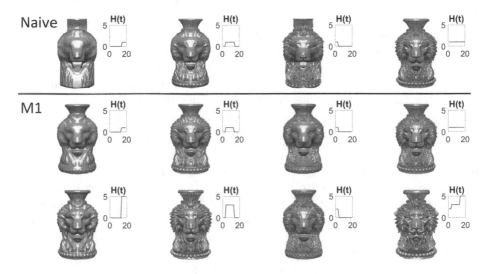

Fig. 2. [Please zoom in] Filtering is applied to the embedded shape, thus magnification and summation of filter-bands magnifies and sums features directly. As t grows, choosing $H(t)$ greater or less than 1, results in amplification or attenuation of finer details. This can be observed in the model's contour and hair-strands. M1 does not cause the shape to be axis-aligned (squaring effect), as in the naive approach.

3.3   Method 2 (M2): Conformalized $\mathcal{P}$-Laplace

Here we modify cMCF [17] to our framework, using our conformalized $\mathcal{P}$-Laplace described below. Our flow inherits cMCF's limb-head smoothing capabilities (Fig. 3), which we then use for shape filtering. The metric of cMCF, $\tilde{g}_t = \sqrt{|g_0^{-1}g_t|}g_0$, is not fixed, and the operator driving the flow, the conformalized Laplace-Beltrami, $\sqrt{\frac{|g_0|}{|g_t|}}\nabla_{g_0} \cdot \nabla_{g_0}$, depends on the evolving shape's metric g_t. To achieve a fixed metric, we re-interpret $|g_t|$ as an operator on a fixed metric g_0. This is valid since the diffused shape defines both the diffused function as well as the evolving metric. This affects homogeneity, as shown below. We define the conformalized $\mathcal{P}$-Laplace as,

$$\tilde{\Delta}_{g,\mathcal{P}}(c) := \sqrt{\frac{|g_0|}{|g_t|}}\nabla_{g_0} \cdot (|\nabla_{g_0}c|^{\mathcal{P}-2}\nabla_{g_0}c). \tag{10}$$

By Eq. (1) we have that $|g_t|$ is absolutely 4-homogeneous, hence $\tilde{\Delta}_{g,\mathcal{P}}$ is $\mathcal{P} - 3$ homogeneous,

$$\tilde{\Delta}_{g,\mathcal{P}}(\alpha c) = \sqrt{\frac{|g_0|}{|\alpha|^4|g_t|}}\nabla_{g_0} \cdot (|\alpha|^{\mathcal{P}-2}|\nabla_{g_0}c|^{\mathcal{P}-2})\alpha\nabla_{g_0}c = \frac{\alpha}{|\alpha|^{4-\mathcal{P}}}\tilde{\Delta}_{g,\mathcal{P}}(c).$$

Thus we choose $\tilde{\Delta}_{g,3}$ as a zero-homogeneous modification of the conformalized Laplace. Once again inter-correlations are accounted for, as in [11], yielding the operator,

$$-p_{M2}(c) := \sqrt{\frac{|g_0|}{|g_t|}}\nabla_{g_0} \cdot \left(\sqrt{\sum_{\tilde{c}=x,y,z}|\nabla_{g_0}\tilde{c}|^2}\nabla_{g_0}c\right) \tag{11}$$

The flow is followed by nonlinear spectral filtering, Eq. (8). Editing extremities, a capability inherited from our conformal 3-Laplace flow, is demonstrated in Fig. 4, where extremities are in the form of human limbs and head.

Fig. 3. M2 conformalized $\mathcal{P}$-Laplace flows. $\mathcal{P} = 2$ is an unscaled version of cMCF. $\mathcal{P} = 1$ is a new conformalized shape TV flow. For $\mathcal{P} = 3$ the flow is zero-homogeneous.

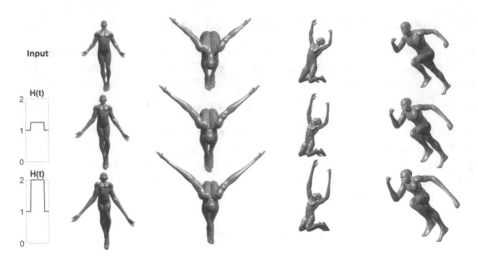

Fig. 4. Shape exaggeration by M2 spectral filtering. The filter inherits its properties from the conformalized 3-Laplace flow (Fig. 3), which translates to interesting limb-head editing. Isometry robustness is demonstrated as well.

3.4   Method 3 (M3): Directional Shape TV

Mesh TV smoothing typically preserve pointy surface points, e.g. tip of chin [12] or ears [11]. Here we propose a method that preserves edges, e.g. muscle contour, similarly to TV processing of images. While M1 and M2 utilized modifications of Heat Flow and cMCF, M3 draws inspiration from MCF.

MCF already has a thoroughly researched fixed-metric zero-homogeneous modification: The TV flow as applied to gray-scale images [18]. For a surface represented as $S = (x, y, f(x, y))$, this modification entails constraining the evolved shape to be of the form $S(t) = (x, y, f(x, y, t))$. This is enforced by constraining each point on the surface to evolve in direction $\hat{z}$ (perpendicular to the x, y plane). We note that unconstrained MCF would necessarily violate this form of $S(t)$, as it theoretically converges to a singular point.

Our third method aims to generalize the above direction-constraint to general shapes, hence the name "directional". The x, y domain is generalized to be an over-smoothed version of the initial shape which we denote $\hat{S}$. Each $p \in S$ is mapped to a $\hat{p} \in \hat{S}$. The direction of evolution is fixed as $\hat{d} = \alpha \frac{S - \hat{S}}{|S - \hat{S}|}$, where α is a sign indicator which makes sure $\hat{d}$ points "outwards". Finally, the evolving initial surface is represented as $f_0 = \alpha |S - \hat{S}|$. Note that $S = \hat{S} + f_0 \hat{d}$. This method is a generalization in the following sense: Consider the form $S = (x, y, f(x, y))$, choosing $\hat{S} = (x, y, 0)$, we have that $\hat{d} = \hat{z}$, and $\alpha |S - \hat{S}| = (0, 0, f(x, y))$.

We advocate the choice of $\hat{S}$ as a cMCF smoothed version of S, because cMCF was shown to provide a conformal mapping from S to $\hat{S}$. By construction - the metric is fixed and inter-correlations are accounted for. The proposed zero-homogeneous operator (acting on a scalar-valued function u) is,

$$- p_{M3}(u) := \nabla_{g_0} \cdot \frac{\nabla_{g_0} u}{|\nabla_{g_0} u|}, \tag{12}$$

where $u(t)$ is the evolution of f_0 at time t, which results in $\frac{\partial S}{\partial t} = \nabla_{g_0} \cdot \frac{\nabla_{g_0} u}{|\nabla_{g_0} u|} \hat{d}$, satisfying the imposed directionality. Finally, f_0 is filtered as in Eq. (8), and a filtered shape is obtained by $\hat{S} + f^{filtered}\hat{d}$. Being closely related to spectral TV on images, this method preserves detail well, as demonstrated in Figs. 1, 5, 6. Though inspired by MCF, the proposed flow is substantially different.

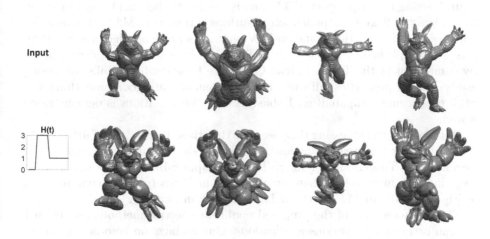

Fig. 5. Shape exaggeration by M3. This filter applies edge preserving smoothing while admitting the requirements for the caricaturiazation task, as posed by [20].

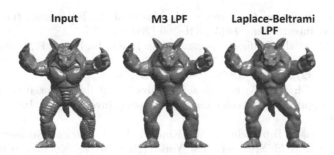

Fig. 6. M3 filtering has distinctive detail-preserving capabilities. All our methods can utilize rough time discretization for runtime efficiency at the expense of reconstruction error - here M3 takes approximately 30 s. The Laplace-Beltrami filtering requires SVD for 2000 eigenvectors of the $[17 \cdot 10^3 \times 17 \cdot 10^3]$ sparse operator matrix, which took us approximately 5 h to obtain (2.5 orders of magnitude slower).

4   Discussion and Conclusion

We propose a general methodology for shape processing based on nonlinear spectral filtering. In this framework we use smoothing flows which satisfy two requirements: zero-homogeneity, and a fixed metric. We process the shape in its embedding space, providing unmediated nonlinear spectral representations and accurate and meaningful filtering capabilities.

To showcase the general concept, three methods are proposed, where spectral processing is done via the same mechanism, Eqs. (6) and (8). M1 is arguably the natural setting for shape spectral TV, and is the most robust and easy to handle. M2 enables the flexible amplification of surface extremities. M3 best entails the well known characteristics of spectral TV on images (such as edge preservation). This is accomplished by restricting the direction of the evolution, generalizing a key component of the TV and Mean Curvature Flow duality. While possessing visibly distinct properties, all three methods demonstrate good smoothing and detail enhancement capabilities. Robustness to pose variations is demonstrated as well.

With respect to processing time, we note that these methods are fairly fast. In order to filter by linear eigenfunctions, one first needs to compute the basis for the specific shape. This entails solving SVD of the Laplace-Beltrami operator, which is significantly more costly than our proposed methods (e.g. the computations for Fig. 6 are 30 s for M3 vs. 5 h for Laplace-Beltrami filtering).

The characteristics of the proposed nonlinear spectral methods are distinct and can carry-over to various applications. Our findings on zero-homogeneous processing are not restricted to shape analysis, and can be implemented even in new neural-network architectures that comply with our requirements. [2]

References

1. Andreu, F., Ballester, C., Caselles, V., Mazón, J.M.: Minimizing total variation flow. Differ. Integral Equ. **14**(3), 321–360 (2001)
2. Benning, M., et al.: Nonlinear spectral image fusion. In: Lauze, F., Dong, Y., Dahl, A.B. (eds.) SSVM 2017. LNCS, vol. 10302, pp. 41–53. Springer, Cham (2017). https://doi.org/10.1007/978-3-319-58771-4_4
3. Bresson, X., Chan, T.F.: Fast dual minimization of the vectorial total variation norm and applications to color image processing. Inverse Probl. Imag. **2**(4), 455–484 (2008)
4. Bronstein, A.M., Bronstein, M.M., Kimmel, R.: Numerical Geometry of Non-Rigid Shapes. MCS. Springer, New York (2009). https://doi.org/10.1007/978-0-387-73301-2
5. Bungert, L., Burger, M., Chambolle, A., Novaga, M.: Nonlinear spectral decompositions by gradient flows of one-homogeneous functionals. Analysis & PDE (2019)

---

[2] Models: Bust of Queen Nefertiti. gyptisches Museum und Papyrussammlung. Model: Trigon art; Stanford armadillo and poses by Belyaev, Yoshizawa, Seidel (2006); Michaels from [4]; various models from LIRIS database.

6. Burger, M., Gilboa, G., Osher, S., Xu, J.: Nonlinear inverse scale space methods. Commun. Math. Sci. **4**(1), 179–212 (2006)
7. Burger, M., Gilboa, G., Moeller, M., Eckardt, L., Cremers, D.: Spectral decompositions using one-homogeneous functionals. SIAM J. Imag. Sci. **9**(3), 1374–1408 (2016)
8. Burger, M., Osher, S.: A guide to the TV Zoo. In: Level Set and PDE Based Reconstruction Methods in Imaging, pp. 1–70. Springer, Cham (2013). https://doi.org/10.1007/978-3-319-01712-9_1
9. Chambolle, A., Caselles, V., Cremers, D., Novaga, M., Pock, T.: An introduction to total variation for image analysis. In: Theoretical Foundations and Numerical Methods for Sparse Recovery, vol. 9, no. 263–340, p. 227 (2010)
10. Do Carmo, M.P.: Differential Geometry of Curves and Surfaces: Revised and Updated Second Edition. Courier Dover Publications, Mineola (2016)
11. Elmoataz, A., Lezoray, O., Bougleux, S.: Nonlocal discrete regularization on weighted graphs: a framework for image and manifold processing. IEEE Trans. Image Process. **17**(7), 1047–1060 (2008)
12. Fumero, M., Möller, M., Rodolà, E.: Nonlinear spectral geometry processing via the TV transform. ACM Trans. Graph. (TOG) **39**(6), 1–16 (2020)
13. Gilboa, G.: A total variation spectral framework for scale and texture analysis. SIAM J. Imag. Sci. **7**(4), 1937–1961 (2014)
14. Gilboa, G.: A spectral approach to total variation. In: Kuijper, A., Bredies, K., Pock, T., Bischof, H. (eds.) SSVM 2013. LNCS, vol. 7893, pp. 36–47. Springer, Heidelberg (2013). https://doi.org/10.1007/978-3-642-38267-3_4
15. Hait, E., Gilboa, G.: Spectral total-variation local scale signatures for image manipulation and fusion. IEEE Trans. Image Process. **28**(2), 880–895 (2019)
16. Jacobson, A., Panozzo, D.: Libigl: prototyping geometry processing research in C++ (2017). https://libigl.github.io/tutorial/#chapter-2-discrete-geometric-quantities-and-operators
17. Kazhdan, M., Solomon, J., Ben-Chen, M.: Can mean-curvature flow be modified to be non-singular? In: Computer Graphics Forum, vol. 31, pp. 1745–1754. Wiley, Hoboken (2012)
18. Kimmel, R., Malladi, R., Sochen, N.: Images as embedded maps and minimal surfaces: movies, color, texture, and volumetric medical images. Int. J. Comput. Vis. **39**(2), 111–129 (2000)
19. Mantegazza, C.: Lecture Notes on Mean Curvature Flow, vol. 290. Springer, Basel (2011). https://doi.org/10.1007/978-3-0348-0145-4
20. Sela, M., Aflalo, Y., Kimmel, R.: Computational caricaturization of surfaces. Comput. Vis. Image Underst. **141**, 1–17 (2015)
21. Sorkine, O., Cohen-Or, D., Lipman, Y., Alexa, M., Rössl, C., Seidel, H.P.: Laplacian surface editing. In: Proceedings of the 2004 Eurographics/ACM SIGGRAPH Symposium on Geometry Processing, pp. 175–184 (2004)
22. Taubin, G.: A signal processing approach to fair surface design. In: Proceedings of 22nd Annual Conference on Computer Graphics and Techniques, pp. 351–358 (1995)
23. Vallet, B., Lévy, B.: Spectral geometry processing with manifold harmonics. In: Computer Graphics Forum, vol. 27, pp. 251–260. Wiley, Hoboken (2008)
24. Wetzler, A., Aflalo, Y., Dubrovina, A., Kimmel, R.: The Laplace-Beltrami operator: a ubiquitous tool for image and shape processing. In: Hendriks, C.L.L., Borgefors, G., Strand, R. (eds.) ISMM 2013. LNCS, vol. 7883, pp. 302–316. Springer, Heidelberg (2013). https://doi.org/10.1007/978-3-642-38294-9_26

Total-Variation Mode Decomposition

Ido Cohen$^{(\boxtimes)}$, Tom Berkov, and Guy Gilboa

Technion - Israel Institute of Technology, 3200003 Haifa, Israel
{idoc,ptom}@campus.technion.ac.il, guy.gilboa@ee.technion.ac.il

Abstract. In this work we analyze the *Total Variation* (TV) flow applied to one dimensional signals. We formulate a relation between *Dynamic Mode Decomposition* (DMD), a dimensionality reduction method based on the Koopman operator, and the spectral TV decomposition. DMD is adapted by time rescaling to fit linearly decaying processes, such as the TV flow. For the flow with finite subgradient transitions, a closed form solution of the rescaled DMD is formulated. In addition, a solution to the TV-flow is presented, which relies only on the initial condition and its corresponding subgradient. A very fast numerical algorithm is obtained which solves the entire flow by elementary subgradient updates.

Keywords: *Total Variation*-flow · *Total Variation*-spectral decomposition · *Dynamic Mode Decomposition* · Time reparametrization

1 Introduction

Finding latent information in high dimensional data is one of the most challenging tasks in data analysis. Often, dimensionality reduction techniques are used to reveal the latent modes. In this work, we bridge between two methods, *Total Variation* (TV) spectral decomposition [13] and *Dynamic Mode Decomposition* (DMD) [17]. These methods analyze gradient descent flows, represented as PDEs, and allow to perform filtering, spectrum analysis, and signal reconstruction.

TV-flow [1], the steepest descent process with respect to the TV functional, decays piecewise linearly and has finite support in time. This was used to compute the spectral TV components and to analyze their properties, see [5,6,12–14]. The two attributes are at odds with *Dynamic Mode Decomposition* (DMD), an approximation of the Koopman operator [16], which is described as an exponential data fitting algorithm [2]. Moreover, it was shown in [10] that DMD converges to highly inaccurate solutions, in certain cases, for flows derived by a γ-homogeneous operator when $\gamma \neq 1$. One way to improve the accuracy of DMD is by adding nonlinear measurements to the state space [20,21]. In [10] a new solution was suggested for homogeneous flows, of applying time-reparametrizing. This allows perfect flow estimations, in certain cases, yielding excellent linearization of the flow. In this work we follow and extend this direction. Understanding the time and manner of transitions of the subgradient are essential for the analysis of the TV flow. This leads both to effective mode decomposition, as well as to an alternative numerical solver. The main contributions of this work are:

1. Formulating a closed form solution of rescaled DMD (R-DMD) for TV-flow.
2. Deriving TV spectral decomposition from R-DMD.
3. Proposing a new algorithm to concisely and rapidly evaluate the TV-flow.
4. Illustrating these methods numerically, showing the alternative solution approaches yield close to identical numerical results.

2 Preliminary

In this section, we summarize the definitions and methods of previous studies relevant to this work.

2.1 *Dynamic Mode Decomposition* (DMD)

Given an observed set of N instances of length M, $\Psi = [\psi_0, \psi_1, \cdots, \psi_N] \in \mathbb{R}^{M \times (N+1)}$, generated by some dynamical system, the DMD algorithm [17] approximates the dynamics using a linear low-dimensional space. It finds the main spatial structures (modes), their amplitude (coefficients), and the respective time changes (eigenvalues). The three main steps of this algorithm are: 1. *Dimensionality reduction*, 2. *Optimal linear mapping*, and 3. *System reconstruction* with *modes, eigenvalues and coefficients*. These steps are detailed below.

Dimensionality Reduction. It is assumed that the data is embedded in a lower dimensional space. To find this space we apply *Singular Vector Decomposition* (SVD) on the data matrix,

$$\Psi = U\Sigma V^*. \tag{1}$$

where V^* is the conjugate transpose of V.

The columns of U span the column space of Ψ, whereas the rows of V^* span the row space of Ψ. The data is projected on the subspace spanned by r columns of U, denoted U_r, related to the most significant eigenvalues,

$$x_k = U_r^* \cdot \psi_k. \tag{2}$$

Thus x_k can be understood as the coordinates of a datum ψ_k with respect to this basis.

Linear Mapping. The linear mapping, F, from x_k to x_{k+1}, is obtained by solving the following optimization problem, $F = \arg\min_F \sum_{k=0}^{N-1} \|F \cdot x_k - x_{k+1}\|^2$, or, in matrix notation,

$$F = \arg\min_F \|F \cdot X - Y\|_{\mathcal{F}}^2, \tag{3}$$

where $\|\cdot\|_{\mathcal{F}}$ denotes the Frobenius norm and

$$X = U_r^* [\psi_0 \cdots \psi_{N-1}], \quad Y = U_r^* [\psi_1 \cdots \psi_N]. \tag{4}$$

The solution to Eq. (3) is

$$F = YX^* \cdot (XX^*)^{-1}. \tag{5}$$

Thus, the relation between two successive data coordinates is given by $x_{k+1} \approx F \cdot x_k$, where $\approx$ is the approximation in the sense of error minimization of Eq. (3). We assume the linear mapping, F, is diagonalizable and therefore can be written as

$$F = WDW^{-1}, \tag{6}$$

where W contains the right eigenvectors of F, and D is a diagonal matrix whose entries are the eigenvalues of F.

Modes, Eigenvalues and Coefficients. Reconstructing the datum ψ_{k+1} from the corresponding coordinates x_{k+1} is formulated as $\tilde{\psi}_{k+1} = U_r x_{k+1}$. Substituting x_{k+1} by $F \cdot x_k$ we get $\tilde{\psi}_{k+1} \approx U_r \cdot F \cdot x_k$. By plugging in the definition of x_k, Eq. (2), we have

$$\tilde{\psi}_{k+1} \approx U_r \cdot F \cdot U_r^* \psi_k = U_r \cdot F \cdot U_r^* \tilde{\psi}_k. \tag{7}$$

Notice that the equality notation is justified as $\tilde{\psi}_k$ is the projection of ψ_k on the subsapce spanned by the columns of U_r. Moreover, by generalizing this relation we can approximate the entire dynamics as

$$\tilde{\psi}_k \approx A^k \cdot \tilde{\psi}_0, \tag{8}$$

where

$$A = U_r \cdot F \cdot U_r^*. \tag{9}$$

Substituting Eqs. (6) and (9) in Eq. (8), we get

$$\tilde{\psi}_k \approx \left(U_r \cdot WDW^{-1} \cdot U_r^*\right)^k \cdot \tilde{\psi}_0 = U_r \cdot WD^kW^{-1} \cdot U_r^* \cdot \tilde{\psi}_0.$$

Now, we can define the modes, $\{\phi_i\}_{i=1}^r$, eigenvalues, $\{\mu_i\}_{i=1}^r$, and coefficients, $\{\alpha_i\}_{i=1}^r$, having the dynamic mode decomposition.

Modes are defined as $\Phi = \begin{bmatrix} \phi_1 \cdots \phi_r \end{bmatrix} = U_r W$. Notice that $\{\phi_i\}_{i=1}^r$ are the right eigenvectors of the matrix A and only them since the rank of A is r.

Eigenvalues are the diagonal entries of the matrix D. These are the eigenvalues of the matrix F and A.

Coefficients are defined by $\alpha = \begin{bmatrix} \alpha_1, \cdots, \alpha_r \end{bmatrix} = W^{-1}U_r^* \tilde{\psi}_0$. We can now reconstruct the approximate dynamics by,

$$\tilde{\psi}_k \approx \Phi D^k \alpha = \sum_{i=1}^r \alpha_i \mu_i^k \phi_i. \tag{10}$$

The DMD algorithm is summarized in Algorithm 1.

2.2 *Total Variation* Spectral Decomposition

Let $\mathcal{H}$ be a Hilbert space with an inner product, $\langle \cdot, \cdot \rangle$, and the corresponding induced norm, $\|\cdot\| = \sqrt{\langle \cdot, \cdot \rangle}$.

Algorithm 1. Standard DMD [17]

1: **Inputs:** Data sequence $\{\psi_k\}_{k=0}^{N}$; reduced dimensionality r.
2: Compute the *Singular Vector Decomposition* (SVD) of Ψ (see [19]) (Eq. (1)).
3: Form the matrices X and Y from the coordinates of the data (Eq. (4)).
4: Find the optimal linear mapping, F, between X and Y (Eq. (5)).
5: Compute the modes and the coordinates as $\Phi \triangleq U_r W$, $\alpha \triangleq W^{-1}U_r^*\psi_0$. The eigen-values of the DMD are the eigenvalues of F, $\{\mu_i\}_{i=1}^{r}$.
6: **Outputs:** $\{\mu_i, \phi_i, \alpha_i\}_1^r$

TV Functional. The TV functional, is defined for smooth functions as,

$$J_{TV}(\psi) = \langle|\nabla\psi|, 1\rangle, \quad \psi \in \mathcal{H}. \tag{11}$$

Precise definitions for functions in BV and various properties of TV can be found in [8]. We denote the subdifferential of J_{TV} at ψ as $\partial J_{TV}(\psi)$ and a subgradient as $-P$. They admit the following relation,

$$P(\psi) \in -\partial J_{TV}(\psi). \tag{12}$$

We assume Neumann boundary conditions. Note that the thorough analysis of the subgradient of the TV flow can be found in [1,3].

Eigenfunctions. The nonlinear eigenfunction, v, of P admits

$$P(v) = \lambda \cdot v, \tag{EF}$$

for some $\lambda \in \mathbb{R}$. It can be shown (see [6]), that with the above definitions of P we have $\lambda \leq 0$.

The gradient descent of TV (TV-flow) is defined by the following PDE,

$$\psi_t = P(\psi), \quad \psi(t = 0) = f, \tag{TV-flow}$$

where ψ_t is the first order time derivative of $\psi(t)$ and f is the initial condition.

TV Spectral Framework [13]. The spectral decomposition of a signal, $f \in \mathcal{H}$, related to the eigenfunctions of P is based on the solution of Eq. (TV-flow). We list below the definitions of the transform, inverse-transform, filtering and spectrum of this framework. We simplify notations, assume all derivatives exist and that the signal has no null-space components of J_{TV}.

   The TV-transform is defined by $\mathcal{G}(t) = t\frac{d^2}{dt^2}\psi(t)$, where $\psi(t)$ is the solution of (TV-flow). The function $\mathcal{G}(t)$ is the spectral component of the signal $f(x)$ at time t.

   The inverse transform is the reconstruction of the original signal f from the spectral components, defined by $\hat{f} = \int_0^\infty \mathcal{G}(t)dt$.

   The filtering of f by the filter $h(t)$ is $f_h = \int_0^\infty \mathcal{G}(t) \cdot h(t)dt$, where $h(t)$ is a real function. Namely, filtering is an amplification (or attenuation) of $\mathcal{G}(t)$ in the transform domain, t.

The spectrum of f at any scale t is defined by $S(t) = \langle f, \mathcal{G}(t) \rangle$.

If the initial condition, f, is an eigenfunction (admits Eq. (EF)), then the solution of Eq. (TV-flow) is,

$$\psi(t) = (1 + \lambda t)^+ \cdot f, \tag{13}$$

where λ is the corresponding eigenvalue and $(a)^+ = \max\{a, 0\}, \forall a \in \mathbb{R}$. The transform of this signal is

$$\mathcal{G}(t) = f \cdot t\lambda^2 \cdot \delta(1 + \lambda \cdot t), \tag{14}$$

where $\delta(\cdot)$ is the Dirac delta function.

Settings: In this work we first note that DMD is fully discrete (time and space) whereas TV-flow and spectral TV are semi-discrete (time-continuous, spatially discrete). Thus, in order to apply DMD on a gradient descent flow we first need to sample (uniformly) with respect to the time variable t. In all cases we use Euclidean inner product and norm. We list below some *Attributes* of the semi discrete one-dimensional TV-flow and the TV spectral components:

1. The subgradient is piecewise constant with respect to t [6].
2. The initial condition can be reconstructed by knowing the subgradient as a function of t (by integration).
3. The flow splits into merging events [4].
4. The average of the subgradient, $-P$, over the spatial variable is zero.
5. The spectrum is a finite set of delta functions, where each delta function represents a spectral component [6].
6. For a given f, the spectral component set is orthogonal [6].
7. Two adjacent points which become equal in value during the flow, will not separate [3, 18].
8. There is a time reparametrization for which the TV decays exponentially [10].

3   DMD of the TV Flow

3.1   Closed Form Solution

Let us formulate *Attribute 1* and *Attribute 2* more formally. The solution of (TV-flow) converges to a steady state in finite time. In this finite time, the solution is divided into L disjoint segments, $[T_i, T_{i+1})$. In each segment, the subgradient is constant, $-p_i \in \partial J(t), t \in [T_i, T_{i+1})$, where for $t > T_L$ it is zero, $p_{L+1} = 0$. The solution can be expressed by (e.g. [6]),

$$\psi(t) = \psi(T_i) + (t - T_i)p_i, \quad t \in [T_i, T_{i+1}). \tag{15}$$

For an initial condition, $\psi(0) = f$, orthogonal to the kernel of J_{TV} (constant functions) the reconstruction of f from the set of subgradients $\{p_i\}$ is

$$f = \psi(0) = \sum_{i=1}^{L} T_i(p_{i+1} - p_i). \tag{16}$$

Proposition 1 (Linear decay). *The solution of* (TV-flow) *is a sum of spectral components decaying linearly. More formally, if the initial condition, f, is orthogonal to the kernel set of P then the solution of Eq.* (TV-flow) *is*

$$\psi(t) = \sum_{i=1}^{L} (1 + \lambda_i t)^+ \, \varphi_i, \text{ where } \lambda_i = -T_i^{-1} \text{ and } \varphi_i = \frac{p_i - p_{i+1}}{\lambda_i}. \tag{17}$$

Proof. Let us reformulate the solution, Eq. (15) for the first time range $t \in [0, T_1]$. Substituting Eq. (16) in Eq. (15), we have

$$\psi(t) = \psi(0) + (t - 0)p_1 = \sum_{i=1}^{L} T_i(p_{i+1} - p_i) + tp_1$$

$$\underset{p_{L+1}=0}{=} \sum_{i=1}^{L} T_i(p_{i+1} - p_i) + t \sum_{i=1}^{L} (p_i - p_{i+1}) = \sum_{i=1}^{L} (T_i - t)(p_{i+1} - p_i).$$

Therefore, $\psi(T_1) = \sum_{i=1}^{L}(T_i - T_1)(p_{i+1} - p_i) = \sum_{i=2}^{L}(T_i - T_1)(p_{i+1} - p_i)$. In a similar manner, we can reformulate the solution for $t \in [T_1, T_2]$ as $\psi(t)|_{t \in [T_1, T_2]} = \sum_{i=2}^{L}(T_i - t)(p_{i+1} - p_i)$. By induction, the general solution is, $\psi(t) = \sum_{i=1}^{L}(T_i - t)^+(p_{i+1} - p_i)$. Denoting $\lambda_i = -T_i^{-1}$, this can be expressed as,

$$\psi(t) = \sum_{i=1}^{L}(-\lambda_i^{-1} - t)^+(p_{i+1} - p_i) = \sum_{i=1}^{L}(1 + \lambda_i t)^+ \frac{p_i - p_{i+1}}{\lambda_i}. \quad \square$$

The spectral decomposition is computed by second order time derivative of $\psi(t)$,

$$\mathcal{G}(t) = \sum_{i=1}^{L} \varphi_i \cdot t\lambda_i^2 \cdot \delta(1 + \lambda_i \cdot t). \tag{18}$$

This coincides with *Attribute 5: a finite set of Dirac delta functions.* The flow can also be defined (without using the operator $(\cdot)^+$) in disjoint time intervals,

$$\psi(t) = \sum_{i=k}^{L} (1 + \lambda_i t) \, \varphi_i, \quad \forall t \in [T_{k-1}, T_k). \tag{19}$$

We will use this formulation later in our analysis.

3.2   Flow Transitions and a Fast TV-flow Algorithm

While there has been on going research on fast methods for TV regularization (e.g. [9,11,15]), few advances were made in fast algorithms of the TV-flow, which is required for computing spectral TV. Our proposed solution, $\psi(t) \in \mathbb{R}^M \times [0, T_L]$, is in a semi-discrete setting. The algorithm is based on the TV-flow attributes listed at the end of Sect. 2.2.

Algorithm 2. Accelerated TV flow and spectral decomposition (1D)

1: **Inputs:** f, $\delta > 0$
2: **Initialize:** $\psi_0 \leftarrow f$, $p_0 \leftarrow P(f)$, $t \leftarrow 0$, $\mathcal{T} = \emptyset$, and $\mathcal{P} = \{p_0\}$
3: **while** $\|p_i\| > \delta$ **do**
4:     Find the next time transition point, T_{i+1} (Eq. (20)).
5:     $\psi_{i+1} \leftarrow \psi_i + (T_{i+1} - T_i) \cdot p_i$.
6:     Find clusters, $\{\mathcal{M}_k\}_{k=1}^r$, where $|\nabla \psi_{i+1}| = 0$
7:     Update the next negative subgradient, p_{i+1}, such that $p_{i+1}(\mathcal{M}_k) = \frac{1}{|\mathcal{M}_k|} \sum_{\mathcal{M}_k} p_i(\mathcal{M}_k)$, $k = 1, \cdots, r$.
8:     Add p_{i+1} to the set $\mathcal{P}$; Add T_{i+1} to the set $\mathcal{T}$. Compute φ_i by (17).
9: **end while**
10: **Outputs:** $\mathcal{T}, \mathcal{P}$

Subgradient transitions are time points where the subgradient is updated, denoted $\mathcal{T} = \{T_i\}_{i=0}^L$ where $T_0 = 0$, and T_L is the extinction time. From *Attributes* 3 and 7 we can find these time points and compute the updated subgradient. These steps are detailed below and concisely formalized in Alogirhtm 2.

Merging occurs when two adjacent pixels become equal. Then, the discrete gradient approximation is $\nabla \psi^{(j)} := \psi^{(j+1)} - \psi^{(j)} = 0$, where $\psi^{(j)}$ is the entry j of the vector ψ. Using Eq. (15), the first merging event after T_i can be calculated by

$$T_{i+1} = T_i + \min_{j \in \mathcal{J}^*} \{-\nabla \psi^{(j)}(T_i)/\nabla P^{(j)}(\psi(T_i))\}, \tag{20}$$

where $\mathcal{J}^* = \{j \, s.t. \, 0 < -\nabla \psi^{(j)}(T_i)/\nabla P^{(j)}(\psi(T_i)) < \infty\}$.

Subgradient Update is necessary after every merging event. According to *Attribute* 7, the merged entries evolve together at the same pace. In addition, the subgradient at other locations is unchanged (see [18]). Since the average of the subgradient is zero (*Attribute* 4), the subgradient of the merged entries is the average of the previous subgradient at these entries.

3.3   Rescaled-DMD

We follow the work of [10] where an analysis of DMD was carried out for flows based on homogeneous operators. The homogeneity order dictates not only the decay profile but also the support in time of the solution. In particular, TV-flow decays linearly and has a finite extinction time, whereas a flow linearization algorithm, such as DMD, can be interpreted as an exponential data fitting algorithm [2]. In [10] it was suggested to solve this problem by time reparametrization. Introducing a new time variable τ, Eq. (TV-flow) is time rescaled by the flow,

$$\psi_\tau = G(\psi) = -\frac{\langle P(\psi), \psi \rangle}{\|P(\psi)\|^2} P(\psi). \tag{R-TV-flow}$$

Note that, $G(a\psi) = aG(\psi), \forall a \in \mathbb{R}$. Therefore, this can be viewed as a flow derived by a one-homogeneous operator, yielding exponential decay. Using (TV-flow) and (R-TV-flow), the relation between t and τ can be derived by,

$$\frac{d}{d\tau}\psi(t(\tau)) = -\frac{\langle P(\psi(t(\tau))), \psi(t(\tau))\rangle}{\|P(\psi(t(\tau)))\|^2}P(\psi(t(\tau))) = -\frac{\langle P(\psi(t(\tau))), \psi(t(\tau))\rangle}{\|P(\psi(t(\tau)))\|^2}\frac{d}{dt}\psi(t(\tau)),$$

yielding,

$$\frac{d}{d\tau}t(\tau) = -\frac{\langle P(\psi(t(\tau))), \psi(t(\tau))\rangle}{\|P(\psi(t(\tau)))\|^2}. \tag{21}$$

This ODE gets a different form in each segment, $[T_{k-1}, T_k)$. Substituting Eq. (19) in Eq. (21), we have

$$\frac{d}{d\tau}t(\tau) = -\frac{\langle \sum_{i=k}^{L}\lambda_i\varphi_i, \sum_{i=k}^{L}(1+\lambda_i t(\tau))\varphi_i\rangle}{\left\|\sum_{i=k}^{L}\lambda_i\varphi_i\right\|^2} = -\frac{\sum_{i=k}^{L}\lambda_i\|\varphi_i\|^2}{\sum_{i=k}^{L}\lambda_i^2\|\varphi_i\|^2} - t(\tau).$$

The solution is,

$$t(\tau) = a_k e^{-\tau} - c_k, \quad c_k = \frac{\sum_{i=k}^{L}\lambda_i\|\varphi_i\|^2}{\sum_{i=k}^{L}\lambda_i^2\|\varphi_i\|^2}, \tag{22}$$

where a_k depends on the initial conditions of every segment such that $t(\tau)$ is continuous (where $t(0) = 0$). Then, the time points $\{T_i\}_{i=1}^{L}$ are mapped to $\{\tau_i\}_{i=1}^{L}$, accordingly.

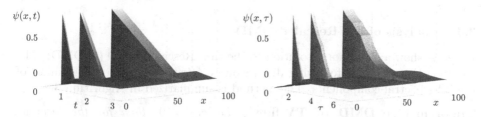

Fig. 1. Time reparametrization - Left, **TV-flow** $\psi(t)$. The TV-flow decays piecewise linearly. With time reparametrization, **R-TV-flow**, $\psi(\tau)$, it is mapped to a piecewise smooth function, depicted in Right. The nonsmooth points represent transitions in the subgradient.

Proposition 2 (Main TV-flow modes). *In every disjoint kth interval, $[\tau_{k-1}, \tau_k)$, the solution of time reparametrizing (TV-flow), Eq. (R-TV-flow), has two main orthogonal modes, ξ_1^k, ξ_2^k, with eigenvalues zero and minus one.*

Proof. Substituting Eq. (22) in Eq. (19), we get

$$\psi(t(\tau)) = \sum_{i=k}^{L} (1 + \lambda_i t(\tau)) \, \varphi_i, \qquad \forall t \in [T_{k-1}, T_k)$$

$$= \sum_{i=k}^{L} \left(1 + \lambda_i \left(a_k e^{-\tau} - c_k\right)\right) \varphi_i, \qquad \forall \tau \in [\tau_{k-1}, \tau_k)$$

$$= \underbrace{\sum_{i=k}^{L} \varphi_i - c_k \sum_{i=k}^{L} \lambda_i \varphi_i}_{\xi_1^k} + \underbrace{e^{-\tau} a_k \sum_{i=k}^{L} \lambda_i \varphi_i}_{\xi_2^k} = \xi_1^k + e^{-\tau} \xi_2^k, \ \forall \tau \in [\tau_{k-1}, \tau_k).$$

By plugging c_k from Eq. (22) into ξ_1^k, ξ_2^k their orthogonality is concluded immediately. ☐

In Fig. 2 we show the TV-modes defined in Proposition 2 with the initial condition Fig. 3a. It contains six disjoint intervals with the corresponding modes $\{\xi_1^k, \xi_2^k\}_{k=1}^{6}$.

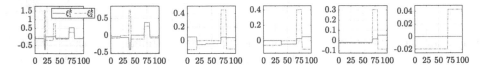

Fig. 2. Modes: ξ_1^k (teal) - constant, ξ_2^k (orange) - exponentially decaying.

3.4 Analysis of the Rescaled-DMD

Here, we show a closed form solution to the time Rescaled-DMD (R-DMD). The common thread in the following discussion is *Attribute 6*, the orthogonality of the TV-spectral components. The method is summarized in Algorithm 3.

Theorem 1 (R-DMD of TV-flow). *Let $\tau_0 = 0$, then for the interval, $[\tau_{k-1}, \tau_k)$, where $k = 1, \cdots, L - 1$, R-DMD reveals two non-zero orthogonal modes that reconstruct accurately the TV-flow in this interval. For the last interval, $[\tau_{L-1}, \tau_L)$, there is only one nonzero mode.*

Proof. According to Proposition 2 and since DMD is an exponential data fitting algorithm, the DMD of the dynamics, Eq. (R-TV-flow), is as follows. The modes are $\phi_1^k = \xi_1^k/\|\xi_1^k\|$, $\phi_2^k = \xi_2^k/\|\xi_2^k\|$, and the coefficients are $\alpha_1^k = \|\xi_1^k\|$ and $\alpha_2^k = \|\xi_2^k\|$. Note that one mode is constant with respect to time and the second decays exponentially. Therefore, the eigenvalues are $\mu_1^k = 1$ for the constant mode and $\mu_2^k = e^{-dt}$ where dt is the sampling step size (see Algorithm 3). ☐

Now we formulate the relation between the TV spectral components φ_k and the R-DMD modes.

Algorithm 3. R-DMD for TV-flow

1: **Inputs:** The initial condition, and sampling step size dt.
2: **Initialize:** Evolve the solution of (**R-TV-flow**) uniformly with a step size dt.
3: Invoke Algo. 2 - the result is $\mathcal{T}$ and $\mathcal{P}$.
4: Map the set of transition time points, $\mathcal{T}$, to a new set $\hat{\mathcal{T}}$ (Eq. (21)).
5: **for** Every time segment $[\tau_i, \tau_{i+1})$, $\tau_i, \tau_{i+1} \in \hat{\mathcal{T}}$ **do**
6:     Invoke Algo. 1 with $r = 2$ (when $i = L - 1$, $r = 1$).
7: **end for**
8: **Outputs:** Modes $\{\phi_1^k, \phi_2^k\}_{k=1}^L$, coefficients $\{\alpha_1^k, \alpha_2^k\}_{k=1}^L$, and eigenvalues $\mu_1^k = 1, \mu_2^k = e^{-dt}$.

Proposition 3 (Revealing TV spectral components from R-DMD).
The kth spectral component, φ_k, admits the following relation

$$c_k \lambda_k \varphi_k = \phi_2^k - \frac{\langle \phi_2^k, \phi_2^{k+1} \rangle}{\left\| \phi_2^{k+1} \right\|^2} \phi_2^{k+1}. \tag{23}$$

Proof.

$$\phi_2^k - \frac{\langle \phi_2^k, \phi_2^{k+1} \rangle}{\left\| \phi_2^{k+1} \right\|^2} \phi_2^{k+1} = c_k \sum_{i=k}^L \lambda_i \varphi_i - \frac{\langle c_k \sum_{i=k}^L \lambda_i \varphi_i, c_{k+1} \sum_{i=k+1}^L \lambda_i \varphi_i \rangle}{\left\| c_{k+1} \sum_{i=k+1}^L \lambda_i \varphi_i \right\|^2} c_{k+1} \sum_{i=k+1}^L \lambda_i \varphi_i$$

$$= c_k \lambda_k \varphi_k + c_k \sum_{i=k+1}^L \lambda_i \varphi_i - c_k \frac{\langle \lambda_k \varphi_k + \sum_{i=k+1}^L \lambda_i \varphi_i, \sum_{i=k+1}^L \lambda_i \varphi_i \rangle}{\left\| \sum_{i=k+1}^L \lambda_i \varphi_i \right\|^2} \sum_{i=k+1}^L \lambda_i \varphi_i = c_k \lambda_k \varphi_k. \qquad \square$$

4   Results and Conclusion

In this section, the theory and algorithms discussed above are illustrated. We use standard first-order discretization of the derivatives and Neumann boundary conditions. We begin with a toy example, depicted in Fig. 3a. In Fig. 1 we show that the solution of (TV-flow) decays linearly and that of Eq. (R-TV-flow) piecewise exponentially. In Fig. 4-top the TV-spectral decomposition, dashed red line (computed in the standard way, see [13]) is compared with two algorithms. First, by Algorithm 3 based on Proposition 3, black dotted line. Second, by Algorithm 2, blue line. The errors between the TV-spectral decomposition and Algorithms 3 and 2 are depicted in Fig. 4-bottom.

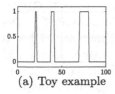
(a) Toy example

(b) Zebra image

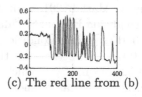
(c) The red line from (b)

Fig. 3. Initial conditions - (a) Three pulses with different widths. (c) The corresponding values of the pixels on the red line in (b). (Color figure online)

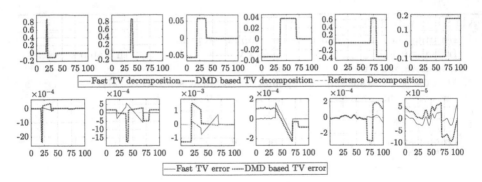

Fig. 4. TV spectral decomposition comparison for toy example - The standard method of spectral decomposition (Dashed red line) vs. fast TV decomposition and R-DMD decomposition in blue and dotted black lines respectively. Bottom row - respecive errors (Color figure online)

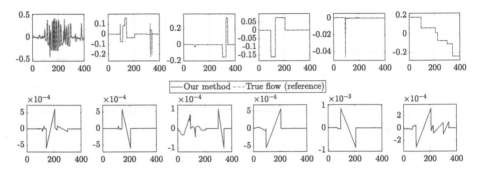

Fig. 5. TV spectral decomposition comparison for initial condition depicted in Fig. 3c - Standard method of spectral decomposition (Dashed red line) vs. fast TV decomposition (Blue line). Bottom row - respective error. (Color figure online)

In Fig. 5 we show results of the fast TV-spectral decomposition, Algorithm 2, applied on a natural signal. We arbitrarily chose the red line from the zebra in Fig. 3b, depicted in Fig. 3c. Bands of standard TV-spectral decomposition and the fast TV-spectral decomposition are shown in Fig. 5-top. One can observe that our proposed fast method recovers the spectral bands faithfully, with negligible error, Fig. 5-bottom.

Rough Performance Comparison. We report the elapsed time in seconds, running in Matlab 2018b on an 8th Gen. Core i7 laptop with 16 GB RAM. Initial condition Fig. 3a: Standard method (iterative application of [7]) - 488.4 s; ours - 0.013 s. Initial condition Fig. 3c (zebra): Standard method - 7.1×10^3 s; ours - 0.15 s. Note that the times specified for our algorithm do not include the initialization of p_0, which takes additional 7.3 s for Fig. 3a, and 27.8 s for Fig. 3c. This time can be shortened further using faster subgradient computation algorithms.

Conclusion. In this paper the modes of one-dimensional TV-flow were analyzed. A popular method for finding modes of flows in fluid dynamics, *Dynamic Mode Decomposition* (DMD) [17], was examined. We have presented an adaptation for the TV case, by a time rescaled version. This was based on the spectral-TV theory, where the spectral components are orthogonal. Obtaining TV modes, or nonlinear spectral TV components, requires a gradient-flow evolution. Since evolving TV-flow is a slow process using optimization techniques, we have proposed a very fast method, based on simple updates of the subgradient.

Acknowledgments. We acknowledge support by grant agreement No. 777826 (NoMADS), by the Israel Science Foundation (Grant No. 534/19) and by the Ollendorff Minerva Center.

References

1. Andreu, F., Ballester, C., Caselles, V., Mazón, J.M.: Minimizing total variation flow. Differ. Integr. Equ. **14**(3), 321–360 (2001)
2. Askham, T., Kutz, J.N.: Variable projection methods for an optimized dynamic mode decomposition. SIAM J. Appl. Dyn. Syst. **17**(1), 380–416 (2018)
3. Bellettini, G., Caselles, V., Novaga, M.: The total variation flow in RN. J. Differ. Equ. **184**(2), 475–525 (2002)
4. Brox, T., Weickert, J.: A TV flow based local scale estimate and its application to texture discrimination. J. Vis. Commun. Image Rep. **17**(5), 1053–1073 (2006)
5. Bungert, L., Burger, M., Chambolle, A., Novaga, M.: Nonlinear spectral decompositions by gradient flows of one-homogeneous functionals. arXiv preprint arXiv:1901.06979 (2019)
6. Burger, M., Gilboa, G., Moeller, M., Eckardt, L., Cremers, D.: Spectral decompositions using one-homogeneous functionals. SIAM J. Imaging Sci. **9**(3), 1374–1408 (2016)
7. Chambolle, A.: An algorithm for total variation minimization and applications. J. Math. Imaging Vis. **20**(1–2), 89–97 (2004)
8. Chambolle, A., Caselles, V., Cremers, D., Novaga, M., Pock, T.: An introduction to total variation for image analysis. Theor. Found. Numer. Meth. Sparse Recovery **9**(263–340), 227 (2010)
9. Cherkaoui, H., Sulam, J., Moreau, T.: Learning to solve TV regularised problems with unrolled algorithms. In: Advances in Neural Information Processing Systems, 33 (2020)
10. Cohen, I., Azencot, O., Lifshits, P., Gilboa, G.: Modes of homogeneous gradient flows. arXiv preprint arXiv:2007.01534 (2020)
11. Darbon, J., Sigelle, M.: Image restoration with discrete constrained total variation Part I: fast and exact optimization. J. Math. Imaging Vis. **26**(3), 261–276 (2006)
12. Gilboa, G.: A spectral approach to total variation. In: Kuijper, A., Bredies, K., Pock, T., Bischof, H. (eds.) SSVM 2013. LNCS, vol. 7893, pp. 36–47. Springer, Heidelberg (2013). https://doi.org/10.1007/978-3-642-38267-3_4
13. Gilboa, G.: A total variation spectral framework for scale and texture analysis. SIAM J. Imaging Sci. **7**(4), 1937–1961 (2014)
14. Gilboa, G., Moeller, M., Burger, M.: Nonlinear spectral analysis via one-homogeneous functionals: overview and future prospects. J. Math. Imaging Vis. **56**(2), 300–319 (2016)

15. Goldfarb, D., Yin, W.: Parametric maximum flow algorithms for fast total variation minimization. SIAM J. Sci. Comput. **31**(5), 3712–3743 (2009)
16. Mezić, I.: Spectral properties of dynamical systems, model reduction and decompositions. Nonlinear Dyn. **41**(1–3), 309–325 (2005)
17. Schmid, P.J.: Dynamic mode decomposition of numerical and experimental data. J. Fluid Mecha. **656**, 5–28 (2010)
18. Steidl, G., Weickert, J., Brox, T., Mrázek, P., Welk, M.: On the equivalence of soft wavelet shrinkage, total variation diffusion, total variation regularization, and sides. SIAM J. Numer. Anal. **42**(2), 686–713 (2004)
19. Trefethen, L.N., Bau III, D.: Numerical Linear Algebra, vol. 50. SIAM, Philadelphia (1997)
20. Williams, M.O., Kevrekidis, I.G., Rowley, C.W.: A data-driven approximation of the Koopman operator: extending dynamic mode decomposition. J. Nonlinear Sci. **25**(6), 1307–1346 (2015)
21. Williams, M.O., Rowley, C.W., Kevrekidis, Y.: A kernel-based method for data-driven Koopman spectral analysis. J. Comput. Dyn. **2**(2), 247–265 (2015)

Fast Morphological Dilation and Erosion for Grey Scale Images Using the Fourier Transform

Marvin Kahra[✉], Vivek Sridhar, and Michael Breuß

Institute for Mathematics, Brandenburg Technical University Cottbus-Senftenberg,
03046 Cottbus, Germany
{marvin.kahra,sridhviv,breuss}@b-tu.de

Abstract. The basic filters in mathematical morphology are dilation
and erosion. They are defined by a flat or non-flat structuring element
that is usually shifted pixel-wise over an image and a comparison process
that takes place within the corresponding mask. Existing fast algorithms
that realise dilation and erosion for grey value images are often limited
with respect to size or shape of the structuring element. Usually their
algorithmic complexity depends on these aspects. Many fast methods
only address flat morphology.

In this paper we propose a novel way to make use of the fast Fourier
transform for the computation of dilation and erosion. Our method is by
design highly flexible, as it can be used with flat and non-flat structuring
elements of any size and shape. Moreover, its complexity does not depend
on size or shape of the structuring element, but only on the number of
pixels in the filtered images. We show experimentally that we obtain
results of very reasonable quality with the proposed method.

Keywords: Mathematical morphology · Fast Fourier Transform ·
Dilation · Erosion · Efficient algorithms

1 Introduction

Mathematical morphology is a theory for the analysis of spatial structures in
images. It has evolved over decades to a very successful field in image processing, see e.g. [4,5,7] for an overview. There are two main building blocks of usual
morphological operators. The first one is the structuring element (SE), charac-
terised by its shape, size and centre location. There are in addition two types of
SEs, flat and non-flat [13]. A flat SE basically defines a neighbourhood of the
centre pixel where morphological operations take place, whereas a non-flat SE
also contains a mask of finite values used as additive offsets. The SE is translated
over an image, and often implemented as a sliding window. The second building
block is a mechanism performing a comparison of values within a SE. The basic
operations in mathematical morphology are dilation and erosion, where a pixel
value is set to the maximum or minimum of the discrete image function within

© Springer Nature Switzerland AG 2021
A. Elmoataz et al. (Eds.): SSVM 2021, LNCS 12679, pp. 65–77, 2021.
https://doi.org/10.1007/978-3-030-75549-2_6

the SE centred upon it, respectively. Many morphological filtering processes of practical interest, like e.g. opening, closing or top hats, can be formulated by combining dilation and erosion. As dilation and erosion are dual operations, it is often sufficient to focus on one of it for algorithm construction.

Part of the success of mathematical morphology is due to the remarkable efficiency that can be obtained in its algorithmic realisation. Many of the fast morphological algorithms may be classified in terms of two families, cf. [2]. One of these families aims to either reduce the size of a SE or to decompose it. The main goal is thereby to reduce the number of basic dilations or erosions needed to generate a desired SE. The second family of algorithms aims to analyse a given image so that the number of redundant comparison operations needed for image filtering can be reduced. Most of these methods refer to specific shape or size of the SE, sometimes also specific hardware is addressed; see e.g. the works [9,11,12] and the references therein. To this end, many approaches explore the use of a GPU [7,10,14].

There are just a few works that deal with fast algorithms for SEs of arbitrary shape and size. Let us mention [17] which employs histogram updates during translation of a structuring element over an image. This classical method may also be employed with non-flat SEs which represents a limitation of many other fast methods. However, as is also the case for the algorithm from [17], the algorithmic complexity of the vast majority of the existing fast algorithms inherently relies on the size (and in practice also on the shape) of the structuring element. Since the structuring element has usually to be processed over a significant part of a given image, this also relates to image size. In view of the constantly increasing image resolution in many applications, we conjecture that morphological algorithms that scale well with image size independently of size and shape of the structuring element, will represent potentially interesting developments.

Considering the algorithms that may scale well with image size, a natural candidate appears to be the Fast Fourier Transform (FFT). Algorithms for FFT [6] were popularised in the 1960s. The fast convolution technique which utilises the FFT has an algorithmic complexity of $\mathcal{O}(n \log n)$ where n is the size of the larger of the two arrays used in convolution. In image processing n usually relates to the number of pixels in an image.

A first attempt to utilise the FFT for morphological operations was made by Tuzikov, Margolin and Grenov [1]. They introduced measures of rotation and reflection symmetries for compact convex sets, which are used to decompose these sets into a Minkowski sum of two sets. In particular, they showed that binary dilation respectively erosion can be represented by a convolution with corresponding characteristic functions, and they expressed these convolutions by Fourier transformations. This results in the possibility of computing binary dilation or erosion using the FFT. The authors of [8] extend this approach in a direct way to grey scale images. They first decompose the image into discrete grey levels. Then they treat each level set as a binary image and apply the technique from [1], combining the processed levels to obtain the final grey value image. The proceeding implies that the method is limited to flat SEs.

Fig. 1. Illustration of dilation and erosion. **From left to right:** Test image, flat dilation and erosion with a 5×5 structuring element, example for non-flat dilation and erosion.

In this paper, we propose a completely different means to realise morphological dilation and erosion with the FFT. We employ a recent analytic relaxation of the maximum function [3] to derive an analytic representation of morphological dilation using the Fourier transform, motivated by the proceeding in [1]. This formulation allows to process a given grey value image directly utilising the FFT. Since the entire formulation is highly flexible, there are no principle restrictions with respect to size or shape of a structuring element, also there is no restriction concerning the choice of flat or non-flat SEs. In experiments we confirm the beneficial properties of our method.

2   Morphological Dilation and Erosion

In order to make this paper self-contained, we briefly recall some fundamentals of mathematical morphology. Thereby we stick to a continuous-scale description which may fit well to the Fourier transform framework.

We start by considering a two dimensional, continuous-scale image domain $\Omega \subset \mathbb{R}^2$. Morphological dilation and erosion operations employ the structuring element $B \subset \mathbb{R}^2$ to work on grey value images represented here as scalar functions $f(x)$ with $x \in \Omega$.

In flat morphology, grey scale dilation $\oplus$ and erosion $\ominus$ with the structuring element B may be defined as

$$(f \oplus B)(x) = \sup\{f(x - x') : x' \in B \text{ and } (x - x') \in \Omega\}$$
$$(f \ominus B)(x) = \inf\{f(x + x') : x' \in B \text{ and } (x + x') \in \Omega\} \tag{1}$$

respectively. On a discrete grid, these operations are performed taking the maximum respectively minimum over a mask representing B. See Fig. 1 for an illustration using a SE of size 5×5 for one filtering step. Let us note that in a lattice theoretic framework for discrete images, one may describe dilation and erosion for grey value images using the umbra and the notion of top surfaces, compare [16].

As indicated, many other morphological operations of practical interest can be composed by dilation and erosion. As examples let us mention here opening $f \circ B = (f \ominus B) \oplus B$ and closing $f \bullet B = (f \oplus B) \ominus B$.

The operations dilation and erosion are dual in the following sense. Let the range of f be given by $L = [0, l]$, where $l > 0$ is the upper limit of the grey values

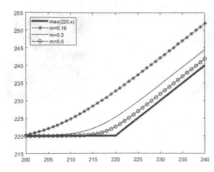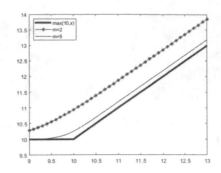

Fig. 2. Visualisation of our smooth relaxation of the maximum function, see (4). **Left:** As observable, for values $0 < m < 1$ we obtain a certain shift of the correct result which quickly approaches the max function when increasing m. **Right:** Already moderate values of m deliver very good approximations. Let us stress that the relaxation will take place in the Fourier domain.

of the pixels in a grey scale image. (The full range is of course $L = [0, 255]$.) We define then the complementary image $-f$ via

$$(-f)(x) = l - f(x). \tag{2}$$

It is evident that the erosion of f may be expressed using dilation as the result of $-(-f \oplus \check{B})$, where $\check{B} = \{-x : x \in B\}$.

In non-flat morphology, the notions of dilation and erosion from above can be extended to

$$(f \oplus S)(x) = \sup\{f(x - x') + S(x') : x' \in B \text{ and } (x - x') \in \Omega\}$$
$$(f \ominus S)(x) = \inf\{f(x + x') - S(x') : x' \in B \text{ and } (x + x') \in \Omega\} \tag{3}$$

where we make use of a structuring element S in order to distinguish from the flat morphology situation. See again Fig. 1 for a visualisation of non-flat filtering. Also for non-flat structuring elements, erosion may be expressed via dilation in the complementary sense.

3 Greyscale Dilation Discretisation with Convolution

Our goal is now to discretise morphological grey scale dilation and erosion with an approximation of the maximum (resp. minimum). We show that this yields a convolution over the image grid. The discrete convolution can be formulated using a Fourier transform. Furthermore we demonstrate how the algorithm works and address some specific aspects of the FFT-based implementation.

3.1 Analytic Motivation of the Algorithm

We consider the maximum function in a general formulation of up to k arguments. The work [3] gives a smooth approximation of the maximum function as

$$\max(x_1,\ldots,x_k) \doteq \frac{1}{m} \log \left(\sum_{i=1}^{k} e^{mx_i} \right) \tag{4}$$

where $x_i \in \mathbb{R}$ for $i = 1,\ldots,k$, compare Fig. 2. In the cited work, the authors employ the equation above in the field of optimal control. In the current article, we consider this equation for creating a new approach to the dilation/erosion operations in mathematical morphology. However, the Eq. (4) is not the only possible approximation, but a simple first approach. Let us note here that in [3], the value of the real number m is considered to be $m \geq 1$ for analytic purposes. In the context of our work we will be interested to find a suitable choice for m for algorithmic realisation. We will elaborate below on this point.

The formula (4) means that the limit for $m \to \infty$ should be $\bar{x} := \max\{x_1,\ldots,x_k\}$, i.e.

$$\lim_{m\to\infty} \frac{1}{m} \log \left(\sum_{i=1}^{k} e^{mx_i} \right) = \bar{x}. \tag{5}$$

For convenience and in order to illustrate the generality of our approach, we now unify the descriptions of flat and non-flat SEs given in (1) and (3). We thus consider the dilation of a grey scale image function $f(x)$ with a general structuring function $b(x)$ (equivalent to a structuring element). The latter maps the Euclidean plane E to $\bar{\mathbb{R}} := \mathbb{R} \cup \{-\infty, \infty\}$, so that we may write:

$$(f \oplus b)(x) := \sup_{y \in E} \{f(y) + b(x - y)\}. \tag{6}$$

f To make the underlying SEs compatible with the process described below, we redefine it as

$$b(x) = \begin{cases} b(x), & x \in B, \\ -\infty, & \text{otherwise} \end{cases}, \quad B \subseteq E. \tag{7}$$

Thus, in order to retrieve a flat structuring element, we set $b(x) = 0$ over B. Let us note, that at hand of $b(x)$ we may realise any non-flat SE over B.

For presentation we focus now on discretisation of dilation only. To this end, we will consider E now as a grid and B as our bounded mask. After switching in this way from continuous to discrete domain, we can change the supremum operator to the maximum. Now we approximate the dilation with the smooth maximum:

$$(f \oplus b)(x) = \lim_{m\to\infty} \frac{1}{m} \log \left(\sum_{y \in E} e^{m\left(f(y)+b(x-y)\right)} \right) \tag{8}$$

$$= \lim_{m\to\infty} \frac{1}{m} \log \left(\sum_{y \in E} e^{mf(y)} e^{mb(x-y)} \right). \tag{9}$$

The sum within the last equation describes the discrete convolution of two functions $f_1, f_2 : E \to \mathbb{C}$ over the grid E:

$$(f_1 \otimes f_2)(x) = \sum_{y \in E} f_1(y) f_2(x - y). \tag{10}$$

This opens up the possibility to use the Fast Fourier Transform for fast convolution, let us refer to the classic work [18] for an exposition on this topic.

Since the actual choice of the parameter m is of interest for definition of the algorithm, let us elaborate a little bit more on the underlying formula (4) for the maximum. We believe that this investigation helps to get an idea about the range of a possible, suitable choice of m.

Let us consider a finite set A of non-negative integers, $A = \{x_1, x_2, \ldots, x_k\}$, and let x_j be the maximum of A. Let us consider the question how to find the least positive real value for m such that

$$\left\lfloor \frac{1}{m} \log \left(\sum_{i=1}^{k} e^{m \cdot x_i} \right) \right\rfloor = x_j \tag{11}$$

where $\lfloor . \rfloor$ is the floor function. Clearly, for all $m \in \mathbb{R}_{>0}$, it holds

$$\frac{1}{m} \log \left(\sum_{i=1}^{k} e^{m \cdot x_i} \right) \geq \frac{1}{m} \log \left(e^{m \cdot x_j} \right) = x_j \tag{12}$$

and $x_j \in \mathbb{Z}$. Therefore, we have

$$\alpha := \left\lfloor \frac{1}{m} \log \left(\sum_{i=1}^{k} e^{m \cdot x_i} \right) \right\rfloor \geq x_j. \tag{13}$$

Let us seek $m > 0$ such that $\alpha \leq x_j$. This means we have

$$\left\lfloor \frac{1}{m} \log \left(\sum_{i=1}^{k} e^{m \cdot x_i} \right) \right\rfloor \leq x_j \quad \Leftrightarrow \quad \frac{1}{m} \log \left(\sum_{i=1}^{k} e^{m \cdot x_i} \right) < x_j + 1. \tag{14}$$

We also have that $x_j = \max A$, so that $e^{m \cdot x_i} \leq e^{m \cdot x_j}$, for each $i \in 1, 2, \ldots, k$. Making use of this we may estimate

$$\frac{1}{m} \log \left(\sum_{i=1}^{k} e^{m \cdot x_i} \right) \leq \frac{1}{m} \log \left(\sum_{i=1}^{k} e^{m \cdot x_j} \right) = \frac{\log \left(k \cdot e^{m \cdot x_j} \right)}{m}. \tag{15}$$

If the latter expression is still smaller than $x_j + 1$, we would obtain $\log(k) < m$. Thus, if $\log(k) < m$, then we would have

$$\left\lfloor \frac{1}{m} \log \left(\sum_{i=1}^{k} e^{m \cdot x_i} \right) \right\rfloor = x_j = \max A. \tag{16}$$

This investigation indicates that a suitable choice for m within an algorithm may not be a large real number as one may think when considering the analytic relation (5). In terms of a suitable approximation, even values close to zero may be convenient, which may appear not intuitive at first glance, given (5). The Fig. 2 illustrates these aspects.

3.2   Details on the Algorithm

Let us first make more precise some notations useful for describing the discrete algorithm. Also in this paragraph we focus on dilation for this purpose.

For clarity of the complete presentation we decided to borrow the meaning of f and b from the analytic derivation. Thus we identify f and b also with the discretised image function $f : F \rightarrow L$ and the discretised filter (flat or non-flat) $b : B \rightarrow L$. Thereby F and B are bounded subsets of $\mathbb{Z}^2$, and L is the set of non-negative integers ≤ 255. After these steps, we have 2-D arrays for image and filter. In practice, Eq. (7) thus becomes

$$b(x) = \begin{cases} b(x), & x \in B, \\ -256, & \text{otherwise} \end{cases}, \quad B \subseteq \mathbb{Z}^2. \tag{17}$$

To fix some important numbers, we denote the size of image by n and size of filter by n_b. Let us first define the function $Expm(\cdot)$ as $Expm(f)(x) = e^{(m \cdot f(x))}$, $\forall x \in F$. The computation of $Expm(f)$ is performed in $\mathcal{O}(n)$. Similarly, we compute $Expm(b)$, with complexity $\mathcal{O}(n_b)$.

The next step of the process is to obtain the fast convolution of $Expm(f)$ and $Expm(b)$. This can be done by the FFT technique, relying on the classic convolution theorem. The corresponding process is performed in $\mathcal{O}(n \log n)$, given that the size of the image is larger than the size of the filter. We employ *scipy.signal.fftconvolve*($\cdot$) from SciPy package [15] to meet this task. Let us note that the package performs zero padding at the image boundary.

Now let $h = Expm(f) \otimes Expm(b)$. Using FFT with large numbers that may arise during the discrete transformations may cause a few overflow errors and errors of form $0/0$ and ∞/∞. We remove these few errors here by ourselves:

$$h(x) = \begin{cases} h(x), & \text{if } h(x) \in \mathbb{R}_{\geq 1} \\ e^{m \cdot 0}, & \text{if } h(x) < 1 \\ e^{m \cdot 255}, & \text{if } h(x) = \infty \\ e^{m \cdot 128}, & \text{otherwise} \end{cases}. \tag{18}$$

In Eq. (18), we have provided the means to deal with all the type of exceptions that might arise at this stage. But in practice, it is observed that, $h(x) \in \mathbb{R}_{\geq 1}$, in 99.99% of times. For example, in Fig. 1, exception occurred for only one pixel (during fast erosion with non-flat SE). No exceptions occurred in computations for Fig. 4.

Now we take the Fourier-based inverse of $Expm(\cdot)$. Let us thus define $InExpm(\cdot)$ as $InExpm(h)(x) = \frac{1}{m} \cdot \log(h(x)) \ \forall x \in F$. This completes the

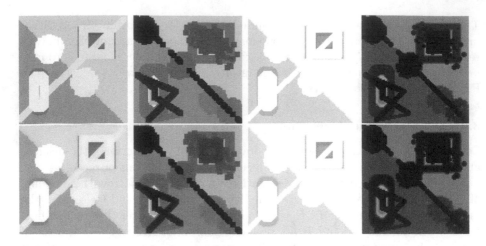

Fig. 3. Comparison of filtering results for dilation and erosion. **From left to right:** Flat dilation and erosion with a 5×5 structuring element, example of non-flat dilation and erosion, compare Fig. 1. **Top row:** Standard method. **Bottom row:** Our new method.

translation of (4) to the discrete setting. This process is again $\mathcal{O}(n \log n)$. The result of the fast dilation is given by $f \oplus b \approx Inte(h)$. Here, $Inte(h)(x) = \lfloor h(x) \rfloor$, $\forall x \in F$. Therefore, the entire process is $\mathcal{O}(n \log n)$ if the filter is smaller than the image, as it is in most cases.

The analogous fast erosion technique follows directly from duality. In theory it is possible to modify the formulae given in previous paragraph for smooth maxima to obtain smooth minima, obtaining $\min\{a, b\} = -\max\{-a, -b\}$. However, in practice, taking the exponential of negative integers causes round-off errors.

Let us finally comment on the choice of the parameter m. As indicated in the previous paragraph, theory seems to motivate to take $m \to \infty$. But, since we are taking the maximum of a finite number of discrete values (in L), taking a finite value of m suffices. Let us note again in this context, that the relation (4) is evaluated in the Fourier domain. Thus, in practice, the value of m relates to the implementation of the FFT. In our experiments we observed that setting $m = 0.16$ helps to get reasonable results and also to minimise computational errors as discussed above while using FFT. Relating to the shift observable in Fig. 2, let us stress that such shifted results still need to be transformed back to the original grey value domain, which has the effect that the shift gets diminished. Still, one may expect a minor shift of certain grey values.

4   Experiments

We demonstrate the efficacy of our fast dilation and erosion methods along with a validation of useful properties in a variety of tests.

Fig. 4. Comparison of filtering results for flat dilation and erosion, employing the standard test image *peppers* (not displayed here). **From left to right:** Standard dilation and fast dilation, standard erosion and fast erosion.

Experiment 1: Visual Quality. The aim of this first experiment is to assess the visual quality of results obtained by our new method, in comparison to standard implementations of morphological dilation/erosion. To this end we go back to our introductory example, see Fig. 1. At hand of Fig. 3, we observe that the results in this test (as well as in general in all our experiments) are for our method very similar to the ones obtained by standard implementations. In particular, we obtain the usual sharp object boundaries as with the standard filtering procedure. By Fig. 4 we show that our method gives high quality filtering results also for more natural images.

While the dilation/erosion results may appear at first glance virtually indistinguishable, a detailed investigation shows that there is as expected a slight shift in a few grey values in the image. As elaborated before, we think that this is related to our first choice of m.

Experiment 2: Algorithmic Complexity. The aim of this experiment is to assess the computational complexity behaviour of the proposed method. By construction, we predicted that the computational scaling of the method is independent of the size of the structuring element, and that it scales well with size of filtered images. The results of related experiments are displayed in Fig. 5.

For the evaluation of the method with respect to increasing image sizes, we have measured the average time to perform morphological operations on images of sizes $n \times n$, with $n = 128, 256, \ldots, 2048$. For this experiment, the size of the structuring element was fixed to 5×5. Let us note that for such simple-shaped flat structuring elements the dilation and erosion can be calculated in an efficient way using separable implementations of horizontal and vertical masks (in this case, 1×3 and 3×1). However, we will not limit ourselves to these here, but focus on more general masks. Let us for example, also refer to Fig. 6, and Fig. 7 (top-left). Results are independent of flat or non-flat structuring element. Most of the test images were obtained from SIPI Image Database [20].

For evaluating dependence on the size of the SE, we perform all operations on the classic *Lena* test image of size 512×512. We vary in this experiment the size of the structuring element only, given here as numbers $n_1 \times n_1$. It is clearly seen that the algorithm is nearly constant time with respect to the size of filter.

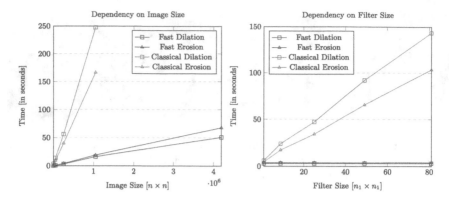

Fig. 5. Evaluation of algorithmic complexity of our method. For comparison only, we relate to a standard implementation of dilation/erosion. Note that fast erosion takes slightly longer than fast dilation, because we use the duality to compute fast erosion. **Left:** We observe the expected FFT behaviour, let us note that the lower axis is given in factors of 10^6. **Right:** We observe the expected behaviour, i.e. no dependence of our method on size of structuring element.

Experiment 3: Shape of Structuring Element. The goal of this experiment is to validate our claim, that the method is by construction independent of the shape of the structuring element. To this end, we have defined an uncommon, non-convex and non-flat SE structuring element, as displayed in Fig. 7. Let us note that the centre of the SE is at $(2,2)$. Since the SE is non-flat, we display here also its umbra [4], see Fig. 6.

In a first test, we employ the *pepper* test image for dilation with the non-convex SE. For comparison, we also display the result of standard filtering. As results after one filtering step are not apparent, we iterate the dilation here for two times. There is no apparent difference in the results between standard method and proposed algorithm with respect to edges, yet we observe again a certain shift of grey values.

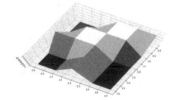

Fig. 6. Umbra of non-flat SE

As a final test we consider again erosion of our synthetic image from Fig. 1. For better assessment of the effect of the non-standard SE, we display also an erosion result with a 5×5 block-like SE. We clearly observe that we obtain with the proposed method very similar results as with standard filtering.

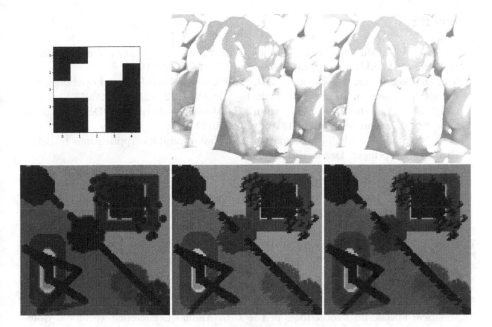

Fig. 7. Comparison of filtering results for dilation and erosion with non-flat, non-convex SE, displayed in top row, left image. **Tow row:** Standard dilation (middle image) and fast dilation with proposed method (right). **Bottom row, from left to right:** Erosion with 5×5 structuring element (for comparison), standard erosion scheme, proposed erosion scheme.

5 Conclusion

We have shown that the proposed algorithm combines many beneficial properties. Most apparently, its complexity does not depend on size or shape of the structuring element but only on the size of an input image, which distinguishes it from other fast methods. We conjecture that it may be particularly suitable for large images or filters. In addition, even non-flat filters can be used.

For future work, we aim to improve the method in terms of the computational range of the relaxation parameter. As we have reported, in some tests we observe for some grey values a shift, so that the current implementation may give slightly different results from standard filtering. While computational results in presented as well as in not reported tests are from our point of view already quite convincing, we conjecture that a further improvement in this point can make the proposed method a good choice for many applications.

Acknowledgements. The work was partly supported by the European Regional Development Fund, EFRE 85037495.

References

1. Tuzikov, A.V., Margolin, G.L., Grenov, A.I.: Convex set symmetry measurement via Minkowski addition. J. Math. Imaging Vis. **7**, 53–68 (1997)
2. Van Droogenbroeck, M., Buckley, M.J.: Morphological erosions and openings: fast algorithms based on anchors. J. Math. Imaging Vis. **22**, 121–142 (2005)
3. Geiger, T., Wachsmuth, D., Wachsmuth, G.: Optimal control of ODEs with state suprema. arXiv preprint arXiv:1810.11402 (2018)
4. Serra, J., Soille, P. (eds.): Mathematical Morphology and its Applications to Image Processing. Springer, Dordrecht (2012)
5. Najman, L., Talbot, H. (eds.): Mathematical Morphology: From Theory to Applications. Wiley-ISTE (2010)
6. Cooley, J.W., Lewis, P.A.W., Welch, P.D.: Historical notes on the fast Fourier transform. Proc. IEEE **55**(10), 1675–1677 (1967). https://doi.org/10.1109/PROC. 1967.5959
7. Roerdink, J.B.T.M.: Mathematical morphology in computer graphics, scientific visualization and visual exploration. In: Soille, P., Pesaresi, M., Ouzounis, G.K. (eds.) ISMM 2011. LNCS, vol. 6671, pp. 367–380. Springer, Heidelberg (2011). https://doi.org/10.1007/978-3-642-21569-8_32
8. Kukal, J., Majerova, D., Procházka, A.: Dilation and erosion of gray images with spherical masks. In: Proceedings of the 15th Annual Conference Technical Computing (2007)
9. Déforges, O., Normand, N., Babel, M.: Fast recursive grayscale morphology operators: from the algorithm to the pipeline architecture. J. Real-Time Image Proc. **8**, 143–152 (2013)
10. Moreaud, M., Itthirad, F.: Fast algorithm for dilation and erosion using arbitrary flat structuring element: improvement of Urbach and Wilkinson's algorithm to GPU computing. In: Proceedings of 2014 International Conference on Multimedia Computing and Systems (ICMCS), pp. 289–294 (2014)
11. Lin, X., Xu, Z.: A fast algorithm for erosion and dilation in mathematical morphology. In: Proceedings of 2009 WRI World Congress on Software Engineering, pp. 185–188 (2009)
12. Van Herk, M.: A fast algorithm for local minimum and maximum filters on rectangular and octagonal kernels. Pattern Recogn. Lett. **13**, 517–521 (1992)
13. Haralick, R., Sternberg, S., Zhuang, X.: Image Analysis Using Mathematical Morphology. IEEE Trans. Pattern Anal. Mach. Intell. **9**, 532–550 (1987)
14. Thurley, M.J., Danell, V.: Fast morphological image processing open-source extensions for GPU processing With CUDA. IEEE J. Sel. Top. Signal Process. **6**(7), 849–855 (2012)
15. Virtanen, P., et al.: SciPy 1.0: Fundamental Algorithms for Scientific Computing in Python. Nat. Methods **17**(3), 261–272 (2020)
16. Haralick, R., Zhuang, X., Lin, C., Lee, J.: The digital morphological sampling theorem. IEEE Trans. Acoust. Speech Signal Process. **37**, 2067–2090 (1990)
17. Van Droogenbroeck, M., Talbot, H.: Fast computation of morphological operations with arbitrary structuring elements. Pattern Recogn. Lett. **17**(14), 1451–1460 (1996)
18. Cooley, J., Lewis, P., Welch, P.: Application of the fast Fourier transform to computation of Fourier integrals, Fourier series, and convolution integrals. IEEE Trans. Audio Electroacoust. **15**(2), 79–84 (1967)

19. Schatzman, J.C.: Accuracy of the discrete Fourier transform and the fast Fourier transform. SIAM J. Sci. Comput. **17**(5), 1150–1166 (1996)
20. SIPI: Image Database, 11:04:00 GMT, October 2017. http://sipi.usc.edu/database/database.php?volume=misc. Accessed 20 Jan 2021

Diffusion, Pre-smoothing and Gradient Descent

Martin Welk[✉]

Institute of Biomedical Image Analysis, UMIT TIROL – Private University
of Health Sciences, Medical Informatics and Technology,
Eduard-Wallnöfer-Zentrum 1, 6060 Hall, Tyrol, Austria
`martin.welk@umit.at`

Abstract. Nonlinear diffusion of images, both isotropic and anisotropic, has become a well-established and well-understood denoising tool during the last three decades. Moreover, it is a component of partial differential equation methods for various further tasks in image analysis. For the analysis of such methods, their understanding as gradient descents of energy functionals often plays an important role. Often the diffusivity or diffusion tensor field for nonlinear diffusion is computed from pre-smoothed image gradients. What was not clear so far was whether nonlinear diffusion with this pre-smoothing step still is the gradient descent for some energy functional. This question is answered to the negative in the present paper. Triggered by this result, possible modifications of the pre-smoothing step to retain the gradient descent property of diffusion are discussed.

1 Introduction

Nonlinear diffusion processes are well-established image denoising methods. They also form indispensable building blocks in numerous image analysis methods involving partial differential equations (PDEs).

The starting point for the modern development of nonlinear diffusion methods in image processing was Perona and Malik's paper [13]. It proposes the isotropic nonlinear diffusion model which is nowadays mostly stated as

$$u_t = \operatorname{div}\left(g(|\boldsymbol{\nabla} u|^2)\,\boldsymbol{\nabla} u\right),\tag{1}$$

with $g : \mathbb{R}_0^+ \to \mathbb{R}_0^+$ being a decreasing diffusivity function. From the two candidates for g proposed in [13], the function

$$g(s^2) = \frac{1}{1 + s^2/\lambda^2}\tag{2}$$

prevails as a widespread standard choice. The parameter $\lambda > 0$ can be interpreted intuitively as a threshold. It separates small gradients $|\boldsymbol{\nabla} u| < \lambda$ that are smoothed out as noise from large ones $|\boldsymbol{\nabla} u| > \lambda$ presumed to represent valuable image structures that should be preserved.

A. Elmoataz et al. (Eds.): SSVM 2021, LNCS 12679, pp. 78–90, 2021.
https://doi.org/10.1007/978-3-030-75549-2_7

Other diffusivities have been proposed, with total variation diffusion [1,2,6] being the most prominent example besides (2).

In semidiscrete (space-discrete, time-continuous) and fully discrete settings well-posedness of nonlinear isotropic diffusion processes has been proved, see [3,4,15].

In the fully continuous setting, Perona-Malik diffusion in its original form is not generally well-posed. Solutions of its initial-boundary value problem in different function spaces have been investigated [7–10,23] with mixed results: Although classical solutions exist in certain conditions, either locally or even globally, this is not generally the case. Weak solutions can be highly non-unique. Stability results are largely limited to extremum principles (L^∞-stability).

A pivotal role in the stability issues of Perona-Malik diffusion is played by the staircasing phenomenon, by which even smooth initial data develop discontinuities within finite time. This is essentially caused by the local appearance of inverse diffusion in regions with large gradients where the flux $g(|\nabla u|^2)|\nabla u|$ decreases with $|\nabla u|$. With the diffusivity (2), for example, this is the case for $|\nabla u| > \lambda$, compare [15].

Regularised Nonlinear Diffusion. An explanation for the discrepancy between the space-continuous and discrete behaviour of Perona-Malik diffusion is given in [17] where it is pointed out that discretisation by itself introduces a regularising effect. However, relying on regularisation by discretisation means that important features of the actual image enhancement process are not part of the space-continuous model. Therefore it is desirable to have an explicit regularisation already in the space-continuous model.

Indeed, explicit regularisations in the continuous setting have been proposed as early as 1992 in [5,11]. The pre-smoothing by Gaussian smoothing of the gradient within the diffusivity expression as introduced by [5] enjoys most popularity till today,

$$u_t = \operatorname{div}\left(g(|G_\sigma * \nabla u|^2)\,\nabla u\right). \tag{3}$$

Well-posedness of (3) has been proven in [5].

Introducing an additional directional dependency of the diffusivity, one arrives at anisotropic diffusion processes. One such process, *edge-enhancing anisotropic diffusion (EED)*, can be stated as [14,15]

$$u_t = \operatorname{div}\left(g(\boldsymbol{J}(u))\,\nabla u\right). \tag{4}$$

Herein, the decreasing diffusivity function g acts on the rank-one symmetric outer-product matrices

$$\boldsymbol{J}(u) = (G_\sigma * \nabla u)(G_\sigma * \nabla u)^{\mathrm{T}} \tag{5}$$

to yield at each image location a diffusion tensor $g(\boldsymbol{J})$. The diffusion tensor possesses one eigenvalue $g(|G_\sigma * \nabla u|^2)$ with an eigenvector parallel to the pre-smoothed gradient $G_\sigma * \nabla u$, and one eigenvalue $g(0) = 1$ for an eigenvector orthogonal to the gradient. Like isotropic nonlinear diffusion, EED is capable of preserving and even enhancing sharp edges because it suppresses, or even reverts, diffusion across edges; in contrast to its isotropic predecessor, however, it allows for undiminished diffusion flow along edges, thus achieving better denoising near edges.

It is worth noting that for grey-value images, the pre-smoothing of the gradient is essential for true anisotropy because the application of g to the rank-one matrices $\nabla u \nabla u^{\mathrm{T}}$ would yield just the exact same flux field as Perona-Malik diffusion, and thereby effectively reproduce isotropic diffusion. For multi-channel (e.g., colour) images, anisotropy can arise even without pre-smoothing if the gradient directions of the colour channels do not coincide.

Diffusion as Gradient Descent. PDE-based methods in image processing, be it for denoising or for other purposes, are often derived from variational approaches, where an image processing task is formulated as the minimisation of some energy functional dependent on the sought image. Variational calculus then allows to derive PDEs either as Euler-Lagrange equations or as gradient descent of the functional. Indeed, the energy functional

$$E[u] = \frac{1}{2} \int_{\Omega} \Psi\big(|\nabla u|^2\big) \, \mathrm{d}\boldsymbol{x} \tag{6}$$

immediately leads to the gradient descent

$$u_t = \mathrm{div}\big(\Psi'(|\nabla u|^2) \, \nabla u\big) , \tag{7}$$

which is exactly the original Perona-Malik equation without pre-smoothing, with diffusivity $g \equiv \Psi'$; compare [12]. For example, the diffusivity (2) arises from $\Psi(s^2) = \lambda^2 \ln(1 + s^2/\lambda^2)$. More general variational models often contain summands of the type (6) which then yield diffusion components in their gradient descent equations.

Similarly, the EED equation without pre-smoothing in the case of multi-channel images can be obtained as gradient descent of an energy functional in which a penalty Ψ as before is applied to the sum of outer products $\nabla u_c \nabla u_c^{\mathrm{T}}$ of the colour channels u_c.

For PDEs arising from gradient descent, it can be attractive to design finite-difference discretisations in a way that they preserve this connection. To this end, the energy functional is discretised, which yields a function of a large finite number of real variables (one per pixel). Gradient descent for this function takes the form of a system of ordinary differential equations which is a discretisation of the PDE, see [19] for an example.

Derivation as a gradient descent from an energy functional can serve as a strong theoretical justification that singles out a particular PDE evolution from similar ones by an optimality criterion. The analysis of energy functionals often provides powerful tools to derive important properties of the models such as convergence to a unique steady state (e.g. for convex energy functionals).

Thus, the gradient descent property makes a strong case for the original Perona-Malik diffusion (1). In practice, however, it is very often the pre-smoothed version (3) which is used for denoising, or as building block within some other

PDE-based image processing method. This raises naturally the question whether (3) and similar diffusion processes are gradient descents for suitable energy functionals, too.

Whereas researchers in the field have pondered about this question time and again, it seems that it has attracted little real effort over the years. Researchers noted that there is no known energy functional yielding this evolution as gradient descent but could not agree whether this would be principally impossible, or whether they had not been inventive enough to write down the proper energy functional, or whether maybe such an energy functional existed but would not admit being stated in a closed form. For example, [19, p. 199, footnote 3] states that "For $\sigma \neq 0$, no energy functional is known that has [isotropic nonlinear diffusion with pre-smoothing] as gradient descent."

However, given the theoretical advantages of PDE methods being derived from variational models, we believe that this question is worth settling, and will do so in this paper. Unfortunately, the answer is negative. We will therefore add some – explorative – discussion of alternatives.

Structure of the Paper. In Sect. 2, we prove that neither 1D nonlinear diffusion, nor, as a consequence, 2D isotropic nonlinear diffusion, nor EED can be stated as gradient descents. In Sect. 3, we discuss how to design regularised diffusion processes that are exact gradient descents. Experimental demonstration of one such process in comparison to Perona-Malik diffusion is provided in Sect. 4. A short summary in Sect. 5 concludes the paper.

2 Integrability Analysis of Diffusion with Pre-smoothing

In order to investigate whether diffusion with pre-smoothing can be stated as a gradient descent, let us recall first the situation in classical vector field analysis.

Classical Integrability Conditions. Consider a continuously differentiable vector field $v : \mathbb{R}^d \to \mathbb{R}^d$, where $v(x) = (v_1(x_1, \ldots, x_d), \ldots, v_d(x_1, \ldots, x_d))^{\mathrm{T}}$. A necessary criterion for v to be the gradient field of a potential V is then given by the *integrability condition*

$$\frac{\partial v_i}{\partial x_j} = \frac{\partial v_j}{\partial x_i} \quad \text{for all } i, j = 1, \ldots, d, i \neq j. \tag{8}$$

In particular, for $d = 2$ or $d = 3$ this boils down to the well-known condition $\mathrm{rot}\, v = 0$ where $\mathrm{rot}\, v = \nabla \wedge v = \partial_1 v_2 - \partial_2 v_1$ (scalar-valued) in two, and $\mathrm{rot}\, v = \nabla \times v$ (vector-valued) in three dimensions.

Coordinate-Free Integrability Conditions. Whereas it is usually assumed in (8) that coordinates x_i, v_i are taken w.r.t. some orthonormal basis, it is obvious that this is not necessary: Note that (8) is trivially fulfilled also for $i = j$. Thus it can easily be extended to linear combinations (with $y = (y_1, \ldots, y_d)^{\mathrm{T}}$ and $z = (z_1, \ldots, z_d)^{\mathrm{T}}$ being unit vectors), yielding

$$\sum_{i=1}^{d} \sum_{j=1}^{d} \alpha_i \beta_j \frac{\partial v_i}{\partial x_j} = \sum_{i=1}^{d} \sum_{j=1}^{d} \beta_j \alpha_i \frac{\partial v_j}{\partial x_i} \tag{9}$$

which by $\langle v, y \rangle = \sum_{i=1}^{d} y_i v_i$, $\partial/\partial z = \sum_{j=1}^{d} z_j \, \partial/\partial x_j$ means that the set of integrability conditions can be re-stated as

$$\frac{\partial \langle v, y \rangle}{\partial z} = \frac{\partial \langle v, z \rangle}{\partial y} \tag{10}$$

for arbitrary unit vectors y, z, i.e. the component of v in direction of y has a directional derivative in direction of z equal to that of the z component in y direction. The virtue of (10) is that it is coordinate-free, i.e. it does not depend on any choice of orthonormal basis.

Moreover, each integrability condition does in fact involve only the projection of v onto a two-dimensional subspace. This is natural since the restriction and projection of a gradient descent to a subspace is again a gradient descent in that subspace. This argument does obviously hold not only in finite, but also in infinite dimensions, which means that the set of necessary conditions (10) remains valid even if $v : V \to V$ with any Hilbert space V.

Transfer to Function Spaces. Assume that $V = V(\Omega)$ is a Hilbert space of sufficiently smooth functions over some domain Ω. We consider a time-dependent partial differential equation

$$u_t = F[u] , \qquad F[u](x) = f\big(u(x), \partial_{\alpha_1} u(x), \partial_{\alpha_2} u(x), \ldots\big) \tag{11}$$

where $\partial_{\alpha_i} u$ are partial derivatives of u w.r.t. spatial coordinates in Ω, and f some sufficiently smooth function combining these. The flux $F[u]$ defined by f can then be considered as a mapping from the space V of functions to the space of perturbations of functions in V, which can be identified with V. Thus, F is a vector field on V.

If (11) is to be the gradient descent of some functional $E[u]$ over V, F needs to be the gradient field of $E[u]$. Translating the integrability conditions (10) to this situation, we obtain as the set of necessary conditions

$$\frac{\partial \langle u_t, v \rangle}{\partial \langle u, w \rangle} = \frac{\partial \langle u_t, w \rangle}{\partial \langle u, v \rangle} \tag{12}$$

where $v, w \in V$ are arbitrary perturbation functions. Using $\partial(\,\cdot\,)/\partial \langle u, v \rangle = \frac{d}{d\varepsilon}(\,\cdot\,)|_{\varepsilon=0}$, this set of conditions can be translated further into

$$\frac{d}{d\varepsilon} \langle F[u + \varepsilon v], w \rangle \bigg|_{\varepsilon=0} = \frac{d}{d\varepsilon} \langle F[u + \varepsilon w], v \rangle \bigg|_{\varepsilon=0} \tag{13}$$

for arbitrary perturbation functions v, w.

Analysis of Pre-smoothed Nonlinear Diffusion in 1D. We turn now to apply (13) to analyse the 1D nonlinear diffusion process with pre-smoothing given by

$$u_t = \partial_x \big(g((G_\sigma * u_x)^2) u_x\big) . \tag{14}$$

Let us assume that u and its perturbation functions come from a suitable Hilbert space of functions over a domain $\Omega \subseteq \mathbb{R}$ with the standard scalar product

$$\langle u, v \rangle = \int_\Omega u(x) v(x) \, dx . \tag{15}$$

(Note that if Ω is not the entire $\mathbb{R}$, the boundary treatment for the convolution must be specified appropriately.) Assume further that the perturbation functions v and w vanish on the boundary of Ω if any (this technical condition could be relaxed but it simplifies the expressions arising from integration by parts later on). Setting for abbreviation $\tilde{u} := G_\sigma * u$, we can calculate

$$F[u] = \partial_x\big(g(\tilde{u}_x^2)\,u_x\big) = g'(\tilde{u}_x^2)\,2\,\tilde{u}_x\,\tilde{u}_{xx}\,u_x + g(\tilde{u}_x^2)\,u_{xx} \tag{16}$$

and thus

$$\frac{\mathrm{d}}{\mathrm{d}\varepsilon}\langle F[u+\varepsilon v], w\rangle\Big|_{\varepsilon=0} = \int_\Omega \frac{\mathrm{d}}{\mathrm{d}\varepsilon}\Big(2\,g'((\tilde{u}_x + \varepsilon\tilde{v}_x)^2)\,(\tilde{u}_x + \varepsilon\tilde{v}_x)\,(\tilde{u}_{xx} + \varepsilon\tilde{v}_{xx})\cdot$$

$$(u_x + \varepsilon v_x) + g((\tilde{u}_x + \varepsilon\tilde{v}_x)^2)(u_{xx} + \varepsilon v_{xx})\Big)\Big|_{\varepsilon=0} w\,\mathrm{d}x$$

$$= \int_\Omega \Big(2\,g''(\tilde{u}_x^2)\cdot 2\,\tilde{u}_x\,\tilde{v}_x\,\tilde{u}_x\,\tilde{u}_{xx}\,u_x + 2\,g'(\tilde{u}_x^2)\,\tilde{v}_x\,\tilde{u}_{xx}\,u_x$$

$$+ 2\,g'(\tilde{u}_x^2)\,\tilde{u}_x\,\tilde{v}_{xx}\,u_x + 2\,g'(\tilde{u}_x^2)\,\tilde{u}_x\,\tilde{u}_{xx}\,v_x$$

$$+ g'(\tilde{u}_x^2)\cdot 2\,\tilde{u}_x\,\tilde{v}_x\,u_{xx} + g(\tilde{u}_x^2)\,v_{xx}\Big)w\,\mathrm{d}x$$

$$= \int_\Omega 4\,g''(\tilde{u}_x^2)\,\tilde{u}_x^2\,u_x\,\tilde{u}_{xx}\,\tilde{v}_x\,w + 2\,g'(\tilde{u}_x^2)\,u_x\,\tilde{u}_{xx}\,\tilde{v}_x\,w$$

$$+ 2\,g'(\tilde{u}_x^2)\,\tilde{u}_x\,u_x\,\tilde{v}_{xx}\,w + 2\,g'(\tilde{u}_x^2)\,\tilde{u}_x\,\tilde{u}_{xx}\,v_x\,w$$

$$+ 2\,g'(\tilde{u}_x^2)\,\tilde{u}_x\,u_{xx}\,\tilde{v}_x\,w + g(\tilde{u}_x^2)\,v_{xx}\,w\,\mathrm{d}x \tag{17}$$

Using integration by parts for the summands involving second derivatives of perturbation functions, most summands cancel, leaving

$$\frac{\mathrm{d}}{\mathrm{d}\varepsilon}\langle F[u+\varepsilon v], w\rangle\Big|_{\varepsilon=0} = \int_\Omega -g(\tilde{u}_x^2)\,v_x\,w_x - 2\,g'(\tilde{u}_x^2)\tilde{u}_x\,u_x\,\tilde{v}_x\,w_x\,\mathrm{d}x . \tag{18}$$

We combine this expression with its counterpart for $\frac{\mathrm{d}}{\mathrm{d}\varepsilon}\langle F[u+\varepsilon v], w\rangle\big|_{\varepsilon=0}$ to obtain, with cancellation of $g(\tilde{u}_x^2)\,v_x\,w_x$,

$$\frac{\mathrm{d}}{\mathrm{d}\varepsilon}\langle F[u+\varepsilon v], w\rangle\Big|_{\varepsilon=0} - \frac{\mathrm{d}}{\mathrm{d}\varepsilon}\langle F[u+\varepsilon v], w\rangle\Big|_{\varepsilon=0}$$

$$= 2\int_\Omega g'(\tilde{u}_x^2)\tilde{u}_x\,u_x\,(v_x\,\tilde{w}_x - \tilde{v}_x\,w_x)\,\mathrm{d}x . \tag{19}$$

Unfortunately, this expression does not vanish identically for non-constant diffusivity g and arbitrary functions u, v, w. We have therefore proven the following statement.

Proposition 1. *The regularised nonlinear 1D diffusion process (14) with a non-constant, twice continuously differentiable diffusivity function g is not the gradient descent for any energy functional on functions $u : \Omega \to \mathbb{R}$, $\Omega \subseteq \mathbb{R}$, w.r.t. the standard metric in function space induced by the scalar product (15).*

Implications for 2D Diffusion Processes. We turn now to consider the nonlinear isotropic diffusion process from [13] with pre-smoothing [5] in 2D (or higher dimension) as given in (3).

We assume that the domain $\Omega \subseteq \mathbb{R}^d$ is of the form $\Omega = \Omega_1 \times \Omega_2$ where $\Omega_1 \subseteq \mathbb{R}$ and $\Omega_2 \subseteq \mathbb{R}^{d-1}$; this is obviously the case e.g. for rectangular images. We assume further that the scalar product of functions on Ω (and thus the Hilbert function space $\mathcal{V}(\Omega)$) is chosen in such a way that the functions that are constant along all but the first coordinate direction and are given by $u(x_1, x_2, \ldots, x_d) = u_1(x_1)$ with $u_1 \in \mathcal{V}(\Omega_1)$ belong to $\mathcal{V}(\Omega)$. This can be ensured, e.g., by taking Ω_2 as an interval or Cartesian product of intervals with periodic boundary conditions, or by equipping $\Omega_2 = \mathbb{R}^{d-1}$ with a weighted scalar product in which the local weights decay quickly enough to ensure finiteness of $\langle 1, 1 \rangle$, e.g.

$$\langle u, v \rangle = \int_{\Omega_2} G_\sigma(\boldsymbol{x}) u(\boldsymbol{x}) v(\boldsymbol{x}) \, \mathrm{d}\boldsymbol{x} . \tag{20}$$

By restriction to functions as described above that depend on the first coordinate only, the process (3) reduces verbatim to (14). As Proposition 1 shows, it cannot be represented as gradient descent in this particular case, and therefore neither in general. Using a suitable limit argument where necessary, the result can be transferred to the function space $\mathcal{V}(\Omega)$ with standard scalar product as stated in the following corollary.

Corollary 1. *The regularised nonlinear isotropic diffusion process* (3) *with a non-constant, twice continuously differentiable diffusivity function g is not the gradient descent for any energy functional on functions $u : \Omega \to \mathbb{R}$, $\Omega \subseteq \mathbb{R}^d$, w.r.t. the standard metric in function space.*

Similar arguments apply to EED, yielding the following statement.

Corollary 2. *The regularised nonlinear anisotropic diffusion process* (4) *with a non-constant, twice continuously differentiable diffusivity function g is not the gradient descent for any energy functional on functions $u : \Omega \to \mathbb{R}$, $\Omega \subseteq \mathbb{R}^d$, w.r.t. the standard metric in function space.*

3   Alternatives

In this section we discuss possible alternatives to the established pre-smoothing in diffusion methods which could be compatible with the gradient descent framework that exists for nonlinear diffusion without pre-smoothing, given that this framework often also inspires applications of the pre-smoothed variants.

We remark first that using the traditional pre-smoothed Perona-Malik diffusion (3) for edge-enhancing denoising, the smoothing of the gradient has a two-fold role. On one hand, it yields an overall smoother diffusivity field, thus supporting edge enhancement in creating a more regular set of edges. On the other hand, it boosts the removal of small-scale structures such as single noise pixels which would otherwise be stabilised longer by their surrounding high gradients and thus low diffusivities.

We also notice that "regularisation by discretisation", as undesirable an intertwining of model and numerics it involves, has the advantage to retain the gradient descent property if an appropriate discretisation is used. However, it does not provide a means to steer the degree of regularisation.

Modified Energy Functional. In order to find diffusion equations with adjustable regularisation parameters that are consistent with gradient descent in the space-continuous setting, we consider modifications of the energy functional (6).

In coherence-enhancing anisotropic diffusion (CED) [16], another anisotropic diffusion process introduced by Weickert that is designed to denoise and enhance line-like structures rather than providing general-purpose denoising such as EED, a smoothed structure tensor is used that involves, besides the smoothing of the gradients ∇u, a second Gaussian convolution that applies to the outer product matrices, leading to

$$J_\varrho(u) = G_\varrho * \left((G_\sigma * \nabla u)(G_\sigma * \nabla u)^\mathrm{T}\right) . \tag{21}$$

Moreover, following [18], the energy functional (6) can be rewritten as

$$E[u] = \frac{1}{2} \int_\Omega \Psi\left(\mathrm{trace}(\nabla u \nabla u^\mathrm{T})\right) \mathrm{d}\boldsymbol{x} . \tag{22}$$

Inspired by this observation, we consider smoothing of $|\nabla u|^2 = \mathrm{trace}(\nabla u \nabla u^\mathrm{T})$. Generalising Gaussian convolutions to linear operators L_1, L_2, we write down the ansatz

$$E[u] = \frac{1}{2} \int_\Omega \Psi\left(L_1(|L_2(\nabla u)|^2)\right) \mathrm{d}\boldsymbol{x} . \tag{23}$$

For the following, we denote by L^* the adjoint operator of a linear operator L, i.e. the linear operator that satisfies

$$\int_\Omega L^*(f) \cdot g \, \mathrm{d}\boldsymbol{x} = \int_\Omega f \cdot L(g) \, \mathrm{d}\boldsymbol{x} \tag{24}$$

for all f, g. Gaussian convolution on $\mathbb{R}^n$ is self-adjoint, i.e. $(G_\sigma *)^* = G_\sigma *$.

Gradient Descent. To determine the gradient descent of (26), we calculate, for some perturbation function v that vanishes on the boundary of Ω,

$$\frac{\mathrm{d}}{\mathrm{d}\varepsilon} E[u + \varepsilon v]\Big|_{\varepsilon=0} = \int_\Omega \Psi'\big(L_1(|L_2(\nabla u)|^2)\big) \cdot L_1\big(\langle L_2(\nabla u), L_2(\nabla v)\rangle\big)\,\mathrm{d}\boldsymbol{x}$$

$$= \int_\Omega L_1^*\Big(\Psi'\big(L_1(|L_2(\nabla u)|^2)\big)\Big) \cdot \langle L_2(\nabla u), L_2(\nabla v)\rangle\,\mathrm{d}\boldsymbol{x}$$

$$= \int_\Omega \Big\langle L_2^*\Big(L_1^*\big(\Psi'(L_1(|L_2(\nabla u)|^2))\big)\cdot L_2(\nabla u)\Big), \nabla v\Big\rangle\,\mathrm{d}\boldsymbol{x}$$

$$= -\int_\Omega \operatorname{div}\Big(L_2^*\Big(L_1^*\big(\Psi'(L_1(|L_2(\nabla u)|^2))\big)\cdot L_2(\nabla u)\Big)\Big)\cdot v\,\mathrm{d}\boldsymbol{x} \tag{25}$$

from which we read off the desired gradient descent as

$$u_t = \operatorname{div}\Big(L_2^*\Big(L_1^*\big(\Psi'(L_1(|L_2(\nabla u)|^2))\big)\Big)\cdot L_2(\nabla u)\Big). \tag{26}$$

Note that in this diffusion-like process the flux is the smoothed version of a vector field which is at each location a scalar multiple of $L_2(\nabla u)$. If L_2 is not the identical operator id, the flux direction can, and will at most locations with non-trivial structure, deviate from that of ∇u. Thus, (26) is an anisotropic process. However, there is an important difference to established anisotropic diffusion processes like EED or CED: In these, the flux is always the product of a positive semidefinite diffusion tensor with ∇u. Therefore, the projection of the flux onto the gradient direction always points in positive gradient direction, ensuring a forward diffusion component in that direction. In contrast, the flux in (26) can even have a negative projection onto ∇u, thus performing inverse diffusion with negative diffusivity. Inverse diffusion is a prototype of an unstable evolution. Although diffusion processes that involve local inverse diffusion can still be stable for entire images, compare [20, 22], it is not obvious whether this is true here. To decide this requires a more detailed stability analysis which is beyond the scope of this paper. At any rate, to devise stable numerical schemes for such a process would be challenging [20–22]. The gradient descent (26) with $L_2 \neq$ id does therefore not lend itself as a promising candidate to replace (3).

With $L_2 \equiv$ id, instead, (26) simplifies into

$$u_t = \operatorname{div}\Big(L_1^*\big(\Psi'(L_1(|\nabla u|^2))\big)\cdot\nabla u\Big). \tag{27}$$

Specifically for L_1 being Gaussian convolution, we have

$$u_t = \operatorname{div}\Big((G_\sigma * \Psi'(G_\sigma * |\nabla u|^2))\nabla u\Big). \tag{28}$$

We will demonstrate the effect of the evolution (28) compared with traditional Perona-Malik diffusion with pre-smoothing (3) by an experiment in the next section. Before we do so, let us shortly discuss what effect can be expected from the modified pre-smoothing in (28). Unlike in (3), where pre-smoothing amounted to a local averaging of (oriented) gradient directions, the Gaussian convolution in (28) locally averages the (non-oriented) gradient flow-line directions (or, equivalently, level line directions).

Regarding the twofold effect of traditional pre-smoothing discussed at the begin of this section, this means that the first effect, creating a more regular set of edges, will still happen in a similar way. The second effect, fast removal of small-scale structures, cannot be expected to the same extent because opposing gradients do no longer cancel, thus leaving a higher average gradient magnitude to be estimated around small-scale structures.

4   Experiments

We use a test image with substantial additive Gaussian noise, Fig. 1a, to compare the gradient-descent-based regularised isotropic nonlinear diffusion (28) with traditional pre-smoothed Perona-Malik diffusion (3). Both PDEs are discretised by essentially the same explicit Euler forward scheme with central spatial differences, see e.g. [22, eq. (10)]. In all experiments, we use the diffusivity (2) with the same threshold, and the same standard deviation for pre-smoothing by Gaussian convolution.

To start with, Fig. 1b, c show the result of (3). As expected, noise is removed quickly, and progressive simplification of image structures, and smoothing of edges takes place.

The next two frames, Fig. 1d, e show a failed evolution: In order to give an indication of the difficulties arising from anisotropy with inverse diffusion that occur in (26) with non-trivial L_2, we show this process with L_2 being Gaussian convolution, and $L_1 \equiv id$, i.e.

$$u_t = \mathrm{div}\Big(G_\sigma * \big(\Psi'(|G_\sigma * \nabla u|^2)(G_\sigma * \nabla u)\big)\Big) . \tag{29}$$

Using the same evolution times as in b, one observes small-scale oscillatory artifacts or ripples, Fig. 1d, that persist even at an evolution time when many meaningful image structures have already been removed, see frame e. Although it remains open whether this process can be stabilised by more advanced numerics, the experiment supports that (29) is not a convincing replacement for (3).

The remaining frames, Fig. 1f–i, demonstrate the evolution (28). Frames f and g show the same evolution times as b and c. As expected, small-scale noise takes longer to be removed but is eventually eliminated. In frame g, still many more small-scale details are preserved than in c although the larger-scale smoothing effect is not that much behind that in c. With doubling the evolution time, frame h, the overall image smoothing is visually comparable to that in c, but still with a stronger tendency to preserve small features. In the long run, of course, (28) converges to a flat homogeneous image, as the final frame Fig. 1i illustrates.

Fig. 1. Comparison of regularised isotropic nonlinear diffusion evolutions. In all experiments, an explicit forward scheme with central spatial differences and time step size $\tau = 0.25$ was employed, and the diffusivity function was fixed to (2) with $\lambda = 1$. All Gaussian convolutions used $\sigma = 2$. **Top left, a** Test image *Cameraman* (256×256 pixels) with Gaussian noise of standard deviation 40. – **Middle horizontal strip (reference):** Pre-smoothed Perona-Malik diffusion (3). **b** diffusion time $T = 125$. – **c** $T = 500$. – **Top right strip:** Unstable evolution (29). **d** Diffusion time $T = 125$ (same as b). – **e** $T = 1000$. – **Lower strip:** Evolution (28). **f** Diffusion time $T = 125$ (same as b). – **g** $T = 500$ (same as c). – **h** $T = 1000$ (visual effect comparable to c). – **i** $T = 10000$.

5 Summary

In this paper we have closed a long-standing – minor, but, in our opinion, relevant – gap in the theoretical framework of diffusion methods in image processing. We have proven that nonlinear isotropic and anisotropic diffusion with the commonly used pre-smoothing of image gradients is not the gradient descent of any energy functional, despite the fact that applications of these components in image processing applications are often justified with energy minimisation arguments.

This result raised the question whether the concepts of diffusion as a gradient descent, and pre-smoothing as an important ingredient for stability in diffusion methods can be reconciliated. By analysing a generalised ansatz for an energy functional with pre-smoothing, we could single out pre-smoothing of the squared gradients as a candidate regularisation procedure for nonlinear diffusion that retains the gradient descent property.

References

1. Andreu-Vaillo, F., Caselles, V., Mazon, J.M.: Parabolic Quasilinear Equations Minimizing Linear Growth Functionals, Progress in Mathematics, vol. 223. Birkhäuser, Basel (2004)
2. Bellettini, G., Caselles, V., Novaga, M.: The total variation flow in R^N. J. Differ. Equ. **184**(2), 475–525 (2002)
3. Bellettini, G., Novaga, M., Paolini, M.: Convergence for long-times of a semidiscrete Perona-Malik equation in one dimension. Math. Models Methods Appl. Sci. **21**(2), 241–265 (2011)
4. Bellettini, G., Novaga, M., Paolini, M., Tornese, C.: Convergence of discrete schemes for the Perona-Malik equation. J. Differ. Equ. **245**, 892–924 (2008)
5. Catté, F., Lions, P.L., Morel, J.M., Coll, T.: Image selective smoothing and edge detection by nonlinear diffusion. SIAM J. Numer. Anal. **32**, 1895–1909 (1992)
6. Dibos, F., Koepfler, G.: Global total variation minimization. SIAM J. Numer. Anal. **37**(2), 646–664 (2000)
7. Ghisi, M., Gobbino, M.: A class of local classical solutions for the one-dimensional Perona-Malik equation. Trans. Am. Math. Soc. **361**(12), 6429–6446 (2009)
8. Ghisi, M., Gobbino, M.: An example of global classical solution for the Perona-Malik equation. Commun. Partial. Differ. Equ. **36**(8), 1318–1352 (2011)
9. Guidotti, P.: Anisotropic diffusions of image processing from Perona-Malik on. In: Ambrosio, L., Giga, Y., Rybka, P., Tonegawa, Y. (eds.) Variational Methods for Evolving Objects. Advanced Studies in Pure Mathematics, vol. 67, pp. 131–156. Mathematical Society of Japan, Tokyo (2015)
10. Kawohl, B., Kutev, N.: Maximum and comparison principle for one-dimensional anisotropic diffusion. Mathematische Annalen **311**, 107–123 (1998)
11. Nitzberg, M., Shiota, T.: Nonlinear image filtering with edge and corner enhancement. IEEE Trans. Pattern Anal. Mach. Intell. **14**, 826–833 (1992)
12. Nordström, N.: Biased anisotropic diffusion - a unified regularization and diffusion approach to edge detection. Image Vis. Comput. **8**, 318–327 (1990)
13. Perona, P., Malik, J.: Scale space and edge detection using anisotropic diffusion. IEEE Trans. Pattern Anal. Mach. Intell. **12**, 629–639 (1990)

14. Weickert, J.: Theoretical foundations of anisotropic diffusion in image processing. Comput. Suppl. **11**, 221–236 (1996)
15. Weickert, J.: Anisotropic Diffusion in Image Processing. Teubner, Stuttgart (1998)
16. Weickert, J.: Coherence-enhancing diffusion filtering. Int. J. Comput. Vis. **31**(2/3), 111–127 (1999)
17. Weickert, J., Benhamouda, B.: A semidiscrete nonlinear scale-space theory and its relation to the Perona-Malik paradox. In: Solina, F., Kropatsch, W.G., Klette, R., Bajcsy, R. (eds.) Advances in Computer Vision, pp. 1–10. Springer, Wien (1997). https://doi.org/10.1007/978-3-7091-6867-7
18. Weickert, J., Schnörr, C.: A theoretical framework for convex regularizers in PDE-based computation of image motion. Int. J. Comput. Vis. **45**(3), 245–264 (2001)
19. Welk, M., Steidl, G., Weickert, J.: Locally analytic schemes: a link between diffusion filtering and wavelet shrinkage. Appl. Comput. Harmon. Anal. **24**, 195–224 (2008)
20. Welk, M., Weickert, J.: PDE evolutions for M-smoothers: from common myths to robust numerics. In: Lellmann, J., Burger, M., Modersitzki, J. (eds.) SSVM 2019. LNCS, vol. 11603, pp. 236–248. Springer, Cham (2019). https://doi.org/10.1007/978-3-030-22368-7_19
21. Welk, M., Weickert, J.: PDE evolutions for M-smoothers in one, two, and three dimensions. J. Math. Imaging Vis. **63**(2), 157–185 (2021)
22. Welk, M., Weickert, J., Gilboa, G.: A discrete theory and efficient algorithms for forward-and-backward diffusion filtering. J. Math. Imaging Vis. **60**(9), 1399–1426 (2018)
23. Zhang, K.: Existence of infinitely many solutions for the one-dimensional Perona-Malik model. Calc. Var. Partial. Differ. Equ. **26**(2), 126–171 (2006)

Local Culprits of Shape Complexity

Mazlum Ferhat Arslan$^{(\boxtimes)}$ and Sibel Tari

Department of Computer Engineering, Middle East Technical University,
06800 Ankara, Turkey
{ferhata,stari}@metu.edu.tr

Abstract. Quantifying shape complexity is useful in several practical problems in addition to being interesting from a theoretical point of view. In this paper, instead of assigning a single global measure of complexity, we propose a distributed coding where each point on the shape domain a measure of its contribution to complexity is assigned. We define the shape *simplicity* as the expressibility of the shape via a prototype shape. To keep discussions concrete we focus on a case where the prototype is a rectangle. Nevertheless, the constructions in the paper is valid in higher dimensions where the prototype is a hyper-cuboid. Thanks to the connection between differential operators and mathematical morphology, the proposed construction naturally extends to the case where diamonds serve as the prototypes.

1 Introduction

Given an 8-connected digital binary pattern representing a digital shape as a mapping $\mathbb{Z}^2 \rightarrow \{0, 1\}$, we are interested in quantifying, at each point on the pattern, the likelihood that the point belongs to a maximal *prototype* shape that fits the digital shape represented by the binary pattern in question. For the prototype shape, the measure is expected to be uniformly zero over the shape. The prototype shape serves as the simplest shape in a certain context.

The practical use of such a measure is two fold: First, if integrated over the pattern, the resulting number can be used as a measure of the tileability of the shape by the maximal prototype shape, which in turn can be used to quantify shape's complexity. Second, directly as a local measure, it can be useful in identifying the locations to cut the shape so that the resulting pieces are tileable by the maximal prototype. A perfectly tileable shape can be digitally represented with maximum compression. By *local culprits*, we mean the points where the measure is significantly low, as they are the points responsible for the failure of tileability.

In this work, we focus on the case where the prototype shape is a rectangle. Nevertheless, as we discuss in Sect. 4, by changing the underlying metric the method naturally extends to the case where the prototype is a diamond. Though we illustrate the method only on 2D shapes (not necessarily simply connected), all discussions are valid in higher dimensions.

Quantifying rectangularity has practical uses in several applications e.g. urban planning and landscape ecology [1]. Rectangularity measures are also used

© Springer Nature Switzerland AG 2021
A. Elmoataz et al. (Eds.): SSVM 2021, LNCS 12679, pp. 91–99, 2021.
https://doi.org/10.1007/978-3-030-75549-2_8

to improve over-segmented images [2]. In the literature, there are several global measures for quantifying the conformity of a shape to a simple prototype [3–5]. These global measures do not convey point-wise information. For circular shapes defined in $\mathbb{R}^2$, the method in [6] provide local information for quantifying conformity to circles.

A related problem of recent interest is quantifying the complexity of high-dimensional datasets for estimating their classification difficulty. Varshney and Willsky [7] measured their classifiers in terms of level sets of the decision hypersurfaces' geometrical complexity using ϵ-entropy. A growing number of works emphasize the role of the shape of the decision hypersurface as a determinant of either how complex the data is or how robust its classification by a certain classifier. An interesting claim by Fawzi et al. [8] is that vulnerability to adversarial attacks is related to positive curvature of the decision boundary. Fawzi et al. further attributed the robustness of the popular deep networks to the flatness of the shape of the produced decision boundaries.

Our construction relies on the connection between differential operators and shape sets, the so-called *structuring elements* of the mathematical morphology. Specifically, we resort to applying morphological derivatives to numerically approximate the infinity-Laplacian as in [9]. Proper numerical realizations of PDEs mimicking morphological process is an important issue. Among the recent works is [10] where the flux-corrected transport scheme to the PDE implementation of erosions and dilations with arbitrary structuring elements is considered.

2   Method

For a shape A, we consider the following PDE with the infinity Laplacian:

$$\left(\Delta_\infty - \frac{1}{\rho^2}\right) f_A = -1 \text{ subject to } f_A\big|_{\partial A} = 0. \tag{1}$$

In numerical solutions to (1), ρ is chosen to be equal to the shape radius, *i.e.* the maximal value of L^∞ (chessboard) distance transform. After obtaining f_A, it is normalized such that the maximum value of the field is 1. These ensure the scale invariance of f_A. To acquire numerical solutions, the approximation to Laplace operator in L^∞ [9],

$$\Delta_\infty f_S(x) \approx \max_{y \in B(x)} f_S(y) + \min_{y \in B(x)} f_S(y) - 2f_S(x) \tag{2}$$

is used where $B(x)$ denotes a unit ball centered at x. The numerical solution to (1) can be acquired by using the scheme proposed in [5].

This equation is favorable for us because the level curves of f_A roughly serve as gradual transformation of the shape boundary ∂A towards a square under the influence of the diffusion governed by the L^∞ metric. The points at a system governed by (1) generate and cumulate the values of the field, f_A. For squares, due to their isotropy in L^∞, the total accumulated values of points equidistant from the boundaries are the same. Therefore, for a square S, the value of the field at x depends only on the minimum distance of x to boundary (equivalently, on its distance to the *shape center*, by which we mean the points attaining the maximum distance transform value). The equidistant points form an equivalence class. As a result, the problem of acquiring an analytic solution reduces to disjoint one dimensional problems over regions of the square, which are continuous on the intersection on the regions.

Consider the points P_1 and P_2 as given in Fig. 1. Since they are equidistant from the boundaries, f_S attains the same values at these two points by the above reasoning. Furthermore, this is true for all points having the same y coordinates in the shaded region R_1. In this region, f_S changes in the y direction only, *i.e.* $\partial f_S / \partial x = 0$. Analogous arguments apply for points in R_2 where instead of y, x coordinates determine the equivalence classes. By the continuity of the field on the intersection of R_1 and R_2, the equivalence classes span both regions, and each is a square by itself.

With these, (1) reduces to

$$\frac{\partial^2 f_S}{\partial y^2} - \frac{1}{\rho^2} f_S = -1, \quad \text{for } |y| \geq |x|$$

$$\frac{\partial^2 f_S}{\partial x^2} - \frac{1}{\rho^2} f_S = -1, \quad \text{for } |y| \leq |x| \tag{3}$$

$$\text{subject to } f_S \big|_{\partial S} = 0.$$

In R_1, for the homogeneous part $f_{S,h} = A \exp\{y/\rho\} + B \exp\{-y/\rho\}$, and for the inhomogeneous part $f_{S,p} = \rho^2$. Due to the symmetry of the boundary conditions, the acquired solution is invariant under $y \mapsto -y$ changes. This dictates $A = B$. Applying the boundary condition we acquire

$$f_S \big|_{R_1} = \rho^2 - \rho^2 \frac{e}{e^2 + 1} \left(\exp\left\{ \frac{y}{\rho} \right\} + \exp\left\{ -\frac{y}{\rho} \right\} \right).$$

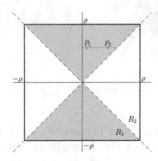

Fig. 1. Square with sides aligned with grid axes

Following the same steps, the solution in R_2 is acquired. The joint solution is given in the closed form as

$$f_S(x,y) = \rho^2 - \rho^2 \frac{e}{e^2+1} \times$$
$$\left(\exp\left\{ \frac{\max\{|x|,|y|\}}{\rho} \right\} + \exp\left\{ \frac{-\max\{|x|,|y|\}}{\rho} \right\} \right). \tag{4}$$

Although this solution is derived for a square, it applies for rectangles as well. This is because the equivalence classes of a square, which are again squares, deform to rectangles: the distances to boundary and the shape center still add up to ρ since the shape center is a line for a rectangle rather than a single point. The validity of the acquired solution for the elongated unit-circle (rectangle in this case) is due to L^∞ norm, and in the general scheme does not hold. For example, in the case of L^2, an analytical solution for which is given in [6], circles are the corresponding equivalence classes, yet, the solution for circles does not apply for ellipses.

In the present form the solution is not translation invariant. To make it so, implicit reference to the origin should be removed. This can be satisfied by reformulating (4) in terms of L^∞ distance transform since $\max\{|x|,|y|\} = \|(x,y)\|_\infty$. We acquire:

$$f_S = \rho^2 - \rho^2 \frac{e}{e^2+1} (\exp\{t'^\infty\} + \exp\{-t'^\infty\}) \tag{5}$$

where $t'^\infty = 1 - t^\infty/\rho$, and t^∞ refers to L^∞ distance transform of S.

In Fig. 2, the difference between the normalized ($i.e.$ has 1 as its maximum value) numerical solution $\hat{f}_{S,numerical}$ and the normalized analytical solution $\hat{f}_{S,analytical}$ for a square of side length 256 is displayed.

Fig. 2. $\hat{f}_{S,numerical}$ (left) and the difference $\hat{f}_{S,numerical} - \hat{f}_{S,analytical}$ (right)

The maxima of the non-normalized fields are 5766.3 and 5797.8, respectively. However, the mean error between the normalized fields is

$$E = \frac{\int |\hat{f}_{S,numerical} - \hat{f}_{S,analytical}|}{|S|} = 0.001$$

This is an acceptable error rate, considering that numerical solution is acquired to a first order approximation.

To construct a measure valid for all shapes, we need (5) to extend beyond rectangles. Thankfully, the solution is in terms of the distance transform of the shape and can be deployed as is as an extension of the field. Then for any shape A,

$$f_{A,assumed} := \rho^2 - \rho^2 \frac{e}{e^2 + 1} \left(\exp\{t'^\infty\} + \exp\{-t'^\infty\} \right).$$

With the choice that we have here, we can assign scores of contribution to complexity to each point in the shape by simply subtracting the assumed extension from the numerically acquired solution. Thus we define the complexity encoding field d_A as

$$d_A := \hat{f}_{A,numerical} - \hat{f}_{A,assumed}.$$

This corresponds to measuring the error in assuming each point is coming from a square of radius ρ in which the point is located $\rho t'^\infty$ away from the center.

Acquired fields, $\hat{f}_{A,numerical}$ and d_A, for a square with an appendage of size 64×128 on one side are shown in Fig. 3.

Fig. 3. A square with a rectangular appendage: $\hat{f}_{A,numerical}$ (left) and d_A (right)

This, being one of the simplest cases of complexity, is informative in understanding the behavior of the proposed field. High negative values occur around the rectangular appendage. Pixels near boundary, be them of base square or appendage, attain smaller values and would be disregarded in a thresholded treatment of the field. The extrema is attained at the two center pixels in the vertical direction along the edge of the square.

3   Illustrative Experiments

In all the experimental results depicting d_As, if the color bar is not shown, the color scale is between 0 and -0.446 where the yellow denotes zero. In the first experiment, we demonstrate the method on composite rectangles of constant width. The size of the square is 64×64, *i.e.*, one quarter of the previous square used in Fig. 2. As shown in Fig. 4, these shapes have no pixels that increase the complexity. Slight fluctuations in d_A are due to the first order approximation to $f_{A,numerical}$.

In the second experiment, we apply the method to floor plans of increasing complexity. Results are depicted in Fig. 5. The first floor plan is composed of

Fig. 4. Composite rectangles of constant width

four identical rooms. This plan has no pixels that increase the complexity. As we introduce missing or extra segments, respective locations start to attain negative d_A values. The last floor plan consists of multiple rooms of varying sizes. The two rooms of the largest size are deemed as the simplest with d_A values near zero. d_A attain higher negative values inside the smaller rooms. Note that the measure d_A is parameterless. As such it implicitly measures the deviation from the rectangle that maximally fits into the shape due to the choice of ρ (Sect. 2). Hence, smaller rooms are identified as parts of the plan that increase the plan's complexity.

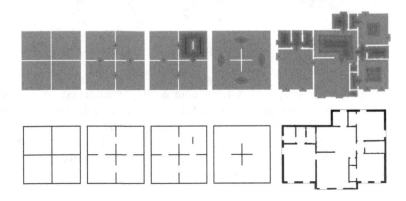

Fig. 5. d_As (top row) for some floor plans (bottom row)

For specific purposes, however, one might be tolerant of the size or interested in identifying complexity with reference to a prototype shape of certain size rather than the maximal size. To this end, one can simply treat ρ as a parameter. In that case positive values for d_A arise in the central parts of larger rectangles. This is illustrated in Fig. 6. The rightmost figure shows the original result from Fig. 5, i.e., $\rho' = \rho$. In the remaining two results, notice that d_As take both negative and positive values as indicated by the color bars. In the leftmost figure, $\rho' = 8\rho/9$ coinciding with the width of the two identical square-shaped rooms on the right-side of the plan. Inside these two rooms d_A is almost uniformly zero as indicated by the color bar. Furthermore, the number of pixels where d_A is negative decreases.

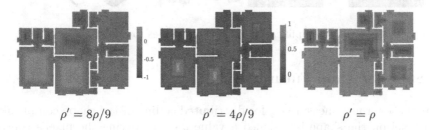

$$\rho' = 8\rho/9 \qquad \rho' = 4\rho/9 \qquad \rho' = \rho$$

Fig. 6. When the critical width ρ' is treated as a parameter

In the third experiment, we explore quantifying the shape complexity with a single value by using $1000 \times \mathrm{mean}(|d_A|)$ rounded to the closest integer. Some illustrative results are shown in Fig. 7.

| 2 | 4 | 16 | 24 | 26 | 31 | 66 | 132 |

Fig. 7. A variety of shapes of increasing complexity

In the final experiment we explore how we can simplify a complex shape towards a rectangle. We had observed that the local extrema of d_A are near the centers of regions that increase complexity. Furthermore, at the extrema, the gradient of $f_{A,numerical}$ indicates the position of such regions relative to the shape body. Thus, the orthogonal direction to the gradient reveals the directions to cut in order to acquire a more rectangular shape.

As a proof of concept, we developed a greedy iterative algorithm. It uses both fields at each iteration step, d_A providing information on the cut location and $f_{A,numerical}$ providing information on the cut direction. For example, for the square with an appendage (Fig. 3), the mean gradient of $f_{A,numerical}$ at the two neighboring extrema of d_A has no component in y direction. Therefore, the shape is cut along the y direction from these extrema, separating the base square and the appendage.

Illustrative cuts are shown in Fig. 8. When determining the direction normal to the gradient, the numerically computed direction is replaced with the direction of closest axis if the angle between them is lesser than 2.86 degrees ($\approx \arctan 0.05$).

Fig. 8. Iterative simplifications of various shapes towards a rectangle

4 Summary and Future Work

We proposed a new measure d_A for distributed coding of the shape complexity. Each pixel on the shape is assigned a value d_A quantifying the pixel's contribution to overall complexity where the simplicity is understood as the expressibility of the shape in terms of an assumed prototype. Throughout the paper we assumed that the prototype is a rectangle (or hyper-cuboid in higher dimensions). We have performed proof of concept experiments to demonstrate the practical applicability of the method.

To first order approximation, the proposed method relies on partial differential equations driven by the morphological Laplacian. An interesting future research based on this would be changing the underlying metric and using a corresponding structuring element that represents the unit circle of the chosen metric. Then, the distance transform appearing in the analytical solution can be computed, for example, in L^1 to acquire a similar shape descriptor addressing rhombicity. In the numerical solution part, this can be imitated by using a diamond structuring element instead of a square one. Sample results for L^1 acquired in this way are displayed in Fig. 9.

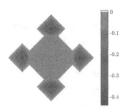

Fig. 9. Extending the method to measure rhombicity

References

1. Moser, D., Zechmeister, H.G., Plutzar, C., Sauberer, N., Wrbka, T., Grabherr, G.: Landscape patch shape complexity as an effective measure for plant species richness in rural landscapes. Landscape Ecol. **17**(7), 657–669 (2002)
2. Ngo, T.-T., Collet, C., Mazet, V.: Automatic rectangular building detection from VHR aerial imagery using shadow and image segmentation. In: IEEE International Conference on Image Processing, pp. 1483–1487 (2015)

3. Maragos, P.: Pattern spectrum and multiscale shape representation. IEEE Trans. Pattern Anal. Mach. Intell. **11**(7), 701–716 (1989)
4. Rosin, P.L., Zunic, J.: Measuring squareness and orientation of shapes. J. Math. Imaging Vis. **39**(1), 13–27 (2011)
5. Arslan, M.F., Tari, S.: Complexity of shapes embedded in $\mathbb{Z}^n$ with a bias towards squares. IEEE Trans. Image Process. **5**(6), 922–937 (2020)
6. Genctav, A., Tari, S.: Discrepancy: local/global shape characterization with a roundness bias. J. Math. Imaging Vis. **61**(1), 160–171 (2019)
7. Varshney, K.R., Willsky, A.S.: Classification using geometric level sets. J. Mach. Learn. Res. **11**, 491–516 (2010)
8. Fawzi, A., Moosavi-Dezfooli, S., Frossard, P.: The robustness of deep networks: a geometrical perspective. IEEE Signal Process. Mag. **34**(6), 50–62 (2017)
9. Oberman, A.: A convergent difference scheme for the infinity Laplacian: construction of absolutely minimizing Lipschitz extensions. Math. Comput. **74**(251), 1217–1230 (2005)
10. Breuß, M., Weickert, J.: Highly accurate PDE-based morphology for general structuring elements. In: Tai, X.-C., Mørken, K., Lysaker, M., Lie, K.-A. (eds.) SSVM 2009. LNCS, vol. 5567, pp. 758–769. Springer, Heidelberg (2009). https://doi.org/10.1007/978-3-642-02256-2_63

Extension of Mathematical Morphology in Riemannian Spaces

El Hadji S. Diop[1]([✉]), Alioune Mbengue[2], Bakary Manga[2], and Diaraf Seck[3]

[1] Department of Mathematics, University Iba Der Thiam, Thies, Senegal
el-hadji.diop@univ-thies.sn
[2] Department of Mathematics and Computer Science, University Cheikh Anta Diop, Dakar, Senegal
[3] LMDAN, NLAGA, University Cheikh Anta Diop, Dakar, Senegal

Abstract. Mathematical morphology remains an efficient image analysis tool due to its morphological scale-spaces capability. It can be formulated using partial differential equations that are in fact a particular case of the first order Hamilton-Jacobi equations for which viscosity solutions exist and are given by Hopf-Lax-Oleinik (HLO) formulas. In this study, we propose an extension of HLO formulas in Riemannian manifolds by considering a general Cauchy problem, and prove the existence of a unique viscosity solution. Some properties are derived, and example on the hyperbolic ball is also provided.

Keywords: Hamilton-Jacobi equation · Hopf-Lax-Oleinik formula · Riemannian manifold · Scale-spaces · Mathematical morphology

1 Introduction

In many image and data processing, for instance, data compression, feature detection, motion analysis/detection, multiband frequency analysis, performing a multiscale analysis is very important, for either at multiple scales or resolutions. Because of its nonlinearity aspects, and good shape and geometry description properties, mathematical morphology (MM) [24] has been for over decades an efficient and powerful multiscale analysis tool [25]. Since then, many theoretical studies on MM have been conducted and plenty algorithms have designed, which contribute to inspiring new breakthroughs in data analysis and processing; for instance, in machine vision or pattern recognition.

Various formulations were proposed for theoretical studies on MM. Indeed, an algebraic formulation based on complete lattice theory was adopted [4,17]. In such an approach, the adjunction property between inf and sup operators was the main ingredient. More precisely, the notion of adjunction was linked two operators (ε, δ) so that for any given dilation δ, there was a unique erosion ε such that (ε, δ) is an adjunction. On another hand, a functional analysis framework was proposed in many other works [1,23]. One of the main benefits of such a formulation is the fact that it yields partial differential equations (PDEs)

A. Elmoataz et al. (Eds.): SSVM 2021, LNCS 12679, pp. 100–111, 2021.
https://doi.org/10.1007/978-3-030-75549-2_9

for morphological scale-spaces for multiscale flat dilations and erosions, which were achieved by meeting some axiomatic principles [3]. In that context, non-linear PDEs were proposed for scale-spaces modeling with convex and concave structuring functions [11,14], on one hand, while edge sets were used as for propagation fronts leading to scale-spaces based on PDEs [10], on the other hand. A particular interest of such a functional analysis-based formulation is the fact that yielded PDEs for multiscale flat dilations and erosions are a particular case of the first order Hamilton-Jacobi equation:

$$\frac{\partial u(x,t)}{\partial t} + H\left(x, \nabla u(x,t)\right) = 0 \text{ on } \mathbb{R}^n \times (0,+\infty), \; u(\cdot,0) = f(\cdot) \text{ on } \mathbb{R}^n. \quad (1)$$

A classical solution of the general Hamilton-Jacobi equation does not exist, it is studied, rather, in the viscosity sense. For more information on viscosity solutions, please see [12]. If the Lagrangian L is convex, then, the solution of the Cauchy problem (8) is given by the Hopf-Lax-Oleinik formula, as [8,19]:

$$h(x,t) = \sup_{y \in \mathbb{R}^n} \left\{ f(y) - tH^*\left(\frac{x-y}{t}\right) \right\}, \quad (2)$$

with H^* being the Legendre-Fenchel transform of H. Hopf-Lax-Oleinik (HLO) formulas for solutions of Hamilton-Jacobi equations have been studied for a while [9,13,15], and are still object of intensive research [7,21]. In this work, we prove an extension of HLO formula in compact Riemannian manifolds (M,g), by considering the general Cauchy problem:

$$\frac{\partial u}{\partial t} + \|\nabla u\|_g^p = 0 \text{ on } M \times (0,T), \; u(x,0) = f(x) \text{ on } M, \quad (3)$$

where $p > 2$, the function f is bounded and $\|.\|_g$ stands for the norm induced by the metric g. Using the concept of minimal action [16], we prove the existence of a unique viscosity solution given as:

$$u(x,t) = \inf_{y \in M} \left\{ f(y) + c_p \frac{d(x,y)^{\frac{p}{p-1}}}{t^{\frac{1}{p-1}}} \right\}, \quad (4)$$

with c_p being a constant as the one obtained in $\mathbb{R}^n$ [20]. Some properties are derived and presented, as well as an example of the hyperbolic ball.

2 Background on MM and HLO Formulas

2.1 Basic Morphological Operators, Scale-Spaces and PDEs

Let $h : \mathbb{R}^2 \to \bar{\mathbb{R}}$ be a concave function usually known as the structuring function (SF). Let $\mathcal{E}$ be a subset of $\mathbb{Z}^2$ and $f : \mathcal{E} \to \bar{\mathbb{R}}$.

Definition 1. *Dilation and erosion are respectively given by:*

$$f \oplus h(x) = \sup_{y \in \mathcal{E}}[f(y) + h(x-y)] \text{ and } f \ominus h(x) = \inf_{y \in \mathcal{E}}[f(y) - h(y-x)]. \quad (5)$$

Let h be a SF and $(h_t)_{t\geq0}$ the family of SFs defined by:

$$h_t(x) = \begin{cases} th(x/t) & \text{for } t > 0 \\ 0 & \text{for } t = 0, \ x = 0 \\ -\infty & \text{otherwise.} \end{cases}$$

Definition 2. *Multiscale dilations and erosions are given by:*

$$(f \oplus h_t)(x) = \sup_{y\in\mathcal{E}}[f(y) + h_t(x - y)] \ and \ (f \ominus h_t)(x) = \inf_{y\in\mathcal{E}}[f(y) - h_t(y - x)]. \quad (6)$$

In image processing, multiscale dilations and erosions can be performed with the PDE [2, 11, 22]:

$$\partial_t u = \pm\|\nabla u\|; \ u(\cdot, 0) = I, \quad (7)$$

I being the processed image. Equation (7) is a particular case of the first order Hamilton-Jacobi (HJ) Eq. [12]:

$$\frac{\partial u(x,t)}{\partial t} + L(x, \nabla u(x,t)) = 0 \text{ on } \mathbb{R}^n \times (0, +\infty); \ u(\cdot, 0) = f \text{ on } \mathbb{R}^n. \quad (8)$$

The Cauchy problem has a viscosity solution if L is convex. The solution is then given by [8, 19]:

$$u(x,t) = \inf_{y\in\mathbb{R}^n} \left\{ f(y) + tL^* \left(\frac{x - y}{t} \right) \right\}, \quad (9)$$

where L^* is the Legendre-Fenchel transform of L.

2.2   HLO Formulas and Hamilton-Jacobi Equations

Lasry and Lion's results on regularization of functions defined on $\mathbb{R}^n$ or on Hilbert spaces by sup-inf convolutions with squares of distances [18] were extended to Riemannian manifolds M of bounded sectional curvature [6]. More precisely, if the sectional curvature K of M satisfies $-K_0 \leq K \leq K_0$ on M for some $K_0 > 0$ and if the injectivity and convexity radius of M are strictly positive, then every bounded and uniformly continuous function $f : M \to \mathbb{R}$ can be uniformly approximated by globally $C^{1,1}$ function defined by:

$$(f_\lambda)^\mu(x) = \sup_{z\in M} \inf_{y\in M} \{f(y) + \frac{1}{2\lambda}d(z, y)^2 - \frac{1}{2\mu}d(x, z)^2\}, \quad (10)$$

as $\lambda, \mu \to 0^+$, with $0 < \mu < \frac{\lambda}{2}$. In [5], formula (10) was used to define a canonic Riemannian opening and closing. In [21], authors used the group structures and considered the Heisenberg group $\mathcal{H}$, and provided contributions for the more general HJ equation:

$$\frac{\partial u}{\partial t} + H(\nabla u) = 0 \text{ on } \mathcal{H} \times \mathbb{R}^+, \ u(x,0) = g(x) \text{ on } \mathcal{H}, \quad (11)$$

where $g : \mathcal{H} \to \mathbb{R}$ is a continuous and bounded function, ∇u denotes the horizontal gradient of u. Thus, if the Hamiltonian H is radial, convex and superlinear, then, the solution of (11) is given by the following HLO formula [21]:

$$u(x,t) = \inf_{y \in \mathcal{H}} \left\{ tL\left(\frac{y^{-1}x}{t}\right) + g(y) \right\}, \tag{12}$$

where L represents the horizontal Legendre transform of H. An extension in the Carnot group combined with a control theoretic approach was proposed [7], by studying the following system:

$$\frac{\partial u}{\partial t} + \Psi(X_1 u, \ldots, X_m u) = 0 \text{ on } \mathbf{G} \times (0, \infty), \ u(x,0) = g(x) \text{ on } \mathbf{G}, \tag{13}$$

with $X_1, \ldots, X_m$ representing a basis of the first layer of the Lie algebra of the group $\mathbf{G}$, and $\Psi : \mathbb{R}^m \to \mathbb{R}$ being a superlinear and convex function. The unique viscosity solution of this HJ equation was given by:

$$u(x,t) = \inf_{y \in \mathbf{G}} \left\{ t\Psi^{\mathbf{G}}\left(\delta_{\frac{1}{t}}(y^{-1} \circ x)\right) + g(y) \right\} \tag{14}$$

where $\Psi^{\mathbf{G}} : \mathbf{G} \to \mathbb{R}$ is the $\mathbf{G}$-Legendre-Fenchel transform of Ψ, which was defined thanks to a theoretical control approach.

3   HLO Formulas in Riemannian Manifolds

Let us first consider the following more general Cauchy problem with $p > 2$ [19]:

$$\frac{\partial u}{\partial t} + \|\nabla u\|^p = 0 \text{ on } \mathbb{R}^n \times (0, \infty), \ u(., x) = f(x) \text{ on } \mathbb{R}^n. \tag{15}$$

Proposition 1. *The viscosity solutions of (15) are given by:*

$$u(x,t) = \inf_{y \in \mathbb{R}^n} \left\{ f(y) + C_p \frac{\|x - y\|^{\frac{p}{p-1}}}{t^{\frac{1}{p-1}}} \right\} \tag{16}$$

This was studied for $p = 2$ [19], without a proof for $p > 2$. Below is a proof:

Proof. The solutions of the Cauchy problem (15) are given as:

$$u(x,t) = \inf_{y \in \mathbb{R}^n} \left\{ tL\left(\frac{x - y}{t}\right) + f(y) \right\}, \tag{17}$$

where $L(v) = \sup_{q \in \mathbb{R}^n} \{q.v - H(q)\}$. Indeed, the Hamiltonian of this Cauchy problem is given as $H(q) = \|q\|^p$. We need then to find the corresponding Lagrangian $L(v)$. For that we consider the function $f(v,q) = q \cdot v - \|q\|^p$. The derivative of f with right to q must vanish, that is $\frac{\partial f}{\partial q} = v - p\frac{q}{\|q\|}\|q\|^{p-1} = 0$. It comes that

$q = \frac{v}{p\|q\|^{p-2}}$ and by taking the norm of both sides we get $\|q\|^{p-1} = \frac{\|v\|}{p}$ and then $\|q\| = \frac{\|v\|^{\frac{1}{p-1}}}{p^{\frac{1}{p-1}}}$. Hence, the Lagrangian, which is the dual function of H, writes:

$$L(v) = \frac{\|v\|^2}{p\|q\|^{p-2}} - \frac{\|v\|^{\frac{p}{p-1}}}{p^{\frac{p}{p-1}}} = \|v\|^{\frac{p}{p-1}} \frac{p^{\frac{p}{p-1}} - p^{\frac{1}{p-1}}}{p^{\frac{p+1}{p-1}}} = \frac{p-1}{p^{\frac{p}{p-1}}} \|v\|^{\frac{p}{p-1}} = c_p \|v\|^{\frac{p}{p-1}}, \quad (18)$$

where $c_p = \frac{p-1}{p^{\frac{p}{p-1}}}$. Hence, one has:

$$u(x,t) = \inf_{y \in \mathbb{R}^n} \left\{ f(y) + tL\left(\frac{x-y}{t}\right) \right\} = \inf_{y \in \mathbb{R}^n} \left\{ f(y) + c_p \frac{\|x-y\|^{\frac{p}{p-1}}}{t^{\frac{1}{p-1}}} \right\}.$$

$\square$

3.1   Viscosity Solutions in Riemannian Manifolds

Let M be a compact and connected Riemannian manifold, TM its tangent bundle and $L : TM \to \mathbb{R}$ a Lagrangian function. For any real number $t \geq 0$ and all points $x, y \in M$, the minimal action from x to y during the time t is the quantity:

$$h_t(x,y) = \inf_{\gamma} \left\{ \int_0^t L(\gamma(s), \gamma'(s)) ds; \gamma : [0,t] \to M \text{ of class } C^1, \gamma(0) = x, \gamma(t) = y \right\}.$$
$$(19)$$

It is well defined since M is assumed to be connected.

Definition 3. *A function $u : V \to \mathbb{R}$ is a viscosity subsolution of $H(x, d_x u) = c$ on the open subset $V \subset M$, if for every C^1 function $\varphi : V \to \mathbb{R}$ and every point $x_0 \in V$ such that $u - \varphi$ has a maximum at x_0, we have $H(x_0, d_{x_0}\varphi) \leq c$.*
A function $u : V \to \mathbb{R}$ is a viscosity super-solution of $H(x, d_x u) = c$ on the open subset $V \subset M$, if for every C^1 function $\Phi : V \to \mathbb{R}$ and every point $y_0 \in V$ such that $u - \Phi$ has a minimum at y_0, we have $H(y_0, d_{y_0}\Phi) \geq c$.
A function $u : V \to \mathbb{R}$ is a viscosity solution of $H(x, d_x u) = c$, if it is both a subsolution and a supersolution.

Theorem 1 ([16]). *The function $u : \mathbb{R}_+ \times M \to \mathbb{R}$ defined by $u(t,x) := \inf_{y \in M} \{u_0(y) + h_t(y,x)\}$ is the solution of the Cauchy problem:*

$$\frac{\partial u}{\partial t}(t,x) + H(x, \nabla u) = 0 \text{ on } M \times (0, +\infty), \quad u(0, \cdot) = u_0(\cdot) \text{ on } M. \quad (20)$$

Proof. The proof is based on two steps.

Step 1. We first prove that u is a super-solution of the system (20). Let (t_0, x_0) be an element of $\mathbb{R}_+ \times M$ and φ a C^1 function s.t. $u - \varphi$ has a local minimum at (t_0, x_0), which is supposed to be null when subtracting from φ a positive constant. Let $y_0 \in M$ s.t. $u(t_0, x_0) = u_0(y_0) + h_{t_0}(y_0, x_0)$, and γ an extremal of

the action from y_0 to x_0 for a duration time from 0 to t_0. For all $t < t_0$, one has:
$(u - \varphi)(t, \gamma(t)) \geq (u - \varphi)(t_0, \gamma(t_0)) - C$, and then, $u(t, \gamma(t)) - \varphi(t, \gamma(t)) \geq C \geq 0$,
$u(t, \gamma(t)) \geq \varphi(t, \gamma(t))$. So, we can write:

$$\varphi(t, \gamma(t)) \leq u(t, \gamma(t)) \leq u_0(y_0) + \int_0^t L(\gamma(s), \gamma'(s)) ds.$$

Then, for all $t < t_0$, one has:

$$\frac{d}{dt} \left(u_0(y_0) + \int_0^t L(\gamma(s), \gamma'(s)) ds - \varphi(t, \gamma(t)) \right)_{t=t_0} \leq 0. \tag{21}$$

Since $u_0(y_0) = u(t_0, x_0) + h_{t_0}(y_0, x_0)$, then, it follows: $\frac{d}{dt} u_0(y_0) = \frac{d}{dt} u(t_0, x_0) = \frac{d}{dt} \varphi(t_0, x_0)$. Equation (21) can be rewritten as:

$$L(\gamma(t_0), \gamma'(t_0)) - \nabla \varphi(t_0, x_0) . \gamma'(t_0) - \varphi_t(t_0, x_0) \leq 0,$$

and then, one gets

$$\varphi_t(t_0, x_0) + \nabla \varphi(t_0, x_0) . \gamma'(t_0) - L(\gamma(t_0), \gamma'(t_0)) \geq 0.$$

So, we have
$$\varphi_t(t_0, x_0) + H(x_0, \nabla \varphi(t_0, x_0)) \geq 0.$$

Step 2: We prove that u is a subsolution of the system.
Let $(t_0, x_0) \in \mathbb{R}_+ \times M$ s.t. $u - \varphi$ has a maximum at (t_0, x_0). Let $x \in M$, $v \in T_x M$, and define $\gamma_v(s) = x_0 + (t_0 - s).v$. Due to the semi-group property of $(f_t)_t$, for all $t < t_0$, one has:

$$u(t_0, x_0) \leq u(t, x_0 + (t_0 - t)v) + h_{t_0-t}(x_0 + (t_0 - t)v, x_0)$$
$$\leq \varphi(t, x_0 + (t_0 - t)v) + \int_{t_0-t}^{t_0} L(\gamma_v(s), \gamma_v'(s)) ds.$$

The last equality becomes an equality for $t = t_0$, and then, considering ideas in step 1 yields: $\varphi_t(t_0, x_0) - L(x, v) + \nabla \varphi(t_0, x_0).v \leq 0$. Thus:

$$\varphi_t(t_0, x_0) + H(x_0, \nabla \varphi(t_0, x_0)) \leq 0.$$

$\square$

Example 31. Let $x \in M$, $H(x, q) = \|q\|^2$, with $q \in T_x^* M$, where $T_x^* M$ is the cotangent bundle. We consider the exponential application $\exp_x : T_x M \to M$, $T_x M$ being the tangent space at x. For all $y \in \exp_x(T_x M)$, there exists a unique geodesic γ parametrized by the unit arc from x to y, where γ is given by:

$$\gamma(t) = \exp_x t(\exp_x^{-1} \| \exp_x^{-1} y \|) \text{ and } \| \exp_x^{-1}(y) \| = d(x, y). \tag{22}$$

This yields:

$$L(x, v) = \frac{1}{4}\|v\|^2 = \frac{d(x, \exp_x(v))^2}{4} = \frac{d(x, y)^2}{4} \tag{23}$$

with $y = \exp_x(v)$. Thus, one gets:

$$f_t(x) = \inf_{v \in T_x M} \left\{ u_0(\exp_x(v)) + \frac{1}{4t}\|v\|^2 \right\}. \tag{24}$$

Thus, we obtain the formula:

$$f_t(x) = \inf_{y \in M} \left\{ u_0(y) + \frac{d(x, y)^2}{4t} \right\}. \tag{25}$$

Definition 4. *A Lagrangian L on a manifold M is called a Tonelli Lagrangian if the following conditions are satisfied:*

1. *$L : TM \to \mathbb{R}$ is of class C^2, at least.*

2. *L is superlinear above compact subset of M; i.e., $\lim_{\|p\| \to \infty} \dfrac{L(p)}{\|p\|} = +\infty$, $\|\cdot\|$ being a norm induced by a Riemannian metric on M.*

3. *For each $(x, v) \in TM$, $\dfrac{\partial^2 L}{\partial^2 v}(x, v)$ is positive definite as a quadratic form.*

Theorem 2 ([16]). *Let M be a compact Riemannian manifold and $L : TM \to \mathbb{R}$ a Tonelli Lagrangian. If $u \in C^0(M, \mathbb{R})$, then, the function $U : [0, +\infty] \times M \to \mathbb{R}$ defined by:*

$$U(t, x) = T_t^- u(x) = f_t(x) = \inf_{y \in M} \{ u(y) + h_t(x, y) \}, \tag{26}$$

is a viscosity solution of the equation:

$$\frac{\partial U}{\partial t}(t, x) + H\left(x, \frac{\partial U}{\partial x}(t, x) \right) = 0 \text{ on }]0, +\infty[\times M, \tag{27}$$

where $H : T^*M \to \mathbb{R}$ is the Hamiltonian associated to L.

Proof. Let us consider the curve $\gamma : [a, b] \to M$. Thanks to the semi-group property, one has $T_b^- u = T_{b-a}^-[T_a^-(u)]$, and then, one gets: $T_b^- u = T_{b-a}^-[T_a^-(u)] \leq T_a^- u + \int_a^b L(\gamma(s), \gamma'(s)) ds$. It follows:

$$U(b, \gamma(b)) - U(a, \gamma(a)) \leq \int_a^b L(\gamma(s), \gamma'(s)) ds. \tag{28}$$

Step 1: We show that U is a viscosity sub-solution.
We suppose that $\varphi \geq u$, with φ in C^1 and $\varphi(t_0, x_0) = U(t_0, x_0)$, with $t_0 > 0$. Let $0 \leq t \leq t_0$, $v \in T_{x_0} M$ be fixed, and $\gamma : [0, t_0] \to M$ be a curve in C^1 s.t. $\gamma(t_0) = x_0$ and $\gamma'(t_0) = v$. Using Eq. (28), one has:

$$U(t_0, \gamma(t_0)) - U(t, \gamma(t)) \leq \int_t^{t_0} L(\gamma(s), \gamma'(s)) ds. \tag{29}$$

Let $\gamma(t_0) = x_0$. Since $\varphi \geq U$, using Eq. 29, one gets: $\varphi(t_0, x_0) - \varphi(t, \gamma(t)) \leq \int_t^{t_0} L(\gamma(s), \gamma'(s))ds$. Next, dividing by $t_0 - t \geq 0$ and letting $t \to t_0$ yield $\forall\, v \in T_{x_0}M$: $\frac{\partial\varphi}{\partial t}(t_0, x_0) + \frac{\partial\varphi}{\partial x}(t_0, x_0)v \leq L(x_0, v)$, so $\frac{\partial\varphi}{\partial x}(t_0, x_0)v - L(x_0, v) \leq -\frac{\partial\varphi}{\partial t}(t_0, x_0)$. Thus, one has

$$\sup_{v \in T_{x_0}M}\left\{\frac{\partial\varphi}{\partial x}(t_0, x_0).v - L(x_0, v)\right\} \leq -\frac{\partial\varphi}{\partial x}(t_0, x_0)$$

then

$$H\left(x_0, \frac{\partial\varphi}{\partial x}(t_0, x_0)\right) + \frac{\partial\varphi}{\partial t}(t_0, x_0) \leq 0.$$

Step 2: We show that U is a viscosity super-solution. A different proof from [16] is proposed here. Let φ be a C^1 function s.t. $\varphi \leq U$, and suppose $U(t_0, x_0) = \varphi(t_0, x_0)$ with $t_0 > 0$. Let $\gamma : [0, t_0] \to M$ s.t. $\gamma(t_0) = x_0$ and $U(t_0, x_0) = T_{t_0}^- u(x_0) = u(\gamma(0)) + \int_0^{t_0} L(\gamma(s), \gamma'(s))ds$. Since $U(0, \gamma(0)) = u(\gamma(0))$, then, we can write: $U(t_0, x_0) - U(0, \gamma(0)) = \int_0^{t_0} L(\gamma(s), \gamma'(s))ds$. It follows: $\int_0^{t_0} L(\gamma(s), \gamma'(s)ds = \int_0^t L(\gamma(s), \gamma'(s))ds + \int_t^{t_0} L(\gamma(s), \gamma'(s))ds$, and then:

$$\int_t^{t_0} L(\gamma(s), \gamma'(s))ds = \int_0^{t_0} L(\gamma(s), \gamma'(s))ds - \int_0^t L(\gamma(s), \gamma'(s))ds$$
$$\leq U(t_0, \gamma(t_0)) - U(0, \gamma(0)) - U(t, \gamma(t)) + U(0, \gamma(0))$$
$$\leq U(t_0, \gamma(t_0)) - U(t, \gamma(t)).$$

Since $\int_t^{t_0} L(\gamma(s), \gamma'(s)ds \geq U(t_0, \gamma(t_0)) - U(t, \gamma(t))$. Thus, $\forall t \in [0, t_0]$, it yields:

$$U(t_0, \gamma(t_0)) - U(t, \gamma(t)) = \int_t^{t_0} L(\gamma(s), \gamma'(s))ds. \tag{30}$$

Since $\varphi \leq U$, and because $\varphi(t_0, x_0) = U(t_0, x_0)$, one has:

$$\varphi(t_0, \gamma(t_0)) - \varphi(t, \gamma(t)) \geq \int_t^{t_0} L(\gamma(s), \gamma'(s))ds. \tag{31}$$

Next, we divide by $t_0 - t$ and let $t \to t_0$ to finally obtain $\forall\, v \in T_{x_0}M$

$$\frac{\partial\varphi}{\partial t}(t_0, x_0) + \frac{\partial\varphi}{\partial x}(t_0, x_0)(v) \geq L(x_0, v).$$

This implies

$$\frac{\partial\varphi}{\partial x}(t_0, x_0)(v) - L(x_0, v) \geq -\frac{\partial\varphi}{\partial t}(t_0, x_0).$$

We obtain then

$$\sup_{v \in T_{x_0}M}\left\{\frac{\partial\varphi}{\partial x}(t_0, x_0)(v) - L(x_0, v)\right\} \geq -\frac{\partial\varphi}{\partial t}(t_0, x_0).$$

Thus, one gets:

$$H\left(x_0, \frac{\partial \varphi}{\partial x}(t_0, x_0)\right) + \frac{\partial \varphi}{\partial t}(t_0, x_0) \geq 0.$$

□

Example 32. Let M be a connected compact Riemannian manifold, and let us consider the Cauchy problem:

$$\begin{cases} \dfrac{\partial u}{\partial t} + \dfrac{1}{2}\|\nabla u\|^2 = 0 \text{ on } M \times [0, T] \\ u(t, x) = f(x) \text{ on } M. \end{cases} \tag{32}$$

In this case, we have $H(x, p) = \dfrac{1}{2}\|p\|^2$. Let us consider $\gamma : [0, t] \rightarrow M$ the geodesic on M joining x and y. Because of the duality of L and H, one has $L(x, v) = \dfrac{1}{2}\|v\|^2$. The definition of $h_t(x, y)$ yields:

$$h_t(x, y) = \int_0^t L(x, v)\mathrm{d}s = \frac{t}{2}\|v\|^2. \tag{33}$$

Since $d(x, y) = \int_0^t \|\gamma'(s)\|\mathrm{d}s = \int_0^t \|v\|\mathrm{d}s = t\|v\|$, one has $\|v\| = \dfrac{d(x, y)}{t}$, and then, $h_t(x, y) = \dfrac{d(x, y)^2}{2t}$. Thanks to Theorem 2, we conclude that: $u(t, x) = \inf_{y \in M}\{f(y) + h_t(x, y)\} = \inf_{y \in M}\{f(y) + \dfrac{1}{2t}d(x, y)^2\}$ is a viscosity solution of the Cauchy problem (32).

Below is a result concerning the general case *i.e.*, any p, where we look at the solutions of the Cauchy problem:

Proposition 2. *Let* $f \in C^0(M, \mathbb{R})$ *and* $p \geq 2$. *Let* $c_p = \dfrac{p-1}{p^{\frac{p}{p-1}}}$. *The viscosity solutions of the general following Cauchy problem:*

$$\begin{cases} \dfrac{\partial u}{\partial t} + \|\nabla u\|^p = 0 \text{ on } M \times (0, \infty) \\ u(\cdot, t) = f(\cdot) \text{ on } M, \end{cases} \tag{34}$$

are given by the HLO formula on Riemanian manifolds:

$$u(x, t) = \inf_{y \in M}\left\{f(y) + c_p \frac{d(x, y)^{\frac{p}{p-1}}}{t^{\frac{1}{p-1}}}\right\}. \tag{35}$$

Proof. Here, we have $H(x, q) = \|q\|^p$. Using previous calculus, one has:

$$L(v) = \frac{p-1}{p^{\frac{p}{p-1}}}\|v\|^{\frac{p}{p-1}} = C_p\|v\|^{\frac{p}{p-1}}. \tag{36}$$

Also, one has

$$d(x, y) = \int_0^t \|\gamma'(s)\| ds = t\|v\|,$$

and then, we get $\|v\| = \frac{d(x,y)}{t}$. It follows that

$$h_t(x, y) = C_p \int_0^t \|v\|^{\frac{p}{p-1}} ds = tC_p\|v\|^{\frac{p}{p-1}},$$

and thus:

$$h_t(x, y) = tC_p\left(\frac{d(x, y)^{\frac{p}{p-1}}}{t^{\frac{p}{p-1}}}\right) = C_p\left(\frac{d(x, y)^{\frac{p}{p-1}}}{t^{\frac{1}{p-1}}}\right). \tag{37}$$

And thus, we conclude that:

$$u(x, t) = \inf_{y \in M}\left\{f(y) + C_p\frac{d(x, y)^{\frac{p}{p-1}}}{t^{\frac{1}{p-1}}}\right\} \tag{38}$$

is the viscosity solution of (34). $\qquad\qquad\square$

3.2   Example and Some Properties

Example 33. Hyperbolic ball

Let us consider the ball: $B^n = \left\{(x_1, \cdots, x_n) \in \mathbb{R}^n \text{ s.t. } \sum_{i=0}^{n} x_i^2 < 1\right\}$ with the metric $g = \dfrac{4(dx_1^2 + \cdots + dx_n^2)}{(1 - \|x\|^2)^2}$, where $\|\cdot\|$ denotes the euclidian norm. The length $\gamma : [a, b] \to B^n$ of all curves is given by:

$$L(\gamma) = \int_a^b \sqrt{g(\gamma'(t), \gamma'(t))} dt = \int_a^b 2\frac{\sqrt{\gamma_1'(t)^2 + \cdots + \gamma_n'(t)^2}}{1 - (\gamma_1(t)^2 + \cdots + \gamma_n(t)^2)} dt, \tag{39}$$

where $\gamma(t) = (\gamma_1(t), \cdots, \gamma_n(t))$. The distance between two points in B^n is the infimum of all curves joining x and y. Thus, the distance between x and y denoted $d_{B^n}(x, y)$ is given by: $\cosh d_{B^n}(x, y) = 1 + \dfrac{2\|x - y\|^2}{(1 - \|x\|^2)(1 - \|y\|^2)}$, and then, one has: $d_{B^n}(x, y) = Arg\cosh\left(1 + \dfrac{2\|x - y\|^2}{(1 - \|x\|^2)(1 - \|y\|^2)}\right)$. For the 2-dimension case, B^2 can be considered as a complex subset defined by $B^2 = \{z \in \mathbb{C} \text{ s.t. } |z| < 1\}$. If z and $z' \in B^2$, then, we have: $d_{B^2}(z, z') = \dfrac{1}{2}\ln\left(\dfrac{1 + |\frac{z' - z}{1 - z'\bar{z}}|}{1 - |\frac{z' - z}{1 - z'\bar{z}}|}\right)$. For all z and z' in B^2, we can consider the parametrization of the geodesic segment joining z and z' given by $\gamma : [0, 1] \to B^2; t \to \gamma(t) = \dfrac{z + t\frac{z' - z}{1 - z'\bar{z}}}{1 + t\bar{z}\frac{z' - z}{1 - z'\bar{z}}}$. Thus:

$$u(t, x) = \inf_{y \in B^2}\left\{f(y) + \frac{d_{B^2}(x, y)^2}{2t}\right\}. \tag{40}$$

Semi-group. Let $f_t(x) = \inf_{y \in M} \{f(y) + h_t(y, x)\}$. Thus, one has: $f_{t+t'}(x) = f_t \circ f_{t'}(x) = f_t(f_{t'}(x))$. Indeed, one gets $f_{t+t'}(x) = \inf_{y \in M} \{f(y) + h_{t+t'}(y, x)\}$, and we also have: $h_{t'+t}(y, x) = \inf_{y \in M} \{h_{t'}(y, z) + h_t(z, x)\}$. Then, it yields:

$$f_{t+t'}(x) = \inf_{y \in M} \{f(y) + \inf_{z \in M} \{h_{t'}(y, z) + h_t(z, x)\}\} \tag{41}$$

$$= \inf_{y \in M} \inf_{z \in M} \{f(y) + h_{t'}(y, z) + h_t(z, x)\} = \inf_{z \in M} \inf_{y \in M} \{f(y) + h_{t'}(y, z) + h_t(z, x)\}$$

$$= \inf_{z \in M} \{ \inf_{y \in M} \{f(y) + h_{t'}(y, z)\} + h_t(z, x)\} = \inf_{z \in M} \{f_{t'}(z) + h_t(z, x)\} \tag{42}$$

$$= f_t(f_{t'}(x)) = f_t \circ f_{t'}(x) \tag{43}$$

Duality. Let $f^t(x) = \sup_{y \in M} \{f(y) - h_t(x, y)\}$. Thanks to the duality between inf and sup operators, we have: $f^t(x) = -(-f)_t$. Indeed, the result follows thanks to the duality property between inf and sup operators. Let $A = f(y) - h_t(x, y)$. Then, one has: $\sup(A) = -\inf(-A)$, it yields then: $f^t(x) = \sup(A) = -\inf(-A) = -\inf(-f(y) + h_t(x, y)) = -(-f)_t$. Thus: $f^t(x) = -(-f)_t$.

Anti-extensivity. Let $(f_t)^s(x) = \sup_{z \in M} \inf_{y \in M} \{f(y) + h_t(z, y) - h_s(x, z)\}$ and $(f^s)_t(x) = \inf_{z \in M} \sup_{y \in M} \{f(y) - h_s(z, y) + h_t(x, z)\}$. Then, one has: $(f_t)^s \leq f \leq (f^s)_t$. Remarking one has sup inf $u \leq u \leq$ inf sup u for any function u yields the result.

4   Conclusion

We have proposed here an extension of HLO formula in Riemannian manifolds by considering a general Cauchy problem, and proved the existence of a unique viscosity solution. Some properties are derived, and example on the hyperbolic ball is also provided. Contributions of this work could be interesting for example for the image processing community, because of the interesting properties of mathematical morphology in multiscale image analysis.

References

1. Akian, M., Quadrat, J.P., Viot, M.: Bellman processes. In: Cohen, G., Quadrat, J.P. (eds.) 11th International Conference on Analysis and Optimization of Systems Discrete Event Systems, pp. 302–311. Springer, Heidelberg (1994). https://doi.org/10.1007/BFb0033561
2. Alvarez, L., Guichard, F., Lions, P.L., Morel, J.M.: Axioms and fundamental equations of image processing. Arch. Rational Mech. Anal. **123**, 199–257 (1993)
3. Alvarez, L., Lions, P.L., Morel, J.M.: Image selective smoothing and edge detection by nonlinear diffusion. II. SIAM J. on Num. Anal. **29**(3), 845–866 (1992)

4. Angulo, J.: Convolution in (max, min)-algebra and its role in mathematical morphology. In: Advances in Imaging and Electron Physics, vol. 203, pp. 1–66. Elsevier (2017)

5. Angulo, J., Velasco-Forero, S.: Riemannian mathematical morphology. Pattern Recogn. Lett. **47**, 93–101 (2014)

6. Azagra, D., Ferrera, J.: Regularization by sup-inf convolutions on riemannian manifolds: an extension of Lasry-Lions theorem to manifolds of bounded curvature. J. Math. Anal. Appl. **423**(2), 994–1024 (2015)

7. Balogh, Z.M., Calogero, A., Pini, R.: The Hopf-Lax formula in carnot groups: a control theoretic approach. Calc. Var. Partial. Differ. Equ. **49**(3–4), 1379–1414 (2013)

8. Bardi, M., Evans, L.: On Hopf's formulas for solutions of Hamilton-Jacobi equations. Nonlinear Anal. Theory Methods Appl. **8**(11), 1373–1381 (1984)

9. Barles, G.: Existence results for first order Hamilton Jacobi equations. Ann. l'Institut Henri Poincare (C) Non Linear Anal. **1**(5), 325–340 (1984)

10. van den Boomgaard, R., Smeulders, A.W.M.: Towards a morphological scale-space theory. In: O, Y.L., Toet, A., Foster, D., Heijmans, H.J.A.M., Meer, P. (eds.) Shape in Picture, pp. 631–640. Springer, Heidelberg (1994). https://doi.org/10.1007/978-3-662-03039-4_46

11. Brockett, R.W., Maragos, P.: Evolution equations for continuous-scale morphological filtering. IEEE Trans. Sig. Proc. **42**(12), 3377–386 (1994)

12. Crandall, M.G., Ishii, H., Lions, P.L.: User's guide to viscosity solutions of second order partial differential equations. Bull. Am. Math. Soc. **27**(1), 1–67 (1992)

13. Crandall, M.G., Lions, P.L.: On existence and uniqueness of solutions of Hamilton-Jacobi equations. Nonlin. Anal. Theo. Meth. Appl. **10**(4), 353–370 (1986)

14. Diop, E.H.S., Angulo, J.: Inhomogeneous morphological PDEs for robust and adaptive image shock filters. IET Image Process. **14**(6), 1035–1046 (2020)

15. Evans, L.C., Souganidis, P.E.: Differential games and representation formulas for solutions of Hamilton-Jacobi-Isaacs equations. Indiana Univ. Math. J. **33**(5), 773–797 (1984)

16. Fathi, A.: The Weak KAM Theorem in Lagrangian Dynamics. Cambridge University Press, Cambridge (2008)

17. Heijmans, H.: Morphological image operators. SIAM Rev. **38**(1), 178–179 (1996)

18. Lasry, J.M., Lions, P.L.: A remark on regularization in hilbert spaces. Israel J. Math. **55**(3), 257–266 (1986)

19. Lions, P.L.: Generalized Solutions of Hamilton-Jacobi Equations. Pitman, Boston (1982)

20. Lions, P.L., Souganidis, P., Vázquez, J.: The relation between the porous medium and the eikonal equations in several space dimensions. Revista Matemática Iberoamericana **3**, 275–310 (1987)

21. Manfredi, J.J., Stroffolini, B.: A version of the HOPF-LAX formula in the Heisenberg group. Commun. Partial Differ. Equ. **27**(5–6), 1139–1159 (2002)

22. Sapiro, G., Kimmel, R., Shaked, D., Kimia, B.B., Bruckstein, A.M.: Implementing continuous-scale morphology via curve evolution. Pattern Recogn. **26**(9), 1363–1372 (1993)

23. Schmidt, M., Weickert, J.: Morphological counterparts of linear shift-invariant scale-spaces. J. Math. Imaging Vis. **56**(2), 352–366 (2016)

24. Serra, J.: Image Analysis and Mathematical Morphology, vol. I. AP, England (1982)

25. Soille, P.: Morphological Image Analysis. Springer, Heidelberg (1999)

Flow, Motion and Registration

Multiscale Registration

Noémie Debroux[1], Carole Le Guyader[2(✉)], and Luminita A. Vese[3]

[1] Université Clermont Auvergne, CNRS, SIGMA Clermont, Institut Pascal,
63000 Clermont-Ferrand, France
`noemie.debroux@uca.fr`

[2] Normandie Univ, Institut National des Sciences Appliquées de Rouen,
Laboratory of Mathematics, 76000 Rouen, France
`carole.le-guyader@insa-rouen.fr`

[3] Department of Mathematics, University of California Los Angeles, Hilgard Avenue,
Los Angeles, CA 90095-1555, USA
`lvese@math.ucla.edu`

Abstract. In the seminal paper E. Tadmor, S. Nezzar and L. Vese, *A multiscale image representation using hierarchical (BV, L^2) decompositions*, Multiscale Model. Simul., 2(4), 554–579, (2004), the authors introduce a multiscale image decomposition model providing a hierarchical decomposition of a given image into the sum of scale-varying components. In line with this framework, we extend the approach to the case of registration, task which consists of mapping salient features of an image onto the corresponding ones in another, the underlying goal being to obtain such a kind of hierarchical decomposition of the deformation relating the two considered images (—from the coarser one that encodes the main structural/geometrical deformation, to the more refined one—). To achieve this goal, we introduce a functional minimisation problem in a hyperelasticity setting by viewing the shapes to be matched as Ogden materials. This approach is complemented by hard constraints on the L^∞-norm of both the Jacobian and its inverse, ensuring that the deformation is a bi-Lipschitz homeomorphism. Theoretical results emphasising the mathematical soundness of the model are provided, among which the existence of minimisers, a Γ-convergence result and an analysis of a suitable numerical algorithm, along with numerical simulations demonstrating the ability of the model to produce accurate hierarchical representations of deformations.

Keywords: Multiscale analysis · Registration · Nonlinear elasticity ·
Ogden materials · Bi-Lipschitz homeomorphisms · Γ-convergence

---

L. Vese acknowledges support from the National Science Foundation under Grant # 2012868. This project was co-financed by the European Union with the European regional development fund (ERDF, 8P03390/18E01750/18P02733) and by the Haute-Normandie Régional Council via the M2SINUM project.

A. Elmoataz et al. (Eds.): SSVM 2021, LNCS 12679, pp. 115–127, 2021.
https://doi.org/10.1007/978-3-030-75549-2_10

1   Introduction

Assuming that the considered grey-level images are L^2-observations that may include scale-varying objects, ranging from large homogeneous regions that are faithfully modelled by the smaller functional space BV, to oscillatory patterns/ texture that require more involved functional spaces, multiscale representation can be viewed as the task which aims to accurately describe these different levels of details. This means representing the noticeable characteristics of a given image into suitable intermediate spaces/subclasses lying in between the rougher space L^2 (or L^p) and the smaller space BV of functions of bounded variation. While, as previously mentioned, this latter space is the proper one to describe homogeneous regions with sharp edges, finer details (whether it be oscillatory patterns or noise) are generally captured by a function or distribution belonging to a larger space such as L^p or a dual space of distributions (such as $\{\operatorname{div} \vec{g} \,|\, \vec{g} \in L^\infty\}$, $\{\operatorname{div} \vec{g} \,|\, \vec{g} \in \mathrm{BMO}$ (Bounded Mean Oscillation)$\}$ or $\{\varDelta g \,|\, g$ Zygmund function$\}$, as suggested by Meyer [11]). To identify such intermediate functional spaces, a standard strategy is interpolation (see [4] or [5]). The theory of interpolation studies the family of spaces Y that are intermediate spaces between Y_0 and Y_1, Banach spaces continuously embedded in the same Hausdorff topological vector space Z, in the sense that $Y_0 \cap Y_1 \subset Y \subset Y_0 + Y_1$. The model investigated in [24] falls within this framework and proves to be a special instance of the canonical form

$$J_p(f, \eta; X, Y) = \inf_{u+v=f} \left\{ \|v\|_X^p + \eta \|u\|_Y \right\}$$

(closely related to the so-called K-functional (cf. [26, Chapter 5])) for which the intermediate spaces $(X, Y)_\theta$, $\theta \in [0, 1]$ with $Y \subset X$, ranging from $(X, Y)_0 = X$—the larger space—to $(X, Y)_1 = Y$—the smaller one—are quantified by the behaviour of $J_p(\cdot, \eta; \cdot, \cdot)$ as η tends to 0. More precisely, in [24], the authors consider the larger functional space X to be L^2, while Y is chosen to be BV, and intend to measure how accurately an L^2-object can be approximated by its BV characteristics by considering the family of J-functionals

$$J(f, \lambda) = J_2(f, \lambda; BV, L^2) = \inf_{u+v=f} \left\{ \lambda \|v\|_{L^2}^2 + \|u\|_{BV} \right\},$$

with increasing λ's. Note that the focus is shifted in comparison to the canonical form, from smaller η's to larger λ's.

The weighting parameter λ serves as a scale level to discriminate properly the two components: u, that captures the structural features of the observation f, while v encodes texture and oscillatory patterns, the discrimination between the two constituents being dictated by the scale parameter λ. This principle is embedded in a dyadic refinement process with λ_0 a given initial scale, that reads as:

$$f = u_0 + v_0, \quad [u_0, v_0] = \arg\min_{u+v=f} J(f, \lambda_0),$$

$$v_j = u_{j+1} + v_{j+1}, \quad [u_{j+1}, v_{j+1}] = \arg\min_{u+v=v_j} J(v_j, \lambda_0 2^{j+1}), \ j = 0, 1, \cdots,$$

producing at the end of the kth step, the following hierarchical decomposition:

$$f = u_0 + v_0 = u_0 + u_1 + v_1 = \cdots = u_0 + u_1 + \cdots + u_k + v_k,$$

the u_k's resolving finer edges. A remarkable result ([24, Theorem 2.2]) states that under the assumption $f \in BV$, the (BV, L^2) hierarchical decomposition of f, $\sum_{j=0}^{k} u_j$, strongly converges to f in L^2.

In accordance with this framework, we propose transposing it to the case of registration, the task which aims to determine an optimal (in a sense to be specified) diffeomorphic deformation φ that aligns the structures visible in an image called Reference into their counterparts in another one—called Template—(see [12,13,23] for a relevant analysis of the registration problem), the underlying goal being twofold: (i) obtaining a hierarchical decomposition of the sought deformation as $\varphi_0 \circ \cdots \varphi_k \circ \cdots \circ \varphi_n$, φ_0 encoding the main geometry-driven/structural deformation, while the $\varphi_0 \circ \cdots \circ \varphi_k$'s capture more refined deformations; (ii) assessing to what extent this hierarchical decomposition conveys the hidden structure of the deformation to be recovered, and how it could be used to derive image statistics, to measure variability inside a population, to retrieve the inherent dynamics of a phenomenon, and so on. The introduced algorithm is patterned after the one in [24] in the sense that it embeds successive applications of a refinement step—the composition of deformations is now a substitute for the sum of the scale-varying constituents of the seminal model [24]—, and can be schematised as follows. Given R (resp. T) the Reference (resp. Template) image and its related multilayered/hierarchical decomposition $\sum_{j=0}^{k} R_j$ (resp. $\sum_{j=0}^{k} T_j$), the algorithm reads as, $\mathcal{F}$ being a general functional that will be specified later on,

$$\varphi_0 = \arg \min_{\varphi} \mathcal{F}(\varphi; R_0, T_0),$$

$$\vdots \tag{$\mathcal{D}$}$$

$$\varphi_k = \arg \min_{\varphi} \mathcal{F}(\varphi_0 \circ \cdots \circ \varphi_{k-1} \circ \varphi; \sum_{j=0}^{k} R_j, \sum_{j=0}^{k} T_j).$$

Before depicting in depth our model, we would like to point out that prior related works [14,17–19] suggest to foster the use of this image multiscale representation in a registration setting.

The work [17] is focused on (rigid) multiscale image registration for the registration of medical images degraded by significant levels of noise. Where conventional methods fail to correctly align the image pair, i.e., to remove the artificial differences while highlighting the real differences due to intrinsic variations of the objects—whether it be ordinary registration techniques alone or sequential treatments including a prior step to denoise the image pair and then the application of a classical registration algorithm —, their proposed joint model achieves accurate registration. Given f an image with hierarchical expansion $\sum_{j=0}^{k} u_j$— recall that for small k, $\sum_{j=0}^{k} u_j$ is a coarse representation of the image f that actually encodes the structural/geometrical information of f whilst removing small details—to be registered with g approximated by $\sum_{j=0}^{k} v_j$, the authors suggest registering the truncated hierarchical representations rather than f and g themselves, to get an accurate estimation of the deformation aligning them. Two main differences can be observed in comparison to our model: (i) first, each hierarchical step k is treated independently of the previous ones, that is, without connecting the optimal deformations $\Phi_0, \cdots, \Phi_{k-1}$ obtained at previous steps and the optimal deformation Φ_k produced at step k, whereas in our approach, motivated by physical arguments stated below, deformation Φ_k is built up by taking advantage of the $\Phi_0, \cdots, \Phi_{k-1}$'s; (ii) second, and this is in some way a corollary of the first point, the optimal deformation Φ meant to bring f and g into spatial alignment is computed as a weighted average of the form $\frac{1}{m} \sum_{l=0}^{m-1} b_l \Phi_l$, the b_l's being weights appropriately chosen, while we promote composition of deformations. This work is then extended to the case of landmark-driven registration [18] in a B-spline setting and then, to the case of non-rigid deformations [19]. The main motivation in our work to promote composition of deformations stems from the following fact stated in [25]. As mapping a point through a first transformation and then through a second one is equivalent to mapping the point through the composition of these two spatial transformations, the most natural and geometrically meaningful operation the space of non-parametric spatial transformations can be endowed with, is the composition. Note, on the contrary, that addition of spatial transformations has no geometric meaning.

More recently in [14], Modin *et al.* propose constructing analogous hierarchical expansions for diffeomorphims, the sum being replaced by composition of maps, in the context of image registration. Their method can be seen as a series of Large Deformation Diffeomorphic Metric Mapping (LDDMM) steps with varying weighting parameters (note that the adjective multiscale applies more to the way these weighting parameters are set at each iteration, *i.e.*, on how strong penalisations on the data fidelity term and on the deviation of the deformation composition from the identity mapping are, since they do not exploit the multiscale expansions of the images contrary to us). Although our method and theirs have in common the composition of deformations to refine the registration process, there are several key differentiating points in addition to the above: among them, the deformation model we embrace to describe the setting in which the objects to be matched are interpreted and viewed. In our case, it originates from physical considerations (*i.e.* elastic models in which the shapes to be matched are considered as the observations of a same body before and after being subjected to constraints) contrary to the model developed by Modin *et al.* which, to our point of view, is rather rooted in pure mathematical considerations and disconnected in some way from the physics of the problem. This choice of devising a model connected to physical considerations and especially to the hyperelasticity framework was motivated by the fact that it proves to be suitable when dealing with large and nonlinear deformations and that many applied problems, *e.g.* biological tissue behaviour, are modelled within this setting.

In [10], the authors propose decomposing the transformation f of the domain $D \subset \mathbb{R}^2$ in which the object is embedded, using quasi-conformal theories. As a desirable one-to-one representation of an orientation-preserving mapping $f : D \to D$ can be achieved via its Beltrami Coefficient (BC) $\mu(f) : D \to \mathbb{C}$ (— measure of non conformality, *i.e.*, to what extent a deformation deviates from a conformal map, and uniquely related to f by $\mu(f)(z) = \left(\frac{\partial f}{\partial \bar{z}} \right) / \left(\frac{\partial f}{\partial z} \right)$ —), the authors suggest applying a wavelet transform to the Beltrami coefficient, yielding a decomposition of the BC into different components of different frequencies compactly supported in different sub-domains. Quasi-conformal mappings related to different components of the BC (different scales) can thus be reconstructed, those yielding a multiscale decomposition of the deformation.

Finally and for the sake of completeness, we refer the reader to [2,9,20] and [22] for alternative approaches.

We now turn to the mathematical formulation and analysis of the proposed physics-based multiscale modelling. We would like to point out that the core of the paper is on the fine theoretical properties exhibited by our modelling. The proofs being long, we have deliberately chosen to focus primarily on the asymptotic result, which, in our view, is the most significant one. A section is dedicated to the design of a suitable algorithm, but this does not constitute the crux of the manuscript. Extensive numerical simulations are currently being carried out to put this mathematical model to the test, and to assess its ability to unveil the hidden structure of the deformation to be retrieved.

2   Mathematical Modelling

2.1   Depiction of the Model

Let Ω be a connected bounded open subset of $\mathbb{R}^2$ of class $\mathcal{C}^1$, thus satisfying the cone property. Let us denote by $R : \bar{\Omega} \to \mathbb{R}$ the reference image and by $T : \bar{\Omega} \to \mathbb{R}$ the Template image, assumed to belong to the functional space $BV(\Omega)$. For theoretical and numerical purposes, we assume that T is compactly supported on Ω to ensure that $T \circ \varphi$ is always well-defined. Of course, in practice, the sought deformation should be with values in $\bar{\Omega}$, but from a mathematical point of view, if we work with such spaces, we lose the structure of vector spaces. Nonetheless, we can show that our model retrieves deformations with values in $\bar{\Omega}$—based on Ball's results [3]. A deformation is a smooth mapping that is orientation-preserving and injective, except possibly on $\partial\Omega$. The deformation gradient is $\nabla\varphi : \bar{\Omega} \to M_2(\mathbb{R})$, the set $M_2(\mathbb{R})$ being the set of real square matrices of order 2. The sought deformation is seen as an argument of minimum of a specifically designed cost function comprising a regularisation on φ prescribing the nature of the deformation, and a term measuring alignment, or how the available data are used to drive the registration process. To allow large deformations, the shapes to be matched are viewed as isotropic (exhibiting the same mechanical properties in each direction), homogeneous (showing the same behaviour everywhere inside the material), hyperelastic materials (they exhibit both nonlinear behaviour and large changes in shape), and more precisely, as Ogden ones. Note that rubber, filled elastomers and biological tissues are often modelled within the hyperelastic framework, which motivates our approach. This perspective dictates the way the regularisation on the deformation φ is devised, based on the stored energy function of an Ogden material. Recall that the general expression of the stored energy function of such a material (see [6]) is given by $W_O(F) = \sum_{i=1}^{K} a_i \|F\|^{\gamma_i} + \Gamma(\det F)$, with $\forall i \in \{1, \cdots, K\}$, $a_i > 0$, $\gamma_i \geq 0$, and $\Gamma :]0, +\infty[\rightarrow \mathbb{R}$, being a convex function satisfying $\lim_{\delta \to 0^+} \Gamma(\delta) = \lim_{\delta \to +\infty} \Gamma(\delta) = +\infty$. The first term penalises changes in length, while the second one restricts the changes in area. The latter one also ensures preservation of topology by enforcing positivity of the Jacobian almost everywhere. In this work, we focus more specifically on the following energy, $\|\cdot\|$ denoting the Frobenius norm,

$$W_{Op}(F) = \begin{cases} a_1\|F\|^4 + a_2(\det F - 1)^2 + \frac{a_3}{(\det F)^{10}} - 4a_1 - a_3 & \text{if} \quad \det F > 0 \\ +\infty & \text{otherwise,} \end{cases}$$

which fulfills the previous assumptions. The two latter terms govern the distribution of the Jacobian determinant. While the middle term promotes Jacobian determinants close to 1, the rightmost one prevents singularities and large contractions by penalising small values of the determinant. The constants are added to comply with the energy property $W_{Op}(I) = 0$, where I stands for the identity matrix. In the following, we will set $W_{Op}(F) = W_{Op}(F, \det F)$.

This stored energy function is complemented by the term $\mathbb{1}_{\{\|\cdot\|_{L^\infty(\Omega, M_2(\mathbb{R}))} \leq \alpha\}}(F) + \mathbb{1}_{\{\|\cdot\|_{L^\infty(\Omega, M_2(\mathbb{R}))} \leq \beta\}}(F^{-1})$, with $\alpha \geq 1$, and $\beta \geq 1$, where $\mathbb{1}_A$ denotes the convex characteristic function of a convex set A, the underlying idea being to recover deformations that are bi-Lipschitz homeomorphisms.

Remark 1. In terms of functional spaces, if $\varphi \in W^{1,\infty}(\Omega, \mathbb{R}^2)$ (suitable space owing to the L^∞ hard constraints), $\det\nabla\varphi$ is automatically an element of $L^\infty(\Omega)$. Penalising the L^∞ norm of $\nabla\varphi$ thus entails control over the Jacobian determinant. This additional term implicitly gives an upper and lower bound on the Jabobian determinant ensuring thus topology preservation.

The aforementioned regulariser is then applied along with a classical L^2-discrepancy measure, yielding the following minimisation problem

$$\inf_{\varphi \in \mathcal{W}} \left\{ F(\varphi) = \frac{\lambda}{2}\|T \circ \varphi - R\|^2_{L^2(\Omega)} + \int_\Omega W_{Op}(\nabla\varphi, \det(\nabla\varphi))\, dx \right. \tag{$\mathcal{P}$}$$

$$\left. + \mathbb{1}_{\{\|\cdot\|_{L^\infty(\Omega, M_2(\mathbb{R}))} \leq \alpha\}}(\nabla\varphi) + \mathbb{1}_{\{\|\cdot\|_{L^\infty(\Omega, M_2(\mathbb{R}))} \leq \beta\}}((\nabla\varphi)^{-1}) \right\},$$

with $\mathcal{W} = \{\psi \in \text{Id} + W_0^{1,\infty}(\Omega, \mathbb{R}^2) \,|\, \|\nabla\varphi\|_{L^\infty(\Omega, M_2(\mathbb{R}))} \leq \alpha, \|(\nabla\varphi)^{-1}\|_{L^\infty(\Omega, M_2(\mathbb{R}))} \leq \beta, \det(\nabla\varphi) > 0 \text{ a.e. in } \Omega\}$, and $\lambda > 0$.

2.2   Theoretical Results

The first theoretical result claims that problem $(\mathcal{P})$ admits at least one minimiser. Due to the limited number of pages, we only sketch the proof here.

Theorem 1. *Problem (P) admits at least one minimiser in W.*

Proof. The proof mainly relies on the following elements. Ball's results [3] enable one to conclude that any minimising sequence is such that $\forall k \in \mathbb{N}$, φ_k is a homeomorphism from Ω to Ω —and even from $\bar{\Omega}$ to $\bar{\Omega}$ owing to [3, Theorem 2]—, and more particularly, a bi-Lipschitz homeomorphism. Also, $\varphi_k \xrightarrow[k \to +\infty]{*}$ $\bar{\varphi}$ in $W^{1,\infty}(\Omega, \mathbb{R}^2)$ and similar arguments to those previously used enable one to get that $\bar{\varphi}$ is a bi-Lipschitz homeomorphism from Ω to Ω and even from $\bar{\Omega}$ to $\bar{\Omega}$. By continuity of the trace operator, one gets that $\bar{\varphi} \in \mathrm{Id} + W_0^{1,\infty}(\Omega, \mathbb{R}^2)$. Now, φ_k^{-1} is uniformly bounded according to k in $W^{1,\infty}(\Omega, \mathbb{R}^2)$ using Poincaré-Wirtinger inequality, thus there exist a subsequence still denoted by φ_k^{-1} and $\bar{u} \in W^{1,\infty}(\Omega, \mathbb{R}^2)$ such that $\varphi_k^{-1} \xrightarrow[k \to +\infty]{*} \bar{u}$ in $W^{1,\infty}(\Omega, \mathbb{R}^2)$. On the one hand, $\varphi_k^{-1} \circ \bar{\varphi} \xrightarrow[k \to +\infty]{} \mathrm{Id}$ in $L^{\infty}(\Omega, \mathbb{R}^2)$ due to the β-Lipschitz property of φ_k^{-1}, thus almost everywhere in Ω up to a subsequence. On the other hand, φ_k^{-1} uniformly converges to $\bar{u}$ in $\mathcal{C}^0(\bar{\Omega}, \mathbb{R}^2)$, yielding $\varphi_k^{-1} \circ \bar{\varphi} \xrightarrow[k \to +\infty]{} \bar{u} \circ \bar{\varphi}$ pointwise in $\bar{\Omega}$. By uniqueness of the limit, $\bar{u} \circ \bar{\varphi} = \mathrm{Id}$ a.e., leading to $\bar{u} = \bar{\varphi}^{-1}$ a.e. in Ω. It remains to address the question of weak lower semicontinuity. Since $T \in BV(\Omega)$ and all φ_k and $\bar{\varphi}$ are bi-Lipschitz homeomorphisms, we get that $\forall k \in \mathbb{N}$, $T \circ \varphi_k \in BV(\Omega) \subset L^2(\Omega)$ and $T \circ \bar{\varphi} \in BV(\Omega)$ (see [1, Theorem 2.6]). We then prove that $\varphi_k \circ \bar{\varphi}^{-1}$ strongly converges to Id in $\mathcal{C}^{0,\alpha}(\bar{\Omega}, \mathbb{R}^2)$, which is a cornerstone step to prove that $T \circ \varphi_k \xrightarrow[k \to +\infty]{} T \circ \bar{\varphi}$ in $L^2(\Omega)$.

We now include a multiscale representation of the deformation in our model, relying on the hierarchical decomposition into the sum of scale-varying components introduced in [24]. Let $(T_j)_j \in BV(\Omega)$ and $(R_j)_j \in BV(\Omega)$ be the sequence of varying scale structural features of respectively T and R coming from the following problems—S standing for either R or T below —:

$$\begin{cases} (S_0, v_0) = \underset{(u,v)\in BV(\Omega)\times L^2(\Omega) \,|\, S=u+v}{\arg\min} \{\lambda_0 \|v\|_2^2 + TV(u)\}, \\ (S_{j+1}, v_{j+1}) = \underset{(u,v)\in BV(\Omega)\times L^2(\Omega) \,|\, v_j=u+v}{\arg\min} \{2^{j+1}\lambda_0 \|v\|_2^2 + TV(u)\}, \, j=1,\dots, \end{cases}$$

λ_0 being an initial scale parameter provided by the user. We refer the reader to [24] for a thorough description and analysis of this multiscale representation of an image.

In this work, we assume that T and R have similar scale structures and that each level of the following hierarchical decomposition of T, $\left(\sum_{j=0}^{k} T_j\right)$, can be matched to the corresponding level of hierarchical decomposition of R, $\left(\sum_{j=0}^{k} R_j\right)$.

We then derive a corresponding hierarchical decomposition of the deformation, going from global structural deformations to more locally refined ones, and based on the composition operator—a more natural and physically meaningful operator than the addition for transformations—. This multiscale decomposition is described as follows:

$$
\begin{cases}
\varphi_0 = \arg\min_{\varphi \in W}\{\mathcal{F}(\varphi, T_0, R_0)\}, \\
\varphi_k = \arg\min_{\varphi \in \mathcal{X}_k}\{\mathcal{F}(\varphi_0 \circ \varphi_1 \circ \ldots \circ \varphi_{k-1} \circ \varphi, \sum_{j=0}^{k} T_j, \sum_{j=0}^{k} R_j)\},
\end{cases} \quad (\mathcal{P}_k)
$$

with

$$
\mathcal{F}(\varphi, T, R) = \frac{\lambda}{2}\|T \circ \varphi - R\|_{L^2(\Omega)}^2 + \int_\Omega \mathcal{W}_{Op}(\nabla\varphi, \det(\nabla\varphi))\, dx
$$
$$
+ \mathbb{1}_{\{\|\cdot\|_{L^\infty(\Omega, M_2(\mathbb{R}))} \le \alpha\}}(\nabla\varphi) + \mathbb{1}_{\{\|\cdot\|_{L^\infty(\Omega, M_2(\mathbb{R}))} \le \beta\}}((\nabla\varphi)^{-1}),
$$

and $\mathcal{X}_k = \{\psi \in \mathrm{Id} + W_0^{1,\infty}(\Omega, \mathbb{R}^2) \,|\, \det\nabla\psi > 0 \, a.e., \, (\nabla(\varphi_0 \circ \varphi_1 \circ \ldots \circ \varphi_{k-1} \circ \psi))^{-1} \in L^\infty(\Omega, M_2(\mathbb{R}))\}$. The next theoretical result is a Γ-convergence result.

Theorem 2. *Problem ($\mathcal{P}_k$) admits at least one minimiser and the associated functional Γ-converges to the functional related to problem ($\mathcal{P}$) according to De Giorgi's definition.*

Proof. For the existence of minimisers, the same arguments as those previously used can be applied since orientation-preserving bi-Lipschitz homeomorphisms form a group stable for the composition, and $BV(\Omega)$ is stable for the sum. The Γ-convergence result mainly relies on [24, Theorem 2.2] stating the strong convergence of hierarchical decompositions to the initial image in $L^2(\Omega)$, as well as Ball's results [3]. These enable one to prove both inequalities required for the Γ-convergence property.

This result ensures that our multiscale decomposition gives a sequence of bi-Lipschitz homeomorphic deformations which model more and more locally refined distortions until it resembles the original deformation that maps T to R.

We now turn to the numerical resolution of our problem ($\mathcal{P}_k$).

3 Numerical Resolution

Due to its non-linearity in both φ and $\nabla\varphi$ and its non-differentiability, problem ($\mathcal{P}_k$) is hard to solve numerically. Inspired by a prior work by Negrón Marrero [15] followed by more related works [7,8,16] in which the proposed numerical method is adapted to the image registration problem in a nonlinear elasticity setting, we introduce auxiliary variables with quadratic penalty terms. Especially, this enables us to decouple the deformation Jacobian from the deformation itself and from its inverse. For the sake of conciseness, we denote by $\bar{T}_k = \sum_{j=0}^{k} T_j$ and by $\bar{R}_k = \sum_{j=0}^{k} R_j$. The decoupled problem is thus defined by :

$$\inf_{\varphi,\psi,\phi,V,W} \mathcal{F}_k = \frac{\lambda}{2} \int_\Omega (\bar{T}_k \circ \phi - \bar{R}_k)^2 \, dx + \int_\Omega W_{Op}(V, \det V) \, dx$$
$$+ \mathbb{1}_{\{\|\cdot\|_{L^\infty(\Omega, M_2(\mathbb{R}))} \leq \alpha\}}(V) + \mathbb{1}_{\{\|\cdot\|_{L^\infty(\Omega, M_2(\mathbb{R}))} \leq \beta\}}(W)$$
$$+ \frac{\gamma_1}{2} \|V - \nabla\phi\|^2_{L^2(\Omega, M_2(\mathbb{R}))} + \frac{\gamma_2}{2} \|W - \nabla\psi\|^2_{L^2(\Omega, M_2(\mathbb{R}))} \qquad (\mathcal{P}_{k,d})$$
$$+ \frac{\gamma_3}{2} \|\zeta_{k-1}^{-1} \circ \phi - \varphi\|^2_{L^2(\Omega, \mathbb{R}^2)} + \frac{\gamma_4}{2} \|\psi \circ \phi - \mathrm{Id}\|^2_{L^2(\Omega, \mathbb{R}^2)}$$

in which ϕ simulates $\varphi_0 \circ \varphi_1 \circ \ldots \circ \varphi_{k-1} \circ \varphi = \zeta_{k-1} \circ \varphi$, V mimics $\nabla\phi$ while W simulates $\nabla(\phi^{-1})$, and ψ approximates ϕ^{-1}. To design the L^∞-bound on W, we used the fact that if u is a homeomorphism of Ω into Ω, and the inverse function u^{-1} belongs to $W^{1,q}(\Omega)$, the matrix of weak derivatives reads $\nabla(u^{-1}) = (\nabla u)^{-1}(u^{-1})$.

We then use a classical alternative scheme which consists of splitting the original problem into sub-problems that are more computationally tractable. The sketch of our numerical method of resolution is summarised in Algorithm 1. We now present numerical simulations validating the accuracy of the multiscale representation of the obtained deformation.

4   Numerical Experiments

We tested our model on 4DMRI[1] acquired during free-breathing of right lobe liver [21]. We chose the images (195×166) corresponding to the liver in full exhalation and the liver in full inhalation to illustrate the capability of our model to handle large deformations. The parameters were chosen as follows: $N = 10$ levels, $\lambda = 1$, $a_1 = 5.0$, $a_2 = 1000.0$, $a_3 = 4.0$, $\gamma_1 = 80000$, $\gamma_2 = 1.0$, $\gamma_3 = 1.0$, $\gamma_4 = 1.0$, $\alpha = 100$, and $\beta = 100$. The coefficients a_1, a_2 and a_3 involved in the Ogden stored energy function affect respectively the averaged local change of length, and the averaged local change of area. This leads to the following phenomenon: the higher the a_i's are, the more rigid the deformation is. The γ_i are chosen rather big to ensure the closeness between the auxiliary variables and the original ones. The λ weighs the fidelity term, and a trade-off has to be met between the accuracy of the registration with high values of λ and physically meaningful smooth deformations with small λ. The obtained results are illustrated in Fig. 1.

We remark that for each level, the deformed Template is well-aligned with the Reference, demonstrating the ability of our model to deal with large deformations and localised small feature movements as the scale grows. The deformation grids do not exhibit overlaps and hence show that the obtained transformations are physically meaningful with a determinant remaining positive everywhere at all scales. By looking at the last column, one can see that the hierarchical decomposition of the deformations obtained with our model behaves as expected, that is to say the deformations on the first scales are global and represent the movements of the main organs and as k grows, the deformation becomes more localised and more refined to model the movements of small features, i.e. blood vessels imaged as small white dots.

---

[1] http://www.vision.ee.ethz.ch/~organmot/chapter_download.shtml.

Algorithm 1 Our Proposed Method (L^∞ constraints applied componentwise)

---

1   **Start from** $\phi_{-1} \leftarrow \mathrm{Id}$, $V_{11,-1} \leftarrow 1$, $V_{12,-1} \leftarrow 0$, $V_{21,-1} \leftarrow 0$,
      $V_{22,-1} \leftarrow 1$, $W_{11,-1} \leftarrow 1$, $W_{12,-1} \leftarrow 0$, $W_{21,-1} \leftarrow 0$, $W_{22,-1} \leftarrow 1$
      $\psi_{-1} \leftarrow \mathrm{Id}$, $\varphi_{-1} \leftarrow \mathrm{Id}$, **and** $\zeta_{-1} \leftarrow \mathrm{Id}$;

2   **Choose** N, **the number of scales.**

3   **Compute** $(T_j)_{j=0,\dots,N}$ **and** $(R_j)_{j=0,\dots,N}$;

4   **for** $k = 0,\dots,N$:

5     $\bar{T}_k \leftarrow \sum_{j=0}^{k} T_j$, **and** $\bar{R}_k \leftarrow \sum_{j=0}^{k} R_j$;

6     $\phi_k \leftarrow \phi_{k-1}$, $V_{11,k} \leftarrow V_{11,k-1}$, $V_{12,k} \leftarrow V_{12,k-1}$, $V_{21,k} \leftarrow V_{21,k-1}$,
      $V_{22,k} \leftarrow V_{22,k-1}$, $W_{11,k} \leftarrow W_{11,k-1}$, $W_{12,k} \leftarrow W_{12,k-1}$,
      $W_{21,k} \leftarrow W_{21,k-1}$, $W_{22,k} \leftarrow W_{22,k-1}$ $\psi_k \leftarrow \psi_{k-1}$, $\varphi_k \leftarrow \mathrm{Id}$,
      and $\zeta_{k-1} \leftarrow \phi_{k-1}$;

7   **for** $l = 1,\dots,nbIter$:

8   **for each pixel** :

9         $V_{11,k} \leftarrow \mathrm{proj}_{\{\|\cdot\|_{L^\infty(\Omega)} \leq \alpha\}}(V_{11,k} - (4\|V_k\|^2 V_{11,k} + 2a_2(\det V_k - 1)$
          $V_{22,k} - \frac{10a_3 V_{22,k}}{\det V_k^{11}} + \gamma_1(V_{11,k} - \frac{\partial \phi_{1,k}}{\partial x})))$;

10        $V_{12,k} \leftarrow \mathrm{proj}_{\{\|\cdot\|_{L^\infty(\Omega)} \leq \alpha\}}(V_{12,k} - (4\|V_k\|^2 V_{12,k} - 2a_2(\det V_k - 1)$
          $V_{21,k} + \frac{10a_3 V_{21,k}}{\det V_k^{11}} + \gamma_1(V_{12,k} - \frac{\partial \phi_{1,k}}{\partial y})))$;

11        $V_{21,k} \leftarrow \mathrm{proj}_{\{\|\cdot\|_{L^\infty(\Omega)} \leq \alpha\}}(V_{21,k} - (4\|V_k\|^2 V_{21,k} - 2a_2(\det V_k - 1)$
          $V_{12,k} + \frac{10a_3 V_{12,k}}{\det V_k^{11}} + \gamma_1(V_{21,k} - \frac{\partial \phi_{2,k}}{\partial x})))$;

12        $V_{22,k} \leftarrow \mathrm{proj}_{\{\|\cdot\|_{L^\infty(\Omega)} \leq \alpha\}}(V_{22,k} - (4\|V_k\|^2 V_{22,k} + 2a_2(\det V_k - 1)$
          $V_{11,k} - \frac{10a_3 V_{11,k}}{\det V_k^{22}} + \gamma_1(V_{22,k} - \frac{\partial \phi_{2,k}}{\partial y})))$;

13        $W_{11,k} \leftarrow \mathrm{proj}_{\{\|\cdot\|_{L^\infty(\Omega)} \leq \beta\}}(\frac{\partial \psi_{1,k}}{\partial x})$;

14        $W_{12,k} \leftarrow \mathrm{proj}_{\{\|\cdot\|_{L^\infty(\Omega)} \leq \beta\}}(\frac{\partial \psi_{1,k}}{\partial y})$;

15        $W_{21,k} \leftarrow \mathrm{proj}_{\{\|\cdot\|_{L^\infty(\Omega)} \leq \beta\}}(\frac{\partial \psi_{2,k}}{\partial x})$;

16        $W_{22,k} \leftarrow \mathrm{proj}_{\{\|\cdot\|_{L^\infty(\Omega)} \leq \beta\}}(\frac{\partial \psi_{2,k}}{\partial y})$;

17   **for each pixel**

18   **Solve the Euler-Lagrange equation with**
        respect to ϕ_k **using an** L^2 **gradient flow**
        with implicit Euler time stepping;

19   **Solve the Euler-Lagrange equation with**
        respect to ψ_k **using an** L^2 **gradient flow**
        with implicit Euler time stepping;

20     $\varphi_k \leftarrow \zeta_{k-1}^{-1} \circ \phi_k$;

21   **return** $\phi_k, \psi_k, V_{11,k}, V_{12,k}, V_{21,k}, V_{22,k}, W_{11,k}, W_{12,k}$,
    $W_{21,k}, W_{22,k}, \varphi_k, \bar{T}_k \circ \phi_k$;

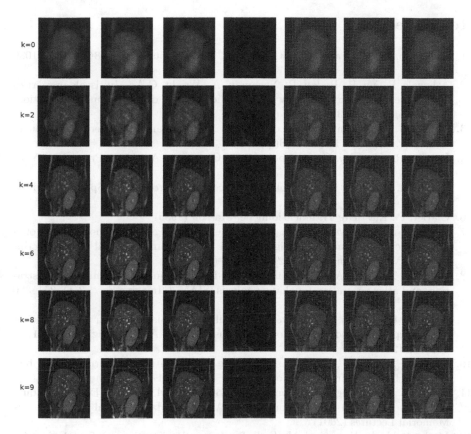

Fig. 1. Multiscale registration results: each row represents a scale of the deformation; the first column displays the Template image at scale k, i.e. $\bar{T}_k$, the second column shows the Reference image at scale k, i.e. $\bar{R}_k$, the third one illustrates the deformed Template obtained at scale k, i.e. $\bar{T}_k \circ \phi_k$, the fourth one exhibits the absolute difference $|\bar{T}_k \circ \varphi_k - \bar{R}_k|$ at scale k, the fifth column presents the inverse deformation at scale k, i.e. $\psi_k \approx \phi_k^{-1} \approx (\varphi_0 \circ \varphi_1 \circ \ldots \circ \varphi_k)^{-1}$, the sixth column represents the composition of deformations at scale k, i.e. $\phi_k \approx \varphi_0 \circ \varphi_1 \circ \ldots \circ \varphi_k$, and finally the last column displays the deformation obtained at scale k, i.e. φ_k.

5   Conclusion

To conclude, we have introduced a mathematically sound and physically relevant multiscale registration model inspired by the hierarchical decomposition presented in [24]. It shows promising results on medical data, producing deformations from global coarse deformations to localised finer transformations at different levels. A theoretical analysis of our numerical scheme (asymptotic result) complemented by additional tests with adapting parameters constitute the next steps of our work, along with an extension to more involved fidelity terms and a discussion on the possible statistics unveiled by this hierarchical decomposition.

References

1. Ambrosio, L., Bertrand, J.: DC calculus. Math. Zeitschrift **288**(3), 1037–1080 (2018)
2. Athavale, P., Xu, R., Radau, P., Nachman, A., Wright, G.A.: Multiscale properties of weighted total variation flow with applications to denoising and registration. Med. Image Anal. **23**(1), 28–42 (2015)
3. Ball, J.M.: Global invertibility of Sobolev functions and the interpenetration of matter. P. Roy. Soc. Edin. A **88**(3–4), 315–328 (1981)
4. Bennet, R., Sharpley, R.: Interpolation of Operators. Academic Press, New York (1988)
5. Bergh, J., Löfström, J.: Interpolation Spaces: An Introduction. Springer, Heidelberg (1976)
6. Ciarlet, P.: Elasticité Tridimensionnelle. Masson (1985)
7. Debroux, N., Le Guyader, C.: A joint segmentation/registration model based on a nonlocal characterization of weighted total variation and nonlocal shape descriptors. SIAM J. Imaging Sci. **11**(2), 957–990 (2018)
8. Debroux, N., et al.: A variational model dedicated to joint segmentation, registration, and atlas generation for shape analysis. SIAM J. Imaging Sci. **13**(1), 351–380 (2020)
9. Gris, B., Durrleman, S., Trouvé, A.: A sub-Riemannian modular framework for diffeomorphism-based analysis of shape ensembles. SIAM J. Imaging Sci. **11**(1), 802–833 (2018)
10. Lam, K.C., Ng, T.C., Lui, L.M.: Multiscale representation of deformation via beltrami coefficients. Multiscale Model. Simul. **15**(2), 864–891 (2017)
11. Meyer, Y.: Oscillating Patterns in Image Processing and Nonlinear Evolution Equations: The Fifteenth Dean Jacqueline B. American Mathematical Society, Lewis Memorial Lectures (2001)
12. Modersitzki, J.: Numerical Methods for Image Registration. Oxford University Press, Oxford (2004)
13. Modersitzki, J.: FAIR: Flexible Algorithms for Image Registration. SIAM (2009)
14. Modin, K., Nachman, A., Rondi, L.: A multiscale theory for image registration and nonlinear inverse problems. Adv. Math. **346**, 1009–1066 (2019)
15. Negrón Marrero, P.: A numerical method for detecting singular minimizers of multidimensional problems in nonlinear elasticity. Numer. Math. **58**, 135–144 (1990)
16. Ozeré, S., Gout, C., Le Guyader, C.: Joint segmentation/registration model by shape alignment via weighted total variation minimization and nonlinear elasticity. SIAM J. Imaging Sci. **8**(3), 1981–2020 (2015)
17. Paquin, D., Levy, D., Schreibmann, E., Xing, L.: Multiscale image registration. Math. Biosci. Eng. **3**(2), 389–418 (2006)
18. Paquin, D., Levy, D., Xing, L.: Hybrid multiscale landmark and deformable image registration. Math. Biosci. Eng. **4**(4), 711–737 (2007)
19. Paquin, D., Levy, D., Xing, L.: Multiscale deformable registration of noisy medical images. Math. Biosci. Eng. **5**(1), 125–144 (2008)
20. Risser, L., Vialard, F.X., Wolz, R., Murgasova, M., Holm, D.D., Rueckert, D.: Simultaneous multi-scale registration using large deformation diffeomorphic metric mapping. IEEE T. Med. Imaging **30**(10), 1746–1759 (2011)
21. von Siebenthal, M., Székely, G., Gamper, U., Boesiger, P., Lomax, A., Cattin, P.: 4D MR imaging of respiratory organ motion and its variability. Phys. Med. Biol. **52**(6), 1547 (2007)

22. Sommer, S., Lauze, F., Nielsen, M., Pennec, X.: Sparse multi-scale diffeomorphic registration: the kernel bundle framework. J. Math. Imaging Vis. **46**, 292–308 (2013)
23. Sotiras, A., Davatzikos, C., Paragios, N.: Deformable medical image registration: a survey. IEEE Trans. Med. Imaging **32**(7), 1153–1190 (2013)
24. Tadmor, E., Nezzar, S., Vese, L.: A multiscale image representation using hierarchical (BV, L^2) decompositions. Multiscale Model. Simul. **2**(4), 554–579 (2004)
25. Vercauteren, T., Pennec, X., Perchant, A., Ayache, N.: Diffeomorphic demons: efficient non-parametric image registration. NeuroImage **45**, S61–72 (2008)
26. Vese, L., Le Guyader, C.: Variational Methods in Image Processing. Chapman & Hall/CRC Mathematical and Computational Imaging Sciences Series, Taylor & Francis (2015)

Challenges for Optical Flow Estimates
in Elastography

Ekaterina Sherina[1]([✉]) [iD], Lisa Krainz[2] [iD], Simon Hubmer[3] [iD],
Wolfgang Drexler[2] [iD], and Otmar Scherzer[1,3] [iD]

[1] Faculty of Mathematics, University of Vienna, Oskar-Morgenstern-Platz 1,
1090 Vienna, Austria
{ekaterina.sherina,otmar.scherzer}@univie.ac.at
[2] Center for Medical Physics and Biomedical Engineering, Medical University of
Vienna, Währinger Gürtel 18-20, 1090 Vienna, Austria
{lisa.krainz,wolfgang.drexler}@meduniwien.ac.at
[3] Johann Radon Institute Linz, Altenbergstrae 69, 4040 Linz, Austria
{simon.hubmer,otmar.scherzer}@ricam.oeaw.ac.at

Abstract. In this paper, we consider visualization of displacement fields
via optical flow methods in elastographic experiments consisting of a
static compression of a sample. We propose an elastographic optical flow
method (EOFM) which takes into account experimental constraints, such
as appropriate boundary conditions, the use of speckle information, as
well as the inclusion of structural information derived from knowledge
of the background material. We present numerical results based on both
simulated and experimental data from an elastography experiment in
order to demonstrate the relevance of our proposed approach.

Keywords: Displacement field estimation · Elastographic optical
flow · Speckle tracking

1 Introduction and Motivation

The ultimate goal of elastography is to reconstruct material parameters of a
sample, such the Lamé parameters λ, μ, the Young's modulus E, or the Possion
ratio ν, by exposing it to external forces. This problem is widely used in Medicine,
in particular for the non-invasive identification of malignant formations inside
the human skin or tissue biopsies during surgeries.

The general strategy has given rise to a number of different elastography
approaches; see e.g. [7,12,16] and the references therein. These use different
external forces (e.g., quasi-static, harmonic, or transient) and measure the resulting deformation either only on the boundary or everywhere inside the sample
(using all kinds of imaging techniques such as e.g., X-ray, ultrasound, magnetic

---

Supported by the Austrian Science Fund (FWF): project F6807-N36 (ES and OS),
project F6805-N36 (SH), and project F6803-N36 (LK and WD).

A. Elmoataz et al. (Eds.): SSVM 2021, LNCS 12679, pp. 128–139, 2021.
https://doi.org/10.1007/978-3-030-75549-2_11

Fig. 1. Example of two tomograms from a compressed sample in a quasi-static OCT elastography experiment with arrows (red) indicating the motion of speckle formations. (Color figure online)

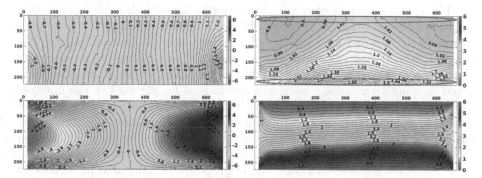

Fig. 2. Lateral (left) and axial (right) components of the displacement fields estimated by standard optical flow (top) and our proposed elastographic optical flow method (bottom) applied to the tomograms depicted in Fig. 1. Circled in red: level lines which approach the boundary perpendicularly due to natural boundary conditions implicit in standard optical flow. This does not agree with the motion occurring in the elastography experiment. (Color figure online)

resonance, or optical imaging, to name but a few). In order to infer elastic material properties from these measurements, computational inversion techniques, assuming suitable material models such as linear, visco, or hyper-elasticity, have to be implemented. The most common strategy for elastography is a two-step approach consisting in first imaging the sample during displacement with the favorite imaging system and secondly by visualizing the displacement field inside the specimen from which the elastic material parameters can be computed. But also all-in-once approaches are used, which aim for visualization of material parameters directly.

In this paper, we focus on the displacement field estimation in two-step approaches to elastography; see e.g. [10, 16, 21, 23] and the references therein. In particular for quantitative results in elastography, accurately estimating this internal displacement field is crucial for obtaining reliable material parameter estimates. However, up to now most of the research on two-step methods for elastography has been concerned with the second step, i.e., the parameter estimation from given displacement fields. Apart from some exceptions discussed below, the first step, i.e., the displacement estimation itself, has been implement with standard optical flow methods. As a result, it can be observed that while many of the proposed methods for material parameter reconstruction work well on simulated displacement fields, their performance with experimental data is limited. In the course of working with different data, see e.g. Fig. 1, we became convinced that this is due to the fact that important physical constraints are not adequately accounted for by standard optical flow techniques: These are a lack of physical assumptions on the motion, an improper treatment of boundary conditions, and peculiarities of the data and the post-processing in scattering-based imaging modalities. From the physical point of view, the motion observed in an experiment with an elastic sample changing its shape due to external forces, see e.g. Fig. 1, is a non-rigid body transformation, where the distance between the neighboring points changes non-linearly. Secondly, the visibility of a sample's internal structure and its geometry affects the motion estimation quality. The standard formulation of optical flow relies on the assumption of the similarity of sequential images, i.e., of their brightness remaining constant. In scattering-based imaging techniques, tomograms are formed by particles in the material which reflect the electromagnetic waves. This means that optical flow can detect motion in a semi-transparent uniform material only in the presence of reflectors inside. The material needs to contain particles or needs to be artificially seeded with reflectors in order to make the deformation accurately visible. Experimental data frequently happens to violate the brightness constancy and physicality in one or another way: the refractive properties of the material change under compression; imaging artifacts; borders of the sample perpendicular to the imaging direction are either invisible or they are removed from the data during post-processing by cutting out to the region of interest inside. As a result, optical flow is "too slow" to follow a depth-varying rate of motion in uniform materials with sparse reflectors, and underestimates the flow near the samples borders by applying the built-in natural boundary conditions which are physically not correct. This can be seen in the level lines of the flow reaching the boundary perpendicularly, see e.g. Fig. 2 (top right, encircled in red).

In this paper, we propose an elastographic optical flow (EOFM) method which takes into account additional experimental and physical side constraints. We concentrate on EOFM based on quasi-static imaging from (a pair of) successive images. Similar techniques can also be used for quasi-static, harmonic, and transient imaging. In particular we consider the proper treatment of boundary conditions, the use of speckle information, as well as structure information derived from knowledge of the background material. Based on both simulated and real experimental data we shall see that by combining the considered techniques we can obtain physically meaningful displacement field.

2   Displacement Field Estimation via Optical Flow

The basis for displacement field estimation typically is the optical flow equation

$$\nabla I \cdot \mathbf{u} + I_t = 0 \,. \tag{1}$$

It connects an image intensity function $I = I(\mathbf{x}, t)$ with a displacement (motion, flow) field $\mathbf{u}(\mathbf{x}) = (u_1(\mathbf{x}), u_2(\mathbf{x}))^T$ for $\mathbf{x} \in \Omega \subset \mathbb{R}^2$ and $t \in \mathbb{R}^+$. Based on this equation, an estimate of the displacement field **u** is typically found by minimizing

$$J(\mathbf{u}) := \int_{\Omega} \left(\nabla I \cdot \mathbf{u} + I_t \right)^2 dx + \alpha \mathcal{R}(\mathbf{u}), \qquad \forall\, t > 0, \tag{2}$$

where $\alpha \geq 0$ is a regularization parameter and $\mathcal{R}$ is some suitably defined regularization functional. In this paper, we focus on the common choice

$$\mathcal{R}(\mathbf{u}) := \|\nabla \mathbf{u}\|^2_{L^2(\Omega)} := \|\nabla u_1\|^2_{L^2(\Omega)} + \|\nabla u_2\|^2_{L^2(\Omega)},$$

which gives rise to the well-known Horn-Schunck method. It enforces certain smoothness constraints on the displacement field **u** and can also be given a physical interpretation [18]. Over the years, a wide variety of advanced motion estimation techniques have been proposed; see e.g. [1–5, 20, 22] and the references therein. Since at their core most of them still follow a similar strategy, we use (2) as our starting point for all further considerations.

2.1 Speckle Tracking

The phenomenon known as speckle is commonly observed in scattering imaging modalities such as ultrasound, optical coherence tomography (OCT), radar-based imaging, and radio-astronomy. Resulting from the constructive and destructive interference of back-scattered waves, it is responsible for granulated images. These speckle patterns are influenced by a number of factors such as the design of the imaging system or optical properties of the imaged samples.

On the one hand, speckle can be seen as a source of noise corrupting the obtained images. On the other hand, they also contain important information, such as on the motion inside the sample; see e.g. [17]. This even lead to the introduction of additional [16] or virtual [8, 15] speckle. Since speckle appears throughout the obtained images, and thus in particular also in otherwise feature-less areas, their movement during an elastography experiment can also be used to estimate the internal displacement field more accurately. A popular method for doing so is the normalized cross-correlation method [6, 16]. Unfortunately, if the size of the correlation area is not carefully chosen, or if the applied strain is too small or too high, then this method is prone to miss-estimations. In addition, its pixel-by-pixel processing is very time-consuming.

Hence, in [19] the authors proposed a novel, heuristics-based image-processing algorithm for the detection and tracking of large speckle formations, termed *bubbles*, which is also used here.

Now, assume that from a given pair of successive images the centers of mass $\hat{\mathbf{x}}^i = (\hat{x}^i_1, \hat{x}^i_2) \in \Omega$ and the directions of motion $\hat{\mathbf{u}}^i = (\hat{u}^i_1, \hat{u}^i_2) \in \mathbb{R}^2$ of a number M of large speckle formations (bubbles) have been extracted. In [19], the authors proposed to complement the optical flow functional $J(u)$ by the functional

$$S_\sigma(\mathbf{u}) := \sum_{i=1}^{M} \int_{\Omega} g_\sigma(\mathbf{x}, \hat{\mathbf{x}}^i) \left| \mathbf{u}(\mathbf{x}) - \hat{\mathbf{u}}^i \right|^2 dx, \tag{3}$$

where the Gaussian-functions $g_\sigma(\mathbf{x}, \hat{\mathbf{x}}^i)$ are defined by

$$g_\sigma(\mathbf{x}, \hat{\mathbf{x}}^i) = \frac{1}{2\pi\sigma^2} e^{-\frac{(x_1 - \hat{x}_1^i)^2 + (x_2 - \hat{x}_2^i)^2}{2\sigma^2}} .$$

This leads to the optical flow method, consisting in minimization of

$$J(\mathbf{u}) + \beta S_\sigma(\mathbf{u}) = \int_\Omega (\nabla I \cdot \mathbf{u} + I_t)^2 \, d\mathbf{x} + \alpha \mathcal{R}(\mathbf{u}) + \beta S_\sigma(\mathbf{u}) , \qquad (4)$$

where $\beta \geq 0$ is another regularization parameter. Depending on the application, the quality of the data, and any given a-priory assumptions on the displacement field $\mathbf{u}$, the values of α, β, and σ can be adjusted to put an emphasis either on the smoothness of the field, or the fit to the given bubble motion $\hat{\mathbf{u}}^i$.

2.2   Boundary Conditions

Consider the minimization of the Horn-Schunck functional $J(\mathbf{u})$ defined in (1) for a fixed time t. Without any additional restrictions, its minimizer $\mathbf{u}$ can be seen to satisfy the so-called natural boundary conditions

$$\nabla u_1 \cdot \mathbf{n} = 0 \qquad \text{and} \qquad \nabla u_2 \cdot \mathbf{n} = 0 , \qquad (5)$$

where $\mathbf{n}$ denotes a unit vector perpendicular to the boundary $\partial\Omega$. The same boundary conditions also hold for the minimizer of the adapted functional defined in (4), i.e., when speckle information is included in the reconstruction.

However, in most cases these natural boundary conditions do not agree with the physical boundary conditions imposed by an actual elastography setup. Consider e.g. a sample which is fixed to a stable surface along a part $\Gamma_1 \subset \partial\Omega$ of its boundary $\partial\Omega$, and which is compressed by an amount g at another part Γ_2 of its boundary. This can be expressed by the Dirichlet boundary conditions

$$\mathbf{u} = 0 \quad \text{on } \Gamma_1 , \qquad \text{and} \qquad \mathbf{u} = g \quad \text{on } \Gamma_2 , \qquad (6)$$

which clearly differ from the natural boundary conditions (5). Hence, in this situation the standard Horn-Schunck optical flow algorithm would yield an estimate of the internal displacement field which is not physically meaningful.

Two possibilities for incorporating known Dirichlet boundary conditions of the form (6) into the displacement field reconstruction suggest themselves. The first is to introduce an additional penalty term of the form

$$\mathcal{B}(\mathbf{u}) := \int_{\Gamma_1 \cup \Gamma_2} |\mathbf{u} - g|^2 \, dS \qquad (7)$$

which can be used to enforce the Dirichlet boundary conditions in a weak form. The other and more natural possibility used in the numerical examples presented below is to restrict the search space in the minimization of either (2) or (4) to contain only those functions which satisfy the required boundary conditions.

Another situation in which the question of proper boundary conditions becomes particularly relevant is when one can only work with measurements of a certain region within the sample. This is e.g. the case if parts of the measurements are corrupted by strong noise or artefacts and thus have to be removed, or if only this region was imaged to begin with. It should be clear that in this case the natural boundary conditions (5) are even less appropriate than before.

In order to deal with this issue, we propose the following strategy: For those boundaries of the measurement region which overlap with the sample boundary, available boundary conditions like (6) can be used as described above. For the remaining boundaries, we propose to use the motion information contained in bubbles to obtain physically meaningful displacement fields. Typically, this can be accomplished by adding the penalty term $\mathcal{S}_\sigma(\mathbf{u})$ defined in (3) and a proper tuning of the corresponding parameters β and σ; compare with (4). In case that only comparatively few bubbles are located near the boundaries without available conditions, it can be advantageous to replace $\mathcal{S}_\sigma(\mathbf{u})$ by the functional

$$\tilde{\mathcal{S}}_\sigma(\mathbf{u}) := \sum_{i=1}^{M} \beta_i \int_\Omega g_\sigma(\mathbf{x}, \hat{\mathbf{x}}^i) \left| \mathbf{u}(\mathbf{x}) - \hat{\mathbf{u}}^i \right|^2 d\mathbf{x},$$

and to emphasize the motion information contained in bubbles close to the boundaries by adapting the corresponding values of the parameters β_i.

2.3   Homogeneous Background Information

General knowledge on the expected structure of the sought for displacement field can be used to enhance the overall quality of the reconstruction methods. Consider e.g. the case that a sample consists of multiple different inclusions in an otherwise homogeneous background material. Then one can write

$$\mathbf{u} = \mathbf{u}^{\mathrm{bg}} + \mathbf{u}^{\mathrm{upd}}, \tag{8}$$

where $\mathbf{u}^{\mathrm{bg}}$ denotes the displacement field which would result from the same elastography experiment carried out on the same sample but without any inclusions, and $\mathbf{u}^{\mathrm{upd}}$ denotes an update which amends this field. A similar idea can also be found in multi-level approaches (see e.g. [9]), where $\mathbf{u}^{\mathrm{bg}}$ mainly plays the role of an initial guess. In contrast, our choice of $\mathbf{u}^{\mathrm{bg}}$ is based on the following physical consideration, which can e.g. also be found in geophysics: Assuming that the material parameters of the background material are known, which is typically the case in elastography experiments and required for uniqueness of parameter reconstruction, then the field $\mathbf{u}^{\mathrm{bg}}$ can be computed by applying a suitable forward model to a homogeneous sample with these parameters. Hence, the task of estimating the displacement field $\mathbf{u}$ reduces to finding the update field $\mathbf{u}^{\mathrm{upd}}$. This can be done by adapting the reconstruction approach outlined above as follows: Since $\mathbf{u} = \mathbf{u}^{\mathrm{bg}} + \mathbf{u}^{\mathrm{upd}}$ should satisfy the optical flow Eq. (1), we can adapt (4) and determine the update field $\mathbf{u}^{\mathrm{upd}}$ as the minimizer of the functional

$$\int_\Omega \left(\nabla I \cdot \left(\mathbf{u}^{\mathrm{bg}} + \mathbf{u}^{\mathrm{upd}} \right) + I_t \right)^2 d\mathbf{x} + \alpha \mathcal{R}(\mathbf{u}^{\mathrm{upd}}) + \beta \tilde{\mathcal{S}}_\sigma(\mathbf{u}^{\mathrm{upd}}),$$

where now the speckle functional $\mathcal{S}_\sigma$ defined in (3) is replaced by the functional

$$\bar{\mathcal{S}}_\sigma(\mathbf{u}^{\mathrm{upd}}) := \sum_{i=1}^{M} \int_\Omega g_\sigma(\mathbf{x}, \hat{\mathbf{x}}^i) \left| \mathbf{u}^{\mathrm{upd}}(\mathbf{x}) - \left(\hat{\mathbf{u}}^i - \mathbf{u}^{\mathrm{bg}}(\hat{\mathbf{x}}^i)\right) \right|^2 dx,$$

which accounts for the relative shifts induced by (8). This also effects the question of appropriate boundary conditions. For example, in the case of the Dirichlet boundary conditions (6), these have to be satisfied for both $\mathbf{u}$ and for $\mathbf{u}^{\mathrm{bg}}$, and thus $\mathbf{u}^{\mathrm{upd}}$ has to satisfy

$$\mathbf{u}^{\mathrm{upd}} = 0 \quad \text{on } \Gamma_1 \cup \Gamma_2.$$

Furthermore, if one assumes that there holds $\mathbf{u} \approx \mathbf{u}^{\mathrm{bg}}$ on $\partial\Omega$, then homogeneous Dirichlet boundary conditions $\mathbf{u}^{\mathrm{upd}} = 0$ can also be applied on those parts of the boundary where no other physically motivated boundary conditions can be used (cf. Sect. 2.2). Even though this may only be a rough approximation, it nevertheless results in a more meaningful condition than the natural boundary condition (5), and together with the speckle information typically helps to improve the overall reconstruction quality.

3   The Elastographic Optical Flow Method

In this section, we formulate the elastographic optical flow method (EOFM) for the problem of determining the internal displacement field of a sample which is subjected to a deformation of the form (6) and imaged with some scattering imaging modality. It takes into account prior knowledge on the homogeneous background material and tracking of bubbles. For this, let $\mathbf{u} = \mathbf{u}^{\mathrm{bg}} + \mathbf{u}^{\mathrm{upd}}$ denote the decomposition of $\mathbf{u}$ described in Sect. 2.3, and the field $\mathbf{u}^{\mathrm{bg}}$ be known. Then EOFM determines $\mathbf{u}^{\mathrm{upd}}$ as the minimizer of

$$F(\mathbf{v}) := \int_\Omega \left(\nabla I \cdot \left(\mathbf{u}^{\mathrm{bg}} + \mathbf{v}\right) + I_t\right)^2 dx + \alpha \mathcal{R}(\mathbf{v}) + \beta \bar{\mathcal{S}}_\sigma(\mathbf{v}), \tag{9}$$

where the space over which this functional is minimized is

$$V := \left\{ \mathbf{v} \in H^1(\Omega)^2 \,|\, \mathbf{v} = 0 \text{ on } \Gamma_1 \cup \Gamma_2 \right\}.$$

In order to analyse this approach and compute a minimizer of (9) we adopt the ideas of Schnörr [18]. That is, we rewrite F in the form

$$F(\mathbf{v}) = \frac{1}{2} a(\mathbf{v}, \mathbf{v}) - b(\mathbf{v}) + c, \tag{10}$$

where the bilinear form $a(\cdot, \cdot)$ and the linear form $b(\cdot)$ are given by

$$a(\mathbf{u}, \mathbf{v}) := 2 \int_\Omega \left((\nabla I \cdot \mathbf{u})(\nabla I \cdot \mathbf{v}) + \alpha \nabla \mathbf{u} : \nabla \mathbf{v} + \beta \sum_{i=1}^{M} g_\sigma(\mathbf{x}, \hat{\mathbf{x}}^i)(\mathbf{u} \cdot \mathbf{v}) \right) dx,$$

$$b(\mathbf{v}) := 2 \int_\Omega \left(\beta \sum_{i=1}^{M} g_\sigma(\mathbf{x}, \hat{\mathbf{x}}^i)(\hat{\mathbf{u}}^i - \mathbf{u}^{\mathrm{bg}}(\hat{\mathbf{x}}^i)) \cdot \mathbf{v} - \left(I_t + \nabla I \cdot \mathbf{u}^{\mathrm{bg}}\right)(\nabla I \cdot \mathbf{v}) \right) dx,$$

and c denotes a constant term. Using this representation (10), we obtain

Theorem 1. *Let $\Omega \subset \mathbb{R}^2$ be a nonempty, bounded, open, and connected set with a Lipschitz continuous boundary $\partial\Omega$ and let $\nabla I \in L^\infty(\Omega)$ and $I_t \in L^2(\Omega)$. Furthermore, let $\alpha > 0$ and let the components of ∇I be linearly independent. Then, the unique minimizer of the problem*

$$\min_{\mathbf{u}\in V} F(\mathbf{u}), \tag{11}$$

is given as the unique solution $\mathbf{u} \in V$ of the linear problem

$$a(\mathbf{u}, \mathbf{v}) = b(\mathbf{v}), \qquad \forall \mathbf{v} \in V. \tag{12}$$

Furthermore, the solution of this equation depends continuously on the right-hand side $b(\cdot)$, but not necessarily on the image intensity function I.

Proof. This follows from the representation (10) in the same way as in [18,19]. □

4 Numerical Results

In this section, we present some numerical results of the application of our proposed EOFM approach based on both simulated and experimental data. The implementation and computational environment is adapted from [19]. We also combine EOFM with a standard multi-scale strategy [9,11,13,14], which can be considered as one step towards image registration. By itself, the multi-scale strategy is not sufficient for satisfactory results, but is seen to provide additional accuracy to our proposed approach.

4.1 Simulated Data

First, we consider a synthetic sample consisting of a circular inclusion within a homogeneous background of different stiffness, a simulated scattering image, which is shown in Fig. 3 (left) with 200 randomly distributed bubbles. The size of one pixel corresponds to $13.34 \cdot 10^{-6}$ m. This sample is assumed to be fixed on top and compressed from the bottom, which corresponds to the Dirichlet boundary conditions (6), and allowed to move freely on the sides. Under the model of linearized elasticity (see [10] or [19]), this results in a displacement field $\mathbf{u}$ with components u_1, u_2 depicted in Fig. 4. The resulting compressed sample is given in Fig. 3 (right). For extracting the motion $\hat{\mathbf{u}}^i$ of added bubbles for (3), we used speckle tracking from Sect. 2.1.

In order to illustrate the effects of different optical flow methods explained in Sects. 2.1, 2.2, 2.3 on the reconstruction of the displacement field, we run tests with various combinations of regularization terms and side constraints. Based on the independent parameter study, we chose $\alpha = 0.8$ for smoothness-regularization and $\beta = 0.5$, $\sigma = 5$ when utilizing the speckle-regularization. For using the multi-scale approach, we took 4 scale-levels (see [19] for details). In

Fig. 3. Simulated sample with randomly distributed speckle formations (bubbles) before (left) and after (right) compression.

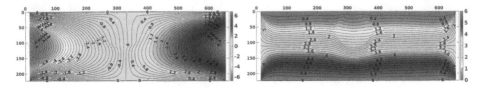

Fig. 4. Components u_1 (left) and u_2 (right) of the simulated displacement field **u**.

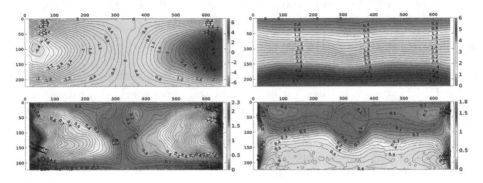

Fig. 5. Test 1. Components u_1 and u_2 (left and right top) of estimated displacement field using (4) with $\alpha = 0.8$, $\beta = 0.5$, multi-scale, and their absolute errors (bottom).

the first test, we solve the minimisation problem for (4) with the bubble motion information. Figure 5 depicts the results of the displacement estimation and absolute error in the field components. The speckle-regularization improves the flow estimate in u_2, with its absolute error being up to 0.6 pixel in the lower half of the sample. However, the lateral motion is underestimated by up to 2.6 pixel in the border area. Next, we minimise (9) with the background field information induced by the same boundary displacement (6) known from the experiment. The relative errors for the resulting estimates are collected in Table 1. Figures 6 and 7 depict the fields reconstructed without ($\beta = 0$) and with ($\beta = 0.5$) the speckle-regularization in addition to the smoothness-regularization and the multi-scale approach. The latter result shows only the maximal error of 0.35 pixel in u_2 and 0.6 pixel in u_1.

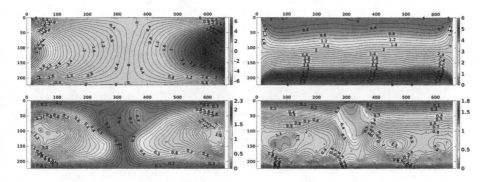

Fig. 6. Test 3. Components u_1 and u_2 (left and right top) of estimated displacement field using (9) with $\alpha = 0.8$, $\beta = 0$, multi-scale, and their absolute errors (bottom).

Table 1. Relative errors of the estimated fields and their components. Note that the settings of the method used in Test No. 1 correspond to the baseline test from [19].

Test No.	Fig. No.	Parameter α	Parameter β	Multi-scale	$e_{rel}(\mathbf{u})$ %	$e_{rel}(u_1)$ %	$e_{rel}(u_2)$ %
1	5	0.8	0.5	yes	19.22	28.07	10.10
2		0.8	0	no	31.69	35.94	28.64
3	6	0.8	0	yes	20.92	23.77	18.87
4		0.8	0.5	no	10.21	12.08	8.81
5	7	0.8	0.5	yes	6.48	7.70	5.56

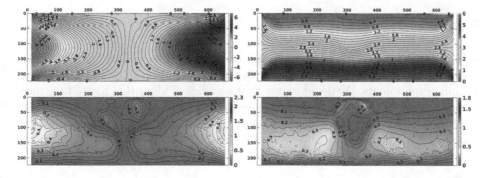

Fig. 7. Test 5. Components u_1 and u_2 (left and right top) of estimated displacement field using (9) with $\alpha = 0.8$, $\beta = 0.5$, multi-scale, and their absolute errors (bottom).

4.2 Experimental Data

Next, we consider the data from an actual quasi-static elastography experiment. The sample has the same structure as the simulated sample, and its OCT tomograms before and after compression are depicted in Fig. 1. As for the simulated data, the size of one pixel corresponds to $13.34 \cdot 10^{-6}$ m. The red arrows

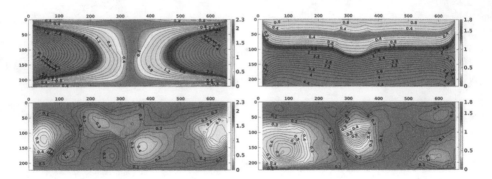

Fig. 8. Approximate absolute errors in u_1 (left) and u_2 (right) of estimated fields with the standard optical flow method (top) and EOFM (bottom).

correspond to the vectors $\hat{u}^i$ obtained by speckle tracking as described in [19]. The resulting field estimate from the standard optical flow is depicted in Fig. 2 (top) and its approximate absolute errors in Fig. 8 (top) in comparison to the expected field for considered sample. The error reaches 5.7 pixels in u_1 and 4.7 pixel in u_2. Figures 2 and 8 (bottom) depict the reconstructed field and its errors using the proposed elastographic optic flow method (9) combining the smoothness and speckle regularization terms together with the background information. For this test, we used the same parameter choice as in Test 5. The resulting displacement from the experimental data features only 0.6 pixel misfit in u_1 and 0.8 pixel in u_2 in certain areas in the sample.

5   Summary

In this paper we introduced an elastographic optical flow method, which allows efficient reconstruction of the displacement field in elastography with scattering imaging experiments. We demonstrate that an efficient algorithm has to take into account information on the physical experiment, such as appropriate boundary conditions, the background medium, and speckle information. The numerical experiments with real data and synthetic data drastically show the necessity of this experimental information.

References

1. Aubert, G., Kornprobst, P.: Mathematical Problems in Image Processing, 2nd edn. Springer, New York (2006)
2. Baker, S., Scharstein, D., Lewis, J.P., Roth, S., Black, M.J., Szeliski, R.: A database and evaluation methodology for optical flow. Int. J. Comput. Vis. **92**(1), 1–31 (2011)
3. Black, M.J., Anandan, P.: The robust estimation of multiple motions: parametric and piecewise-smooth flow fields. Comput. Vis. Image Und. **63**, 75–104 (1996)

4. Brox, T., Malik, J.: Large displacement optical flow: descriptor matching in variational motion estimation. IEEE Trans. Pattern Anal. Mach. Intell. **33**(3), 500–513 (2011)
5. Chen Z., Jin H., Lin Z., Cohen S., Wu Y.: Large displacement optical flow from nearest neighbor fields. In: IEEE Conference on Computer Vision and Pattern Recognition, Portland, OR, pp. 2443–2450 (2013)
6. Duncan, D.D., Kirkpatrick, S.J.: Processing algorithms for tracking speckle shifts in optical elastography of biological tissues. J. Biomed. Opt. **6**(4), 418 (2001)
7. Doyley, M.M.: Model-based elastography: a survey of approaches to the inverse elasticity problem. Phys. Med. Biol. **57**(3), R35–R73 (2012)
8. Glatz, T., Scherzer, O., Widlak, T.: Texture generation for photoacoustic elastography. J. Math. Imaging Vision **52**(3), 369–384 (2015)
9. Haber, E., Modersitzki, J.: A multilevel method for image registration. SIAM J. Sci. Comput. **27**(5), 1594–1607 (2006)
10. Hubmer, S., Sherina, E., Neubauer, A., Scherzer, O.: Lamé parameter estimation from static displacement field measurements in the framework of nonlinear inverse problems. SIAM J. Imag. Sci. **11**(2), 1268–1293 (2018)
11. Lauze, F., Kornprobst, P., Memin, E.: A Coarse to fine multiscale approach for linear least squares optical flow estimation. In: British Machine Vision Conference, pp. 767–776 (2010)
12. Manduca, A., et al.: Magnetic resonance elastography: non-invasive mapping of tissue elasticity. Med. Image Anal. **5**, 237–354 (2001)
13. Meinhardt-Llopis, E., Sánchez, P.J., Kondermann, D.: Horn-Schunck optical flow with a multi-scale strategy. Image Proc. On Line **3**, 151–172 (2013)
14. Modersitzki, J.: FAIR: Flexible Algorithms for Image Registration. Society for Industrial and Applied Mathematics (SIAM), Philadelphia, PA (2009)
15. Schmid J., et al.: Texture generation in compressional photoacoustic elastography. In: Photons Plus Ultrasound: Imaging and Sensing 2015, Proceedings of SPIE, p. 93232S (2015)
16. Schmitt, J.M.: OCT elastography: imaging microscopic deformation and strain of tissue. Opt. Express **3**(6), 199–211 (1998)
17. Schmitt, J.M., Xiang, S.H., Yung, K.M.: Differential absorption imaging with optical coherence tomography. J. Opt. Soc. Amer. A **15**, 2288–2296 (1998)
18. Schnörr, C.: Determining optical flow for irregular domains by minimizing quadratic functionals of a certain class. Int. J. Comput. Vision **6**, 25–38 (1991)
19. Sherina, E., Krainz, L., Hubmer, S., Drexler, W., Scherzer, O.: Displacement field estimation from OCT images utilizing speckle information with applications in quantitative elastography. Inverse Prob. **36**(12), 124003 (2020)
20. Sun, D., Roth, S., Black, M.J.: A quantitative qnalysis of current practices in optical flow estimation and the principles behind them. Int. J. Comput. Vision **106**(2), 115–137 (2013)
21. Wang, S., Larin, K.V.: Optical coherence elastography for tissue characterization: a review. J. Biophotonics **8**(4), 279–302 (2015)
22. Weickert, J., Bruhn, A., Brox, T., Papenberg, N.: A survey on variational optic flow methods for small displacements. In: Scherzer, O. (ed.) Mathematical Models for Registration and Applications to Medical Imaging, vol. 10, pp. 103–136. Springer, Heidelberg (2006). https://doi.org/10.1007/978-3-540-34767-5_5
23. Wijesinghe, P., Kennedy, B.F., Sampson, D.D.: Chapter 9 - Optical elastography on the microscale. In: Alam, S.K., Garra, B.S. (eds.) Tissue Elasticity Imaging, pp. 185–229. Elsevier, Amsterdam (2020)

An Anisotropic Selection Scheme for Variational Optical Flow Methods with Order-Adaptive Regularisation

Lukas Mehl[1]([✉]), Cedric Beschle[2], Andrea Barth[2], and Andrés Bruhn[1]

[1] Institute for Visualization and Interactive Systems,
University of Stuttgart, Stuttgart, Germany
{lukas.mehl,andres.bruhn}@vis.uni-stuttgart.de
[2] Institute for Applied Analysis and Numerical Simulation,
University of Stuttgart, Stuttgart, Germany
{cedric.beschle,andrea.barth}@mathematik.uni-stuttgart.de

Abstract. Approaches based on order-adaptive regularisation belong to the most accurate variational methods for computing the optical flow. By locally deciding between first- and second-order regularisation, they are applicable to scenes with both fronto-parallel and ego-motion. So far, however, existing order-adaptive methods have a decisive drawback. While the involved first- and second-order smoothness terms already make use of anisotropic concepts, the underlying selection process itself is still isotropic in that sense that it locally chooses the same regularisation order for all directions. In our paper, we address this shortcoming. We propose a generalised order-adaptive approach that allows to select the local regularisation order for each direction individually. To this end, we split the order-adaptive regularisation across and along the locally dominant direction and perform an energy competition for each direction separately. This in turn offers another advantage. Since the parameters can be chosen differently for both directions, the approach allows for a better adaption to the underlying scene. Experiments for MPI Sintel and KITTI 2015 demonstrate the usefulness of our approach. They not only show improvements compared to an isotropic selection scheme. They also make explicit that our approach is able to improve the results from state-of-the-art learning-based approaches, if applied as a final refinement step – thereby achieving top results in both benchmarks.

Keywords: Higher-order regularisation · Anisotropic selection scheme · Variational methods · Optical flow · Variational refinement

1 Introduction

With applications such as structure from motion, video processing or autonomous driving, the estimation of the displacement vector field between two consecutive frames of an image sequence, the so-called optical flow, is one

© Springer Nature Switzerland AG 2021
A. Elmoataz et al. (Eds.): SSVM 2021, LNCS 12679, pp. 140–152, 2021.
https://doi.org/10.1007/978-3-030-75549-2_12

of the key problems in computer vision. Since the early approach of Horn and Schunck [8] variational methods have proven to be a well understood and successful tool to solve this task. At first, they have been used as stand-alone approaches often relying on some form of coarse-to-fine estimation [1,3,12,17,22]. More recently, however, such methods have frequently been applied to refine the solution of other approaches allowing variational methods to benefit from recent progress in randomised patch matching [10,11,18] and convolutional neural networks [6].

From a modelling viewpoint, variational methods compute the optical flow as minimiser of suitable energy functionals with two terms. While a data term penalises deviations from constancy assumptions on image features, a smoothness term penalises deviations from spatial regularity constraints. In the context of refinement approaches, in particular the spatial regularity turns out to be vital. It yields smooth transitions and allows for (improved) sub-pixel accuracy [13].

Unfortunately, the diversity of motion that may occur in a scene makes it difficult to come up with a generic regulariser that works in most cases. Hence, it is not surprising that since the seminal work of Horn and Schunck [8] significant progress has been made to develop suitable regularisers for different scenarios and to improve their degree of adaptivity to the underlying scene. These efforts can be classified into two main research directions: higher-order regularisation and anisotropy. While earlier approaches penalise first-order derivatives of the optical flow and thus favour piecewise constant flow fields [1,3,18], more recent methods propose to employ higher-order regularisation to handle more complex scenarios including piecewise affine motion [5,14,17]. In the latter context, indirect second-order regularisation became rather popular [7,14,17], since the underlying coupling model allows to preserve both kinks and jumps in the solution [2,7]. Orthogonal to this research, the use of directional information has been proposed in the literature to render the underlying regularisation anisotropic. Such ideas, in turn, date back to [16] and, in the meantime, led to a number of sophisticated anisotropic regularisers [7,12,14,22].

Recently, there have also been efforts to combine both research directions, i.e. to model anisotropic regularisers of higher order; see e.g. [14] and the references therein. Most remarkable in this context is the order-adaptive regulariser of Maurer *et al.* [12] that does not rely on a single regularisation order but locally decides between an anisotropic first-order and an anisotropic second-order regulariser based on a patch-wise energy competition. Compared to direct [5,22] or indirect single-order regularisers [2,14,17] it allows to generalise much better across different scenes [10,12]. Moreover, it also allows to improve the results substantially when refining flow fields [10,11,13]. However, this regulariser still has a decisive drawback: While it is based on anisotropic regularisers of first and second order, its order selection scheme itself is *isotropic*. Thus, despite of treating local directions differently in terms of regularisation strength, the selected regularisation order must be the same for all directions.

Contributions. In this paper we address the aforementioned drawback. We propose a generalised order-adaptive approach that not only combines anisotropic smoothness terms, but also incorporates an anisotropic selection scheme. In this context, our contributions are twofold: (i) On the one hand, by performing a separate energy competition along and across the locally dominant direction, we are able to model a novel anisotropic selection scheme that is capable of selecting the most appropriate regularisation order for each direction individually. Evidently, this strategy also allows to select different regularisation orders for different directions – a generalised regularisation concept that has not been considered in the optical flow literature so far. (ii) On the other hand, by using our novel selection scheme within a single-level refinement approach, we demonstrate that variational methods are able to improve the flow fields of recent learning-based methods. More precisely, we show that our approach allows to improve the results of the currently most accurate method in the literature: RAFT [20].

Related Work. Regarding the order-adaptive regularisation in the context of optical flow, there only exists the method of Maurer *et al.* [12] and its refinement variant [13] which both form the basis of our approach. As outlined before, our approach generalises their underlying isotropic order selection scheme to an anisotropic setting thus allowing different regularisation orders in different directions. Regarding the adaptive refinement of previously computed correspondence fields, in addition to the method of Maurer *et al.* [13], two learning-based approaches have recently been proposed in the literature. On the one hand, in the context of optical flow, Wannenwetsch and Roth [21] suggested a CNN-based refinement method which makes use of a probabilistic pixel-adaptive refinement network. On the other hand, in the context of stereo, Knöbelreiter and Pock [9] proposed a variational network for the joint refinement of image data, depth estimate and confidence map that is based on unrolling iterates of a proximal gradient method. Both methods, however, rely on the availability of a confidence map or a cost volume, respectively, provided by the initial method to be refined. Since for recent optical methods, this additional information is often not available, we propose a rather general refinement method that only uses an initial flow estimate and the image data itself.

Organisation. In Sect. 2 we introduce a generic variational optical flow model that serves as basis for our novel approach. Afterwards, in Sect. 3 we review the order-adaptive regulariser with isotropic selection scheme by Maurer *et al.* [12]. Based on its drawbacks we then propose our novel generalised order-adaptive regulariser in Sect. 4 and detail on its minimisation in Sect. 5. Finally, we evaluate our novel approach both as stand-alone and refinement method in Sect. 6. We conclude with a summary in Sect. 7.

2   Variational Optical Flow

Let us start by reviewing a general variational optical flow approach that serves as basis for our novel generalised order-adaptive method. To this end, we consider an image sequence $f(\mathbf{x}, t)$ where $\mathbf{x}$ denotes the location within the rectangular image domain $\Omega \in \mathbb{R}^2$ and t denotes time. Then the optical flow $\mathbf{u}(\mathbf{x}) = (u(\mathbf{x}), v(\mathbf{x}))^\top$ between two frames at time t and $t+1$ of that sequence can be computed as minimiser of an energy functional of the form

$$E(\mathbf{u}) = D(\mathbf{u}) + \alpha \cdot R(\mathbf{u}). \tag{1}$$

While the data term $D(\mathbf{u})$ models the constancy of image features along the path of motion, the regularisation term $R(\mathbf{u})$ assumes the resulting flow field to be piecewise smooth. Both assumptions are balanced using the parameter $\alpha > 0$.

Data Term. As data term we make use of the illumination-aware term proposed by Demetz et al. [5]. In the spirit of Brox et al. [3] it features the well-known assumptions of brightness and gradient constancy which are separately robustified using a subquadratic penaliser function Ψ_D and balanced using a weight $\gamma > 0$:

$$D(\mathbf{u}) = \inf_{\mathbf{c}} \Bigg\{ \int_\Omega \quad \Psi_D \left((f(\mathbf{x} + \mathbf{u}, t+1) - \Theta(f(\mathbf{x}, t), \mathbf{c}))^2 \right)$$
$$+ \gamma \cdot \Psi_D \left(\|\nabla f(\mathbf{x} + \mathbf{u}, t+1) - \nabla \Theta(f(\mathbf{x}, t), \mathbf{c})\|_2^2 \right)$$
$$+ \mu \cdot \sum_{l=1}^{2} \Psi_D^l \left(\sum_{k=1}^{N} \left(\mathbf{r}_l^\top \nabla c_k \right)^2 \right) d\mathbf{x} \Bigg\}. \tag{2}$$

Moreover, by jointly estimating the local coefficients $\mathbf{c}(\mathbf{x}) = (c_1(\mathbf{x}), \ldots, c_N(\mathbf{x}))^\top$ of a parametrised brightness transfer function

$$\Theta(f, \mathbf{c}) = f + \sum_{i=1}^{N} c_i \cdot \theta_i(f), \tag{3}$$

with suitable basis functions θ_i, the data term explicitly models illumination changes from frame t to frame $t+1$. Finally, the data term comprises a first-order anisotropic regulariser weighted by the parameter $\mu > 0$, in order to prevent arbitrary adaptations of $\mathbf{c}$. This regularisation makes use of different penaliser functions Ψ_D^1 and Ψ_D^2 in the local directions $\mathbf{r}_1$ and $\mathbf{r}_2 = \mathbf{r}_1^\perp$.

While we will detail on the choice of $\mathbf{r}_1$ and $\mathbf{r}_2$ in Sect. 3, the corresponding penaliser functions are chosen such that Ψ_D^1 is the Perona-Malik penaliser

$$\Psi_{\mathrm{PM}}(s^2) = \epsilon^2 \log \left(1 + s^2/\epsilon^2 \right), \tag{4}$$

and Ψ_D^2 is the Charbonnier penaliser

$$\Psi_{\mathrm{CH}}(s^2) = 2\epsilon^2 \sqrt{1 + s^2/\epsilon^2}, \tag{5}$$

both equipped with a small constant ϵ. Moreover, for the robust function Ψ_D in the data term we employ the Charbonnier penaliser. Finally, for the illumination basis functions θ_i, we use the normalised affine functions; see [13].

3   Regularisation with Isotropic Selection Scheme

Regarding the choice of the regularisation term, approaches from the literature typically rely either on first-order smoothness terms which favour fronto-parallel motion or on second-order smoothness terms which model affine motion. The main idea of the order-adaptive regulariser of Maurer *et al.* [12] is to locally select the most appropriate regularisation order from those two classes. To this end, it adaptively combines two anisotropic smoothness terms which make use of image structures to steer the local smoothing behaviour.

As *first-order* smoothness term it considers the anisotropic complementary regulariser of Zimmer *et al.* [22], which reads

$$S_1(\mathbf{u}) = \sum_{l=1}^{2} \Psi_S^l \left((\mathbf{r}_l^\top \nabla u)^2 + (\mathbf{r}_l^\top \nabla v)^2 \right). \tag{6}$$

Here, the local directions $\mathbf{r}_1$ and $\mathbf{r}_2$ correspond to the eigenvectors of the *regularisation tensor*, a generalisation of the structure tensor to arbitrary constancy assumptions [22]. While $\mathbf{r}_1$ points in the direction of the most dominant change, i.e. across edges, $\mathbf{r}_2$ points orthogonal to it, i.e. along edges. To achieve the desired anisotropic behaviour, the two directions are penalised separately using the Perona-Malik and Charbonnier penaliser for Ψ_S^1 and Ψ_S^2, respectively.

As *second-order* smoothness term it employs the anisotropic coupling-based model of Hafner *et al.* [7]. First introduced in the context of focus fusion it was later adapted by Maurer *et al.* [14] to optical flow estimation. The coupling-based approach introduces two additional vector-valued variables $\mathbf{a}$ and $\mathbf{b}$ and consists of two terms. The first term enforces similarity between the auxiliary variables and the gradient of the flow field. It is given by

$$S_2(\mathbf{u}, \mathbf{a}, \mathbf{b}) = \sum_{l=1}^{2} \Psi_S^l \left((\mathbf{r}_l^\top (\nabla u - \mathbf{a}))^2 + (\mathbf{r}_l^\top (\nabla v - \mathbf{b}))^2 \right). \tag{7}$$

The second term realises a first-order regulariser on the auxiliary variables, and thus performs an indirect second-order regularisation on the flow. It reads

$$S_{\text{aux}}(\mathbf{a}, \mathbf{b}) = \sum_{l=1}^{2} \Psi_{\text{aux}}^l \left(\sum_{k=1}^{2} (\mathbf{r}_k^\top \mathcal{J} \mathbf{a} \, \mathbf{r}_l)^2 + (\mathbf{r}_k^\top \mathcal{J} \mathbf{b} \, \mathbf{r}_l)^2 \right), \tag{8}$$

where $\mathcal{J}$ is the Jacobian and Ψ_{aux}^1 and Ψ_{aux}^2 are defined according to Ψ_S^1 and Ψ_S^2.

The *order-adaptive* regulariser finally combines both smoothness terms via

$$R(\mathbf{u}) = \inf_{\mathbf{a}, \mathbf{b}, o} \left\{ \int_\Omega \bar{o} \cdot S_1(\mathbf{u}) + (1 - \bar{o}) \cdot (S_2(\mathbf{u}, \mathbf{a}, \mathbf{b}) + T) + \beta \cdot S_{\text{aux}}(\mathbf{a}, \mathbf{b}) + \lambda \cdot \phi(o) d\mathbf{x} \right\}, \tag{9}$$

with

$$\bar{o}(\mathbf{x}) = \frac{1}{|\mathcal{N}_w(\mathbf{x})|} \int_{\mathcal{N}_w(\mathbf{x})} o(\mathbf{y}) \, d\mathbf{y}. \tag{10}$$

It includes a convex combination of S_1 and S_2 using the spatially-variant order selection function $o(\mathbf{x})$. Additionally, a penalty T is introduced, which requires S_2 to have a minimum average benefit over S_1 to be selected. Furthermore, the parameter $\beta > 0$ is used to weight the term S_{aux}. Finally, the order-adaptive regulariser is complemented by a specifically tailored order selection term

$$\phi(o) = o \cdot \ln(o) + (1 - o) \cdot \ln(1 - o), \tag{11}$$

with weight $\lambda > 0$ which controls the selection behaviour. As Maurer *et al.* [12] have shown it leads to the following closed form solution with respect to $o(\mathbf{x})$:

$$o(\mathbf{x}) = \frac{1}{1 + e^{-\Delta(\mathbf{x})/\lambda}} \quad \text{with} \quad \Delta(\mathbf{x}) = \int_{\mathcal{N}_w(\mathbf{x})} \frac{1}{|\mathcal{N}_w(\mathbf{y})|} \cdot (T + S_2 - S_1) \, d\mathbf{y}. \tag{12}$$

As can be seen, this approach yields an energy competition between the costs of the first-order regulariser S_1 and the costs of the coupling term S_2 of the second-order regulariser (with penalty T). Its result is integrated in a box-shaped neighbourhood $\mathcal{N}$ of size w and fed into a sigmoid function with slope λ, leading to values between 0 and 1 depending on which cost supersedes the other.

4   Regularisation with Anisotropic Selection Scheme

Although the order-adaptive regulariser of Maurer *et al.* [12] allows to adapt well to local image structures due to the anisotropy of the underlying smoothness terms, it is not capable of individually selecting the regularisation order for the two local directions $\mathbf{r}_1$ and $\mathbf{r}_2$. In the following, we will address this problem and derive a novel generalised order-adaptive regulariser that relies on an anisotropic selection scheme. To this end, we propose to perform an energy competition for each of the two local directions $\mathbf{r}_1$ and $\mathbf{r}_2$ separately. Thereby we make use of the fact that the two terms S_1 and S_2 involved in the energy competition are already formulated in an anisotropic manner. Hence, we split up these terms into their directional components S_1^l and S_2^l for $l \in \{1, 2\}$:

$$S_1(\mathbf{u}) = \sum_{l=1}^{2} \Psi_S^l \left((\mathbf{r}_l^\top \nabla u)^2 + (\mathbf{r}_l^\top \nabla v)^2 \right) = \sum_{l=1}^{2} S_1^l(\mathbf{u}), \tag{13}$$

$$S_2(\mathbf{u}, \mathbf{a}, \mathbf{b}) = \sum_{l=1}^{2} \Psi_S^l \left((\mathbf{r}_l^\top (\nabla u - \mathbf{a}))^2 + (\mathbf{r}_l^\top (\nabla v - \mathbf{b}))^2 \right) = \sum_{l=1}^{2} S_2^l(\mathbf{u}, \mathbf{a}, \mathbf{b}). \tag{14}$$

Now, we propose a component-wise competition between S_1^1 and S_2^1 as well as S_1^2 and S_2^2. To this end, we introduce two penalties T_1, T_2 and two slope parameters λ_1, λ_2. Also, we allow the window size to be different for each competition, leading to w_1, w_2. Finally, we extend our selection function to be vector-valued, i.e. $\mathbf{o}(\mathbf{x}) = (o_1(\mathbf{x}), o_2(\mathbf{x}))^\top$. The resulting order-adaptive regulariser then reads

$$
R(\mathbf{u}) = \inf_{\mathbf{a},\mathbf{b},\mathbf{o}} \left\{ \int_\Omega \sum_{l=1}^2 \left(\bar{o}_l \cdot S_1^l(\mathbf{u}) + (1 - \bar{o}_l) \cdot \left(S_2^l(\mathbf{u},\mathbf{a},\mathbf{b}) + T_l \right) \right) \right.
$$

$$
\left. + \beta \cdot S_{\text{aux}}(\mathbf{a},\mathbf{b}) + \sum_{l=1}^2 \lambda_l \cdot \phi(o_l) \, d\mathbf{x} \right\}, \tag{15}
$$

with

$$
\bar{o}_l(\mathbf{x}) = \frac{1}{|\mathcal{N}_{w_l}(\mathbf{x})|} \int_{\mathcal{N}_{w_l}(\mathbf{x})} o_l(\mathbf{y}) \, d\mathbf{y} \tag{16}
$$

for $l = 1, 2$ and where the selection term ϕ is defined as in Eq. (11). When determining the infimum in Eq. (15) with respect to the selection function $\mathbf{o}$, one finally obtains the desired anisotropic selection behaviour, i.e.

$$
o_l(\mathbf{x}) = \frac{1}{1 + e^{-\Delta_l(\mathbf{x})/\lambda_l}} \quad \text{with} \quad \Delta_l(\mathbf{x}) = \int_{\mathcal{N}_{w_l}(\mathbf{x})} \frac{1}{|\mathcal{N}_{w_l}(\mathbf{y})|} \cdot (T_l + S_2^l - S_1^l) \, d\mathbf{y} \tag{17}
$$

for $l = 1, 2$. Please note that by choosing window sizes of $w_1 = w_2 = 1$, we are also able to model a fully local anisotropic selection scheme without spatial integration – analogous to the local variant in [12]. However, larger window sizes not only lead to more stable decisions, in case of our anisotropic scheme, the size can also be chosen differently for each direction to better adapt to the scene.

5   Minimisation

The energy functional including the order-adaptive regulariser with anisotropic selection scheme as given in Eq. (15) is minimised by determining the corresponding Euler-Lagrange equations. They are discretised and then solved using the incremental coarse-to-fine fixed point iteration introduced by Brox et al. [3]. Since the minimising order-selection function $\mathbf{o}$ can be computed in closed form, we introduce an alternating minimisation between computing $\mathbf{o}$ and solving the variational scheme for $\mathbf{u}$, $\mathbf{a}$ and $\mathbf{b}$. In order to do the latter, we linearise the constancy assumptions and apply constraint normalisation [22]. Finally, as in [3] we make use of the lagged nonlinearity approach to then solve the resulting linear system of equations with a multicolour SOR scheme.

6   Evaluation

To evaluate the performance of our order-adaptive variational method based on the novel anisotropic selection scheme, we performed several experiments on the data sets of MPI Sintel [4] and KITTI 2015 [15]. Thereby we made use of our method both as a stand-alone approach and as a refinement technique. For both approaches we used the optimisation framework from [19] to learn the free model parameters individually for each benchmark from the respective training data.

6.1 Stand-Alone Approach

Isotropic vs. Anisotropic Order Selection. In our first experiment, we compare our generalised variational approach with anisotropic order selection to the original model of Maurer *et al.* [12] that relies on isotropic order selection. We thereby resorted to an own implementation of both methods, since in [12] some results are missing (MPI Sintel, *final* pass), while others are obsolete (KITTI 2015, due to an updated ground truth). Moreover, to allow for a fair comparison, we replaced the brightness transfer function Θ in the data term of our model by the identity function. Hence, both methods use the same data term.

Table 1. Comparison of isotropic and anisotropic order selection for the *training* data set of KITTI 2015 [15] using the average endpoint error (AEE) and the percentage of bad pixels (*Fl*) as well as both passes of the Sintel dataset [4] in terms of the AEE.

Order selection	Sintel *clean*	Sintel *final*	KITTI 2015	
	AEE [px]	AEE [px]	Fl [%]	AEE [px]
Isotropic [12]	4.168	6.116	20.352	8.816
Anisotropic (with data term from [12])	**4.099**	**6.038**	**20.154**	**8.654**

Fig. 1. Isotropic vs. anisotropic order selection (KITTI 2015, training data set, Sequence #49). *First row*: Input frames. *Second row*: Computed flow fields. Isotropic order selection [12], anisotropic order selection (our method), sparse ground truth. *Third Row*: Corresponding error visualisation (red: large, blue: small). *Fourth Row*: Corresponding order selection maps (black: 1st order, white: 2nd order). Isotropic order selection map o, anisotropic order selection maps o_1 and o_2. (Color figure online)

Table 2. Comparison of different window sizes (KITTI 2015 training dataset).

w_2	1×1		3×3		5×5		7×7		9×9	
w_1	Fl	AEE	Fl	AEE	Fl	AEE	Fl	AEE	Fl	AEE
1×1	23.79	10.04	20.57	8.75	20.47	8.69	20.57	8.92	20.47	8.78
3×3	23.98	10.12	20.34	8.90	20.23	8.85	20.28	8.66	20.32	8.80
5×5	23.82	10.12	20.33	8.77	20.24	8.67	**20.15**	**8.65**	20.37	8.93
7×7	23.76	10.07	20.29	8.84	20.41	8.77	20.22	8.67	20.32	8.76
9×9	23.65	10.00	20.30	8.72	20.43	8.73	20.37	8.88	20.29	8.73

As can be seen from Table 1, our novel anisotropic order selection scheme consistently outperforms the original isotropic selection scheme. In particular, the results show similar improvements independent of the type of motion (MPI Sintel: fronto-parallel, KITTI 2015: affine). These improvements can also be seen in Fig. 1. While the anisotropic selection scheme is able to capture the moving car much better than its isotropic counterpart, the corresponding order maps show that this improvement is in fact due to the underlying anisotropic order selection behaviour: c_1 decides for first-order smoothness (black) across object edges (which allows to better preserve jumps), while c_2 decides for second-order smoothness (white) along object edges (affine motion of the car). Please note that the ground truth is sparse and does not include the motion of the truck.

Window Sizes. In a second experiment, we investigate the influence of the window sizes w_1 and w_2 which model a nonlocal decision process when determining the regularisation order across and along object boundaries in our anisotropic selection scheme. To this end, we used the KITTI 2015 training data set and computed results for different combinations of these window sizes.

The outcome in Table 2 shows that choosing a moderate window size is beneficial for deciding on the regularisation order in both directions, i.e. across and along object boundaries. In particular along object boundaries, the robustness provided by a nonlocal decision process, i.e. by choosing $w_2 > 1$, turns out to be crucial. This intuitively makes sense, since most of the smoothing takes place along object boundaries and hence the estimation of the appropriate regularisation order in this direction is particularly important. Accordingly, we choose $w_1 = 5$ and $w_2 = 7$ for KITTI. For Sintel $w_1 = 3$ and $w_2 = 5$ worked best.

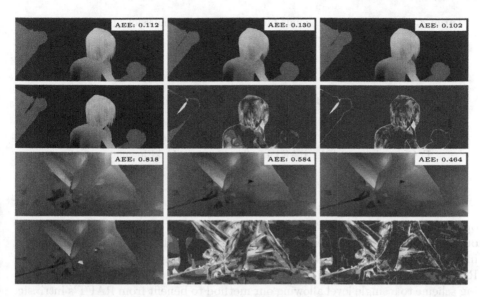

Fig. 2. Refinement examples (MPI Sintel, training data set, *clean* pass, Sequences *alley_1* and *cave_4*). *First and third row*: Computed flow fields. Stand-alone result (our method), RAFT result (CTSKH+warm-start [20]), refined RAFT result (our method). *Second and fourth row*: Ground truth, error visualisation (red: large, blue: small) for the RAFT result and the refined result (our method). (Color figure online)

Table 3. Variational refinement on the results of the RAFT [20] method for the MPI Sintel and the KITTI 15 dataset in terms of the average endpoint error (AEE) and the percentage of bad pixels (Fl). RAFT results with n/a are not provided in [20].

Method	Sintel *train*		Sintel *test*		KITTI 15 *train*		*test*
	Clean	Final	Clean	Final		occ	occ
	AEE	AEE	AEE	AEE	Fl	AEE	Fl
RAFT(CT) [20]	1.43	2.71	n/a	n/a	17.4	5.04	n/a
+ refinement (our method)	**1.35**	**2.58**	–	–	**16.3**	**4.82**	–
RAFT(CTSKH), 2-view [20]	0.76	1.22	1.94	3.18	**1.5**	**0.63**	5.10
+ refinement (our method)	**0.74**	**1.20**	**1.85**	**3.17**	1.5	0.63	**5.07**
RAFT(CTSKH), warm-start [20]	0.77	1.27	1.61	2.86	n/a	n/a	n/a
+ refinement (our method)	**0.74**	**1.23**	**1.54**	**2.81**	–	–	–

6.2   Refinement Approach

Refinement. Let us now consider the performance of our anisotropic selection scheme when refining the results of a recent state-of-the-art method. In our third experiment, we hence used the outcome of the currently best performing

Table 4. Top five non-anonymous pure optical flow submissions (test data sets).

Sintel *clean*		Sintel *final*		KITTI 15	
Method	AEE	Method	AEE	Method	Fl
Our method	**1.544**	**Our method**	**2.813**	**Our method**	**5.07**
RAFT [20]	1.609	RAFT [20]	2.855	RAFT [20]	5.10
DICL-Flow+	1.863	DICL-Flow+	3.317	PPAC-HD3 [21]	6.06
DICL_update	2.121	RAFT+LCV	3.365	MaskFlownet	6.11
RAFT+LCV	2.489	DICL_update	3.438	VCN+LCV	6.25

optical flow method – the RNN-based RAFT method [20] – as initialisation for our approach. More precisely, we used results from three RAFT variants with increasing accuracy: CT, CTSKH and CTSKH+warm-start; see [20] for details. Regarding the minimisation of our energy, we restricted the coarse-to-fine warping scheme to a single level allowing our method to benefit from RAFT's intrinsic large-displacement capabilities. Moreover we optimised the model parameters for each variant and for each benchmark separately, since, not surprisingly, a generic refinement setting cannot keep up with the partly over-fitted RAFT models.

The results of our refinement approach can be seen in Table 3. Regarding the performance on the training data sets, it shows that our approach is not only able to consistently improve the results of the CT variant (which has not been fine-tuned on Sintel or KITTI), but is also partially capable of improving the highly over-fitted and hence extremely accurate results of the CTSKH variant (which has been fine-tuned on Sintel and KITTI). Even more impressively, in the case of the top performing CTSKH+warm-start variant, which itself uses results from previous frames as initialisation, our refinement shows significant improvements up to 4%. This is also confirmed in Fig. 2, where the refined flow fields are compared to the ones of the CTSKH+warm-start variant and the results of our stand-alone approach. The refined flow fields not only show more details (e.g. hair) but also smoother transitions compared to RAFT (error plots: darker blue). Moreover, in terms of precision, they also show a higher accuracy.

Comparison to State of the Art. In our fourth experiment we check, if the excellent performance of our order-adaptive variational refinement method with anisotropic selection scheme generalises to the test data sets of MPI Sintel and KITTI 2015. The corresponding results in Table 3 not only show smaller errors in all cases, they also demonstrate that variational methods are able to improve the highly precise results of recent learning-based techniques even further. Finally, Table 4 compares our method to the state of the art in terms of the top five non-anonymous two-frame optical flow methods for MPI Sintel and KITTI 2015. As one can see, our refinement method takes the lead in all benchmarks.

7   Conclusion

In this paper we proposed a generalised order-adaptive variational optical flow method based on an anisotropic order-selection scheme. In contrast to previous order-adaptive approaches that were purely isotropic in their order selection behaviour, our novel method offers the advantage that it allows to select the local regularisation order individually for each direction and hence to adapt better to the underlying scene. Experiments confirmed our theoretical considerations. They did not only demonstrate that our novel anisotropic order selection scheme is superior to a purely isotropic selection strategy, they also made explicit that our novel method is capable of improving the highly accurate results of recent RNN-based methods when applied as a subsequent refinement step.

Acknowledgements. Funded by the Deutsche Forschungsgemeinschaft (DFG, German Research Foundation) – Project-ID 251654672 – TRR 161 (B04, B07).

References

1. Black, M.J., Anandan, P.: The robust estimation of multiple motions: parametric and piecewise-smooth flow fields. Comput. Vis. Image Underst. **63**(1), 75–104 (1996)
2. Bredies, K., Kunisch, K., Pock, T.: Total generalized variation. SIAM J. Imag. Sci. **3**(3), 492–526 (2010)
3. Brox, T., Bruhn, A., Papenberg, N., Weickert, J.: High accuracy optical flow estimation based on a theory for warping. In: Pajdla, T., Matas, J. (eds.) ECCV 2004. LNCS, vol. 3024, pp. 25–36. Springer, Heidelberg (2004). https://doi.org/10.1007/978-3-540-24673-2_3
4. Butler, D.J., Wulff, J., Stanley, G.B., Black, M.J.: A naturalistic open source movie for optical flow evaluation. In: Fitzgibbon, A., Lazebnik, S., Perona, P., Sato, Y., Schmid, C. (eds.) ECCV 2012. LNCS, vol. 7577, pp. 611–625. Springer, Heidelberg (2012). https://doi.org/10.1007/978-3-642-33783-3_44
5. Demetz, O., Stoll, M., Volz, S., Weickert, J., Bruhn, A.: Learning brightness transfer functions for the joint recovery of illumination changes and optical flow. In: Fleet, D., Pajdla, T., Schiele, B., Tuytelaars, T. (eds.) ECCV 2014. LNCS, vol. 8689, pp. 455–471. Springer, Cham (2014). https://doi.org/10.1007/978-3-319-10590-1_30
6. Dosovitskiy, A., et al.: Flownet: learning optical flow with convolutional networks. In: Proceedings of International Conference on Computer Vision (ICCV), pp. 2758–2766 (2015)
7. Hafner, D., Schroers, C., Weickert, J.: Introducing maximal anisotropy into second order coupling models. In: Gall, J., Gehler, P., Leibe, B. (eds.) GCPR 2015. LNCS, vol. 9358, pp. 79–90. Springer, Cham (2015). https://doi.org/10.1007/978-3-319-24947-6_7
8. Horn, B., Schunck, B.G.: Determining optical flow. AI **17**, 185–203 (1981)
9. Knöbelreiter, P., Pock, T.: Learned collaborative stereo refinement. In: Fink, G.A., Frintrop, S., Jiang, X. (eds.) DAGM GCPR 2019. LNCS, vol. 11824, pp. 3–17. Springer, Cham (2019). https://doi.org/10.1007/978-3-030-33676-9_1

10. Maurer, D., Bruhn, A.: ProFlow: learning to predict optical flow. In: Proceedings of British Machine Vision Conference (BMVC). BMVA Press (2018)

11. Maurer, D., Marniok, N., Goldluecke, B., Bruhn, A.: Structure-from-motion-aware PatchMatch for adaptive optical flow estimation. In: Ferrari, V., Hebert, M., Sminchisescu, C., Weiss, Y. (eds.) ECCV 2018. LNCS, vol. 11212, pp. 575–592. Springer, Cham (2018). https://doi.org/10.1007/978-3-030-01237-3_35

12. Maurer, D., Stoll, M., Bruhn, A.: Order-adaptive regularisation for variational optical flow: global, local and in between. In: Lauze, F., Dong, Y., Dahl, A.B. (eds.) SSVM 2017. LNCS, vol. 10302, pp. 550–562. Springer, Cham (2017). https://doi.org/10.1007/978-3-319-58771-4_44

13. Maurer, D., Stoll, M., Bruhn, A.: Order-adaptive and illumination-aware variational optical flow refinement. In: Proceedings of British Machine Vision Conference (BMVC), pp. 150.1–150.13 (2017)

14. Maurer, D., Stoll, M., Volz, S., Gairing, P., Bruhn, A.: A comparison of isotropic and anisotropic second order regularisers for optical flow. In: Lauze, F., Dong, Y., Dahl, A.B. (eds.) SSVM 2017. LNCS, vol. 10302, pp. 537–549. Springer, Cham (2017). https://doi.org/10.1007/978-3-319-58771-4_43

15. Menze, M., Geiger, A.: Object scene flow for autonomous vehicles. In: Proceedings of Conference on Computer Vision and Pattern Recognition (CVPR), pp. 3061–3070 (2015)

16. Nagel, H., Enkelmann, W.: An investigation of smoothness constraints for the estimation of displacement vector fields from image sequences. IEEE Trans. Pattern Anal. Mach. Intell. **8**(5), 565–593 (1986)

17. Ranftl, R., Bredies, K., Pock, T.: Non-local total generalized variation for optical flow estimation. In: Fleet, D., Pajdla, T., Schiele, B., Tuytelaars, T. (eds.) ECCV 2014. LNCS, vol. 8689, pp. 439–454. Springer, Cham (2014). https://doi.org/10.1007/978-3-319-10590-1_29

18. Revaud, J., Weinzaepfel, P., Harchaoui, Z., Schmid, C.: Epicflow: edge-preserving interpolation of correspondences for optical flow. In: Proceedings of Conference on Computer Vision and Pattern Recognition (CVPR), pp. 1164–1172 (2015)

19. Stoll, M., Volz, S., Maurer, D., Bruhn, A.: A time-efficient optimisation framework for parameters of optical flow methods. In: Sharma, P., Bianchi, F.M. (eds.) SCIA 2017. LNCS, vol. 10269, pp. 41–53. Springer, Cham (2017). https://doi.org/10.1007/978-3-319-59126-1_4

20. Teed, Z., Deng, J.: RAFT: recurrent all-pairs field transforms for optical flow. In: Vedaldi, A., Bischof, H., Brox, T., Frahm, J.-M. (eds.) ECCV 2020. LNCS, vol. 12347, pp. 402–419. Springer, Cham (2020). https://doi.org/10.1007/978-3-030-58536-5_24

21. Wannenwetsch, A.S., Roth, S.: Probabilistic pixel-adaptive refinement networks. In: Proceedings of Conference on Computer Vision and Pattern Recognition (CVPR), pp. 11642–11651 (2020)

22. Zimmer, H., Bruhn, A., Weickert, J.: Optic flow in harmony. Int. J. Comput. Vision **93**(3), 368–388 (2011)

Low-Rank Registration of Images Captured Under Unknown, Varying Lighting

Matthieu Pizenberg[(✉)], Yvain Quéau, and Abderrahim Elmoataz

Normandie Univ, UNICAEN, ENSICAEN, CNRS, GREYC, 14000 Caen, France
{matthieu.pizenberg,yvain.queau,abderrahim.elmoataz-billah}@unicaen.fr

Abstract. Photometric stereo infers the 3D-shape of a surface from a sequence of images captured under moving lighting and a static camera. However, in real-world scenarios the viewing angle may slightly vary, due to vibrations induced by the camera shutter, or when the camera is hand-held. In this paper, we put forward a low-rank affine registration technique for images captured under unknown, varying lighting. Optimization is carried out using convex relaxation and the alternating direction method of multipliers. The proposed method is shown to significantly improve 3D-reconstruction by photometric stereo on unaligned real-world data, and an open-source implementation is made available.

Keywords: Photometric stereo · Registration · Shape-from-X

1 Introduction

Photometric stereo is a 3D-reconstruction technique pioneered by Woodham [17]. It infers the geometry of a surface from a set of images captured under the *same viewing angle*, but varying illumination. The camera is thus assumed to be perfectly still, an assumption which may be violated when the camera is hand-held or when it is manually triggered (which induces slight vibrations of the sensor). Consequently, 3D-reconstruction by photometric stereo may be blurry (cf. Fig. 1), and geometric information may even be hallucinated (cf. Fig. 2). To prevent this, the image sequence could be registered prior to 3D-reconstruction.

We focus in this work on quasi-planar scenes, such that the geometric transformation between images can be assumed affine. Despite this simplifying assumption, registration remains arduous due to lighting variations, which may induce cast-shadows or specularities. In the present work, we propose an open-source[1], robust affine registration technique for images captured under unknown, varying lighting, based on low-rank approximation and convex relaxation.

The rest of this paper is organized as follows. After reviewing related works in Sect. 2, we introduce in Sect. 3 a variational approach to image registration

---

[1] https://github.com/mpizenberg/lowrr

© Springer Nature Switzerland AG 2021
A. Elmoataz et al. (Eds.): SSVM 2021, LNCS 12679, pp. 153–164, 2021.
https://doi.org/10.1007/978-3-030-75549-2_13

under varying lighting. Then, we describe in Sect. 4 an augmented Lagrangian approach for numerically solving this inverse problem. The implementation of this scheme is discussed in Sect. 5, along with experimental evaluation. Our conclusions and perspectives are eventually drawn in Sect. 6.

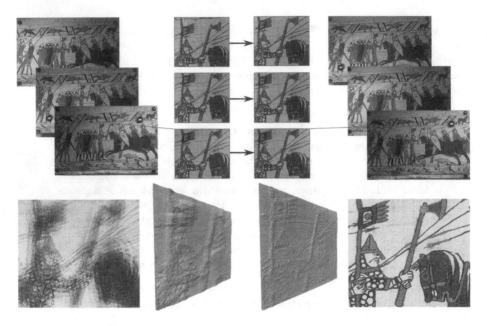

Fig. 1. Top, left: three (out of $m = 13$) images of the Bayeux Tapestry, acquired under varying lighting, using a hand-held camera. Top, right: images registered using the proposed low-rank technique. Bottom: estimated reflectance and 3D-shape, using the original (left) or the registered (right) images. Registering images allows one to use a hand-held camera while obtaining a 3D-reconstruction comparable to that obtained using a high-quality tripod (cf. Fig. 3).

2   Related Works

Low-Rank Techniques in Photometric Stereo. Early works on photometric stereo have focused on the ideal case of a Lambertian surface lit by directional sources [17]. In practical setups, these assumptions are however rarely met: low-rank approximation techniques have thus been developed to solve photometric stereo under less restrictive assumptions. Based on earlier investigations on the linear image subspaces spanned by Lambertian surfaces [2], Basri et al. have shown in [3] how to handle general, unknown lighting using spherical harmonics approximation. Non-Lambertian phenomena such as specularities or cast shadows can also be viewed as sparse outliers, and removed from the photometric stereo images by seeking a low-rank observation map sparsely deviating from the input one [18]. Such low-rank approximation techniques for outliers handling also relate to sparsity-promoting estimators used in robust calibrated photometric stereo, see e.g. [11,13]. Low-rank approximation was also considered for solving the uncalibrated case by jointly minimizing the rank and enforcing integrability [15].

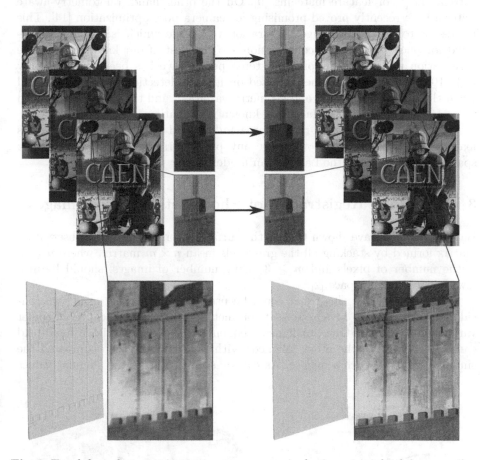

Fig. 2. Top, left: a photometric stereo sequence acquired using a tripod, while manually triggering the camera. Manual trigger induces slight image transformations, which are compensated with the proposed registration technique (top, right). Bottom: shape and reflectance estimated by photometric stereo, using either the unregistered (left) or the registered sequence (right). The misalignment causes the estimated reflectance to appear blurry, and induces geometry hallucination (the pictured surface is a perfectly planar book cover).

Image Registration in Photometric Stereo. All the works mentioned above assume that the input images have been registered beforehand. When this is not the case, the geometry recovered by using photometric stereo might appear blurry. In some sense, the effect would be similar to observing a translucent object where scattering is observed. In the latter case, the estimated geometry could be deconvolved a posteriori, as suggested in [12]. Yet, pre-registering the images seems more relevant in the absence of scattering. Registration is a long-standing problem in imaging, which has been addressed using e.g., phase correlation [8] or feature matching [9]. On the other hand, photometry-aware criteria have recently proved promising for camera pose optimization [14]. This invites us to design an image registration technique which is specifically tailored for photometric 3D-reconstruction. To the best of our knowledge, image registration under the photometric stereo perspective has been explored only in [4,10]. However, the former is based on feature detection, hence it may fail when the data lacks texture and geometry variation; and the latter is restricted to perfectly Lambertian surfaces and known illumination.

On the contrary, the low-rank approach discussed in the next section avoids feature detection; it does not require any pre-calibration; and it is robust to sparse deviations from the Lambertian model.

3   Low-Rank Registration of Photometric Stereo Images

Basri and Jacob have shown that, if the surface is Lambertian, the observation matrix formed by stacking all the graylevels in an $n \times m$ matrix, where $n \geq m$ is the number of pixels and $m \geq 3$ is the number of images, should lie in a low-dimensional subspace [2].

To verify whether this result could be used for image registration, we considered a set of $m = 13$ real-world photometric stereo images, captured under varying illumination. One (unaligned) sequence was acquired with a hand-held camera (cf. Fig. 1), the other (aligned) with a tripod. We then compared the singular values of each sequence. As can be seen in Fig. 3, the singular values

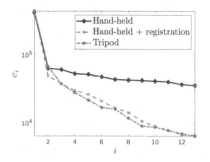

Fig. 3. Left: singular values of the observation matrix for the data in Fig. 1, using either a hand-held camera (without, or with the proposed registration technique) or a tripod ensuring perfect alignment. The singular values of the aligned sequences are smaller than those of the unaligned sequence. Right: reflectance and 3D-shape recovered from the (aligned) tripod sequence, which are comparable with those obtained from the registered, hand-held sequence, cf. Fig. 1-right.

of the aligned sequence are smaller. This invites us to recast photometric stereo images registration as a low-rank optimization problem.

3.1 Displacement Parameterization

Let us denote by $u^1, \ldots, u^m \in \mathbb{R}^n$, $n \gg m \geq 3$ the vectorized (graylevel) images (m is the number of images and n is the number of pixels in each image), by $u_j^i := u^i(x_j)$ the graylevel of the i-th image at pixel $x_j \in \mathbb{R}^2$, and by $u = \left[u^{1\top}, \ldots, u^{m\top} \right]^\top \in \mathbb{R}^{mn}$ the full observation set in vector form.

We are looking for a set $W(u; \theta) \in \mathbb{R}^{mn}$ of registered images, where W is a warping function and $\theta := \left[\theta^{1\top}, \ldots, \theta^{m\top} \right]^\top \in \mathbb{R}^{mp}$ is the unknown set of warp parameters. Therein, the i-th transformation, $i \in \{1, \ldots, m\}$, is characterized by the subset of parameters $\theta^i \in \mathbb{R}^p$. We will focus in this paper on invertible parametric image registration i.e., a closed-form expression for the warping function W and its inverse W^{-1} will be assumed, which both will depend on a "small" number of parameters ($p \leq 2n$).

In particular, in our experiments we focus on affine transformations, which combine translation, rotation, scaling and shearing. In this case, $p = 6$, and the warping depends on parameters $\theta^i := \left[\theta_1^i, \ldots, \theta_6^i \right]^\top \in \mathbb{R}^6$ according to:

$$W(u; \theta) = \text{vec} \begin{bmatrix} u^1 \left(w^1(x_1; \theta^1) \right) & \cdots & u^m \left(w^m(x_1; \theta^m) \right) \\ \vdots & & \vdots \\ u^1 \left(w^1(x_n; \theta^1) \right) & \cdots & u^m \left(w^m(x_1; \theta^m) \right) \end{bmatrix},$$

where the i-th warping is locally defined as

$$w^i(\cdot; \theta^i) : \mathbb{R}^2 \to \mathbb{R}^2$$
$$x_j \mapsto \begin{bmatrix} 1 + \theta_1^i & \theta_3^i \\ \theta_2^i & 1 + \theta_4^i \end{bmatrix} x_j + \begin{bmatrix} \theta_5^i \\ \theta_6^i \end{bmatrix} \tag{1}$$

and its inverse by

$$w^{i^{-1}}(\cdot; \theta^i) : \mathbb{R}^2 \to \mathbb{R}^2$$
$$x_j \mapsto \begin{bmatrix} 1 + \theta_1^i & \theta_3^i \\ \theta_2^i & 1 + \theta_4^i \end{bmatrix}^{-1} \left(x_j - \begin{bmatrix} \theta_5^i \\ \theta_6^i \end{bmatrix} \right). \tag{2}$$

3.2 Low-Rank Formulation and Convex Relaxation

Our approach consists in seeking the warping function which minimizes the rank of the warped observation matrix. Formally, this would come down to solving

$$\min_{\theta \in \mathbb{R}^{mp}} \text{rank} \left(\text{Mat}(W(u; \theta)) \right), \tag{3}$$

where operator Mat is such that $\text{vec} \circ \text{Mat} = \text{id}$: it reorganizes the elements of a vector into a matrix according to

$$\text{Mat} : \qquad \mathbb{R}^{mn} \qquad \to \mathbb{R}^{n \times m}$$
$$u := \left[u^{1\top}, \ldots, u^{m\top} \right]^\top \mapsto \text{Mat}(u) = \left[u^1, \ldots, u^m \right].$$

However, both the camera small displacements and specular spots in the image may break the low-rank assumption, by generating occlusions and outliers to the Lambertian model. To take such outliers into account, we simultaneously estimate the displacement parameters θ and a restored observation matrix A, which would sparsely deviate from $\text{Mat}\,(W(u;\theta))$, and have minimal rank:

$$\min_{\substack{A \in \mathbb{R}^{n \times m} \\ \theta \in \mathbb{R}^{mp}}} \quad \text{rank}\,(A) + \lambda\,\|\text{vec}(A) - W(u;\theta)\|_0 \tag{4}$$

with $\lambda \geq 0$ some tuning parameter and $\|\cdot\|_0$ the number of nonzero elements in a vector. Let us remark that if W is the identity transform, then the proposed model comes down to the low-rank image correction technique advocated in [18].

The optimization problem (4) is difficult to solve since both the rank and the 0-"norm" are nonconvex. So we rather optimize a convex relaxation of (4):

$$\min_{\substack{A \in \mathbb{R}^{n \times m} \\ \theta \in \mathbb{R}^{mp}}} \quad \|A\|_* + \lambda\,\|\text{vec}(A) - W(u;\theta)\|_1 \tag{5}$$

where $\|\cdot\|_*$ denotes the sum of singular values (nuclear norm) and $\|\cdot\|_1$ the sum of absolute values, which are the tightest convex relaxation of rank and $\|\cdot\|_0$, respectively. Next, we propose an algorithm for numerically solving (5).

4   Augmented Lagrangian Framework

In this section, we describe an alternating direction method of multipliers (see, e.g., [6] for a thorough presentation of this algorithm) for solving Problem (5).

4.1   Splitting

Both terms in the objective function of (5) are convex, yet simultaneously optimizing them remains challenging. We therefore split the optimization problem over both terms by turning it into the following equivalent, constrained one:

$$\min_{\substack{A \in \mathbb{R}^{n \times m} \\ e \in \mathbb{R}^{mn} \\ \theta \in \mathbb{R}^{mp}}} \|A\|_* + \lambda\,\|e\|_1\,,$$
$$\text{s.t.} \quad \text{vec}(A) = W(u;\theta) + e. \tag{6}$$

Let us consider the augmented Lagrangian associated with this constrained optimization problem:

$$\mathcal{L}_\rho^\sharp(A, e, \theta, y) := \|A\|_* + \lambda\,\|e\|_1 + \langle y | W(u;\theta) + e - \text{vec}(A)\rangle + \frac{\rho}{2}\,\|W(u;\theta) + e - \text{vec}(A)\|^2, \tag{7}$$

where $y \in \mathbb{R}^{mn}$ is the set of Lagrange multipliers, and $\rho > 0$ is a penalty parameter arbitrarily set to 0.1 in all our experiments. Problem (6) can be solved iteratively, by alternating minimizations of (7) wrt the primal variables A, e and θ, and a dual ascent step over the dual variable y, yielding the following sequence:

$$A^{(k+1)} = \operatorname*{argmin}_{A} \mathcal{L}_{\rho}^{\sharp}(A, e^{(k)}, \theta^{(k)}, y^{(k)}), \tag{8}$$

$$e^{(k+1)} = \operatorname*{argmin}_{e} \mathcal{L}_{\rho}^{\sharp}(A^{(k+1)}, e, \theta^{(k)}, y^{(k)}), \tag{9}$$

$$\theta^{(k+1)} = \operatorname*{argmin}_{\theta} \mathcal{L}_{\rho}^{\sharp}(A^{(k+1)}, e^{(k+1)}, \theta, y^{(k)}), \tag{10}$$

$$y^{(k+1)} = y^{(k)} + \rho\left(W(u; \theta^{(k+1)}) + e^{(k+1)} - \text{vec}(A^{(k+1)})\right),$$

starting from the initial estimate $A^{(0)} = \text{Mat}(u)$, $e^{(0)}, \theta^{(0)}, y^{(0)} \equiv 0$ (i.e., the initial solution is set to the unregistered data), and until the relative residual $\left\|\text{vec}\left(A^{(k+1)} - A^{(k)}\right)\right\| / \left\|\text{vec}\left(A^{(k)}\right)\right\|$ falls below some threshold (this threshold is set to 10^{-3} in our experiments).

Next, we detail the resolution of each sub-problem in the above ADMM algorithm.

4.2 Updating the Corrected Observation Matrix A

Update (8) can be computed in closed-form. Indeed:

$$\begin{aligned}
A^{(k+1)} &= \operatorname*{argmin}_{A} \mathcal{L}_{\rho}^{\sharp}(A, e^{(k)}, \theta^{(k)}, y^{(k)}) \\
&= \operatorname*{argmin}_{A} \|A\|_{*} + \frac{\rho}{2}\left\|W(u; \theta^{(k)}) + e^{(k)} - \text{vec}(A) + y^{(k)}/\rho\right\|^{2} \\
&= \operatorname*{argmin}_{A} \|A\|_{*} + \frac{\rho}{2}\left\|A - \text{Mat}(W(u; \theta^{(k)}) + e^{(k)} + y^{(k)}/\rho)\right\|_{F}^{2} \\
&= \operatorname*{prox}_{\frac{1}{\rho}\|\cdot\|_{*}} (\text{Mat}(W(u; \theta^{(k)}) + e^{(k)} + y^{(k)}/\rho)),
\end{aligned}$$

where $W(u; \theta^{(k)})$ can be evaluated by interpolation (we used linear interpolation in our experiments) of the unregistered images, and the proximity operator of the nuclear norm admits a closed-form expression as shrinkage of the singular values. More precisely, denoting by $B = U \operatorname{diag}(\sigma^{1}, \ldots, \sigma^{m}) V^{\top}$ the singular value decomposition of some matrix $B \in \mathbb{R}^{n \times m}$, $m \leq n$:

$$\operatorname{prox}_{\frac{1}{\rho}\|\cdot\|_{*}}(B) = U \operatorname{diag}\left(\text{shrink}(\sigma^{1}, 1/\rho), \ldots, \text{shrink}(\sigma^{m}, 1/\rho)\right) V^{\top},$$

using the following definition of the shrinkage (soft-thresholding) operator:

$$\text{shrink}(\cdot, 1/\rho) = \text{sign}(\cdot) \max\left\{|\cdot| - 1/\rho, 0\right\}.$$

4.3 Updating the Error Vector e

Update (9) can also be computed in closed-form:

$$e^{(k+1)} = \operatorname*{argmin}_{e} \mathcal{L}_\rho^\sharp(A^{(k+1)}, e, \theta^{(k)}, y^{(k)})$$

$$= \operatorname*{argmin}_{e} \lambda \|e\|_1 + \frac{\rho}{2} \left\| e - \left(\operatorname{vec}(A^{(k+1)}) - W(u; \theta^{(k)}) - y^{(k)}/\rho\right)\right\|^2$$

$$\overset{d}{=} \operatorname{prox}_{\frac{\lambda}{\rho}\|\cdot\|_1} \left(\operatorname{vec}(A^{(k+1)}) - W(u; \theta^{(k)}) - y^{(k)}/\rho\right)$$

$$= \operatorname{shrink}(\operatorname{vec}(A^{(k+1)}) - W(u; \theta^{(k)}) - y^{(k)}/\rho, \lambda/\rho)$$

where shrinkage is to be understood component-wise.

4.4 Updating the Displacement Coefficients θ

Let us now consider the update (10), which writes as

$$\theta^{(k+1)} = \operatorname*{argmin}_{\theta} \left\| W(u; \theta) - \left(\operatorname{vec}(A^{(k+1)}) - e^{(k+1)} - y^{(k)}/\rho\right)\right\|^2. \qquad (11)$$

This is a classic least-squares warping problem resembling the Kanade-Lucas-Tomasi (KLT) problem, and we refer the interested reader to [1] for an in-depth presentation. Therein, several iterative algorithms are discussed, which have different preconditions but are shown to be equivalent if the set of warps forms a group and if warps are differentiable - as it is the case for affine warping. Among the various possibilities, we chose the forward compositional algorithm for the simpler expression of the Jacobian in general.

Let us denote $v^{(k)} := \operatorname{vec}(A^{(k+1)}) - e^{(k+1)} - y^{(k)}/\rho$. Instead of solving (11), one step of the forwards compositional algorithm computes $\theta^{(k+1)}$ as follows:

$$\delta\theta^{(k+1)} = \operatorname*{argmin}_{\delta\theta} \left\| W(W(u; \delta\theta); \theta^{(k)}) - v^{(k)}\right\|^2, \qquad (12)$$

$$W(\cdot; \theta^{(k+1)}) = W(\cdot; \theta^{(k)}) \circ W(\cdot; \delta\theta^{(k+1)}).$$

This update can be carried out independently for each image u^i. We denote $\theta^{i(k)} \in \mathbb{R}^p$ the warp parameters for image i at step k, such that $\theta^{(k)} \in \mathbb{R}^{mp}$ is the concatenation of all $\theta^{i(k)}$ at step k. Similarly, we denote $v^{i(k)} \in \mathbb{R}^n$ the part of $v^{(k)} \in \mathbb{R}^{mn}$ corresponding to the i-th image. As explained in [1], performing a first-order Taylor expansion of the expression in (12) gives

$$\delta\theta^{i(k+1)} = \operatorname*{argmin}_{\delta\theta} \sum_j \left[W(u; \theta^{(k)})_j^i + \nabla W(u; \theta^{(k)})_j^i \frac{\partial W}{\partial\theta}\bigg|_{x_j} \delta\theta - v_j^{i(k)}\right]^2, \qquad (13)$$

where $(\nabla W(u; \theta^{(k)})_j^i)^\top \in \mathbb{R}^2$ is the warped image gradient at pixel j and $\frac{\partial W}{\partial\theta}\big|_{x_j} \in \mathbb{R}^{2\times p}$ are the partial derivatives of the warp function relative to warp parameters. The solution to this least-squares optimization problem is

$$\delta\theta^{i(k+1)} = H^{-1} \left(J^\top [v^{i(k)} - W(u; \theta^{(k)})^i]\right),$$

where $J \in \mathbb{R}^{n \times p}$ is the Jacobian matrix and $H = J^\top J$ is the Gauss-Newton approximation of the Hessian matrix. The rows of J are computed as

$$J_j = \nabla W(u; \theta^{(k)})_j^i \frac{\partial W}{\partial \theta}\Big|_{x_j}.$$

Therein, for any point $x_j := [x_{j,1}, x_{j,2}]^\top \in \mathbb{R}^2$:

$$\frac{\partial W}{\partial \theta}\Big|_{x_j} = \begin{bmatrix} x_{j,1} & 0 & x_{j,2} & 0 & 1 & 0 \\ 0 & x_{j,1} & 0 & x_{j,2} & 0 & 1 \end{bmatrix},$$

since we focus on affine transformations as defined in (1),

We now have all the ingredients for solving the low-rank registration problem (5). In the next section, we discuss the practical implementation of the proposed scheme, and evaluate it on synthetic and real-world data.

5   Implementation

The numerical solution comes down to iterating steps (8) to (10). Yet, convergence of these steps can be established only towards a locally optimum solution, which may be far away from the global optimum if the displacements are important. In order to ease convergence towards a reasonable solution, as well as to accelerate the algorithm, we embed this algorithm inside a multi-scale scheme.

5.1   Multi-scale Scheme

The first-order Taylor expansion done in (13) is only viable in the case of small displacements, when the image gradients and the residuals $v^{(k)} - W(u; \theta^{(k)})$ are correlated. Yet at full resolution, warps may generate displacements of dozens of pixels. One workaround, commonly used in KLT-related problems [5], is to consider a multi-scale approach, by generating a pyramid of images as in Fig. 4. Each level halves/doubles the resolution of the previous one, and the only parameter to tune is the number of levels of the pyramid. At the lowest resolution, every pixel covers a much larger area, therefore artificially reducing the image displacement, and improving the conditions for the first-order Taylor approximation. Once the algorithm has converged at one level, we adapt the resulting registration parameters to the next level of the pyramid and use those as initialization. In the case of the affine warp, θ_1^i to θ_4^i stay unchanged, and θ_5^i and θ_6^i are doubled at a level transition.

5.2   Using a Sparse Subset of Pixels

Engel et al. showed in [7] that direct image alignment is possible with a sparse subset of points as long as they are well distributed in the image, and located at pixels with higher gradient magnitude. This can be used as a trick to accelerate the proposed algorithm.

If the scene is highly textured, as in the datasets of Figs. 1 and 2, it is possible to restrict our attention to pixels where the gradient magnitude is maximal. At the coarsest level, all pixels can be considered. Then, for each selected pixel, only one or two of the four subpixels can be selected at the next pyramid level, as illustrated in Fig. 4. Let us emphasize, however, that this feature-based speed up is only a way to accelerate the process, unlike e.g. in [10] where the whole registration builds upon feature detection, hence cannot be employed for unsufficiently textured scenes.

Fig. 4. Top row: pyramid of resolutions for one image of the sequence in Fig. 1. The resolution doubles at each level. Bottom row: sparse subset of pixels (indicated in red) which can be selected at each level, in order to accelerate the process. (Color figure online)

5.3   Empirical Evaluation

We first quantitatively evaluated our method on simulated image transformations. We considered two challenging real-world photometric stereo datasets from [16], namely the "Buddha" and "Harvest" sequences. Both consist of $m = 96$ images of a shiny surface (exhibiting significant deviations from the Lambertian assumption) taken from the same viewing angle under varying lighting. For each object, we generated random synthetic translations, with the parameters uniformly taken between 0 and 1% of the image width. We then evaluated the accuracy of different algorithms by measuring their mean displacement error. This process was repeated 40 times per object image sequence. The results in Fig. 5 show that the proposed method estimates the correct displacement in most cases. It outperforms standard image registration algorithms which are not designed for handling illumination changes. This can be seen by comparing with Matlab's implementation of similarity optimization (`imregtform`) or phase correlation (`imregcorr`).

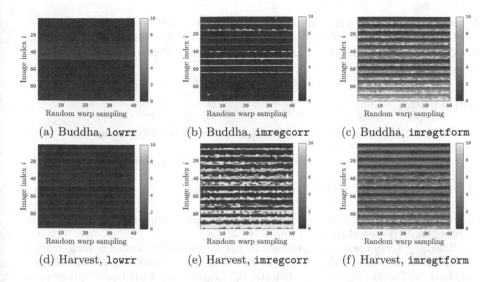

Fig. 5. Mean displacement error (in pixels) of every warp estimation for our approach (`lowrr`), and the `imregtform` and `imregcorr` Matlab's algorithms, on two sequences ("Buddha" and "Harvest") of the DiLiGenT dataset [16].

Next, we considered a real-world hand-held sequence of $m = 13$ images of a medieval embroidery (cf. Fig. 1), exhibiting fine-scale geometric variations. We registered the images with the proposed method, and then observed the impact of image registration on the results of a recent photometric stereo algorithm [13]. The proposed approach allows to reach results comparable with those obtained from perfectly aligned images acquired with a tripod (compare with Fig. 3). Lastly, we carried out a similar experiment, on a perfectly planar surface with spatially-varying reflectance (cf. Fig. 2). In this experiment, the camera stands on a tripod, yet the camera is manually triggered, inducing tiny displacements (a few pixels). Taking into account these displacements improves the sharpness of the estimated reflectance, and avoids the hallucination of geometric details.

6   Conclusions and Perspectives

We have presented an affine registration technique for images captured under unknown, varying lighting. The proposed approach seeks the registration parameters which minimize the rank of the warped observations, while robustness is ensured by sparsity enforcement. An efficient solution based on the augmented Lagrangian has also been introduced. Experiments on real-world data show that the proposed approach significantly improves 3D-reconstruction by photometric stereo on unaligned data, and even allows using hand-held images. In future work, we will extend the proposed formulation so as to handle non-affine transformations, by inserting optical flow constraints inside the variational problem.

Acknowledgements. This work was supported by the RIN project "Guide Muséal", and by the ANR grant "Inclusive Museum Guide" (ANR-20-CE38-0007). The authors would like to thank C. Berthelot at the Bayeux Tapestry Museum for supervising the image acquisition campaign.

References

1. Baker, S., Matthews, I.: Lucas-Kanade 20 years on: a unifying framework. IJCV **56**(3), 221–255 (2004)
2. Basri, R., Jacobs, D.W.: Lambertian reflectance and linear subspaces. PAMI **25**(2), 218–233 (2003)
3. Basri, R., Jacobs, D.W., Kemelmacher, I.: Photometric stereo with general, unknown lighting. IJCV **72**, 239–257 (2007)
4. Berkiten, S., Rusinkiewicz, S.: Alignment of images captured under different light directions. Technical report, TR-974-14 (2014)
5. Bouguet, J.Y.: Pyramidal implementation of the affine Lucas-Kanade feature tracker: description of the algorithm. Technical report, Intel corporation (2001)
6. Boyd, S., Parikh, N., Chu, E., Peleato, B., Eckstein, J.: Distributed optimization and statistical learning via the alternating direction method of multipliers. Found. Trends Mach. Learn. **3**, 1–122 (2010)
7. Engel, J., Koltun, V., Cremers, D.: Direct sparse odometry. PAMI **40**(3), 611–625 (2017)
8. Foroosh, H., Zerubia, J., Berthod, M.: Extension of phase correlation to subpixel registration. TIP **11**(3), 188–200 (2002)
9. Förstner, W.: A feature based correspondence algorithm for image matching. In: ISPRS ComIII, pp. 150–166 (1986)
10. Harrison, A.P., Joseph, D.: Translational photometric alignment of single-view image sequences. CVIU **116**(6), 765–776 (2012)
11. Ikehata, S., Wipf, D., Matsushita, Y., Aizawa, K.: Robust photometric stereo using sparse regression. In: CVPR, pp. 318–325 (2012)
12. Inoshita, C., Mukaigawa, Y., Matsushita, Y., Yagi, Y.: Surface normal deconvolution: photometric stereo for optically thick translucent objects. In: Fleet, D., Pajdla, T., Schiele, B., Tuytelaars, T. (eds.) ECCV 2014. LNCS, vol. 8690, pp. 346–359. Springer, Cham (2014). https://doi.org/10.1007/978-3-319-10605-2_23
13. Quéau, Y., Wu, T., Lauze, F., Durou, J.D., Cremers, D.: A non-convex variational approach to photometric stereo under inaccurate lighting. In: CVPR, pp. 99–108 (2017)
14. Schmitt, C., Donne, S., Riegler, G., Koltun, V., Geiger, A.: On joint estimation of pose, geometry and svBRDF from a handheld scanner. In: CVPR, pp. 3493–3503 (2020)
15. Sengupta, S., Zhou, H., Forkel, W., Basri, R., Goldstein, T., Jacobs, D.: Solving uncalibrated photometric stereo using fewer images by jointly optimizing low-rank matrix completion and integrability. JMIV **60**(4), 563–575 (2018)
16. Shi, B., Mo, Z., Wu, Z., Duan, D., Yeung, S., Tan, P.: A benchmark dataset and evaluation for non-Lambertian and uncalibrated photometric stereo. PAMI **41**(2), 271–284 (2019)
17. Woodham, R.J.: Photometric method for determining surface orientation from multiple images. Opt. Eng. **19**(1), 134–144 (1980)
18. Wu, L., Ganesh, A., Shi, B., Matsushita, Y., Wang, Y., Ma, Y.: Robust photometric stereo via low-rank matrix completion and recovery. In: ACCV, pp. 703–717 (2010)

Towards Efficient Time Stepping for Numerical Shape Correspondence

Alexander Köhler[(✉)] and Michael Breuß

Institute for Mathematics, Brandenburg Technical University,
Platz der Deutschen Einheit 1, 03046 Cottbus, Germany
{koehlale,breuss}@b-tu.de

Abstract. The computation of correspondences between shapes is a principal task in shape analysis. To this end, methods based on partial differential equations (PDEs) have been established, encompassing e.g. the classic heat kernel signature as well as numerical solution schemes for geometric PDEs. In this work we focus on the latter approach.

We consider here several time stepping schemes. The goal of this investigation is to assess, if one may identify a useful property of methods for time integration for the shape analysis context. Thereby we investigate the dependence on time step size, since the class of implicit schemes that are useful candidates in this context should ideally yield an invariant behaviour with respect to this parameter.

To this end we study integration of heat and wave equation on a manifold. In order to facilitate this study, we propose an efficient, unified model order reduction framework for these models. We show that specific l_0 stable schemes are favourable for numerical shape analysis. We give an experimental evaluation of the methods at hand of classical TOSCA data sets.

Keywords: Shape analysis · Second order time integration · Heat equation · Wave equation · Model order reduction

1 Introduction

The computation of shape correspondences is a fundamental task in computer vision with many potential applications, cf. [4,8]. In the setting of three-dimensional shape analysis, the underlying problem amounts to identify an explicit relation between the surface elements of two or more shapes. The variety of possible shape correspondence mappings that is of interest in applications includes non-rigid transformations where shapes are just almost isometric, allowing e.g. to match different poses of human or animal shapes.

An important solution strategy is to achieve a pointwise shape correspondence using so called descriptor based methods. For this, a feature descriptor has to be computed that characterizes each point on a shape by describing the surrounding shape surface geometry. A mathematically sound approach to compute such shape signatures is to make use of the spectral decomposition of the

© Springer Nature Switzerland AG 2021
A. Elmoataz et al. (Eds.): SSVM 2021, LNCS 12679, pp. 165–176, 2021.
https://doi.org/10.1007/978-3-030-75549-2_14

Laplace-Beltrami operator, see e.g. [4,14,15]. To this end, the Laplace-Beltrami operator may be incorporated in a certain variety of partial differential equations (PDEs) that are potentially useful as models for feature computation. The arguably most important classic signature is the heat kernel signature (HKS) [15] which relies on the heat equation, however also versions of the Schrödinger equation leading to the wave kernel signature (WKS) [1] as well as the hyperbolic wave equation [6] have already been proposed.

For the computation of such PDE-based signatures, the basic task amounts to resolve the underlying PDEs on a manifold representing a shapes' boundary. The HKS and WKS both rely on the eigenfunction expansion of the Laplace-Beltrami operator to tackle this task and for achieving efficient algorithms. An alternative to the spectral approach is to consider the numerical integration of the underlying PDEs as proposed in [2,5,6]. However, when following this path it is highly advocated to employ very efficient computational means such as the model order reduction framework presented in [2,3] in order to avoid high computational times.

In previous works based on numerical integration, first order implicit time integration has been studied in detail [5]. For the wave equation model, it has been shown that backward differencing in time may yield favorable results over simple central differences [6]. Especially, as has been illustrated in [6], the classic wave equation may yield in some experiments results of higher correspondence quality compared to the other mentioned models. Let us note that this holds again in particular when assessing the models using numerical integration with implicit first order time stepping.

Our Contributions. In this paper we give an account of ongoing work on improving numerical PDE-based shape descriptors with respect to computational efficiency. The underlying question addressed in this paper is, if the class of implicit schemes has particularly useful key properties that should be respected in the shape analysis context. We present a study of three implicit time stepping schemes for heat and wave equation on manifolds, respectively. We show that standard finite differences may not be appropriate since the typical initial condition used for the construction of shape signatures, which is a discrete Dirac delta function, may yield oscillatory artefacts that may spoil correspondence quality. In order to resolve this issue, we propose to adopt l_0-stable methods. For performing the study, we have developed a unified numerical model order reduction framework based on [2]. We give an experimental account of our investigations at hand of selected shape data sets.

2   Modelling of the Shape Correspondence Framework

The shape of three-dimensional geometric object can be described by its bounding surface, given as a compact two-dimensional Riemannian manifold $\mathcal{M} \subset \mathbb{R}^3$. Two such shapes $\mathcal{M}$ and $\tilde{\mathcal{M}}$ are considered to be isometric if there is a smooth homeomorphism between the corresponding object surfaces that preserves the intrinsic distances between surface points. For many applications, isometry may only hold approximately, leading to the notion of almost isometric shapes.

As indicated, a widely used descriptor class that can handle almost isometric transformations relies on PDEs on manifolds. We will consider in this paper the corresponding heat and wave equation, respectively. These PDEs may be put together in a more general form as described below.

All the mentioned PDEs rely on the Laplace-Beltrami operator as a fundamental building block. This operator is the geometric version of the Laplace operator, meaning that it takes into account the local curvature of a smooth manifold in 3D. Applied to a scalar-valued function u, it can be written as:

$$\Delta_{\mathcal{M}} u = \frac{1}{\sqrt{|g|}} \sum_{i,j=1}^{2} \partial_i \left(\sqrt{|g|} g^{ij} \partial_j u \right) \tag{1}$$

where $|g|$ is the determinant of the metric tensor $g \in \mathbb{R}^{2 \times 2}$ that describes locally the geometry, and g^{ij} are the entries of its inverse, see e.g. [7] for more details on the differential geometric notions.

Heat Equation. The heat equation on a manifold is a proper shape descriptor [15] and reads as

$$\partial_t u(x,t) = \Delta_{\mathcal{M}} u(x,t). \quad x \in \mathcal{M}, \ t \in I \tag{2}$$

This PDE basically describes heat flow along a surface $\mathcal{M}$.

Wave Equation. Proposed in [6] for computing shape correspondences, this equation can be written as

$$\partial_{tt} u(x,t) = \Delta_{\mathcal{M}} u(x,t). \quad x \in \mathcal{M}, \ t \in I \tag{3}$$

Meta PDE. We can combine the PDEs from above into a single equation as

$$\phi \partial_{tt} u(x,t) + \psi \partial_t u(x,t) = \Delta_{\mathcal{M}} u(x,t). \quad x \in \mathcal{M}, \ t \in I \tag{4}$$

We see if we choose $\phi = 0$ and $\psi = 1$ we receive heat Eq. (2) and for $\phi = 1$ and $\psi = 0$ it will lead us to the wave Eq. (3). The PDE formulation (4) is useful for us in this paper, since it allows us to describe in a compact way numerical developments in later sections.

Initial Condition for Feature Computation. Both equations need initial conditions to be solved accurately. The classic initial condition used for PDE-based feature generation is the Dirac Delta function $u(x,0) = u_0(x) = u_{x_i}$ centred around a point $x_i \in \mathcal{M}$, cf. [15]. We also adopt this initial condition in the numerical integration framework. Because Eq. (3) is a PDE of second order in time we need an additional condition at the initial velocity. A canonical choice is to consider the zero initial velocity condition $\partial_t u(x,0) = 0$.

Feature Descriptor. For establishing shape correspondence we consider for each point on an objects' surface a feature descriptor, which is a computational object that contains geometric shape information. Within our framework, a simple way to construct it is to restrict the spatial component of $u(x,t)$ to

$$f_{x_i}(t) := u(x,t)|_{x=x_i} \quad \text{with} \quad u(x,0)|_{x=x_i} = u_{x_i} \tag{5}$$

and call f_{x_i} the feature descriptor at the location $x_i \in \mathcal{M}$.

With this definition we may adopt a physical interpretation of (5). The feature descriptor f_{x_i} describes heat transferred away from x_i respectively the motion amplitudes of a wave front emitted and observed again at x_i. Thereby the processes run on the surface $\mathcal{M}$, as indicated via the Laplace-Beltrami operator.

Shape Correspondence. To compare the feature descriptors for different locations $x_i \in \mathcal{M}$ and $\tilde{x}_j \in \widetilde{\mathcal{M}}$ on two different shapes $\mathcal{M}$ and $\widetilde{\mathcal{M}}$, we define a distance $d_f(x_i, \tilde{x}_j)$ via the L_1 norm

$$d_f(x_i, \tilde{x}_j) = \int_I |f_{x_i} - f_{\tilde{x}_j}| \mathrm{d}t \tag{6}$$

The tuple of locations $(x_i, \tilde{x}_j) \in \mathcal{M} \times \widetilde{\mathcal{M}}$ with the smallest distance belong together.

The discrete surface representation is given by a set of points $P := (x_1, \ldots, x_N)$ containing coordinates of each point x_i and one for triangles T. T stores triples of points forming a triangle. We denote this data set tuple as $\mathcal{M}_d = (P, T)$. Furthermore we denote by Ω_i the barycentric cell volume surround the point x_i. The discrete shape is given by non-uniform linear triangles. Compare also Fig. 1 (figure adopted from [2]).

Fig. 1. Continuous & discrete shape

3   Spatial Discretisation

We employ the Eq. (4) with a finite volume method for discretisation. First, we consider (4) over a cell Ω_i and a time interval I_k so that the integration in time and space will lead to

$$\int_{I_k} \int_{\Omega_i} \phi \partial_{tt} u(x,t) + \psi \partial_t u(x,t) \, \mathrm{d}x \, \mathrm{d}t = \int_{I_k} \int_{\Omega_i} \Delta_{\mathcal{M}} u(x,t) \, \mathrm{d}x \, \mathrm{d}t \tag{7}$$

The definition of the cell average

$$u_i(t) = u(\bar{x}_i, t) = \frac{1}{|\Omega_i|} \int_{\Omega_i} u(x,t) \, \mathrm{d}x \tag{8}$$

where $|\Omega_i|$ denotes the area of Ω_i, is used to define the averaged Laplacian as

$$Lu_i(t) = \frac{1}{|\Omega_i|} \int_{\Omega_i} \Delta_{\mathcal{M}} u(x,t) \, dx \tag{9}$$

Evaluation of this integral making use of the divergence theorem will lead to a line integral over the boundary of each cell. Application of the cotangent weight scheme [10] turns the resulting integrals over a full shape into the ODE system

$$\phi \ddot{\mathbf{u}}(t) + \psi \dot{\mathbf{u}}(t) = L\mathbf{u}(t) \tag{10}$$

The cotangent weight scheme will give us the discrete Laplace-Beltrami operator $L \in \mathbb{R}^{N \times N}$ via the sparse matrix representation $L = D^{-1}W$. The matrix W represents spatial weights computed from the cotangent formulae, encoding spatial connectivity and local geometry information. The matrix $D = \mathrm{diag}(|\Omega_1|, \ldots, |\Omega_i|, \ldots, |\Omega_N|)$ contains the local cell areas and all functions $u_i(t)$ are stored in the N-dimensional vector $\mathbf{u}(t) = (u_1(t), \ldots, u_N(t))^{\top}$.

Revisiting Eq. (10), we may notice that for $\phi = 0$ we have an ODE system of first order. Additionally we want to transform the equation for $\phi \neq 0$ into an ODE system of first order as well, since this is advantageous for the proceeding. We find

$$\dot{\mathbf{q}}(t) = H\mathbf{q}(t) \tag{11}$$

with $\mathbf{q}(t) = \left(\mathbf{u}(t), \dot{\mathbf{u}}(t)\right)^{\top} \in \mathbb{R}^{2N}$,

$$H = \begin{pmatrix} 0 & I \\ \frac{1}{\phi}L & -\frac{\psi}{\phi}I \end{pmatrix} \overset{\phi=1}{=} \begin{pmatrix} 0 & I \\ L & -\psi I \end{pmatrix} \in \mathbb{R}^{2N \times 2N} \tag{12}$$

and $I \in \mathbb{R}^{N \times N}$ the identity matrix.

Let us turn to the discrete initial conditions. Choosing $\dot{\mathbf{u}}(0) = \mathbf{0}$ let us only need to consider the initial spatial conditions $\mathbf{u}(0)$. Again using cell averages, the normalised initial condition at the location x_i can be written as

$$\mathbf{u}_{\mathbf{x_i}} = \mathbf{u}(x_i, 0) = (0, \ldots, 0, |\Omega_i|^{-1}, 0, \ldots, 0)^{\top} \tag{13}$$

Eigenproblem and Modal Coordinate Reduction. Systems like (10) have to deal with large sparse matrices and consequently high cost of computation time. With the Model Order Reduction (MOR) we can approximate our high dimensional system with lower dimensional one. In [12,13] one may find an overview about the MOR topic. For this work we are interested in the specific Modal Coordinate Reduction (MCR) technique as presented in [3], which we briefly recall now.

Solving Eq. (10) via MCR, it is essential to calculate the eigenvalues from the Laplace-Beltrami Operator $L\mathbf{v} = \lambda \mathbf{v}$. With $L = D^{-1}W$ we can transform the eigenvalue problem to the generalised eigenvalue Problem

$$W\mathbf{v} = \lambda D\mathbf{v} \tag{14}$$

By the properties of W and D all eigenvalues are real and the eigenvectors are linear independent. Additionally they are D-orthogonal with $\mathbf{v}_i^\top D\mathbf{v}_j = \delta_{ij}$. Thus we obtain the equalities

$$I = V^\top DV, \qquad L = V\Lambda V^\top D, \qquad \Lambda = V^\top WV \tag{15}$$

with V being the right eigenvector matrix of L and Λ the diagonal matrix of eigenvalues.

To execute the MCR transformation we take advantage of these considerations and substitute Eq. (10) with $\mathbf{u} = V\mathbf{w}$ to obtain:

$$\phi V\ddot{\mathbf{w}} + \psi V\dot{\mathbf{w}} = LV\mathbf{w} \tag{16}$$

Multiplication with $V^\top D$ from the left results in

$$\phi\ddot{\mathbf{w}} + \psi\dot{\mathbf{w}} = \Lambda\mathbf{w} \tag{17}$$

Equation (17) is still the full system. To reduce it we are only interested in the first $r \ll N$ ordered eigenvalues $0 = \lambda_1 < \lambda_2 \leq \ldots \leq \lambda_r$. This gives us the reduced matrix $\Lambda_r \in \mathbb{R}^{r \times r}$ extracted from Λ and $V_r \in \mathbb{R}^{N \times r}$. Consequently we obtain the reduced model

$$\phi\ddot{\mathbf{w}}_r + \psi\dot{\mathbf{w}}_r = \Lambda_r\mathbf{w}_r \quad \text{where} \quad \mathbf{w}_r = V_r^\top D\mathbf{u} \tag{18}$$

Analogously to the procedure by which we obtained Eq. (11) from (10), we can build

$$\dot{\mathbf{p}}_r(t) = H_r\mathbf{p}_r(t) \quad \text{with} \quad H_r = \begin{pmatrix} 0 & I_r \\ \Lambda_r & -\psi I_r \end{pmatrix} \in \mathbb{R}^{2r \times 2r} \tag{19}$$

and $\mathbf{p}_r = (\mathbf{w}_r, \dot{\mathbf{w}}_r)^\top \in \mathbb{R}^{2r}$.

Now, we want to discuss the transformation of the initial conditions. Remember that $\mathbf{u}(x_i, 0) = (0, \ldots, 0, |\Omega_i|^{-1}, 0, \ldots, 0)^\top$ and together with the second part of (18) we obtain $\mathbf{w}_r(x_i, 0) = V_r^\top\mathbf{e}_i$. Nothing will change for the initial velocity $\dot{\mathbf{u}}(0) = \mathbf{0}$, because we set it zero and so will $\dot{\mathbf{w}}_r(0) = \mathbf{0}$.

4 Time Discretisation

For discretisation of the explored time interval we divide $[0, t_M]$ into intervals $I_k = [t_{k-1}, t_k]$ with $\tau = t_k - t_{k-1}$ as time increment, and we set $t_0 = 0$.

Let us now turn to time derivatives. The classic method of second order in time accuracy is the Crank-Nicolson method [11]. We conjecture at this point that this method may not be beneficial in our shape analysis framework. Illustrating this, we calculate the first two iterations of a heat equation with the (l_0 unstable) Crank-Nicolson method in comparison to the l_0-stable implicit Euler method, cf. Fig. 2. For the solution of the heat equation we expect a smooth propagation away from the peak. However, the Crank-Nicolson method causes significant oscillations in the solution, even negative values appear. We conjecture at this point that such oscillations may spoil correspondence quality, when

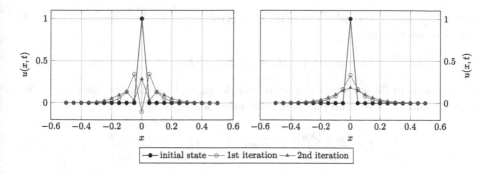

Fig. 2. This first two iterations of a heat equation solved with the Crank-Nicolson method (left) and with the l_0-stable implicit Euler scheme (right), respectively.

performing numerical shape correspondence. This example illustrates why we aim to explore l_0 stability of schemes.

Second Order l_0-stable Scheme. In this section, we briefly explain the idea and approach of the method of [16]. We have chosen this method because it does not depend on complicated Butcher tableaus, but offers an intuitive approach.

Looking at Eq. (11) or (19), we find these are of the form

$$\dot{\mathbf{x}}(t) = A\mathbf{x}(t) \tag{20}$$

The solution of this is $\mathbf{x}(t) = \exp(tA)$ or for a time step further

$$\mathbf{x}(t + \tau) = \exp((t + \tau)A) = \exp(\tau A)\mathbf{x}(t) \tag{21}$$

Calculating $\exp(\tau A)$ is not a trivial task, so we use an approximation

$$R(\tau A) \approx \exp(\tau A) \tag{22}$$

instead. With the approximation $R(\tau H) = (I - a\tau H)^{-1}(I + (1-a)\tau H)$ we obtain a first-order method. Choosing a will lead to different well-known schemes. For $a = 0$ we receive the explicit method and for $a = 1$ we obtain the implicit method, like it is used in [2,3] for time discretisation. The Crank-Nicolson scheme can be obtained by setting $a = 0.5$.

However, two out of three are not of second order and, in case of the explicit and Crank-Nicolson method, are not l_0-stable. Twizell and co-authors [16] present with

$$R(\tau A) = (I - r_1\tau A)^{-1}(I - r_2\tau A)^{-1}(I + (1-a)\tau A) \tag{23}$$

with $r_{1,2} = \frac{1}{2}\left(a \mp \sqrt{a^2 - 4a + 2}\right)$ a method of second order and choosing $a = 2 - \sqrt{2} - \varepsilon$, with ε an arbitrarily small positive number, will provide an l_0-stable method.

These considerations lead to the following discrete procedure for the wave equation:

$$\mathbf{p}^{k+1} = (I_{2r} - r_1\tau H_r)^{-1}(I_{2r} - r_2\tau H_r)^{-1}(I_{2r} + (1-a)\tau H_r)\mathbf{p}^k \tag{24}$$

with $\mathbf{p}^k = \mathbf{p}(t_k)$ and $\mathbf{p}^{k+1} = \mathbf{p}(t_{k+1}) = \mathbf{p}(t + \tau)$. Analogously for the heat equation, we receive

$$\mathbf{w}^{k+1} = (I_r - r_1\tau\Lambda_r)^{-1}(I_r - r_2\tau\Lambda_r)^{-1}(I_r + (1 - a)\tau\Lambda_r)\mathbf{w}^k \tag{25}$$

As discussed in [3] we adopt the temporal domain $[0, t_M]$ for the heat equation to $[0, t_h^*]$ and for wave and damped wave equation to $[0, t_w^*]$ by

$$t_h^*(\lambda_r) = \frac{t_M\sqrt{|\lambda_N|}}{\sqrt{|\lambda_r|}} \qquad \text{resp.} \qquad t_w^*(\lambda_r) = \frac{t_M\sqrt[4]{|\lambda_N|}}{\sqrt[4]{|\lambda_r|}} \tag{26}$$

The computational parameters are thus $\tau = \frac{t^*}{M}$, with M being the number of iterations.

5   Experiments

Hit Rate and Geodesic Error. The percentage Hit Rate is defined as $TP/(TP + FP)$, where TP is the number of true positives and FP is the number of false positives events.

The Princeton benchmark protocol [9] delivers the important key points for the evaluation of the correspondence quality. It defines the quality of a matching by the normalised intrinsic distance $d_\mathcal{M}(x_i, x^*)/\sqrt{A_\mathcal{M}}$ from the calculated matching x_i to the ground-truth correspondence x^*. A matching is accepted to be true if the normalised intrinsic distance is smaller then the threshold 0.25.

Dataset. For the experiments we used a selection of shapes from the TOSCA dataset [4] containing a total of 80 shapes of animals and humans in different poses. The range of shape resolution in TOSCA goes from 4 344 points for the wolf shape to 52 565 points for the david and michael shape.

Experimental Settings. As discussed, we use three methods to solve the heat and wave equation on different shapes. As a reference method we employ the l_0-stable implicit Euler method of first order in time [3]. We compare it with the discussed second order methods, i.e. Crank-Nicolson and the l_0-stable method (23) from [16].

Let us recall that we aim to explore the possibility to increase the time step size τ. The point with this is, that an implicit scheme should theoretically perform stable independently on the time step size [11]. Thus we expect that a rescaling of time step size may reveal if the l_0 stability is consistently a useful property.

The time step τ and the number of iterations depend directly on each other through $\tau M = t^*$. Since we fixed the value for t^* with the calculations from (26) an increase of τ leads to a decrease of M. We denote the factor by which we have increased the time step size τ by c. We have chosen $c = \{1, 5, 10\}$ for our experiments. With these values for c the number of iterations changed to $M = \{100, 20, 10\}$.

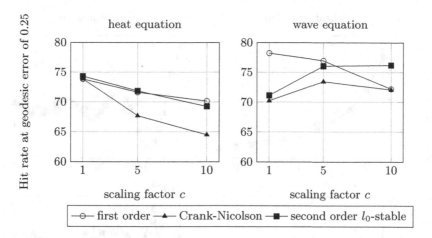

Fig. 3. The figure displays mean hit rates in % at geodesic error of 0.25 for different shapes and schemes in dependence of the scaling factor c. The shape sets over which we average are cat, centaur, dog, horse, and wolf, taken from the TOSCA Dataset [4].

Results for the Heat Equation. A plot of results can be found in the left figure within Fig. 3. We see that all three methods lose some accuracy when the time step size is increased. The Crank-Nicolson method gives the worst hit rates. For $c = 1$ it can still keep up with the other methods (73.84%), but it loses significantly more accuracy (64.56%) when the time step size is increased, compared to the other two methods. The second order l_0 stable method and the implicit Euler method are very similar. Generally speaking, if we compare the hit rates from the l_0-stable methods (first and second order), we observe nearly no differences.

A look at the left plots of Fig. 4 shows that the feature descriptor for the individual methods is not very different. However, the beneficial aspect of an l_0-stable method becomes evident, since for an increase of the time step the Crank-Nicolson method produces clear oscillations, like in the simple motivation example in Fig. 2.

The results for the heat equation emphasize the usefulness of the l_0 stability as an important notion of practical importance in numerical shape analysis.

Results for Wave Equation. The right figure within Fig. 3 is providing the plots of the mean hit rates. Interesting is the fact, that with choosing a second order method we lose about 8% accuracy in the hit rate for $c = 1$ compared to the Euler integrator. At this point, further investigation is needed. However, we can notice an increase in hit rate again by increasing the time step size. Again we observe here that l_0 stability is a beneficial property.

Looking at the right plots in Fig. 4, we see that a much more detailed feature descriptor can be produced by the second order methods. However, this does not increase the hit rate.

Summarising the findings, we observe again that the l_0 stable schemes perform better than the classical Crank-Nicolson method.

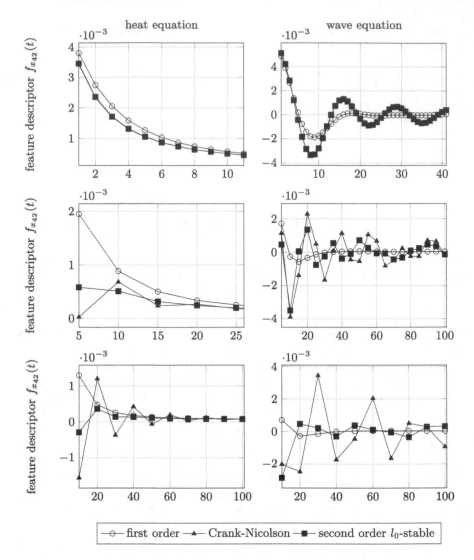

Fig. 4. The feature descriptors of the wolf shape in dependencies of the time. The feature descriptor is received from the heat (left) or wave (right) equation solved with a first order method [3], the Crank-Nicolson method and a l_0-stable second order method [16]. From Top to Bottom we see the different time scaling factor c. Starting with $c = 1$ (top) going to $c = 5$ (middle) and end up with $c = 10$ (bottom). In the top plots the second-order method are close to each other so they appear as one curve.

6   Conclusion and Further Work

Summarising the experiments, we have validated that l_0 stability is an important property for numerical integration in shape analysis. This is surely a useful aspect for possible future developments.

There are some other indirect findings, e.g. that the use of higher order schemes for time integration may not be of highly beneficial impact, since the l_0-stable second order time integrator did not perform significantly better than the implicit Euler scheme. However, this may be partly due to the fact of the observed drop in accuracy in case of the wave equation, a subject we aim to resolve in future work.

References

1. Aubry, M., Schlickewei, U., Cremers, D.: The wave kernel signature: a quantum mechanical approach to shape analysis. In: 2011 IEEE International Conference on Computer Vision Workshops (ICCV Workshops). IEEE, November 2011
2. Bähr, M., Breuß, M., Dachsel, R.: Fast Solvers for Solving Shape Matching by Time Integration (2018)
3. Bähr, M., Breuß, M., Dachsel, R.: Model Order Reduction for Shape Correspondence arXiv: 1910.00914v2 (2020)
4. Bronstein, A.M., Bronstein, M.M., Kimmel, R.: Numerical Geometry of Non-Rigid Shapes. MCS. Springer, New York (2009). https://doi.org/10.1007/978-0-387-73301-2
5. Dachsel, R., Breuß, M., Hoeltgen, L.: Shape matching by time integration of partial differential equations. In: Lauze, F., Dong, Y., Dahl, A.B. (eds.) SSVM 2017. LNCS, vol. 10302, pp. 669–680. Springer, Cham (2017). https://doi.org/10.1007/978-3-319-58771-4_53
6. Dachsel, R., Breuß, M., Hoeltgen, L.: The classic wave equation can do shape correspondence. In: Felsberg, M., Heyden, A., Krüger, N. (eds.) CAIP 2017. LNCS, vol. 10424, pp. 264–275. Springer, Cham (2017). https://doi.org/10.1007/978-3-319-64689-3_22
7. Do Carmo, M.P.: Differential Geometry of Curves and Surfaces: Revised and Updated, 2nd edn. Dover Publications, Mineola (2016)
8. van Kaick, O., Zhang, H., Hamarneh, G., Cohen-Or, D.: A survey on shape correspondence. Comput. Graph. Forum **30**(6), 1681–1707 (2011)
9. Kim, V.G., Lipman, Y., Funkhouser, T.: Blended intrinsic maps. In: ACM SIGGRAPH 2011 papers on - SIGGRAPH 2011. ACM Press (2011)
10. Meyer, M., Desbrun, M., Schröder, P., Barr, A.H.: Discrete differential-geometry operators for triangulated 2-manifolds. In: Hege, H.C., Polthier, K. (eds.) Mathematics and Visualization, pp. 35–57. Springer, Heidelberg (2003). https://doi.org/10.1007/978-3-662-05105-4_2
11. Morton, K.W., Mayers, D.F.: Numerical Solution of Partial Differential Equations. Cambridge University Press, Cambridge (2005)
12. Nouri, S.: Advanced model-order reduction techniques for large scale dynamical systems. Ph.D. thesis, Ottawa-Carleton (2014)
13. Qu, Z.Q.: Model Order Reduction Techniques with Applications in Finite Element Analysis. Springer, London (2004). https://doi.org/10.1007/978-1-4471-3827-3
14. Rustamov, R.M.: Laplace-beltrami eigenfunctions for deformation invariant shape representation, functional map. In: Proceedings of the Fifth Eurographics Symposium on Geometry Processing, SGP 2007, pp. 225–233. Eurographics Association, Goslar, DEU (2007)

15. Sun, J., Ovsjanikov, M., Guibas, L.: A concise and provably informative multi-scale signature based on heat diffusion. Comput. Graph. Forum **28**(5), 1383–1392 (2009)
16. Twizell, E.H., Gumel, A.B., Arigu, M.A.: Second-order, L_0-stable methods for the heat equation with time-dependent boundary conditions. Adv. Comput. Math. **6**(1), 333–352 (1996)

First Order Locally Orderless Registration

Sune Darkner$^{(\boxtimes)}$, José D. T. Vidarte, and François Lauze

Department of Computer Science, University of Copenhagen, Copenhagen, Denmark
{darkner,jota,francois}@di.ku.dk

Abstract. First Order Locally Orderless Registration (FLOR) is a scale-space framework for image density estimation used for defining image similarity, mainly for Image Registration. The Locally Orderless Registration framework was designed in principle to use zeroth-order information, providing image density estimates over three scales: image scale, intensity scale, and integration scale. We extend it to take first-order information into account and hint at higher-order information. We show how standard similarity measures extend into the framework. We study especially Sum of Squared Differences (SSD) and Normalized Cross-Correlation (NCC) but present the theory of how Normalised Mutual Information (NMI) can be included.

Keywords: Image registration · Locally Orderless Images · First order information

1 Introduction

Image similarity is generally based on zeroth-order information by a scalar to scalar comparison, e.g. Sum of Squared Differences (SSD), Normalised Cross-Correlation (NCC) or Mutual Information (MI) [9]. However, images have structure and they encode information that extends beyond zeroth-order, they do not look like random noise. MI and NCC do incorporate more than just pixel intensity but very weakly and indirectly. Higher-order information is seldom used with a few exceptions, notably the normalized gradient fields [5]. We aim to integrate high-order information for registration based on Locally Orderless Images (LOI) [10] and Locally Orderlles Registration LOR [4].

LOI defines three fundamental scales for estimating a density from an image: the *spatial* scale, which is the "classical" scale-space one, the *intensity or information* scale, as "bin scale" and the *integration* scale, which define the localisation of the density estimates of intensity distributions. The key is to 'marginalize over the geometry' and leave only the correspondence of information. The locally orderless registration gives us a theoretical platform to perform this marginalization for scalar-valued images.

Locally Orderless Registration (LOR) [8] explored its application for Magnetic Resonance Diffusion-Weighted Imaging (DWI), which are images containing complicated geometries. Indeed, DWI images can be seen as functions

© Springer Nature Switzerland AG 2021
A. Elmoataz et al. (Eds.): SSVM 2021, LNCS 12679, pp. 177–188, 2021.
https://doi.org/10.1007/978-3-030-75549-2_15

$\mathbf{I} : \Omega \times \mathbb{S}^2 \to \mathbb{R}$, with Ω an open subset of $\mathbb{R}^3$, where $\mathbb{S}^2$ is seen as the space of directions (with orientation) in $\mathbb{R}^3$. An extra directional scale is added before building and localizing densities.

In this work we extend the LOI and LOR [3,4,8] framework for images $\mathbf{I} : \Omega \to \mathbb{R}$, by lifting these images to images $\mathbf{I} : \Omega \times \mathbb{S}^2 \to \mathbb{R}$, where $\mathbb{S}^2$ this time parametrizes the local orientations of the image $\mathbf{I}$. This is performed through directional responses of derivatives of Gaussian. Other kernels could be used, for instance, non-symmetric ones. Lifts to second or higher-order structures can similarly be defined via higher kernel derivatives. Once the lifting has been performed, ideas similar to DWI image registration can be used. However, as opposed to the DWI case, this lifting comes already with its scale parameter. This lifting idea is not new, with especially works in the context of image smoothing and disentangling of directions ([7] and references therein). Tools and end goals in this work are different: classical Gaussian filters and image registration.

Given two images I and J, the registration problem is to find the transformation $\varphi : \mathbb{R}^3 \to \mathbb{R}^3$ that maps I onto J such that some similarity/dissimilarity $M(I \circ \varphi, J)$ is optimized. Registration is an ill-posed problem. Therefore, the deformation φ requires regularization. Typical regularizations use constraints on the family of admissible transformations e.g. diffeomorphisms. Other alternatives are to enforced local constraints by using additional smoothing (enforcing scale to the transformation). The LOI and LOR framework provides building blocks for similarity measures, and do not impose regularisers forms. We use a very simple ones here.

Organisation and Contributions. The paper is organized as follows. First we review previous work in Sect. 2 and recall the Locally Orderless Imaging and Locally Orderless Registration frameworks in Sect. 3. Our main contribution, the extension of the LOI and LOR frameworks to first order information, is presented in Sect. 4. Registration objective functions are also discussed in this section. We illustrate the effects first order extensions for SSD and NCC similarities on the quality, and convergence of the registration in Sect. 5. Finally, we summarise and discuss perspectives in Sect. 5.3.

2   Related Work

LOI was originally proposed by Koenderink and van Dorn [10] and describes the three inherent scales of images: spatial scale, intensity scale, and integration scale. This notion of images was used to describe image similarity in a variational framework [6] and formalized into a generalized framework for image registration and the image similarity measures as LOR in [4]. Some of the groundwork for LOR as well as the properties of the density estimators used for images in image registration where investigated in [3], revealing a 'scale imbalance' in the partial volume density estimator. The idea of marginalizing over more complex geometries than $\mathbb{R}^n$ was proposed in [8].

The idea of using higher order information for estimation of similarity between images is not new and normalized gradient fields NGF [5] were one of the first. In [14] an extension to the LDDMM using higher order information was presented.

There are few recent implementations of registration algorithms with NGF. The most noticeable uses NGF and a Gauss-Newton optimization scheme with locally rigid constraints [12]. This work was further evaluated on pelvis CT/CBCT images [11]. A recent first-order information approach adds another metric based on gradients to the registration cost function with NGF [15]. This metric is defined as the sum of three gradients norms, i.e. the transformed moving image, the fixed image, and the difference between moving and fixed while offering a small increase in registration accuracy.

3   Background on Locally Orderless Image Information

3.1   Notations

$\Omega \subset \mathbb{R}^3$ is the spatial domain of the images we use in the sequel. A scalar image is a function $f : \Omega \to \mathbb{R}$. We assume that images can be extended out of Ω to $\mathbb{R}^3$ – typically by 0 – as it is necessary for convolution. Convolution of two images $I, J : \mathbb{R}^3$ is defined by $I * J(\boldsymbol{x}) = \int_{\mathbb{R}^3} I(\boldsymbol{y}) J(\boldsymbol{x} - \boldsymbol{y}) \, d\boldsymbol{y}$. This actually extends to the case where one of the images is vector-valued directly. G_σ denotes a 3D isotropic Gaussian of standard deviation σ.

3.2   Lebesgue Integration and Histograms

Consider a function integrable $I : \mathbb{R}^n \to \mathbb{R}$. Its integral $\int I \, d\mu$ with respect to the Lebesgue measure μ of $\mathbb{R}^n$, denoted in the sequel as $\int_{\mathbb{R}^n} I(\boldsymbol{x}) \, d\boldsymbol{x}$, can be computed as the limit over all subdivisions $0 \le i_0 < \cdots < i_N$, $\sum_{n=0}^{N-1} i_n \mu(I^{-1}([i_n, i_{n+1}]))$. At the limit, when $i_{n+1} - i_n \to 0$, this can be rewritten as $\int_{\mathbb{R}} i h_I(i) \, di$ where $h_I(i)$ is the length of isophote $I^{-1}(i)$. The function $i \mapsto h_I(i)$ is a generalized histogram of the values of I. Many standard integrals can be rewritten using this form. For instance $\int_{\mathbb{R}^n} I(x)^2 \, dx = \int_{\mathbb{R}} i^2 h_I(i) \, di$. This generalizes to joint histograms: given two images $I, J : \mathbb{R}^n \to \mathbb{R}$, $(I, J) : \mathbb{R}^n \to \mathbb{R}^2$, $\boldsymbol{x} \mapsto (I(\boldsymbol{x}), J(\boldsymbol{x}))$ and its integral can be written as $\int_{\mathbb{R}^2} (i, j) h_{I,J} \, di \, dj$ where $h_{I,J}$ is the joint histogram of I and J. Classical similarities can be rewritten using histograms, for instance, Sum of Square Differences (SSD): $\int_{\mathbb{R}^n} (I(x) - J(x))^2 \, dx = \int_{\mathbb{R}^2} (i - j)^2 h_{I,J}(i, j) \, di \, dj$. Normalised Cross-Correlation, (Normalised) Mutual information, etc. can be written in terms of image histograms and their normalisations.

3.3   LOI and LOR Framework

LOI is a way to map images into local histograms, with three inherent scales: the spatial or image scale, the intensity scale, and integration scale. The image or spatial scale σ is used to smooth input images I and obtain $I_\sigma = I * G_\sigma$. A localised histogram over the values of I_σ is computed as

$$h_{I,\sigma\beta\alpha}(i|\boldsymbol{x}) := \int_\Omega P_\beta(I_\sigma(\boldsymbol{y}) - i)\, W_\alpha(\boldsymbol{y} - \boldsymbol{x})\, d\boldsymbol{y} \tag{3.1}$$

where P_β is a Parzen window of scale β, which provides the *intensity* scale and $W_\alpha(x)$ is an integration window which provides the *integration* scale α. The histogram $h_{I,\sigma\beta\alpha}(\cdot|\boldsymbol{x})$ is defined over $\mathbb{R}$ or at least over an interval Λ containing the range of values of I_σ. Normalising it, we obtain the image density

$$p_{I,\sigma\beta\alpha}(i|\boldsymbol{x}) = \frac{h_{I,\sigma\beta\alpha}(i|\boldsymbol{x})}{\int_\Lambda h_{I,\sigma\beta\alpha}(j|\boldsymbol{x})\, dj}. \tag{3.2}$$

By letting the integration scale $\alpha \to \infty$, we obtain *global* histograms an densities $h_{I,\sigma\beta}(i) := \int_\Omega P_\beta(I_\sigma(\boldsymbol{x}) - i)\, d\boldsymbol{x}$ and $p_{I,\sigma\beta}(i)$. This will be the case in this paper. This construction extends to the definition of joint histograms and densities, at the heart of Locally Orderless Registration by

$$h_{I,J,\sigma\beta\alpha}(i,j|\boldsymbol{x}) := \int_\Omega P_\beta(I_\sigma(\boldsymbol{y}) - i)P_\beta(J_\sigma(\boldsymbol{y}) - j)\, W_\alpha(\boldsymbol{y} - \boldsymbol{x})\, d\boldsymbol{y} \tag{3.3}$$

$$p_{I,J,\sigma\beta\alpha}(i,j|\boldsymbol{x}) = \frac{h_{I,J,\sigma\beta\alpha}(i,j|\boldsymbol{x})}{\int_\Lambda h_{I,J,\sigma\beta\alpha}(u,v|\boldsymbol{x})\, du\, dv} \tag{3.4}$$

and similar formulas in the global case. Single histograms and densities can also be obtained from them by marginalisation. LOR Image similarities are defined through single and joint density estimates Eq. (3.2) and Eq. (3.4). Similarity measures are defined as

$$M_L(I, J) = \int_\Omega \int_{\Lambda^2} f(i, j, p_{I,J,\sigma\beta\alpha}(i,j|\boldsymbol{x}))\, di\, dj\, d\boldsymbol{x}, \tag{3.5}$$

$$M_G(I, J) = \int_{\Lambda^2} f(i, j, p_{I,J,\sigma\beta}(i,j))\, di\, dj \tag{3.6}$$

with M_L built from localised densities and M_G from global ones. Among them, $p-$linear ones are characterized by $f(i, j, p) = g(i, j)p$, while nonlinear ones take more complex forms. We already mentioned in the previous section how SSD can be simply written using joint histograms. By normalising it, it can be written via densities (3.4). Another classical similarity, normalised cross-correlation (NCC), can also easily be written in term of histograms and densities.

$$NCC(I, J) = \frac{\langle I - \bar{I}, J - \bar{J} \rangle}{\|I - \bar{I}\|\|J - \bar{J}\|}$$

where $\bar{I}$ and $\bar{J}$ are the average values of I and J on Ω and the inner product and norms are L^2 ones. The inner product $\langle I, J \rangle$ is $\int_{\mathbb{R}^2} ij h_{I,J}(i,j)\, di dj$. Replacing $h_{I,J}$ by $h_{I,J;\sigma\beta\alpha}$ provides its LOI counterpart expression. The average $\bar{I}$ is $\int_\mathbb{R} ih_I(i)\, di / (\int_\mathbb{R} h(i)\, di)$. Again, we replace h_I by $h_{I;\sigma\beta\alpha}$ to obtain its LOI counterpart expression.

To use it in registration, the setting is typically the following. One chooses a hold on domain $D \subset \mathbb{R}^3$ large enough, with $\Omega \subset D$ and mappings $\varphi : \mathbb{R}^3 \to \mathbb{R}^3$, with $\varphi \equiv \mathrm{id}_3$ out of D, where id_3 is the identity transform. Here, we assume $D = \Omega$. These transformations are usually of class C^k, $k \geq 1$, often more. They are often, but not always, constrained to be diffeomorphic. A goodness of fit functional is obtained by evaluation the (dis)similarity $\varphi \mapsto M(I \circ \varphi, J)$.

4   Extension of LOI and LOR to Higher Information

In this section, we introduce a straightforward way to extend the LOI to incorporate higher order image information in histogram and density formulations. We focus on first order, as higher order may be limited in practice because of the complexity and resulting memory footprint.

4.1   First Order Locally Orderless Registration (FLOR)

In this paper we probe and use first order differential information of an image $I : \mathbb{R}^3 \to \mathbb{R}$ (with effective spatial domain Ω). It is obtained by lifting it to image $I_\sigma : \mathbb{R}^3 \times \mathbb{S}^2 \to \mathbb{R}$ which encodes gradient responses at different directions in a straightforward way. The differential $d_x G_\sigma : \mathbb{R}^3 \to \mathbb{R}$ is, for each x linear, it is enough to know it on $\mathbb{S}^2 \subset \mathbb{R}^3$.

$$I_\sigma(x, v) = \left(\int_{\mathbb{R}^3} I(y) d_{(y-x)} G_\sigma \, dy \right) v = d_x (I * G_\sigma) v \qquad (4.1)$$

This can of course be rewritten as $I_\sigma(x, v) = \nabla I_\sigma(x)^T v$. Note that $I_\sigma(x, -v) = -I_\sigma(x, v)$ due to our lifting choice. Using a higher order operator, such as, for instance, the Hessian of Gaussian $\nabla^2 G_\sigma$ would allow us to probe second order structure as a $\tilde{I}_\sigma(x, v) = \mathrm{Hess}\, I_\sigma(x)(v, v)$.

Once the lifting is performed, we can now define local histograms and densities. They are spatially localised, not directionally.

$$h_{I;\sigma\beta\alpha}(i|x) = \int_{\mathbb{R}^3 \times \mathbb{S}^2} P_\beta(I_\sigma(y, v) - i) W_\alpha(x - y) \, dv \, dy \qquad (4.2)$$

$$p_{I;\sigma\beta\alpha}(i|x) = \frac{h_{I;\sigma\beta\alpha}(i|x)}{\int_\Lambda h_{I;\sigma\beta\alpha}(j|x) \, dj} \qquad (4.3)$$

where this time Λ is an interval containing the range of I_σ. As in the zeroth order case, global histograms and densities can be obtained by letting $\alpha \to \infty$. Given two images $I, J : \mathbb{R}^3 \to \mathbb{R}$, we can lift them to I_σ and J_σ and define joint histograms and densities

$$h_{I,J;\sigma\beta\alpha}(i, j|x) = \int_{\mathbb{R}^3 \times \mathbb{S}^2} P_\beta(I_\sigma(y, v) - i) P_\beta(J_\sigma(y, v) - i) W_\alpha(x - y) \, dv \, dy \quad (4.4)$$

$$p_{I,J;\sigma\beta\alpha}(i, j|x) = \frac{h_{I;J\sigma\beta\alpha}(i, j|x)}{\int_{\Lambda^2} h_{I;\sigma\beta\alpha}(u, v|x) \, du \, dv} \qquad (4.5)$$

Here again, by letting $\alpha \to 0$, we obtain global histograms and densities.

4.2   First Order Deformation Model

Let $\varphi : \mathbb{R}^3 \to \mathbb{R}^3$ a deformation. By the chain rule, $\frac{d}{dt}\big|_{t=0} I_\sigma(\varphi(\boldsymbol{x} + t\boldsymbol{v})) = \nabla I_\sigma(\varphi(\boldsymbol{x}))^T J_{\boldsymbol{x}}\varphi(\boldsymbol{v})$, with $J_{\boldsymbol{x}}\varphi$ the Jacobian of φ. This implies of course that φ acts on the first order information via its differential. Here comes the problem that, as we have limited the directional probing space to $\mathbb{S}^2$, there is no guarantee that $J_{\boldsymbol{x}}\varphi(\boldsymbol{v}) \in \mathbb{S}^2$, let alone non zero. This is however the case if we restrict φ to be a diffeormorphism, and this is what we assume from now. From its very definition, the mapping of directions at $\boldsymbol{x} \in \Omega$ is given by

$$\psi_{\boldsymbol{x}} : \mathbb{S}^2 \to \mathbb{S}^2, \quad v \mapsto \frac{J_{\boldsymbol{x}}\varphi(\boldsymbol{v})}{|J_{\boldsymbol{x}}\varphi(\boldsymbol{v})|}, \tag{4.6}$$

This lead to define the action of φ^1 on the lifted image $\boldsymbol{I}_\sigma(\boldsymbol{x}, \boldsymbol{v})$ as

$$(\varphi.\boldsymbol{I}_\sigma)(\boldsymbol{x}, \boldsymbol{v}) = |J_{\boldsymbol{x}}\varphi(\boldsymbol{v})| \boldsymbol{I}_\sigma(\varphi(\boldsymbol{x}), \psi_{\boldsymbol{x}}(\boldsymbol{v})). \tag{4.7}$$

It clearly satisfies $(\varphi.\boldsymbol{I}_\sigma)(\boldsymbol{x}, -\boldsymbol{v}) = -(\varphi.\boldsymbol{I})(\boldsymbol{x}, \boldsymbol{v})$, thus respecting the structure of lifted images. Alternatively, one could consider another first order deformation model, where the Jacobian scaling factor is ignored, i.e.

$$(\varphi.\boldsymbol{I}_\sigma)(\boldsymbol{x}, \boldsymbol{v}) = \boldsymbol{I}_\sigma(\varphi(\boldsymbol{x}), \psi_{\boldsymbol{x}}(\boldsymbol{v})). \tag{4.8}$$

This may apply to images of more categorical nature. This can be the case for two images showing similar anatomical structures, with same tissue density, but which cannot be registered by a (local) rigid motion.

By using either the local or global histograms and densities, higher order similarities $M(\boldsymbol{I}, \boldsymbol{J})$ are obtained exactly the same way as discussed in the previous section. Finally one may combine zeroth and first order to get new similarity measures, and use them in a registration framework via

$$\varphi \mapsto M(I \circ \varphi, J) + \lambda M(\varphi.\boldsymbol{I}_\sigma, \boldsymbol{J}_\sigma). \tag{4.9}$$

4.3   Registration Objectives and Deformations

The similarities used in this paper are 1) SSD for zeroth and first order information, and 2) NCC for zeroth and first order information. Free-form B-spline deformation models [13] are used, with simple control point grid motion limitation as regularisation. We also use simple translation deformations on some experiments.

4.4   Implementation

The implementation has been made in PyTorch 1.7.1 and the basis consists of a Cubic B-spline from which both the image interpolation and deformation field can be estimated. Analytical Jacobians of both image and deformation has been

---

[1] Note that this is not stricto senso a group action here.

implemented which allow us to use the backpropagation of PyTorch and optimizer for finding the solution. The action of the Jacobian on the directional
derivative have 2 implementations, given by Eq. (4.7) and Eq. (4.8). The implementations ensures that all scales are consistent and to change image scale we
simply blur the images prior to registration with the desired kernel. Objectives
are optimised using PyTorch Adam implementation. The code runs both on
CPU and GPU; a full 3D registration takes 2–3 minutes on a laptop and around
1 min on an RTX3090.

5   Experiments and Results

We conduct 2 main experiments. First we investigate the properties of the 1st-
order information compared to the 0th-order using translation only. Secondly
we show that we can perform 3D non-rigid registration with convincing results.
We have used two 3D T1 weighted magnetic resonance images (MRI) from two
separate individuals for our proof of principle. The images are shown in Fig. 1.

5.1   The Similarity Properties

To illustrate the effects of including higher-order (1st-order) information in the
similarity-measure, we map the 0th-order and 1st-order information as a function
of translation in 2D (x, y)-plane.

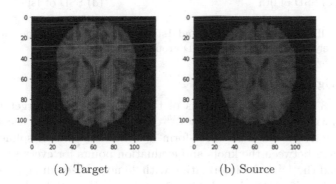

(a) Target   (b) Source

Fig. 1. A Slice of the target and source image used for our experiments

Our first experiment shows how the information from the images using SSD
and NCC respectively appears in the simple case where the deformation φ is a
pure translation in 2D, for an MRI compared with itself. As Fig. 2 illustrates, the
similarity in the 1st-order information has a significantly steeper slope close to the
optimum, compared to that of the 0th-order information for both SSD and NCC.
This indicates that including 1st-order information may improve registration close
to the optimum. However, when comparing 2 different images in Fig. 3 we observe

that multiple minima exist with 1st-order only, and that 0th-order has a better and a wider basin of attraction. Furthermore NCC seems to be more suitable compared to SSD. Therefore a combination of 0th-order, 1st-order information and NCC seems more appropriate for image registration applications.

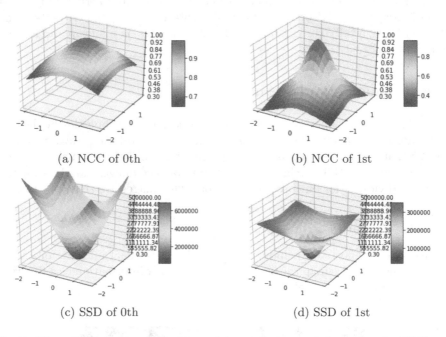

(a) NCC of 0th　　　　　　　　　　　(b) NCC of 1st

(c) SSD of 0th　　　　　　　　　　　(d) SSD of 1st

Fig. 2. The self similarity of the 0th and 1st order image information for NCC and SSD respectively under translation in 2D around the identity.

5.2　Non-rigid Registration

We perform 3 non-rigid registrations of the source and the target using only 1st-order information and using only 0th-order information and a combination of both respectively. We used a free-form deformation cubic B-spline [13] with 5 voxel spacing between the knots and evaluation points for every second voxel. We discretized the 1st-order information with 26 normalized directions, pointing to each neighbouring voxel in a $3 \times 3 \times 3$ local grid. We weighted 0th-order and 1st-order terms by the ratio between the number evaluation of the 0th-order and the 1st-order ($\frac{1}{26}$). As can be seen from the convergence plots (Fig. 4), this ratio will align both gradient information and intensity information, in contrast to optimizing only the 0th-order or the 1st-order information.

The final registration results are shown in Fig. 5. We have used both formulations from Eq. (4.7), with Jacobian normalisation, and Eq. (4.8), without Jacobian normalisation. As Fig. 5 shows, the results are quite convincing, and the difference between the two is very small. However, in this MRI registration case, the version not including scaling seems to be more suitable compared to SSD.

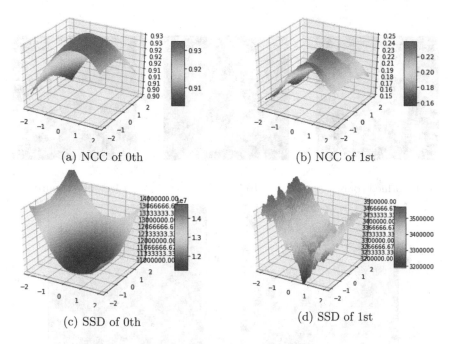

Fig. 3. The similarity of the 0th- and 1st-order image information for NCC and SSD respectively under translation in 2D around the identity for the source and target image. Clearly, multiple local minima exist in the 1st-order information, in contrast to 0th-order that only has one.

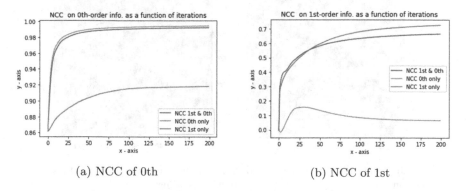

Fig. 4. Convergence plot of NCC for 0th- and 1st-order similarities, separately, as function of iterations. The experiment was performed optimizing only for 0th-order information, only 1st-order information and both with weight between 0th- and 1st-order terms as the ratio of the number of evaluation-orientations ($\frac{1}{26}$). Note how the 1st-order appears to also maximize the 0th-order and how 1st-order fails to maximize the gradient information.

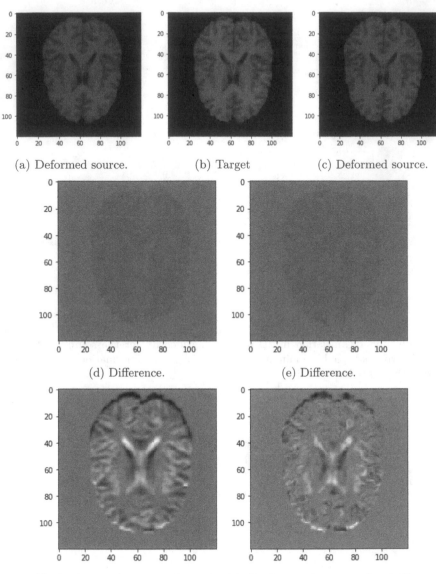

(a) Deformed source.   (b) Target   (c) Deformed source.

(d) Difference.   (e) Difference.

(f) Gradient information for 1 di-   (g) Gradient information for 1 direction, no scaling   rection, scaling

Fig. 5. (a) and (c) are the result of the registration to (b), where (a) has been registered with Eq. (4.7) and (c) with Eq. (4.8). (d) and (e) are the difference images of (a) and (b), and (c) and (b) respectively. (f) and (g) are examples of the 1st order information in (c) matched to (b) in the frame of (c)

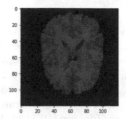

Fig. 6. An example of the deformation from pure 1st-order information.

5.3   Discussion and Limitations

The experiments presented in the previous section illustrate the gain obtained by including 1st-order information, but also some limitations. First, it is clear from our experiments that 1st-order information itself is not sufficient to make a proper registration in our setup. Figure 6 shows it clearly, where the registration of the image is inferior and suffers from undesirable deformations. Another point is that experiment here only serve as a proof of principle, and future work will include a thorough comparison over multiple large data sets similar to [9]. Furthermore, we will extend the work to use diffeomorphisms like previous methods, such as: Symmetric Normalization [1] or the Collocation for Diffeormorphic Deformations [2]. Finally, we will include an evaluation of information theory metrics and the exploration of all the scales presented in the formulation.

6   Conclusion

We introduced a framework for including higher-order information into image similarity and illustrated the application in image registration. We have shown that the method is able to match both $0^{th}-$ and $1^{st}-order$ information using SSD and NCC. The framework allows us to use all admissible measures from the LOR framework. We have shown that the framework is able to deliver high-quality non-rigid registration and that it has the potential to improve the accuracy of image registration in general.

Acknowledgements. This project has received funding from the European Union's Horizon 2020 research and innovation programme under the Marie Skłodowska-Curie grant agreement No. 764644. This paper only contains the author's views and the Research Executive Agency and the Commission are not responsible for any use that may be made of the information it.

References

1. Avants, B.B., Epstein, C.L., Grossman, M., Gee, J.C.: Symmetric diffeomorphic image registration with cross-correlation: evaluating automated labeling of elderly and neurodegenerative brain. Med. Image Anal. **12**(1), 26–41 (2008)
2. Darkner, S., Pai, A., Liptrot, M., Sporring, J.: Collocation for diffeomorphic deformations in medical image registration. IEEE Trans. Pattern Anal. Mach. Intell. **40**(7), 1570–1583 (2018). https://doi.org/10.1109/TPAMI.2017.2730205
3. Darkner, S., Sporring, J.: Generalized partial volume: an inferior density estimator to Parzen windows for normalized mutual information. In: Székely, G., Hahn, H.K. (eds.) IPMI 2011. LNCS, vol. 6801, pp. 436–447. Springer, Heidelberg (2011). https://doi.org/10.1007/978-3-642-22092-0_36
4. Darkner, S., Sporring, J.: Locally orderless registration. IEEE Trans. Pattern Anal. Mach. Intell. **35**(6), 1437–1450 (2013)
5. Haber, E., Modersitzki, J.: Intensity gradient based registration and fusion of multi-modal images. In: Larsen, R., Nielsen, M., Sporring, J. (eds.) MICCAI 2006. LNCS, vol. 4191, pp. 726–733. Springer, Heidelberg (2006). https://doi.org/10.1007/11866763_89
6. Hermosillo, G., Chefd'Hotel, C., Faugeras, O.: Variational methods for multimodal image matching. Int. J. Comput. Vision **50**(3), 329–343 (2002)
7. Janssen, M.H., Janssen, A.J., Bekkers, E.J., Bescos, J.O., Duits, R.: Design and processing of 3D invertible orientation scores of 3D images. J. Math. Imaging Vis. **60**, 1427–1458 (2018)
8. Jensen, H.G., Lauze, F., Nielsen, M., Darkner, S.: Locally orderless registration for diffusion weighted images. In: Navab, N., Hornegger, J., Wells, W.M., Frangi, A.F. (eds.) MICCAI 2015. LNCS, vol. 9350, pp. 305–312. Springer, Cham (2015). https://doi.org/10.1007/978-3-319-24571-3_37
9. Klein, A., et al.: Evaluation of 14 nonlinear deformation algorithms applied to human brain MRI registration. Neuroimage **46**(3), 786–802 (2009)
10. Koenderink, J.J., Van Doorn, A.J.: The structure of locally orderless images. Int. J. Comput. Vision **31**(2), 159–168 (1999)
11. König, L., Derksen, A., Papenberg, N., Haas, B.: Deformable image registration for adaptive radiotherapy with guaranteed local rigidity constraints. Radiat. Oncol. **11**(1), 1–9 (2016)
12. König, L., Rühaak, J.: A fast and accurate parallel algorithm for non-linear image registration using normalized gradient fields. In: 2014 IEEE 11th International Symposium on Biomedical Imaging (ISBI), pp. 580–583. IEEE (2014)
13. Rueckert, D., Sonoda, L.I., Hayes, C., Hill, D.L., Leach, M.O., Hawkes, D.J.: Nonrigid registration using free-form deformations: application to breast MR images. IEEE Trans. Med. Imaging **18**(8), 712–721 (1999)
14. Sommer, S., Nielsen, M., Darkner, S., Pennec, X.: Higher-order momentum distributions and locally affine LDDMM registration. SIAM J. Imag. Sci. **6**(1), 341–367 (2013)
15. Theljani, A., Chen, K.: An augmented lagrangian method for solving a new variational model based on gradients similarity measures and high order regularization for multimodality registration. Inverse Probl. Imaging **13**(2), 309–335 (2019)

Optimization Theory and Methods
in Imaging

First-Order Geometric Multilevel Optimization for Discrete Tomography

Jan Plier[1,3]([✉]), Fabrizio Savarino[2]([iD]), Michal Kočvara[4,5], and Stefania Petra[3]([iD])

[1] Heidelberg Institute for Theoretical Studies, Heidelberg, Germany
jan.plier@h-its.org
[2] Image and Pattern Analysis Group, Heidelberg University, Heidelberg, Germany
[3] Mathematical Imaging Group, Heidelberg University, Heidelberg, Germany
[4] School of Mathematics, University of Birmingham, Birmingham, UK
[5] Institute of Information Theory and Automation, Prague, Czech Republic

Abstract. Discrete tomography (DT) naturally leads to a hierarchy of models of varying discretization levels. We employ *multilevel optimization (MLO)* to take advantage of this hierarchy: while working at the fine level we compute the search direction based on a coarse model. Importing concepts from information geometry to the n-orthotope, we propose a smoothing operator that only uses first-order information and incorporates constraints smoothly. We show that the proposed algorithm is well suited to the ill-posed reconstruction problem in DT, compare it to a recent MLO method that nonsmoothly incorporates box constraints and demonstrate its efficiency on several large-scale examples.

1 Introduction

This paper introduces a *geometric* multilevel optimization approach for solving

$$\min_{x \in \mathbb{R}^n} f(x), \qquad f(x) = \mathrm{KL}(Ax, b) + \lambda \|Dx\|_{1,\tau} + \delta_C(x) \tag{1}$$

to recover a discretized function x on a spatial domain from linear projection measurements $b = Ax$ by minimizing the Kullback-Leibler (KL) divergence and a sparsity promoting prior subject to box constraints $C = [l, u] \subset \mathbb{R}^n_+$ – see Fig. 1 for an illustration. We aim at exploiting 'geometry' in a twofold way. On one hand, multiple grid sizes are used for discretizing the domain with different resolutions, which mitigates the ill-posedness of the inverse recovery problem at coarser levels. On the other hand, by turning the bounded interior of the convex feasible set into a Riemannian manifold, the geometry of the space makes first-order updates of the iterate x more efficient. Our approach combines these design aspects in a principled manner using problem (1) and discrete tomography [1] as a scenario that is representative for a range of approaches to inverse problems using constrained convex optimization.

Related Work. Seminal work on unconstrained smooth multilevel optimization includes [2–4]. These ideas were elaborated for nonsmooth convex optimization

© Springer Nature Switzerland AG 2021
A. Elmoataz et al. (Eds.): SSVM 2021, LNCS 12679, pp. 191–203, 2021.
https://doi.org/10.1007/978-3-030-75549-2_16

in [5,6]. Our approach is applicable to such problems but essentially relies on smoothness induced by changing the geometry of the feasible set. Regarding discrete tomography, multiresolution approaches include [7,8] with a focus on filtered backprojection and heuristics for acceleration, whereas our approach solves a constrained optimization problem.

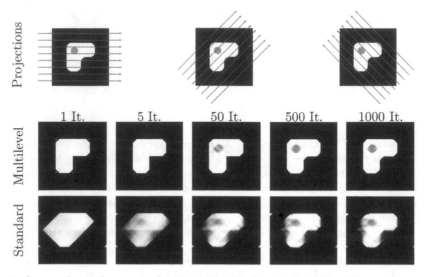

Fig. 1. Scenario and approach. Top row: In discrete tomography, we reconstruct *finite* range functions from a finite set of parallel projections. The incidence relation of projection rays and the discretized domain (pixels in 2D, voxels in 3D) is represented by a matrix A. All line integrals are collected in a vector b. **Bottom row:** Comparing to a standard iterative reconstruction scheme for solving (1), our novel multilevel approach recovers more efficiently the large scale structure and subsequently the fine scale structure of the unknown function.

Contribution and Organization. Section 2 introduces essential concepts of multilevel optimization. Our geometric multilevel optimization approach is introduced in Sect. 3, building on [9]. In Sect. 4, we show results of single- and multilevel optimization and compare them to [5]. Our large scale experiments show that our approach is on par with the state of the art and holds potential for further elaboration.

Basic Notation. $\langle \cdot, \cdot \rangle$ denotes the standard inner product on $\mathbb{R}^n$, ∇f the gradient and $\nabla^2 f$ the Hessian of a sufficiently differentiable function $f : \mathbb{R}^n \to \mathbb{R}$. We denote componentwise multiplication of vectors by $uv = (u_1 v_1, \ldots, u_n v_n)^\top$ and, for strictly positive vectors $v \in \mathbb{R}^n_{++}$, componentwise division by $\frac{u}{v}$. Likewise, the functions e^x and $\log x$ apply componentwise to a vector x. For a smooth Riemannian manifold $(\mathcal{M}, g)$ with metric g, $T_x \mathcal{M}$ denotes the tangent space at $x \in \mathcal{M}$ and $\mathrm{d}_x f : T_x \mathcal{M} \to \mathbb{R}$ the differential (aka tangent map) of a smooth function

$f : \mathcal{M} \to \mathbb{R}$. The Riemannian gradient $\nabla_{\mathcal{M}} f(x) \in T_x \mathcal{M}$ of f is uniquely defined by $d_x f[\xi] = g_x \left(\nabla_{\mathcal{M}} f(x), \xi \right), \ \forall \xi \in T_x \mathcal{M}$.

2   Multilevel Optimization in Euclidean Space

Two Level Optimization. We next describe a two grid cycle, that is computing an update x^+ at a fine grid from the current iterate x. This is done either by a search direction obtained from a model defined on a coarse grid using a much smaller number of variables (*coarse correction*) or, whenever coarse correction is not effective, by a standard local approximation defined on the fine grid (*fine correction*). The general approach is summarized in Algorithm 2.1.

Algorithm 2.1: Two Level Optimization

---

1 **initialization:** Set $i = 0$ and choose initial point x, two grids, transfer operators R and P and a coarse representation $\overline{f}$ of the objective.
2 **repeat**
3   **if** *condition to use coarse model is satisfied at x* **then**
4   Define coarse model $\overline{\psi}(\overline{y}; \overline{x}, \overline{f}, R\nabla f(x))$.   /* coarse model */
5   Find a descent direction $\overline{d}$ w.r.t. the fine objective f at x using $\overline{\psi}$.
6   Set $d = P\overline{d}$.
7   Find $\alpha > 0$ such that $f(x + \alpha d) < f(x)$.   /* line search */
8          $x \leftarrow x + \alpha d$.
9   **else**
10   Apply one iteration of the monotone fine level algorithm to find x^+ with $f(x^+) < f(x)$ and update $x \leftarrow x^+$.
11   Increment $i \leftarrow i + 1$.
12 **until** *a stopping rule is met.*

The key question is how to represent the problem on a coarser grid. The starting point is the coarse discretization of the fine grid objective $f \colon \mathbb{R}^n \to \mathbb{R}$, denoted by $\overline{f} \colon \mathbb{R}^{\overline{n}} \to \mathbb{R}$, that represents f on the coarse grid in a meaningful way. We use the following notation

$y \in \mathbb{R}^n$: fine grid variable, $\overline{y} \in \mathbb{R}^{\overline{n}}$: coarse grid variable,

$d \in \mathbb{R}^n$: search direction on fine grid, $\overline{d} \in \mathbb{R}^{\overline{n}}$: search direction on coarse grid.

We assume linear maps $R \colon \mathbb{R}^n \to \mathbb{R}^{\overline{n}}$ and $P \colon \mathbb{R}^{\overline{n}} \to \mathbb{R}^n$, called *restriction* and *prolongation* to be given, for translating various quantities between the coarse and fine grid level. Typically, information transfer between levels is done via linear interpolation or simple injection as in classical multigrid techniques [10].

Coarse Model. The central principle is defining the *coarse grid model*, first proposed in [2], given by

$$\overline{\phi}_x(\overline{y}) := \overline{f}(\overline{y}) - \langle \overline{v}_x, \overline{y} \rangle, \tag{2a}$$

$$\overline{v}_x := \nabla \overline{f}(\overline{x}) - R\nabla f(x), \quad \overline{x} := Rx, \tag{2b}$$

which is based on a linear modification of the coarse grid objective $\overline{f}$. In the following, we drop the explicit dependence of $\overline{\phi}$ and $\overline{v}$ on x for simplifying notation. The objective of the coarse grid model is to determine a gradient-like descent direction in an efficient way using a much smaller number of coarse grid variables. For the *initial* iterate of the coarse grid $\overline{x}$ defined in (2b), we have

$$\nabla \overline{\phi}(\overline{x}) = R\nabla f(x). \tag{3}$$

This property, also known as the *first order coherence condition*, ensures that a critical point of the objective function on the fine grid is also a critical point of the coarse model when transferred to the coarse grid. Note, that at this stage we have *not* imposed a relation between the intergrid transfer operators P and R. The update is defined as

$$x^+ = x + \alpha d, \tag{4a}$$

$$d = P(\overline{y}_* - \overline{x}), \qquad \overline{y}_* = \arg\min_{\overline{y}} \overline{\phi}(\overline{y}). \tag{4b}$$

Remark 1. We should underline that $\overline{y}_*$ in (4b) is typically replaced by an *inexact* solution of the coarse model (2) obtained by some iterative method.

Relation to the FAS. The coarse model $\overline{\phi}$ is closely related to the coarse grid correction equation of the FAS (full approximation scheme) in the context of multigrid methods for nonlinear equation [10, Chap. 5.3]. Applying FAS for solving the nonlinear critical point equation $\nabla f(y) = 0$ at the approximation x gives the *coarse grid correction*, see [10, Eq. 5.3.13]

$$\nabla \overline{f}(\overline{x} + \overline{y}) - \nabla \overline{f}(\overline{x}) = \overline{r}, \quad \overline{r} := 0 - R\nabla f(x), \tag{5}$$

that needs to be solved for $\overline{y}$. In FAS *both* the current approximation x and the residual, here $r := 0 - \nabla f(x)$, are transferred to the coarse grid. The coarse grid correction is defined in terms of $\overline{x}$, $\overline{r}$ and the coarse representation of the nonlinear equation. A solution of the coarse grid correction Eq. (5) is a critical point of $\overline{y} \mapsto \overline{\phi}(\overline{x} + \overline{y})$ in (2).

Bregman Gap, Coarse Model-Based Descent Direction. The following notion will be used for evaluating the coarse grid model.

Definition 1 (Coarse model, Bregman gap). *Given a differentiable function* $\overline{f} \colon \mathbb{R}^{\overline{n}} \to \mathbb{R}$ *and* $\overline{x} \in \mathbb{R}^{\overline{n}}$ *define the coarse model by*

$$\overline{\psi}_{x,\overline{x}}(\overline{y}) := B_{\overline{f}}(\overline{x} + \overline{y}, \overline{x}) + \langle \nabla f(x), P\overline{y} \rangle, \tag{6a}$$

with Bregman gap

$$B_{\overline{f}}(\overline{x} + \overline{y}, \overline{x}) := \overline{f}(\overline{x} + \overline{y}) - \overline{f}(\overline{x}) - \langle \nabla \overline{f}(\overline{x}), \overline{y} \rangle. \tag{6b}$$

Again we drop the explicit dependence of $\overline{\psi}$ on x and $\overline{x}$ for simplifying notation. The rational behind this definition is that it allows to efficiently obtain a descent direction, as the gap function is always nonnegative for any convex function $\overline{f}$.

Lemma 1. *Assume that $\overline{f}$ is convex and $\overline{\psi}(\overline{d}) < 0$ holds. Then $d := P\overline{d}$ is a descent direction satisfying $\langle \nabla f(x), d \rangle < 0$.*

Proof. Since $\overline{f}$ is convex, the statement follows from $B_{\overline{f}}(\overline{x} + \overline{y}, \overline{x}) \geq 0$ for all $\overline{y}$. $\square$

Remark 2. Whenever $R = P^\top$ holds (a standard assumption[1] in multigrid literature [10]), the coarse model $\overline{\psi}$ and the 'shifted' coarse model $\overline{y} \mapsto \overline{\phi}(\overline{x} + \overline{y})$ only differ by a constant that depends on x and $\overline{x} = Rx$. Indeed, using $\overline{v}_x$ from (2b) we rewrite

$$\overline{\phi}(\overline{x} + \overline{y}) = \overline{f}(\overline{x} + \overline{y}) - \langle \nabla \overline{f}(\overline{x}) - R\nabla f(x), \overline{x} + \overline{y} \rangle \tag{7a}$$
$$= B_{\overline{f}}(\overline{x} + \overline{y}, \overline{x}) + \langle R\nabla f(x), \overline{y} \rangle + const \tag{7b}$$
$$\overset{R=P^\top}{=} \overline{\psi}(\overline{y}) + const. \tag{7c}$$

In the following we disregard constant terms in (7) and consider (the simplified) coarse model $\overline{\psi}$ from (6a).

Remark 3. The first-order coherence applied to $\overline{\psi}$ now reads $\nabla \overline{\psi}(0) = P^\top \nabla f(x)$.

Remark 4. The coarse model $\overline{\psi}$ incorporates both first order information of the fine objective and second order information of the coarse objective. Indeed, for $\overline{f} \in C^2$ we can write

$$\overline{\psi}(\overline{y}) = \langle \nabla f(x), P\overline{y} \rangle + B_{\overline{f}}(\overline{x} + \overline{y}, \overline{x}) = \langle \nabla f(x), P\overline{y} \rangle + \frac{1}{2} \langle \overline{y}, \nabla^2 \overline{f}(\overline{z}), \overline{y} \rangle$$

for some $\overline{z} \in \{(1 - t)(\overline{x} + \overline{y}) + t\overline{x}\}_{t \in [0,1]}$. We interpret the first term in $\overline{\psi}$ as the first-order Taylor expansion of $f(x + P\overline{y})$ at the current iterate x on the fine grid and ignore $f(x)$ as it is a constant with respect to $\overline{y}$. Hence, coarse model $\overline{\psi}$ resembles the quadratic approximation model in *single level* optimization

$$q_x(y) := f(x) + \langle \nabla f(x), y \rangle + \frac{1}{2} \langle y, H_x y \rangle, \tag{8}$$

where H_x is a symmetric positive definite approximation of $\nabla^2 f(x)$.

Coarse Correction Condition. We adopt the following criteria from [4]

$$\|P^\top \nabla f(x)\| \geq \kappa \|\nabla f(x)\| \quad \text{and} \quad \|P^\top \nabla f(x)\| > \varepsilon, \tag{9}$$

---

[1] Our model does not require this assumption.

where $\kappa \in (0, \min(1, \|P\|))$ and $\varepsilon \in (0, 1)$. The above criteria prevent us from using the coarse model for computing a descent direction when $\bar{x} \approx \overline{x + d}$, i.e. the coarse correction direction $\bar{d}$ is close to 0.

Box Constrained Coarse Model. We now extend the coarse model $\overline{\psi}$ to box constraints in order to approach (1) by MLO. We introduce

$$\min_{\overline{y}} \overline{\psi}(\overline{y}) \qquad \text{subject to} \quad \overline{l}_{x,\overline{x},P} \leq \overline{y} \leq \overline{u}_{x,\overline{x},P}, \tag{10}$$

where the bounds at the coarse level are defined as

$$(\overline{l}_{x,\overline{x},P})_j = \overline{x}_j + \frac{1}{\|P\|_\infty} \max_{i=1,\dots,n} \begin{cases} (l - x)_i, & \text{if } P_{ij} > 0, \\ (x - u)_i, & \text{if } P_{ij} < 0, \end{cases} \tag{11a}$$

$$(\overline{u}_{x,\overline{x},P})_j = \overline{x}_j + \frac{1}{\|P\|_\infty} \min_{i=1,\dots,n} \begin{cases} (u - x)_i, & \text{if } P_{ij} > 0, \\ (x - l)_i, & \text{if } P_{ij} < 0, \end{cases} \tag{11b}$$

and adopted from [11]. A closely related coarse model was considered in [5]. In our notation, we drop the dependency of these bounds on x, $\overline{x}$ and P. The above definitions also handle negative elements in P (as in e.g. cubic interpolation). The next result states that box constraints are preserved by prolongation.

Lemma 2 ([11, Lemma 4.3]) *Let $x, l, u \in \mathbb{R}^n$ with $l < u$, $P : \mathbb{R}^{\overline{n}} \to \mathbb{R}^n$ and $\overline{l}$ and $\overline{u}$ be defined as in (11). Consider any $\overline{d} \in [\overline{l}, \overline{u}]$. Then $l \leq x + P\overline{d} \leq u$ holds.*

In the unconstrained case, it suffices to test whether $\|P^\top \nabla f(x)\|$ is large enough compared to $\|\nabla f(x)\|$, see (9). However, this criterion is inadequate for the box-constrained problem. Instead, we use the scaled gradient [12],

$$G(x) = S(x)\nabla f(x), \qquad S(x) = \operatorname{diag}(s_1(x), \dots, s_n(x)), \tag{12a}$$

$$s_i(x) = \begin{cases} \min\{1, x_i - l_i\}, & \text{if } (\nabla f(x))_i > 0, \\ \min\{1, u_i - x_i\}, & \text{if } (\nabla f(x))_i < 0, \\ \min\{1, x_i - l_i, u_i - x_i\}, & \text{if } (\nabla f(x))_i = 0, \end{cases} \tag{12b}$$

and replace $\nabla f(x)$ by $G(x)$ in (9). This gives

$$\|P^\top G(x)\| \geq \kappa \|G(x)\| \quad \text{and} \quad \|P^\top G(x)\| > \varepsilon, \tag{13}$$

where $\kappa \in (0, \min(1, \|P\|))$ and $\varepsilon \in (0, 1)$. One can show that

$$s_i(x) \begin{cases} = 0, & \text{if } x_i = l_i \quad \text{and} \quad (\nabla f(x))_i > 0, \\ = 0, & \text{if } x_i = u_i \quad \text{and} \quad (\nabla f(x))_i < 0, \\ \geq 0, & \text{if } x_i \in \{l_i, u_i\} \quad \text{and} \quad (\nabla f(x))_i = 0, \\ > 0, & \text{otherwise.} \end{cases}$$

Thus any x with $G(x) = 0$ is a stationary point of the box-constrained fine level problem.

Application to Discrete Tomography. We now represent problem (1) on a coarser grid and evaluate the coarse grid model $\overline{\psi}$ from (6a). We assume that images are discretized on $n = N \times N$ grid points in a two dimensional domain in $\mathbb{R}^2$. Using the one-dimensional discrete derivative operator $\partial_d \colon \mathbb{R}^d \to \mathbb{R}^d$, $(\partial_d)_{ij} = -1$, if $i = j < d$, $(\partial_d)_{ij} = +1$ if $j = i + 1 \leq d$ and $(\partial_d)_{ij} = 0$ otherwise, along each spatial direction, we define the discrete gradient matrix of an $N \times N$ discrete image by $D := \left(\begin{smallmatrix} D_1 \\ D_2 \end{smallmatrix} \right) = \left(\begin{smallmatrix} \partial_N \otimes I_N \\ I_N \otimes \partial_N \end{smallmatrix} \right)$, where $\otimes$ stands for the Kronecker product and I_N is the identity matrix of dimension N. Analogously, we define the discrete gradient on a coarse $\overline{n} = \overline{N} \times \overline{N}$ grid and denote it by $\overline{D}$.

We denote the projection matrix at the coarser level by $\overline{A}$. Next, we show that the specific ray geometry corresponding to the coarse grid can be selected independently of the ray geometry at the fine grid, as we do not need to transfer the projection information between levels. To this end, we evaluate the Bregman gap in (6b), as only this term involves the coarse objective $\overline{f}$.

Lemma 3. *Denote the data term in f from (1) by* $p(y) := \mathrm{KL}(Ay, b)$ *and the regularizer with* $q(y) := \|Dy\|_{1,\tau} := \langle \rho_\tau(Dy), \mathbb{1} \rangle$, *where* ρ_τ *is the Huber function applied component-wise. Assume* $\overline{A}$ *and* $\overline{D}$ *are given on the coarse grid. Then*

$$B_{\overline{f}}(\overline{y}, \overline{x}) = KL(\overline{A}\overline{y}, \overline{A}\overline{x}) + \lambda B_{\overline{q}}(\overline{y}, \overline{x}). \tag{14}$$

Proof. A simple calculation shows $B_{\overline{p}}(\overline{y}, \overline{x}) = KL(\overline{A}\overline{y}, \overline{A}\overline{x})$. Then the result follows from linearity of the Bregman gap $B_{p+\lambda q}(y, x) = B_p(y, x) + \lambda B_q(y, x)$.

Remark 5. One can show that the observation above applies to all variational models that involve a data term formulated by means of a Bregman divergence.

Final Algorithm. We now particularize the steps of the general framework in Algorithm 2.1.

- Line 3: We choose the coarse correction condition as in (13).
- Line 4: We use the box constrained coarse model in (10).
- Line 5: We obtain $\overline{d}$ with a few iterations of the projected gradient method with inexact line search [13] until $\overline{\psi}(\overline{d}) < 0$ holds.
- Line 6: We employ a full weighting operator [10].
- Line 7: This line search may be omitted due to our choice of the restricted box (11), see Lemma 2.
- Line 10: As f is not gradient Lipschitz continuous, we do fine corrections via the projected gradient with inexact (rather than fixed) line search.

The algorithm can be implemented also recursively using multiple levels.

3   Geometric Approach

We focus on the minimization of f subject to box constraints in a smooth setting.

Riemannian Geometry of the Box. Following [14] we turn the open box into a manifold

$$(\mathcal{M}, g), \qquad \mathcal{M} := (l, u), \quad l, u \in \mathbb{R}^n, \quad l < u \tag{15}$$

with the Hessian Riemannian metric $g_x(v, w) = \langle v, H_x w \rangle$, $v, w \in T_x \mathcal{M} = \mathbb{R}^n$, induced by $h(x) = \langle \mathbb{1}, (x - l) \log(x - l) + (u - x) \log(u - x) \rangle$ (a convex Legendre function [15, Chapter 26]) and its Hessian given by $H_x := \nabla^2 h(x) = \text{Diag}\left(\frac{u-l}{(x-l)(u-x)}\right)$. The Riemannian gradient is now given by

$$\nabla_{\mathcal{M}} f(x) = H_x^{-1} \nabla f(x) = \frac{(x-l)(u-x)}{u-l} \nabla f(x). \tag{16}$$

Though the choice of h may appear arbitrary at this point, it will prove beneficial in connection with the constructed retraction below.

Retraction. Conceptually, any reasonable numerical first-order update for the minimization of f has to map the Riemannian gradient $\nabla_{\mathcal{M}} f(x)$ from the tangent space $T_x \mathcal{M}$ at a current point $x \in \mathcal{M}$ onto the manifold in a meaningful way in order to produce an update $x^+ \in \mathcal{M}$. On a Riemannian manifold, the natural candidate for this purpose is given by the exponential map with respect to the Levi-Civita connection. However, in our case it can be shown that this map is only defined around a small neighborhood of $0 \in T_x \mathcal{M}$ and does not extend onto all of $T_x \mathcal{M}$. To overcome this limitation, we consider a *retraction map* $\mathscr{R}_x : T_x \mathcal{M} \to \mathcal{M}$, smoothly varying in $x \in \mathcal{M}$, which is required to fulfill

$$(i) \quad \mathscr{R}_x(0) = x, \quad 0 \in T_x \mathcal{M} \quad \text{and} \quad (ii) \quad d\mathscr{R}_x(0) = \text{id}_{T_x \mathcal{M}}, \quad \forall x \in \mathcal{M}. \tag{17}$$

These conditions ensure that the curve $\gamma(t) := \mathscr{R}_x(tv)$ realizes the tangent vector $v \in T_x \mathcal{M}$ at $x \in \mathcal{M}$ by satisfying $\gamma(0) = x$ and $\dot{\gamma}(0) = v$. See [16, Section 4] for more background and details of retraction maps.

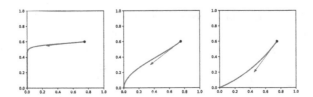

Fig. 2. Behavior of the retraction $\widetilde{\exp}_x$ from (18) on $\mathcal{M} = (0, 1)^2$ in terms of the curve $\gamma(t) = \widetilde{\exp}_x(tv)$ for various choices of $v \in T_x \mathcal{M} = \mathbb{R}^2$. This illustrates that $\gamma(t)$ realizes the tangent vector v near $t = 0$ and never leaves the manifold $\mathcal{M}$.

Proposition 1. *Let $(\mathcal{M}, g)$ be given by* (15). *Then the map*

$$\widetilde{\exp} : T\mathcal{M} \to \mathcal{M}, \qquad \widetilde{\exp}_x(v) := l + \frac{(u-l)(x-l)e^{\frac{u-l}{(x-l)(u-x)}v}}{u-x+(x-l)e^{\frac{u-l}{(x-l)(u-x)}v}} \tag{18}$$

is a proper retraction map.

Proof. The relative interior of the probability 2-simplex is given by $\mathrm{relint}(\Delta_2) = \{p \in \mathbb{R}^2 | p > 0 \text{ and } p_1 + p_2 = 1\} =: S_2$. For any index $i \in [n]$ we can identify the interval (l_i, u_i) with S_2 via $F_i \colon (l_i, u_i) \to S_2$, by sending a point $x_i \in (l_i, u_i)$ to $F_i(x_i) := \frac{1}{u_i - l_i}\left(\begin{smallmatrix} u_i - x_i \\ x_i - l_i \end{smallmatrix}\right)$. The manifold S_2 possesses an exponential map with respect to the so called *e-connection* from information geometry [17,18], which is defined on all of $T_p S_2$ and given by $\exp_p(v) = \langle p, e^{\frac{v}{p}}\rangle^{-1} p e^{\frac{v}{p}}$ at any $p \in S_2$ and $v \in T_p S_2$. Since exponential maps always fulfill condition (ii) of (17), the map $\exp_p \colon T_p S_2 \to S_2$ and therefore also the pullback onto (l_i, u_i) under F_i

$$F_i^{-1}\big(\exp_{F_i(x_i)}(dF_{i,x}(v))\big) = \big(\widetilde{\exp}_x(v)\big)_i$$

are both proper retractions. Applying this argument to each coordinate $i \in [n]$ proves the statement. $\square$

The retraction in (18) allows us to compute updates on the manifold based on numerical operation in the tangent space (Fig. 2). Due to the simple structure of the constraints, this can be done separately for each coordinate. Furthermore, as a consequence of the choice for h, the corresponding Hessian H_x defined before Eq. (16) exactly matches the exponent in the expression for $\widetilde{\exp}_x$ in (18). Thus, applying $\widetilde{\exp}_x$ to the Riemannian gradient (16) simplifies to

$$\widetilde{\exp}_x\big(-\alpha\nabla_{\mathcal{M}}f(x)\big) = l + \frac{(u-l)(x-l)e^{-\alpha\nabla f(x)}}{u - x + (x-l)e^{-\alpha\nabla f(x)}}. \tag{19}$$

Coarse Grid Model, Coarse Grid Correction. For $x \in \mathcal{M}$ and $\overline{x} = Rx$ define $\overline{\mathcal{M}} := (\overline{l}, \overline{u}) \subset \mathbb{R}^{\overline{n}}$ endowed with the Riemannian geometry from (15). We consider $\overline{\psi}(\overline{y}) = B_{\overline{f}}(\overline{y}, \overline{x}) + g_x(\nabla_{\mathcal{M}}f(x), P(\overline{y} - \overline{x}))$. To find a $\overline{y} \in \overline{\mathcal{M}}$ such that $\overline{\psi}(\overline{y}) < 0$ we employ the Riemannian gradient method, see [16, Alg. 1].

Final Algorithm. Algorithm 3.1 summarizes a multilevel implementation of the two grid general framework of Algorithm 2.1 specified to our geometric setting. Note that in Algorithm 3.1, just as $\overline{f}$ represents f on the first coarse level, $\overline{\overline{f}}$ represents f on the second coarse level.

Remark 6. Strictly speaking, tangent vectors from different vector spaces $T_x\mathcal{M}$ and $T_{x'}\mathcal{M}$ are incompatible unless parallel transport is used. This issue arises in Algorithm 3.1, line 5, when the tangent vector field $\nabla_{\mathcal{M}}f$ is evaluated at a fine grid point x, as part of the coarse grid model. Since $\mathcal{M}$ is an open subset of an ambient Euclidean space with a trivial tangent bundle, however, this problem is merely a *formal* one, and we deliberately ignore it throughout this paper.

Algorithm 3.1: Multilevel Optimization (ML RG)

```
1  Function MLO(f, f̄, x, P)
2  │  i ← 0
3  │  while x is not optimal and i < i_max do
4  │  │  if ‖RG(x)‖ ≥ κ‖G(x)‖ and ‖RG(x)‖ ≥ ε and f̄ defined then
5  │  │  │  ψ̄(ȳ) = D_f̄(ȳ, x̄) + g_x(∇_M f(x), P(ȳ − x̄))    /* coarse model */
6  │  │  │  if f̿ defined then
7  │  │  │  │  ȳ ← MLO(ψ̄, f̿, x̄, P̄)                       /* recursive call */
8  │  │  │  else
9  │  │  │  │  find ȳ with ψ̄(ȳ) < 0
10 │  │  │  d ← Pd̄,   d̄ = ȳ − x̄                          /* descent direction */
11 │  │  │  x ← exp̃_x(αd)          /* α > 0 such that f(exp̃_x(αd)) < f(x) */
12 │  │  x ← RiemannianGradientDescent(f, x)
13 │  │  i ← i + 1
14 │  return x
```

Fig. 3. Phantoms that exhibit both fine and large scale structures.

4   Experiments

To illustrate our approach, we compare it to a state-of-the-art first-order multi-level approach [5] (capable of handling box constraints in a Euclidean setting) which we adapt as described in Sect. 2 and denote it as *multilevel projected gradi-ent (ML PG)*. We denote its single-level counterpart as *projected gradient (PG)*. Similarly, we denote our proposed geometric multilevel approach in Sect. 3 with *multilevel Riemmanian gradient (ML RG)* and its single-level version as *Rie-mannian gradient (RG)*. We summarize our results in Fig. 4.

Data Setup. We consider the phantoms ($n = 1024 \times 1024$) in Fig. 3. We gener-ated the projection matrices using the ASTRA-toolbox[2]. We used parallel beam projections along equidistant angles between 0 and π. The undersampling rate at the fine grid is 20%. Entry a_{ij} of projection matrix A holds the length of the line segment of the i-th projection ray passing through the j-th pixel. At every level the width of the sensor-array was set to the grid size, so that at each scale every pixel intersects with at least one projection ray. For the information

---

[2] https://www.astra-toolbox.com/.

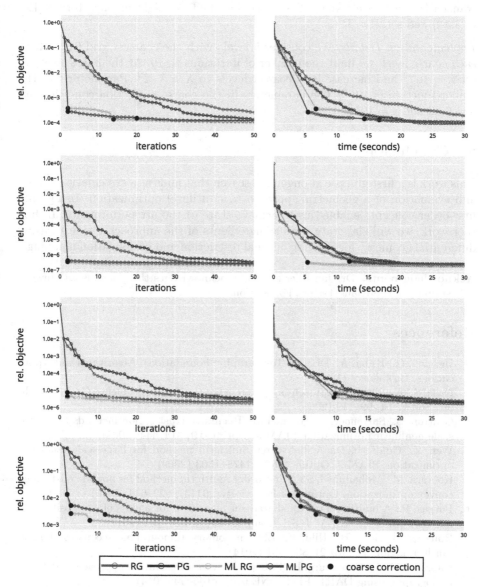

Fig. 4. Comparison of decreasing objective function values (left column) and runtime (right column) for single-level resp. multilevel (ML) versions of projected gradient descent (PG, ML PG) and Riemannian gradient descent (RG, ML RG). The i-th row corresponds to the i-th image in Fig. 3. Black dots indicate when descent directions were computed on coarser grids. The multilevel schemes aggressively minimize the objective. The computational overhead (checking feasibility of coarse grid descent directions, grid transfer) takes some computation time. Yet, in view of the objective function decrease, multilevel iterations can be terminated earlier than the single-level schemes.

transfer between levels we used as restriction the full weighting operator [10, Eq. (2.3.3.)] and set $P = R^\top$.

Implementation Details. We consider 5 levels with the coarsest grid 64×64. In each coarse level, we limit the number of iterations to 10. At the finest level, we set $\lambda = 10^{-3}$ and increase it at coarser levels to $\overline{\lambda} = \lambda \cdot 2^2$. Parameter τ of the Huber function is 10^{-4}. The parameters in the coarse correction condition are $\kappa = 0.49$ and $\varepsilon = 10^{-3}$.

5   Conclusion

This work is a first glimpse at ongoing research that aims at a systematic analysis and evaluation of a geometric approach to multilevel optimization. Using only first-order concepts enabled us to achieve state of the art performance. In further work, we will elaborate various ingredients of the approach like, e.g., using differential geometry for deriving optimal restriction and prolongation mappings.

Acknowledgments. Dr. Jan Plier gratefully acknowledges the generous and invaluable support of the Klaus Tschira Foundation.

References

1. Herman, G., Kuba, A.: Discrete Tomography: Foundations, Algorithms and Applications. Birkhäuser (1999)
2. Nash, S.: A multigrid approach to discretized optimization problems. Optim. Methods Softw. **14**, 99–119 (2000)
3. Gratton, S., Sartenaer, A., Toint, P.L.: Recursive trust-region methods for multiscale nonlinear optimization. SIAM J. Optim. **19**, 414–444 (2008)
4. Wen, Z., Goldfarb, D.: A line search multigrid method for large-scale nonlinear optimization. SIAM J. Optim. **20**(3), 1478–1503 (2009)
5. Kočvara, M., Mohammed, S.: A first-order multigrid method for bound-constrained convex optimization. Optim. Methods Softw. **31**(3), 622–644 (2016)
6. Parpas, P.: A multilevel proximal gradient algorithm for a class of composite optimization problems. SIAM J. Sci. Comput. **39**(5), 681–701 (2017)
7. Roux, S., Leclerc, H., Hild, F.: Efficient binary tomographic reconstruction. J. Math. Imaging Vis. **49**(2), 335–351 (2014)
8. Dabravolski, A., Batenburg, K., Sijbers, J.: A multiresolution approach to discrete tomography using DART. PLoS ONE **9**(9), e106090 (2014)
9. Plier, J.: Theoretical and numerical approaches to co-/sparse recovery in discrete tomography. Ph.D. thesis, Heidelberg University (2020)
10. Trottenberg, U., Oosterlee, C., Schüller, A.: Multigrid. Academic Press, Cambridge (2001)
11. Gratton, S., Mouffe, M., Toint, P.L., Weber-Mendonca, M.: A recursive formula-trust-region method for bound-constrained nonlinear optimization. IMA J. Numer. Anal. **28**(4), 827–861 (2008)
12. Ulbrich, M.: Nonmonotone trust-region methods for bound-constrained semismooth equations with applications to nonlinear mixed complementarity problems. SIAM J. Optim. **11**, 889–916 (2001)

13. Iusem, A.N.: On the convergence properties of the projected gradient method for convex optimization. Comput. Appl. Math. **22**, 37–52 (2003)
14. Alvarez, F., Bolte, J., Brahic, O.: Hessian Riemannian gradient flows in convex programming. SIAM J. Control. Optim. **43**(2), 477–501 (2004)
15. Rockafellar, R.T.: Convex Analysis. Princeton University Press, Princeton (1997)
16. Absil, P.A., Mahony, R., Sepulchre, R.: Optimization Algorithms on Matrix Manifolds. Princeton University Press, Princeton (2008)
17. Amari, S.I., Nagaoka, H.: Methods of Information Geometry. American Mathematical Society and Oxford University Press (2000)
18. Ay, N., Jost, J., Lê, H.V., Schwachhöfer, L.: Information Geometry. EMG-FASMSM, vol. 64. Springer, Cham (2017)

Bregman Proximal Gradient Algorithms for Deep Matrix Factorization

Mahesh Chandra Mukkamala[1](✉), Felix Westerkamp[2], Emanuel Laude[2], Daniel Cremers[2], and Peter Ochs[1]

[1] University of Tübingen, Tübingen, Germany
{mamu,ochs}@math.uni-tuebingen.de
[2] Technical University of Munich, Munich, Germany
{felix.westerkamp,cremers}@tum.de, emanuel.laude@in.tum.de

Abstract. A typical assumption for the convergence of first order optimization methods is the Lipschitz continuity of the gradient of the objective function. However, for many practical applications this assumption is violated. To overcome this issue extensions based on generalized proximity measures, known as Bregman distances, were introduced. This initiated the development of the Bregman Proximal Gradient (BPG) algorithms, which, however, rely on problem dependent Bregman distances. In this paper, we develop Bregman distances for deep matrix factorization problems, which yields a BPG algorithm with theoretical convergence guarantees, while allowing for a constant step size strategy. Moreover, we demonstrate that the algorithms based on the developed Bregman distance outperform their Euclidean counterparts as well as alternating minimization based approaches.

1 Introduction

We consider the deep matrix factorization problem or equivalently training a so-called deep linear neural network model, which involves solving the following optimization problem:

$$\min_{\mathbf{W}_i \in \mathcal{W}_i, \forall i \in \{1,\dots,N\}} \left\{ g(\mathbf{W}) := \frac{1}{2} \|\mathbf{W}_1 \mathbf{W}_2 \cdots \mathbf{W}_N \mathbf{X} - \mathbf{Y}\|_F^2 \right\}, \tag{1}$$

where N denotes the number of layers, $d_i \in \mathbb{N}$ for all $i \in \{1,\dots,N\}$, $\mathcal{W}_i = \mathbb{R}^{d_i \times d_{i+1}}$ for all $i \in \{1,\dots,N\}$, $\mathbf{X} \in \mathbb{R}^{d \times n_T}$ be fixed with $n_T \in \mathbb{N}$ corresponding to the number of training samples and set $d =: d_{N+1}$. Similarly we have fixed $\mathbf{Y} \in \mathbb{R}^{d_1 \times n_T}$, which typically corresponds to the labels of the inputs in $\mathbf{X}$. We denote $\mathbf{W} := (\mathbf{W}_1, \dots, \mathbf{W}_N)$, meaning $\mathbf{W}$ lies in the product space $\mathcal{W} := \mathcal{W}_1 \times \cdots \times \mathcal{W}_N$, equipped with the norm $\|\mathbf{W}\|_F^2 := \sum_{i=1}^N \|\mathbf{W}_i\|_F^2$. We focus

---

Electronic supplementary material The online version of this chapter (https://doi.org/10.1007/978-3-030-75549-2_17) contains supplementary material, which is available to authorized users.

© Springer Nature Switzerland AG 2021
A. Elmoataz et al. (Eds.): SSVM 2021, LNCS 12679, pp. 204–215, 2021.
https://doi.org/10.1007/978-3-030-75549-2_17

on $N \geq 2$ in this paper. The main goal of this paper is to propose globally convergent first-order algorithms in order to solve variants of (1). The analysis and convergence of many first-order optimization methods relies on the Lipschitz continuous gradient property for the involved objective. Let $\tilde{g}$ be a continuously differentiable function over $\mathbb{R}^d$. Then $\tilde{g}$ is said to be (classically) L-smooth (has Lipschitz continuous gradient), if there exists $L > 0$, such that for all $x, y \in \mathbb{R}^d$, we have

$$\|\nabla \tilde{g}(x) - \nabla \tilde{g}(y)\|_2 \leq L \|x - y\|_2 .$$

This implies that the functions $(L/2) \|\cdot\|^2 - \tilde{g}$ and $(L/2) \|\cdot\|^2 + \tilde{g}$ are convex on $\mathbb{R}^d$, an equivalence to the well-known descent lemma (cf. Lemma 1.2.3 of [27]). However, the L-smoothness assumption can be restrictive and does not hold for the deep matrix factorization objective in (1). The L-smoothness assumption is also not satisfied for various simple examples, such as the two dimensional function $g(x, y) = (x^2 + y^2)^2$. This means that, in general, Proximal Gradient (PG) algorithms (with constant step size) cannot be used for optimization. Notably, this issue persists even if we resort to alternating, Gauss–Seidel like algorithms such as PALM [8], iPALM [29] which rely on the L-smoothness of the objective with respect to one (block) variable. In the above example, even if we fix $y = c$, for some constant $c \in \mathbb{R}$, the function $g_1(x) = (x^2 + c^2)^2$ fails to be L-smooth. Likewise, quadratic inverse problems, matrix factorization problems and many other practical problems, lack L-smoothness. To overcome this limitation, recent works [4,9,23] consider an extension of L-smooth functions called L-smooth adaptable functions, which relies on the concept of a Bregman distance. Such distances played a crucial role in the development of the Bregman Proximal Gradient (BPG) algorithms [9,26]. However, in order to use BPG methods, an appropriate *problem dependent Bregman distance* must be developed, which is non-trivial in general and particularly for the objective in (1).

Key Contribution. Our contributions are detailed below.

– We propose a novel class of Bregman distances for deep matrix factorization problems with a squared loss, as in (1). Our proposed Bregman distance (Sect. 4.1) is generated by a so-called kernel generating distance of the following form:

$$C_1 \left(\|\mathbf{W}\|_F^2\right)^N + C_2 \left(\|\mathbf{W}\|_F^2\right)^{\frac{N}{2}} , \quad \text{for some } C_1, \ C_2 > 0 .$$

– Such distances are key for the applicability and also the transfer of global convergence (to a stationary point) results of BPG algorithm (Theorem 2).
– The developed Bregman distances yield a base algorithm (BPG) that allows for modifications to alternating, stochastic, inertial variants (cf. Sect. 4.4).
– We propose closed form analytic expressions for the BPG update step under various practical settings (cf. Sect. 4.2).
– We empirically illustrate that BPG based algorithms are superior to Euclidean distance based algorithms such as PG and PALM methods on matrix completion problem based on (1) (cf. Sect. 5).

2　Related Work

Extensions of Lipschitz Continuity of the Gradient. For many practical problems including Poisson inverse problems [4], structured low-rank matrix factorization problems [12,25], quadratic inverse problems [9], the corresponding objective functions are not L-smooth. This hinders us from a straight application of proximal gradient related schemes. To overcome this limitation, in [4,9] the notion of a L-smooth adaptable (L-smad) function is introduced which extends the classical L-smoothness property by means of a problem-dependent Bregman distance. However, the choice of the Bregman distance is typically non-trivial.

Bregman Proximal Minimization. The L-smad property can be characterized in terms of a generalized non-Euclidean descent lemma [9]. This yields generalized proximal gradient method called Bregman proximal gradient (BPG), see [4,9]. Inertial BPG was explored in [26]. Mirror descent algorithm (a special case of BPG) has been extended to a stochastic setting in [15].

Matrix Factorization. Bregman distances for matrix factorization problems has become an active research area. In [12,25] various Bregman distances suitable for matrix factorization problems were proposed. However, such Bregman distances are not valid for deep matrix factorization.

Deep Matrix Factorization. The main contribution of this work is to derive Bregman distances suitable for performing deep matrix factorization with a quadratic loss, or equivalently, the training of a deep neural network with linear activations. As remarked by [14] and in view of [10,17,32,33], it is well justified to study the theoretically more tractable deep linear neural networks instead of the more challenging deep nonlinear networks. A popular application of Deep matrix factorization is matrix completion [1,19]. Various related state of the art results were also explored in [6,24,31].

3　Bregman Proximal Minimization

We revisit required concepts from related works [4,9]. Bregman distances are constructed from a kernel generating distance, defined below. The rest of the section introduces the concepts of [9], specialized to our unconstrained setting.

Definition 1. *A function $h : \mathbb{R}^d \to (-\infty, +\infty]$ is a kernel generating distance if h is C^1 on $\mathbb{R}^d$ and convex. The class of such functions is denoted by $\mathcal{G}(\mathbb{R}^d)$.*

For $h \in \mathcal{G}(\mathbb{R}^d)$, the associated Bregman distance for $(x, y) \in \mathbb{R}^d \times \mathbb{R}^d$ is given by

$$D_h(x, y) = h(x) - h(y) - \langle \nabla h(y), x - y \rangle . \tag{2}$$

Henceforth, we assume the following.

Assumption A. $h \in \mathcal{G}(\mathbb{R}^d)$ and $g : \mathbb{R}^d \to \mathbb{R}$ is continuously differentiable.

For nonconvex functions, the extension of Lipschitz continuity of the gradient is referred to as the L-smad property which we record below.

Definition 2. *A pair (g, h) is L-smooth adaptable (L-smad) on $\mathbb{R}^d$ if there exists $L > 0$ such that $Lh - g$ and $Lh + g$ are convex on $\mathbb{R}^d$.*

The L-smad property can be reformulated in terms of Bregman distances, which yields the following extended descent lemma. For $h = (1/2) \|\cdot\|^2$, the notion of L-smoothness and the classical descent lemma are recovered.

Lemma 1 (Extended Descent Lemma [9, Lemma 2.1]). *The pair of functions (g, h) is L-smad on $\mathbb{R}^d$ if and only if the following holds:*

$$|g(x) - g(y) - \langle \nabla g(y), x - y \rangle| \leq L D_h(x, y), \text{ for all } x, y \in \mathbb{R}^d. \tag{3}$$

3.1 Bregman Proximal Gradient

In analogy to the Euclidean case, the extended descent lemma leads us to consider the following iterative majorization–minimization scheme, which minimizes the following upper bound (majorizer) at each iteration k. Let $x^k \in \mathbb{R}^d$. Under the L-smad assumption, the extended descent lemma yields:

$$g(x) \leq M_k(x) := g(x^k) + \langle \nabla g(x^k), x - x^k \rangle + L D_h(x, x^k), \tag{4}$$

and $M_k(x^k) = g(x^k)$. Then clearly for x^{k+1} given as

$$x^{k+1} \in \operatorname*{argmin}_{x \in \mathbb{R}^d} M_k(x), \tag{5}$$

we have $g(x^{k+1}) \leq M_k(x^{k+1}) \leq M_k(x^k) = g(x^k)$, i.e. objective function is non-increasing. The update in (5) is the mirror descent algorithm [5] which for $h = (1/2)\| \cdot \|^2$ specializes to the classical gradient descent method. Like in the classical proximal gradient method, the majorization property of M_k still holds if we add a second convex nonsmooth term f to both sides of the inequality (4). Minimization of $M_k + f$ then yields the BPG scheme for nonconvex additive composite problems given as

$$(\mathcal{P}) \qquad \inf \{\Psi(x) := f(x) + g(x) : x \in \mathbb{R}^d\}, \tag{6}$$

where g, h satisfy Assumption A. The complete BPG algorithm is given in Algorithm 1. It is formulated in terms of the BPG mapping given by

$$T_\lambda(x) := \operatorname*{argmin}_{u \in \mathbb{R}^d} \{f(u) + \langle \nabla g(x), u \rangle + \frac{1}{\lambda} D_h(u, x)\}. \tag{7}$$

The BPG mapping generalizes the proximal gradient mapping, by replacing the Euclidean distance with a Bregman distance. For convergence and well-definedness we require the following standard assumption (see [9]).

Algorithm 1. BPG: Bregman Proximal Gradient [9]

Input. Choose $h \in \mathcal{G}(\mathbb{R}^d)$ such that (g, h) is L-smooth adaptable on $\mathbb{R}^d$.
Initialization. $x^1 \in \mathbb{R}^d$ and $0 < \lambda < (1/L)$.
General Step. For $k \geq 1$, compute $x^{k+1} \in T_\lambda(x^k)$.

Assumption B. (i) $f : \mathbb{R}^d \rightarrow (-\infty, +\infty]$ is a proper, lower semicontinuous, convex function.

(ii) $v(\mathcal{P}) := \inf \{ \Psi(x) : x \in \mathbb{R}^d \} > -\infty$.

(iii) h is σ-strongly convex on $\mathbb{R}^d$.

(iv) For all $\lambda > 0$, $h + \lambda f$ is supercoercive, i.e., $\lim_{\|x\| \rightarrow \infty} \frac{h(x) + \lambda f(x)}{\|x\|} = \infty$.

Assumption B(iv) ensures the well-definedness of the T_λ, in the sense that $T_\lambda(x)$ is non-empty and compact, for any $x \in \mathbb{R}^d$. We provide below the condensed global convergence result from [9], which states that the full sequence of iterates $\{x^k\}_{k \in \mathbb{N}}$ generated by BPG converges to a stationary point. For generic classes of nonsmooth nonconvex optimization problems, this result is currently the best that we can expect. The global convergence of Bregman proximal algorithms relies on the nonsmooth Kurdyka–Lojasiewicz (KL) property [2, 7]. The KL property is typically satisfied for many practical objectives, for example, for semi-algebraic functions (cf. [3]).

Theorem 1 (Global Convergence of BPG [9, **Theorem 4.1]).** *Let Assumptions A, B hold and let g be L-smad w.r.t. h. Assume $\nabla g, \nabla h$ to be Lipschitz continuous on any bounded subset of $\mathbb{R}^d$. Let $\{x^k\}_{k \in \mathbb{N}}$ be a bounded sequence generated by BPG with $0 < \lambda L < 1$, and suppose Ψ satisfies the KL property, then, $\{x^k\}_{k \in \mathbb{N}}$ has finite length, and converges to a critical point of Ψ.*

By a critical point, following the standard in nonsmooth optimization, we mean a point for which the limiting subdifferential of the objective contains zero, i.e., Fermat's rule holds (Theorem 10.1 of [30]). We remark that additional assumptions mentioned in Theorem 1 are typically satisfied in practice. For example, the boundedness assumption in the statement is automatically satisfied, if the objective is coercive (lower level-bounded). The rest of the assumptions are standard in the nonconvex nonsmooth optimization literature. In this context, we refer the reader to [3,8]. Even though the sequence is bounded and hence the Lipschitz constant of the gradient exists in theory on compact sets, we insist on the fact that this set is unknown before running the algorithm and thus cannot be exploited algorithmically.

4 BPG for Deep Matrix Factorization

This section is the main part of our paper. We specialize g to be a quadratic loss function with a deep matrix factorization model or equivalently with a deep linear neural network (DLNN) model, as in (1). Henceforth, we use the abbreviation "DLNN" for simplicity. Deep matrix factorization model also has applications in matrix completion (c.f. Sect. 5). Note that g in (1) is not classically L-smooth and therefore lacks a quadratic upper bound even for the two layer case. Therefore, our main goal is to derive a novel kernel generating distance h that allows us to apply Theorem 1 in our setting and, therefore, to prove global convergence of BPG for certain DLNN based problems. Thus, we now focus on the kernel generating distances that are suitable for DLNN's, using L-smad property.

4.1   Smooth Adaptable Property for DLNN

To prove the L-smad property we consider its characterization via the Hessian. More precisely, if h and g are twice continuously differentiable, $Lh - g$ and $g + Lh$ are convex if and only if $L\nabla^2 h(x) \succeq \nabla^2 g(x)$ and $-L\nabla^2 h(x) \preceq \nabla^2 g(x)$, i.e., the eigenvalues of the Hessian of g are bounded by the eigenvalues of the Hessian of Lh. Our analysis suggests that the odd and the even case have to be considered separately. Let $N \geq 2$ and define the following functions:

$$H_1(\mathbf{W}) = \left(\frac{\|\mathbf{W}\|_F^2}{N} \right)^N, H_2(\mathbf{W}) = \left(\frac{\|\mathbf{W}\|_F^2}{N} \right)^{\frac{N}{2}}, H_3(\mathbf{W}) = \left(\frac{\|\mathbf{W}\|_F^2 + 1}{N + 1} \right)^{\frac{N+1}{2}}.$$

Then, we have the following result, which shows that for an appropriate linear combination of H_1 and H_2 we obtain the L-smad property for g in (1).

Proposition 1. *Let H_1, H_2, H_3 be as defined above and let g be as in (1). Then, for $L = 1$, the function g satisfies the L-smad property with respect to the following kernel generating distance*

$$H_a(\mathbf{W}) = c_1(N)H_1(\mathbf{W}) + c_2(N)H_2(\mathbf{W}), \qquad \textit{if } N \textit{ is even,} \qquad (8)$$
$$H_b(\mathbf{W}) = c_1(N)H_1(\mathbf{W}) + c_2(N+1)H_3(\mathbf{W}), \qquad \textit{if } N \textit{ is odd,} \qquad (9)$$

where

$$c_1(N) = \frac{(2N-1)N^N}{2N^{N-1}} \|\mathbf{X}\|_F^2, \quad c_2(N) = \frac{\|\mathbf{Y}\|_F \|\mathbf{X}\|_F (N-1)N^{\frac{N-2}{2}}}{(N-2)^{\frac{N-2}{2}}}.$$

The proof is given in Section A.4 in the supplementary material. The resulting Bregman distances are data-dependent, as the coefficients $c_1(N)$ and $c_2(N)$ are dependent on the number of layers, $\mathbf{X}$ and $\mathbf{Y}$. We remark that for $N = 2$ and $\|\mathbf{X}\|_F = 1$, our results match with [25]. An independent work in [21] is closely related to our work, however, the explicit evaluation of the exact kernel generating distance for DLNN was not considered there. Moreover, the value of L is much larger compared to ours. For example, set $\mathbf{X} = 1$, $N = 2$, h in (8), then the value of L is $O(1 + \|\mathbf{Y}\|_F)$ in [21], whereas in our case we require only that $L \geq 1$, while the kernel generating distances coincide. In the context of matrix factorization in 2D, we compare Bregman distances from [9,21] with our method (Proposition 1) in Section B in the supplementary material, and show that our method results in tightest L such that L-smad property is satisfied.

Strong Convexity. The global convergence result of BPG in Theorem 1 (also Theorem 2) relies on the strong convexity of h. In this regard, we use:

$$h(\mathbf{W}) = H_a(\mathbf{W}), \quad \text{for } N = 2, \tag{10}$$

$$h(\mathbf{W}) = H_a(\mathbf{W}) + \rho H_4(\mathbf{W}), \quad \text{for even } N > 2 \text{ for some } \rho > 0, \tag{11}$$

$$h(\mathbf{W}) = H_b(\mathbf{W}), \quad \text{for odd } N > 2, \tag{12}$$

where $H_4(\mathbf{W}) := \|\mathbf{W}\|_F^2/N$. For even $N > 2$, the strong convexity parameter is $\frac{2\rho}{N}$. We fix ρ in the initialization phase of the algorithms. Note that for $N = 2$ and for odd $N > 2$, the function h is strongly convex.

4.2 Closed Form Updates for BPG

In practice, in order to make use of kernel generating distances proposed in Proposition 1 with BPG, we require efficient update steps. It is in general difficult to compute the Bregman proximal mapping (T_λ in (7)) in closed form, even for common f. Typically this involves the computation of the convex conjugate function of the problem-dependent h which can be hard to derive. In our case we show in Proposition 2, that the computation of the BPG map (7) can be reduced to a simple projection problem and a simple one-dimensional nonlinear equation, more precisely a polynomial equation with a unique real root. Such a closed form solution is also valid for any other Bregman proximal algorithm including, stochastic BPG [11]. We denote $g = \Psi$ from (1) and $f := 0$ and we set h as in Sect. 4.

Proposition 2. *In BPG, with above defined g, f, h, denoting $\mathbf{P}_i^k :=$ $\lambda \nabla_{\mathbf{W}_i} g\left(\mathbf{W}^k\right) - \nabla_{\mathbf{W}_i} h(\mathbf{W}^k)$, the update steps in each iteration are given by*

$$\mathbf{W}_i^{k+1} = -r \frac{\sqrt{N}\,\mathbf{P}_i^k}{\|\mathbf{P}^k\|_F}, \quad \text{for all } i \in \{1, \ldots, N\},$$

with $\|\mathbf{P}^k\|_F^2 = \sum_{i=1}^N \|\mathbf{P}_i^k\|_F^2$. Set $\rho = 0$ for $N = 2$. For even $N \geq 2$, $r \geq 0$ satisfies

$$2c_1(N)r^{2N-1} + c_2(N)r^{N-1} + \frac{2\rho}{N}r - \frac{\|\mathbf{P}^k\|_F}{\sqrt{N}} = 0, \tag{13}$$

and, for odd $N > 2$, $r \geq 0$ satisfies

$$2c_1(N)r^{2N-1} + \frac{c_2(N+1)(Nr^2+1)^{\frac{N-1}{2}}}{(N+1)^{\frac{N-1}{2}}} - \frac{\|\mathbf{P}^k\|_F}{\sqrt{N}} = 0. \tag{14}$$

The proof is given in Section A.7 in the supplementary material. With a slight abuse of denotation, we are referring that Proposition 2 provides closed form solution, even though r must be found by solving one-dimensional nonlinear equation. We now consider non-zero f. The squared L2-regularizer is given by

$$f(\mathbf{W}) := \frac{\lambda_0}{2} \sum_{i=1}^N \|\mathbf{W}_i\|_F^2, \quad \text{with } \lambda_0 > 0. \tag{15}$$

In this setting, to obtain closed forms for even $N \geq 2$ replace $\frac{2\rho}{N}$ with $\left(\frac{2\rho}{N} + \lambda\lambda_0\right)$ in Proposition 2 and for odd $N \geq 2$, $\lambda\lambda_0 r$ must be added to left hand side of (14). The L1-regularizer is given by

$$f(\mathbf{W}) := \sum_{i=1}^{N} \mu_i \|\mathbf{W}_i\|_1 , \tag{16}$$

with $\mu_i > 0$ for all $i \in \{1, \ldots, N\}$. Using the element wise soft-thresholding operator $\mathcal{S}_\theta(x) = \max\{|x| - \theta, 0\}\mathrm{sgn}(x)$, the closed form updates are obtained by replacing $-\mathbf{P}_i^k$ with $\mathcal{S}_{\lambda\mu_i}(-\mathbf{P}_i^k)$ in Proposition 2. The proof is similar to the proof of Proposition 2, hence we skip for brevity.

4.3   Global Convergence of BPG for Regularized DLNN

We prove the global convergence of BPG applied to minimize $\Psi := f + g$ by invoking Theorem 1, where f being either L2-regularizer or L1-regularizer, and g be as defined in (1). The proof is provided in Section A.1 in the supplementary material. We remark that our theory does not provide global convergence guarantees for no regularization ($f := 0$).

Theorem 2 (Global Convergence of BPG for Regularized DLNN). *Let g be defined as in (1) with $N > 1$, and f be either L2-regularization as in (15) or L1-regularization as in (16). If $N = 2$, choose the kernel generating distance function $h = H_a$ as in (8). If $N > 2$ and even, then choose h as in (11), otherwise, if $N > 2$ and odd, then choose h as in (12). Then, Ψ has the KL-property and g is L-smad w.r.t. h. Moreover, the sequence $\{x^k\}_{k \in \mathbb{N}}$ generated by BPG is bounded, has finite length, and converges to a critical point of Ψ.*

4.4   Discussion of BPG Variants

The Base Algorithm BPG. The key advantage of BPG for DLNN compared to its Euclidean variant, the Proximal Gradient (PG) method, is the guaranteed convergence when a constant step size rule is used. This is enabled by validity of L-smad property (Proposition 1). On the contrary, PG, which requires a classical L-smoothness can only be used by the following trick. Under a coercivity assumption, all iterates generated by PG lie in a compact set, on which a global Lipschitz constant for the objective's gradient can be found. However, the compact set is usually unknown (and cannot be determined before running the algorithm), and can potentially be extremely large which makes the practical computation of such a global Lipschitz constant difficult or computationally intractable. A good heuristic guess may result in PG being more efficient than BPG. Therefore, BPG based methods render promising alternatives to PG when line search must be avoided due to a prohibitively expensive function evaluation.

BPG with Backtracking. BPG with backtracking (BPG-WB) involves the same setting as BPG with the update $x^{k+1} \in T_{\tau_k}(x^k)$ with the step-size τ_k in $(0, 1/\bar{L}_k)$, where $\bar{L}_k > 0$ is chosen such that:

$$D_g(x^{k+1}, x^k) \leq \bar{L}_k D_h\left(x^{k+1}, x^k\right),$$

holds true and $\bar{L}_k \geq \bar{L}_{k-1}$ holds true for $k > 1$. The value of $\bar{L}_k$ can be found by taking a guess and then increasing it by a constant multiplicative factor iteratively such that above condition holds true. BPG-WB is helpful if the value of L in BPG is not known. BPG-WB can be seen as a special case of inertial BPG algorithm, known as CoCaIn BPG [26], with zero inertia. If backtracking line search variants are affordable for solving the given optimization problem, then BPG and their Euclidean variants PG and iPiano provide the same convergence guarantees. Intuitively, from a global perspective, the adapted upper and lower bounds given by the Bregman distance for BPG should more tightly approximate to the objective function than quadratic functions as required for L-smoothness. However, this situation can change when backtracking line search is used and only locally tight approximations are required.

Alternating vs Non-alternating Strategies. We would like to stress two important advantages of non-alternating schemes such as BPG over alternating minimization strategies like PALM or iPALM. Firstly, BPG allows for block-wise parallelization, and, secondly, there are interesting settings for which alternating minimization is not applicable. The obvious example is symmetric Matrix Factorization, for which BPG is studied in [12]. In the context of DLNN ($N > 2$ in (1)) requiring $\mathbf{W}_1 = \mathbf{W}_2 = \ldots = \mathbf{W}_N$ (upto a transpose) can be considered as a prototype for an unrolled recurrent neural network architecture, where weights are shared across layers. Here, there is no natural way to apply alternating minimization schemes and the objective is not classically L-smooth.

BPG vs PALM. Proximal Alternating Linearized Minimization (PALM) [8] has a clear bias towards the first block of coordinates, if the update direction points into a narrow valley. This effect may be compensated by its inertial variant iPALM [29]. Additionally, in PALM based methods, the block wise step-size computation can be computationally intensive, unlike in BPG.

Stochastic Setting Extensions. A stochastic version of BPG was developed recently in [11], for which our Bregman distances are valid to train DLNN. Several popular stochastic variants such as Adam [18], Adagrad [13] can potentially be extended with a Bregman proximal framework.

5   Experiments

For the purpose of practical illustration, we consider the matrix completion problem [20] that uses the following function:

$$g(\mathbf{W}) := \frac{1}{2}\left\|P_\Omega(\mathbf{W}_1\mathbf{W}_2\cdots\mathbf{W}_N\mathbf{X} - \mathbf{Y})\right\|_F^2,$$

where P_Ω is a masking operator over a given set of indices Ω, which sets the elements at indices that are not in Ω to zero, while retaining other elements. Note that we recover g in (1) when Ω contains all indices.

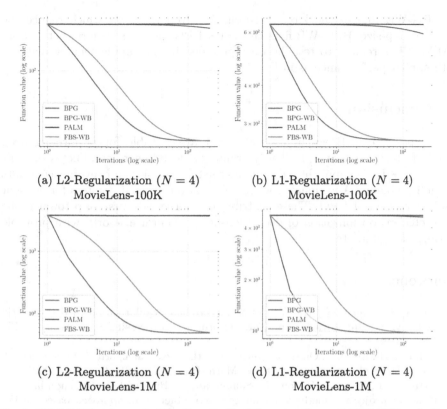

(a) L2-Regularization ($N = 4$)
MovieLens-100K

(b) L1-Regularization ($N = 4$)
MovieLens-100K

(c) L2-Regularization ($N = 4$)
MovieLens-1M

(d) L1-Regularization ($N = 4$)
MovieLens-1M

Fig. 1. BPG-WB performs better than its Euclidean counterpart FBS-WB and alternating updates based PALM, when function value vs iterations is considered. The regularization setting and the dataset name is provided below each figure.

The changes incurred in the theory are the replacement of $\|\mathbf{Y}\|_F$ by $\|P_\Omega(\mathbf{Y})\|_F$ in Proposition 1, and the replacement of the term $(\mathbf{W_1W_2}\ldots\mathbf{W_N X} - \mathbf{Y})$ to $P_\Omega(\mathbf{W_1W_2}\ldots\mathbf{W_N X} - \mathbf{Y})$ in the gradient expression given in Proposition 3 in the supplementary material. We provide experiments for matrix completion problem with squared L2-regularizer and L1-regularizer. Note that Theorem 2 provides global convergence guarantees for BPG. We compare BPG (Algorithm 1), BPG-WB, Forward–Backward Splitting with backtracking (FBS-WB, also known as Proximal Gradient Method with backtracking) [22,28], PALM [8] algorithms. We use MovieLens-100K and MovieLens-1M datasets [16] with $N = 4$ and $\mathbf{X}$ is a scalar with $\mathbf{X} = 1$. We use 80% of the data here and we use 20% of the data to test the performance of the model. The weights $\mathbf{W_1} \in \mathbb{R}^{943 \times 5}, \mathbf{W_2} \in \mathbb{R}^{5 \times 5}, \mathbf{W_3} \in \mathbb{R}^{5 \times 5}, \mathbf{W_4} \in \mathbb{R}^{5 \times 1682}$ are initialized with 0.1. The convergence plots are given in Fig. 1 which illustrate the superior performance of BPG-WB compared to FBS-WB and PALM. BPG is slow compared to BPG-WB due to the large constants involved in c_1, c_2, which reduce the effective step-size. BPG-WB fixes this issue. BPG methods perform better than PALM. Plots related to test data are provided in the supplementary material with similar performance as in Fig. 1.

6   Conclusion

We proposed new Bregman distances that are suitable for performing deep matrix factorization, or equivalently, training a deep linear neural network. This result makes BPG applicable and enables the transfer of their convergence results to such problems. Additionally, we provide various crucial pointers for efficient implementation of BPG. These contributions serve as a first step towards the more efficient optimization of deep (non-linear) neural networks with provable convergence guarantees.

References

1. Arora, S., Cohen, N., Hu, W., Luo, Y.: Implicit regularization in deep matrix factorization. In: Advances in Neural Information Processing Systems, pp. 7413–7424 (2019)
2. Attouch, H., Bolte, J.: On the convergence of the proximal algorithm for nonsmooth functions involving analytic features. Math. Program. **116**(1–2), 5–16 (2009)
3. Attouch, H., Bolte, J., Redont, P., Soubeyran, A.: Proximal alternating minimization and projection methods for nonconvex problems: an approach based on the Kurdyka-Łojasiewicz inequality. Math. Oper. Res. **35**(2), 438–457 (2010)
4. Bauschke, H.H., Bolte, J., Teboulle, M.: A descent lemma beyond Lipschitz gradient continuity: first-order methods revisited and applications. Math. Oper. Res. **42**(2), 330–348 (2017)
5. Beck, A., Teboulle, M.: Mirror descent and nonlinear projected subgradient methods for convex optimization. Oper. Res. Lett. **31**(3), 167–175 (2003)
6. Berg, R.V.D., Kipf, T.N., Welling, M.: Graph convolutional matrix completion. arXiv preprint arXiv:1706.02263 (2017)
7. Bolte, J., Daniilidis, A., Lewis, A., Shiota, M.: Clarke subgradients of stratifiable functions. SIAM J. Optim. **18**(2), 556–572 (2007)
8. Bolte, J., Sabach, S., Teboulle, M.: Proximal alternating linearized minimization for nonconvex and nonsmooth problems. Math. Program. **146**, 459–494 (2013). https://doi.org/10.1007/s10107-013-0701-9
9. Bolte, J., Sabach, S., Teboulle, M., Vaisbourd, Y.: First order methods beyond convexity and Lipschitz gradient continuity with applications to quadratic inverse problems. SIAM J. Optim. **28**(3), 2131–2151 (2018)
10. Choromanska, A., Henaff, M., Mathieu, M., Arous, G.B., LeCun, Y.: The loss surfaces of multilayer networks. In: Artificial Intelligence and Statistics, pp. 192–204 (2015)
11. Davis, D., Drusvyatskiy, D., MacPhee, K.J.: Stochastic model-based minimization under high-order growth. arxiv preprint arXiv:1807.00255 (2018)
12. Dragomir, R.A., d'Aspremont, A., Bolte, J.: Quartic first-order methods for low rank minimization. arxiv preprint arXiv:1901.10791 (2019)

13. Duchi, J., Hazan, E., Singer, Y.: Adaptive subgradient methods for online learning and stochastic optimization. J. Mach. Learn. Res. **12**(Jul), 2121–2159 (2011)
14. Goodfellow, I., Bengio, Y., Courville, A.: Deep Learning. MIT Press, Cambridge (2016)
15. Hanzely, F., Richtárik, P.: Fastest rates for stochastic mirror descent methods. arxiv preprint arXiv:1803.07374 (2018)
16. Harper, F.M., Konstan, J.A.: The movielens datasets: history and context. ACM Trans. Interact. Intell. Syst. (TIIS) **5**(4), 19 (2016)
17. Kawaguchi, K.: Deep learning without poor local minima. In: Advances in Neural Information Processing Systems, pp. 586–594 (2016)
18. Kingma, D.P., Ba, J.: Adam: a method for stochastic optimization. arxiv preprint arXiv:1412.6980 (2014)
19. Kolda, T.G., Bader, B.W.: Tensor decompositions and applications. SIAM Rev. **51**(3), 455–500 (2009)
20. Koren, Y., Bell, R., Volinsky, C.: Matrix factorization techniques for recommender systems. Computer **42**(8), 30–37 (2009)
21. Li, Q., Zhu, Z., Tang, G., Wakin, M.B.: Provable Bregman-divergence based methods for nonconvex and non-lipschitz problems. arXiv preprint arXiv:1904.09712 (2019)
22. Lions, P.L., Mercier, B.: Splitting algorithms for the sum of two nonlinear operators. SIAM J. Numer. Anal. **16**(6), 964–979 (1979)
23. Lu, H., Freund, R.M., Nesterov, Y.: Relatively smooth convex optimization by first-order methods, and applications. SIAM J. Optim. **28**(1), 333–354 (2018)
24. Monti, F., Bronstein, M.M., Bresson, X.: Geometric matrix completion with recurrent multi-graph neural networks. In: Proceedings of the 31st International Conference on Neural Information Processing Systems, pp. 3700–3710 (2017)
25. Mukkamala, M.C., Ochs, P.: Beyond alternating updates for matrix factorization with inertial Bregman proximal gradient algorithms. In: Advances in Neural Information Processing Systems, pp. 4266–4276 (2019)
26. Mukkamala, M.C., Ochs, P., Pock, T., Sabach, S.: Convex-Concave backtracking for inertial Bregman proximal gradient algorithms in nonconvex optimization. SIAM J. Math. Data Sci. **2**(3), 658–682 (2020)
27. Nesterov, Y.: Introductory lectures on convex optimization: a basic course (2004)
28. Ochs, P., Chen, Y., Brox, T., Pock, T.: iPiano: inertial proximal algorithm for nonconvex optimization. SIAM J. Imaging Sci. **7**(2), 1388–1419 (2014)
29. Pock, T., Sabach, S.: Inertial proximal alternating linearized minimization (iPALM) for nonconvex and nonsmooth problems. SIAM J. Imaging Sci. **9**(4), 1756–1787 (2016)
30. Rockafellar, R.T., Wets, R.J.B.: Variational Analysis, Fundamental Principles of Mathematical Sciences, vol. 317. Springer, Berlin (1998)
31. Wang, X., He, X., Wang, M., Feng, F., Chua, T.S.: Neural graph collaborative filtering. In: Proceedings of the 42nd International ACM SIGIR Conference on Research and Development in Information Retrieval, pp. 165–174 (2019)
32. Wu, Y., Poczos, B., Singh, A.: Towards understanding the generalization bias of two layer convolutional linear classifiers with gradient descent. In: The 22nd International Conference on Artificial Intelligence and Statistics, pp. 1070–1078. PMLR (2019)
33. Yun, C., Sra, S., Jadbabaie, A.: Global optimality conditions for deep neural networks. In: International Conference on Learning Representations (2018)

Hessian Initialization Strategies for ℓ-BFGS Solving Non-linear Inverse Problems

Hari Om Aggrawal[1(✉)] and Jan Modersitzki[1,2]

[1] Institute of Mathematics and Image Computing, University of Lübeck,
Lübeck, Germany
hariom85@gmail.com, modersitzki@mic.uni-luebeck.de
[2] Fraunhofer Institute for Digital Medicine MEVIS, Lübeck, Germany

Abstract. ℓ-BFGS is the state-of-the-art optimization method for many large scale inverse problems. It has a small memory footprint and achieves superlinear convergence. The method approximates Hessian based on an initial approximation and an update rule that models current local curvature information. The initial approximation greatly affects the scaling of a search direction and the overall convergence of the method.

We propose a novel, simple, and effective way to initialize the Hessian. Typically, the objective function is a sum of a data-fidelity term and a regularizer. Often, the Hessian of the data-fidelity is computationally challenging, but the regularizer's Hessian is easy to compute. We replace the Hessian of the data-fidelity with a scalar and keep the Hessian of the regularizer to initialize the Hessian approximation at every iteration. The scalar satisfies the secant equation in the sense of ordinary and total least squares and geometric mean regression.

Our new strategy not only leads to faster convergence, but the quality of the numerical solutions is generally superior to simple scaling based strategies. Specifically, the proposed schemes based on ordinary least squares formulation and geometric mean regression outperform the state-of-the-art schemes.

The implementation of our strategy requires only a small change of a standard ℓ-BFGS code. Our experiments on convex quadratic problems and non-convex image registration problems confirm the effectiveness of the proposed approach.

Keywords: Inverse problem · Optimization · Quasi-newton · ℓ-BFGS · Hessian initialization

1 Introduction

Many real-life problems fit the framework of an inverse problem. Fluorescence optical tomography [17], ultrasound tomography [2], and photoacoustic tomography [18] are just a few non-invasive imaging techniques that image a human body's internal structure by solving inverse problems.

© Springer Nature Switzerland AG 2021
A. Elmoataz et al. (Eds.): SSVM 2021, LNCS 12679, pp. 216–228, 2021.
https://doi.org/10.1007/978-3-030-75549-2_18

Inverse problems are typically ill-posed in nature [8]. The solution may not be unique and unstable with variations in the data due to unavoidable factors such as physical noise. Regularizing the problem with prior information, we obtain a solution by minimizing an objective function

$$J : \mathbb{R}^n \to \mathbb{R}, \quad J(x) = D(x) + S(x),$$

where D denotes a data-fitting term and S a regularizer. For many non-linear problems, the objective function is non-convex, and the main limitation is computationally demanding operations. Hence, an efficient optimization method to be designed that requires fewer evaluations of an objective function, its gradient and Hessian, and more occasional calls to a linear solver.

Numerous optimization schemes exist to solve these problems. Still, schemes that do not require more than first-order information are generally preferable. Hessian computation is usually expensive.

Steepest-descent (SD) and quasi-newton methods are the most popular first-order methods. SD converges only linearly; hence super-linear convergent quasi-Newton methods such as Gauss-Newton (GN) schemes or the Broyden-class are preferable. The quasi-Newton method's key idea is to replace Hessian with an approximation that models the local curvature information. It leads to not only faster convergence but as well higher solution accuracy than simple gradient descent methods; see, e.g. [9, 17] and references therein.

The Hessian approximation in the GN method is based on linearizing a function that involves a matrix-vector product with a Jacobian matrix. For applications such as optical tomography [17], the Jacobians are generally dense, and hence per iteration costs can be very high. Therefore, Broyden-class methods are preferred in practice. The most popular member is the limited-memory version of BFGS scheme (ℓ-BFGS) for large scale inverse problems; see its application to recent work in ultrasound tomography [2] and image registration [11].

The Broyden-class works with approximations of the Hessian that are based on an initial approximation and an update rule that is typically based on current curvature information derived from a secant Eq. (1). Setting $x = x_k$, $x' = x_{k+1}$, $p = x' - x$ and introduce $y := \nabla J(x') - \nabla J(x)$, a Taylor expansion $y = \nabla J(x') - \nabla J(x) \approx \nabla^2 J(x) p$ motivates the so-called secant-equations [15, Chapter 2] for Hessian B,

$$B' p = y \quad \text{or, for the inverse of } B, \quad p = H' y. \tag{1}$$

Based on an initial choice H_0, BFGS-schemes update the current approximation $H = H_k$ using a constrained and weighted least squares fit,

$$H_{k+1} \in \text{argmin}\{|M - H_k|_F, \ M = M^\top, \quad M y_k = p_k\},$$

where a weighted Frobenius-norm $|A|_F^2 = \text{trace}(W A W A^\top)$ is used. If the weight matrix satisfies secant equation $W p = y$, one obtains a unique and scale-invariant solution for H_{k+1} as a rank two update of the H_k,

$$H_{k+1} = V_k^\top H_k V_k + \alpha_k \, p_k p_k^\top, \quad \text{with} \quad \alpha_k := (y_k^\top p_k)^{-1}, \quad V_k := I - \alpha \, y_k p_k^\top. \tag{2}$$

For large-scale problems, a limited-memory version of BFGS (ℓ-BFGS) is used [12]. In ℓ-BFGS, at most the last ℓ pairs are used. More precisely, only pairs (y_j, p_j) with $k_\ell := \max\{1, k - \ell - 1\} \leq j \leq k$ are used. Formally, $H_{k+1} = M_{k+1}$ results from the modified recursion

$$M_{j+1} := V_j^\top M_j V_j + \alpha_j \, p_j p_j^\top, \quad j = k_\ell, \ldots, k. \tag{3}$$

The convergence depends on the quality of Hessian approximation which generally can not be controlled. It has been observed numerically that a "good" initial guess of the Hessian greatly affects the scaling of a search direction and convergence of the overall scheme [1,5,7,13]. Note that ℓ-BFGS-method allows to re-initialize Hessian at every iteration. This opportunity provides a window to rescale the search direction and infuse more information in the scheme.

The state-of-the-art strategy initialize the Hessian (or it's inverse) with a scaled identity matrix

$$H_0^k = \tau_k I.$$

The scalar τ_k is computed at each iteration to satisfy the secant Eq. (1) in a ordinary least square sense following the Oren–Luenberger scaling strategy [16]. This results in two choices for scaling factor τ_k, i.e.,

$$\tau_k^{\text{LSy}} = (y^\top p)/(y^\top y) \quad \text{and} \quad \tau_k^{\text{LSp}} = (p^\top p)/(y^\top p). \tag{4}$$

In practice, it has been observed that the factor τ_k^{LSy} ensures a well-scaled search direction and as a result, most of the iterations accept a steplength of one [7].

In this paper, we suggest to improve the quality of the initial Hessian approximation by including computationally manageable parts from the regularizer. More precisely, we suggest to use

$$B_0^k = \tau_k I + \nabla^2 S,$$

where $\nabla^2 S$ denotes the Hessian of a not necessarily quadratic regularizer. We derive four options for scaling factor τ_k based on ordinary and total-least squares formulations, and geometric mean regression.

Our work is motivated by ideas in image registration [9,14] and molecular energy minimization [10]. In [9,14], the Hessian is initialized by a positive-definite matrix $B_0 = \tau I + A$, where A is the Hessian of a quadratic regularizer and constant. The parameter $\tau > 0$ is chosen manually. In [9], it is reported that this strategy outperforms the simple scaling approach. In [10], the proposed strategy is similar to ours but requires expensive incomplete Cholesky factorization at each iteration to ensure positive-definiteness of Hessian approximation. Moreover, the value for τ is heuristically defined. But, in this paper, we show that it satisfies secant equation in a sense of geometric mean regression.

We assume the regularization part to be computationally manageable. Typical examples include L_2-norm based Tikhonov regularizers [8,14], smooth total-variation norm [19], or, more generally, quadratic forms of derivative based regularization. Here, $R(x) = \|Bx\|_{L_2}$ and B is a linear differential operator. Non-quadratic forms such as the hyperelastic regularizer [3] also fit into this class.

Our new strategy is easy to integrate into an ℓ-BFGS code. Only the Hessian initialization routine needs to be changed, all other parts remain unchanged.

In this paper, we also demonstrate on various test cases that the proposed approach achieves fast convergence and improves the solution accuracy compared to the standard scaling based approaches. Our test cases include convex quadratic problems and non-convex image registration problems with both, quadratic and non-quadratic regularization. Due to the page limitation, the theoretical investigation will be a part of the extended version of this paper.

In Sect. 2, we derive four scaling factors for the proposed initialization strategy and present a practical algorithm. We report the numerical experiments with results in Sect. 3. In Sect. 4, we conclude our findings.

2   Proposed Hessian Initialization Strategy and Algorithm

As common for inverse problems, we assume that the objective function J is a sum of a data fitting term D and a regularizer S. Hence

$$\nabla^2 J = \nabla^2 D + \nabla^2 S, \tag{5}$$

where $A_k := \nabla^2 S(x_k)$ is symmetric positive semidefinite (SPSD). We assume that A_k is "easy", i.e. has low memory requirements and $A_k x = b$ can be solved efficiently. Problems may occur from the data fitting part $\nabla^2 D(x_k)$, which might be computational complex and potentially ill-conditioned.

In the proposed strategy, we suggest to approximate $\nabla^2 D(x_k) \approx \tau_k I$, where τ_k is a tuning parameter to be determined. Hence

$$B_0^k = \tau_k I + A_k. \tag{6}$$

The role of B_0^k is to mimic the Hessian at least for the current update. In this regard, we aim to satisfy the secant Eq. (1) in some least squares sense, where now

$$y = B_0^k\, p = \tau_k\, p + A_k p \iff \tau_k p - z = 0, \quad z := y - A_k p. \tag{7}$$

Since B_0^k is required to be symmetric positive definite (SPD), we have $\tau_k \geq \tau_{\min} := \varepsilon - \mu_{\min}$, where $\varepsilon > 0$ is a small tolerance, typically $\varepsilon = 10^{-6}$ and $\mu_{\min}$ is the smallest eigenvalue of A_k. This adds a constraint to the least squares problems. First we summarize the ordinary least square approach; cf. Lemma 1. We removed subscript k for clarity.

Lemma 1. *Let $p, z \in \mathbb{R}^n$ with $p^\top z \neq 0$ and $\tau_{\min} \in \mathbb{R}$. Then*

$$\tau^{Dp} := \max\{(p^\top z)/(p^\top p), \tau_{\min}\} \text{ and } \tau^{Dz} := \max\{(z^\top z)/(p^\top z), \tau_{\min}\}$$

are optimal scaling parameters resulting from a minimization of $|\xi u - v|$ subject to $\xi \geq \tau_{\min}$, where $(u, v) = (p, z)$ for (Dp) and $(u, v) = (z, p)$ for (Dz).
Moreover, it holds $|\tau^{Dp}| \leq |\tau^{Dz}|$.

Proof. For (Dp): the unique minimizer ξ of the unconstrained problem follows from the basic calculus. If $\xi < \tau_{\min}$, the minimum is attained on the boundary. For (Dz): the result follows from rescaling. The inequality follows the Cauchy-Schwarz-inequality $|p^\top z|^2 \le (p^\top p)(z^\top z)$.

The above choices have a preference either for the p or z direction. A total least squares approach can be used for an unbiased approach; cf. Lemma 2.

Lemma 2. *Let* $p, z \in \mathbb{R}^n$ *with* $\delta := p^\top z \ne 0$, $\tau_{\min} \in \mathbb{R}$, *Then*

$$\tau^{\mathrm{Du}} := \max\{(|z|^2 - \lambda)/\delta, \tau_{\min}\}, \quad \lambda = (|p|^2 + |z|^2 - \sqrt{(|p|^2 - |z|^2)^2 + 4\delta^2})/2,$$

is an optimal scaling parameter from the rescaling of minimizer $\eta = [\eta_1, \eta_2]$ *of the total least squares formulation* $|\eta_1 p - \eta_2 z|$ *subject to* $|\eta| = 1$. *With* τ^{Dp} *and* τ^{Dz} *as in Lemma 1, it holds* $|\tau^{\mathrm{Dp}}| \le |\tau^{\mathrm{Du}}| \le |\tau^{\mathrm{Dz}}|$.

Proof. We have the necessary condition of first order $(U^\top U - \mu I)\eta = 0$ subject to $|\eta| = 1$, where $U = [p, -z]$ and μ denotes the Lagrange-multiplier. This indicates that μ is the smallest eigenvalue of the symmetric 2-by-2 matrix $U^\top U$ with diagonal elements $|p|^2$ and $|z|^2$ and off diagonal $-\delta$. Hence, $\mu = \lambda$ and $\eta := v/|v|$ is a normalized version of the associated eigenvector $v = (\delta, |p|^2 - \lambda)$. The value for τ^{Du} follows from proper scaling.

To show the inequality, we use the relationship $(|z|^2 - \lambda)/\delta = \delta/(|p|^2 - \lambda)$ derived from $\det(U^\top U - \mu I) = 0$. Since $U^\top U$ is SPSD, we know $\lambda \ge 0$. Hence, the inequalities $|(|z|^2 - \lambda)/\delta| \le |z|^2/|\delta|$ and $|\delta|/|p|^2 \le |\delta/(|p|^2 - \lambda)|$ satisfies. It leads to $|\tau^{\mathrm{Dp}}| \le |\tau^{\mathrm{Du}}| \le |\tau^{\mathrm{Dz}}|$ following the definition of τ^{Dp}, τ^{Dz}, and τ^{Du}.

Geometric mean regression is an another unbiased approach; see [6] for details. The optimal scaling parameter is defined as the geometric mean of scaling parameters obtained from ordinary least squares problems in Lemma 1, *i.e.*,

$$\tau^{\mathrm{GM}} = \max\{(\tau^{\mathrm{Dp}}\tau^{\mathrm{Dz}})^{1/2}, \tau_{\min}\} = \max\{(|z|^2/|p|^2)^{1/2}, \tau_{\min}\} \tag{8}$$

and follows $|\tau^{\mathrm{Dp}}| \le \tau^{\mathrm{GM}} \le |\tau^{\mathrm{Dz}}|$.

Remarks on Scaling Parameters: Note that, the tuning of parameter τ changes both the angle and length of search direction, whereas the simple scaling based schemes τ^{LSy} and τ^{LSp} majorly changes the length of the search direction.

To achieve fast convergence, we aim to reduce the number of iterations and the line-search steps at every iteration. For that, we seek a search direction that is closer to the Newton direction and take fewer iterations to convergence. Furthermore, we seek well-scaled search directions that satisfy steplength equal to one and avoid any line-search steps for reducing the total run-time.

For a simple quadratic problem, we observe that the search directions with the proposed choices for τ behave almost in a similar fashion with respect to the Newton direction. Hence, all options are practically equivalent.

But, in practice, we observe that the length of a search direction is inversely proportional to the value of τ for our schemes. Hence, a small τ leads to a

---

Algorithm 1: Standard ℓ-BFGS algorithm with the proposed Hessian initialization strategy

---

1 Initialize a starting guess x_0, integer $\ell > 0$, and $\varepsilon > 0$;
2 $k \longleftarrow 0$;
3 repeat
4   Compute B_0^k following steps in Algo. 2;
5   Compute search direction $d_k \longleftarrow -H_k \nabla J(x_k)$ using B_0^k in the two-step recursive algorithm based on (3); see details in [12];
6   Compute $x_{k+1} \longleftarrow x_k + \alpha_k d_k$ where α_k is obtained with a line-search algorithm;
7   **if** $k > \ell$ **then**
8 | Discard the vector pair $\{p_{k-\ell}, y_{k-\ell}\}$ from the storage;
9   **end**
10   Compute and save $p_k \longleftarrow x_{k+1} - x_k$ and $y_k = \nabla J(x_{k+1}) - \nabla J(x_k)$;
11 $k \longleftarrow k + 1$;
12 until *convergence*;

Algorithm 2: Secant equation based Hessian initialization strategies

---

1 Compute A_k, Hessian of regularizer at x_k;
2 Set $\tau_{\min} \longleftarrow \varepsilon$;
3 if *first iteration* $(k = 0)$ **then**
4 | Set $\tau_k \longleftarrow \tau_{\min}$;
5 else
6   Compute $z_k \longleftarrow y_k - A_k p_k$;
7   Set τ_k to either τ_k^{Dp}, τ_k^{Du}, τ_k^{Dz}, or τ_k^{GM};
8 end
9 Initialize $B_0^k \longleftarrow \tau_k I + A_k$;

---

long step. Although it is desirable, but an overestimated length leads to many line-search steps. On the other hand, with a large τ, we take small steps and, as results, require many iterations for convergence; see results in Sect. 3. These facts suggest an optimal scaling factor, but the exact criterion are so far unknown to us. Nevertheless, we provide four choices for τ covering a wide range and describe their inter-relationship in Lemma 2 and (8).

Practical Algorithm: Now, we are ready to present pseudo code for the standard ℓ-BFGS algorithm with the proposed Hessian initialization strategies where we motivate to initialize Hessian B_0^k at every iteration with (6); see Algorithm 1 [15]. In the standard ℓ-BFGS code, we only need to change the Hessian initialization routine; see Line 4 in Algorithm 1; with a few lines of code described in Algorithm 2.

Table 1. Optimization methods used for evaluations. A_k be the Hessian of regularizer.

No.	Optimization methods	Hessian initialization	References
1	Steepest descent (SD)	–	see [15]
2	ℓ-BFGS	$H_k^0 = \tau_k I$	$\tau_k = I$, τ_k^{LSy}, or τ_k^{LSp}; see (4)
3	ℓ-BFGS (FAIR scheme)	$B_k^0 = \tau I + A_k$	τ is set manually; see [14]
4.	ℓ-BFGS (proposed)	$B_k^0 = \tau_k I + A_k$	$\tau_k = \tau_k^{\mathrm{Dp}}$, τ_k^{Dz}, τ_k^{Du}, or τ_k^{GM}; see Lemma 1, Lemma 2, and (8)
5	Gauss-Newton (GN)	–	see [15]

To initialize B_0^k, we start with setting $\tau_{\min}$, computing the Hessian of regularizer, and evaluating z_k. The parameter τ_k can be set with either τ_k^{Du}, τ_k^{Dp}, τ_k^{Dz}, or τ_k^{GM}.

In the first iteration, we can not compute z_k due to the lack of information on the required iterates. Hence, initially, we set $\tau = \varepsilon$ in our experiments. Other initialization options, *e.g.*, based on the norm of a gradient [15], are also possible.

Recall that the parameter τ_k should be greater than $\tau_{\min} = \varepsilon - \mu_{\min}(A_k)$ to ensure the positive-definiteness of Hessian B_0^k. To determine $\tau_{\min}$, we need the smallest eigenvalue of A_k that could be computationally expensive operation for large scale problems. Hence, we avoid the eigenvalue computation in practice and set $\tau_{\min} = \varepsilon = 10^{-6}$ in our experiments.

3   Numerical Experiments and Results

We report on the performance of the proposed Hessian initialization strategies for typical inverse problems: **a) Strictly convex quadratic problems:** This class is chosen to validate the convergence properties of the proposed strategy numerically. **b) Non-convex image registration problems:** This class is chosen to show the effectiveness of the proposed strategy on a few challenging real-world problems.

We investigate in total eight Hessian initialization strategies for ℓ-BFGS method; see Table 1 for details. Along with ℓ-BFGS, we also report results with Gauss-Newton (GN) and steepest descent (SD) method. GN is a widely used method in the field of image registration. GN may achieve quadratic convergence close to the solution. Even though, GN may converge to a local optimal point in a few iterations, but for large scale problems, per iteration cost for GN could be very high due to additional matrix-vector products with Jacobian; see run-time for GN in Table 4 for image registration problems. On the other hand, SD follows a linear and ℓ-BFGS a superlinear convergence.

Note that, the methods Dp, Du, Dz, GM, FAIR, and GN solve a linear system at each iteration to compute the search direction. For that, we use Jacobi preconditioned conjugate gradient (PCG) method. Moreover, the associated system matrices are not stored, rather matrix-vector product has been computed

Table 2. Optimization results for quadratic problem with eight Hessian initialization strategies (S). Iteration counts and average line-searches (LS) per iteration are mentioned for weakly ($\alpha = 10^{-5}$), mildly ($\alpha = 10^{-3}$), and strongly ($\alpha = 10^{-1}$) regularized problems with $\ell = 1, 5, 10$, and ∞.

S./ℓ		$\alpha = 10^{-5}$				$\alpha = 10^{-3}$				$\alpha = 10^{-1}$			
		1	5	10	∞	1	5	10	∞	1	5	10	∞
Iterations	Id	5000	3858	3893	171	5000	754	628	128	567	137	97	47
	LSp	4108	908	383	37	539	145	65	28	79	37	31	23
	LSy	2705	1460	870	90	558	230	115	39	69	39	30	23
	FAIR	5000	869	389	18	168	93	31	15	18	7	7	7
	Dp	4400	817	262	30	565	79	47	20	23	12	10	10
	Dz	3716	1128	633	84	564	228	163	50	49	27	19	17
	Du	5000	760	274	30	578	88	36	20	23	11	10	10
	GM	2880	588	269	61	356	125	56	30	30	13	11	11
Avg. LS per iter.	Id	1.00	1.00	1.00	1.00	1.00	1.00	1.00	1.00	1.00	1.00	1.00	1.00
	LSp	2.73	7.05	8.13	1.16	2.43	3.78	3.08	1.18	1.76	1.62	1.48	1.43
	LSy	1.15	1.10	1.11	1.00	1.15	1.15	1.17	1.00	1.07	1.05	1.10	1.04
	FAIR	14.74	12.92	12.04	5.56	8.08	6.41	3.74	2.67	2.11	1.43	1.43	1.43
	Dp	2.68	6.64	6.86	1.80	2.32	3.62	2.79	1.65	1.35	1.17	1.20	1.20
	Dz	1.15	1.10	1.11	1.18	1.10	1.06	1.09	1.18	1.04	1.07	1.11	1.12
	Du	2.71	6.49	7.55	1.80	2.38	3.10	2.58	1.65	1.26	1.27	1.20	1.20
	GM	1.77	3.04	3.54	1.28	1.63	1.88	1.89	1.33	1.10	1.15	1.18	1.18

directly. We run PCG until the relative residual is less than 10^{-6} or the maximum iterations reach to 100. The iteration count is set to low with the purpose of reducing extra computational time at each iteration due to the linear solver.

The Hessian initialization in the FAIR [14] is similar to ours. But, they set manually the parameter $\tau = 10^{-3}c$, where c is the first diagonal element of A.

We use Armijo backtracking line-search algorithm to estimate the step-size. As noted in [15], if curvature condition is not satisfied at any iteration, we skip the Hessian update. For stopping criteria, we follow [14, p. 78] and set $\varepsilon_J = 10^{-5}$, $\varepsilon_W = 10^{-1}$, and $\varepsilon_G = 10^{-2}$. For ℓ-BFGS, we use the standard choice $\ell = 5$.

Run-time and solution accuracy are our main criteria to evaluate the performance of optimization methods.

3.1   Quadratic Problem

We minimize a strictly convex quadratic function $0.5(x - c)^\top (D + \alpha R)(x - c)$ that has the unique minimizer at $x^* = c \in \mathbb{R}^n$ with D and R be a symmetric and positive-definite matrix and $\alpha > 0$. In experiments, D is a diagonal matrix with exponentially decaying eigenvalues, i.e., $D_{ii} = \exp(-i)$. It is a highly ill-conditioned matrix with condition number of order 10^6 reflecting Hessian of a typical data-fidelity term in inverse problems. The regularization matrix R be a well-known Laplacian matrix with zero boundary conditions. The regularization parameter α controls the ill-conditioning of the quadratic function. Here, we

Table 3. Image registration test problems (TP) for the performance evaluation of optimization strategies. For TP-4, the initial TRE is not available (N.A.).

TP	Dataset	Problem size	Data-Fidelity	Regularizer	Parameters	Initial TRE
1	Hand (2D)	$2 \times 128 \times 128$	SSD	Curvature	$\alpha = 1.5 \times 10^3$	1.04 (0.62)
2	Hand (2D)	$2 \times 128 \times 128$	MI	Elastic	$\alpha = 5 \times 10^{-3}$	1.04 (0.62)
3	Lung (3D)	$3 \times 64 \times 64 \times 24$	NGF	Curvature	$\alpha = 10^2$	3.89 (2.78)
4	Disc-C (2D)	$2 \times 16 \times 16$	SSD	Hyperelastic	$\alpha = (100, 20)$	N.A.

investigate weakly ($\alpha = 10^{-5}$), mildly ($\alpha = 10^{-3}$), and strongly ($\alpha = 10^{-1}$) regularized problems; see results in Table 2.

The iterations start with x be a zero vector. The iterations stop when either the relative error $\|x - x^*\|/\|x^*\| \leq 10^{-5}$ or the iteration count reaches 5000.

As expected, the highly ill-conditioned problem, *i.e.*, weakly regularized, requires many iterations for convergence. Especially, if the local curvature is not well-estimated; see results for $\ell = 1$ in Table 2. The iteration counts are decreasing with increasing regularization levels and with improving Hessian approximation that is increasing ℓ. Note that this behavior is consistent across all Hessian initialization strategies taken into consideration in this work.

Identity initialized Hessian scheme converges very slow. But mostly, it satisfies step-length equal to one, which means it takes tiny steps at each iteration.

In most cases, the Hessian initialization schemes equipped with regularization require fewer iterations than the simple scaling based LSy and LSp schemes. In particular, the FAIR scheme takes the lowest iterations, but search-directions are badly scaled. Hence, line-searches (LS) per iteration are much higher than the other schemes. Moreover, LS steps highly depend on the regularization level.

But, for the proposed four schemes, the LS steps depend on the goodness of Hessian approximation, *i.e.*, the value of ℓ rather than the regularization level. In practice, we generally work with a fixed ℓ and adjust the regularization level as per the need. Hence, the proposed scheme suits better for such a scenario. In particular, the Dz scheme generally take 1.15 LS steps per iteration and does not depend much on ℓ. The Dp and Du schemes require higher LS steps than Dz whereas the GM between the Dp and the Dz. In terms of iterations, we observe an almost inverse relationship; *e.g.*, the Dp and Du scheme take fewer iterations than Dz; follow the discussion in Sect. 2 for the underlying reason.

3.2   Image Registration

Now, we show effectiveness on four real-life large-scale problems from image registration. The registration problems are generally highly non-convex and ill-posed in nature; see [14] for details. Here, given a pair of images T and R, the goal is to find a transformation field ϕ such that the transformed image $T(\phi)$ is similar to R, *i.e.*, $T(\phi) \approx R$. To determine ϕ, we solve an unconstrained optimization problem

$$J(\phi) = D(T(\phi), R) + \alpha S(\phi) \xrightarrow{\phi} \min$$

where D measures the similarity between the transformed image $T(\phi)$ and R. The regularizer S enforces smoothness in the field. Curvature, elastic, and hyperelastic are a few commonly used regularizers. The typical choices for similarity measures are the sum of squared difference (SSD), normalized gradient fields (NGF), and mutual information (MI).

Our four test problems (TP) represent a big class of registration models; see Table 3. The popular X-ray hand images are from [14], lung CT images from the well-known DIR dataset [4,11], and the academic Disc-C images from [3].

Note that our strategy works even when the Hessian of regularizer is available only partially. For that, we consider hyperelastic regularizer; see [3] for details.

Table 4. Optimization results for four image registration test problems (TP). The iteration counts (iter), the function evaluations (feval), the reduction in objective function $\frac{J(\phi)}{J(\phi_0)}$, the average run-time in seconds, and the mean and standard deviation of TRE are reported. The gray-colored cell denotes the ℓ-BFGS method that achieve either the smallest TRE (higher accuracy) or lowest run-time (faster convergence).

M.	\multicolumn{5}{c}{TP-1: Hand, SSD, Curvature}					\multicolumn{5}{c}{TP-2: Hand, MI, Elastic}				
	iter	feval	$\frac{J(\phi)}{J(\phi_0)}$	time (sec.)	TRE mean (std.)	iter	feval	$\frac{J(\phi)}{J(\phi_0)}$	time (sec.)	TRE mean (std)
SD	999	1600	86.27	23.31	1.04 (0.62)	169	270	84.80	8.23	0.99 (0.63)
LSp	1000	4157	28.07	40.46	0.67 (0.52)	154	446	73.53	9.15	0.64 (0.37)
LSy	1000	1029	24.48	20.44	0.52 (0.29)	135	137	73.64	5.74	0.64 (0.36)
FAIR	1000	1001	21.62	57.10	0.36 (0.18)	170	171	72.78	12.27	0.56 (0.30)
Dp	53	59	21.94	8.96	0.38 (0.17)	43	67	72.76	7.63	0.58 (0.34)
Dz	444	445	20.49	71.63	0.37 (0.16)	72	94	72.84	6.73	0.58 (0.32)
Du	444	445	20.49	71.79	0.37 (0.16)	43	67	72.76	7.70	0.58 (0.34)
GM	78	80	20.84	13.13	0.35 (0.17)	64	66	72.77	8.83	0.57 (0.31)
GN	18	19	28.45	4.58	0.69 (0.59)				$\gg 360$	

M.	\multicolumn{5}{c}{TP-3: Lung, NGF, Curvature}					\multicolumn{4}{c}{TP-4: Disc-C, SSD, Hyperelastic}			
SD	145	251	97.17	145.83	3.64 (2.70)	999	1585	25.96	10.15
LSp	75	204	95.93	87.44	2.94 (2.20)	778	3520	16.90	11.33
LSy	184	185	94.84	163.00	1.70 (0.99)	671	829	16.93	6.03
FAIR	127	128	94.67	170.31	1.59 (0.79)	214	1083	6.03	7.99
Dp	58	87	94.79	128.58	1.61 (0.83)	117	695	6.02	4.22
Dz	131	132	94.79	161.70	1.64 (0.88)	148	656	6.02	3.90
Du	148	149	94.74	183.07	1.61 (0.80)	134	535	6.07	3.31
GM	77	79	94.76	123.45	1.62 (0.83)	213	384	16.85	4.35
GN	68	254	94.90	2616.43	1.68 (0.99)	60	100	8.30	12.08

The ground truth transformation fields are not available for real-world problems. Hence, we compute the target registration error (TRE), defined as the Euclidean distance between the ground truth landmarks and the estimated landmarks after registration. To accumulate the TRE for each landmark position, we compute the mean and standard deviation (std.) of TRE. The regularization parameter is set to achieve the lowest TRE without foldings in the field.

The field ϕ is initialized with an identity map, *i.e.*, $\phi_0(x) = x$ in all experiments. The open-source FAIR image registration toolbox [14] is the backbone of our implementations. We follow FAIR matrix-free approach.

In all the experiments, the regularization-equipped initialization schemes achieve higher accuracy than the simple scaling based approaches, *i.e.*, LSp, LSy, and Id. Moreover, these simple scaling schemes converge to a higher value of the objective function; see TRE and reduction factor column in Table 4.

In terms of TRE, the FAIR scheme is almost similar to the proposed schemes, but it converges much slower than others; see the run-time column in Table 4. The proposed schemes are faster than others in all the experiments but TP-2. Here, LSy converges faster but at the cost of lower accuracy. It is important to note that, even though the regularization-equipped schemes' per-iteration cost is higher due to the linear solver, they converge faster. It is mainly because of the lower iteration counts, as also seen for quadratic problems; see Table 2.

Among the four proposed choices, the Dp and the GM turn out to be the best performing schemes. Although, we notice that the performance of a particular scheme greatly depends on the minimizing objective function at hand.

As expected, the steepest descent method is one of the slowest and inaccurate among all. The GN method generally needs fewer iterations, but the per-iteration cost is much higher due to the Jacobian computation; hence the run-time is high.

4   Conclusion

We have proposed a Hessian initialization strategy particularly suited for large-scale non-linear inverse problems. Typically, the objective function is the sum of a data-fidelity term and a regularizer. Often, the Hessian of the data-fidelity is computationally expensive. But not the Hessian of the regularizer.

We propose to replace the Hessian of the data-fidelity with a scalar and keep the Hessian of regularizer to initialize the Hessian approximation at every iteration. The scalar satisfies the well-known secant equation in the sense of ordinary and total least squares, and geometric mean regression. In total, we have proposed four choices for the scalar that leads to well-scaled search directions. We also established the inter-relationship between the derived scalars and discussed the consequences of a scalar choice on the convergence in terms of iteration counts and line-search steps. The implementation of our strategy requires only a small change of a standard ℓ-BFGS code.

Our experiments on highly non-convex image registration problems indicate that the proposed schemes converge faster and achieve higher accuracy than the simple scaling based approaches. The Dp, based on ordinary least squares, and GM, based on geometric mean regression, are best-performing schemes.

Under suitable assumptions, we can also show that the proposed parameters are the eigenvalue's estimates of the Hessian of a data-fidelity term. The theoretical investigation will be a part of the extended version of this paper. Future work also addresses the application to inverse problems, *e.g.*, ultrasound tomography [2], and optical tomography [18].

References

1. Andrei, N.: A new accelerated diagonal Quasi-Newton updating method with scaled forward finite differences directional derivative for unconstrained optimization. Optimization **70**, 1–16 (2020). https://doi.org/10.1080/02331934.2020.1712391
2. Bernhardt, M., Vishnevskiy, V., Rau, R., Goksel, O.: Training variational networks with multidomain simulations: speed-of-sound image reconstruction. IEEE Trans. Ultrason. Ferroelectr. Freq. Control **67**(12), 2584–2594 (2020). https://doi.org/10.1109/tuffc.2020.3010186
3. Burger, M., Modersitzki, J., Ruthotto, L.: A hyperelastic regularization energy for image registration. SIAM J. Sci. Comput. **35**(1), B132–B148 (2013). https://doi.org/10.1137/110835955
4. Castillo, R., et al.: A framework for evaluation of deformable image registration spatial accuracy using large landmark point sets. Phys. Med. Biol. **54**(7), 1849–1870 (2009). https://doi.org/10.1088/0031-9155/54/7/001
5. Dener, A., Munson, T.: Accelerating limited-memory quasi-newton convergence for large-scale optimization. In: RodriguesRodrigues, J.M.F., et al. (eds.) ICCS 2019. LNCS, vol. 11538, pp. 495–507. Springer, Cham (2019). https://doi.org/10.1007/978-3-030-22744-9_39
6. Draper, N.R.: Straight line regression when both variables are subject to error. In: Conference on Applied Statistics in Agriculture (1991). https://doi.org/10.4148/2475-7772.1414
7. Gilbert, J.C., Lemaréchal, C.: Some numerical experiments with variable-storage Quasi-Newton algorithms. Math. Program. **45**(1–3), 407–435 (1989). https://doi.org/10.1007/bf01589113
8. Hansen, P.C.: Discrete Inverse Problems. Society for Industrial and Applied Mathematics (2010). https://doi.org/10.1137/1.9780898718836
9. Heldmann, S.: Non-linear Registration Based on Mutual Information Theory, Numerics, and Application. Logos-Verl, Berlin (2006)
10. Jiang, L., Byrd, R.H., Eskow, E., Schnabel, R.B.: Preconditioned L-BFGS algorithm with application to molecular energy minimization. Technical report, Colorado University at Boulder, Department of Computer Science (2004)
11. König, L., Rühaak, J., Derksen, A., Lellmann, J.: A matrix-free approach to parallel and memory-efficient deformable image registration. SIAM J. Sci. Comput. **40**(3), B858–B888 (2018). https://doi.org/10.1137/17m1125522
12. Liu, D.C., Nocedal, J.: On the limited memory BFGS method for large scale optimization. Math. Program. **45**(1–3), 503–528 (1989). https://doi.org/10.1007/bf01589116
13. Marjugi, S.M., Leong, W.J.: Diagonal Hessian approximation for limited memory Quasi-Newton via variational principle. J. Appl. Math. **2013**, 1–8 (2013). https://doi.org/10.1155/2013/523476
14. Modersitzki, J.: FAIR: Flexible Algorithms for Image Registration. SIAM, Philadelphia (2009)
15. Nocedal, J., Wright, S.J.: Numerical Optimization. Springer, Heidelberg (2006). https://doi.org/10.1007/978-0-387-40065-5
16. Oren, S.S.: Perspectives on self-scaling variable metric algorithms. J. Optim. Theory Appl. **37**(2), 137–147 (1982). https://doi.org/10.1007/bf00934764
17. Patil, N., Naik, N.: Second-order adjoint sensitivities for fluorescence optical tomography based on the SPN approximation. J. Opt. Soc. Am. A **36**(6), 1003 (2019). https://doi.org/10.1364/josaa.36.001003

18. Saratoon, T., Tarvainen, T., Cox, B.T., Arridge, S.R.: A gradient-based method for quantitative photoacoustic tomography using the radiative transfer equation. Inverse Probl. **29**(7), 075006 (2013). https://doi.org/10.1088/0266-5611/29/7/075006
19. Vogel, C.R.: Computational Methods for Inverse Problems. Society for Industrial and Applied Mathematics (2002). https://doi.org/10.1137/1.9780898717570

Inverse Scale Space Iterations for Non-convex Variational Problems Using Functional Lifting

Danielle Bednarski$^{(\boxtimes)}$ and Jan Lellmann

Institute of Mathematics and Image Computing, University of Lübeck,
Lübeck, Germany
{bednarski,lellmann}@mic.uni-luebeck.de

Abstract. Non-linear filtering approaches allow to obtain decompositions of images with respect to a non-classical notion of scale. The associated inverse scale space flow can be obtained using the classical Bregman iteration applied to a convex, absolutely one-homogeneous regularizer. In order to extend these approaches to general energies with non-convex data term, we apply the Bregman iteration to a lifted version of the functional with sublabel-accurate discretization. We provide a condition for the subgradients of the regularizer under which this lifted iteration reduces to the standard Bregman iteration. We show experimental results for the convex and non-convex case.

1 Motivation and Introduction

We consider variational image processing problems with energies of the form

$$F(u) := \underbrace{\int_\Omega \rho(x, u(x)) \, \mathrm{d}x}_{H(u)} + \underbrace{\int_\Omega \eta(\nabla u(x)) \, \mathrm{d}x}_{J(u)}, \tag{1}$$

where the integrand $\eta : \mathrm{I\!R}^d \mapsto \mathrm{I\!R}$ of the *regularizer* is non-negative and convex, and the integrand $\rho : \Omega \times \Gamma \mapsto \overline{\mathrm{I\!R}}$ of the *data term* H is proper, non-negative and possibly non-convex with respect to u. We assume that the domain $\Omega \subset \mathrm{I\!R}^d$ is open and bounded and that the range, or *label space*, $\Gamma \subset \mathrm{I\!R}$ is compact.

Such problems are common in image reconstruction, segmentation, and motion estimation [1,28]. We are mainly concerned with three distinct problem classes. Whenever we are working with the *total variation* regularizer, we use the abbreviation TV-(1). If the data term is furthermore given by

$$\rho(x, u(x)) = \frac{\lambda}{2}(u(x) - f(x))^2 \tag{2}$$

for some input f and $\lambda > 0$ we use the abbreviation ROF-(1). For data term (2) and arbitrary convex, absolute one-homogeneous regularizer η we write OH-(1).

A. Elmoataz et al. (Eds.): SSVM 2021, LNCS 12679, pp. 229–241, 2021.
https://doi.org/10.1007/978-3-030-75549-2_19

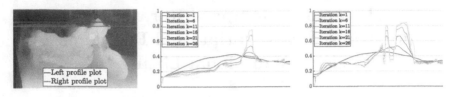

Fig. 1. Scale-space of solutions for non-convex depth estimation. Applying the sublabel-accurate lifting approach [18] to the non-convex problem of depth estimation results in a convex problem to which the Bregman iteration [21] can be applied. In addition to the final depth map **(left)**, the Bregman iteration generates a scale space of solutions with increasing spatial detail, as can be seen from the two horizontal sections **(center, right)**.

Consider the so-called *inverse scale space flow* (ISS) [3,4,21] equation

$$\partial_s p(s) = f - u(s, \cdot), \quad p(s) \in \partial J(u(s, \cdot)), \quad p(0) = 0, \tag{3}$$

where J is assumed to be convex and absolutely one-homogeneous. The evolution $u : [0, T] \times \Omega \to \mathbb{R}$ starts at $u(0, \cdot) = \text{mean}(f)$ and $p(s)$ is forced to lie in the subdifferential of the term J. E.g., for total variation regularization $J = \text{TV}$, the flow $u(s, \cdot)$ progressively incorporates details of finer scales contained in the input image f as s increases; for $s \to \infty$ the flow converges to the input image.

By considering the derivative u_s, one can even define a non-linear decomposition of the input f [5,12] based on the solution u of the inverse scale space flow and derive non-linear filters. Similar ideas have been developed for variational models of the form OH-(1) and *gradient flow* formulations [2–4,10,11].

For problems in the class OH-(1), the inverse scale space flow can be understood [4] as a continuous limit of the so-called *Bregman iteration* [21]. For both the data term H and regularizer J being non-negative and convex (!) the Bregman iteration is defined as:

Algorithm 1: Bregman iteration

Initialize $p_0 = 0$ and repeat for $k = 1, 2, \ldots$

$$u_k \in \arg\min_u \{H(u) + J(u) - \langle p_{k-1}, u \rangle\}, \tag{4}$$

$$p_k \in \partial J(u_k). \tag{5}$$

In case of the ROF-(1) problem the subgradient p_k can be chosen explicitly as $p_k = p_{k-1} - \lambda(u_k - f)$. Further extensions include the *split Bregman method* for ℓ_1-regularized problems [13] and the *linearized Bregman iteration* for compressive sensing and sparse denoising [6,22].

However, applying the Bregman iteration to variational problems with non-convex data term H is not trivial since the well-definedness of the iterations as well as the convergence results in [21] rely on the convexity of the data term. In [14], the Bregman iteration was used to solve a non-convex *optical flow* problem, however, the approach relies on an iterative reduction to a convex problem using first-order Taylor approximations.

In this work, we aim to apply the Bregman iteration to energies with a non-convex data term such as the non-convex stereo matching problem (Fig. 1 and Fig. 3). In order to do so, we follow a *lifting approach:* Instead of minimizing the non-convex problem

$$\inf_{u \in U} \{H(u) + J(u)\} \tag{6}$$

over some suitable (discrete or function) space U we solve a *lifted problem*

$$\inf_{\boldsymbol{u} \in \boldsymbol{U}} \{\boldsymbol{H}(\boldsymbol{u}) + \boldsymbol{J}(\boldsymbol{u})\} \tag{7}$$

over a larger space $\boldsymbol{U}$ but with *convex* energies $\boldsymbol{H}, \boldsymbol{J}$. The Bregman iteration can then be performed on the convex problem (7):

Algorithm 2: Lifted Bregman iteration

Initialize $\boldsymbol{p}_0 = 0$ and repeat for $k = 1, 2, \ldots$

$$\boldsymbol{u}_k \in \arg\min_{\boldsymbol{u} \in \boldsymbol{U}} \{\boldsymbol{H}(\boldsymbol{u}) + \boldsymbol{J}(\boldsymbol{u}) - \langle \boldsymbol{p}_{k-1}, \boldsymbol{u} \rangle\}, \tag{8}$$

$$\boldsymbol{p}_k \in \partial \boldsymbol{J}(\boldsymbol{u}_k). \tag{9}$$

This allows to extend the Bregman iteration to non-convex data terms. Of course it raises the question whether the iterates of Algorithm 1 and Algorithm 2 are related, and whether the lifted method still generates a scale space in practice. In the following, we will investigate these questions.

Outline and Contribution. In Sect. 2 we summarize the sublabel-accurate relaxation approach for problems of the form TV-(1) as presented in [18]. In Sect. 3 we derive conditions under which the original and lifted Bregman iteration are equivalent. These conditions are in particular met by the anisotropic TV. In Sect. 4 we validate these findings experimentally by comparing the original and lifted iteration on the convex ROF-(1) problem and present first numerical results on the non-convex stereo matching problem.

Related Work. In a fully discrete setting with discretized domain and finite range Γ, Ishikawa and Geiger proposed first lifting strategies for the labeling problem [15,16]. Later the relaxation of the labeling problem was studied in a spatially continuous setting with binary [8,9] and multiple labels [17,30].

Our work is based on methods for scalar but continuous range Γ with first-order regularization in the spatially continuous setting [23,24]: The feasible set of scalar-valued functions $u : \Omega \to \Gamma$ is embedded into the convex set of functions $v : \Omega \times \Gamma \to [0, 1]$ by associating each function u with the characteristic function of the subgraph, i.e., $\mathbf{1}_u(x, z) := 1$ if $u(x) > z$ and 0 otherwise. To extend the energy F in (1) for $\Gamma = \mathbb{R}$ onto this larger space, a lifted convex functional $\mathcal{F}$ is defined:

$$\mathcal{F}(v) := \sup_{\phi \in \mathcal{K}} \int_{\Omega \times \Gamma} \langle \phi, Dv \rangle, \tag{10}$$

where Dv denotes the distributional derivative of v. With η^* denoting the point-wise conjugate of the regularizer, the admissible dual vector fields are given by

$$\mathcal{K} := \{(\phi_x, \phi_t) \in C_0(\Omega \times \mathbb{R}; \mathbb{R}^d \times \mathbb{R}) :$$
$$\phi_t(x,t) + \rho(x,t) \geq \eta^*(\phi_x(x,t)), \qquad \forall (x,t) \in \Omega \times \mathbb{R}\}. \tag{11}$$

In [23] the authors show that $F(u) = \mathcal{F}(1_u)$ holds for any $u \in W^{1,1}$. Moreover, if the non-convex set $\{1_u : u \in W^{1,1}\}$ is relaxed to the convex set

$$C := \{v \in BV_{loc}(\Omega \times \mathbb{R}, [0,1]) :$$
$$v(x,t) = 1 \; \forall t \leq \min(\Gamma), \qquad v(x,t) = 0 \; \forall t > \max(\Gamma)\}, \tag{12}$$

any minimizer of the lifted problem $\inf_{v \in C} \mathcal{F}(v)$ can be transformed into a global minimizer of the original nonconvex problem $\inf_{u \in W^{1,1}} \mathcal{F}(1_u)$ by thresholding.

In practice, the discretization of the label space Γ during the implementation process leads to artifacts and the quality of the solution strongly depends on the number and positioning of the chosen discrete labels. Therefore, it is advisable to employ a *sublabel-accurate* discretization [18], which allows to preserve information about the data term in between discretization points, resulting in smaller problems. In [19] the authors point out that this approach is closely linked to the approach in [23] when a combination of piecewise linear and piecewise constant basis functions is used for discretization.

More recent developments in the field of functional lifting include an extension to the sublabel-accurate lifting approach to arbitrary convex regularizers [20] and a connection to Dynamical Optimal Transport and the Benamou-Brenier formulation that also allows to incorporate higher-order regularization [29].

Notation. We denote the *extended real line* as $\overline{\mathbb{R}} := \mathbb{R} \cup \{\pm\infty\}$. Given a function $f : \mathbb{R}^n \mapsto \overline{\mathbb{R}}$ the conjugate $f^* : \mathbb{R}^n \mapsto \overline{\mathbb{R}}$ is defined as [26, Ch. 11]

$$f^*(u^*) := \sup_{u \in \mathbb{R}^n} \{\langle u^*, u \rangle - f(u)\}. \tag{13}$$

If f has a proper convex hull, both the conjugate and biconjugate are proper, lower semi-continuous and convex. The *indicator function* of a set C is defined as $\delta_C(x) := 0$ if $x \in C$ and $+\infty$ otherwise. Whenever u denotes a vector, we use subscripts u_k to indicate an iteration or sequence, and superscripts u^k to indicate the k-th value of the vector.

2   Sublabel-Accurate Lifting Approach

For reference, we provide a short summary of the lifting approach with sublabel-accurate discretization for TV-(1) problems using the notation from [18]. The approach comprises three steps:

Lifting of the Label Space. First, we choose L labels $\gamma_1 < \gamma_2 < ... < \gamma_L$ such that $\Gamma = [\gamma_1, \gamma_L]$. These labels decomposese the label space Γ into $l := L - 1$ *sublabel spaces* $\Gamma_i := [\gamma_i, \gamma_{i+1}]$. Any value in Γ can be written as

$$\gamma_i^\alpha := \gamma_i + \alpha(\gamma_{i+1} - \gamma_i), \tag{14}$$

for some $i \in \{1, 2, ..., l\}$ and $\alpha \in [0, 1]$. The lifted representation of such a value in $\mathbb{R}^l$ is defined as

$$\mathbf{1}_i^\alpha := \alpha \mathbf{1}_i + (1 - \alpha)\mathbf{1}_{i-1}, \tag{15}$$

where $\mathbf{1}_i \in \mathbb{R}^l$ is the vector of i ones followed by $l-i$ zeroes. The *lifted label space* – which is non-convex – is given as $\boldsymbol{\Gamma} := \{\mathbf{1}_i^\alpha \in \mathbb{R}^l | i \in \{1, 2, ..., l\}, \alpha \in [0, 1]\}$. Any lifted value $\boldsymbol{u}(x) = \mathbf{1}_i^\alpha \in \boldsymbol{\Gamma}$ can be mapped uniquely to the equivalent value in the unlifted label space by applying

$$u(x) = \gamma_1 + \sum_{i=1}^{l} \boldsymbol{u}^i(x)(\gamma_{i+1} - \gamma_i). \tag{16}$$

We refer to such functions $\boldsymbol{u}$ as *sublabel-integral*.

Lifting of the Data Term. Next, a lifted formulation of the data term is derived that in effect approximates the energy locally convex between neighboring labels. For the possibly non-convex data term of (1), the lifted – yet still non-convex – representation for fixed $x \in \Omega$ is defined as $\boldsymbol{\rho} : \mathbb{R}^l \mapsto \overline{\mathbb{R}}$,

$$\boldsymbol{\rho}(\boldsymbol{u}) := \inf_{i \in \{1,...,l\}, \alpha \in [0,1]} \{\rho(\gamma_i^\alpha) + \delta_{\mathbf{1}_i^\alpha}(\boldsymbol{u})\}. \tag{17}$$

Note that the domain is $\mathbb{R}^l$ and not just $\boldsymbol{\Gamma}$. Outside of the lifted label space $\boldsymbol{\Gamma}$ the lifted representation $\boldsymbol{\rho}$ is set to ∞. Applying the definition of Legendre-Fenchel conjugates twice to the integrand of the data term results in a relaxed – and convex – data term:

$$\boldsymbol{H}(\boldsymbol{u}) = \int_\Omega \boldsymbol{\rho}^{**}(x, \boldsymbol{u}(x)) dx. \tag{18}$$

For explicit expressions of $\boldsymbol{\rho}^{**}$ in the linear and non-linear case we refer to [18, Prop. 1, Prop. 2].

Lifting of the Total Variation Regularizer. Lastly, a lifted representation of the (isotropic) total variation regularizer is established, building on the theory developed in the context of multiclass labeling approaches [7,17]. For fixed $x \in \Omega$ the lifted – and non-convex – integrand $\boldsymbol{\phi} : \mathbb{R}^{l \times d} \mapsto \overline{\mathbb{R}}$ is defined:

$$\boldsymbol{\phi}(\boldsymbol{g}) := \inf_{1 \leq i \leq j \leq l, \alpha, \beta \in [0,1]} |\gamma_i^\alpha - \gamma_j^\beta| \cdot \|v\|_2 + \delta_{(\mathbf{1}_i^\alpha - \mathbf{1}_j^\beta)v^\top}(\boldsymbol{g}). \tag{19}$$

Applying the definition of Legendre-Fenchel conjugates twice to the lifted integrand of the regularizer results in a relaxed – and convex – regularization term:

$$\boldsymbol{TV}(\boldsymbol{u}) := \int_\Omega \boldsymbol{\phi}^{**}(D\boldsymbol{u}), \tag{20}$$

where Du is the distributional derivative in the form of a Radon measure. For isotropic TV, it can be shown that for $g \in \mathbb{R}^{l \times d}$,

$$\phi^{**}(g) = \sup_{q \in \mathcal{K}_{\text{iso}}} \langle q, g \rangle, \tag{21}$$

$$\mathcal{K}_{\text{iso}} = \left\{ q \in \mathbb{R}^{l \times d} \ \middle| \ \|q_i\|_2 \leq \gamma_{i+1} - \gamma_i, \quad \forall i = 1, ..., l \right\}. \tag{22}$$

For more details we refer to [18, Prop. 4] and [7]. Unfortunately isotropic TV in general does not allow to prove global optimality for the discretized system. Therefore we also consider the lifted anisotropic (L^1) TV, by replacing (22) with

$$\mathcal{K}_{\text{an}} = \left\{ q \in \mathbb{R}^{l \times d} \ \middle| \ \|q_i\|_\infty \leq \gamma_{i+1} - \gamma_i, \quad \forall i = 1, ..., l \right\} \tag{23}$$

$$= \bigcap_{j=1,...,d} \left\{ q \in \mathbb{R}^{l \times d} \ \middle| \ \|q_{i,j}\|_2 \leq \gamma_{i+1} - \gamma_i, \quad \forall i = 1, ..., l \right\}. \tag{24}$$

Together, the previous three sections allow us to formulate a version of the problem of minimizing the lifted energy (10) over the relaxed set (12) that is discretized in the label space Γ:

$$\inf_{u \in BV(\Omega, \Gamma)} \int_\Omega \rho^{**}(x, u(x)) + \int_\Omega \phi^{**}(Du). \tag{25}$$

Once the non-convex set Γ is relaxed to its convex hull, we obtain a fully convex lifting of problem TV-(1) similar to (7), which can now be spatially discretized.

3   Equivalency of the Lifted Bregman Iteration

This chapter addresses the question under which conditions Algorithm 1 and Algorithm 2 are equivalent. We stipulate a sufficient condition on the subgradients used in the Bregman iteration and prove in Chap. 4 that this condition is met in case of the anisotropic TV regularizer. The key idea is to note that the Bregman iteration amounts to extending the data term by a linear term, and that the sum of the separately relaxed terms is point-wise equal to the relaxation of their sum. Note that this additivity does not hold for general sums.

The following considerations are formal due to the mostly pointwise arguments; we leave a rigorous investigation in the function space to future work. However, they can equally be understood in the spatially discrete setting with finite Ω, where arguments are more straightforward. For readability, we consider a fixed $x \in \Omega$ and omit x in the arguments.

Proposition 1. *Assume* $\rho_1, \rho_2, h : \Gamma \mapsto \overline{\mathbb{R}}$ *with*

$$\rho_2(u) := \rho_1(u) - h(u), \qquad h(u) := pu, \qquad p \in \mathbb{R}, \tag{26}$$

where ρ_1 and ρ_2 should be understood as two different data terms in (1). *Define*

$$\tilde{\gamma} := \left(\gamma_2 - \gamma_1, \ldots, \gamma_L - \gamma_l\right)^{\top} \tag{27}$$

Then, for the lifted representations $\rho_1, \rho_2, h : \mathbb{R}^l \mapsto \overline{\mathbb{R}}$ in (17), *it holds*

$$\rho_2^{**}(u) = \rho_1^{**}(u) - h^{**}(u) = \rho_1^{**}(u) - \langle p\tilde{\gamma}, u \rangle. \tag{28}$$

Proof of Proposition 1. The proof is slightly technical and we only sketch it. By definition of the Fenchel conjugate and after some transformations, ρ_2^* becomes

$$\rho_2^*(v) = \sup_{j \in \{1,\ldots,l\}, \beta \in [0,1]} \left\{ \langle 1_j^\beta, v + p\tilde{\gamma} \rangle - \rho_1(\gamma_j^\beta) \right\}. \tag{29}$$

Applying the definition of the Fenchel conjugate once again eventually leads to

$$\rho_2^{**}(u) = \rho_1^{**}(u) - \langle p\tilde{\gamma}, u \rangle. \tag{30}$$

Comparing this to [18, Prop. 2] we see that $\langle p\tilde{\gamma}, u \rangle = h^{**}(u)$.   □

The following proposition shows that Algorithm 1 and Algorithm 2 are equivalent as long as we base the iteration on subgradients p_{k-1} and p_{k-1} in the subdifferential of $J(u_{k-1})$ and $J(u_{k-1})$ that are linked in a particular way.

Proposition 2. *Assume that the minimization problems* (4) *in the original Bregman iteration have unique solutions. Moreover, assume that in the lifted iteration, the solutions u_k of* (8) *in each step satisfy $u(x) \in \Gamma$, i.e., are sublabel-integral. If at every point x the chosen subgradients $p_{k-1} \in \partial J(u_{k-1})$ and $p_{k-1} \in \partial J(u_{k-1})$ satisfy*

$$p_{k-1}(x) = p_{k-1}(x)\tilde{\gamma} \tag{31}$$

with $\tilde{\gamma}$ as in (27), *then the lifted iterates u_k correspond to the iterates u_k of the classical Bregman iteration* (4) *according to* (16).

Proof of Proposition 2. We define the *extended data term*

$$\tilde{H}(u) := \int_\Omega \rho(x, u(x)) - p(x)u(x) \, \mathrm{d}x, \tag{32}$$

which incorporates the linear term of the Bregman iteration. Using Proposition 1, we reach the following lifted representation:

$$\tilde{H}(u) = \int_\Omega \rho^{**}(x, u(x)) - \langle p(x)\tilde{\gamma}, u(x) \rangle \, \mathrm{d}x. \tag{33}$$

Hence the lifted version of (4) is

$$\arg\min_{u \in U} \left\{ H(u) + J(u) - \langle p_{k-1}\tilde{\gamma}, u \rangle \right\}. \tag{34}$$

Comparing this to (8) shows that the minimization problem in the lifted iteration is the lifted version of (4) if the subgradients $p_{k-1} \in \partial J(u_{k-1})$ and $\boldsymbol{p}_{k-1} \in \partial \boldsymbol{J}(\boldsymbol{u}_{k-1})$ satisfy $\boldsymbol{p}_{k-1} = p_{k-1}\tilde{\gamma}$. In this case, since we have assumed that the solution of the lifted problem (8) is sublabel-integral, it can be associated via (16) with the solution of the original problem (4), which is unique by assumption.
□

Thus, under the condition of the proposition, the lifted and unlifted Bregman iterations are equivalent.

4   Numerical Discussion and Results

In this section, we consider the spatially discretized problem on a finite discretized domain Ω^h with grid spacing h. In particular, we will see that the subgradient condition in Proposition 2 can be met in case of anisotropic TV and how such subgradients can be obtained in practice.

Finding a Subgradient. The discretized, sublabel-accurate relaxed total variation is of the form

$$J^h(\nabla \boldsymbol{u}^h) = \max_{q^h:\Omega^h \to \mathbb{R}^{k \times d}} \left\{ \sum_{x \in \Omega^h} \langle \boldsymbol{q}^h(x), \nabla \boldsymbol{u}^h(x) \rangle - \delta_{\mathcal{K}}(\boldsymbol{q}^h(x)) \right\}, \quad (35)$$

with $\mathcal{K}$ defined by (21) or (23) and ∇ denoting the discretized forward-difference operator. By standard convex analysis ([25, Thm. 23.9], [26, Cor. 10.9], [26, Prop. 11.3]) we can show that if q^h is a maximizer of (35), then $p^h := \nabla^\top q^h$ is a subgradient of $J^h(\nabla \boldsymbol{u}^h)$. Thus, the step of choosing a subgradient (9) boils down to $\boldsymbol{p}_k^h = \nabla^\top \boldsymbol{q}_k^h$ and for the dual maximizer $\boldsymbol{q}_{k-1}^h$ of the last iteration we implement (8) as:

$$\boldsymbol{u}_k^h = \arg \min_{u^h:\Omega^h \mapsto \mathbb{R}^l} \max_{q_k^h:\Omega^h \mapsto \mathcal{K}} \sum_{x \in \Omega^h} (\rho^h)^{**}(x, \boldsymbol{u}^h(x)) + \langle \boldsymbol{q}_k^h - \boldsymbol{q}_{k-1}^h, \nabla \boldsymbol{u}^h \rangle. \quad (36)$$

Transforming the Subgradient. In Proposition 2 we formulated a constraint on the subgradients for which the original and lifted Bregman iteration are equivalent. While this property is not necessarily satisfied if the subgradient $\boldsymbol{p}_{k-1}^h$ is chosen according to the previous paragraph, we will now show that any such subgradient can be transformed into another valid subgradient that satisfies condition (31).

Consider a pointwise sublabel-integral solution $\boldsymbol{u}_k^h$ with subgradient $\boldsymbol{p}_k^h := \nabla^\top \boldsymbol{q}_k^h \in \partial \boldsymbol{J}^h(\boldsymbol{u}_k^h)$ for $\boldsymbol{q}_k^h(\cdot) \in \mathcal{K}$ being a maximizer of (35). We define a pointwise transformation: For fixed $x^m \in \Omega^h$ and $\boldsymbol{u}_k^h(x^m) = 1_i^\alpha$, let $(\boldsymbol{q}_k^h(x^m))^i \in \mathbb{R}^d$ denote the i-th row of $\boldsymbol{q}_k^h(x^m)$ corresponding to the i-th label as prescribed by $\boldsymbol{u}_k^h(x^m) = 1_i^\alpha$. Both in the isotropic and anisotropic case the transformation

$$\tilde{\boldsymbol{q}}_k^h(x^m) := \frac{(\boldsymbol{q}_k^h(x^m))^i}{\gamma_{i+1} - \gamma_i} \tilde{\gamma} \quad (37)$$

returns an element of the set $\mathcal{K}$, i.e., $\mathcal{K}_{\mathrm{iso}}$ or $\mathcal{K}_{\mathrm{an}}$. In the anisotropic case we can furthermore show that $\tilde{q}_k^h$ also maximizes (35) and therefore the transformation gives a subgradient $\tilde{p}_k^h := \nabla^\top \tilde{q}_k^h \in \partial J^h(u_k^h)$ of the desired form (31):

Proposition 3. *Consider the anisotropic TV-regularized case (23). Assume that the iterate u_k^h is sublabel-integral. Moreover, assume that $p_k^h := \nabla^\top q_k^h$ is a subgradient in $\partial J^h(u_k^h)$ and define $\tilde{q}_k^h$ pointwise as in (37). Then $\tilde{p}_k^h := \nabla^\top \tilde{q}_k^h$ is also a subgradient and furthermore of the form*

$$\tilde{p}_k^h = p_k^h \tilde{\gamma}^h, \tag{38}$$

where p_k^h is a subgradient in the unlifted case, i.e., $p_k^h \in \partial J^h(u_k^h)$.

Proof of Proposition 3. In the anisotropic case the spatial dimensions are uncoupled, therefore w.l.o.g. assume $d = 1$. Consider two neighboring points x^m and x^{m+1} with $u_k^h(x^m) = 1_i^\alpha$ and $u_k^h(x^{m+1}) = 1_j^\beta$. Applying the forward difference operator gives

$$\nabla u_k^h(x^m) = \frac{1}{h} \begin{cases} (\mathbf{0}_{i-1}, \quad 1-\alpha, \quad 1_{j-i-2}, \quad \beta, \quad \mathbf{0}_{l-j})^\top, & i < j, \\ (\mathbf{0}_{i-1}, \quad \beta - \alpha, \quad \mathbf{0}_{l-i})^\top, & i = j, \\ (\mathbf{0}_{j-1}, \quad \beta - 1, \quad -1_{i-j-2}, \quad -\alpha, \quad \mathbf{0}_{l-j})^\top, & i > j. \end{cases} \tag{39}$$

Maximizers $q_k^h(x^m) \in \mathcal{K}_{an}$ of the dual problem (35) are exactly all vectors

$$q_k^h(x^m) = \begin{cases} (\ast\ast\ast, \quad \gamma_{i+1} - \gamma_i, \quad ..., \quad \gamma_{j+1} - \gamma_j, \quad \ast\ast\ast)^\top, & i < j, \\ (\ast\ast\ast, \quad \mathrm{sgn}(\beta - \alpha)(\gamma_{i+1} - \gamma_i), \quad \ast\ast\ast)^\top, & i = j, \\ (\ast\ast\ast, \quad \gamma_j - \gamma_{j+1}, \quad ..., \quad \gamma_i - \gamma_{i+1}, \quad \ast\ast\ast)^\top, & i > j. \end{cases} \tag{40}$$

The elements marked with $\ast$ can be chosen arbitrarily as long as $q_k^h(x^m) \in \mathcal{K}_{an}$. Due to this special form, the transformation (37) leads to $\tilde{q}_k^h(x^m) = \pm\tilde{\gamma}$ depending on the case. Crucially, this transformed vector is another equally valid choice in (40) and therefore (37) returns another valid subgradient $\tilde{p}_k^h = \nabla^\top \tilde{q}_k^h$.

In order to show that $p_k^h = \nabla^\top q_k^h$ for $q_k^h(\cdot) = \pm 1$ is a subgradient in the unlifted setting we use the same arguments. To this end, we use the sublabel-accurate notation with $L = 2$. The "lifted" label space is $\Gamma = [0,1]$, independently of the actual $\Gamma \subset \mathbb{R}$; see [18, Prop. 3]. Then with $u_k^h(x^m) = \gamma_i^\alpha$ and $u_k^h(x^{m+1}) = \gamma_j^\beta$ (corresponding to 1_i^α and 1_j^β from before), applying the forward difference operator $\nabla u_k^h(x^m) = \frac{1}{h}(\gamma_j^\beta - \gamma_i^\alpha)$ shows that dual maximizers are $q_k^h(x^m) = \mathrm{sgn}(\gamma_j^\beta - \gamma_i^\alpha)|\Gamma| = \pm 1$. It can be seen that the algebraic signs coincide pointwise in the lifted and unlifted setting. Thus p_k^h in (38) is of the form $p_k^h = \nabla^\top q_k^h$ and in particular a subgradient in the unlifted setting. □

Convex Energy with Artificial Data. We compare the results of the original and lifted Bregman iteration for the ROF-(1) problem with $\lambda = 20$, synthetic input data and anisotropic TV regularizer. In the lifted setting, we compare implementations with and without transforming the subgradients as in (37). The results shown in Fig. 2 clearly support the theory: Once subgradients are transformed as in Proposition 2, the iterates agree with the classical, unlifted iteration.

A subtle issue concerns points where the minimizer of the lifted energy is non-sublabel-integral, i.e., cannot be easily identified with a solution of the original problem. This impedes the recovery of a suitable subgradient as in (16), which leads to diverging Bregman iterations. We found this issue to occur in particular with isotropic TV discretization, which does not satisfy a discrete version of the coarea formula – which is used to prove in the continuous setting that solutions of the original problem can be recovered by thresholding – but is also visible to a smaller extent around the boundaries of the objects in Fig. 2.

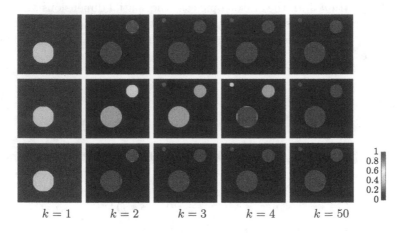

$$k = 1 \qquad k = 2 \qquad k = 3 \qquad k = 4 \qquad k = 50$$

Fig. 2. Equivalency of classical and lifted Bregman on a convex problem. On the *convex* ROF-(1) problem with anisotropic TV, with a naïve implementation, the classical Bregman iteration as in Algorithm 1 **(top row)** and the lifted generalization as in Algorithm 2 **(middle row)** show clear differences. If the lifted subgradients are transformed as in Proposition 2, the lifted iterates **(bottom row)** are visually indistinguishable from the classical iteration. However, the lifted version also allows to transparently handle nonconvex energies (Fig. 3).

Non-convex Stereo Matching with Real-World Data. Let us demonstrate the applicability of the lifted Bregman iteration on a non-convex stereo-matching problem for depth estimation. We use TV-(1) with data term

$$\rho(x, u(x)) = \int_{W(x)} \sum_{d=1,2} h(\partial_{x_d} I_1((y_1, y_2 + u(x))) - \partial_{x_d} I_2((y_1, y_2))), \qquad (41)$$

where $W(x)$ denotes a patch around x and $h(\alpha) := \min\{\alpha, \beta\}$ is a truncation with threshold $\beta > 0$. This data term is non-convex and non-linear in u. We apply the lifted Bregman iteration on three data sets [27] with $L = 5$ labels,

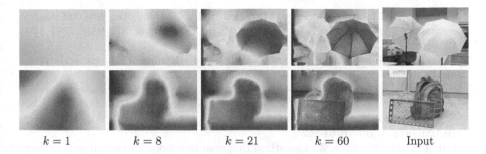

$k = 1$ $k = 8$ $k = 21$ $k = 60$ Input

Fig. 3. Lifted Bregman on stereo matching problem with isotropic TV. TV-(1) problem with data term (41). This problem is non-convex and non-linear in u. At $k = 1$ the solution is a coarse approximation of the depth field. As the iteration advances, details are progressively recovered. Although the problem is not of the form OH-(1) classically associated with the ISS flow, the results show a qualitative similarity to a nonlinear scale space for this non-convex problem.

isotropic TV regularizer and untransformed subgradients. For results see Fig. 1, 3 (Motorbike: $\lambda = 20$, $k = 30$; Umbrella: $\lambda = 10$; Backpack: $\lambda = 25$). We also ran the experiment with an anisotropic TV regularizer as well as transformed subgradients. Overall, the behavior was similar, but transforming the subgradients led to more pronounced jumps. Interestingly, even in this non-convex case the solution of the lifted Bregman iteration also strongly reminds of an ISS flow: The first solution is a smooth estimation; as the iteration continues, finer structures are added. This behavior is also visible in the progression of the profiles in Fig. 1.

Conclusion. We proposed a combination of the Bregman iteration and a lifting approach with sublabel-accurate discretization in order to extend the Bregman iteration to non-convex energies. If a certain form of the subgradients can be ensured – which can be shown to be the case with the convex ROF-(1) problem and anisotropic TV – the iterates agree in theory and in practice. In the future, it will be interesting to see if such methods can lead to the development of scale space transformations and nonlinear filters for arbitrary nonconvex data terms.

Acknowledgments. The authors acknowledge support through DFG grant LE 4064/1-1 "Functional Lifting 2.0: Efficient Convexifications for Imaging and Vision" and NVIDIA Corporation.

References

1. Aubert, G., Kornprobst, P.: Mathematical Problems in Image Processing: Partial Differential Equations and the Calculus of Variations, vol. 147. Springer, New York (2006). https://doi.org/10.1007/978-0-387-44588-5
2. Benning, M., Burger, M.: Ground states and singular vectors of convex variational regularization methods. arXiv preprint arXiv:1211.2057 (2012)

3. Burger, M., Gilboa, G., Moeller, M., Eckardt, L., Cremers, D.: Spectral decompositions using one-homogeneous functionals. SIAM J. Imag. Sci. **9**(3), 1374–1408 (2016)

4. Burger, M., Gilboa, G., Osher, S., Xu, J., et al.: Nonlinear inverse scale space methods. Commun. Math. Sci. **4**(1), 179–212 (2006)

5. Burger, M., Eckardt, L., Gilboa, G., Moeller, M.: Spectral representations of one-homogeneous functionals. In: Aujol, J.-F., Nikolova, M., Papadakis, N. (eds.) SSVM 2015. LNCS, vol. 9087, pp. 16–27. Springer, Cham (2015). https://doi.org/10.1007/978-3-319-18461-6_2

6. Cai, J., Osher, S., Shen, Z.: Linearized Bregman iterations for compressed sensing. Math. Comput. **78**(267), 1515–1536 (2009)

7. Chambolle, A., Cremers, D., Pock, T.: A convex approach for computing minimal partitions (2008)

8. Chan, T.F., Esedoglu, S., Nikolova, M.: Algorithms for finding global minimizers of image segmentation and denoising models. SIAM J. Appl. Math. **66**(5), 1632–1648 (2006)

9. Chan, T.F., Vese, L.A.: Active contours without edges. IEEE Trans. Image Proc. **10**(2), 266–277 (2001)

10. Gilboa, G.: Semi-inner-products for convex functionals and their use in image decomposition. J. Math. Imag. Vis. **57**(1), 26–42 (2017)

11. Gilboa, G.: A spectral approach to total variation. In: Kuijper, A., Bredies, K., Pock, T., Bischof, H. (eds.) SSVM 2013. LNCS, vol. 7893, pp. 36–47. Springer, Heidelberg (2013). https://doi.org/10.1007/978-3-642-38267-3_4

12. Gilboa, G., Moeller, M., Burger, M.: Nonlinear spectral analysis via one-homogeneous functionals: overview and future prospects. J. Math. Imag. Vis. **56**(2), 300–319 (2016)

13. Goldstein, T., Osher, S.: The split Bregman method for L1-regularized problems. SIAM J. Imag. Sci. **2**(2), 323–343 (2009)

14. Hoeltgen, L., Breuß, M.: Bregman iteration for correspondence problems: A study of optical flow. arXiv preprint arXiv:1510.01130 (2015)

15. Ishikawa, H.: Exact optimization for Markov random fields with convex priors. Patt. Anal. Mach. Intell. **25**(10), 1333–1336 (2003)

16. Ishikawa, H., Geiger, D.: Segmentation by grouping junctions. In: CVPR, vol. 98, p. 125. Citeseer (1998)

17. Lellmann, J., Schnörr, C.: Continuous multiclass labeling approaches and algorithms. SIAM J. Imag. Sci. **4**(4), 1049–1096 (2011)

18. Möllenhoff, T., Laude, E., Möller, M., Lellmann, J., Cremers, D.: Sublabel-accurate relaxation of nonconvex energies. CoRR abs/1512.01383 (2015)

19. Mollenhoff, T., Cremers, D.: Sublabel-accurate discretization of nonconvex free-discontinuity problems. In: Proceedings of the IEEE International Conference on Computer Vision, pp. 1183–1191 (2017)

20. Mollenhoff, T., Cremers, D.: Lifting vectorial variational problems: a natural formulation based on geometric measure theory and discrete exterior calculus. In: Proceedings of the IEEE Conference on Computer Vision and Pattern Recognition, pp. 11117–11126 (2019)

21. Osher, S., Burger, M., Goldfarb, D., Xu, J., Yin, W.: An iterative regularization method for total variation-based image restoration. Multiscale Model. Simul. **4**(22), 460–489 (2005)

22. Osher, S., Mao, Y., Dong, B., Yin, W.: Fast linearized Bregman iteration for compressive sensing and sparse denoising. arXiv preprint arXiv:1104.0262 (2011)

23. Pock, T., Cremers, D., Bischof, H., Chambolle, A.: Global solutions of variational models with convex regularization. SIAM J. Imag. Sci. **3**(4), 1122–1145 (2010)
24. Pock, T., Schoenemann, T., Graber, G., Bischof, H., Cremers, D.: A convex formulation of continuous multi-label problems, pp. 792–805 (2008)
25. Rockafellar, R.T.: Convex Analysis, vol. 28. Princeton University Press, Princeton (1970)
26. Rockafellar, R.T., Wets, R.J.: Variational Analysis, vol. 317. Springer, Heidelberg (2009)
27. Scharstein, D., et al.: High-resolution stereo datasets with subpixel-accurate ground truth. In: Jiang, X., Hornegger, J., Koch, R. (eds.) GCPR 2014. LNCS, vol. 8753, pp. 31–42. Springer, Cham (2014). https://doi.org/10.1007/978-3-319-11752-2_3
28. Scherzer, O., Grasmair, M., Grossauer, H., Haltmeier, M., Lenzen, F.: Variational Methods in Imaging, Applied Mathematical Sciences, vol. 167. Springer, New York (2009). https://doi.org/10.1007/978-0-387-69277-7
29. Vogt, T., Haase, R., Bednarski, D., Lellmann, J.: On the connection between dynamical optimal transport and functional lifting. arXiv preprint arXiv:2007.02587 (2020)
30. Zach, C., Gallup, D., Frahm, J.M., Niethammer, M.: Fast global labeling for real-time stereo using multiple plane sweeps. In: Vis. Mod. Vis., pp. 243–252 (2008)

A Scaled and Adaptive FISTA Algorithm for Signal-Dependent Sparse Image Super-Resolution Problems

Marta Lazzaretti[1]([⊠]), Simone Rebegoldi[2], Luca Calatroni[3],
and Claudio Estatico[1]

[1] Dip. di Matematica, Università di Genova, Via Dodecaneso 35, 16146 Genoa, Italy
marta.lazzaretti@edu.unige.it, estatico@dima.unige.it
[2] Dip. di Ingegneria Industriale, Università di Firenze,
Via di S. Marta 3, 50139 Florence, Italy
simone.rebegoldi@unifi.it
[3] CNRS, UCA, Inria, Lab. I3S, 2000 Route des Lucioles,
06903 Sophia-Antipolis, France
calatroni@i3s.unice.fr

Abstract. We propose a scaled adaptive version of the Fast Iterative Soft-Thresholding Algorithm, named S-FISTA, for the efficient solution of convex optimization problems with sparsity-enforcing regularization. S-FISTA couples a non-monotone backtracking procedure with a scaling strategy for the proximal–gradient step, which is particularly effective in situations where signal-dependent noise is present in the data. The proposed algorithm is tested on some image super-resolution problems where a sparsity-promoting regularization term is coupled with a weighted-ℓ_2 data fidelity. Our numerical experiments show that S-FISTA allows for faster convergence in function values with respect to standard FISTA, as well as being an efficient inner solver for iteratively reweighted ℓ_1 algorithms, thus reducing the overall computational times.

Keywords: Inertial forward-backward splitting · Variable metric · Sparse super-resolution · Scaled FISTA · Sparse optimization

1 Introduction

Among the plethora of first-order convex optimization schemes proposed over the last decades to solve image reconstruction problems (see [6] for a review), forward-backward (FB) algorithms have attracted a significant attention due to their easy applicability in many scenarios and their fast convergence properties whenever endowed with appropriate inertial updates, as in the case of the celebrated FISTA algorithm [1]. Their successful application relies, in practice, on the accurate estimation of the 'steepness' of the smooth component of the functional to minimize; whenever such estimate is coarse, the practical effectiveness of FB methods may be limited. One remedy to solve this issue consists in the

A. Elmoataz et al. (Eds.): SSVM 2021, LNCS 12679, pp. 242–253, 2021.
https://doi.org/10.1007/978-3-030-75549-2_20

adoption of adaptive backtracking procedures, see e.g. [5,16]. Moreover, as noted in several works in the literature [3,7,10], FB algorithms can also benefit from suitable scaling approaches aimed at capturing some second-order information of the smooth component. In particular, the use of scaling approaches has been shown to render particularly effective in the context of signal-dependent image reconstruction problems in a variety of applications ranging from biological to astronomical imaging, see e.g., [2]. Recently, in [14] the authors proposed SAGE-FISTA, a general FISTA-type algorithm designed for (possibly strongly) convex composite problems where a suitable scaling strategy is combined with an adaptive backtracking technique. Upon suitable conditions on the scaling updates, accelerated convergence rates are rigorously proved for SAGE-FISTA.

In this work we consider a particular instance of the SAGE-FISTA algorithm (denominated S-FISTA) for efficiently solving sparse reconstruction problems arising in the field of image super-resolution with signal-dependent Poisson noise. The use of composite functionals promoting signal sparsity by means of both convex and non-convex regularizers has been extensively explored in recent years, see, e.g. [18]. However, as preliminary results showed in [11], when such models are used in correspondence with data terms explicitly incorporating signal-dependence, the numerical efficiency of standard FISTA-type algorithms may suffer due to the use of excessively small step-sizes. By contrast, our numerical experience shows that S-FISTA significantly improves computational convergence when used both as global solver in convex optimization regimes and as inner solver for iterative non-convex optimization schemes such as iteratively reweighted ℓ_1 algorithm [13]. The main contribution of this work is the use of a tailored efficient algorithm for improving overall computational efficiency.

2   Scaled and Adaptive FISTA

We consider the following convex composite problem

$$\min_{\mathbf{x} \in \mathbb{R}^N} F(\mathbf{x}) := f(\mathbf{x}) + g(\mathbf{x}), \tag{1}$$

where $f : Y \to \mathbb{R}$ is a convex, continuously differentiable function with $\mathcal{L}$–Lipschitz continuous gradient on a closed, convex set $Y \subseteq \mathbb{R}^N$ and $g : \mathbb{R}^N \to \bar{\mathbb{R}}$ is convex and possibly non-smooth.

Forward–backward (FB) algorithms are classical optimization tools for solving (1), whose main iteration consists in alternating a gradient step on the differentiable part f with a proximal step on the term g [8]. The following sequence of iterates is thus generated:

$$\begin{aligned}
\mathbf{x}^{k+1} &= \text{prox}_{\tau_{k+1} g}(\mathbf{x}^k - \tau_{k+1}\nabla f(\mathbf{x}^k)) \\
&= \underset{\mathbf{z} \in \mathbb{R}^N}{\text{argmin}}\, g(\mathbf{z}) + \frac{1}{2\tau_{k+1}}\|\mathbf{z} - \mathbf{x}^k + \tau_{k+1}\nabla f(\mathbf{x}^k)\|^2, \quad k \geq 0, \tag{2}
\end{aligned}$$

where $\tau_{k+1} > 0$ is the step-size parameter. The standard scheme (2) is usually accelerated by introducing an inertial term in the proximal–gradient step, as in the popular Fast Iterative Soft-Thresholding Algorithm (FISTA) [1], where each iterate is obtained by extrapolating the two previous iterates as follows:

$$\mathbf{y}^{k+1} = \mathbf{x}^k + \left(\frac{t_k - 1}{t_{k+1}}\right)(\mathbf{x}^k - \mathbf{x}^{k-1}), \quad \mathbf{x}^{k+1} = \text{prox}_{\tau_{k+1}g}(\mathbf{y}^{k+1} - \tau_{k+1}\nabla f(\mathbf{y}^{k+1})),$$

where $t_0 = 1$ and $t_{k+1} = \frac{1+\sqrt{1+4t_k^2}}{2}$. FISTA enjoys a favorable $\mathcal{O}(1/k^2)$ convergence rate, which is optimal for first-order methods, see [1,12]. However, it is also well-known that the practical performance of FISTA may not be as satisfactory as its theoretical convergence rate suggests. This is mainly due to the restricted range of admissible step-sizes τ_{k+1} discussed in [1] to ensure convergence, which requires computing τ_{k+1} either as $1/\mathcal{L}$ for all $k \geq 0$ or using a backtracking procedure that monotonically reduces τ_{k+1} until an estimate of $\mathcal{L}$ is found. Such an approach may negatively affect the practical convergence rate of FISTA, especially when only a large (pessimistic) estimate of $\mathcal{L}$ is known.

In [14] the authors propose the so-called Scaled Adaptive GEneralized FISTA (SAGE-FISTA), a novel variant of FISTA aimed at improving its practical convergence rate. SAGE-FISTA employs an adaptive backtracking strategy allowing the adaptive (increasing/decreasing) adjustment of τ_{k+1} at each iteration, which comes in handy when the initial τ_0 is extremely small due to a large estimate of $\mathcal{L}$. The adaptive backtracking is further combined with a symmetric and positive definite scaling matrix $\mathbf{D}_{k+1}$ in the proximal–gradient step, by which some second-order information of the function f around the current iterate $\mathbf{x}^k$ may be incorporated while keeping the iteration computationally inexpensive, e.g. adopting Majorization-Minimization techniques or split-gradient strategies [4,7]. The algorithm in [14] also takes into account the inexact computation of the proximal

---

Algorithm 1. S-FISTA($\mathbf{x}^0$,τ_0,f,g)

Parameters: $\rho \in (0,1)$, $\delta \in (0,1]$, $\{\eta_{inf}^k\}_k$, $\{\eta_{sup}^k\}_k$ s.t. $0 < \eta_{inf} \leq \eta_{inf}^k \leq \eta_{sup}^k \leq \eta_{sup}$.

Initialization: $\mathbf{x}^{-1} = \mathbf{x}^0$, $t_0 \geq 1$, $\mathbf{D}_0 \in \mathcal{D}_{\eta_{inf}^0}^{\eta_{sup}^0}$.

FOR $k = 0, 1, \ldots$ REPEAT

Choose $\mathbf{D}_{k+1} \in \mathcal{D}_{\eta_{inf}^k}^{\eta_{sup}^k}$ and set $\tau_{k+1}^0 = \frac{\tau_k}{\delta}$.

FOR $i = 0, 1, \ldots$ REPEAT

STEP 1. Set $\tau_{k+1} = \rho^i \tau_{k+1}^0$.

STEP 2. Set $t_{k+1} = \frac{1+\sqrt{1+4\frac{\tau_k}{\tau_{k+1}}t_k^2}}{2}$.

STEP 3. Set $\mathbf{y}^{k+1} = \text{proj}_{Y,\mathbf{D}_{k+1}}\left(\mathbf{x}^k + \left(\frac{t_k-1}{t_{k+1}}\right)(\mathbf{x}^k - \mathbf{x}^{k-1})\right)$.

STEP 4. Compute the next iterate as

$$\mathbf{x}^{k+1} = \text{prox}_{\tau_{k+1}g}^{\mathbf{D}_{k+1}}(\mathbf{y}^{k+1} - \tau_{k+1}\mathbf{D}_{k+1}^{-1}\nabla f(\mathbf{y}^{k+1})).$$

UNTIL $f(\mathbf{x}^{k+1}) \leq f(\mathbf{y}^{k+1}) + \nabla f(\mathbf{y}^k)^T(\mathbf{x}^{k+1} - \mathbf{y}^{k+1}) + \frac{1}{2\tau_{k+1}}\|\mathbf{x}^{k+1} - \mathbf{y}^{k+1}\|_{\mathbf{D}_{k+1}}^2$

UNTIL stopping criterion

---

operator and the possible introduction of the strong convexity moduli of either f or g. Here, we consider a simplified, convex version, denominated Scaled FISTA (S-FISTA), where the proximal operator of g is assumed to be computable in closed form. We report S-FISTA in Algorithm 1.

Let $\mathcal{SP}(\mathbb{R}^{N \times N})$ denote the set of $N \times N$ symmetric, positive definite real matrices and $\mathcal{D}_{\eta_{inf}^k}^{\eta_{sup}^k} = \{\mathbf{D} \in \mathcal{SP}(\mathbb{R}^{N \times N}) : \eta_{inf}^k \|x\|^2 \le \|x\|_{\mathbf{D}}^2 \le \eta_{sup}^k \|x\|^2, \ \forall \ x\}$, where $\|x\|_{\mathbf{D}} = \sqrt{x^T \mathbf{D} x}$ denotes the norm induced by the matrix $\mathbf{D}$.

For all iterations $k \ge 0$, we first select a scaling matrix $\mathbf{D}_{k+1} \in \mathcal{D}_{\eta_{inf}^k}^{\eta_{sup}^k}$ and a tentative step-size $\tau_{k+1}^0 = \tau_k/\delta$. Note that if $\delta < 1$, this corresponds to an increasing of τ_k, whereas when $\delta = 1$, nothing happens at this step. Then, the computation of $\mathbf{x}^{k+1}$ is performed by means of an inner backtracking procedure depending on the scaling matrix $\mathbf{D}_{k+1}$. At each inner backtracking iteration $i \ge 0$, we set the parameter t_{k+1} and the projected inertial point $\mathbf{y}^{k+1}$, where $\text{proj}_{Y, \mathbf{D}_{k+1}}(\mathbf{x}) = \text{argmin}_{\mathbf{z} \in Y} \|\mathbf{z} - \mathbf{x}\|_{\mathbf{D}_{k+1}}^2$ denotes the projection operator onto Y w.r.t. the norm induced by $\mathbf{D}_{k+1}$. The proximal–gradient point $\mathbf{x}^{k+1}$ is then computed by pre-multiplying $\nabla f(\mathbf{y}^{k+1})$ with $\mathbf{D}_{k+1}^{-1}$ and performing a scaled proximal step, where $\text{prox}_{\tau_{k+1}g}^{\mathbf{D}_{k+1}}(\mathbf{x}) = \text{argmin}_{\mathbf{z} \in \mathbb{R}^N} \ g(\mathbf{z}) + \frac{1}{2}\|\mathbf{z} - \mathbf{x}\|_{\mathbf{D}_{k+1}}^2$ is the proximal operator in the norm induced by $\mathbf{D}_{k+1}$. Finally, a check is made on a sufficient decrease condition based on a scaled version of the classical descent lemma [1, Lemma 2.1]; if such a condition is not met, the step-size is reduced by a factor $\rho \in (0, 1)$ and the backtracking steps are performed again until the condition holds. Note that this procedure ends in a finite number of steps, since the decrease condition is verified for sufficiently small step-sizes [14, Lemma 2.2].

If the bounds $\eta_{inf}^k, \eta_{sup}^k$ converge sufficiently fast to the same constant value, then it is possible to prove an optimal $\mathcal{O}(1/k^2)$ convergence rate for S-FISTA.

Theorem 1. [14, Corollary 3.3] Let $\mathbf{x}^*$ be the solution of problem (1) and suppose the parameters $\eta_{inf}^k, \eta_{sup}^k$ are computed as

$$\eta_{inf}^k = \eta - \nu_{inf}^k, \quad \eta_{sup}^k = \eta + \nu_{sup}^k, \tag{3}$$

where $\nu_{inf}^k, \nu_{sup}^k \ge 0$ and $\sum_{k=0}^{\infty} \nu_{inf}^k < \infty$, $\sum_{k=0}^{\infty} \nu_{sup}^k < \infty$. Then, for all $k \ge 0$

$$F(\mathbf{x}^{k+1}) - F(\mathbf{x}^*) \le \frac{C}{(k + 1 + t_0)^2}, \tag{4}$$

where $C = C(\mathbf{x}^*, \mathbf{x}^0, \tau_0, \mathbf{D}_0, t_0, \rho, \eta_{inf}, \eta_{sup}) > 0$ is a constant depending on $\mathbf{x}^*$ and the parameters of Algorithm 1.

We refer the reader to Sect. 4 for a practical implementation of (3).

S-FISTA has already proven to be effective on several imaging problems (see [14, Section 4]) thanks to the combination of the scaling procedure with the adaptive backtracking, which allows to select bigger step-sizes, thus improving the practical convergence speed towards the solution of the problem. We now show the effectiveness of S-FISTA on image super-resolution problems.

3   Application to Sparse Weighted Models for Super-Resolution Microscopy

In several works (see, e.g., [3,10]) it has been shown that variable metric first-order methods perform particularly well in image reconstruction/restoration problems where data are corrupted by signal-dependent Poisson noise, as it is common, for instance, in the field of image microscopy. In the following, we focus on the problem of image super-resolution.

3.1   Sparse Image Super-Resolution with Poisson Data

The goal of super-resolution fluorescence image microscopy is to estimate molecules' intensities and positions from blurred and noisy acquisitions of sparse samples of molecules. Namely, for a given Low Resolution (LR), blurred and noisy (vectorized) image $\mathbf{y} \in \mathbb{R}_{>0}^{M}$, the ill-posed inverse problem of reconstructing a Super-Resolved image $\mathbf{x} \in \mathbb{R}_{\geq 0}^{N}$, where $N = L^2 M$ and $L > 0$ denotes the Super-Resolution (SR) factor, consists in solving:

$$\text{find} \quad \mathbf{x} \quad \text{s.t.} \quad \mathbf{y} = \mathcal{P}(\mathbf{R_L H x}), \tag{5}$$

where $\mathbf{H} \in \mathbb{R}^{N \times N}$ is the blurring operator corresponding to the microscope PSF $\mathbf{h}$, $\mathbf{R_L} \in \mathbb{R}^{M \times N}$ is the under-sampling operator and $\mathcal{P}(\mathbf{z})$ denotes the realization of a Poisson-distributed random vector with mean values $(\mathbf{z})_i \geq 0$, for $i = 1, \ldots, M$ which are assumed to be strictly positive for simplicity. To simplify notations, we will denote in the following by $\mathbf{A} := \mathbf{R_L H} \in \mathbb{R}^{M \times N}$ the whole forward operator. Following a standard Maximum A Posteriori estimation yields to having the Kullback-Leibler (KL) divergence $D_{KL}(\mathbf{Ax}; \mathbf{y}) := \sum_{j=1}^{M} (\mathbf{Ax})_j - y_j \log(\mathbf{Ax})_j$ as data fidelity term. Several approximations of D_{KL} have been considered in the literature (see [15]); in the following, we consider the second order Taylor expansion around $\mathbf{y}$ which results in a weighted-ℓ_2 fidelity term:

$$\frac{1}{2} \sum_{i=1}^{M} \frac{((\mathbf{Ax})_i - y_i)^2}{y_i} = \frac{1}{2} \langle \mathbf{Ax} - \mathbf{y}, \mathbf{W}(\mathbf{Ax} - \mathbf{y}) \rangle =: \frac{1}{2} \|\mathbf{Ax} - \mathbf{y}\|_{\mathbf{W}}^2, \tag{6}$$

where the weighted norm is defined in terms of the diagonal positive definite matrix $\mathbf{W} := \text{diag}(1/\mathbf{y}) \in \mathbb{R}^{M \times M}$ where division is intended element-wise. Upon this choice and consistently with the signal-dependent modeling considered, the weighted-ℓ_2 norm measures the relative discrepancy between the underlying noiseless signal $\mathbf{Ax}$ and the noisy signal $\mathbf{y}$ component-wise, so as to have a low data-fidelity whenever intense signal (hence high noise) is measured.

The natural choice for promoting sparsity in the desired signal consists in considering the ℓ_0 pseudo-norm or its continuous relaxations, see, e.g. [18]. Coupling such regularization with the fidelity term (6), we can thus consider the following minimization problem:

$$\hat{\mathbf{x}} \in \arg\min_{\mathbf{x} \in \mathbb{R}^N} G_{w\ell_0}(\mathbf{x}) \equiv \frac{1}{2} \|\mathbf{Ax} - \mathbf{y}\|_{\mathbf{W}}^2 + \lambda \|\mathbf{x}\|_0 + i_{\geq 0}(\mathbf{x}), \tag{7}$$

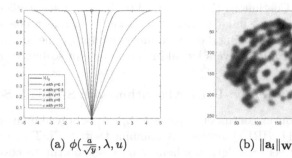

(a) $\phi(\frac{a}{\sqrt{y}}, \lambda, u)$ (b) $\|\mathbf{a_i}\|_\mathbf{W}$

Fig. 1. Left: Dependence of the penalty in (9) on the data y for fixed $\lambda = 1$ and $a = 0.7$. Right: plot of parameters used for the test image shown in Fig. 2(b).

where $\lambda > 0$ is the regularization parameter and $i_{\geq 0}(\cdot)$ denotes the indicator function of $Y := \{x \in \mathbb{R}^N \mid x_i \geq 0 \; \forall i = 1, \ldots, N\}$ which forces the desired solution $\mathbf{x}$ to be non-negative. Note that the functional $G_{w\ell_0}$ is not continuous nor convex, due to the presence of the ℓ_0 pseudo-norm which further makes the problem NP-hard. A common strategy to mitigate such numerical difficulties is to consider a continuous relaxation of the ℓ_0 pseudo-norm, in order to make use of standard optimization algorithms, such as, for instance, the Iterative Reweighted ℓ_1 (IRL1) algorithm [13]. We consider here the weighted-Continuous Exact ℓ_0 (wCEL0) relaxation recently proposed in [11]. Inspired by [17], the relaxed model computed by double Fenchel biconjugation applied to $G_{w\ell_0}$ reads:

$$\hat{\mathbf{x}} \in \underset{\mathbf{x} \subset \mathbb{R}^N}{\arg \min} \; G_{\text{wCEL0}}(\mathbf{x}) \equiv \frac{1}{2}\|\mathbf{Ax} - \mathbf{y}\|^2_\mathbf{W} + \Phi_{\text{wCEL0}}(\mathbf{x}; \lambda, \mathbf{A}, \mathbf{W}) + \iota_{\geq 0}(\mathbf{x}), \quad (8)$$

where the continuous and non-convex functional $\Phi_{\text{wCEL0}}(\cdot; \lambda, \mathbf{A}, \mathbf{W})$ is defined by

$$\Phi_{\text{wCEL0}}(\mathbf{x}; \lambda, \mathbf{A}, \mathbf{W}) = \sum_{i=1}^N \phi(\|\mathbf{a}_i\|_\mathbf{W}, \lambda, |x_i|), \quad (9)$$

for $\lambda > 0$, with $\mathbf{a_i} \in \mathbb{R}^M$ being i-th column of the matrix $\mathbf{A}$ for $i = 1, \ldots, N$ and

$$\phi(\bar{a}, \lambda, u) = \lambda - \frac{\bar{a}^2}{2}\left(u - \frac{\sqrt{2\lambda}}{\bar{a}}\right)^2 \mathbb{1}_{\{u < \frac{\sqrt{2\lambda}}{\bar{a}}\}}. \quad (10)$$

We remark that (9) depends both on the forward model and on the observed data and it is a sum of N one-dimensional functions. The expression of (10) is obtained by a direct analytical computation of the convex envelope of the corresponding 1D restriction of the $G_{w\ell_0}$ functional (7) and $\bar{a}$ encodes the data-dependence of the 1D penalty as $\bar{a} = \frac{a}{\sqrt{y}}$, with $a > 0$ being the 1D forward model and $y > 0$ the observed data with background. In Φ_{wCEL0}, the dependence on the data is reflected in the weighted column norms $\bar{a} = \|\mathbf{a}_i\|_\mathbf{W}$, which depend on the data $\mathbf{y}$, as shown in Fig. 1.

Compared to the Continuous Exact ℓ_0 (CEL0) Φ_{CEL0} penalty studied in [17], the Φ_{wCEL0} penalty is more suited to deal with signal-dependent noise models. Moreover, the data term (6) offers the opportunity to improve the computational performance of optimization solvers by taking advantage of scaling techniques.

3.2   Solving (8) Efficiently: IRL1 Algorithm with S-FISTA Solver

The composite non-convex continuous problem (8) can be solved by means of an iteratively reweighted ℓ_1 (IRL1) scheme [13] combined with S-FISTA Algorithm 1 as inner solver. Iteratively reweigthed schemes are based on the successive minimization of a sequence of convex weighted ℓ_1 subproblems, each obtained by appropriately linearizing the (possibly non-convex) penalty term. In the case of (8), following [13, Algorithm 3], one can replace each term $\phi(\|\mathbf{a}_i\|_{\mathbf{W}}, \lambda, |x_i|)$ in (9) with the linear function $x_i \mapsto \omega_i^{\mathbf{x}^k}|x_i|$, where

$$\omega_i^{\mathbf{x}^k} = \left(\sqrt{2\lambda} - \|\mathbf{a}_i\|_{\mathbf{W}}^2 |x_i^k|\right) \mathbb{1}_{\{|x_i^k| < \frac{\sqrt{2\lambda}}{\|\mathbf{a}_i\|_{\mathbf{W}}}\}} \in \partial_{x_i} \phi(\|\mathbf{a}_i\|_{\mathbf{W}}, \lambda, |x_i|^k)$$

is the generalized derivative of $\phi(\|\mathbf{a}_i\|_{\mathbf{W}}, \lambda, \cdot)$ at point $|x_i|^k$. This generates the following sequence of convex subproblems:

$$\mathbf{x}^{k+1} = \arg\min_{\mathbf{x} \in \mathbb{R}^N} \frac{1}{2}\|\mathbf{A}\mathbf{x} - \mathbf{y}\|_{\mathbf{W}}^2 + \lambda \sum_{i=1}^{N} \omega_i^{\mathbf{x}^k}|x_i| + \iota_{\geq 0}(\mathbf{x}), \quad k \geq 0, \qquad (11)$$

which do not usually admit a closed-form solution and hence require to be solved approximately via an iterative optimization solver able to deal with weighted ℓ_1 terms, such as ISTA or FISTA. As highlighted in Sect. 2, these solvers may be extremely inefficient when only a pessimistic estimate of the Lipschitz constant $\mathcal{L}$ of the gradient is at our disposal. In particular, this is true for problem (11) for which, by denoting by $\mathcal{F}(\cdot)$ the 2D Fourier transform operator, we get

$$\mathcal{L} = \|\mathbf{A}^T\mathbf{W}\mathbf{A}\|_2 \leq \|\mathbf{A}\|^2\|\mathbf{W}\| = L^2 \max(\mathcal{F}(\mathbf{h}))^2 \min(\mathbf{y})^{-1}, \qquad (12)$$

which may be extremely large due to the possible values close to zero assumed by $\mathbf{y}$. Such a large estimate leads to a sequence a small step-sizes slowing down the convergence speed of the inner solver, which in turn worsen the progress of the outer IRL1 scheme towards a stationary point of (8).

In order to solve problem (8) more efficiently, we propose to employ the S-FISTA Algorithm 1 for computing inexactly the IRL1 iterate (11). The resulting scheme is reported in Algorithm 2.

Note that a warm-start strategy is adopted inside Algorithm 2, as the previous outer iterate $\mathbf{x}^k$ and step-size τ_k are used as initial guesses for the S-FISTA inner solver. We also remark that the convergence of the IRL1 scheme to a stationary point of (8) has been proved so far only when the subproblems (11) are solved exactly (see [13,17]), whereas the convergence of Algorithm 2 still needs to be addressed. Nonetheless, the numerical performance of Algorithm 2 is promising, as we show in the next section.

4   Numerical Results

In this section we report two tests confirming the effectiveness of S-FISTA when used both in convex optimization regimes as a global solver (Algorithm 1) and in non-convex scenarios as an efficient inner routine for the IRL1 in Algorithm 2. In both cases, the weighted-ℓ_2 data term (6) is used as the smooth component f in the splitting (1). Concerning the choice of the scaling matrices $\{\mathbf{D}_k\}_k$, we consider the split-gradient strategy proposed in [10] where the decomposition $-\nabla f(\mathbf{x}) = U(\mathbf{x}) - V(\mathbf{x})$ with $U(\mathbf{x}) \geq 0$ and $V(\mathbf{x}) > 0$ and the choice $\mathbf{D}_k = \text{diag}\left(\mathbf{y}^{(k)}/V(\mathbf{y}^{(k)})\right)^{-1}$ is made. In order to ensure condition (3), we further introduce an appropriate sequence of thresholding parameters $\{\gamma_k\}_k$ in the definition of $\mathbf{D}_k$, thus obtaining

---

Algorithm 2. Inexact IRL1 algorithm for Weighted CEL0

Input: $\mathbf{y} \in \mathbb{R}^M$, $\mathbf{x}^0 \in \mathbb{R}^N$, $\tau_0 > 0$, $\lambda > 0$
FOR $k = 0, 1, \dots$ REPEAT

    STEP 1. Update the weights $\omega_i^{\mathbf{x}^k} \in \partial_{x_i} \phi(\|a_i\|_{\mathbf{W}}, \lambda, |x_i|^k)$.
    STEP 2. Set $f(\mathbf{x}) = \frac{1}{2}\|\mathbf{Ax} - \mathbf{y}\|_{\mathbf{W}}^2$, $g^k(\mathbf{x}) := \lambda \sum_{i=1}^{N} \omega_i^{\mathbf{x}^k} |x_i| + \iota_{\geq 0}(\mathbf{x})$ and compute

$$\mathbf{x}^{k+1} = \text{S-FISTA}(\mathbf{x}^k, \tau_k, f, g^k).$$

UNTIL stopping criterion
Output: x

$$\mathbf{D}_k = \text{diag}\left(\max\left(\frac{1}{\gamma_k}, \min\left(\gamma_k, \frac{\mathbf{y}^{(k)}}{V(\mathbf{y}^{(k)})}\right)\right)\right)^{-1}, \quad \gamma_k = \sqrt{1 + \frac{s_1}{(k+1)^{s_2}}}, \quad (13)$$

where $s_1 > 0$, $s_2 > 0$. Note that when $s_1 = 0$, then $\mathbf{D}_k \equiv \mathbf{I}$, hence the standard Euclidean metric is recovered. In general, large values of s_1 promote the use of the variable metric in the early iterations, while the behavior for $k \gg 1$ is driven by s_2 which determines the speed of convergence of $\mathbf{D}_k$ towards $\mathbf{I}$. For the term (6), simple calculations show that $V(\mathbf{y}^{(k)}) = \tilde{\mathbf{A}}^T \tilde{\mathbf{A}} \mathbf{y}^{(k)}$, with $\tilde{\mathbf{A}} := \sqrt{\mathbf{W}} \mathbf{A}$.

4.1   Weighted-ℓ_2-ℓ_1 Minimization

As a preliminary test, we consider the S-FISTA Algorithm 1 endowed with both Armijo ($\delta = 1$) and adaptive ($\delta < 1$) backtracking to solve the weighted-$\ell_2 - \ell_1$ convex optimization problem:

$$\hat{\mathbf{x}} \in \arg\min_{\mathbf{x} \in \mathbb{R}^N} F(\mathbf{x}) \equiv \frac{1}{2}\|\mathbf{Ax} - \mathbf{y}\|_{\mathbf{W}}^2 + \lambda\|\mathbf{x}\|_1 + \iota_{\geq 0}(\mathbf{x}). \quad (14)$$

For this test we considered fluorescent microscopy data $\mathbf{y} \in \mathbb{R}_{>0}^M$ with $M = 128^2$ simulated by generating a random density of points uniformly distributed over the image domain Ω, then corrupted by Gaussian blur with standard deviation $\sigma_h = 150$ nm and Poisson noise by using the inbuilt `imnoise` MATLAB routine, see Fig. 2(a). As SR factor, we set $L = 4$.

S-FISTA Algorithm 1 is applied to solve problem (14) with the following parameters: $\lambda = 0.85$, $x^0 = A^T y$ maximum number of outer iterations `maxiter` $= 500$, maximum number of backtracking iterations `max_bt` $= 10$, $\rho = 0.8$ (decreasing backtracking factor), $L_0 = 1/\tau_0 = 100$ and $t_0 = 1.01$. In order to compare the monotone backtracking strategy with the adaptive one, we then choose $\delta = 0.99$ and $\delta = 1$, respectively. Note that STEP 4 can be here computed in closed-form by rescaled soft-thresholding.

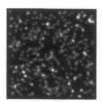 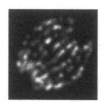

(a) `density`   (b) ISBI 2013, single frame (c) ISBI 2013, $\bar{y}$ (d) IRL1+S-FISTA adapt.

Fig. 2. Test images used for experiments and results obtained by applying the IRL1+S-FISTA Algorithm 1 on the whole ISBI 2013 image dataset.

In Figs. 3 we compare the performance of standard (i.e. non-scaled) FISTA with its scaled variant S-FISTA used for solving (14) for different choices of the scaling matrices $\{D_k\}_k$ in (13) and for monotone (first row) and adaptive (second row) backtracking. To do so, we compute the relative error $(F(x^k) - F(x^*))/F(x^*)$ (where x^* has been pre-computed) calculated both along iterations (first column) and for a maximum run of 15 s (second column) of CPU time. In the third column we report the values of the estimated Lipschitz constant L_k along iterations. We observe that a *sage* combination of scaling (for selected values of s_1 and s_2 in (13)) and adaptive backtracking significantly improves convergence speed and computational performance.

4.2   Weighted-CEL0 Minimization

We now test the effectiveness of S-FISTA when used as inner solver for computing solutions of the non-convex sparse optimization problem (8) via the inexact IRL1 Algorithm 2. For the following experiments, we consider the ISBI SMLM 2013 data, composed of $K = 361$ images representing 8 tubes of 30 nm diameter, see Fig. 2(b) for an exemplar frame and Fig. 2(c) for the average (wide-field) image $\bar{y} = \frac{1}{K} \sum_{i=1}^K y_i$ where each y_i, $i = 1,\ldots,K$ is a LR image $y_i \in \mathbb{R}_{>0}^M$ with $M = 64^2$ and SR factor $L = 4$. The Gaussian PSF h is fully-defined in terms of its Full Width at Half Maximum which for these simulations is FWHM = 258.2 nm, and the noise is signal-dependent, so the choice of a Poisson-type fidelity appears natural, see [11].

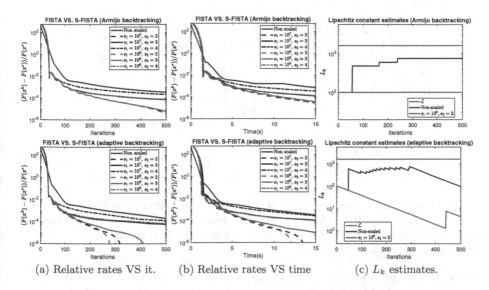

(a) Relative rates VS it.   (b) Relative rates VS time   (c) L_k estimates.

Fig. 3. FISTA VS. S-FISTA with Armijo (top row) and adaptive (bottom row) backtracking with $L_0 = 100$ for different choices of s_1 and s_2 for problem (14) on density image.

Due to the large number of data in the image stack, the use of fast optimization solvers is crucial in these applications to compute solutions of (8) in limited time. Note that for CEL0, in [9], a standard FISTA routine with fixed time-step $\tau = 1/\mathcal{L}_{\text{CEL0}}$ with $\mathcal{L}_{\text{CEL0}} = L^2 \max(\mathcal{F}(\mathbf{h}))^2$ was used. However, in the case of weighted-ℓ_2 fidelity, the estimation (12) may possibly lead to very pessimistic estimates. As a consequence, the convergence speed of the inner FISTA routine and of the outer IRL1 algorithm may suffer. Motivated by the previous example in Sect. 4.1, we thus tested Algorithm 2 endowed with standard (non-scaled) FISTA and S-FISTA (for selected parameters scaling parameters $s_1 = 10^9, s_2 = 2$ in (13)) as inner routines and compare their computational efficiency when used to solve (8) on the ISBI 2013 test image in Fig. 2b.

The parameters are set as follows: $\lambda = 6.25$, $\mathbf{x}^0 = \mathbf{A}^T \mathbf{y}$, maximum number of IRL1 iterations maxIt_IRL1 $= 15$, maximum number of inner S-FISTA iterations maxIt_SFISTA $= 1000$, $L_0 = 1/\tau_0 = 0.01$, maximum number of backtracking iterations maxIt_bt $= 10$, $\rho = 0.9$, $\delta \in \{0.99, 1\}$ and $t_0 = 1.01$. As a stopping criterion for the inner S-FISTA solver, used a control on the relative cost (11) for between two successive inner iterates with tolerance of tol $= 10^{-7}$.

Our computational results are reported in Fig. 4. In particular, Fig. 4(a) shows that differently from non-scaled and non-adaptive FISTA which in the early iterations of IRL1 requires a large number of inner iterations to reach convergence, the use of scaled and adaptive algorithms significantly reduces this drawback. The same behavior is clearly shown in Fig. 4(b) where the decreasing of the non-convex G_{wCELO} cost is shown as a function of the computational time. In Fig. 2(d) the reconstruction obtained with the scaled and adaptive inner routine is shown and in Table 1 a comparison of the quantitative assessment of

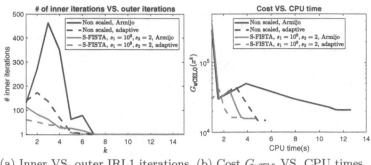

(a) Inner VS. outer IRL1 iterations   (b) Cost G_{wCELO} VS. CPU times

Fig. 4. Computational improvements of using IRL1 Algorithm 2 combined with non-scaled FISTA and S-FISTA Algorithm 1. Left: number of inner iterations for convergence of the inner routine VS. outer iteration number $k \geq 1$. Right: decrease of cost functional G_{wCELO} along CPU times.

Table 1. Jaccard index $\mathbf{J}_\beta := \frac{\text{CD}}{\text{CD+FN+FP}} \in [0,1]$, with CD correct detections, FN false negatives and FP false positives up to tolerance radius of $\beta \in \mathbb{N}$ pixels. Values obtained as average over the frames.

	J_0	J_2	J_4		J_0	J_2	J_4
FISTA Armijo	0.0379	0.3108	0.3515	S-FISTA Armijo	0.0264	0.3401	0.4811
FISTA adapt.	0.0344	0.3071	0.3442	S-FISTA adapt.	0.0233	0.2901	0.3785

localisation precision in terms of Jaccard index is reported. We observe that the use of adaptive S-FISTA favors a much faster decreasing of the cost without undermining the localisation precision, making its use amenable for the time-consuming processing of the entire image stack.

5   Conclusions

We considered a scaled and adaptive variant of the well-known FISTA algorithm for the efficient numerical solution of composite convex optimization problems. The combination of a variable metric scaling with the adaptive adjustment of the algorithmic step-size has recently been shown in [14] to favor faster convergence in function values for problems with signal-dependent (Poisson-type) data terms. We applied the proposed S-FISTA to two image super-resolution problems where sparsity is promoted either via standard ℓ_1 penalty or via an exact continuous relaxation of the ℓ_0 norm (see [11,17]). Our results show that S-FISTA significantly improves upon the computational performance of standard FISTA when used both in convex scenarios and as inner solver for non-convex iteratively reweighted ℓ_1 schemes.

Acknowledgments. LC and SR acknowledge the support of the IEA CNRS grant VaMOS. LC acknowledges the support of the EU H2020 RISE project NoMADS, GA

777826. All authors acknowledge the support of the Italian INdAM group on scientific calculus GNCS.

References

1. Beck, A., Teboulle, M.: A fast iterative shrinkage-thresholding algorithm for linear inverse problems. SIAM J. Imaging Sci. **2**, 183–202 (2009)
2. Bertero, M., Boccacci, P., Desiderà, G., Vicidomini, G.: Image deblurring with Poisson data: from cells to galaxies. Inverse Problems **25**(12) (2009)
3. Bonettini, S., Porta, F., Prato, M., Rebegoldi, S., Ruggiero, V., Zanni, L.: Recent advances in variable metric first-order methods. In: Donatelli, M., Serra-Capizzano, S. (eds.) Computational Methods for Inverse Problems in Imaging. SIS, vol. 36, pp. 1–31. Springer, Cham (2019). https://doi.org/10.1007/978-3-030-32882-5_1
4. Bonettini, S., Rebegoldi, S., Ruggiero, V.: Inertial variable metric techniques for the inexact forward-backward algorithm. SIAM J. Sci. Comput. **40**(5), A3180–A3210 (2018)
5. Calatroni, L., Chambolle, A.: Backtracking strategies for accelerated descent methods with smooth composite objectives. SIAM J. Optim. **29**(3), 1772–1798 (2019)
6. Chambolle, A., Pock, T.: An introduction to continuous optimization for imaging. Acta Numer. **25**, 161–319 (2016)
7. Chouzenoux, E., Pesquet, J.C., Repetti, A.: Variable metric forward-backward algorithm for minimizing the sum of a differentiable function and a convex function. J. Optim. Theory Appl. **162**, 107–132 (2014)
8. Combettes, P., Wajs, V.R.: Signal recovery by proximal forward-backward splitting. Multiscale Model. Sim. **4**(4), 1168–1200 (2005)
9. Gazagnes, S., Soubies, E., Blanc-Féraud, L.: High density molecule localization for super-resolution microscopy using CEL0 based sparse approximation. In: IEEE ISBI 2017 (2017)
10. Lantéri, H., Roche, M., Cuevas, O., Aime, C.: A general method to devise maximum likelihood signal restoration multiplicative algorithms with non-negativity constraints. Signal Process. **81**(5), 945–974 (2001)
11. Lazzaretti, M., Calatroni, L., Estatico, C.: Weighted-CEL0 sparse regularisation for molecule localisation in super-resolution microscopy with Poisson data (2020). https://arxiv.org/abs/2010.13173
12. Nesterov, Y.: Introductory Lectures on Convex Optimization: A Basic Course. Applied Optimization. Kluwer Academic Publication, Boston (2004)
13. Ochs, P., Dosovitskiy, A., Brox, T., Pock, T.: On iteratively reweighted algorithms for nonsmooth nonconvex optimization in computer vision. SIAM J. Imaging Sci. **8**(1), 331–372 (2015)
14. Rebegoldi, S., Calatroni, L.: Scaled, inexact and adaptive generalized FISTA for strongly convex optimization (2021). https://arxiv.org/abs/2101.03915
15. Sawatzky, A., Brune, C., Müller, J., Burger, M.: Total variation processing of images with poisson statistics. In: Jiang, X., Petkov, N. (eds.) CAIP 2009. LNCS, vol. 5702, pp. 533–540. Springer, Heidelberg (2009). https://doi.org/10.1007/978-3-642-03767-2_65
16. Scheinberg, K., Goldfarb, D., Bai, X.: Fast first-order methods for composite convex optimization with backtracking. Found. Comput. Math. **14**, 389–417 (2014)
17. Soubies, E., Blanc-Féraud, L., Aubert, G.: A continuous exact ℓ^0 penalty (CEL0) for least squares regularized problem. SIAM J. Imaging Sci. **8**(3), 1607–1639 (2015)
18. Soubies, E., Blanc-Féraud, L., Aubert, G.: A unified view of exact continuous penalties for ℓ_2-ℓ_0 minimization. SIAM J. Optim. **27**(3), 2034–2060 (2017)

Convergence Properties of a Randomized Primal-Dual Algorithm with Applications to Parallel MRI

Eric B. Gutiérrez[1(✉)], Claire Delplancke[1(✉)], and Matthias J. Ehrhardt[1,2(✉)]

[1] Department of Mathematical Sciences, University of Bath, Bath, UK
{ebgc20,cd902,M.Ehrhardt}@bath.ac.uk
[2] Institute for Mathematical Innovation, University of Bath, Bath, UK

Abstract. The Stochastic Primal-Dual Hybrid Gradient (SPDHG) was proposed by Chambolle et al. (2018) and is an efficient algorithm to solve some nonsmooth large-scale optimization problems. In this paper we prove its almost sure convergence for convex but not necessarily strongly convex functionals. We also look into its application to parallel Magnetic Resonance Imaging reconstruction in order to test performance of SPDHG. Our numerical results show that for a range of settings SPDHG converges significantly faster than its deterministic counterpart.

Keywords: Inverse problems · Parallel magnetic resonance imaging · Stochastic optimization · Primal-dual algorithm · Convex optimization

1 Introduction

Optimization problems have numerous applications among many fields such as imaging, data science or machine learning, to name a few. Many such problems in these areas are often formulated as

$$\hat{x} \in \arg\min_{x \in X} \sum_{i=1}^{n} f_i(A_i x) + g(x) \tag{1}$$

where $f_i : Y_i \to \mathbb{R} \cup \{\infty\}$ and $g : X \to \mathbb{R} \cup \{\infty\}$ are convex functionals, and $A_i : X \to Y_i$ are linear operators between finite-dimensional real Hilbert spaces.

Examples of such problems are total variation regularized image reconstruction [13,21] such as image denoising [7] or PET reconstruction [10]; regularized empirical risk minimization [22,23] such as support vector machine (SVM) [2] or least absolute shrinkage and selection operator (LASSO) [4]; and optimization with large number of constraints [12,17], among others.

MJE and CD acknowledge support from the EPSRC (EP/S026045/1). MJE is also supported by EPSRC (EP/T026693/1), the Faraday Institution (EP/T007745/1) and the Leverhulme Trust (ECF-2019-478). EBG acknowledges the Mexican Council of Science and Technology (CONACyT).

© Springer Nature Switzerland AG 2021
A. Elmoataz et al. (Eds.): SSVM 2021, LNCS 12679, pp. 254–266, 2021.
https://doi.org/10.1007/978-3-030-75549-2_21

While some classical approaches such as gradient descent are not applicable when the functionals f_i or g are not smooth [7], primal-dual methods are able to find solutions to (1) without assuming differentiability. For convex, proper and lower-semicontinuous functionals f_i, g, a primal-dual formulation for (1) reads

$$\hat{x}, \hat{y} \in \arg\min_{x \in X} \max_{y \in Y} \sum_{i=1}^{n} \langle A_i x, y_i \rangle - f_i^*(y_i) + g(x) \tag{2}$$

where f^* is the *convex conjugate* of f and $Y = \Pi_{i=1}^{n} Y_i$. We refer to any solution $\hat{w} = (\hat{x}, \hat{y})$ of (2) as a *saddle point*.

A well-known example of primal-dual methods that solve (2) is the Primal-Dual Hybrid Gradient (PDHG) [6,11,19], as presented by Chambolle and Pock (2011). It naturally breaks down the complexity of (1) into separate optimization problems by doing separate updates for the primal and dual variables x, y, as shown in (3). PDHG is proven to converge to a solution of (2), however its iterations become very costly for large-scale problems, e.g. when $n \gg 1$ [5].

More recently, Chambolle et al. proposed the Stochastic Primal-Dual Hybrid Gradient (SPDHG) [5] which reduces the per-iteration computational cost of PDHG by randomly sampling the dual variable: at each step, instead of the full dual variable y, only a random subset of its coordinates y_i gets updated. This offers significantly better performance than the deterministic PDHG for large-scale problems [5]. Examples of similar random primal-dual algorithms are found in [13,15,16,23].

We are interested in the convergence of SPDHG. In [5], it is shown that, for arbitrary convex functionals f_i and g, SPDHG converges in the sense of *Bregman distances*, which does not imply convergence in the norm. In this paper we present a proof for the almost sure convergence of SPDHG for convex but not necessarily strongly convex functionals, using alternative arguments to the recently proposed proof by Alacaoglu et al. ([1], Theorem 4.4). In contrast to [1], where they represent SPDHG by using projections of a single iterative operator, we present a more intuitive representation through a random sequence of operators (Lemma 6). A summary of the proof is laid out in Sect. 4 and the complete proof is detailed in Sect. 5. A comparison with the proof of Alacaouglu et al. and other related work is discussed in Sect. 6.

Finally, in Sect. 7 we look into the application of parallel Magnetic Resonance Imaging (MRI) in order to compare the performance of SPDHG with that of the deterministic PDHG.

2   Algorithm

In order to solve (2), the deterministic PDHG with dual extrapolation [6] reads

$$\begin{aligned}
x^{k+1} &= \text{prox}_{\tau g}(x^k - \tau A^T \bar{y}^k) \\
y^{k+1} &= \text{prox}_{\sigma f^*}(y^k + \sigma A x^{k+1})
\end{aligned} \tag{3}$$

where $\bar{y}^k = 2y^k - y^{k-1}$ is an extrapolation on the previous iterates, A and f are given by $Ax = (A_1x, ..., A_nx)$ and $f(y) = \sum_i f_i(y_i)$, and the *proximity operator* of any functional f is defined as $\text{prox}_{\sigma f}(v) := \arg\min_{y \in Y} \frac{\|v-y\|^2}{2} + \sigma f(y)$.

SPDHG, in contrast, reduces the cost of iterations by only partially updating the dual variable $y = (y_i)_{i=1}^n$: at every iteration k, choose $j \in \{1, ..., n\}$ at random with probability $p_i = \mathbb{P}(j = i) > 0$, so that only the variable y_j^{k+1} is updated, while the rest remain unchanged, i.e. $y_i^{k+1} = y_i^k$ for $i \neq j$.

Algorithm 1 (SPDHG)

Choose $\tau, \sigma_i > 0$ and $x^0 \in X$. Set $y^0 = \mathbf{0} \in Y$ and $z^0 = \bar{z}^0 = \mathbf{0} \in X$.
For $k \geq 0$ do

$$\text{select } j^k \in \{1, ..., n\} \text{ at random}$$

$$x^{k+1} = \text{prox}_{\tau g}(x^k - \tau \bar{z}^k)$$

$$y_i^{k+1} = \begin{cases} \text{prox}_{\sigma_i f_i^*}(y_i^k + \sigma_i A_i x^{k+1}) & \text{if } i = j^k \\ y_i^k & \text{else} \end{cases}$$

$$\delta^k = A_{j^k}^T(y_{j^k}^{k+1} - y_{j^k}^k)$$

$$z^{k+1} = z^k + \delta^k$$

$$\bar{z}^{k+1} = z^{k+1} + p_{j^k}^{-1}\delta^k$$

3   Main Result

We establish the almost sure convergence of SPDHG for any convex functionals, under the same step size conditions as in [5]:

Assumption 2. *We assume the following to hold:*

1. *The set of solutions to (2) is nonempty.*
2. *The functionals g, f_i are convex, proper and lower-semicontinuous.*
3. *The step sizes $\tau, \sigma_i > 0$ satisfy*

$$\tau \sigma_i \|A_i\|^2 < p_i \quad \text{for every } i. \tag{4}$$

Theorem 3 (Convergence of SPDHG). *Let $(w^k)_{k \in \mathbb{N}} = (x^k, y^k)_{k \in \mathbb{N}}$ be a random sequence generated by Algorithm 1. Under Assumption 2, the sequence $(w^k)_{k \in \mathbb{N}}$ converges almost surely to a solution of (2).*

4   Sketch of the Proof

The following results lay out the proof of Theorem 3. The complete proof is detailed in Sect. 5. We use the notation $\|x\|_T^2 = \langle Tx, x \rangle$, as well as the block diagonal operators $Q, S : Y \to Y$ given by $Q = \text{diag}(p_1^{-1}, ..., p_n^{-1})$ and $S = \text{diag}(\sigma_1, ..., \sigma_n)$. The conditional expectation at time $k + 1$ is denoted, for any functional φ, by $\mathbb{E}^{k+1}(\varphi(w^{k+1})) = \mathbb{E}(\varphi(w^{k+1})|w^k)$.

The proof of Theorem 3 uses the following important inequality from SPDHG, which is a consequence of ([5], Lemma 4.4). This inequality is best summarized in ([1], Lemma 4.1), which we have further simplified by using the fact that Bregman distances of convex functionals are nonnegative ([5], Sect. 4).

Lemma 1 ([1], **Lemma 4.1**). *Let $(w^k)_{k \in \mathbb{N}}$ be a random sequence generated by Algorithm 1 under Assumption 2. Then for every saddle point $\hat{w}$,*

$$V^k(w^k - \hat{w}) \geq \mathbb{E}^{k+1}(V^{k+1}(w^{k+1} - \hat{w})) + V(x^{k+1} - x^k, y^k - y^{k-1}) \quad (5)$$

where V and V^k are given by $V(x, y) = \|x\|_{\tau^{-1}}^2 + 2\langle QAx, y \rangle + \|y\|_{QS^{-1}}^2$ and

$$V^k(x, y) = \|x\|_{\tau^{-1}}^2 - 2\langle QAx, y^k - y^{k-1} \rangle + \|y^k - y^{k-1}\|_{QS^{-1}}^2 + \|y\|_{QS^{-1}}^2.$$

The following result is the central argument of our proof. It makes use of inequality (5) and a classical result from Robbins & Siegmund (Lemma 3) to establish an important convergence result. Its proof is detailed in Sect. 5.

Proposition 1. *Let $(w^k)_{k \in \mathbb{N}}$ be a random sequence generated by Algorithm 1 under Assumption 2 and let $\hat{w}$ be a saddle point. Then:*

   i) *The sequence $(w^k)_{k \in \mathbb{N}}$ is a.s. bounded.*
  ii) *The sequence $(V^k(w^k - \hat{w}))_{k \in \mathbb{N}}$ converges a.s.*
 iii) *The sequence $(\|w^k - \hat{w}\|)_{k \in \mathbb{N}}$ converges a.s.*
 iv) *If every cluster point of $(w^k)_{k \in \mathbb{N}}$ is a.s. a saddle point, the sequence $(w^k)_{k \in \mathbb{N}}$ converges a.s. to a saddle point.*

Lastly, we prove that every cluster point of $(w^k)_{k \in \mathbb{N}}$ is almost surely a solution to (2). This is also explained in Sect. 5.

Proposition 2. *Let $(w^k)_{k \in \mathbb{N}}$ be a random sequence generated by Algorithm 1 under Assumption 2. Then every cluster point of $(w^k)_{k \in \mathbb{N}}$ is almost surely a saddle point.*

Proof (Proof of Theorem 3). By Proposition 2, every cluster point of $(w^k)_{k \in \mathbb{N}}$ is almost surely a saddle point and, by Proposition 1 iv), the sequence $(w^k)_{k \in \mathbb{N}}$ converges almost surely to a saddle point.

5   Proof of Convergence

This section contains detailed proofs for our two main arguments, Propositions 1 and 2. The proof of Proposition 1 follows a similar strategy to that of Combettes and Pesquet in ([9], Proposition 2.3), and we have divided it into three sections.

5.1 Proof of Proposition 1 i)

To show this first part, we borrow the following lemma from [5] (Lemma 4.2 with $c = 1$, $v_i = \tau\sigma_i\|A_i\|^2$ and $\gamma^2 = \max_i \frac{v_i}{p_i}$):

Lemma 2 ([5], Lemma 4.2). *Let $p_i^{-1}\tau\sigma_i\|A_i\|^2 \leq \gamma^2 < 1$ for every i and let y^k be defined as in Algorithm 1. Then for every $x \in X$,*

$$\mathbb{E}^k(V(x, y^k - y^{k-1})) \geq (1 - \gamma)\mathbb{E}^k\big(\|x\|^2_{\tau^{-1}} + \|y^k - y^{k-1}\|^2_{QS^{-1}}\big).$$

Proof (Proof of Proposition 1 i)). By ([1], Lemma 4.1), for any saddle point $\hat{w}$ we have

$$\Delta^k \geq \mathbb{E}^{k+1}(\Delta^{k+1}) + V(x^{k+1} - x^k, y^k - y^{k-1}) \tag{6}$$

where $\Delta^k = V^k(w^k - \hat{w})$. By Lemma 2,

$$\mathbb{E}^k(V(x^{k+1} - x^k, y^k - y^{k-1})) \geq (1 - \gamma)\mathbb{E}^k\Big\{\|x^{k+1} - x^k\|^2_{\tau^{-1}} + \|y^k - y^{k-1}\|^2_{QS^{-1}}\Big\}.$$

Hence, taking the full expectation in (6) yields

$$\mathbb{E}(\Delta^k) \geq \mathbb{E}(\Delta^{k+1}) + (1 - \gamma)\mathbb{E}\big(\|x^{k+1} - x^k\|^2_{\tau^{-1}} + \|y^k - y^{k-1}\|^2_{QS^{-1}}\big).$$

Taking the sum from $k = 0$ to $k = N - 1$ gives

$$\Delta^0 \geq \mathbb{E}(\Delta^N) + (1 - \gamma)\mathbb{E}\Big\{\sum_{k=0}^{N-1}\|x^{k+1} - x^k\|^2_{\tau^{-1}} + \|y^k - y^{k-1}\|^2_{QS^{-1}}\Big\} \tag{7}$$

where $y^{-1} = y^0$. This implies $\Delta^0 \geq \mathbb{E}(\Delta^N)$ and, by Lemma 2 we have

$$\mathbb{E}(\Delta^N) \geq \mathbb{E}\big\{(1 - \gamma)(\|x^N - \hat{x}\|^2_{\tau^{-1}} + \|y^N - y^{N-1}\|^2_{QS^{-1}}) + \|y^N - \hat{y}\|^2_{QS^{-1}}\big\}. \tag{8}$$

It follows that $\Delta^0 \geq (1 - \gamma)\|x^N - \hat{x}\|^2_{\tau^{-1}} + \|y^N - \hat{y}\|^2_{QS^{-1}}$ a.s., from where it is clear that the sequence $(w^N)_{N\in\mathbb{N}}$ is bounded almost surely.

5.2 Proof of Proposition 1 ii)–iii)

As in ([9], Proposition 2.3), we use a classical result from Robbins & Siegmund:

Lemma 3 ([20], Theorem 1). *Let $\mathcal{F}_k$ be a sequence of sub-σ-algebras such that $\mathcal{F}_k \subset \mathcal{F}_{k+1}$ for every k, and let α_k, η_k be nonnegative $\mathcal{F}_k$-measurable random variables such that $\sum_{k=1}^{\infty} \eta_k < \infty$ almost surely and*

$$\mathbb{E}(\alpha_{k+1} \mid \mathcal{F}_k) \leq \alpha_k + \eta_k \quad a.s.$$

for every k. Then α_k converges almost surely to a random variable in $[0, \infty)$.

Proof (Proof of Proposition 1 ii)–iii)). From (8) we have $\mathbb{E}(\Delta^N) \geq 0$. Thus taking the limit as $N \to \infty$ in (7) yields

$$\mathbb{E}\left\{ \sum_{k=0}^{\infty} \|x^{k+1} - x^k\|_{\tau^{-1}}^2 + \|y^k - y^{k-1}\|_{QS^{-1}}^2 \right\} < \infty \tag{9}$$

which implies

$$\sum_{k=0}^{\infty} \|x^{k+1} - x^k\|_{\tau^{-1}}^2 + \|y^k - y^{k-1}\|_{QS^{-1}}^2 < \infty \quad \text{a.s.} \tag{10}$$

and, in particular,

$$\|y^k - y^{k-1}\|_{QS^{-1}} \to 0 \quad \text{a.s.} \tag{11}$$

Since $(w^k)_{k \in \mathbb{N}}$ is bounded a.s., so is $(x^k)_{k \in \mathbb{N}}$ and, since the operators Q, A and S are also bounded, there exists $M > 0$ such that, for every k,

$$|\langle QA(x^k - \hat{x}), y^k - y^{k-1} \rangle| \leq \|QA\| \|x^k - \hat{x}\| \|y^k - y^{k-1}\| \leq M \|y^k - y^{k-1}\|_{QS^{-1}}$$

a.s. and therefore, by (11),

$$\langle QA(x^k - \hat{x}), y^k - y^{k-1} \rangle \to 0 \quad \text{a.s.} \tag{12}$$

The fact that $(w^k)_{k \in \mathbb{N}}$ is a.s. bounded, together with (11) and (12) imply the sequence $(\Delta^k)_{k \in \mathbb{N}}$ is also a.s. bounded. Thus there exists $\hat{M} \geq 0$ such that $\Delta^k + \hat{M} \geq 0$ for every k. Let $\alpha_k = \Delta^k + \hat{M}$ and $\eta_k = 2|\langle QA(x^{k+1} - x^k), y^k - y^k^{-1} \rangle|$. From (6) we deduce

$$\alpha_k + \eta_k \geq \mathbb{E}^{k+1}(\alpha_{k+1}) \quad \text{a.s. for every } k, \tag{13}$$

where all the terms are nonnegative and, for some $\tilde{M} > 0$,

$$\begin{aligned} \eta_k = 2|\langle QA(x^{k+1} - x^k), y^k - y^{k-1} \rangle| &\leq 2\|QA\| \|x^{k+1} - x^k\| \|y^k - y^{k-1}\| \\ &\leq 2\tilde{M} \|x^{k+1} - x^k\|_{\tau^{-1}} \|y^k - y^{k-1}\|_{QS^{-1}} \\ &\leq \tilde{M} \left(\|x^{k+1} - x^k\|_{\tau^{-1}}^2 + \|y^k - y^{k-1}\|_{QS^{-1}}^2 \right) \end{aligned}$$

which implies, by (10), $\sum_{k=1}^{\infty} \eta_k < \infty$ a.s.. Thus (13) satisfies all the assumptions of Lemma 3 and it yields $\Delta^k \to \alpha$ a.s. for some $\alpha \in [-\hat{M}, \infty)$. Furthermore, from (11) and (12) we know some of the terms in Δ^k converge to 0 a.s., namely

$$-2\langle QA(x^k - \hat{x}), y^k - y^{k-1} \rangle + \|y^k - y^{k-1}\|_{QS^{-1}}^2 \to 0 \quad \text{a.s.}$$

hence $\|x^k - \hat{x}\|_{\tau^{-1}}^2 + \|y^k - \hat{y}\|_{QS^{-1}}^2 \to \alpha$ a.s. Finally, the norm $\|w\|_R^2 := \|x\|_{\tau^{-1}}^2 + \|y\|_{QS^{-1}}^2$ is equivalent to the norm in $X \times Y$. Since the sequence $(\|w^k - \hat{w}\|_R)_{k \in \mathbb{N}}$ converges a.s., so does $(\|w^k - \hat{w}\|)_{k \in \mathbb{N}}$.

5.3 Proof of Proposition 1 iv)

The two following lemmas are consequence of ([9], Proposition 2.3) and their proofs are not included here. We use the standard notation $(\Omega, \mathcal{F}, P)$ for the probability space corresponding to the random iterations w^k.

Lemma 4 ([9], **Proposition 2.3 iii)).** *Let* $\mathbf{F}$ *be a closed subset of a separable Hilbert space and let* $(w^k)_{k \in \mathbb{N}}$ *be a sequence of random variables such that the sequence* $(\|w^k - w\|)_{k \in \mathbb{N}}$ *converges almost surely for every* $w \in \mathbf{F}$. *Then there exists* $\Omega \in \mathcal{F}$ *such that* $\mathbb{P}(\Omega) = 1$ *and the sequence* $(\|w^k(\omega) - w\|)_{k \in \mathbb{N}}$ *converges for all* $\omega \in \Omega$ *and* $w \in \mathbf{F}$.

Lemma 5 ([9], **Proposition 2.3 iv)).** *Let* $\mathbf{G}(w^k)$ *be the set of cluster points of a random sequence* $(w^k)_{k \in \mathbb{N}}$. *Assume there exists* $\Omega \in \mathcal{F}$ *such that* $\mathbb{P}(\Omega) = 1$ *and for every* $\omega \in \Omega$, $\mathbf{G}(w^k(\omega))$ *is nonempty and the sequence* $(\|w^k(\omega) - w\|)_{k \in \mathbb{N}}$ *converges for all* $w \in \mathbf{G}(w^k(\omega))$. *Then* $(w^k)_{k \in \mathbb{N}}$ *converges almost surely to an element of* $\mathbf{G}(w^k)$.

Proof (**Proof of Proposition 1 iv)**). Let $\mathbf{F}$ be the set of solutions to the saddle point problem (2). By Proposition 1 iii) and Lemma 4, there exists $\Omega \in \mathcal{F}$ such that the sequence $(\|w^k(\omega) - w\|)_{k \in \mathbb{N}}$ converges for every $w \in \mathbf{F}$ and $\omega \in \Omega$. This implies, since $\mathbf{F}$ is nonempty, that $(w^k(\omega))_{k \in \mathbb{N}}$ is bounded and thus $\mathbf{G}(w^k(\omega))$ is nonempty for all $\omega \in \Omega$. By assumption, there exists $\tilde{\Omega} \in \mathcal{F}$ such that $\mathbf{G}(w^k(\omega)) \subset \mathbf{F}$ for every $\omega \in \tilde{\Omega}$. Let $\omega \in \Omega \cap \tilde{\Omega}$, then $(\|w^k(\omega) - w\|)_{k \in \mathbb{N}}$ converges for every $w \in \mathbf{G}(w^k(\omega)) \neq \emptyset$. By Lemma 5, we get the result.

5.4 Proof of Proposition 2

The following lemma describes Algorithm 1 as a random sequence $(T_{j^k})_{k \in \mathbb{N}}$ of continuous operators on the primal-dual space $X \times Y$, such that the fixed points of the operators are saddle points of (2):

Lemma 6. *Denote* $w = (w_i)_{i=0}^n = (x, y_1, ..., y_n)$ *and for every* $j \in \{1, ..., n\}$ *let the operator* $T_j : X \times Y \to X \times Y$ *be defined by*

$$(T_j w)_0 = \mathrm{prox}_{\tau g}\left(x - \tau A^T y - \left(1 + \frac{1}{p_j}\right)\tau A_j^T((T_j w)_j - y_j)\right)$$

$$(T_j w)_i = \begin{cases} \mathrm{prox}_{\sigma_i f_i^*}(y_i + \sigma_i A_i x) & \text{if } i = j \\ y_i & \text{else} \end{cases} \qquad \text{for } 1 \leq i \leq n.$$

Then the iterations w^k *generated by Algorithm 1 satisfy*

$$T_{j^k}(x^{k+1}, y^k) = (x^{k+2}, y^{k+1}). \tag{14}$$

Furthermore, $\hat{w}$ *is a solution to the saddle point problem (2) if and only if it is a fixed point of* T_j *for each* $j \in \{1, ..., n\}$.

Proof. By definition of the iterates in Algorithm 1, $(T_{j^k}(x^{k+1}, y^k))_i = y_i^{k+1}$ for every $i \in \{1, ..., n\}$. By induction it is easy to check that $z^k = A^T y^k$, and thus

$$\bar{z}^{k+1} = z^k + (1 + \frac{1}{p_{j^k}})\delta^k$$

$$= A^T y^k + (1 + \frac{1}{p_{j^k}})A_{j^k}^T(y_{j^k}^{k+1} - y_{j^k}^k)$$

$$= A^T y^k + (1 + \frac{1}{p_{j^k}})A_{j^k}^T((T_{j^k}(x^{k+1}, y^k))_{j^k} - y_{j^k}^k).$$

Thus $(T_{j^k}(x^{k+1}, y^k))_0 = \text{prox}_{\tau g}(x^{k+1} - \tau \bar{z}^{k+1}) = x^{k+2}$, which proves (14). Now let w be a fixed point of T_j for every j. Then, for any j,

$$y_j = w_j = (T_j w)_j = \text{prox}_{\sigma_i f_j^*}(y_j + \sigma_j A_j x),$$

from where it follows that, for any j,

$$x = w_0 = (T_j w)_0 = \text{prox}_{\tau g}(x - \tau A^T y - (1 + \frac{1}{p_j})\tau A_j^T((T_j w)_j - y_j)$$

$$= \text{prox}_{\tau g}(x - \tau A^T y).$$

These conditions on x and y define a saddle point ([3], 6.4.2). The converse result is direct.

Proof (Proof of Proposition 2). Let j^k be the sampling generated by the algorithm and let $z^k = (x^{k+1}, y^k)$. By Lemma 6 we have $z^{k+1} = T_{j^k} z^k$ and, by (9),

$$\mathbb{E}(\|z^k - z^{k-1}\|^2) = \mathbb{E}(\|x^{k+1} - x^k\|^2 + \|y^k - y^{k-1}\|^2) \to 0. \qquad (15)$$

Furthermore, by the properties of the conditional expectation,

$$\mathbb{E}(\|z^{k+1} - z^k\|^2) = \mathbb{E}(\mathbb{E}^k(\|z^{k+1} - z^k\|^2)) = \mathbb{E}\left(\sum_{j=1}^n \mathbb{P}(j^k = j)\|T_j z^k - z^k\|^2\right)$$

$$= \sum_{j=1}^n \mathbb{P}(j^k = j)\mathbb{E}(\|T_j z^k - z^k\|^2).$$

By assumption $p_j = \mathbb{P}(j^k = j) > 0$, thus by (15) we have $\mathbb{E}(\|T_j z^k - z^k\|^2) \to 0$ for every j and therefore

$$T_j z^k - z^k \to 0 \quad \text{a.s. for every } j \in \{1, ..., n\}. \qquad (16)$$

Assume now a convergent subsequence $w^{\ell_k} \to w^*$. From (15), $y^k - y^{k-1} \to 0$ a.s. and so z^{ℓ_k} also converges to w^*. By (16) and the continuity of T_j, there holds

$$w^* = \lim_{k \to \infty} z^{\ell_k} = \lim_{k \to \infty} T_j z^{\ell_k} = T_j(\lim_{k \to \infty} z^{\ell_k}) = T_j w^* \quad \text{a.s.}$$

for every j. Hence w^* is almost surely a fixed point of T_j for each j and, by Lemma 6, w^* is a saddle point.

6   Relation to Other Work

6.1   Chambolle et al. (2018)

In the original paper for SPDHG [5], it is shown that, under Assumption 2, the *Bregman distance* to any solution $\hat{x}, \hat{y}$ of (2) converges to zero, i.e. the iterates x^k, y^k of Algorithm 1 satisfy

$$D_g^{-A^T \hat{y}}(x^k, \hat{x}) + D_{f^*}^{A\hat{x}}(y^k, \hat{y}) \to 0 \quad \text{a.s.,} \tag{17}$$

where the Bregman distance is defined by $D_h^q(u, v) = h(u) - h(v) - \langle q, u - v \rangle$ for any functional h and any point $q \in \partial h(v)$ in the subdifferential of h. In [5], it is shown that (17) implies $(x^k, y^k) \to (\hat{x}, \hat{y})$ a.s. if f_i or g are strongly convex.

6.2   Combettes and Pesquet (2014)

In [9], Combettes and Pesquet look into the convergence of random sequences $(w^k)_{k \in \mathbb{N}}$ of the form

$$w_i^{k+1} = \begin{cases} (Tw^k)_i & \text{if } i \in \mathbb{S}^k \\ w_i^k & \text{else} \end{cases} \tag{18}$$

where $\mathbb{S}^k \subset \{1, ..., n\}$ is chosen at random. They use the Robbins-Siegmund lemma (Lemma 3) to prove that, for a nonexpansive operator T, the sequence $(w^k)_{k \in \mathbb{N}}$ converges to a fixed point of T. Later, Pesquet and Repetti [18] used this to prove convergence for a wide class of random algorithms of the form (18), where $T = (I + B)^{-1}$ is the resolvent operator of a monotone operator B.

6.3   Alacaoglu et al. (2019)

Recently Alacaoglu et al. also proposed a proof for the almost sure convergence of SPDHG ([1], Theorem 4.4) using a strategy similar to ours. They describe SPDHG as a special case of a sequence $(\mathbf{w}^k)_{k \in \mathbb{N}} \subset \mathbb{R}^{dn+n^2}$ of the form (18) for an operator $\mathbf{T} : \mathbb{R}^{dn+n^2} \to \mathbb{R}^{dn+n^2}$, where $X = \mathbb{R}^d$ and $Y = \mathbb{R}^n$. They show that, under the step size condition (4), the iterations $\mathbf{w}^k \in \mathbb{R}^{dn+n^2}$ converge to a fixed point of $\mathbf{T}$. In contrast, we describe SPDHG using a random sequence $(T^k)_{k \in \mathbb{N}}$ of more intuitive operators $T^k : X \times Y \to X \times Y$, and then prove the sequence $(w^k)_{k \in \mathbb{N}}$ converges to a fixed point $w \in \bigcap_{k \in \mathbb{N}} \text{Fix} T^k$.

7   Numerical Examples

In this section we use MRI reconstruction as a case study to illustrate the performance of SPDHG in comparison to PDHG. In parallel MRI reconstruction [14], a signal x is reconstructed from multiple data samples $b_1, ..., b_n$ of the form $b_i = A_i x + \eta_i$, where each $A_i : \mathbb{C}^d \to \mathbb{C}^m$ is an encoder operator from the

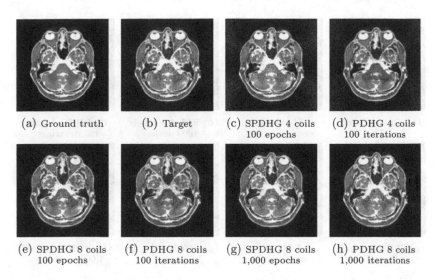

(a) Ground truth   (b) Target   (c) SPDHG 4 coils
100 epochs   (d) PDHG 4 coils
100 iterations

(e) SPDHG 8 coils
100 epochs   (f) PDHG 8 coils
100 iterations   (g) SPDHG 8 coils
1,000 epochs   (h) PDHG 8 coils
1,000 iterations

Fig. 1. Images reconstructed using different algorithms. Subfigure (a) shows the original ground truth $x^\dagger$ from which the noisy data samples have been synthetically generated, while subfigure (b) shows a reliable reconstruction x^* to which the other figures should be compared.

signal space to the sample space, and $\eta_i \in \mathbb{C}^m$ represents noise added to the measurements. A least-squares solution is given by

$$\hat{x} \in \arg\min_x \sum_{i=1}^n \|A_i x - b_i\|^2 + g(x) \tag{19}$$

where g acts as a regularizer. We recover our convex minimization template (1) by identifying $\mathbb{C}^d$ with $\mathbb{R}^{2d}$ and setting $X = \mathbb{R}^{2d}$, $Y_i = \mathbb{R}^{2m}$ and $f_i(y) = \|y - b_i\|^2$.

Here, we consider undersampled data with n coils, i.e. $A_i = S \circ F \circ C_i$, where $S : \mathbb{C}^d \to \mathbb{C}^m$ is a subsampling operator, $F : \mathbb{C}^d \to \mathbb{C}^d$ represents the discrete Fourier transform and $C_i x = c_i \cdot x$ is the element-wise multiplication of x and the i-th coil-sensitivity map $c_i \in \mathbb{C}^d$. Data samples $b_i = A_i x^\dagger + \eta_i$ have been synthetically generated from an original reconstruction $x^\dagger$ from the BrainWave database [8].

In order to choose step-size parameters σ_i and τ which comply with step-size condition (4), we consider $\sigma_i = \gamma(p_i/\|A_i\|)$ and $\tau = \gamma^{-1}(0.99/\max_i \|A_i\|)$, and then choose the value of γ (among orders of magnitude $\gamma = 10^{-5}, 10^{-4}, ..., 10^5$) that gives the lowest objective $\Phi(x^k) = \sum_{i=1}^n f_i(Ax^k) + g(x^k)$ after 100 epochs. Step sizes for PDHG have been optimized in the same way by setting $\sigma = \gamma/\|A\|$ and $\tau = \gamma^{-1}(0.99/\|A\|)$, which satisfies its step-size condition $\tau\sigma\|A\|^2 < 1$ [6].

Figure 1 shows reconstructions obtained through SPDHG and PDHG. Here, we considered model (19) with squared 2-norm regularizer $g(x) = 10^{-4}\|x\|^2$. The *target* solution x^* in (b) has been computed by running SPDHG for large number of epochs ($> 10^4$). Comparing subfigures (e) and (f), the solution for

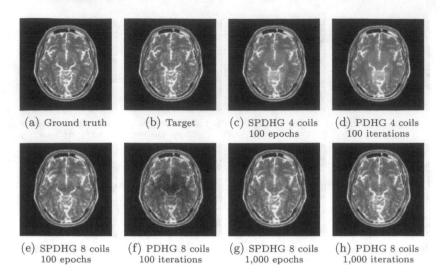

Fig. 2. Images reconstructed using noisy data samples synthetically generated from the ground truth (a). Subfigure (b) shows the target reconstruction for comparison.

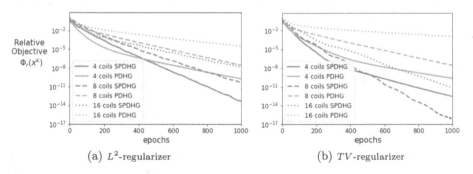

Fig. 3. Relative objective $\Phi_r(x^k) = \frac{\Phi(x^k) - \Phi(x^*)}{\Phi(x^0) - \Phi(x^*)}$ at every epoch for image reconstruction, including the examples from Figs. 1 and 2.

SPDHG seems to be closer to the target (b) than that of PDHG at 100 epochs. At 1,000 epochs, solutions (g) and (h) appear visually similar to each other.

Figure 2 shows reconstructions using model (19) with total-variation regularizer $g(x) = 10^{-4}\|\nabla x\|_1$. As before, solutions by SPDHG appear closer to the target than those by PDHG. Notice as well how increasing the number of coils seems to yield more detailed although noisier reconstructions.

Figure 3 summarizes the performance of both algorithms for all of these reconstructions and some more.

8   Conclusions

In this paper we showed the almost sure convergence of SPDHG by using alternative arguments as provided by Alacaoglu et al. [1]. We also investigated SPDHG in the context of parallel MRI where we observed a significant speed-up, thereby supporting the practical use of variational regularization methods for this application.

References

1. Alacaoglu, A., Fercoq, O., Cevher, V.: On the convergence of stochastic primal-dual hybrid gradient. arXiv preprint arXiv:1911.00799 (2019)
2. Boser, B.E., Guyon, I.M., Vapnik, V.N.: A training algorithm for optimal margin classifiers. In: Proceedings of the Fifth Annual Workshop on Computational Learning Theory, pp. 144–152 (1992)
3. Bredies, K., Lorenz, D.: Mathematical Image Processing. Springer, Cham (2018). https://doi.org/10.1007/978-3-030-01458-2_6
4. Cevher, V., Becker, S., Schmidt, M.: Convex optimization for big data: scalable, randomized, and parallel algorithms for big data analytics. IEEE Signal Process. Mag. **31**(5), 32–43 (2014)
5. Chambolle, A., Ehrhardt, M.J., Richtárik, P., Schönlieb, C.B.: Stochastic primal-dual hybrid gradient algorithm with arbitrary sampling and imaging applications. SIAM J. Optim. **28**(4), 2783–2808 (2018)
6. Chambolle, A., Pock, T.: A first-order primal-dual algorithm for convex problems with applications to imaging. J. Math. Imaging Vis. **40**(1), 120–145 (2011)
7. Chambolle, A., Pock, T.: An introduction to continuous optimization for imaging. Acta Numer. **25**, 161–319 (2016)
8. Cocosco, C., Kollokian, V., Kwan, R.S., Evans, A.: BrainWeb: online interface to a 3D MRI simulated brain database. Neuroimage **5**(4), 425 (1997)
9. Combettes, P.L., Pesquet, J.C.: Stochastic quasi-Fejér block-coordinate fixed point iterations. SIAM J. Optim. **25**(2), 1221–1248 (2015)
10. Ehrhardt, M.J., Markiewicz, P., Schönlieb, C.B.: Faster PET reconstruction with non-smooth priors by randomization and preconditioning. Phys. Med. Biol. **64**(22), 225019 (2019)
11. Esser, E., Zhang, X., Chan, T.F.: A general framework for a class of first order primal-dual algorithms for convex optimization in imaging science. SIAM J. Imaging Sci. **3**(4), 1015–1046 (2010)
12. Fercoq, O., Alacaoglu, A., Necoara, I., Cevher, V.: Almost surely constrained convex optimization. arXiv preprint arXiv:1902.00126 (2019)
13. Fercoq, O., Bianchi, P.: A coordinate-descent primal-dual algorithm with large step size and possibly nonseparable functions. SIAM J. Optim. **29**(1), 100–134 (2019)
14. Fessler, J.A.: Optimization methods for magnetic resonance image reconstruction: key models and optimization algorithms. IEEE Signal Process. Mag. **37**(1), 33–40 (2020)
15. Gao, X., Xu, Y.Y., Zhang, S.Z.: Randomized primal-dual proximal block coordinate updates. J. Oper. Res. Soc. China **7**(2), 205–250 (2019). https://doi.org/10.1007/s40305-018-0232-4

16. Latafat, P., Freris, N.M., Patrinos, P.: A new randomized block-coordinate primal-dual proximal algorithm for distributed optimization. IEEE Trans. Autom. Control **64**(10), 4050–4065 (2019)
17. Patrascu, A., Necoara, I.: Nonasymptotic convergence of stochastic proximal point methods for constrained convex optimization. J. Mach. Learn. Res. **18**(1), 7204–7245 (2017)
18. Pesquet, J.C., Repetti, A.: A class of randomized primal-dual algorithms for distributed optimization. arXiv preprint arXiv:1406.6404 (2014)
19. Pock, T., Cremers, D., Bischof, H., Chambolle, A.: A algorithm for minimizing the Mumford-Shah functional. In: 2009 IEEE 12th International Conference on Computer Vision, pp. 1133–1140 (2009)
20. Robbins, H., Siegmund, D.: A convergence theorem for non negative almost super-martingales and some applications. In: Optimizing Methods in Statistics, pp. 233–257. Elsevier (1971)
21. Rudin, L.I., Osher, S., Fatemi, E.: Nonlinear total variation based noise removal algorithms. Physica D **60**(1–4), 259–268 (1992)
22. Shalev-Shwartz, S., Zhang, T.: Stochastic dual coordinate ascent methods for regularized loss minimization. J. Mach. Learn. Res. **14**(Feb), 567–599 (2013)
23. Zhang, Y., Xiao, L.: Stochastic primal-dual coordinate method for regularized empirical risk minimization. J. Mach. Learn. Res. **18**(1), 2939–2980 (2017)

Machine Learning in Imaging

Wasserstein Generative Models for Patch-Based Texture Synthesis

Antoine Houdard[1]([✉]), Arthur Leclaire[1], Nicolas Papadakis[1], and Julien Rabin[2]

[1] Univ. Bordeaux, Bordeaux INP, CNRS, IMB, UMR 5251, 33400 Talence, France
`antoine.houdard@u-bordeaux.fr`
[2] Normandie Univ., UniCaen, ENSICAEN, CNRS, GREYC, UMR 607, Caen, France

Abstract. This work addresses texture synthesis by relying on the local representation of images through their patch distributions. The main contribution is a framework that imposes the patch distributions at several scales using optimal transport. This leads to two formulations. First, a pixel-based optimization method is proposed, based on discrete optimal transport. We show that it generalizes a well-known texture optimization method that uses iterated patch nearest-neighbor projections, while avoiding some of its shortcomings. Second, in a semi-discrete setting, we exploit differential properties of Wasserstein distances to learn a fully convolutional network for texture generation. Once estimated, this network produces realistic and arbitrarily large texture samples in real time. By directly dealing with the patch distribution of synthesized images, we also overcome limitations of state-of-the-art techniques, such as patch aggregation issues that usually lead to low frequency artifacts (e.g. blurring) in traditional patch-based approaches, or statistical inconsistencies (e.g. color or patterns) in machine learning approaches.

1 Introduction

General Context. Among the boiling topic of image synthesis, exemplar-based texture synthesis aims at generating large textures from a small sample. This is of great interest in various fields ranging from computer graphics for cinema or video games to image or artwork restoration. The exemplar texture is often assumed to be perceptually stationary, *i.e.* with no large geometric deformations nor lighting changes.

Representing an image with its set of patches, which are small sub-images of size $s \times s$, is particularly well suited in this stationary setting. Patches take profit of the self-similarities contained in natural images, and are at the core of efficient image restoration methods [2,8,13]. Patch representation has proven to be fruitful for designing texture synthesis methods ranging from simple iterative copy/paste or nearest-neighbor procedures [3,12] to methods that aim at imposing the patch distribution using optimal transport [4,7,14].

This study has been carried out with financial support from the French Research Agency through the GOTMI project (ANR-16-CE33-0010-01). The authors also acknowledge the French GdR ISIS through the support of the REMOGA project.

These patch-based approaches generally suffer from three main practical limitations. First, the patches are often processed independently and then combined to form a recomposed image [4,14]. The overlap between patches leads to low frequency artifacts such as blurring. Second, the optimization has to be performed sequentially in a coarse-to-fine manner (both in image resolution and patch size) starting from a good initial guess. Last, global patch statistics must be controlled along the optimization to prevent strong visual artifacts [7,10].

In this work, we circumvent these limitations and introduce a framework for synthesizing a texture such that its patch distributions at different scales are close, in an optimal transport sense, to the one of the exemplar image. We demonstrate that patches are sufficient features for texture synthesis when used in an appropriate optimal transport framework, as illustrated in Fig. 1. Actually, even if deep features representations have shown impressive performances for image synthesis [5,9], patch-based methods are still competitive when only a single image is available for training [1,17], both considering the computational cost and the visual performance. Moreover, deep learning methods are still difficult to interpret, whereas patch-based models offer a better understanding of the synthesis process and its cases of success and failure.

Our framework is naturally adapted to the training of a generative network with the semi-discrete formulation of optimal transport. To this end, we rely on feed-forward convolutional neural networks proposed in [1,17,19] for texture and image generation. Training a texture generator with our optimal transport approach allows to generate on-the-fly arbitrarily large textures. Finally, an interesting by-product of the proposed method is that it defines a discrepancy measure between two textures by computing the mean of the optimal transport cost between patch distributions at various scales.

Overview of the Paper. The goal of the method is to synthesize a texture whose multi-scale patch distributions are close in an optimal transport sense to the ones of the target texture. Consider a target texture image $v \in \mathcal{K}^m$ with m pixels taking values in $\mathcal{K} \subset \mathbf{R}^d$ bounded[1]. We define the collection of its patches $\{P_1 v, \dots, P_m v\}$ as the set of all sub-images of size $p = s \times s$ extracted from v. For the sake of simplicity, we use periodic boundary conditions so that the number of patches is exactly m. Then we define the empirical patch distribution of v by $\nu = \frac{1}{m} \sum_{i=1}^{m} \delta_{P_i v}$. Our objective is therefore to prescribe the statistics of the patch distribution of the synthesized textures, in order to match the target distribution ν. To do so, we propose one framework for synthesizing a single texture image u (Sect. 3) and one for learning a generative model $g_\theta \sharp \zeta$ (Sect. 4), where the push-forward of a measure ζ is defined by $g_\theta \sharp \zeta(B) = \zeta(g_\theta^{-1}(B))$ for any borel set B. In both cases, we force the patch distribution μ_t of the synthesized textures (depending on a variable t which is either the image u or the parameters θ) to be close to the empirical patch distribution ν for the optimal transport cost

$$\mathrm{OT}_c(\mu_t, \nu) = \inf_{\pi \in \Pi(\mu_t, \nu)} \int c(x, y) d\pi(x, y), \tag{1}$$

---

[1] For color images we generally have $d = 3$ and $\mathcal{K} = [0, 1]^3$.

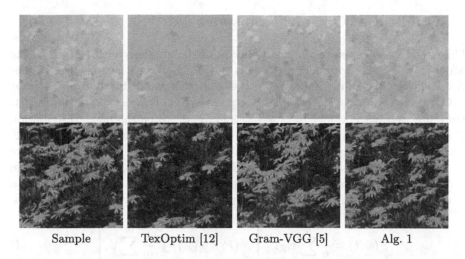

Sample   TexOptim [12]   Gram-VGG [5]   Alg. 1

Fig. 1. Two synthesis examples illustrating the drawbacks that are overcome with our optimal transport framework on patches (Algorithm 1). Contrary to the method TexOptim from [12], our method respects the statistics from the exemplar image and unlike the method of Gram-VGG from [5], it does not suffers from color inconsistency artifacts.

where $c : \mathcal{K}^p \times \mathcal{K}^p \to \mathbf{R}$ is a Lipschitz cost function and $\Pi(\mu_t, \nu)$ is the set of probability distributions on $\mathcal{K}^p \times \mathcal{K}^p$ having marginals μ_t and ν. When using $c(x,y) = \|x - y\|^2$, as done for experiments in this work, OT_c corresponds to the square of the Wasserstein-2 distance. Minimizing with respect to t the optimal transport cost from Eq. (1) appears to be a hard task. Therefore, considering the discrete nature of ν, we propose to take advantage of the semi-discrete formulation of the optimal transport cost (see [16])

$$\mathrm{OT}_c(\mu_t, \nu) = \max_{\psi \in \mathbf{R}^m} F(\psi, t) := \int \psi^c(x) d\mu_t(x) + \frac{1}{m} \sum_{j=1}^m \psi_j \qquad (2)$$

where the c-transform of ψ writes in this case $\psi^c(x) = \min_j [c(x, P_j v) - \psi_j]$. Estimating the variable t, *i.e.* the image u or the generator's parameters θ, amounts to solving the following minimax problem

$$\min_t \max_\psi F(\psi, t). \qquad (3)$$

For fixed t, the function $\psi \mapsto F(\psi, t)$ is concave and an optimal ψ^* can be approximated with an averaged stochastic gradient ascent as proposed in [6].

When the variable t is an image, we propose in Sect. 3 to perform a gradient descent, which outcome is illustrated in Fig. 1. A stochastic gradient-based algorithm is then proposed in Sect. 4 to learn a generative model using a convolutional neural network. Both approaches exploit the property, demonstrated in Sect. 2, that the gradient of the optimal transport $\nabla_t \mathrm{OT}_c(\mu_t, \nu)$ coincides with the gradient $\nabla_t F(\psi^*, t)$ of the function F at an optimal value ψ^*.

2   Gradient of Optimal Transport

This section is devoted to the computation of the gradient with respect to the parameter t of the optimal transport cost $\mathrm{OT}_c(\mu_t, \nu)$ in the considered semi-discrete case. We show, under hypotheses on the patch distribution μ_t, that $\nabla_t \mathrm{OT}_c(\mu_t, \nu) = \nabla_t F(\psi^*, t)$ with $\psi^* \in \arg\max_\psi F(\psi, t)$, when both terms exist. More precisely, we consider $\mu_t = \frac{1}{n} \sum_{i=1}^n (P_i \circ g_t) \sharp \zeta$ to be the patch distribution of a generative model $g_t \sharp \zeta$, where ζ is a fixed probability measure on $\mathcal{Z} \subset \mathbf{R}^r$ and $g_t(\cdot) = g(t, \cdot)$ is defined with a measurable function $g : \mathcal{T} \times \mathcal{Z} \to \mathcal{K}^n$ where $\mathcal{T} \subset \mathbf{R}^q$ is the open set of parameters. For a given z, $g_t(z)$ is an image whose patches are $P_i g_t(z) = (P_i \circ g_t)(z)$. This encompasses the image optimization problem of Sect. 3 using $g_t \sharp \zeta = \delta_t$. The function F defined in (2) then becomes

$$F(\psi, t) = \frac{1}{n} \sum_{i=1}^n \mathbf{E}_{Z \sim \zeta} \left[\psi^c(P_i \circ g_t(Z)) + \frac{1}{m} \sum_{j=1}^m \psi_j \right]. \tag{4}$$

In order to solve the related minimax problem (3), we need the gradients of F with respect to ψ and t. The function F is concave in ψ and its gradient has been studied in [6]. To compute the gradient with respect to t, one has to deal with the points of non-differentiability of $\psi^c(x)$, that reads $\psi^c(x) = \min_j [c(x, P_j v) - \psi_j]$ in the semi-discrete case. To that end, we introduce the open Laguerre cells

$$L_j(\psi) = \{ x \mid \forall k \neq j, \ c(x, P_j v) - \psi_j < c(x, P_k v) - \psi_k \}. \tag{5}$$

The set of points where ψ^c is differentiable coincides with $\cup_j L_j(\psi)$, whose complement is negligible if c is a ℓ_p cost. In order to avoid points living in this complementary set, we introduce the following hypothesis.

Hypothesis 1. *g satisfies Hypothesis 1 at (t, ψ) if $\zeta \left((P_i \circ g_t)^{-1} \{\cup_j L_j(\psi)\} \right) = 1$ for any patch position i, that is, for a given variable t, the generated patches are almost surely within the Laguerre cells defined by ψ.*

Note that Hypothesis 1 is satisfied for any ψ if $c(x, y) = \|x - y\|_p^p$ and if, for any i, $(P_i \circ g_t) \sharp \zeta$ is absolutely continuous with respect to the Lebesgue measure.

   We also introduce a regularity hypothesis for the generative model g_t to control its variations and differentiate under expectation.

Hypothesis 2. *There exists $K : \mathcal{T} \times \mathcal{Z} \to \mathbf{R}_+$ such that for all t, there exists a neighborhood V of t such that $\forall \ t' \in V$ and $\forall \ z \in \mathcal{Z}$*

$$\|g(t, z) - g(t', z)\| \le K(t, z)\|t - t'\| \tag{6}$$

with K verifying for all t, $\mathbf{E}_{Z \sim \zeta}[K(t, Z)] < \infty$.

Now, we show the following theorem that ensures the differentiability of F with respect to the parameter t.

Theorem 1. *Assume c to be $\mathscr{C}^1$. Let g satisfy Hypothesis 2. Let t_0 be a point where $t \to g_t(z)$ is differentiable $\zeta(z)$-a.e. and let g satisfy Hypothesis 1 at (t_0, ψ). Then $t \mapsto F(\psi, t)$ is differentiable at t_0 and*

$$\nabla_t F(\psi, t_0) = \frac{1}{n} \sum_{i=1}^{n} \mathbf{E}_{Z \sim \varsigma} \left[(\partial_t g(t_0, Z))^T \nabla \psi^c(P_i g(t_0, Z)) \right] \tag{7}$$

with $\nabla \psi^c(P_i g(t_0, z)) = \nabla_x c(P_i g(t_0, z), P_{\sigma(i)} v)$ where $\sigma(i)$ is the unique index such that $P_i g(t_0, z) \in L_{\sigma(i)}(\psi)$ (which exists $\zeta(z)$-almost surely).

Proof. Since ψ^c is differentiable on $\cup_j L_j(\psi)$, Hypothesis 1 implies that for any i, $\psi^c(P_i g_t(z))$ is differentiable at t_0 for ζ-almost every z with gradient $\partial_t g(t_0, z))^T \nabla \psi^c(P_i g(t_0, z))$. Since the derivatives of c are bounded by a constant C and since g satisfies Hypothesis 2, we get a neighborhood V of t_0 such that for any $t \in V$ and any z, $\|(\partial_t g(t, z))^T \nabla \psi^c(P_i g(t, z))\| \leq K(t_0, z)C$ with $\mathbf{E}[K(t_0, Z)] < \infty$. Differentiating under the expectation yields the final result. □

Finally, we relate the gradient of F to the gradient of the optimal transport.

Theorem 2. *Let t_0 such that $t \mapsto OT_c(\mu_t, \nu)$ and $t \mapsto F(\psi^*, t)$ are differentiable at t_0 with $\psi^* \in \arg\max_\psi F(\psi, t_0)$ then*

$$\nabla_t OT_c(\mu_{t_0}, \nu) = \nabla_t F(\psi^*, t_0) \tag{8}$$

Proof. Let fix $\psi^* \in \arg\max_\psi F(\psi, t_0)$. The function $h(t) = F(\psi^*, t) - OT_c(\mu_t, \nu)$ is differentiable at t_0 and maximized at t_0, therefore we get $\nabla_t h(t_0) = 0$. □

3 Image Optimization

In this section we introduce a pixelwise optimization algorithm that minimizes a optimal transport cost between patch distributions. This discrete optimal transport problem is covered by the framework of Sect. 2 by taking $g_u(z) = z - u$ for all z and $\zeta = \delta_0$, so that $g_u \sharp \delta_0 = \delta_u$. Then we relate this algorithm to the texture optimization framework of [12] and propose a multi-scale extension to account for different scales in the sample texture to synthetize. Finally, we illustrate the experimental stability and convergence of the proposed framework.

3.1 Mono-Scale Texture Synthesis Algorithm

Let $u \in \mathbf{R}^n$ be the image to synthesize with n pixels and $\mu_u = \frac{1}{n} \sum_{i=1}^{n} \delta_{P_i u}$ its patch distribution. In order to prescribe to u the patch distribution ν of the exemplar image v, we aim at solving

$$\min_{u \in \mathbf{R}^n} OT_c(\mu_u, \nu) = \min_{u \in \mathbf{R}^n} \max_{\psi \in \mathbf{R}^m} F(\psi, u), \tag{9}$$

where

$$F(\psi, u) = \frac{1}{n} \sum_{i=1}^{n} \psi^c(P_i u) + \frac{1}{m} \sum_{j=1}^{m} \psi_j \text{ and } \psi^c(P_i u) = \min_j \left[c(P_i u, P_j v) - \psi_j \right].$$

Note that the problem is quite similar to [7] where the primal formulation of optimal transport between discrete patch distributions is however considered. As in [12], the authors resort to an alternative minimization scheme requiring to solve an optimal assignment problem at each step, which turns out to be computationally prohibitive. To solve the minimax problem (9), we also propose an iterative alternate scheme in u and ψ, starting with an initial image u^0. However, for the discrete case of optimal transport being considered, the authors of [6] show that an optimal potential ψ^* can be estimated with a gradient ascent on ψ. Therefore, for a fixed u^k, we perform a gradient ascent with respect to ψ to obtain an approximation ψ^{k+1} of $\psi^* \in \arg\max_\psi F(\psi, u^k)$. A gradient descent step is then realized on u, using the gradient of F with respect to u given in Theorem 1. Note that if ψ^* is an optimal potential and if we are at a point of differentiability of the OT_c, Theorem 2 relates this gradient of F to the gradient of OT_c. Realizing a gradient step in this direction thus corresponds to performing a gradient descent step for the optimal transport.

Using the quadratic cost $c(x, y) = \frac{1}{2}\|x - y\|^2$, as in the experiments, we have

$$\nabla_u F(\psi^{k+1}, u^k) = \frac{1}{n} \left(\sum_{i=1}^n P_i^T P_i u^k - \sum_{i=1}^n P_i^T P_{\sigma^{k+1}(i)} v \right), \tag{10}$$

at a point (ψ^{k+1}, u^k) where we can uniquely define

$$\sigma^{k+1}(i) = \arg\min_j \frac{1}{2}\|P_i u^k - P_j v\|^2 - \psi_j^{k+1}. \tag{11}$$

Notice that P_j is a linear operator whose adjoint P_j^T maps a given patch q to an image whose j-patch is q and is zero elsewhere. Therefore $\sum_{i=1}^n P_i^T$ corresponds to an uniform patch aggregation. To simplify, we consider periodic conditions for patch extraction, so that $\sum_{i=1}^n P_i^T P_i = pI$, where $p = s \times s$ denotes the number of pixels in the patches. Hence, from (10) and considering a step size $\eta\frac{n}{p}$, $\eta > 0$, the update of u through gradient descent writes

$$u^{k+1} = (1 - \eta)u^k + \eta v^k, \tag{12}$$

where $v^k = \frac{1}{p}\sum_{i=1}^n P_i^T P_{\sigma^{k+1}(i)} v$ is the image formed with patches from the exemplar image v which are the nearest neighbors to the patches of u^k in the sense of (11). The gradient step then mixes the current image u^k with v^k. In the case $\psi = 0$, the minimum in (11) is reached by associating to each patch of u^k its ℓ_2 nearest neighbor in the set $\{P_1 v, \ldots, P_n v\}$, as similarly done in [12].

This image synthesis process is illustrated in Fig. 2, with a comparison of image synthesis for various patch sizes. As expected, this method cannot take into account variations that may occur at scales larger than s. We therefore propose a multi-scale extension in the next section.

3.2 Multi-scale Texture Synthesis

In order to deal with various texture scales, we extend our method in a multi-scale fashion. In [7], a coarse-to-fine greedy strategy is used, where the optimization

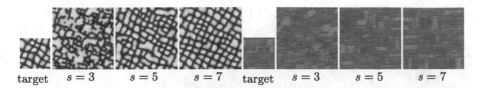

target $s = 3$ $s = 5$ $s = 7$ target $s = 3$ $s = 5$ $s = 7$

Fig. 2. Influence of patch-size s for the pixel optimization method Alg. 1 with $L = 1$.

Algorithm 1. Multi-scale Texture synthesis

Input: target image v, initial image u_0, learning rates η_u and η_ψ, number of iterations N_u and N_ψ, number of scales L

Output: image u

$u \leftarrow u_0$ and $\psi_l = 0$ for $l = 1 \ldots L$

for $k = 1$ **to** N_u **do**

    for $l = 1$ **to** L **do**

        estimate ψ_l^k with N_ψ iterations of gradient ascent of learning rate η_u

        $G_l(u^k, \psi_l^k) \leftarrow S_l^T \nabla_u F_{v_l}(\psi_l^k, S_l(u^k))$

    end for

    $u^{k+1} \leftarrow u^k - \eta_u(k) \sum_{l=1}^{L} G_l(u^k, \psi_l^k)$

end for

---

is performed iteratively at a smaller resolution, before upscaling the solution to the next scale. This strategy is employed on both image resolution and patch size in [12]. In this work, we propose to solve the optimal transport problem at different scales *simultaneously*.

We first create a pyramid of down-sampled and blurred images. For each scale $l - 1, \ldots, L$, we use a linear blurring and down-sampling operator S_l that computes a reduced version $u_l = S_l u$ of u of size $n/2^{l-1} \times n/2^{l-1}$. The multi-scale texture synthesis is obtained by minimizing

$$\mathcal{L}(u) = \sum_{l=1}^{L} \max_{\psi_l} F_{v_l}(u_l, \psi_l), \tag{13}$$

where $F_{v_l}(u_l, \psi_l) = \frac{1}{n} \sum_{j=1}^{n} \min_i \left[c(P_j u_l, P_i v_l) - (\psi_l)_i \right] + \frac{1}{m} \sum_{i=1}^{m} (\psi_l)_i$. As for the single-scale case, an alternate scheme is considered to minimize $\mathcal{L}$. The gradient descent update of u combines gradient at multiple scales: $\nabla_u \mathcal{L}(u) = \sum_{l=1}^{L} S_l^T \nabla_u F_{v_l}(u_l, \psi_l)$. The multi-scale process is summarized in Algorithm 1.

3.3 Experiments

In all experiments, we consider $L = 4$ scales and patches of size $s = 4$. We use auto-differentiation from the Pytorch package and gradient descent is performed with the Adam optimizer [11] with learning rate 0.01. We use $N_\psi = 10$ iterations for the estimation of ψ at each step. The process takes approximately 3 min to run 500 iterations for synthesizing a 256×256 image on a GPU Nvidia K40m.

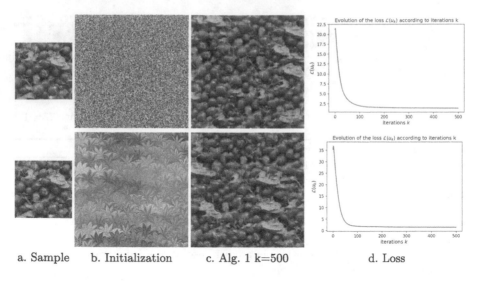

a. Sample   b. Initialization   c. Alg. 1 k=500   d. Loss

Fig. 3. Algorithm 1 is run for the same 100×100 sample (a) with two initial images (b). Both results in faithful 200×200 synthesis (c) in $k = 500$ iterations and the loss $\mathcal{L}(u_k)$ (13) shows a monotone convergence behaviour (d).

Figure 1 shows examples of synthesized textures with Algorithm 1 and comparisons with a patch-based method [12] and the state-of-the art method [5] prescribing deep neural features from VGG-19 [18]. While it is already known that the approach of [5] might have color inconsistencies [15], it mostly suffers here from the small resolution of the input, which makes difficult the extraction of deep features.

Contrary to our method and [5], the approach of [12] does not rely on statistics and does not respect the distribution of features from the original sample. Therefore, it must be initialized with a good guess (permutation of patch) instead of any random image. Additionally, it requires to sample large patches (from $s = 32$ to $s = 8$) on a sub-grid to enforce local copy and avoid blurring. We illustrate in Fig. 3 the stability of our method with respect to the initialization. Faithful textures are obtained from any initialization (column b): random image (first row) or another texture (second row).

With our Algorithm 1, the optimization nevertheless has to be done each time a new image is synthesized. In order to define a versatile algorithm that generates new samples on-the-fly, we rely on generative models in the next section.

4   Training a Convolutional Generative Network

In this section, we consider the problem of training a network to generate images that have a prescribed patch distribution at multiple scales. Then we present some visual results together with a comparison with existing methods. Finally we discuss quantitative evaluation of texture synthesis methods and we propose a framework to derive a multi-scale optimal transport loss between patch distributions that can be used as an evaluation score for texture synthesis.

4.1 Proposed Algorithm for Semi-discrete Formulation

We now consider a generator g_θ defined through a function g that is assumed to satisfy Hypothesis 2, which guarantees the existence of gradients in Theorem 1. The optimal transport formulation is now semi-discrete (in comparison with the previous discrete case). We then propose a stochastic alternate algorithm for training the generator g_θ.

From Theorem 2, the gradient of the optimal transport can be expressed with the gradient of the function F. Since this gradient writes as an expectation from Theorem 1, we perform a stochastic gradient descent considering the term inside the expectation in (7) as a stochastic gradient. In the semi-discrete case, an optimal potential ψ^* can also be approximated with an averaged stochastic gradient ascent [6]. This leads us to propose Algorithm 2 for minimizing the following loss w.r.t. parameters θ:

$$L(\theta) = \sum_{l=1}^{L} \max_{\psi_l} \mathbf{E}_{z \sim \zeta} \left[F_{v_l}(\psi_l, (g_\theta(z))_l) \right]. \tag{14}$$

In practice, for each iteration k and at each layer l we first update the corresponding potential ψ_l with an averaged stochastic gradient ascent as proposed in [6]. Then we sample an image and perform a stochastic gradient step in θ. In order to test our framework, the function g_θ has the same convolutional architecture as the one used for texture generation in [19]. This network has been designed to synthesize textures by minimizing the Gram-VGG loss introduced in [5]. We next demonstrate that we are able to learn the parameters of such a generative network by only enforcing the patch distributions at various scales.

In our PyTorch implementation , we use the Adam optimizer [11] to estimate the parameters θ. We run the algorithm for 10000 iterations with a learning-rate $\eta_\theta = 0.01$. An averaged stochastic gradient ascent with 100 inner iterations is used for computing ψ^*. In this setting, 30 min are required to train our generator with a GPU Nvidia K40m.

Algorithm 2. Learning a texture generator with stochastic gradient descent

---

Input: target image v, initial weight θ_0, learning rate η_θ, number of iterations N_u and N_ψ, number of scales L
Output: generator parameters θ
for $k = 1$ **to** N_u **do**
  for $l = 1$ **to** L **do**
    estimate ψ_l^k with N_ψ iterations of averaged stochastic gradient ascent
    sample z from ζ
    update $G_l(\theta_k)$ with Adam [11] step using $\nabla_\theta F_{v_l}(\psi_l^k, (g_{\theta^k}(z))_l)$
  end for
  $\theta^{k+1} \leftarrow \theta^k - \eta_\theta(k) \sum_{l=1}^{L} G_l(\theta^k)$
end for

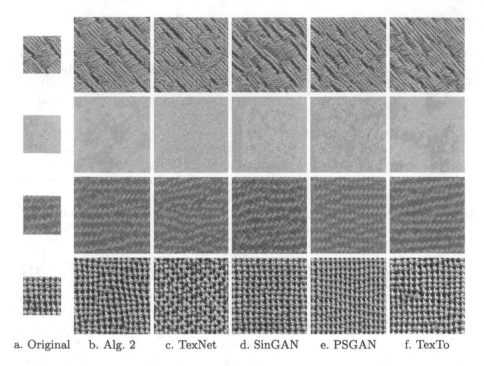

a. Original   b. Alg. 2   c. TexNet   d. SinGAN   e. PSGAN   f. TexTo

Fig. 4. Texture synthesis from a generative network trained on a single 256×256 sample (a). Comparison of our multi-scale approach (b) using 4×4 patches (see Alg. 2) with TexNet [19] (using VGG-19 features), SinGAN [17], PSGAN [1] and TexTo [14].

4.2   Experimental Results and Discussions

Figure 4 gives a comparison of our results with four relevant synthesis methods from the literature. We first consider the Texture Networks (TexNet) method [19], which trains a generative network using VGG-19 feature maps computed on a sample texture. Note that the very same convolutional architecture has been used for our model. We also compare to SinGAN [17] (a recent generative adversarial newtork (GAN) technique generating images from a single example relying on patch sampling), to PSGAN [1] (a previous approach that similarly adapts the GAN framework to the training of a single image) and to TexTo [14] (which also constrains patch distributions with optimal transport but in an indirect way). We used Pytorch implementations of SinGAN, PSGAN, and TexNet, with their default parameters.

The results obtained with our Alg. 2 are visually close to the ones from TexTo [14]. However, the patch-aggregation step from TexTo makes the results blurrier than our method which inherently deals with the aggregation issue. Although TexNet [19] produces textures that look sharper than our results, it may fail to reconstruct larger structures as in the fourth image. Observe that patch-based networks TexTo and SINGAN create less visual artifacts (checkerboard patterns due to VGG pooling, false colors, etc.). Dealing with patch

Table 1. Evaluation of texture synthesis from Fig. 4 for various discrepancy measures, emphasizing **best result** and second best (lower is better). The average score (*Avg*) is computed over all images. **SIFID** [17] is computed from first max pooling Inception features. **VGG Gram norm** [5] is computed from cross-correlation of VGG features as used by [19]. The proposed distance is based on multi-scale patch optimal transport.

	SIFID				*Avg*	VGG Gram norm				*Avg*	Muti-scale patch OT				*Avg*
Alg. 1	0.43	**0.02**	**0.08**	0.71	_0.31_	122	6	_141_	865	283	**0.45**	**0.15**	**0.09**	0.69	**0.35**
Alg. 2	1.13	_0.06_	0.18	1.82	0.80	233	19	151	922	331	_0.48_	_0.16_	_0.10_	0.78	_0.38_
TEXNET	**0.11**	0.08	0.18	**0.17**	**0.14**	_218_	9	**54**	**190**	**118**	0.65	0.24	0.17	1.22	0.57
SINGAN	0.93	0.10	_0.17_	_0.37_	0.39	299	_8_	207	_394_	_227_	0.54	0.24	0.26	0.79	0.46
PSGAN	_0.27_	0.91	1.14	0.49	0.70	224	512	753	1366	714	0.68	0.43	0.34	1.19	0.66
TEXTO	1.22	0.07	0.18	1.67	0.79	260	24	152	1030	367	0.49	_0.16_	0.11	_0.75_	_0.38_

distributions can lead to local copy of large-scale structures as we can easily see in the second row of Fig. 4. Although similar large scale structures are copied, they are not exact copy/pastes of the same area from the exemplar texture and local changes can be observed within these similar patterns. Moreover, this phenomenon also appears for other methods, particularly in SINGAN and PSGAN.

4.3 Evaluation of Texture Synthesis Methods

Evaluating texture synthesis methods is a complex and open question. The visual quality is subjective and for now, there is no widely accepted perceptual metric. For quantitative evaluation, several metrics have been proposed, such as SIFID [17] (Single Image Fréchet Inception Distance) or the metric given by the feature correlations of VGG [5]. Using our framework, we also propose a new evaluation metric that measures the Wasserstein distance between patch distributions at each scale. Table 1 presents the scores for these three metrics for textures from Fig. 4.

As expected, each method performs better for its associated metric. Our two algorithms and TexTo present the lowest scores for the proposed optimal transport loss, whereas TexNet [19] obtains the best results with the metric based on VGG features or the inception network. Our algorithms reach competitive scores for all these metrics and achieve the best results for two of the four proposed textures with Alg. 1. Let us mention however that the considered metrics are not always directly correlated to perception: in the last texture of Fig. 4, TexNet presents smaller values for both SIFID and VGG scores, whereas the synthesized texture does not match the input one in term of large-scale coherence.

References

1. Bergmann, U., Jetchev, N., Vollgraf, R.: Learning texture manifolds with the periodic spatial gan. In: Proceedings of the 34th International Conference on Machine Learning, vol. 70, pp. 469–477. JMLR.org (2017)
2. Buades, A., Coll, B., Morel, J.M.: A non-local algorithm for image denoising. In: 2005 IEEE Computer Society Conference on Computer Vision and Pattern Recognition (CVPR 2005), vol. 2, pp. 60–65. IEEE (2005)
3. Efros, A.A., Leung, T.K.: Texture synthesis by non-parametric sampling. In: IEEE International Conference on Computer Vision, p. 1033 (1999)
4. Galerne, B., Leclaire, A., Rabin, J.: A texture synthesis model based on semi-discrete optimal transport in patch space. SIAM J. Imaging Sci. **11**(4), 2456–2493 (2018)
5. Gatys, L., Ecker, A.S., Bethge, M.: Texture synthesis using convolutional neural networks. In: NIPS, pp. 262–270 (2015)
6. Genevay, A., Cuturi, M., Peyré, G., Bach, F.: Stochastic optimization for large-scale optimal transport. In: Advances in Neural Information Processing Systems, pp. 3440–3448 (2016)
7. Gutierrez, J., Galerne, B., Rabin, J., Hurtut, T.: Optimal patch assignment for statistically constrained texture synthesis. In: Scale-Space and Variational Methods in Computer Vision (2017)
8. Houdard, A., Bouveyron, C., Delon, J.: High-dimensional mixture models for unsupervised image denoising (HDMI). SIAM J. Imaging Sci. **11**(4), 2815–2846 (2018)
9. Karras, T., Laine, S., Aila, T.: A style-based generator architecture for generative adversarial networks. In: Proceedings of the IEEE Conference on Computer Vision and Pattern Recognition, pp. 4401–4410 (2019)
10. Kaspar, A., Neubert, B., Lischinski, D., Pauly, M., Kopf, J.: Self tuning texture optimization. Comput. Graph. Forum **34**, 349–359 (2015)
11. Kingma, D.P., Ba, J.: Adam: a method for stochastic optimization. In: ICLR (2014)
12. Kwatra, V., Essa, I., Bobick, A., Kwatra, N.: Texture optimization for example-based synthesis. In: ACM SIGGRAPH 2005 Papers, pp. 795–802 (2005)
13. Lebrun, M., Buades, A., Morel, J.M.: A nonlocal bayesian image denoising algorithm. SIAM J. Imaging Sci. **6**(3), 1665–1688 (2013)
14. Leclaire, A., Rabin, J.: A fast multi-layer approximation to semi-discrete optimal transport. In: Lellmann, J., Burger, M., Modersitzki, J. (eds.) SSVM 2019. LNCS, vol. 11603, pp. 341–353. Springer, Cham (2019). https://doi.org/10.1007/978-3-030-22368-7_27
15. Liu, G., Gousseau, Y., Xia, G.: Texture synthesis through convolutional neural networks and spectrum constraints. In: International Conference on Pattern Recognition (ICPR), pp. 3234–3239. IEEE (2016)
16. Santambrogio, F.: Optimal transport for applied mathematicians. Progr. Nonlinear Differ. Equ. Appl. **87** (2015)
17. Shaham, T.R., Dekel, T., Michaeli, T.: Singan: learning a generative model from a single natural image. In: Proceedings of the IEEE International Conference on Computer Vision, pp. 4570–4580 (2019)
18. Simonyan, K., Zisserman, A.: Very deep convolutional networks for large-scale image recognition. arXiv preprint arXiv:1409.1556 (2014)
19. Ulyanov, D., Lebedev, V., Vedaldi, A., Lempitsky, V.: Texture networks: feed-forward synthesis of textures and stylized images. In: Proceedings of the International Conference on Machine Learning, vol. 48, pp. 1349–1357 (2016)

Sketched Learning for Image Denoising

Hui Shi[✉], Yann Traonmilin, and Jean-François Aujol

Univ. Bordeaux, Bordeaux INP, CNRS, IMB, UMR 5251, 33400 Talence, France
`hui.shi@u-bordeaux.fr`

Abstract. The Expected Patch Log-Likelihood algorithm (EPLL) and its extensions have shown good performances for image denoising. It estimates a Gaussian mixture model (GMM) from a training database of image patches and it uses the GMM as a prior for denoising. In this work, we adapt the *sketching* framework to carry out the compressive estimation of Gaussian mixture models with low rank covariances for image patches. With this method, we estimate models from a compressive representation of the training data with a learning cost that does not depend on the number of items in the database. Our method adds another dimension reduction technique (low-rank modeling of covariances) to the existing sketching methods in order to reduce the dimension of model parameters and to add flexibility to the modeling. We test our model on synthetic data and real large-scale data for patch-based image denoising. We show that we can produce denoising performance close to the models estimated from the original training database, opening the way for the study of denoising strategies using huge patch databases.

Keywords: Image denoising · Sketching · Optimisation · Machine learning

1 Introduction

In image processing, non-local patch-based models have been producing state-of-the art results for classic image denoising problems [2,18,24]. Patch-based methods are also beneficial to other image inverse problems such as superresolution [8,12], inpainting [7] and deblurring [16]. Among these various non-local methods, the Expected Patch Log-Likelihood algorithm (EPLL) [26] shows very good restoration performances.

The EPLL method uses Gaussian mixture models (GMMs) as a prior model for natural images. In order to maximize the redundancy of structural information to estimate the best possible model parameter, we would want to use a very large training database. However, estimating parameters from a large database can be impractical for classic parameter estimation techniques such as Expectation-Maximization (EM), as their memory consumption and computation time depend on the database size.

This work was partly funded by ANR project EFFIREG - ANR-20-CE40-0001.

Recent works [4, 13, 17] propose a scalable technique to learn model parameters from a compressive representation: a *sketch* of the training data collection. It leverages ideas from compressive sensing [11] and streaming algorithms [6] to compress a large database into a size-fixed representation. Thus, space and time complexity of the algorithm for the estimation of the model no longer depends on the original database size, but only on the size of compressed data and on the dimensionality of the model. Sketching has been used successfully for clustering [4] and GMM estimation with diagonal covariances [17] using the greedy Continuous Orthogonal Matching Pursuit (COMP) algorithm. Sketching produces accurate estimates while requiring fewer memory space and calculations. Sketching also has the advantage to be suitable for distributed computing.

Estimating GMMs on image patches is a complex large-scale learning task. The objective of this paper is to explore the sketching method in this context. In this work, we estimate GMMs priors with non-diagonal covariances which is an extension of previous works. Moreover, for a denoising task it has been shown that the rank of covariance matrices can be reduced [20]. We implement a model using low-rank modeling for GMMs covariances in order to manage the modeling of the image patches in the most possible flexible way.

Contributions: The main contributions of this work are the following.

- We describe how we can learn a GMM prior from a compressed database of patches in the context of image denoising.
- We extend the Continuous Orthogonal Matching Pursuit algorithm to be able to estimate GMM models with non-diagonal and possibly low rank covariances.
- We demonstrate the potential of the approach on real large-scale data (over 4 millions training samples) for the task of patch-based image denoising. We show that we can obtain denoising performances with models trained with the compressed database close to the performance of the denoising with the model obtained with the classical EM algorithm. To the best of our knowledge, this is also the first time that the sketching framework has been applied for such high dimensional GMM (GMM in dimension 25).

Outline: The article is organized as follows. In Sect. 2, we recall the EPLL framework for image denoising. In Sect. 3, we describe how sketching can be implemented within the specific setting of patch based denoising and we give an implementation with a low-rank technique for GMM estimation. In Sect. 4, we provide experimental results both on synthetic data and real images showing that our method has denoising performances close to the EM framework. Finally, we discuss future works in the conclusion.

2   Model Estimation and Denoising with EPLL

2.1   Denoising with EPLL

Expected Patch Log-Likelihood (EPLL) algorithm is a patch-based image restoration algorithm introduced by Zoran and Weiss [26]. It uses priors learned on patches extracted from a database of clean images. We consider the problem of recovering an image $u \in \mathbb{R}^N$ with N the number of pixels from a noisy version $v = u + w$, where $w \sim \mathcal{N}(0, \sigma^2 I_N)$ is a white Gaussian noise component. The EPLL framework restores an image u by using the following maximum a posteriori (MAP) estimation:

$$u^* = \arg\min_{u \in \mathbb{R}^N} \frac{P}{2\sigma^2} \|u - v\|^2 - \sum_{i=1}^{N} log(p(\mathcal{P}_i u)) \tag{1}$$

where $\mathcal{P}_i : \mathbb{R}^N \rightarrow \mathbb{R}^P$ is the linear operator that extracts a patch with P pixels centered at the position i and $p(.)$ is the density of the prior probability distribution of the patches.

Problem (1) is a large non-convex optimization problem as $p(.)$ is chosen as the density of a GMM prior. It can be extended to generalized Gaussian mixture model (GGMM) [10] for a better performance. In the following we keep the GMM model to simplify the description of the model and we leave the extension to GGMM to future work. In the case of GMM, the denoising can be performed with a simple patch by patch Wiener filter with the denoising parameter β.

$$\hat{u} = (I + \frac{\beta\sigma^2}{P} \sum_{i=1}^{N} \mathcal{P}_i^T \mathcal{P}_i)^{-1} (v + \frac{\beta\sigma^2}{P} \sum_{i=1}^{N} \mathcal{P}_i^T \hat{x}_i) \tag{2}$$

where the

$$\hat{x}_i = (\Sigma_{k_i^*} + \frac{1}{\beta} I_P)^{-1} \Sigma_{k_i^*} \tilde{x}_i \tag{3}$$

are denoised patches estimated from noisy patches $\tilde{x}_i$ which are attributed to a Gaussian prior k_i^*, where k_i^* is the component of the GMM that maximizes the likelihood for the given patch $\tilde{x}_i$, i.e. $k_i^* = \arg\max_{1 \leq k_i \leq K} p(k_i | \tilde{x}_i)$ (see e.g. [10]). Note that these operations are applied a few times with increasing β for best denoising performance.

2.2   EM

A classical technique to estimate the GMM is the Expectation-Maximization (EM) algorithm. It is an iterative algorithm to find estimates of GMM parameters, which carries out at each iteration two steps: the expectation step (E-Step), which creates a function for the expectation of the log-likelihood evaluated using the current estimate for the parameters; and the maximization step (M-Step), which computes parameters maximizing the expected log-likelihood found on the E-Step. These estimated parameters are then used to determine the distribution of the latent variables in the next E-Step.

The EM algorithm's average time complexity is $\mathcal{O}(K^2 n)$ when estimating a K-components model on a database of n elements. Learning parameters using EM technique face computational issues linked to the size of the data and the number of parameters to estimate, which would make the use of (very) large image patches databases impractical. Moreover, the EM algorithm is not guaranteed to lead us to the global optimum, it typically converges to a local one [1,25]. It may be arbitrarily poor in high dimensions [9].

3 Compressive GMM Learning from Large Image Patches Database with Sketches

We begin by recalling the sketching method, then we show how to extend previous works to manage the case of GMM prior on image patches.

3.1 Compressive Mixture Estimation

In the sketching framework [14,15], a measure $f \in \mathcal{D}$ ($\mathcal{D}$ is the set of probability measures over $\mathbb{R}^d$) is encoded with a linear sketching operator $\mathcal{S} : \mathcal{D} \to \mathbb{C}^m$ into a compressed representation $z \in \mathbb{R}^m$:

$$z = \mathcal{S}f \qquad (4)$$

We call z a sketch of f. In practice we only have access to the empirical probability distribution $y = \frac{1}{n} \sum_{i=1}^{n} \delta_{x_i}$ where $\chi = \{x_1, ..., x_n\} \subset \mathbb{R}^d$ is the training database (δ_{x_i} is a unit mass at x_i), which we compress into a sketched database $\tilde{y} = \frac{1}{n} \mathcal{S} \sum_{i=1}^{n} \delta_{x_i}$. The goal of the sketching framework is to recover f from $\tilde{y}$.

For some finite $K \in \mathbb{N}^*$, we define a K-sparse model in $\mathcal{D}$ with the parameters $\Theta = \{\theta_1, ..., \theta_K\}$ and the weights $\alpha = \{\alpha_1, ..., \alpha_K\}$:

$$f_{\Theta,\alpha} = \sum_{k=1}^{K} \alpha_k f_{\theta_k} \qquad (5)$$

where $f_{\theta_k} \in \mathcal{D}$ are measures parametrized by θ_k, $\alpha_k \in \mathbb{R}^+$ for all components and $\sum_{k=1}^{K} \alpha_k = 1$. The vector z can then be expressed as

$$z = \mathcal{S}f_{\Theta,\alpha} = \sum_{k=1}^{K} \alpha_k \mathcal{S} f_{\theta_k} \qquad (6)$$

The objective of sketched learning algorithms is to minimize the energy between the compressed database and the sketch of the estimation. It corresponds to the traditional parametric optimization Generalized Method of Moments. We estimate the parameters with the following minimization

$$(\hat{\Theta}, \hat{\alpha}) = \underset{\substack{\Theta \in \mathbb{R}^K \\ \alpha \in \mathbb{R}^K, \alpha_k > 0, \sum_{k=1}^{K} \alpha_k = 1}}{\arg\min} \|\mathcal{S}f_{\Theta,\alpha} - \tilde{y}\|_2^2, \qquad (7)$$

i.e. our aim is to find the probability distribution (the parameters α, Θ) whose sketch is closest to the empirical sketch $\tilde{y}$. It was shown in [15] that we can theoretically guarantee the success of this estimation with a condition on the sketch size. In particular, sketching uses the "lower restricted isometry property" (LRIP) for the recovery guarantee. This property, is verified, for GMM with sufficiently separated means and random Fourier sketching with high probability as long as $m \geq O(k^2 d \text{polylog}(k, d))$, i.e. when the size of the sketch essentially depends on the number of parameters k, d (empirical results seem to indicate that for Γ the number of parameters, a database size of the order of Γ is sufficient). The excess risk of the GMM learning task is then controlled by the sum of an empirical error term and a modeling error term. This guarantees that the estimated GMM approximates well the distribution of the data.

In our case, the sketched GMM learning problem reduces to the estimation of the sum of k zero-mean Gaussians with covariances $\Theta = (\Sigma_k)_{k=1}^{K}$, i.e. $f_{\Theta,\alpha} = \sum_{k=1}^{K} \alpha_k g_{\Sigma_k}$ where g_Σ is the zero mean Gaussian measure with covariance Σ. The mean is not needed in the denoising process and it is removed from the patches before sketching and denoising. In this context, the notion of separation used to prove guarantees in [15] does not hold. We still show empirically that the sketching process is successful without this separation assumption.

Examples on synthetic data illustrate that a different notion of separation might be more suitable, which opens interesting new theoretical questions.

3.2 Design of Sketching Operator: Randomly Sampling the Characteristic Function

In [17], the sketch is a sampling of the characteristic function (*i.e.* the Fourier transform of the probability distribution f). The characteristic function ψ_f of a distribution f is defined as:

$$\psi_f(\omega) = \int_{\mathbb{R}^d} e^{-i\omega^T x} df(x) \quad \forall \omega \in \mathbb{R}^d \tag{8}$$

The sketching operator is therefore expressed as:

$$\mathcal{S}f = \frac{1}{\sqrt{m}}[\psi(\omega_1), ..., \psi(\omega_m)]^T \tag{9}$$

where $\Omega = (\omega_1, ..., \omega_m)$ is a set of well chosen frequencies.

In the context of images, given a training set of n centered patches $\chi = \{x_1, ..., x_n\} \subset \mathbb{R}^P$, we define the empirical characteristic function with $\tilde{\psi}(w) = \frac{1}{n}\sum_{i=1}^{n} e^{-i\omega^T x_i}$. Thus the empirical sketch is:

$$\tilde{y} = \frac{1}{\sqrt{m}}[\tilde{\psi}(\omega_1), ..., \tilde{\psi}(\omega_m)]^T \tag{10}$$

In other words, a sample of the sketched database is a P-dimensional frequency component calculated by averaging over patches (not to be mixed with usual 2D Fourier components of images)

$$\tilde{\psi}(\omega_l) = \frac{1}{n}\sum_{i=1}^{n} e^{-i\omega_l^T x_i} \tag{11}$$

Thanks to the properties of the Fourier transform of Gaussians, the sketch of a single zero-mean Gaussian component g_Σ is

$$(\mathcal{S}(g_\Sigma))_l = e^{-\frac{1}{2}\omega_l^T \Sigma \omega_l}. \tag{12}$$

The choice of frequencies is essential to the success of sketching. Theoretical estimation results are given with random Gaussian frequencies. In practice we generate a Gaussian profile of the amplitude of the frequency using a small sample of the database and we generate randomly the angle of the frequency [17].

3.3   Extension to Low Rank Covariances

Bayesian MAP theory permits to use degenerate covariances as a denoising prior. As we perform Wiener filtering, this is useful as we can reduce the number of parameters by just truncating the component of noisy patches supported on the lowest eigenvalues of Σ. A Gaussian covariance Σ_k is low-rank if there exists a rank r such that we can write $\Sigma_k = X_k X_k^T$ with X_k a $P \times r$ matrix. Our goal is to estimate covariances Θ^* close to the optimal $\hat{\Theta}$. Remark that:

$$\|\mathcal{S}f_{\Theta^*} - \mathcal{S}f_{\hat{\Theta}}\|^2 = \|\sum_{k=1}^K \alpha_k \mathcal{S}(f_{\Sigma_k^*} - f_{\hat{\Sigma}_k})\|^2 \tag{13}$$
$$= \sum_{l=1}^m e_l^2$$

where

$$e_l := \left| \sum_{k=1}^K \alpha_k (e^{-\frac{1}{2}\omega_l^T \Sigma_k^* \omega_l} - e^{-\frac{1}{2}\omega_l^T \hat{\Sigma}_k \omega_l}) \right| \tag{14}$$

We have, using the Taylor expansion of the exponential,

$$\left| e^{-\frac{1}{2}\omega_l^T \Sigma_k^* \omega_l} - e^{-\frac{1}{2}\omega_l^T \hat{\Sigma}_k \omega_l} \right| = \left| e^{-\frac{1}{2}\omega_l^T \Sigma_k^* \omega_l} \left(1 - e^{-\frac{1}{2}\omega_l^T (\hat{\Sigma}_k - \Sigma_k^*)\omega_l}\right) \right|$$
$$= e^{-\frac{1}{2}\omega_l^T \Sigma_k^* \omega_l} \mathcal{O}(\|\hat{\Sigma}_k - \Sigma_k^*\|_F) \tag{15}$$
$$\leq C_{\Theta,\Omega} \|\Sigma_k^* - \hat{\Sigma}_k\|_F.$$

Close to the minimizer, the energy (7) is close to the weighted sum of the Frobenius distance between covariance matrices.

Following classical ideas in low-rank matrix estimation we parametrize Σ_k by its factors X_k: $\Sigma_k = X_k X_k^T$. This is often referred as the Burer-Monteiro method [3,5]. Assume that

$$X_k^* \in \underset{X \in \mathbb{R}^{P \times r}}{\arg\min} \|X_k X_k^T - \Sigma_k\|_F^2. \tag{16}$$

A classical result is that $X_k^* X_k^{*,T} = U_k \Lambda_k U_k^T$ with $\Lambda_k = \text{diag}(\lambda_1, ..., \lambda_P)$ and $\lambda_1 \geq \lambda_2... \geq \lambda_P$ are the ordered eigenvalues of Σ_k (Eckart and Young theorem). Hence, minimizing the Frobenius distance with a reduced rank recovers the largest components of Σ_k. Using this qualitative argument, we approximate minimization (7) by

$$(\hat{X}, \hat{\alpha}) = \underset{\substack{X_k \in \mathbb{R}^{P \times r}, \forall k \\ \alpha \in \mathbb{R}^K, \alpha_k > 0, \sum_{k=1}^{K} \alpha_k = 1}}{\arg\min} \frac{1}{\sqrt{m}} \sum_{l=1}^{m} \left| \frac{1}{n} \sum_{i=1}^{n} e^{-i\omega_l^T x_i} - \sum_{k=1}^{K} \alpha_k e^{-\frac{1}{2}\omega_l^T X_k X_k^T \omega_l} \right|^2$$

(17)

where $\hat{X} = (\hat{X}_1, ..., \hat{X}_K)$ is the collection of factorized rank reduced covariances.

3.4 An Algorithm for Patch Prior Learning from Sketch: LR-COMP (Low Rank Continuous Orthogonal Matching Pursuit)

Problem (17) can be solved approximately using the greedy Continuous Orthogonal Matching Pursuit (COMP) algorithm (also called CL-OMP) [17]. We adapt this algorithm in the GMMs context with our low-rank approximation (Algorithm 1).

Algorithm 1: LR-COMP: Compressive GMM estimation with low-rank covariances.

Data: Empirical sketch $\tilde{y}$, sketching operator $\mathcal{S}$, sparsity K, number of iterations $T \geq K$

Result: Support Θ, weights α

$\hat{r} \leftarrow \tilde{y}$; $\Theta \leftarrow \emptyset$;

for $t - 1$ **to** T **do**

> **Step 1:** Find a X such that: $X \leftarrow \arg\max_X Re \left\langle \frac{\mathcal{S}f_X}{\|\mathcal{S}f_X\|_2}, \hat{r} \right\rangle_2$, init = rand;
>
> **Step 2:** $\Theta \leftarrow \Theta \cup \{X\}$;
>
> **Step 3:** Enforce sparsity by Hard Thresholding if needed;
>
> **if** $|\Theta| > K$ **then**
>
>> $\eta \leftarrow \arg\min_{\eta \geq 0} \left\| \tilde{y} - \sum_{k=1}^{|\Theta|} \eta_k \frac{\mathcal{S}f_{X_k}}{\|\mathcal{S}f_{X_k}\|_2} \right\|_2^2$;
>>
>> Select K largest entries $\eta_{i_1}, ..., \eta_{i_K}$;
>>
>> Reduce the support $\Theta \leftarrow \{X_{i_1}, ..., X_{i_K}\}$;
>
> **Step 4:** Project to find weights;
>
> $\alpha \leftarrow \arg\min_{\alpha \geq 0} \left\| \tilde{y} - \sum_{k=1}^{|\Theta|} \alpha_k \mathcal{S}f_{X_k} \right\|_2^2$;
>
> **Step 5:** Perform a gradient descent initialized with current parameters;
>
> $\Theta, \alpha \leftarrow \arg\min_{\Theta, \alpha \geq 0} \left\| \tilde{y} - \sum_{k=1}^{|\Theta|} \alpha_k \mathcal{S}f_{X_k} \right\|_2^2$, init = (Θ, α);
>
> **Step 6:** Update residual: $\hat{r} \leftarrow \tilde{y} - \sum_{k=1}^{|\Theta|} \alpha_k \mathcal{S}f_{X_k}$;

Normalize α such that $\sum_{k=1}^{K} \alpha_k = 1$.

The main tool for the implementation of Algorithm 1 is to compute the gradients necessary to perform the gradients in Steps 1, 4 and 5. We define the vector $v(X) = [Re(\mathcal{S}f_X); Im(\mathcal{S}f_X)] \in \mathbb{R}^{2m}$ with

$$v(X) = \begin{bmatrix} [Re(\frac{1}{\sqrt{m}}\psi_X(\omega_l))]_{l=1,...,m} \\ [Im(\frac{1}{\sqrt{m}}\psi_X(\omega_l))]_{l=1,...,m} \end{bmatrix} = \begin{bmatrix} [\frac{1}{\sqrt{m}}e^{-\frac{1}{2}(\omega_l^T X X^T \omega_l)}]_{l=1,...,m} \\ 0 \end{bmatrix} \quad (18)$$

To calculate the gradient, we only need to be able to calculate, for a given vector $y \in \mathbb{R}^{2m}$, the scalar products

$$\langle \nabla_X v(X), y \rangle = -B(v(X)_{1:m} * y_{1:m}) \quad (19)$$

where $B \in M_{J,m}(\mathbb{R})$, $J = P \times r$ is a block matrix with

$$B(j,:) = X(:,q)^T W * W(s,:), \quad \forall j = (q-1)P + s \quad (20)$$

where $W = [\omega_1, ..., \omega_m] \in M_{P,m}(\mathbb{R})$ the frequency matrix and $*$ the multiplication element by element.

4   Results and Analysis

4.1   Experiments with Synthetic Data

We generate data with the following settings: $n = 10^5$ items, dimension $d = 4$, the sparsity level of GMM $K = 8$. The parameters of sketching are: the size of sketch $m = 500$, the rank $r = 2$. We compare the estimation from sketch with the estimation of the GMM with EM algorithm. Figure 1 shows the reconstruction performance (projected on the first 2 dimensions). We see that we are able to estimate an accurate GMM model from the sketch of the data and the energy (7) of the 2 models are closed. This figure also illustrates that, although Gaussians have zero mean, they have an angular separation instead of a separation of the means (used to give estimation guarantees in [14]). This opens the question of establishing recovery guarantees for zero mean Gaussians using a different notion of separation.

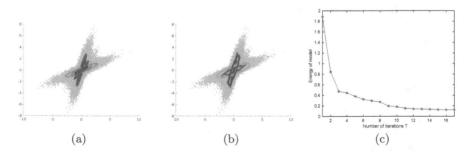

(a)   (b)   (c)

Fig. 1. Modeling on synthetic data: (a) with EM, (b) with sketching, (c) the model (b) is obtained by minimising energy in (17).

Fig. 2. Denoising results: (a) original, (b) noisy images with $\sigma^2 = 15$, (c) results with truncated EM model, (d) results LR-COMP model with PSNR/SSIM. Similar denoising performances are obtained with LR-COMP with a compressed database 250 times smaller.

Fig. 3. Denoising results: (a) original, (b) noisy images with $\sigma^2 = 20$, (c) results with truncated EM model, (d) results LR-COMP model with PSNR/SSIM. Similar denoising performances are obtained with LR-COMP with a compressed database 250 times smaller.

4.2 Results with Real Images

We extract randomly $n = 4 \times 10^6$ patches of size $P = 5 \times 5$ from the training images of Berkeley Segmentation Database (BSDS) [19]. We show the result of denoising with a prior model estimated with EM (with covariances truncated to have rank r) and with LR-COMP. We use $K = 20$ to demonstrate the capability of our algorithm. Our experiments show that we cannot reduce the rank too much to keep good denoising performance. Setting a rank $r = 20$ shows no loss of performance (for both EM and LR-COMP). We set $m = 40K(P \times r + 1) \approx 4 \times 10^5$, i.e. the compressed database is 250 times smaller than the original patch database. The gains in terms of memory is approximately $\frac{n}{m}$ times compared to the EM approach (most of memory is used to store the frequency matrix). We show results for noise levels $\sigma^2 = 15$ (Fig. 2) and $\sigma^2 = 20$ (Fig. 3). We observe that we obtain similar denoising performances for most images, the worst case being with the "barbara" image which has high contrast and high frequency content. Better results are obtained for the satellite image.

5 Conclusion

In this work, we provide an implementation of the sketching method to estimate a prior model from a compressed database for image denoising. It is shown that a high-dimensional Gaussian mixture model can be learned from a compressed database of patches, and then used for patch-based denoising. We achieve performance close to state-of-the art model based methods.

This work opens several perspectives. We saw that performance is degraded for a particular type of image. One possible explanation is that the sketching (i.e. the choice of frequencies) "missed" these particular images as we used frequencies from previous sketching literature. Adapting this choice to the case of zero mean GMM is still an open question (theoretically and practically). We demonstrated the feasibility of image denoising with sketches. Even if its complexity does not depend on the size of the original database, the LR-COMP algorithm still has computational issues. A possible direction is to extend algorithms proposed in [21,22] for the estimation of sums of Diracs to the case of GMM with potential performance guarantees [23].

References

1. Balakrishnan, S., Wainwright, M.J., Yu, B., et al.: Statistical guarantees for the EM algorithm: from population to sample-based analysis. Ann. Stat. **45**(1), 77–120 (2017)
2. Buades, A., Coll, B., Morel, J.M.: A review of image denoising algorithms, with a new one. Multiscale Model. Simul. **4**(2), 490–530 (2005)
3. Burer, S., Monteiro, R.D.: Local minima and convergence in low-rank semidefinite programming. Math. Program. **103**(3), 427–444 (2005)

4. Chatalic, A., Gribonval, R., Keriven, N.: Large-scale high-dimensional clustering with fast sketching. In: 2018 International Conference on Acoustics, Speech and Signal Processing (ICASSP), pp. 4714–4718. IEEE (2018)
5. Chi, Y., Lu, Y.M., Chen, Y.: Nonconvex optimization meets low-rank matrix factorization: an overview. IEEE Trans. Signal Process. **67**(20), 5239–5269 (2019)
6. Cormode, G., Muthukrishnan, S.: An improved data stream summary: the count-min sketch and its applications. J. Algorithms **55**(1), 58–75 (2005)
7. Criminisi, A., Pérez, P., Toyama, K.: Region filling and object removal by exemplar-based image inpainting. IEEE Trans. Image Process. **13**(9), 1200–1212 (2004)
8. Danielyan, A., Foi, A., Katkovnik, V., Egiazarian, K.: Image upsampling via spatially adaptive block-matching filtering. In: 2008 16th European Signal Processing Conference, pp. 1–5. IEEE (2008)
9. Dasgupta, S., Schulman, L.J.: A probabilistic analysis of EM for mixtures of separated, spherical gaussians. J. Mach. Learn. Res. **8**, 203–226 (2007)
10. Deledalle, C.A., Parameswaran, S., Nguyen, T.Q.: Image denoising with generalized gaussian mixture model patch priors. SIAM J. Imag. Sci. **11**(4), 2568–2609 (2018)
11. Foucart, S., Rauhut, H.: A Mathematical Introduction to Compressive Sensing. Springer, New York (2013). https://doi.org/10.1007/978-0-8176-4948-7
12. Glasner, D., Bagon, S., Irani, M.: Super-resolution from a single image. In: 2009 12th International Conference on Computer Vision, pp. 349–356. IEEE (2009)
13. Gribonval, R., Chatalic, A., Keriven, N., Schellekens, V., Jacques, L., Schniter, P.: Sketching datasets for large-scale learning (long version). arXiv preprint arXiv:2008.01839 (2020)
14. Gribonval, R., Blanchard, G., Keriven, N., Traonmilin, Y.: Compressive statistical learning with random feature moments (2020)
15. Gribonval, R., Blanchard, G., Keriven, N., Traonmilin, Y.: Statistical learning guarantees for compressive clustering and compressive mixture modeling (2020)
16. Katkovnik, V., Egiazarian, K.: Nonlocal image deblurring: variational formulation with nonlocal collaborative l 0-norm prior. In: 2009 International Workshop on Local and Non-Local Approximation in Image Processing, pp. 46–53. IEEE (2009)
17. Keriven, N., Bourrier, A., Gribonval, R., Pérez, P.: Sketching for large-scale learning of mixture models. Inf. Inf. J. IMA **7**(3), 447–508 (2018)
18. Lebrun, M., Buades, A., Morel, J.M.: A nonlocal Bayesian image denoising algorithm. SIAM J. Imag. Sci. **6**(3), 1665–1688 (2013)
19. Martin, D., Fowlkes, C., Tal, D., Malik, J.: A database of human segmented natural images and its application to evaluating segmentation algorithms and measuring ecological statistics. In: Proceedings 8th International Conference on Computer Vision. ICCV 2001, vol. 2, pp. 416–423. IEEE (2001)
20. Parameswaran, S., Deledalle, C.A., Denis, L., Nguyen, T.Q.: Accelerating GMM-based patch priors for image restoration: three ingredients for a 100× speed-up. IEEE Trans. Image Process. **28**(2), 687–698 (2018)
21. Traonmilin, Y., Aujol, J.F.: The basins of attraction of the global minimizers of the non-convex sparse spike estimation problem. Inverse Probl. **36**(4), 045003 (2020)
22. Traonmilin, Y., Aujol, J.F., Leclaire, A.: Projected gradient descent for non-convex sparse spike estimation. IEEE Signal Process. Lett. **27**, 1110–1114 (2020)
23. Traonmilin, Y., Aujol, J.F., Leclaire, A.: The basins of attraction of the global minimizers of non-convex inverse problems with low-dimensional models in infinite dimension (2020)
24. Wang, Y.Q., Morel, J.M.: Sure guided gaussian mixture image denoising. SIAM J. Imag. Sci. **6**(2), 999–1034 (2013)

25. Wu, C.J.: On the convergence properties of the EM algorithm. Ann. Stat. 95–103 (1983)
26. Zoran, D., Weiss, Y.: From learning models of natural image patches to whole image restoration. In: 2011 International Conference on Computer Vision, pp. 479–486. IEEE (2011)

Translating Numerical Concepts for PDEs into Neural Architectures

Tobias Alt$^{(\boxtimes)}$, Pascal Peter, Joachim Weickert, and Karl Schrader

Mathematical Image Analysis Group, Faculty of Mathematics and Computer Science,
Campus E1.7, Saarland University, 66041 Saarbrücken, Germany
{alt,peter,weickert,schrader}@mia.uni-saarland.de

Abstract. We investigate what can be learned from translating numerical algorithms into neural networks. On the numerical side, we consider explicit, accelerated explicit, and implicit schemes for a general higher order nonlinear diffusion equation in 1D, as well as linear multigrid methods. On the neural network side, we identify corresponding concepts in terms of residual networks (ResNets), recurrent networks, and U-nets. These connections guarantee Euclidean stability of specific ResNets with a transposed convolution layer structure in each block. We present three numerical justifications for skip connections: as time discretisations in explicit schemes, as extrapolation mechanisms for accelerating those methods, and as recurrent connections in fixed point solvers for implicit schemes. Last but not least, we also motivate uncommon design choices such as nonmonotone activation functions. Our findings give a numerical perspective on the success of modern neural network architectures, and they provide design criteria for stable networks.

Keywords: Numerical algorithms · Partial differential equations · Convolutional neural networks · Nonlinear diffusion · Stability

1 Introduction

The remarkable success of convolutional neural networks (CNNs) has triggered many researchers to analyse their behaviour and to come up with mathematical foundations and stability guarantees. One strategy consists of interpreting networks as approximations of evolution equations; see e.g. [3,22,23]. Then training a network comes down to parameter identification of ordinary or partial differential equations (PDEs). This can be challenging, since it requires various model assumptions: Without additional smoothness assumptions, the models may become very complicated involving millions of parameters. Moreover, connecting a discrete network to a continuous evolution equation involves ambiguous

This work has received funding from the European Research Council (ERC) under the European Union's Horizon 2020 research and innovation programme (grant agreement no. 741215, ERC Advanced Grant INCOVID).

A. Elmoataz et al. (Eds.): SSVM 2021, LNCS 12679, pp. 294–306, 2021.
https://doi.org/10.1007/978-3-030-75549-2_24

limit assumptions: The same discrete model can approximate multiple evolution equations with different orders of consistency.

We address these problems by following two guiding principles:

1. We refrain from the *analytic* strategy of translating a complex neural network into a compact model, since it involves the discussed problems and only analyses how a network is, but not how it should be. Instead we pursue a *synthetic* approach: We translate successful concepts into networks. This is easier, and it allows to understand how a network should be to guarantee desirable qualities such as stability and efficiency.
2. Our concepts of choice are numerical algorithms rather than continuous models. This avoids ambiguities in the limit assumptions. Similar to neural architectures, numerical algorithms can be applied to a multitude of models. We believe that the design principles of modern neural networks realise a small but powerful set of numerical strategies as a basis of their success.

Thus, we want to justify key components of neural architectures and derive novel design principles by translating popular numerical algorithms into networks.

Our Contributions. As an exemplary starting point and a basis for exploring different numerical algorithms, we consider a general evolution equation for higher order nonlinear diffusion in 1D.

First we show that an explicit finite difference discretisation can be interpreted as a residual network (ResNet) [11]. This gives two central insights: The diffusion flux function determines the activation function of the ResNet, and the two convolutional filters follow a transposed structure. This motivates the use of nonmonotone activation functions and allows us to guarantee stability in the Euclidean norm for specific ResNets. Moreover, we identify the skip connections in the ResNet as discrete time derivatives.

Additional interpretations of skip connections are obtained with alternative numerical methods based on fast semi-iterative (FSI) accelerations of explicit schemes [9] and on fixed point algorithms for fully implicit schemes. We show that the latter ones can be regarded as recurrent neural networks [12].

Since multigrid methods [2] are efficient numerical methods for PDEs, it is worthwhile to analyse them as well. We demonstrate that they have structural connections to U-nets [20], shedding some light on their efficiency.

Our results do not only inspire general design criteria for neural networks as well as stable architectures. They also provide structural insights into ResNets, RNNs and U-nets from the perspective of numerical algorithms.

Related Work. Our philosophy to translate numerics into neural networks is shared in [14,18]. Both works motivate additional skip connections in ResNets from multistep schemes for ordinary differential equations. Our paper provides alternative motivations via time discretisations, acceleration via extrapolation, as well as fixed point schemes for implicit discretisations.

The stability of ResNets is studied from a differential equations perspective in [21,22,27]. Particularly, Ruthotto and Haber [22] show stability in the Euclidean norm for a specific form of residual networks. However, they require the activation function to be monotone. In contrast to their approach, our theory allows nonmonotone activation functions.

Several works connect multigrid ideas and CNNs, e.g. to learn the restriction and prolongation operators [8], or to couple feature channels for parameter reduction [6]. He and Xu [10] present a CNN architecture implementing multigrid approaches, however without connecting it to a U-net. We on the other hand present a direct correspondence between a simple multigrid solver and a U-net.

Organisation of the Paper. In Sect. 2, we translate different numerical approximations for higher order diffusion into CNNs and analyse the resulting architectures. Covering multi-resolution approaches, we show that a multigrid solver can be cast into a U-net form in Sect. 3. Finally, we present our conclusions and an outlook on future work in Sect. 4.

2 Networks from Algorithms for Evolution Equations

In this section, we start with generalised diffusion filters in 1D, and we translate three numerical algorithms for them into neural network architectures. This gives novel insights into the value of skip connections, stable network design, and the potential of nonmonotone activation functions.

2.1 Generalised Nonlinear Diffusion

We consider a generalised higher order nonlinear diffusion model in 1D. It creates filtered signals $u(x,t) : (a,b) \times [0,\infty) \to \mathbb{R}$ from an initial signal $f(x)$ on a domain $(a,b) \subset \mathbb{R}$ according to the PDE

$$\partial_t u = -\mathcal{D}^* \big(g(|\mathcal{D}u|^2)\, \mathcal{D}u \big) \tag{1}$$

which is the gradient flow that minimises the energy $E(u) = \int_a^b \Psi(|\mathcal{D}u|^2)\, dx$ with $g = \Psi'$. We use a general differential operator $\mathcal{D} = \sum_{m=0}^M \alpha_m \partial_x^m$ and its adjoint version $\mathcal{D}^* = \sum_{m=0}^M (-1)^m \alpha_m \partial_x^m$, both consisting of weighted derivatives of up to order M. Thus, the corresponding PDE is of order $2M$. The evolution is initialised at time $t = 0$ by $u(x,0) = f(x)$, and we impose reflecting boundary conditions at $x = a$ and $x = b$. Equation (1) creates gradually simplified versions of f. The scalar *diffusivity* function $g(s^2)$ controls the amount of smoothing depending on the local structure of the evolving signal. We consider diffusivities that are smooth, nonnegative, nonincreasing, and bounded from above.

Depending on the operator $\mathcal{D}$ and the choice of the diffusivity, the evolution describes different models. For $\mathcal{D} = \partial_x$, one obtains a 1D version of the nonlinear diffusion filter of Perona and Malik [19]. For this model the exponential diffusivity $g(s^2) = \exp(-s^2/(2\lambda^2))$ inhibits smoothing near discontinuities where $|\partial_x u|$ exceeds a contrast parameter λ. This allows discontinuity-preserving smoothing. A higher order choice of $\mathcal{D} = \partial_x^2$ yields a 1D version of the fourth order PDE of You and Kaveh [26].

2.2   Residual Networks

Residual networks [11] are a popular CNN architecture as they are easy to train, even for a high number of network layers. They consist of chained residual blocks. A residual block is made up of two convolutional layers with biases and nonlinear activation functions after each layer. Each block computes a discrete output u from an input f by

$$u = \sigma_2(f + W_2\,\sigma_1(W_1 f + b_1) + b_2)\,, \tag{2}$$

with discrete convolution matrices W_1, W_2, activation functions σ_1, σ_2 and bias vectors b_1, b_2. The main difference to feed-forward CNNs lies in the skip connection which adds the original input signal f to the result of the inner activation function σ_1. This facilitates training of very deep networks.

2.3   Expressing Explicit Schemes as Residual Networks

In the following, we derive a direct correspondence between an explicit scheme for the generalised diffusion equation (1) and a ResNet. With the help of the *flux function* $\Phi(s) = g(s^2)\,s$, we rewrite (1) as

$$\partial_t u = -\mathcal{D}^*(\Phi(\mathcal{D}u))\,. \tag{3}$$

We now discretise this equation with an explicit finite difference scheme. To obtain discrete vectors $u, f \in \mathbb{R}^N$, we sample the continuous functions u, f with distance h. We employ a forward difference with time step size τ for the time derivative. Moreover, we represent a discretisation of the operator $\mathcal{D}$ by a convolution matrix K. Thus, the adjoint operator $\mathcal{D}^*$ is represented by $K^\top$.

Starting with an initial signal $u^0 = f$, the evolving signal u^{k+1} at time step $k + 1$ arises from the previous one by

$$u^{k+1} = u^k - \tau K^\top \Phi(K u^k)\,. \tag{4}$$

In this notation, the connection between an explicit diffusion step and a ResNet block becomes apparent:

Theorem 1 (Diffusion-inspired ResNets). *A higher order diffusion step* (4) *is equivalent to a residual block* (2) *if*

$$\sigma_1 = \tau\,\Phi, \quad \sigma_2 = \mathrm{Id}, \quad W_1 = K, \quad W_2 = -K^\top, \tag{5}$$

and the bias vectors $b_1,\ b_2$ *are set to* 0.

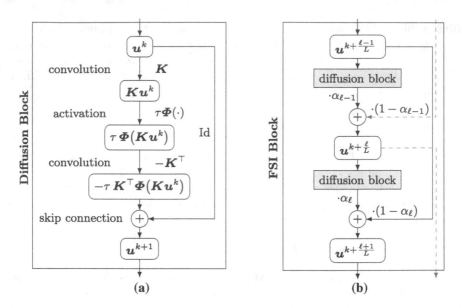

Fig. 1. (a) Diffusion block for an explicit diffusion step (4) with flux function Φ, time step size τ, and a discrete derivative operator K. (b) FSI block.

Interestingly, the inner activation function σ_1 corresponds to a scaled version of the flux function Φ. The effect of the skip connection in the residual block also becomes clear now: It is the central component to realise a time discretisation. We call a ResNet block of this form a *diffusion block*. It is visualised in Fig. 1(a). Graph nodes contain the current state of the signal, while edges describe operations which are applied to proceed from one node to the next.

We observe that the convolution matrices satisfy $W_2 = -W_1^\top$. This is a direct consequence of the gradient flow structure of the diffusion process. In the following, we prove stability and well-posedness for this specific form of ResNets.

2.4   Criteria for Well-Posed and Stable Residual Networks

Now we are able to transfer stability [5] and well-posedness [24] results for diffusion to a residual network consisting of diffusion blocks. We show Euclidean stability, which states that the Euclidean norm of the signal is nonincreasing in each iteration, i.e. $\|u^{k+1}\|_2 \leq \|u^k\|_2$. Well-posedness guarantees that the network output is a continuous function of the input data.

Theorem 2 (Euclidean Stability of ResNets with Diffusion Blocks).
Consider a residual network chaining any number of diffusion blocks (4) with convolutions represented by a convolution matrix K and activation function $\tau\Phi$. Moreover, assume that the activation function arises from a diffusion flux function $\Phi(s) = g(s^2)\, s$ with finite Lipschitz constant L. Then the residual network is well-posed and stable in the Euclidean norm if $\tau \leq 2\left(L\|K\|_2^2\right)^{-1}$. Here, $\|\cdot\|_2$ denotes the spectral norm which is induced by the Euclidean norm.

Proof. We first notice that since $\Phi(s) = g(s^2)\, s$, applying the flux function leads to a rescaling with a diagonal matrix $G(u^k)$ with $g((Ku^k)_i^2)$ as i-th diagonal element. Therefore, we can write (4) as

$$u^{k+1} = \left(I - \tau K^\top G(u^k)K\right) u^k. \tag{6}$$

At this point, well-posedness follows directly from the continuity of the operator $I - \tau K^\top G(u^k)K$, as the diffusivity g is assumed to be smooth [24].

We now show that the time step size restriction guarantees that the eigenvalues of the operator always lie in the interval $[-1, 1]$. As the spectral norm is sub-multiplicative, we can estimate the eigenvalues of $K^\top G(u^k)K$ for each matrix separately. Since g is nonnegative, the diagonal matrix G is positive semi-definite. The maximal eigenvalue of G is the given by the supremum of g. As g is non-increasing and bounded, this value is bounded by the Lipschitz constant L of Φ. Thus, the eigenvalues of $K^\top G(u^k)K$ lie in the interval $[0, \tau L\|K\|_2^2]$. Consequently, the operator $I - \tau K^\top G(u^k)K$ has eigenvalues in $[1 - \tau L\|K\|_2^2, 1]$, and the condition $1 - \tau L\|K\|_2^2 \geq -1$ leads to the bound $\tau \leq 2 \left(L\|K\|_2^2\right)^{-1}$. □

How General Is This Result? Theorem 2 is of fairly general nature and applies to a broad class of ResNets. The fact that K represents a discrete differential operator is no restriction on the convolution, since any convolution kernel can be seen as a discretisation of a suitable differential operator $\mathcal{D} = \sum_{m=0}^{M} \alpha_m \partial_x^m$.

Interestingly, our proof does not require the matrix K to have a convolution structure: It can be any arbitrary matrix. This even includes neural networks beyond CNNs, since the weights within a layer may differ from node to node.

> **The key requirement for network stability is the transposed convolution structure $W_2 = -W_1^T$.**

While this requirement is not fulfilled by the original ResNet [11], several works employ the transposed structure [3,22,27] as it is justified from a PDE perspective, requires less parameters, and provides stability guarantees.

In contrast to Ruthotto and Haber [22], our stability result does not require activation functions to be monotone. Let us now see that widely used diffusivities naturally lead to nonmonotone activation functions.

2.5   Nonmonotone Activation Functions

The connection between diffusivity $g(s^2)$ and activation function $\sigma(s) = \tau \Phi(s)$ with the diffusion flux $\Phi(s) = g(s^2)\, s$ revitalises an old idea of neural network design [4,15]. As an example, we translate the exponential Perona–Malik diffusivity $g(s^2) = \exp\left(-\frac{s^2}{2\lambda^2}\right)$ into its corresponding activation $\sigma(s) = \tau s \exp\left(-\frac{s^2}{2\lambda^2}\right)$. Interestingly, this activation function is *antisymmetric* and *nonmonotone*.

Antisymmetry is very natural in the diffusion case with $\mathcal{D} = \sum_{m=1}^{M} a_m \partial_x^m$, where the argument of the flux function consists of signal derivatives. It reflects the invariance axiom that signal negation and filtering are commutative. Nonmonotone flux functions were considered somewhat problematic for continuous diffusion PDEs. However, it has been shown that their discretisations are well-posed [25], in spite of the fact that they may act contrast enhancing.

The concept of a nonmonotone activation function is unusual in the CNN world. Although there have been a few early proposals in the neural network literature arguing in favour of nonmonotone activations [4,15], they are rarely used in modern CNNs. In practice, CNNs often fix the activation to simple functions such as the rectified linear unit (ReLU). From a PDE perspective, this appears restrictive. The diffusion interpretation suggests that activation functions should be learned in the same manner as convolution weights and biases. In practice, this hardly happens apart from a few notable exceptions such as [3,7,17]. As nonmonotone flux functions outperform monotone ones in the diffusion setting, it appears promising to incorporate them into CNNs. For more examples of diffusion-inspired activation functions, we refer to [1].

2.6 FSI Schemes and Additional Skip Connections

In the following, we show that an acceleration strategy of the explicit scheme induces a natural modification for the skip connections of the corresponding ResNet architecture. To speed up explicit schemes, Hafner et al. [9] proposed *fast semi-iterative* (FSI) schemes. They perform a cycle of extrapolated explicit steps. For our diffusion scheme (4), an FSI acceleration with cycle length L reads

$$\boldsymbol{u}^{k+\frac{\ell+1}{L}} = \alpha_\ell \left(\boldsymbol{I} - \tau \boldsymbol{K}^\top \boldsymbol{\Phi}\left(\boldsymbol{K} \boldsymbol{u}^{k+\frac{\ell}{L}} \right) \right) + (1 - \alpha_\ell)\, \boldsymbol{u}^{k+\frac{\ell-1}{L}} \tag{7}$$

with $\ell = 0, \ldots, L-1$ and extrapolation weights $\alpha_\ell := (4\ell + 2)/(2\ell + 3)$. One formally initialises with $\boldsymbol{u}^{k-\frac{1}{L}} := \boldsymbol{u}^k$. This cycle realises a super time step of size $\frac{L(L+1)}{3}\tau$. Thus, with one cycle involving L explicit steps, one reaches a super step size of $\mathcal{O}(L^2)$ rather than $\mathcal{O}(L)$. This explains its remarkable efficiency [9].

We see that FSI extrapolates the diffusion result at time step $k + \frac{\ell}{L}$ with the previous time step $k + \frac{\ell-1}{L}$ and the weight α_ℓ. This can be realised with a small change in the original diffusion block from Fig. 1(a) by adding an additional skip connection. The two skip connections are weighted by α_ℓ and $(1 - \alpha_\ell)$, respectively. This gives the architecture in Fig. 1(b).

We observe a different benefit of skip connections: Additional and more general skip connections constitute a whole class of acceleration strategies, which is in line with observations in the CNN literature; see e.g. [13,14].

2.7 Implicit Schemes and Recurrent Neural Networks

So far, we have connected variants of explicit schemes to ResNets. However, implicit discretisations are another important class of solvers. We now show that such a discretisation of our diffusion equation leads to a recurrent neural network (RNN). RNNs are classical neural network architectures; see e.g. [12]. The fully implicit discretisation of (1) is given by

$$u^{k+1} = u^k - \tau K^\top \Phi(K u^{k+1}).$$ (8)

We solve the resulting nonlinear system of equations by L fixed point iterations:

$$u^{k+\frac{\ell+1}{L}} = u^k - \tau K^\top \Phi\left(K u^{k+\frac{\ell}{L}}\right),$$ (9)

where $\ell = 0, \ldots, L-1$, and where we assume that τ is sufficiently small to yield a contraction mapping. For $L = 1$, we obtain the explicit scheme (4) with its ResNet interpretation. For larger L, however, different skip connections arise. They connect the layer at time step k with all subsequent layers at steps $k + \frac{\ell}{L}$ with $\ell = 0, \ldots, L-1$. This feedback can be seen as an RNN architecture.

In the context of variational models, Chen and Pock [3] have obtained a similar architecture. However, they explicitly supplement the diffusion process with an additional reaction term which results from the data term of the energy. Our feedback term is a pure numerical phenomenon of the fixed point solver.

We see that skip connections can implement a number of successful numerical concepts: forward difference approximations of the time derivative in explicit schemes, extrapolation steps to accelerate them e.g. via FSI, and recurrent connections within fixed point solvers for implicit schemes.

3   Multigrid Solvers and U-Nets

Multigrid methods [2] are very efficient numerical strategies for solving PDE-based problems. Neural networks, on the other hand, have benefitted from multiscale ideas as well, as can be seen e.g. from the high popularity of U-nets [20]. In this section we shed some light on their structural connections. For simplicity we restrict ourselves to a linear multigrid setting with two levels.

3.1   U-Net Architectures

The U-net [20] has proven useful in applications such as segmentation [20] or pose estimation [16], where features on multiple scales need to be extracted

As its name suggests, the U-net has a symmetric shape: On the left half of the architecture, convolutions extract features while repeated downsampling operators reduce the resolution. On the right half, features are successively upsampled, combined and convolved, starting with the coarsest resolution. The original U-net [20] combines features by concatenation, while other works such as [16] use addition. In the following, we focus on the latter design choice.

For our purposes, it is sufficient to consider a U-net with only two resolutions and a constant number of channels. We use superscripts h and H to denote computations on the fine and coarse grid, respectively. The following six steps capture the essential structure of such a U-net:

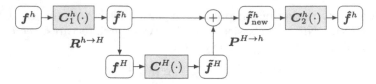

(a) U-net architecture for an input f^h.

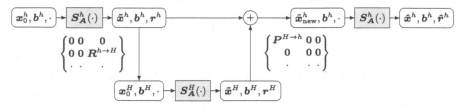

(b) Two-level V-Cycle in the form of a U-net utilising three-channel signals containing the iteration variable x, a right hand side b, and the residual r.

Fig. 2. Architectures for a general U-net and a multigrid V-cycle.

1. One applies a number of CNN layers to the input f^h, yielding a modified signal $\tilde{f}^h$. We denote this general operation by a function $C_1^h(\cdot)$.
2. To provide a coarse input $f^H = R^{h \to H} \tilde{f}^h$ to the next level, a restriction operator $R^{h \to H}$ brings the modified signal $\tilde{f}^h$ to a coarse resolution H. For example, the restriction can consist of an averaging or max-pooling.
3. On the coarse grid, the downsampled signal f^H is again modified by a series of layers to obtain $\tilde{f}^H = C^H(f^H)$.
4. One upsamples the coarse result $\tilde{f}^H$ with a prolongation operator $P^{H \to h}$.
5. On the fine grid, one adds the modified fine grid signal $\tilde{f}^h$ and the upsampled one $P^{H \to h} f^H$ and obtains $\tilde{f}^h_{\text{new}}$.
6. Lastly, applying more layers $C_2^h(\cdot)$ yields the final solution $\hat{f} = C_2^h(\tilde{f}^h_{\text{new}})$.

Figure 2(a) visualises this architecture. In the following, we express a multigrid V-cycle in this form by utilising multiple network channels.

3.2 Expressing a Multigrid V-Cycle Within a U-Net

Multigrid methods [2] allow for the efficient solution of equation systems that result from the numerical approximation of PDEs. For simplicity we consider a linear system of equations given by $Ax = b$. For classical iterative solvers such as the Jacobi or the Gauss–Seidel method, one observes that low-frequent error components are attenuated only very slowly. Hence, their convergence is slow. Multigrid methods transfer the low-frequent error components to a coarser scale, where the iterative solvers work more efficiently. The coarse scale solution is then used to correct the fine scale approximation.

To connect multigrid ideas to U-nets, we consider a V-cycle on two levels with grid sizes h and H for the fine and coarse grid. For our U-net, we use three channels. They contain the iteration variable x of the solver, the right hand side b of the equation system, and the current residual r. Even though we do not always need all channels, we keep the channel number constant for simplicity. A two-level V-cycle solves the linear system by repeating the following steps:

1. The inputs are a fine grid initialisation $x_0^h = 0$ and the given right hand side b^h. The residual at this point is ignored, as it is not relevant to the solver input. We assume that we are given a solver $S_A^h(\cdot)$ for the operator A^h. It produces a three-channel signal containing an approximate solution $\tilde{x}^h$, the right hand side b^h, and a residual $r^h = b^h - A^h\tilde{x}^h$.
2. While the true error e^h of the approximation is unknown, the residual r^h can be computed. This leads to the residual equation $A^h e^h = r^h$ which can be solved efficiently on a coarser grid. To this end, one uses a restriction operator $R^{h \to H}$. As the downsampling is now explicitly concerned with three channels, the corresponding operator in the CNN is a 3×3 block matrix. The coarse initialisation $x_0^H = 0$ does not require any information from the fine scale. Crucially, the new right hand side b^H is the downsampled fine residual, i.e. $b^H = R^{h \to H} r^h$. Lastly, an input residual is not required for the coarse solver. Thus, one obtains the coarse grid residual equation $A^H x^H = b^H$.
3. The coarse grid solver $S_A^H(\cdot)$ now solves the residual equation. It outputs a coarse approximation $\tilde{x}^H$ to the residual error, the right hand side b^H, and a new coarse residual r^H. The latter two would be required if one wants to add another level to the cycle.
4. In the upsampling step, we prepare the coarse scale outputs for the following addition. Similar to the downsampling, we upsample only the coarse error approximation $\tilde{x}^H$ by a prolongation operator $P^{H \to h}$. The coarse right hand side b^H is set to 0 as to not interfere with the fine right hand side.
5. On the fine grid, one adds the three signal channels. The initial fine grid approximation is updated with the upsampled error on the coarse grid by $\tilde{x}_{\text{new}}^h := \tilde{x}^h + P^{H \to h}\tilde{x}^H$. As we have applied the prolongation only to the coarse solution, the fine grid right hand side b^h is propagated.
6. Another instance of the fine grid solver $S_A^h(\cdot)$ takes the corrected solution x_{new}^h and the original right hand side b^h, yielding a new approximation $\hat{x}^h$.

We visualise this architecture in Fig. 2(b). Restriction and prolongation operators are applied only to certain channels of the solver output instead of all channels. In the downsampling phase, the restriction is applied to the residual, while in the upsampling phase, it is applied to the approximated error. This enables the coarse solver to work on the residual equation instead of only a coarse version of the original equation, which is the crucial idea of multigrid methods. The architecture utilises yet another form of skip connection: The fine scale approximation is corrected by adding an upsampled error approximation.

Our two-level setting can be generalised to more levels. Deeper V-cycles are constructed by stacking the two-level V-cycle recursively, and so-called W-cycles are built by concatenating two V-cycles. On the CNN side, this leads to U-nets with more levels, as well as concatenations thereof. This idea is also used in practice: Successful U-nets work on multiple resolutions [20], and so-called stacked hourglass models [16] arise by concatenating multiple V-cycle architectures. It shows that multigrid architectures share essential structural properties with U-nets.

4 Conclusions

Our paper is based on the philosophy of regarding a trained neural network as a numerical algorithm. To substantiate this claim, we have translated a number of efficient numerical algorithms for PDEs into popular building blocks for network architectures. Apart from a few notable exceptions such as [14], this strategy has rarely been pursued in its full consequence. We have shown that valuable structural insights can be gained from such a direct translation, and we have derived systematic design principles for well-founded network components.

More specifically, we have shown the value of skip connections from three different numerical perspectives: as time discretisations in explicit schemes, as extrapolation terms to increase their efficiency, and as recurrent connections in implicit schemes with fixed point structure. By connecting multigrid methods to U-nets, we provide a basis for explaining for their remarkable efficiency. Numerical schemes for generalised diffusion processes suggest that nonmonotone activation functions are permissible and can be advantageous. Last but not least, we have seen that a ResNet block with a transposed structure of both convolution layers can guarantee Euclidean stability in a simple and elegant way.

Our contributions can serve as a blueprint for translating a larger class of successful numerical concepts for PDEs to CNNs. This is part of our ongoing work. It is our hope that this will lead to a closer connection of both worlds and to hybrid methods that unite the stability and efficiency of modern numerical algorithms with the performance of neural networks.

Acknowlegdements. We thank Matthias Augustin and Michael Ertel for fruitful discussions and feedback on our manuscript.

References

1. Alt, T., Weickert, J., Peter, P.: Translating diffusion, wavelets, and regularisation into residual networks. arXiv:2002.02753v3 [cs.LG] (Jun 2020)
2. Briggs, W.L., Henson, V.E., McCormick, S.F.: A Multigrid Tutorial, 2nd edn. SIAM, Philadelphia (2000)
3. Chen, Y., Pock, T.: Trainable nonlinear reaction diffusion: a flexible framework for fast and effective image restoration. IEEE Trans. Pattern Anal. Mach. Intell. **39**(6), 1256–1272 (2016)
4. De Felice, P., Marangi, C., Nardulli, G., Pasquariello, G., Tedesco, L.: Dynamics of neural networks with non-monotone activation function. Netw. Comput. Neural Syst. **4**(1), 1–9 (1993)

5. Didas, S., Weickert, J., Burgeth, B.: Properties of higher order nonlinear diffusion filtering. J. Math. Imaging Vis. **35**, 208–226 (2009)
6. Eliasof, M., Ephrath, J., Ruthotto, L., Treister, E.: Multigrid-in-Channels neural network architectures. arXiv:2011.09128v2 [cs.CV] (Nov 2020)
7. Goodfellow, I., Warde-Farley, D., Mirza, M., Courville, A., Bengio, Y.: Maxout networks. In: Dasgupta, S., McAllester, D. (eds.) Proceedings of the 30th International Conference on Machine Learning. Proceedings of Machine Learning Research, Atlanta, GA, vol. 28, pp. 1319–1327, June 2013
8. Greenfeld, D., Galun, M., Kimmel, R., Yavneh, I., Basri, R.: Learning to optimize multigrid PDE solvers. In: Chaudhuri, K., Salakhutdinov, R. (eds.) Proceedings of the 36th International Conference on Machine Learning. Proceedings of Machine Learning Research, Long Beach, CA, vol. 97, pp. 2415–2423, June 2019
9. Hafner, D., Ochs, P., Weickert, J., Reißel, M., Grewenig, S.: FSI schemes: fast semi-iterative solvers for PDEs and optimisation methods. In: Rosenhahn, B., Andres, B. (eds.) GCPR 2016. LNCS, vol. 9796, pp. 91–102. Springer, Cham (2016). https://doi.org/10.1007/978-3-319-45886-1_8
10. He, J., Xu, J.: MgNet: a unified framework of multigrid and convolutional neural network. Sci. China Math. **62**(7), 1331–1354 (2019). https://doi.org/10.1007/s11425-019-9547-2
11. He, K., Zhang, X., Ren, S., Sun, J.: Deep residual learning for image recognition. In: Proceedings of the 2016 IEEE Conference on Computer Vision and Pattern Recognition, pp. 770–778. IEEE Computer Society Press, Las Vegas, June 2016
12. Hopfield, J.J.: Neural networks and physical systems with emergent collective computational abilities. Proc. Natl. Acad. Sci. **79**(8), 2554–2558 (1982)
13. Huang, G., Liu, Z., van der Maaten, L., Weinberger, K.Q.: Densely connected convolutional networks. In: Proceedings of the 2017 IEEE Conference on Computer Vision and Pattern Recognition, pp. 4700–4708. IEEE Computer Society Press, Honolulu, July 2017
14. Lu, Y., Zhong, A., Li, Q., Dong, B.: Beyond finite layer neural networks: bridging deep architectures and numerical differential equations. In: Dy, J., Krause, A. (eds.) Proceedings of the 35th International Conference on Machine Learning. Proceedings of Machine Learning Research, Stockholm, Sweden, vol. 80, pp. 3276–3285, Jul 2018
15. Meilijson, I., Ruppin, E.: Optimal signalling in attractor neural networks. In: Tesauro, G., Touretzky, D., Leen, T. (eds.) Proceedings of the 7th Annual Conference on Neural Information Processing Systems. Advances in Neural Information Processing Systems, Denver, CO, vol. 7, pp. 485–492, December 1994
16. Newell, A., Yang, K., Deng, J.: Stacked hourglass networks for human pose estimation. In: Leibe, B., Matas, J., Sebe, N., Welling, M. (eds.) ECCV 2016. LNCS, vol. 9912, pp. 483–499. Springer, Cham (2016). https://doi.org/10.1007/978-3-319-46484-8_29
17. Ochs, P., Meinhardt, T., Leal-Taixe, L., Moeller, M.: Lifting layers: analysis and applications. In: Ferrari, V., Hebert, M., Sminchisescu, C., Weiss, Y. (eds.) ECCV 2018. LNCS, vol. 11205, pp. 53–68. Springer, Cham (2018). https://doi.org/10.1007/978-3-030-01246-5_4
18. Ouala, S., Pascual, A., Fablet, R.: Residual integration neural network. In: Proceedings of the 2019 IEEE International Conference on Acoustics, Speech and Signal Processing. pp. 3622–3626. IEEE Computer Society Press, Brighton, May 2019
19. Perona, P., Malik, J.: Scale space and edge detection using anisotropic diffusion. IEEE Trans. Pattern Anal. Mach. Intell. **12**, 629–639 (1990)

20. Ronneberger, O., Fischer, P., Brox, T.: U-Net: convolutional networks for biomedical image segmentation. In: Navab, N., Hornegger, J., Wells, W.M., Frangi, A.F. (eds.) MICCAI 2015. LNCS, vol. 9351, pp. 234–241. Springer, Cham (2015). https://doi.org/10.1007/978-3-319-24574-4_28

21. Rousseau, F., Drumetz, L., Fablet, R.: Residual networks as flows of diffeomorphisms. J. Math. Imaging Vis. **62**, 365–375 (2020)

22. Ruthotto, L., Haber, E.: Deep neural networks motivated by partial differential equations. J. Math. Imaging Vis. **62**, 352–364 (2020)

23. Smets, B., Portegies, J., Bekkers, E., Duits, R.: PDE-based group equivariant convolutional neural networks. arXiv:2001.09046v2 [cs.LG], March 2020

24. Weickert, J.: Anisotropic Diffusion in Image Processing. Teubner, Stuttgart (1998)

25. Weickert, J., Benhamouda, B.: A semidiscrete nonlinear scale-space theory and its relation to the Perona-Malik paradox. In: Solina, F., Kropatsch, W.G., Klette, R., Bajcsy, R. (eds.) Advances in Computer Vision, pp. 1–10. Springer, Wien (1997)

26. You, Y.L., Kaveh, M.: Fourth-order partial differential equations for noise removal. IEEE Trans. Image Process. **9**(10), 1723–1730 (2000)

27. Zhang, L., Schaeffer, H.: Forward stability of ResNet and its variants. J. Math. Imaging Vis. **62**, 328–351 (2020)

CLIP: Cheap Lipschitz Training of Neural Networks

Leon Bungert[1(✉)], René Raab[2], Tim Roith[1], Leo Schwinn[2],
and Daniel Tenbrinck[1]

[1] Department Mathematics,
Friedrich-Alexander University Erlangen-Nürnberg,
Cauerstraße 11, 91058 Erlangen, Germany
{leon.bungert,tim.roith,daniel.tenbrinck}@fau.de
[2] Machine Learning and Data Analytics Lab, Friedrich-Alexander University
Erlangen-Nürnberg, Carl-Thiersch-Straße 2b, 91052 Erlangen, Germany
{rene.raab,leo.schwinn}@fau.de

Abstract. Despite the large success of deep neural networks (DNN) in
recent years, most neural networks still lack mathematical guarantees in
terms of stability. For instance, DNNs are vulnerable to small or even
imperceptible input perturbations, so called adversarial examples, that
can cause false predictions. This instability can have severe consequences
in applications which influence the health and safety of humans, e.g.,
biomedical imaging or autonomous driving. While bounding the Lipschitz constant of a neural network improves stability, most methods rely
on restricting the Lipschitz constants of each layer which gives a poor
bound for the actual Lipschitz constant.

In this paper we investigate a variational regularization method named
CLIP for controlling the Lipschitz constant of a neural network, which
can easily be integrated into the training procedure. We mathematically
analyze the proposed model, in particular discussing the impact of the
chosen regularization parameter on the output of the network. Finally,
we numerically evaluate our method on both a nonlinear regression problem and the MNIST and Fashion-MNIST classification databases, and
compare our results with a weight regularization approach.

Keywords: Deep neural network · Machine learning · Lipschitz
constant · Variational regularization · Stability · Adversarial attack

1 Introduction

Deep neural networks (DNNs) have led to astonishing results in various fields,
such as computer vision [13] and language processing [18]. Despite its large suc-

This work was supported by the European Union's Horizon 2020 research and innovation programme under the Marie Skłodowska-Curie grant agreement No. 777826
(NoMADS) and by the German Ministry of Science and Technology (BMBF) under
grant agreement No. 05M2020 (DELETO).

cess, deep learning and the resulting neural network architectures also bear certain drawbacks. On the one hand, their behaviour is mathematically not yet fully understood and there exist not many rigorous analytical results. On the other hand, most trained networks are vulnerable to adversarial attacks [9]. In image processing, adversarial examples are small, typically imperceptible perturbations to the input that cause misclassifications. In domains like autonomous driving or healthcare this can potentially have fatal consequences. To mitigate these weaknesses, many methods were proposed to make neural networks more robust and reliable. One straight-forward approach is to regularize the norms of the weight matrices [10]. Another idea is adversarial training [16,23,24] which uses adversarial examples generated from the training data to increase robustness locally around the training samples. In this context, also the Lipschitz constant of neural networks has attracted a lot of attention (e.g., [7,8,26,29]) since it constitutes a worst-case bound for their stability, and Lipschitz-regular networks are reported to have superior generalization properties [17]. In [27] Lipschitz regularization around training samples has been related to adversarial training.

Unfortunately, the Lipschitz constant of a neural network is NP-hard to compute [22], hence several methods in the literature aim to achieve more stability of neural networks by bounding the Lipschitz constant of each individual layer [1,10,12,20] or the activation functions [3]. However, this strategy is a very imprecise approximation to the real Lipschitz constant and hence the trained networks may suffer from inferior expressivity and can even be less robust [15]. The following example demonstrates a representation of the absolute value function, which clearly has Lipschitz constant 1, using a neural network whose individual layers have larger Lipschitz constants (cf. [11] for details).

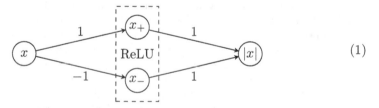

$$(1)$$

In this paper we propose a new variational regularization method for training neural networks, while simultaneously controlling their Lipschitz constant. Since the additional computations can easily be embedded in the standard training process and are fully parallelizable, we name our method *Cheap Lipschitz Training of Neural Networks (CLIP)*. Instead of bounding the Lipschitz constant of each layer individually as in prior approaches, we aim to minimize the global Lipschitz constant of the neural network via an additional regularization term.

The **main contributions** of this paper are as follows. First, we motivate and introduce the proposed variational regularization method for Lipschitz training of neural networks, which leads to a min-max problem. For optimizing the proposed model we formulate a stochastic gradient descent-ascent method, the CLIP algorithm. Subsequently, we perform mathematical analysis of the proposed variational regularization method by discussing existence of solutions. Here we use techniques, developed for the analysis of variational regularization methods of inverse problems [4,6]. We analyse the two limit cases, namely the regularization parameter tending to zero and infinity, where we prove convergence to a Lipschitz minimal fit of the data and a generalized barycenter, respectively. Finally, we evaluate the proposed approach by performing numerical experiments on a regression example and the MNIST and Fashion-MNIST classification databases [14,28]. We show that the introduced variational regularization term leads to improved stability compared to networks without additional regularization or with layerwise Lipschitz regularization. For this we investigate the impact of noise and adversarial attacks on the trained neural networks.

2   Model and Algorithms

Given a finite training set $\mathcal{T} \subset \mathcal{X} \times \mathcal{Y}$, where $\mathcal{X}$ and $\mathcal{Y}$ denote the input and output space, we propose to determine parameters $\theta \in \Theta$ of a neural network $f_\theta : \mathcal{X} \to \mathcal{Y}$ by solving the empirical risk minimization problem with an additional Lipschitz regularization term

$$\theta_\lambda \in \arg\min_{\theta \in \Theta} \frac{1}{|\mathcal{T}|} \sum_{(x,y) \in \mathcal{T}} \ell(f_\theta(x), y) + \lambda \operatorname{Lip}(f_\theta). \tag{2}$$

Here, $\ell : \mathcal{Y} \times \mathcal{Y} \to \mathbb{R}$ is a loss function, $\lambda > 0$ is a regularization parameter and the Lipschitz constant of the network $f_\theta : \mathcal{X} \to \mathcal{Y}$ is defined as

$$\operatorname{Lip}(f_\theta) := \sup_{x,x' \in \mathcal{X}} \frac{\|f_\theta(x) - f_\theta(x')\|}{\|x - x'\|}. \tag{3}$$

The norms involved in this definition can be chosen freely, however, for our CLIP algorithm we assume differentiability. The fundamental difference of the proposed model (2) compared to existing approaches in the literature [1,10,12,20] is that we use the actual Lipschitz constant (3) as regularizer and do not rely on the layer-based upper bound

$$\operatorname{Lip}(f_\theta) \leq \prod_{l=1}^{L} \operatorname{Lip}(\Phi_l) \tag{4}$$

for neural networks of the form $f_\theta = \Phi_L \circ \ldots \Phi_1$. By plugging (3) into (2) this becomes a min-max problem, which we solve numerically by using a modification of the stochastic batch gradient descent algorithm with momentum $\operatorname{SGDM}_{\eta,\gamma}$ with learning rate $\eta > 0$ and momentum $\gamma \geq 0$, see, e.g. [21].

Since evaluating the Lipschitz constant of a neural network is NP-hard [22], we approximate it on a finite subset $\mathcal{X}_{\operatorname{Lip}} \subset \mathcal{X} \times \mathcal{X}$ by setting

$$\operatorname{Lip}(f_\theta, \mathcal{X}_{\operatorname{Lip}}) := \max_{(x,x') \in \mathcal{X}_{\operatorname{Lip}}} \frac{\|f_\theta(x) - f_\theta(x')\|}{\|x - x'\|}. \tag{5}$$

Algorithm 1: AdversarialUpdate$_\tau$ (with step size $\tau > 0$)

$L(x, x') := \| f_\theta(x) - f_\theta(x') \| / \| x - x' \|, \quad x, x' \in \mathcal{X}$

for $(x, x') \in \mathcal{X}_{\text{Lip}}$ **do**

$\quad \mid \quad x \leftarrow x + \tau L(x, x') \nabla_x L(x, x')$

$\quad \mid \quad x' \leftarrow x' + \tau L(x, x') \nabla_{x'} L(x, x')$

Algorithm 2: CLIP (Cheap Lipschitz Training)

---

for *epoch* $e = 1$ **to** E **do**

$\quad$ **for** *minibatch* $B \subset \mathcal{T}$ **do**

$\qquad \mathcal{X}_{\text{Lip}} \leftarrow \text{AdversarialUpdate}_\tau(\mathcal{X}_{\text{Lip}})$

$\qquad \theta \leftarrow \text{SGDM}_{\eta, \gamma} \left(\frac{1}{|B|} \sum_{(x,y) \in B} \ell(f_\theta(x), y) + \lambda \, \text{Lip}(f_\theta, \mathcal{X}_{\text{Lip}}) \right)$

$\qquad$ **if** $\mathcal{A}(f_\theta; \mathcal{T}) > \alpha$ **then**

$\qquad \quad \llcorner \lambda \leftarrow \lambda + d\lambda$

$\qquad$ **else**

$\qquad \quad \llcorner \lambda \leftarrow \lambda - d\lambda$

To make sure that the tuples in $\mathcal{X}_{\text{Lip}}$ correspond to points with a high Lipschitz constant during the whole training process, we update $\mathcal{X}_{\text{Lip}}$ using gradient ascent of the difference quotient in (5) with adaptive step size, see Algorithm 1. We call this *adversarial update*, since the tuples in $\mathcal{X}_{\text{Lip}}$ approach points which have a small distance to each other while still getting classified differently by the network.

To illustrate the effect of adversarial updates, Fig. 1 depicts a pair of images from the Fashion-MNIST database before and after applying Algorithm 1, using a neural network trained with CLIP (see Sect. 4.2 for details). The second pair

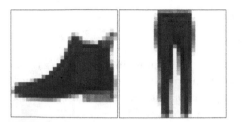 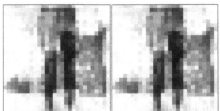

Fig. 1. Effect of Algorithm 1. The initial tuple gets classified correctly with high confidence (>90%). After adversarial updates the pair gets misclassified as the respective other class with low confidence (<30%).

realizes a large Lipschitz constant and hence lies close to the decision boundary of the neural network.

The regularization parameter λ in (2) is chosen by a discrepancy principle from inverse problems [2,5]: We compare an accuracy measure $\mathcal{A}(f_\theta; \mathcal{T})$ for the neural network f_θ, evaluated on the training data, to a target accuracy α. Combining the adversarial update from Algorithm 1 with stochastic gradient descent with momentum, we propose the CLIP algorithm, which is given in Algorithm 2. Similarly to [24] one can reuse computations of the backward pass in the gradient step of the Lipschitz regularizer with respect to θ in Algorithm 2 to compute the gradient with respect to the input x, needed in Algorithm 1, which indeed makes CLIP a 'cheap' extension to conventional training. We would like to emphasize that the set $\mathcal{X}_{\mathrm{Lip}}$, on which we approximate the Lipschitz constant of the neural network, is not fixed but is updated in an optimal way in every minibatch. Hence, it plays the role of an adaptive 'training set' for the Lipschitz constant. Future analysis will investigate the approximation quality of this approach in the framework of stochastic gradient descent methods.

3 Analysis

In this section we prove analytical results on the model (2), which are inspired by analogous statements for variational regularization methods for inverse problems (see, e.g., [4,6]). While the CLIP Algorithm 1 is stochastic in nature, our analysis focuses on the deterministic model (2), whose solutions are approximated by CLIP. For our results we have to pose some mild assumptions on the loss function ℓ and the neural network f_θ which are fulfilled for most loss functions and network architectures, such as mean squared error or cross entropy loss and architectures like feed-forward, convolutional, or residual networks with continuous activation functions. Furthermore, we assume that $\mathcal{X}, \mathcal{Y}$, and Θ are finite-dimensional spaces, which is the case in most applications and lets us state our theory more compactly than using Banach space topologies.

Assumption 1. We assume that the loss function $\ell : \mathcal{Y} \times \mathcal{Y} \to \mathbb{R} \cup \{\infty\}$ satisfies:

(a) $\ell(y, y') \geq 0$ for all $y, y' \in \mathcal{Y}$,
(b) $y \mapsto \ell(y, y')$ is lower semi-continuous for all $y' \in \mathcal{Y}$.

Assumption 2. We assume that the map $\theta \mapsto f_\theta(x)$ is continuous for all $x \in \mathcal{X}$.

Assumption 3. We assume that there exists $\theta \in \Theta$ such that

$$\frac{1}{|\mathcal{T}|} \sum_{(x,y) \in \mathcal{T}} \ell(f_\theta(x), y) + \lambda \operatorname{Lip}(f_\theta) < \infty.$$

We will also need the following lemma, which states that the Lipschitz constant of f_θ is lower semi-continuous with respect to the parameters θ.

Lemma 1. *Under Assumption 2 the functional* $\theta \mapsto \operatorname{Lip}(f_\theta)$ *is lower semi-continuous.*

Proof. Let $(\theta_i)_{i\in\mathbb{N}} \subset \Theta$ converge to $\theta \in \Theta$. Using the continuity of $\theta \mapsto f_\theta(x)$ and the lower semi-continuity of the norm $\|\cdot\|$, one can compute

$$\text{Lip}(f_\theta) = \sup_{x,x'\in\mathcal{X}} \frac{\|f_\theta(x) - f_\theta(x')\|}{\|x - x'\|} \leq \sup_{x,x'\in\mathcal{X}} \liminf_{i\to\infty} \frac{\|f_{\theta_i}(x) - f_{\theta_i}(x')\|}{\|x - x'\|}$$

$$\leq \liminf_{i\to\infty} \sup_{x,x'\in\mathcal{X}} \frac{\|f_{\theta_i}(x) - f_{\theta_i}(x')\|}{\|x - x'\|} = \liminf_{i\to\infty} \text{Lip}(f_{\theta_i}).$$

3.1   Existence of Solutions

We start with an existence statement for a slight modification of (2), which also regularizes the network parameters and guarantees coercivity of the objective.

Proposition 1. *Under Assumptions 1–3 the problem*

$$\min_{\theta\in\Theta} \frac{1}{|T|} \sum_{(x,y)\in T} \ell(f_\theta(x), y) + \lambda\, \text{Lip}(f_\theta) + \mu\, \|\theta\|_\Theta \tag{6}$$

has a solution for all values $\lambda, \mu > 0$. Here, $\|\cdot\|_\Theta$ denotes a norm on Θ.

Proof. We let $(\theta_i)_{i\in\mathbb{N}} \subset \Theta$ be a minimizing sequence, whose existence is assured by Assumption 3. Since $\mu > 0$ in (6), the norms $\|\theta_i\|_\Theta$ for $i \in \mathbb{N}$ are uniformly bounded and hence, up to a subsequence which we do not relabel, $\theta_i \to \theta^* \in \Theta$ as $i \to \infty$. The lower semi-continuity of ℓ, Lip, and $\|\cdot\|_\Theta$ together with the continuity of $\theta \mapsto f_\theta(x)$ then shows that θ^* solves (6).

Remark 1. If Θ is compact or even finite, existence for (2) is assured since any minimizing sequence in Θ is compact. The reason that in the general case we can only show existence for the regularized problem (6) is that it assures boundedness of minimizing sequences. Even for the unregularized empirical risk minimization problem existence is typically assumed in the *realizability assumption* [25].

3.2   Dependency on the Regularization Parameter

We start with the statement that the Lipschitz constant in (2) decreases and the empirical risk increases as the regularization parameter λ grows.

Proposition 2. *Let θ_λ solve (2) for $\lambda \geq 0$. Then it holds*

$$\lambda \longmapsto \frac{1}{|T|} \sum_{(x,y)\in T} \ell(f_{\theta_\lambda}(x), y) \quad \text{is non-decreasing,} \tag{7}$$

$$\lambda \longmapsto \text{Lip}(f_{\theta_\lambda}) \qquad\qquad \text{is non-increasing.} \tag{8}$$

Proof. The proof works precisely as in [6].

Now we will study the limit cases $\lambda \searrow 0$ and $\lambda \to \infty$ in (2). As in Sect. 3.1, because of lack of coercivity, we have to assume that the corresponding sequences of optimal network parameters converge.

Our first statement deals with the behavior as the regularization parameter λ tends to zero. We show that in this case the learned neural networks converge to one which fits the training data with the smallest Lipschitz constant.

Proposition 3. *Let Assumptions 1–3 and the realizability assumption [25]*

$$\min_{\theta \in \Theta} \frac{1}{|\mathcal{T}|} \sum_{(x,y) \in \mathcal{T}} \ell(f_\theta(x), y) = 0 \tag{9}$$

be satisfied. Let θ_λ be a solution of (2) for $\lambda > 0$. If $\theta_\lambda \to \theta^\dagger \in \Theta$ as $\lambda \searrow 0$, then

$$\theta^\dagger \in \arg\min \left\{ \mathrm{Lip}(f_\theta) \,:\, \theta \in \Theta, \, \frac{1}{|\mathcal{T}|} \sum_{(x,y) \in \mathcal{T}} \ell(f_\theta(x), y) = 0 \right\}. \tag{10}$$

Proof. Using the lower semi-continuity of the loss ℓ and the continuity of the map $\theta \mapsto f_\theta(x)$ for all $x \in \mathcal{X}$, we infer

$$\ell(f_{\theta^\dagger}(x), y) \le \liminf_{\lambda \searrow 0} \ell(f_{\theta_\lambda}(x), y), \quad \forall (x, y) \in \mathcal{T}.$$

Furthermore, using Lemma 1 we get

$$\mathrm{Lip}(f_{\theta^\dagger}) \le \liminf_{\lambda \searrow 0} \mathrm{Lip}(f_{\theta_\lambda}).$$

Using these two estimates together with the optimality of θ_λ we infer

$$\frac{1}{|\mathcal{T}|} \sum_{(x,y) \in \mathcal{T}} \ell(f_\theta(x), y) + \lambda \, \mathrm{Lip}(f_{\theta^\dagger})$$

$$\le \liminf_{\lambda \searrow 0} \frac{1}{|\mathcal{T}|} \sum_{(x,y) \in \mathcal{T}} \ell(f_{\theta_\lambda}(x), y) + \lambda \, \mathrm{Lip}(f_{\theta_\lambda}) \le \lambda \, \mathrm{Lip}(f_\theta)$$

for all $\theta \in \Theta$ which satisfy (9). Letting $\lambda \searrow 0$ and using that ℓ is non-negative shows that $\theta^\dagger$ solves (10). $\qquad\square$

Note that if the realizability assumption (9) is not satisfied, e.g., for noisy data or small network architectures, the previous statement has to be refined, which is subject to future work. The next proposition deals with the case in which the parameter λ approaches infinity. In this case the learned neural networks approach a constant map, coinciding with a generalized barycenter of the data. We denote by $\mathcal{T}_{\mathcal{Y}} = \{y : (x, y) \in \mathcal{T}\}$ the set of the data in the output space.

Proposition 4. *Let Assumptions 1–3 be satisfied and assume that*

$$\mathcal{M} := \{y \in \mathcal{Y} \,:\, \exists \theta \in \Theta, \, f_\theta(x) = y, \, \forall x \in \mathcal{X}\} \ne \emptyset. \tag{11}$$

Let θ_λ denote a solution of (2) for $\lambda > 0$. If $\theta_\lambda \to \theta_\infty \in \Theta$ as $\lambda \to \infty$, then $f_{\theta_\infty}(x) = \hat{y}$ for all $x \in \mathcal{X}$ where

$$\hat{y} \in \arg\min \left\{ \frac{1}{|\mathcal{T}|} \sum_{y \in \mathcal{T}_y} \ell(y', y) : y' \in \mathcal{M} \right\}. \tag{12}$$

Proof. From the optimality of θ_λ we deduce

$$\sum_{(x,y) \in \mathcal{T}} \ell(f_{\theta_\lambda}(x), y) + \lambda \operatorname{Lip}(f_{\theta_\lambda}) \leq \sum_{y \in \mathcal{T}_y} \ell(y', y), \quad \forall y' \in \mathcal{M}.$$

Hence, by letting $\lambda \to \infty$ we obtain $\lim_{\lambda \to \infty} \operatorname{Lip}(f_{\theta_\lambda}) = 0$. If we now assume that $\theta_\lambda \to \theta_\infty$ as $\lambda \to \infty$, we can use the lower semi-continuity of the Lipschitz constant to obtain

$$\operatorname{Lip}(f_{\theta_\infty}) \leq \liminf_{\lambda \to \infty} \operatorname{Lip}(f_{\theta_\lambda}) = 0.$$

Hence, $f_{\theta_\infty} \equiv \hat{y}$ for some element $\hat{y} \in \mathcal{Y}$. Using the lower semi-continuity of the loss function, we can conclude the proof with

$$\sum_{y \in \mathcal{T}_y} \ell(\hat{y}, y) = \sum_{(x,y) \in \mathcal{T}} \ell(f_{\theta_\infty}(x), y) \leq \liminf_{\lambda \to \infty} \sum_{(x,y) \in \mathcal{T}} \ell(f_{\theta_\lambda}(x), y)$$

$$\leq \sum_{y \in \mathcal{T}_y} \ell(y', y), \quad \forall y' \in \mathcal{M}.$$

Remark 2. For a Euclidean loss function, i.e., $\ell(y', y) = \|y' - y\|^2$, the quantity (12) coincides with a projection of the barycenter of the training samples in $\mathcal{T}_y$ onto the manifold of constant networks $\mathcal{M}$, given by (11). This can be seen as follows: Let $b := \frac{1}{|\mathcal{T}|} \sum_{y \in \mathcal{T}_y} y$ be the barycenter of the training samples. Then

$$\frac{1}{|\mathcal{T}|} \sum_{y \in \mathcal{T}_y} \|y' - y\|^2 = \|y'\|^2 - 2 \langle y', b \rangle + \frac{1}{|\mathcal{T}|} \sum_{y \in \mathcal{T}_y} \|y\|^2$$

$$= \|y' - b\|^2 - \|b\|^2 + \frac{1}{|\mathcal{T}|} \sum_{y \in \mathcal{T}_y} \|y\|^2, \quad \forall y' \in \mathcal{Y}.$$

Using this equality for general $y' \in \mathcal{M}$ and for $y' = \hat{y}$ given by (12) we obtain

$$0 \leq \frac{1}{|\mathcal{T}|} \sum_{y \in \mathcal{T}_y} \|y' - y\|^2 - \frac{1}{|\mathcal{T}|} \sum_{y \in \mathcal{T}_y} \|\hat{y} - y\|^2 = \|y' - b\|^2 - \|\hat{y} - b\|^2.$$

This in particular implies that $\|\hat{y} - b\| \leq \|y' - b\|$ for all $y' \in \mathcal{M}$ as desired.

4   Experiments

After having studied theoretical properties of the proposed variational Lipschitz regularization method (2), we will apply the CLIP algorithm to regression and classification tasks in this section. In a first experiment we train a neural network to approximate a nonlinear function given by noisy samples, where we illustrate that CLIP prevents strong oscillations, which appear in unregularized networks. In a second experiment we use CLIP for training two different classification networks on the MNIST and Fashion-MNIST databases. Here, we quantitatively show that CLIP yields superior robustness to noise and adversarial attacks compared to unregularized and weight-regularized neural networks. In all experiments we used Euclidean norms for the Lipschitz constant (3).

Our implementation of CLIP is available on `github`.[1]

4.1   Nonlinear Regression

In this experiment we train a neural network to approximate the nonlinear function $f(x) = \frac{1}{2} \max(|x| - 3, 0)$ with three hidden layers consisting of 500, 200 and 100 neurons using the *sigmoid* activation function. We construct 100 noisy training pairs by sampling values of f on the set $[-4, -3] \cup [-0.3, 0.3] \cup [3, 4]$. The set of Lipschitz training samples is randomly chosen in every epoch and consists of tuples (x, x') where $x \in [-4, 4]$ and x' is a noise perturbation of x. Figure 2 shows the learned networks after applying the CLIP Algorithm 2 with different values of the regularization parameter λ. Note that for this experiment we do not use a target accuracy but choose a fixed value of λ. For $\lambda = 10$ the learned network is very smooth, which yields a poor approximation of the ground truth especially at the boundary of the domain. For decreasing values of λ the networks approach the ground truth solution steadily, showing no instabilities. In contrast, the unregularized network with $\lambda = 0$ shows strong oscillations in regions of missing data, which implies a poor generalization capability. Both for $\lambda = 10^{-10}$ and $\lambda = 0$ the trained networks do not fit the noisy data. A reason for this is that some of the data points are negative and we use non-negative sigmoid activation in the last layer. Also, using stochastic gradient descent implicitly regularizes the problem and avoids convergence to spurious critical points of the loss.

Note that we warmstart the computation for $\lambda = 10$ with the pretrained unregularized network. The computations for $\lambda = 1$ and $\lambda = 10^{-10}$ are warmstarted with the respective larger value of λ. The latter successive reduction of the regularization parameter is necessary in order to "select" the desired Lipschitz-minimal solution as $\lambda \to 0$, which resembles Bregman iterations [19].

4.2   Classification on MNIST and Fashion-MNIST

We evaluate the robustness of neural networks trained with the proposed variational CLIP algorithm on the popular MNIST [14] and Fashion-MNIST [28] databases, which contain 28×28 grayscale images of handwritten digits and fashion articles, respectively. They are split into 60,000 samples for training and 10,000 samples for evaluation. For MNIST we train a neural network with two

---

[1] https://github.com/TimRoith/CLIP.

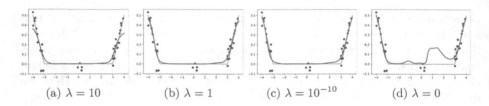

(a) $\lambda = 10$ (b) $\lambda = 1$ (c) $\lambda = 10^{-10}$ (d) $\lambda = 0$

Fig. 2. Lipschitz regularized networks for different values of λ. **Red:** ground truth. **Blue:** noisy ground truth samples. **Green:** trained neural network. (Color figure online)

Table 1. Accuracies [%] for different training setups, tested on train, test, noisy, and adversarial **(Fashion-)MNIST** data. The target accuracies are given by subscript. The regularization parameters μ and λ after training are also shown.

	Training type	Train	Test	Noise	PGD	μ, λ
MNIST	Standard	**95.6**	89.8	71.8	25.3	0
	Weight Reg.$_{95}$	93.1	89.0	71.4	25.4	0.0003
	Weight Reg.$_{90}$	89.8	89.7	71.9	24.8	0.0015
	Weight Reg.$_{85}$	84.8	87.2	72.3	24.9	0.0031
	CLIP$_{95}$	95.3	90.7	70.3	24.4	0.01
	CLIP$_{90}$	91.8	**91.6**	**78.0**	36.2	3.91
	CLIP$_{85}$	87.6	87.2	74.5	**37.7**	9.82
Fashion-MNIST	Standard	**98.5**	**92.0**	25.7	1.9	0
	Weight Reg.$_{95}$	94.8	91.2	26.9	3.5	0.0011
	Weight Reg.$_{90}$	89.8	89.6	28.1	6.0	0.0023
	Weight Reg.$_{85}$	85.3	85.4	**34.2**	10.2	0.0091
	CLIP$_{95}$	94.6	**91.9**	19.8	9.1	0.18
	CLIP$_{90}$	90.6	88.9	23.3	13.0	0.74
	CLIP$_{85}$	85.9	83.1	31.9	**14.9**	4.18

hidden layers with sigmoid activation function containing 64 neurons each. For Fashion-MNIST we use a simple convolutional neural network from [16]. We train non-regularized networks, CLIP regularized networks, and networks where we perform a L_2 regularization of the weights. The latter corresponds to (6) with $\lambda = 0$ and $\mu > 0$, where we update μ with the same discrepancy principle as for CLIP. Note that the weight regularization implies a layerwise Lipschitz bound of the type (4). The quantity $\mathcal{A}(f_\theta; \mathcal{T})$ in Algorithm 2 is chosen as the percentage of correctly classified training samples. We set the target accuracies for L_2 regularized and CLIP regularized networks to 85%, 90%, and 95%. The Lipschitz training set consists of 6,000 images removed from the training set and their noisy versions. We compare the robustness of all trained networks to adversarial perturbations $x + \delta$ of a sample $x \in \mathcal{X}$ in the L_2 norm, such that $\|\delta\| \leq \varepsilon$ with $\varepsilon = 2$, calculated with the Projected Gradient Descent (PGD) attack [16] as

$$x_{\text{adv}}^{t+1} = \Pi \left(x_{\text{adv}}^t + \alpha \, \text{sign}(\nabla_x \ell(f_\theta(x_{\text{adv}}^t), y)) \right), \tag{13}$$

where $\alpha > 0$ is the step size and x_{adv}^t describes the adversarial example at iteration t. Here, $\Pi : \mathcal{X} \to \mathcal{X}$ is a projection operator onto the L_2-ball with radius ε, and sign is the componentwise signum operator. We set the step size $\alpha = 0.25$ and the total number of iterations to 100. Furthermore, we use Gaussian noise with zero mean and unit variance for an additional robustness evaluation. We track the performance against the PGD attack of each model during training and use the model checkpoint with the highest robustness for testing.

Table 1 shows the networks' performance for the MNIST and Fashion-MNIST databases. The CLIP algorithm increases robustness with respect to Gaussian noise and PGD attacks depending on the target accuracy. One observes an inversely proportional relationship between the target accuracy and the magnitude of the regularization parameter λ as expected from Proposition 2. The results for MNIST indicate a higher robustness of the CLIP regularized neural networks with respect to noise perturbations and adversarial attacks in comparison to unregularized or weight-regularized neural networks with similar accuracy on unperturbed test sets. Similar observations can be made for the Fashion-MNIST database. While our method is also more robust under adversarial attacks, it is slightly more prone to noise than the weight-regularized neural networks.

5 Conclusion and Outlook

We have proposed a variational regularization method to limit the Lipschitz constant of a neural network during training and derived *CLIP*, a stochastic gradient descent-ascent training method. Furthermore, we have studied the dependency of the trained neural networks on the regularization parameter and investigated the limiting behavior as the parameter tends to zero and infinity. Here, we have proved convergence to Lipschitz-minimal data fits or constant networks, respectively. We have evaluated the CLIP algorithm on regression and classification tasks and showed that our method effectively increases the stability of the learned neural networks compared to weight regularization methods and unregularized neural networks. In future work we will analyze convergence and theoretical guarantees of CLIP using techniques from stochastic analysis.

References

1. Anil, C., Lucas, J., Grosse, R.B.: Sorting out Lipschitz function approximation. In: ICML, vol. 97, pp. 291–301. PMLR (2019)
2. Anzengruber, S.W., Ramlau, R.: Morozov's discrepancy principle for Tikhonov-type functionals with nonlinear operators. Inverse Probl. **26**(2), 025001 (2009)

3. Aziznejad, S., Gupta, H., Campos, J., Unser, M.: Deep neural networks with trainable activations and controlled Lipschitz constant. IEEE Trans. Signal Process. **68**, 4688–4699 (2020)

4. Bungert, L., Burger, M.: Solution paths of variational regularization methods for inverse problems. Inverse Probl. **35**(10), 105012 (2019)

5. Bungert, L., Burger, M., Korolev, Y., Schönlieb, C.B.: Variational regularisation for inverse problems with imperfect forward operators and general noise models. Inverse Probl. **36**(12), 125014 (2020)

6. Burger, M., Osher, S.: A guide to the TV zoo. In: Level Set and PDE Based Reconstruction Methods in Imaging, vol. 2090, pp. 1–70. Springer, Cham (2013). https://doi.org/10.1007/978-3-319-01712-9_1

7. Combettes, P.L., Pesquet, J.C.: Lipschitz certificates for layered network structures driven by averaged activation operators. SIAM J. Math. Data Sci. **2**(2), 529–557 (2020)

8. Fazlyab, M., Robey, A., Hassani, H., Morari, M., Pappas, G.: Efficient and accurate estimation of Lipschitz constants for deep neural networks. In: NeurIPS (2019)

9. Goodfellow, I.J., Shlens, J., Szegedy, C.: Explaining and harnessing adversarial examples. In: ICLR (2015)

10. Gouk, H., Frank, E., Pfahringer, B., Cree, M.J.: Regularisation of neural networks by enforcing Lipschitz continuity. Mach. Learn. **110**, 1–24 (2020). https://doi.org/10.1007/s10994-020-05929-w

11. Huster, T., Chiang, C.-Y.J., Chadha, R.: Limitations of the Lipschitz constant as a defense against adversarial examples. In: Alzate, C., et al. (eds.) ECML PKDD 2018. LNCS (LNAI), vol. 11329, pp. 16–29. Springer, Cham (2019). https://doi.org/10.1007/978-3-030-13453-2_2

12. Krishnan, V., Makdah, A.A.A., Pasqualetti, F.: Lipschitz bounds and provably robust training by Laplacian smoothing. arXiv preprint arXiv:2006.03712 (2020)

13. Krizhevsky, A.: Learning multiple layers of features from tiny images. Technical report (2009)

14. LeCun, Y., Bottou, L., Bengio, Y., Haffner, P., et al.: Gradient-based learning applied to document recognition. Proc. IEEE **86**(11), 2278–2324 (1998)

15. Liang, Y., Huang, D.: Large norms of CNN layers do not hurt adversarial robustness. arXiv preprint arXiv:2009.08435 (2020)

16. Madry, A., Makelov, A., Schmidt, L., Tsipras, D., Vladu, A.: Towards deep learning models resistant to adversarial attacks. In: ICLR (2018)

17. Oberman, A.M., Calder, J.: Lipschitz regularized deep neural networks converge and generalize. arXiv preprint arXiv:1808.09540 (2018)

18. van den Oord, A., et al.: WaveNet: a generative model for raw audio. In: The 9th ISCA Speech Synthesis Workshop, p. 125 (2016)

19. Osher, S., Burger, M., Goldfarb, D., Xu, J., Yin, W.: An iterative regularization method for total variation-based image restoration. Multiscale Model Sim. **4**(2), 460–489 (2005)

20. Roth, K., Kilcher, Y., Hofmann, T.: Adversarial training is a form of data-dependent operator norm regularization. In: NeurIPS (2019)

21. Ruder, S.: An overview of gradient descent optimization algorithms. arXiv preprint arXiv:1609.04747 (2016)

22. Scaman, K., Virmaux, A.: Lipschitz regularity of deep neural networks: analysis and efficient estimation. In: NeurIPS (2018)

23. Schwinn, L., Raab, R., Eskofier, B.: Towards rapid and robust adversarial training with one-step attacks. arXiv preprint arXiv:2002.10097 (2020)

24. Shafahi, A., et al.: Adversarial training for free! In: NeurIPS, pp. 3353–3364 (2019)
25. Shalev-Shwartz, S., Ben-David, S.: Understanding Machine Learning: From Theory to Algorithms. Cambridge University Press, New York (2014)
26. Szegedy, C., et al.: Intriguing properties of neural networks. In: International Conference on Learning Representations (2014)
27. Terjék, D.: Adversarial Lipschitz regularization. arXiv preprint arXiv:1907.05681 (2019)
28. Xiao, H., Rasul, K., Vollgraf, R.: Fashion-MNIST: a novel image dataset for benchmarking machine learning algorithms (2017)
29. Zou, D., Balan, R., Singh, M.: On Lipschitz bounds of general convolutional neural networks. IEEE Trans. Inf. Theory **66**(3), 1738–1759 (2019)

Variational Models for Signal Processing with Graph Neural Networks

Amitoz Azad[✉], Julien Rabin, and Abderrahim Elmoataz

Normandie Univ, UNICAEN, ENSICAEN, CNRS, GREYC, 14000 Caen, France
{amitoz.azad,julien.rabin,abderrahim.elmoataz}@unicaen.fr

Abstract. This paper is devoted to signal processing on point-clouds by means of neural networks. Nowadays, state-of-the-art in image processing and computer vision is mostly based on training deep convolutional neural networks on large datasets. While it is also the case for the processing of point-clouds with Graph Neural Networks (GNN), the focus has been largely given to high-level tasks such as classification and segmentation using supervised learning on labeled datasets such as ShapeNet. Yet, such datasets are scarce and time-consuming to build depending on the target application. In this work, we investigate the use of variational models for such GNN to process signals on graphs for unsupervised learning.

Our contributions are two-fold. We first show that some existing variational - based algorithms for signals on graphs can be formulated as Message Passing Networks (MPN), a particular instance of GNN, making them computationally efficient in practice when compared to standard gradient-based machine learning algorithms. Secondly, we investigate the unsupervised learning of feed-forward GNN, either by direct optimization of an inverse problem or by model distillation from variational-based MPN.

Keywords: Graph processing · Neural network · Total variation · Variational methods · Message passing network · Unsupervised learning

1 Introduction

Variational methods have been a popular framework for signal processing since the last few decades. Although, with the advent of deep Convolutional Neural Networks (CNN), most computer vision tasks and image processing problems are nowadays addressed using state-of-the-art machine learning techniques. Recently, there has been a growing interest in hybrid approaches combining machine learning techniques with variational modeling have seen a (see *e.g.* [1, 4, 11, 15, 19]).

Motivations. In this paper, we focus on such hybrid approaches for processing of point-clouds and signals on graphs. Graph representation provides a unified framework to work with regular and irregular data, such as images, text, sound,

© Springer Nature Switzerland AG 2021
A. Elmoataz et al. (Eds.): SSVM 2021, LNCS 12679, pp. 320–332, 2021.
https://doi.org/10.1007/978-3-030-75549-2_26

manifold, social network and arbitrary high dimensional data. Recently, many machine learning approaches have also been proposed to deal with high-level computer vision tasks, such as classification and segmentation [21,22,28], point-cloud generation [31], surface reconstruction [29], point cloud registration [27] and up-sampling [16] to name a few. These approaches are built upon Graph Neural Networks (GNN) which are inspired by powerful techniques employed in CNN, such as Multi-Linear Perceptron (MLP), pooling, convolution, *etc*. The main caveat is that transposing CNN architectures for graph-like inputs is not a simple task. Compared to images which are defined on a standard, regular grid that can be easily sampled, data with graph structures can have a large structural variability, which makes the design of a universal architecture difficult. This is likely the main reason why GNN are still hardly being used for signal processing on graph for which variational methods are still very popular. In this work, we investigate the use of data-driven machine learning techniques combined with variational models to solve inverse problems on point clouds.

1.1 Related Work

Graph Neural Networks. GNN are types of Neural Networks which directly operate on the graph structure. GNN are emerging as a strong tool to do representation learning on various types of data which, similarly to CNN, can hierarchically learn complex features.

PointNet [21] was one of the first few successful attempts to build GNN. Rather than preprocessing the input point-cloud by casting it into a generic structure, Qi et al. proposed to tackle the problem of structure irregularity by only considering permutation invariant operators that are applied to each vertex independently. Further works such as [22] have built upon this framework to replicate other successful NN techniques, such as pooling and interpolation in auto-encoder, which enables to process neighborhood of points.

Most successful attempts to build Graph *Convolutional* Networks (GCN) were based upon graph spectral theory (see for instance [2,5,14]). In particular, Kipf and Welling [14] proposed a simple, linear propagation rule which can be seen as a first-order approximation of local spectral filters on graphs. More recently, Wang et al. introduced an edge convolution technique [28] which maintains the permutation invariance property. Interestingly, [30] showed that GCN may act as low pass filters on internal features.

Message Passing Networks. At the core of the most the aforementioned GNNs, is *Message Passing* (MP). The term was first coined in [10], in which authors proposed a common framework that reformulated many existing GNNs, such as [2,14,21]. In the simplest terms, MP is a generalized convolution operator on irregular domains which typically expresses some neighborhood aggregation mechanism consisting of a message function, a permutation invariant function (*e.g.* max) and an update function.

Inverse Problems on Graphs. PDEs and variational methods are two major frameworks to address inverse problems on a regular grid structure like images. These methods have also been extended to weighted graphs in non-local form by defining p-Laplace operators [7]. Typical but not limited applications of these operators are filtering, classification, segmentation, or inpainting [20]. Specifically, we are interested in this work in non-local Total-Variation on a graph, which has been thoroughly studied, *e.g.* for non-local signal processing [6,9,25], multi-scale decomposition of signals [12], point-cloud segmentation [18] and sparsification [26], using various optimization schemes (such as finite difference schemes [6], primal-dual [3], forward-backward [23], or cut pursuit [24]).

Contributions and Outline. The paper is organised as follows. Section 2 focus on variational methods to solve inverse problems for graph signal processing. After a short overview of graph representation and non-local regularization, we demonstrate that two popular optimization algorithms can be interpreted as a specific instance of Message Passing Networks (MPNs). This allows using efficient GPU-based machine learning libraries to solve the inverse problem on graphs. Experiments on point cloud compression and color processing show that such MPN optimization is more efficient than standard machine learning optimization techniques. Section 3 is devoted to the use of a trainable GNN to perform signal processing on graphs in a *feed-forward* fashion. We investigate two techniques for unsupervised training, *i.e.* in absence of ground-truth data, using variational models as prior knowledge to drive the optimization. We show that either model distillation from a variational MPN or optimizing the inverse problem directly with GNN gives satisfactory approximate solutions and allows for faster processing.

2   Solving Inverse Problem with MPN Optimization

2.1   Notations and Definitions

A graph $\mathcal{G} = (\mathcal{V}, \mathcal{E}, \omega)$ is composed of an ensemble of nodes (vertices) $\mathcal{V}$, an ensemble of edges $\mathcal{E} = \mathcal{V} \times \mathcal{V}$ and weights $\omega : \mathcal{E} \mapsto \mathbb{R}_+$, where $\mathbb{R}_+$ indicates the set of non-negative values $[0, \infty)$. Let $f_i \in \mathbb{R}^d$ represents the feature vector (signal) on the node i of $\mathcal{G}$, $w_{i,j}$ the scalar weight on the edge from the node i to j, and $j \in \mathcal{N}(i)$ indicates a node j in the neighborhood of node i, such that $\omega_{i,j} \neq 0$. For sake of simplicity, we only consider non-directed graph, *i.e.* defined with symmetric weights $\omega_{i,j} = \omega_{j,i}$.

We refer to the weighted difference operator as $\nabla_\omega \colon (\nabla_\omega f)_{i,j} = \sqrt{\omega_{i,j}}\,(f_j - f_i) \in \mathbb{R}^d$, $\forall\,(i,j) \in \mathcal{E}$. Assuming symmetry of ω, the adjoint of this linear operator is $(\nabla_\omega^* g)_i = \sum_{j \in \mathcal{V}} \sqrt{\omega_{i,j}}\,(g_{j,i} - g_{i,j}) \in \mathbb{R}^d$, $\forall\,i \in \mathcal{V}$. We denote as $\|.\|_p$ the ℓ_p norm and $\|.\|_{1,p}$ the composed norm for parameter $p \in \mathbb{R}_+ \cup \{\infty\}$:

$$\forall\,g \in \mathbb{R}^{|\mathcal{V}| \times |\mathcal{V}| \times d}, \quad \|g\|_{1,p} = \sum_{i \in \mathcal{V}} \|g_{i,\cdot}\|_p = \sum_{i \in \mathcal{V}} \left(\sum_{j \in \mathcal{V}, 1 \le k \le d} |g_{i,j,k}|^p \right)^{1/p}.$$

2.2 Non-local Regularization on Graph

In this work, we focus on inverse problems for signal processing on graph using the non-local regularization framework [6]. The optimization problem is formulated generically using ℓ_2 fidelity term and a non-local regularization defined from the ℓ^p norm, parametrized by weights ω and power coefficient $q \in \mathbb{R}_+$, and penalized by $\lambda \in \mathbb{R}_+$

$$\inf_f \left\{ J(f) := \tfrac{1}{2}\|f - f_0\|^2 + \tfrac{\lambda}{q}\mathrm{R}_p^q(f) \right\} \quad \text{where } \mathrm{R}_p^q(f) = \|\nabla_\omega f\|_{1,p}^q. \quad (1)$$

Here f_0 is the given signal which requires processing. In experiments, we will mainly focus on two cases: when $p = q = 2$, the smooth regularization term relates to the Tikhonov regularization and boils down to Laplacian diffusion on a graph [6]. For $q = 1$, the regularization term is the Non-Local Total Variation (NL-TV), referred to as *isotropic* when $p = 2$ and *anisotropic* when $p = 1$. Other choice of norm might be useful: using $p \leq 1$, as studied for instance for $p = 0$ in [26], results in sparsification of signals; using $p = \infty$ [25] is also useful when considering unbiased symmetric schemes.

As the functional J is not smooth, an ε-approximation J_ε is often considered to circumvent numerical problems (see *e.g.* [6]); this is for instance achieved by substituting the term R_p^q with (respectively for the case $p = 2$ and $p = 1$)

$$\mathrm{R}_{2,\varepsilon}^q(f) = \sum_{i \in \mathcal{V}} \left(\|\nabla_\omega f_{i,.}\|_p^2 + \varepsilon^2\right)^{\frac{q}{2}} \quad \text{and} \quad \mathrm{R}_{1,\varepsilon}^q(f) = \sum_{i,j \in \mathcal{V}} \omega_{i,j}^{\frac{q}{2}} \left(|f_i - f_j| + \varepsilon\right)^q. \quad (2)$$

2.3 Variational Optimization

We consider now two popular algorithms to solve inverse problem (1), described in Algorithm 1 and 2 with update rules on $f^{(t)}$, where (t) is the iteration number.

Algorithm 1: Gauss-Jacobi [6]	**Algorithm 2:** Primal-Dual [3]
Initialization: $f^{(0)} = f_0$, set $\varepsilon > 0$ compute $\gamma_{i,j}^{(t)}$ using Eq.(3) $\tilde{f}_i^{(t+1)} = (f_0)_i + \lambda \sum_{j \in \mathcal{N}_i} \gamma_{i,j}^{(t)} f_j^{(t)}$ $g_i^{(t+1)} = 1 + \lambda \sum_{j \in \mathcal{N}_i} \gamma_{i,j}^{(t)}$ $f_i^{(t+1)} = \tilde{f}_i^{(t+1)}/g_i^{(t+1)}$	Parameters: $\tau, \beta > 0$ and $\theta \in [0,1]$ Initialization: $f^{(0)} = \bar{f}^{(0)} = f_0$ $g^{(t+1)} = \mathrm{Prox}_{\beta A^*}(g^{(t)} + \beta K \bar{f}^{(t)})$ $f^{(t+1)} = \mathrm{Prox}_{\tau B}(f^{(t)} - \tau K^* g^{(t+1)})$ $\bar{f}^{(t+1)} = f^{(t+1)} + \theta(f^{(t+1)} - f^{(t)})$

For any $p, q > 0$, one can turn to the Gauss-Jacobi (GS) iterated filter [6] described in Algorithm 1. It relies on ε approximation to derive the objective function, giving for instance the following update rule when using Eq. (2)

$$\gamma_{i,j}^{(t)} = \begin{cases} w_{i,j} \left((\|(\nabla_w f)_{i,.}\|^2 + \varepsilon^2)^{\frac{q-2}{2}} + (\|(\nabla_w f)_{j,.}\|^2 + \varepsilon^2)^{\frac{q-2}{2}} \right) & \text{if } p = 2 \\ 2 w_{i,j}^{\frac{q}{2}} \left(|f_i - f_j| + \varepsilon\right)^{q-2} & \text{if } p = 1 \end{cases}. \quad (3)$$

When considering the specific case of NL-TV ($q = 1$), one can turn to the primal-dual algorithm [3] described in Algorithm 2, as done in [12,18]. To do so, we recast the problem (1) as $\min_f \max_g \langle Kf, g \rangle - A^*(g) + B(f)$ where $K \equiv \nabla_\omega$, $A \equiv \|.\|_{1,p}$, $B \equiv \frac{1}{2\lambda}\|. - f_0\|_2^2$. Parameters must be chosen such that $\tau\beta\|K\|^2 < 1$ to ensure convergence. For numerical experiments, we have used $\theta = 1$ and $\tau = \beta = (4\max_{i \in \mathcal{V}} \sum_{j \in \mathcal{V}} \omega_{i,j})^{-1}$. Note that other methods might be used, such as pre-conditioning [23], and cut pursuit [24]. Proximal operators corresponds to

$$\text{Prox}_{\tau B}(f) = \frac{\lambda f + \tau f_0}{\lambda + \tau}, \quad \text{and } \text{Prox}_{\beta A^*}(g) = \text{Proj}_{\mathcal{B}_{\infty,p'}}(g) \quad (4)$$

where the dual norm parameter $p' > 0$ verifies $1/p' + 1/p = 1$. The projection on the unit ball $\mathcal{B}_{\infty,p'}$ for $p = 2$ and $p = 1$ is given respectively by, $\forall \, g \in \mathbb{R}^{|\mathcal{V}| \times |\mathcal{V}| \times d}$

$$\text{Proj}_{\mathcal{B}_{\infty,2}}(g)_{i,j,k} = \frac{g_{i,j,k}}{\max\{1, \|g_{i,.}\|_2\}}, \text{Proj}_{\mathcal{B}_{\infty,\infty}}(g)_{i,j,k} = \frac{g_{i,j,k}}{\max\{1, |g_{i,j,k}|\}}. \quad (5)$$

2.4   Message Passing Network Optimization

Message Passing Networks [10] can be formulated as follows

$$f_i^{(n+1)} = \psi^{(n)}\left(f_i^{(n)}, \square_{j \in \mathcal{N}(i)} \, \phi^{(n)}\left(f_i^{(n)}, f_j^{(n)}, \omega_{i,j}\right)\right) \quad \forall \, i \in \mathcal{V}, \quad (6)$$

where (n) indicates the depth in the network. As already shown in [10], many existing GNN can be recast in this generic framework. Typically, $\square$ is the summation operator but can be any differentiable permutation invariant functions (such as max, mean); $\psi^{(n)}$ (update rule) and $\phi^{(n)}$ (message rule) are the differentiable operators, such as MLPs (*i.e.* affine functions combined with simple non-linear functions such as RELU). Note that they are independent of the input vertex index to preserve permutation invariance.

We now show that Algorithm 1 and 2 can be also formulated within this framework.

G-J algorithm. This is quite straightforward for the anisotropic case $(p = 1)$ using two MPNs to compute the update of f in Algorithm 1. A first MPN is used for the numerator $\bar{f}$, identifying $\psi^{(n)}(f_i^{(n)}, F_i^{(n)}) = (f_0)_i + \lambda F_i^{(n)}$, $\square = \sum$ and $\phi^{(n)}(f_i^{(n)}, f_j^{(n)}, \omega_{ij}) = \gamma_{i,j}^{(n)} f_j^{(n)}$ where $\gamma_{i,j}^{(n)}$, as defined in Eq. (3), is a function of the triplet $(f_i^{(n)}, f_j^{(n)}, \omega_{i,j})$. A second MPN is used to compute the denominator g using $\psi^{(n)}(f_i^{(n)}, F_i^{(n)}) = 1 + \lambda F_i^{(n)}$ and $\phi^{(n)}(f_i^{(n)}, f_j^{(n)}, \omega_{ij}) = \gamma_{i,j}^{(n)}$. For the isotropic case $p = 2$, an additional MPN is ultimately required to compute the norm of the gradient $(Kf^{(n)})_i$ at each vertex i, as detailed in the next paragraph.

P-D Algorithm. Two *nested* MPNs are now required to compute the update of the primal variable f and the dual variable g in Algorithm 2. For f, the MPN reads as the affine operator $\psi^{(n)}(f_i^{(n)}, F_i^{(n)}) = \frac{\tau}{\tau+\lambda}(f_0)_i + \frac{\tau}{\tau+\lambda}(f_i^{(n)} - \tau F_i^{(n)})$. Similarly, for the update of g one computes the projection: $\psi^{(n)}(g_i^{(n)}, G_i^{(n)}) =$

$\text{Proj}_{\mathcal{B}_{\infty,p'}}(g_i^{(n)} + \beta G_i^{(n)})$. The gradient operator K and its adjoint K^* (see definition in Sect. 2.1) can be applied using the processing functions ϕ and ϕ_j, respectively

$$G_i^{(n)} = (K\bar{f}^{(n)})_i = \phi^{(n)}(\bar{f}_i^{(n)}, \bar{f}_j^{(n)}, \omega_{i,j}) = \omega_{i,j}(\bar{f}_j^{(n)} - \bar{f}_i^{(n)})$$

$$F_i^{(n)} = (K^*g^{(n+1)}))_i = \Box\phi_j^{(n)}(g_i^{(n+1)}, g_j^{(n+1)}, \omega_{i,j}) = \sum_{j \in N_i} \omega_{i,j}(g_{j,i}^{(n+1)} - g_{i,j}^{(n+1)})$$

Observe that ϕ_j is a slightly modified message passing function using index j to incorporate the edge features from not only source i to target j (*i.e.* $g_{i,j}$ and $\omega_{i,j}$) but also from target j to source i ($g_{j,i}$ and $\omega_{j,i}$).

These MPN formulations of Algorithm 1 and Algorithm 2 allows to use machine learning libraries to solve the inverse problem Eq. 1 and to take full advantage of fast GPU optimization. In the following experiments, we illustrate the advantage of such MPN optimization in comparison with standard gradient-based machine learning optimization techniques relying on auto-differentiation.

2.5   Experiments on Point-Clouds and Comparison with Auto-Diff

Experimental Setting. Edges on point-clouds are defined with weights $\omega_{i,j} \in \{0,1\}$ using the indicator function of a k-nearest neighbor search ($k = 4$) on points coordinates using ℓ_2 norm, then imposing symmetry by setting $\omega_{i,j} \leftarrow \omega_{i,j} \vee \omega_{j,i}$. Two types of features f are tested for $q = 1$, as illustrated in Fig. 1: 3D point coordinates x for point-cloud simplification (Fig. 2), and colors $c \in [0,1]^{|\mathcal{V}| \times 3}$ for denoising (Fig. 3). We also consider the non-convex case where $q = 0.1$ for point cloud sparsification in Fig. 2.

Auto-differentiation. As a baseline, we compare these MPN-based algorithms with algorithmic auto-differentiation. In this setting, the update of features at iteration (t) is defined by gradient descent on the loss function $J(f^{(t)})$ defined in Eq. (1), *e.g.* $f^{(t+1)} = f^{(t)} - \rho^{(t)}D^{(t)}$. Since the NL-TV regularization term is not smooth, we use the ε approximation of Eq. (2) to compute the gradient $D^{(t)}$. We consider here various standard gradient descent techniques used for NN training: Gradient Descent (GD), ADAM [13] and LBFGS [17].

Implementation Details. All algorithms are implemented using the *Pytorch Geometric* library [8] and tested with a Nvidia GPU GTX1080Ti with 11 GB of memory. Note that reported computation time is only indicative, as implementation greatly influence the efficiency of memory and GPU cores allocation. Different learning rates are used for each method in the experiments. The learning rate of ADAM algorithm is always set to $lr = 0.001$. GD is also used with $lr = 0.001$ for point-cloud simplification but with $lr = 0.1$ for color processing. LBFGS is set with $lr = 0.01$ for $q = 1$, and with $lr = 0.1$ for $q = 0.1$.

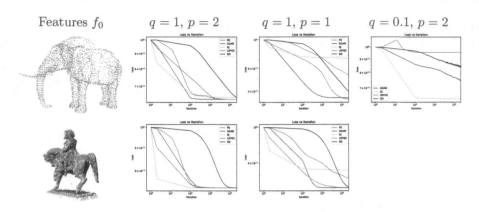

Fig. 1. (left) Raw data f_0 used for experiments in Fig. 2 and Fig. 3. (right) Comparison of the evolution of the objective function $J(f^{(t)})$ depending on iteration (t), in logarithmic scale, for various setting of (p, q) and for different algorithms.

Results. Observe that the noise on the raw color point cloud in Fig. 1 results from the registration of multiple scans with different lighting conditions, for which we do not have any ground-truth. In such a case, variational models like in (1) are useful to remove color artifacts without more specific prior knowledge.

Figures 2 and 3 show the visual results of the aforementioned algorithms after 20k iterations. The objective function $J(f^{(t)})$ is plotted for each one in Fig. 1. Since the computation time *per* iteration is roughly same for all methods (about 0.5 ms for MPN-PD and 1 ms for others on 'elephant' and 'Napoléon' pointclouds with $|\mathcal{V}| = 3k$ and $11k$ respectively), curves are here simply displayed versus the number of iterations. As one can observe from these experiments, MPN based algorithms perform better than gradient based algorithm when it comes to precision. Indeed, for $q = 1$, Algorithm 2 is the only algorithm that can be used without approximation (*i.e.* $\varepsilon = 0$, which is numerically intractable for other methods) and that gives ultimately an optimal solution. As for $q = 0.1$, Algorithm 1 reaches an (approximated) solution much faster than gradient based methods. Among these, it is interesting to notice that ADAM is quite consistent and performs quite well. For this reason, we will consider this method for the next part devoted to unsupervised training of GNN.

We have shown that the two algorithms can be implemented efficiently as MPN, a special instance of GNN. In the next section, we demonstrate the interest of training such MPNs to design feed-forward GNNs allowing fast processing.

3   Variational Training of GNN

One of the most attractive aspect of machine learning is the ability to learn the model itself from the data. As already mentioned in the introduction, most approaches rely on *supervised* training techniques which require datasets with ground-truth, potentially obtained by means of data augmentation (*i.e.* artificially degrading the data), which might be a simple task for some applications such as Gaussian denoising on images. However, it is challenging and time consuming to achieve this for signal processing tasks on pointclouds, as most of the available datasets are only devoted to high-level tasks like segmentation and classification on shapes.

Inspired from the previous results, we investigate the *unsupervised* training of a GNN to reproduce the behavior of variational models on point-clouds while reducing the computation time. To achieve such a goal, two different settings are considered in this section: Unsupervised training using the variational energy as a loss function (Sect. 3.1), and model distillation using exemplars from the variational model defined as an MPN (Sect. 3.2).

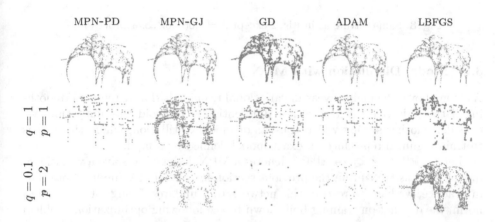

Fig. 2. Comparison of numerical results of MPN implementations of Algorithms 1 (MPN-GJ) and 2 (MPN-PD), with GD, ADAM and LBFGS after $20k$ iterations. Point-cloud simplification is defined by problem (1) with $\lambda = 0.2$ and various settings of p and q. For all methods except MPN-PD, the approximated formulation (2) with $\epsilon = 10^{-8}$ is used.

3.1 Unsupervised Training

We denote by $G_\theta : \mathbb{R}^{|\mathcal{V}| \times d} \times \mathbb{R}_+^{\mathcal{E}} \to \mathbb{R}^{|\mathcal{V}| \times d}$ a trainable GNN parametrized by θ, processing a graph defined by the vertex features f and the edge weights ω. The idea here is simply to train a GNN in such a way that it minimizes in expectation the objective function (1) of the variational model for graphs from a dataset

$$\inf_\theta \mathbb{E}_{X \sim P_{\text{data}}} J\left(G_\theta(X, \omega(X))\right). \tag{7}$$

where X is a random point-cloud (and its associated weights $\omega(X)$) from the training dataset with probability distribution P_{data}. Once again, in practice, we need to consider the ε-approximation J_ε (2) to overcome numerical issues.

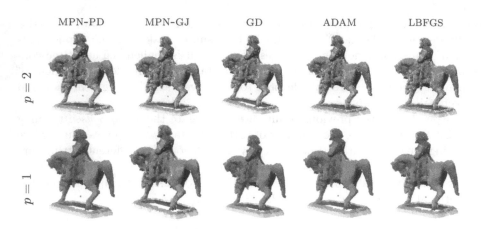

Fig. 3. Same setting as in Fig. 2 except $\lambda = 0.05$, for color processing.

3.2   Model Distillation with MPN

As seen above, the main caveat in variational training is that it requires smoothing if the objective function is not differentiable. to avoid this limitation, we consider another setting which consists of model distillation, where the goal is to train a gnn to reproduce an exact model defined by a mpn.

Let $F : \mathbb{R}^{|\mathcal{V}| \times d} \times \mathbb{R}^{\mathcal{E}}_{+} \mapsto \mathbb{R}^{|\mathcal{V}| \times d}$ denotes a MPN designed to solve a variational model, such as MPN-PD in the previous section for the NL-TV model. Note that in practice, we have to restrict this network to n updates. Using for instance ℓ_2 norm, the distillation training boils down to the following optimization problem

$$\inf_{\theta} \mathbb{E}_{X \sim P_{\text{data}}} \|G_\theta(X, \omega(X)) - F(X, \omega(X))\|^2 \tag{8}$$

Other choices of distance could be as well considered, such as ℓ_1 norm or optimal transport cost for instance [29].

3.3   Experiments

Dataset Setting. We use the same experimental setting as in the previous section, except that k nearest-neighbor parameters $\omega_{i,j} \in \{0, 1\}$ are now weighted using features proximity: $\omega_{i,j} \leftarrow \omega_{i,j} e^{-\kappa \|(f_0)_i - (f_0)_j\|^2}$ where $\kappa = 10^4$. For point-cloud processing, we use the ShapeNet part dataset [32]. Point-clouds are pre-processed by normalizing their features to 0 and 1 using minmax normalization; 15 classes out of 16 are used for training and a hold out class for testing. For each training class, 20% of shapes are also hold-out for validation. In total, 16.5k point clouds are used during training, composed of an average of 2.5k nodes.

GNN Architecture. The GNN is based on MPNs. The architecture is shallow and composed of 3 MPN layers followed by a linear layer. Inspired from MPNs formulated in Sect. 2.4 from Algorithm 1 and Algorithm 2, we define the message function (ϕ) as a MLP acting on the tuple $(f_i, f_j, \sqrt{w_{i,j}}(f_j - f_i))$. Note that, this is quite similar to edge-convolution operation introduced in [28], in which the MLP layers act on features $(f_i, f_j - f_i, w_{i,j})$. The $\square$ operator is $\sum$ and there is no update rule (ψ).

The MLP layers have respectively 4.8k, 20.6k, and 20.6k parameters. These layers operate on 64 dimensional features, which are then fed to a linear layer to return 3-dimensional features (color or coordinates).

Training Settings. Using the dataset and the architecture described above, two GNNs are trained using each of the techniques described in Sect. 3.1 and Sect. 3.2, with $k = 4$, $p = 2$, $q = 1$, $\lambda = 0.05$ and $\varepsilon = 10^{-8}$. Training is performed using ADAM algorithm with 60 epochs and a fixed learning rate of 0.01. Regarding the model distillation, rather than re-evaluating the MPN F in (8) for every sample X at each epoch, we precompute $F(X, \omega(X))$ to save the computation time.

Results. Figure 4 gives a visual comparison of results of trained GNNs with the exact MPN-PD optimization on test data, showing that both training methods give satisfactory results, close to MPN-PD model. Figure 5 displays the objective loss function in (7) during training, on the training and the validation set. The average error relative to mpn-pd on the test set is similar for both methods (2.2% for distillation and 1.2% for unsupervised training). One can observe that model distillation is more stable on the validation set, but need extra computation time to pre-process the MPN-PD on the dataset. Last, but not least, the average

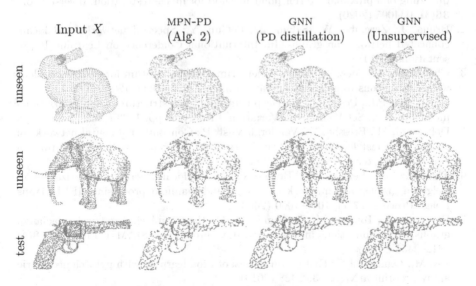

Fig. 4. Comparison of NL-TV regularization (1) of point-clouds X (for $p = 2$, $q = 1$, $\lambda = 0.05$) with the MPN algorithm MPN-PD and trained GNN, using model distillation (8) or the unsupervised setting with variational optimization (7). Input here is either *unseen* point-clouds during training, or from the *test* category, hold out during training.

Fig. 5. (left) Objective loss function (7) during GNN training versus epochs for the training and the validation sets. (right) Comparison of the average objective function J in logarithmic scale versus *average computation time* for GNN, ADAM and MPN-PD.

computation time for MPN-PD on the test set is compared in Fig. 5 to the feed-forward GNNs, showing several order of magnitude speed-up for the latter, which makes such approaches very interesting for practical use.

Acknowledgments. This work has been carried out with financial support from the French Research Agency through the SUMUM project (ANR-17-CE38-0004)

References

1. Bertocchi, C., Chouzenoux, E., Corbineau, M.C., Pesquet, J.C., Prato, M.: Deep-unfolding of a proximal interior point method for image restoration. Inverse Probl. **36**(3), 034005 (2020)
2. Bruna, J., Zaremba, W., Szlam, A., LeCun, Y.: Spectral networks and locally connected networks on graphs. In: International Conference on Learning Representations (2014)
3. Chambolle, A., Pock, T.: A first-order primal-dual algorithm for convex problems with applications to imaging. J. Math. Imaging Vis. **40**(1), 120–145 (2011)
4. Combettes, P.L., Pesquet, J.C.: Deep neural network structures solving variational inequalities. In: Set-Valued and Variational Analysis, pp. 1–28 (2020)
5. Defferrard, M., Bresson, X., Vandergheynst, P.: Convolutional neural networks on graphs with fast localized spectral filtering. In: Advances in Neural Information Processing Systems, pp. 3844–3852 (2016)
6. Elmoataz, A., Lezoray, O., Bougleux, S.: Nonlocal discrete regularization on weighted graphs: a framework for image and manifold processing. IEEE Trans. Image Process. **17**(7), 1047–1060 (2008)
7. Elmoataz, A., Toutain, M., Tenbrinck, D.: On the p-Laplacian and ∞-Laplacian on graphs with applications in image and data processing. SIAM J. Imag. Sci. **8**(4), 2412–2451 (2015)
8. Fey, M., Lenssen, J.E.: Fast graph representation learning with pytorch geometric. arXiv preprint arXiv:1903.02428 (2019)
9. Gilboa, G., Osher, S.: Nonlocal operators with applications to image processing. Multiscale Model. Simul. **7**(3), 1005–1028 (2009)
10. Gilmer, J., Schoenholz, S.S., Riley, P.F., Vinyals, O., Dahl, G.E.: Neural message passing for quantum chemistry. In: ICML (2017)

11. Hasannasab, M., Hertrich, J., Neumayer, S., Plonka, G., Setzer, S., Steidl, G.: Parseval proximal neural networks. J. Fourier Anal. Appl. **26**(4), 1–31 (2020)
12. Hidane, M., Lézoray, O., Elmoataz, A.: Nonlinear multilayered representation of graph-signals. J. Math. Imaging Vis. **45**(2), 114–137 (2013)
13. Kingma, D.P., Ba, J.: Adam: a method for stochastic optimization. In: Proceedings of the 3rd International Conference on Learning Representations (ICLR) (2014)
14. Kipf, T.N., Welling, M.: Semi-supervised classification with graph convolutional networks. In: International Conference on Learning Representations (2017)
15. Kobler, E., Klatzer, T., Hammernik, K., Pock, T.: Variational networks: connecting variational methods and deep learning. In: Roth, V., Vetter, T. (eds.) GCPR 2017. LNCS, vol. 10496, pp. 281–293. Springer, Cham (2017). https://doi.org/10.1007/978-3-319-66709-6_23
16. Li, R., Li, X., Fu, C.W., Cohen-Or, D., Heng, P.A.: PU-GAN: a point cloud upsampling adversarial network. In: Proceedings of the IEEE/CVF International Conference on Computer Vision, pp. 7203–7212 (2019)
17. Liu, D.C., Nocedal, J.: On the limited memory BFGS method for large scale optimization. Math. Program. **45**(1), 503–528 (1989)
18. Lozes, F., Hidane, M., Elmoataz, A., Lézoray, O.: Nonlocal segmentation of point clouds with graphs. In: 2013 IEEE Global Conference on Signal and Information Processing, pp. 459–462. IEEE (2013)
19. Meinhardt, T., Moller, M., Hazirbas, C., Cremers, D.: Learning proximal operators: using denoising networks for regularizing inverse imaging problems. In: Proceedings of the IEEE ICCV, pp. 1781–1790 (2017)
20. Peyré, G., Bougleux, S., Cohen, L.: Non-local regularization of inverse problems. Inverse Probl. Imag. **5**(2), 511 (2011)
21. Qi, C.R., Su, H., Mo, K., Guibas, L.J.: Pointnet: Deep learning on point sets for 3d classification and segmentation. In: Proceedings of the IEEE Conference on Computer Vision and Pattern Recognition, pp. 652–660 (2017)
22. Qi, C.R., Yi, L., Su, H., Guibas, L.J.: PointNet++: Deep hierarchical feature learning on point sets in a metric space. In: Advances in Neural Information Processing Systems, vol. 30 (2017)
23. Raguet, H., Landrieu, L.: Preconditioning of a generalized forward-backward splitting and application to optimization on graphs. SIAM J. Imag. Sci. **8**(4), 2706–2739 (2015)
24. Raguet, H., Landrieu, L.: Cut-pursuit algorithm for regularizing nonsmooth functionals with graph total variation. In: International Conference on Machine Learning, pp. 4247–4256 (2018)
25. Tabti, S., Rabin, J., Elmoataz, A.: Symmetric upwind scheme for discrete weighted total variation. In: IEEE International Conference on Acoustics, Speech and Signal Processing (ICASSP), pp. 1827–1831 (2018)
26. Tenbrinck, D., Gaede, F., Burger, M.: Variational graph methods for efficient point cloud sparsification. arXiv preprint arXiv:1903.02858 (2019)
27. Wang, Y., Solomon, J.M.: Deep closest point: learning representations for point cloud registration. In: Proceedings of the IEEE International Conference on Computer Vision, pp. 3523–3532 (2019)
28. Wang, Y., Sun, Y., Liu, Z., Sarma, S.E., Bronstein, M.M., Solomon, J.M.: Dynamic graph CNN for learning on point clouds. ACM Trans. Graph. (TOG) **38**(5), 1–12 (2019)
29. Williams, F., Schneider, T., Silva, C., Zorin, D., Bruna, J., Panozzo, D.: Deep geometric prior for surface reconstruction. In: Proceedings of the IEEE Conference on Computer Vision and Pattern Recognition, pp. 10130–10139 (2019)

30. Wu, F., Souza, A., Zhang, T., Fifty, C., Yu, T., Weinberger, K.: Simplifying graph convolutional networks. In: Proceedings of the 36th International Conference on Machine Learning, vol. 97, pp. 6861–6871. PMLR (2019)
31. Yang, G., Huang, X., Hao, Z., Liu, M.Y., Belongie, S., Hariharan, B.: PointFlow: 3D point cloud generation with continuous normalizing flows. In: Proceedings of the IEEE International Conference on Computer Vision, pp. 4541–4550 (2019)
32. Yi, L., et al.: A scalable active framework for region annotation in 3D shape collections. ACM Trans. Graph. (ToG) 35(6), 1–12 (2016)

Synthetic Images as a Regularity Prior for Image Restoration Neural Networks

Raphaël Achddou$^{(\boxtimes)}$, Yann Gousseau, and Saïd Ladjal

LTCI, Telecom Paris, Institut Polytechnique de Paris, Paris, France
`raphael.achddou@telecom-paris.fr`

Abstract. Deep neural networks have recently surpassed other image restoration methods which rely on hand-crafted priors. However, such networks usually require large databases and need to be retrained for each new modality. In this paper, we show that we can reach near-optimal performances by training them on a synthetic dataset made of realizations of a dead leaves model, both for image denoising and super-resolution. The simplicity of this model makes it possible to create large databases with only a few parameters. We also show that training a network with a mix of natural and synthetic images does not affect results on natural images while improving the results on dead leaves images, which are classically used for evaluating the preservation of textures. We thoroughly describe the image model and its implementation, before giving experimental results.

Keywords: Image restoration · Deep learning · Natural image models

1 Introduction

A key ingredient of image restoration methods is an *a priori* hypothesis about image regularity. Methods based on total variation regularization terms [24] assume a Laplacian distribution for the image gradient. Methods involving wavelet shrinkage [11] are optimal in some Besov spaces. Non-local methods [4] rely on an auto-similarity hypothesis. More recently, deep neural networks have achieved impressive results in all fields of image restoration: denoising, single image super-resolution, deconvolution, etc. Although these models do not usually incorporate image regularity assumptions, it has been shown that networks themselves can be seen as regularization terms [27]. Instead of hand-crafted mathematical priors, these models necessitate a training on voluminous databases and necessitate to be retrained for each new modality or specific imaging device [8]. In this contribution, we show that these models can be efficiently trained from synthetic image databases based on a mathematical model grounded in physical priors and depending on few parameters. To the best of our knowledge, this is the first such result for networks aimed at restoring natural images.

Among available mathematical models for natural images, we show that an occlusion-based dead leaves model, equipped with a scaling size distribution, is sufficient to reach near state-of-the-art restoration performances, for both tasks

© Springer Nature Switzerland AG 2021
A. Elmoataz et al. (Eds.): SSVM 2021, LNCS 12679, pp. 333–345, 2021.
https://doi.org/10.1007/978-3-030-75549-2_27

of denoising and single image super-resolution. Moreover, we show that this model can be efficiently combined with natural image databases to enhance the capacity of deep neural networks to preserve details, without impairing their classical performance evaluation.

We believe that such a study both sheds light on the way convolutional neural networks can address restoration problems and opens interesting perspectives. First, this result shows that the mere structure of such networks is adapted to image restoration tasks and that despite their huge number of parameters they can be made near-optimal from just a few principles and hyper-parameters. This result, and the fact that simpler, less structured models cannot achieve satisfying restoration performance, also highlights the type of geometric structures a neural networks needs to be efficiently trained. Second, the proposed learning database has the potential to be modified according to specific acquisition devices and in particular to their point spread function, dynamic range, noise modality, etc. This opens the way to flexible, generic and relatively light learning schemes.

2   Related Works

Image Regularity and Restoration Priors. A classical principled approach to image restauration is to assume some closed form statistical prior on the images distribution. Given an observation, one seeks the best explanation according to the model. Depending on the method the "best" may be the maximum a posteriori or the risk minimiser. Among the models that fall in this general framework let us cite the Wiener model, for which the distribution of images is assumed to be Gaussian and translation invariant and the total variation [24] for which the log-likelihood is the l^1 norm of the gradient vector field. In [11] the authors derive an algorithm for restoring signals under the assumption that the targeted signals are well approximated by a sparse representation in some wavelet decomposition. Later, a consequent body of literature discussed the implications of the sparsity assumption and a variety of algorithms where proposed to take advantage of this particular form of regularity [5,10]. The total variation model itself can be viewed as a form of sparsity.

Other fruitful models do not translate easily into a restoration algorithm. Among these, the autosimilarity assumption assumes that the image possesses repeats of the same pattern or patch [4,9]. Methods derived from this regularity assumption are algorithms that typically average similar patches in order to co-denoise those that are likely to represent the same ground truth.

On the other hand deep learning methods for restoration seek directly to build a machinery (the trained network) that minimises the reconstruction error. Typically, the loss function is taken to be the mean square error between the perfect image and the output of the network. Here the prior on images is represented by the training dataset which is believed to convey sufficient information about the distribution of images [30,31]. The effort goes into the careful design of the network and its training.

In turn, a trained denoising network can be used as an implicit prior on the image distribution. This idea, named plug and play, consists in using the denoising network in place of a proximal operator during an iterative optimisation of a variational model [21]. Another approach to using a network as a prior is presented in [27] in which it is showed that the architecture of a network can serve as a regulariser.

Synthesis Models. Some statistical priors on natural images have been turned into generative models that have the ability to synthesize images, mostly for the task of texture synthesis: Gaussian models [12], wavelet-based models [15,22], Markov fields, dead leaves model [2,18], neural networks [13]. In this work, we investigate the use of such generative models as a way to train restoration neural networks. We turn to the dead leaves model, because of its simplicity, limited number of parameters and ability to generate complex images with details at all scales.

Let us mention that the use of synthetic models to train neural networks is not new and has been extensively used for image analysis tasks such as segmentation, recognition, or detection, see e.g. [26]. To the best of our knowledge, no such strategy has been used for restoration tasks.

3   Dead Leaves Image Generation

3.1   The Continuous Dead Leaves Model

Informally, the dead leaves model is a random field obtained as the sequential superimposition of random shapes. It is defined (see [3]) from a set of random positions, times and shapes $\{(x_i, t_i, X_i)_{i \in \mathbb{N}}$, where $\mathcal{P} = \sum \delta_{x_i, t_i}$ is a homogeneous Poisson process on $\mathbb{R}^2 \times (-\infty, 0]$ and the X_i are random sets of $\mathbb{R}^2$ that are independent of $\mathcal{P}$. The sets $x_i + X_i$ are called *leaves* and for each i, the *visible part* of the leaf is defined as

$$V_i = (x_i + X_i) \setminus \bigcup_{t_j \in (t_i, 0)} (x_j + X_j),$$

that is, the visible part of the leaf (x_i, t_i, X_i) is obtained by removing from this leaf all leaves that are indexed by a time greater than t_i (that falls after it). The dead leaves model is then defined as the collection of all visible parts. A random image can be obtained by assigning a random gray level (or color) to each visible part. An example of a dead leaves model where the leaves are disks with a constant radius can be seen in Fig. 2a.

A particular type of dead leaves model, where the leaves have a size with scaling properties, has been shown to reproduce many statistical properties of natural images [2,18]. Such models are obtained by considering random leaves $R.X$, where X is a given shape and R is a real random variable with density $f(r) = C.r^{-\alpha}$, with C a normalizing constant. The case $\alpha = 3$ corresponds to a scale invariant model [18]. In order for such models to be well defined, values of R have to be restricted to values in (r_{min}, r_{max}), see [14]. The resulting model therefore depends on 3 parameters: r_{min}, r_{max} and α. This model is especially appealing for natural images, because it incorporates two of their most fundamental property, non Gaussianity (as a result of edges) and scaling properties [20], in a very simple setting. Because this model contains details and edges at all scales, potentially of arbitrary contrast, it has been proposed as a tool for the evaluation of the ability of imaging devices to respect textures [6,7] and was recently retained as a standard for quality evaluation.

3.2 Implementation and Discussion on the Parameters

Algorithm 1: Dead leaves image generation algorithm

Parameters: $(r_{min}, r_{max}, \alpha, w)$, color_image
Output   : X
mask = ones(w, w);
X = zeros$(w, w, 3)$;
while $\| mask \| > 0$ **do**
    tmp $= r_{max}^{1-\alpha} + (r_{min}^{1-\alpha} - r_{max}^{1-\alpha}) \times$random();
    r = tmp$^{-\frac{1}{\alpha-1}}$;
    x,y = randint(0,w), randint(0,w);
    color = color_image(randint(0,w), randint(0,w));
    new_disk = disk(r, color);
    X = update_image(X, x, y, new_disk);
    mask = update_mask(mask, r, x, y);
end
X = downscale(X,5).

We now detail how to generate digital versions of the dead leaves model, following the procedure summarized in Algorithm 1. At each step, a random discrete disk of radius r and center (x, y) is generated as the set of discrete positions satisfying the corresponding disk equation. Centers are uniformly distributed in the image domain and radiuses are distributed according to a power law density with exponent α, as discussed in the previous paragraph. Radiuses are limited between r_{min} and r_{max}. To generate the image, we rely on a *perfect simulation* technique [17] and sequentially put the disks *below* the previously drawn disks, until the image domain has been fully covered. That is, at each step, pixels which have not been colored yet are given the color of the disk added at this step. The choice of the disk color will be shortly discussed. The used definition of discrete disk is crude and in particular does not include any anti-aliasing scheme. Therefore we first generate a large image that is then down-sampled by a factor 5 after convolution with a Gaussian filter with $\sigma = 5/3$. This step is a critical component of our algorithm. It allows for sub-pixel sized objects and for more natural boundaries. In Fig. 1a, we display a full size (2000, 2000) dead leaves image before down-scaling. A (20, 20) crop on that same image (see Fig. 1b) exhibit very sharp boundaries and piecewise constant zones. A (20, 20) crop on the down-scaled image has a much more realistic aspect (see Fig. 1b). The whole procedure can be seen as a very simple simulation of the camera acquisition of a dead leaves model with tiny objects.

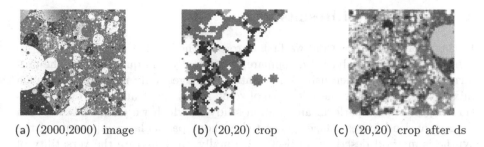

(a) (2000,2000) image   (b) (20,20) crop   (c) (20,20) crop after ds

Fig. 1. Visual example of the downscaling step (ds)

Color Sampling. Colors of disks are picked randomly from the color histogram of a given natural image (different for each image generation). Another simpler option is to consider colors that are uniformly sampled from the RGB cube, but this lead to unnatural colors and no spatial color coherency, as well, as we shall see, to a loss in performance. We can see in Fig. 2 that Fig. 2c,d have more realistic colors than Fig. 2b.

r_{min}, r_{max}, α. Those parameters control the distribution of the disks size in the image. Images generated with different parameters can be seen in Fig. 2. As we can see, at fixed $\alpha = 3$ and $r_{max} = 2000$, the type of image strongly depends on the value of r_{min}. A large r_{min} yields images with clear edges and homogeneous zones, whereas smaller r_{min} yields micro-textures (see Fig. 2e,f). Similar observations can be made when varying α (see Fig. 2g,h). In the rest of the paper, we chose to keep $\alpha = 3.0$ (the scale invariant case) and to vary r_{min}.

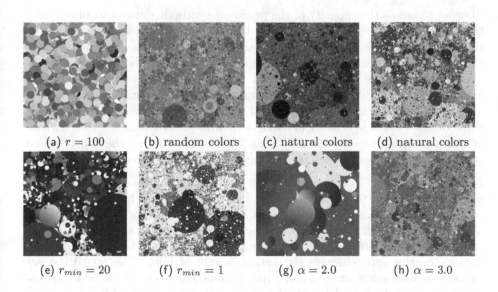

(a) $r = 100$ (b) random colors   (c) natural colors   (d) natural colors

(e) $r_{min} = 20$ (f) $r_{min} = 1$ (g) $\alpha = 2.0$ (h) $\alpha = 3.0$

Fig. 2. Dead leaves images generated with different parameters.

4   Experimental Results

In this experimental section, we first introduce the synthetic image dataset we consider. We then analyze and compare, numerically and qualitatively, the performance obtained when using only synthetic images, only natural images or a mix of both. To assess the relevance of some important features of our generation algorithm, we perform an ablation study in which we train FFDNet on a dataset of dead leaves images generated without particular components of the synthesis method described in Sect. 3.2. Finally, we illustrate the versatility of the proposed dataset by training the super-resolution network RDN [32].

Dead Leaves Dataset. In order to account for both homogenous areas and micro-textures, we build a dataset made of images generated with either $r_{min} = 1$ or $r_{min} = 16$, in both cases combined with parameters $\alpha = 3.0$ and $r_{max} = 2000$. Micro-textures being harder to restore than homogeneous areas, we chose to have a 2 to 1 ratio between the two possible r_{min} values. The color distribution of the disks is given by the histograms of the natural images from the Waterloo database [19]. As shown previously, this leads to a more coherent color distribution than randomly sampling the RGB cube. Finally, we decided to apply a Gaussian blur to a 10th of the dataset, with a standard deviation uniformly sampled between 1 and 3. Indeed, most natural images tend to contain blurry zones due to the depth-of-field of cameras. By adding a very simple blur model to some of the images of the dataset, we expect blurry areas in natural images to be better restored.

4.1   Denoising Results

In order to assess the capacity of the proposed synthetic dataset to successfully train a denoising network, we consider the network FFDNet. It is a state-of-the-art image denoising CNN, which was introduced by Zhang et al. [31] and thoroughly examined in [25]. Its main specificity relies in the first layer of the network: to increase the receptive field and to handle a wide range of noise levels, the image is divided in four sub-images which are concatenated to a noise map indicating the local noise standard deviation. This tensor is then passed through a more classic network of batch normalized convolutional layers, with an architecture similar to that of DNCNN's [30]. It then outputs the four denoised sub-images, which are reassembled to create the final denoised image.

FFDNet Results. To compare different trainings fairly, we use the same optimization algorithm for all trainings. It consists of 80 epochs with the Adam optimizer and the L2 loss, starting with a 10^{-3} learning rate. There is a decay of factor 10 at epoch 50, and another decay of factor 100 at epoch 60. For each training, we used 350k $(50, 50, 3)$ patches, extracted from either the dead leaves dataset, or the natural image dataset, or a mix of both. The mixed dataset contains $\frac{1}{3}$ dead leaves images, and $\frac{2}{3}$ natural images. To show that scaling properties are needed to model natural images, we also trained FFDNet on dead leaves images generated from disks with a fixed radius of 100. In addition, we also consider two alternative training schemes from datasets of synthetic images: white noise images and Gaussian random fields [12]. Numerical evaluation is performed on 2 test sets of natural images (CBSD68, Kodak24) and one set of 24 dead leaves images, generated from the colors of Kodak24. For each test, we compute the average PSNR, SSIM [28] and PieAPP metric [23], a recent perceptual metric based on human annotation, which tends to fit very well with human perception.

Table 1. Numerical comparisons of the different trainings of FFDNet. We evaluated the results on two benchmark datasets for image denoising (CBSD68 and Kodak24), and our dead leaves testset, at two noise levels. Each cell contains the triplet PSNR/SSIM/PieAPP. The best results are in blue, the second in red.

σ	Dataset	CBSD68	Kodak24	Dead leaves testset
$\sigma = 25$	White Noise	19.52/0.416/2.386	19.68/0.365/2.502	20.36/0.607/2.043
	Gaussian field	29.63/0.845/1.402	30.24/0.835/1.471	26.23/0.826/1.254
	DL $r = 100$	29.56/0.820/1.218	30.49/0.819/1.024	26.13/0.799/1.263
	Dead leaves	30.58/0.867/0.711	31.27/0.859/0.739	27.46/0.865/0.573
	Mix	31.07/0.881/0.639	31.98/0.876/0.603	27.33/0.860/0.567
	Natural Images	31.09/0.882/0.629	32.00/0.878/0.599	27.05/0.851/0.576
$\sigma = 50$	White Noise	15.58/0.247/4.682	15.71/0.209/4.785	16.24/0.387/2.932
	Gaussian field	26.68/0.738/2.203	27.41/0.737/2.353	23.31/0.694/2.158
	DL $r = 100$	26.85/0.720/1.563	27.91/0.739/1.314	23.24/0.654/2.005
	Dead leaves	27.40/0.762/1.088	28.21/0.765/1.154	24.21/0.737/1.020
	Mix	27.86/0.782/0.997	28.86/0.789/0.985	24.12/0.732/1.015
	Natural Images	27.87/0.786/0.991	28.89/0.792/0.978	23.90/0.722/1.053

On both natural image testsets and on the dead leaves testset, we observe that the model trained on dead leaves outperforms by a large margin all other models trained on alternative synthetic image datasets (0.9 dB for the Gaussian model and, without surprise, 11 dB for the white noise model), see Table 1. Visually, the Gaussian field model leads to denoised images still containing noise and grid-like artifacts, which severely impact the PieAPP metric. Observe that for both image models of white noise and Gaussian noise, the optimal solution is known and given by the Wiener filter (multiplication by a constant in the first case and linear filtering in the second). It is interesting to note that the network did not learn to apply this theoretical optimal solution to natural images in either cases. Confirming our intuition that an image model with scaling properties is needed, the dead leaves model with a fixed radius tends to strongly over-smooth the image, thus losing all texture information. This amounts to a loss of 0.65 dB on natural image testsets, and 1.2 dB on dead leaves images.

More surprisingly, the model trained exclusively on dead leaves images performs only 0.6 dB lower than the model classically trained on natural images. Visually, the results are still almost as good, despite some limitations. In particular, the synthetically trained model has some difficulties with thin and low contrast lines, and occasionally creates dot artifacts. In other situations, the syn-

Fig. 3. Denoising comparison with different FFDNet trainings. Top row from left to right: clean image, noisy image with $\sigma = 25$, model trained on white noise, model trained on Gaussian fields. Bottom row from left to right: model trained on dead leaves images with fixed radius r = 100, model trained on the dead leaves dataset, model trained on the mixed dataset, model trained on natural images.

thetic training improves the results, as can be seen in Fig. 3, where the texture of the rusty artwork is quite well restored, with a better preservation of fine details than with the model trained on natural images. Another very interesting result is the fact that training on a mix of dead leaves images and natural images does not affect the result of the denoising model on testsets of natural images, the difference in PSNR being less than 0.02 dB. Visually, the results are almost identical, with a slight advantage for the mixed trained model on texture areas. On the dead leaves test set, the mixed trained model clearly outperforms the natural image trained model by 0.25 dB. This result suggests that jointly optimizing the response to this kind of mixed datasets has the ability to increase some aspects on which imaging devices are evaluated. Indeed the scaling dead leaves model is classically used to evaluate the ability of imaging devices to preserve texture areas [6,7] and the corresponding scale-invariant test chart has recently become an ISO standard (ISO/TS 19567-2:2019).

Ablation Study. To confirm the choices made to build the synthetic dataset, we compare different trainings performed with different parameters or design choices, both visually and numerically.

We first illustrate the impact of r_{min} on the denoising results. As shown in Fig. 4, the smaller the r_{min}, the better micro-textures are restored. Conversely, they are smoothed when r_{min} gets larger. On the other hand, homogeneous zones contain artifacts when r_{min} is too small, and are well restored when r_{min} is larger. This behaviour is expected since a large r_{min} leads to dead leaves images with homogeneous zones, and a small r_{min} to more micro-textures zones. Referring

Table 2. Impact of the parameters and ablation study. In the first 5 columns(DL1 to DL16), we fix the parameters to $r_{max} = 2000$, $\alpha = 3.0$, with natural colors and the downscaling step. From column 6 to 8, we keep the same parameters as in the final dataset, but we remove some important features of the generation. The last column corresponds to the final result.

σ	DL-1	DL-2	DL-4	DL-8	DL-16	Rand. col	No sub	No blur	Final
25	31.03	31.03	31.09	31.07	30.98	29.99	30.79	31.25	31.27
50	27.98	27.96	28.04	28.06	28.05	27.16	27.74	28.20	28.21

to Table 2, the optimal r_{min} seems to be between 4 and 8. However, by mixing images generated with $r_{min} = 1$ and $r_{min} = 16$, we get a noticeable improvement in PSNR (0.17 dB) and in image quality, as can be seen in Fig. 4.

Other important features of our algorithm are: the color distribution, the downscaling step, and the blur. As we can see in Fig. 4, when we sample the disks colors uniformally in the RGB cube, the denoised images show many color artifacts. The additive Gaussian noise creates unnatural colors that the network doesn't identify as such, since it has not been trained on images with natural colors. This leads to a performance gap of more than 1 dB in PSNR. The downscaling step is also critical, as it allows sub-pixel sized objects and more natural boundaries. In Fig. 4, we can see that the network trained on the dead leaves dataset without subsampling tends to over-smooth texture areas, and to produce stair-casing artifacts. We can also identify some disk-like objects with hard boundaries in the images, creating an unnatural aspect. In terms of PSNR, this amounts to a loss of 0.5 dB compared to the training on our final dead leaves dataset. For the final synthetic dataset, we decided to blur 10 % of images. As we can see in Table 2, removing this step has almost no impact on the PSNR. Nonetheless, we observe in Fig. 4 that removing this step makes blurry zones look sharper than they really are. Overall, the final dead leaves dataset yields better results, numerically and visually, than the ablated datasets.

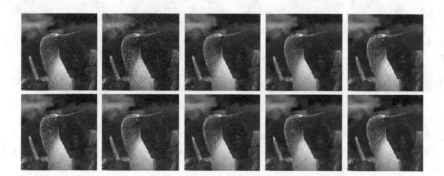

Fig. 4. Visual illustration of the ablation study. From left to right: top line: clean image, noisy image, $r_{min} = 1$, $r_{min} = 2$, $r_{min} = 4$, / bottom line: $r_{min} = 16$, No subsampling, Uniform color distribution, No blur, Final result.

4.2   Single-Image Super-Resolution

To assess our dataset's versatility, we consider the task of single-image super-resolution (SISR). We chose to retrain the Residual Dense Network (RDN) [32], a state-of-the-art super-resolution network. Its architecture is based on residual dense blocks, a combination of dense blocks introduced in [16] and residual connections.

Table 3. Numerical evaluation of our Super-resolution results. We report the PSNR of RDN trained either on the dead leaves dataset or on the DIV2K dataset [1].

Dataset	Set 5		Set 14	
scale	×2	×3	×2	×3
Dead leaves	36.76	33.82	32.93	30.42
Natural Images	38.18	34.71	33.88	30.73

The numerical evaluation shows a similar behaviour to the one observed in image denoising in Sect. 4.1. The gaps are of 1.2 dB and 0.6 dB for a super-resolution of scale 2 and 3 respectively. The results on the Set5 and Set14 datasets, which are common benchmarks for super-resolution, are given in Table 3. Visually, the super-resolution results are very similar as we can see

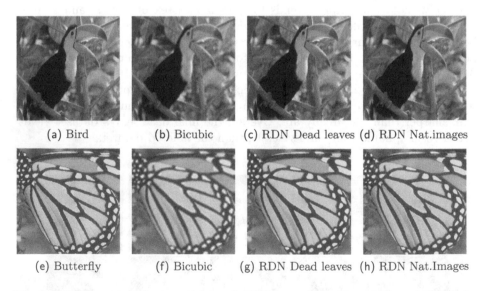

(a) Bird   (b) Bicubic   (c) RDN Dead leaves (d) RDN Nat.images

(e) Butterfly   (f) Bicubic   (g) RDN Dead leaves (h) RDN Nat.Images

Fig. 5. Visual results of our trainings of RDN at scale 3. From left to right: High resolution image, Bicubic interpolation, RDN trained on dead leaves images, and RDN trained on natural images.

in Fig. 5. However, if we look closely to Fig. 5g and h, we see in Fig. 5g that some small white spots in the top black region are better restored. Conversely, thin lines in yellow regions have a "dotted" aspect in Fig. 5g, which disappears in Fig. 5h. The dead leaves model does not contain any straight and thin lines, making it harder for this model to retrieve them. A lead to tackle this problem would be to complete our database with patches generated from a sinusoïdal basis, as was done in [29].

5   Conclusion and Future Works

To the best of our knowledge, this work is the first effort to train an image restoration network on synthetic images. After introducing the dead leaves model and its digital implementation, we carefully studied the role of each component of the image generation method, and their impact on the restoration perfor- mances. Both for denoising and super-resolution, models trained on our dead leaves dataset are surprisingly close to those trained on natural images. When mixing natural and synthetic images in the training, the results reach perfor- mances on par with the model trained on natural images only. Both results indicate that the dead leaves model with scaling properties is a good candidate to replace natural images for training, with only a few parameters. Indeed, the geometry of the model only depends on three parameters: α, r_{min}, r_{max}. Even though the color parameters are still relatively numerous, we plan to investigate simple samplings of the horse-shoe color space in order to avoid unnatural colors. Another perspective would be to complement our dataset with sinusoidal patches to better restore oscillating patches and straight lines. Eventually we believe, as already explained, that a synthetic dataset can be a simple way to avoid retrain- ing new imaging devices with relatively heavy acquisition campaigns [8] and we plan to investigate this ability further.

References

1. Agustsson, E., Timofte, R.: Ntire 2017 challenge on single image super-resolution: dataset and study. In: The IEEE Conference on Computer Vision and Pattern Recognition (CVPR) Workshops, July 2017
2. Alvarez, L., Gousseau, Y., Morel, J.M.: The size of objects in natural and artificial images. In: Advances in Imaging and Electron Physics, vol. 111, pp. 167–242. Elsevier (1999)
3. Bordenave, C., Gousseau, Y., Roueff, F.: The dead leaves model: a general tessel- lation modeling occlusion. Adv. Appl. Probab. **38**(1), 31–46 (2006)
4. Buades, A., Coll, B., Morel, J.M.: A review of image denoising algorithms, with a new one. Multiscale Model. Simul. **4**(2), 490–530 (2005)
5. Candes, E.J., Tao, T.: Decoding by linear programming. IEEE Trans. Inf. Theory **51**(12), 4203–4215 (2005). https://doi.org/10.1109/TIT.2005.858979
6. Cao, F., Guichard, F., Hornung, H.: Measuring texture sharpness of a digital cam- era. In: Digital Photography V, vol. 7250, p. 72500H. International Society for Optics and Photonics (2009)

7. Cao, F., Guichard, F., Hornung, H.: Dead leaves model for measuring texture quality on a digital camera. In: Digital Photography VI, vol. 7537, p. 75370E. International Society for Optics and Photonics (2010)

8. Chen, C., Chen, Q., Xu, J., Koltun, V.: Learning to see in the dark. In: Proceedings of the IEEE Conference on Computer Vision and Pattern Recognition, pp. 3291–3300 (2018)

9. Dabov, K., Foi, A., Katkovnik, V., Egiazarian, K.: Image restoration by sparse 3D transform-domain collaborative filtering. In: Astola, J.T., Egiazarian, K.O., Dougherty, E.R. (eds.) Image Processing: Algorithms and Systems VI, vol. 6812, pp. 62–73. International Society for Optics and Photonics, SPIE (2008). https://doi.org/10.1117/12.766355

10. Donoho, D.L.: Compressed sensing. IEEE Trans. Inf. Theory **52**(4), 1289–1306 (2006). https://doi.org/10.1109/TIT.2006.871582

11. Donoho, D.L., Johnstone, I.M., et al.: Minimax estimation via wavelet shrinkage. Ann. Stat. **26**(3), 879–921 (1998)

12. Galerne, B., Gousseau, Y., Morel, J.M.: Micro-texture synthesis by phase randomization. Image Process. Line **1**, 213–237 (2011)

13. Gatys, L.A., Ecker, A.S., Bethge, M.: Texture synthesis using convolutional neural networks. arXiv preprint arXiv:1505.07376 (2015)

14. Gousseau, Y., Roueff, F.: Modeling occlusion and scaling in natural images. Multiscale Model. Simul. **6**(1), 105–134 (2007)

15. Heeger, D.J., Bergen, J.R.: Pyramid-based texture analysis/synthesis. In: Proceedings of the 22nd Annual Conference on Computer Graphics and Interactive Techniques, pp. 229–238 (1995)

16. Huang, G., Liu, Z., Van Der Maaten, L., Weinberger, K.Q.: Densely connected convolutional networks. In: Proceedings of the IEEE Conference on Computer Vision and Pattern Recognition, pp. 4700–4708 (2017)

17. Kendall, W.S., Thönnes, E.: Perfect simulation in stochastic geometry. Pattern Recogn. **32**(9), 1569–1586 (1999)

18. Lee, A.B., Mumford, D., Huang, J.: Occlusion models for natural images: a statistical study of a scale-invariant dead leaves model. Int. J. Comput. Vis. **41**(1–2), 35–59 (2001)

19. Ma, K., et al.: Waterloo exploration database: new challenges for image quality assessment models. IEEE Trans. Image Process. **26**(2), 1004–1016 (2016)

20. Mumford, D., Gidas, B.: Stochastic models for generic images. Q. Appl. Math. **59**(1), 85–111 (2001)

21. Ono, S.: Primal-dual plug-and-play image restoration. IEEE Signal Process. Lett. **24**(8), 1108–1112 (2017). https://doi.org/10.1109/LSP.2017.2710233

22. Portilla, J., Simoncelli, E.P.: A parametric texture model based on joint statistics of complex wavelet coefficients. Int. J. Comput. Vis. **40**(1), 49–70 (2000)

23. Prashnani, E., Cai, H., Mostofi, Y., Sen, P.: PieAPP: perceptual image-error assessment through pairwise preference. In: Proceedings of the IEEE Conference on Computer Vision and Pattern Recognition, pp. 1808–1817 (2018)

24. Rudin, L.I., Osher, S., Fatemi, E.: Nonlinear total variation based noise removal algorithms. Physica D **60**(1–4), 259–268 (1992)

25. Tassano, M., Delon, J., Veit, T.: An analysis and implementation of the FFDNet image denoising method. Image Process. Line **9**, 1–25 (2019)

26. Tremblay, J., et al.: Training deep networks with synthetic data: bridging the reality gap by domain randomization. In: Proceedings of the IEEE Conference on Computer Vision and Pattern Recognition Workshops, pp. 969–977 (2018)

27. Ulyanov, D., Vedaldi, A., Lempitsky, V.: Deep image prior. In: Proceedings of the IEEE Conference on Computer Vision and Pattern Recognition, pp. 9446–9454 (2018)
28. Wang, Z., Bovik, A.C., Sheikh, H.R., Simoncelli, E.P.: Image quality assessment: from error visibility to structural similarity. IEEE Trans. Image Process. **13**(4), 600–612 (2004)
29. Yu, G., Sapiro, G., Mallat, S.: Solving inverse problems with piecewise linear estimators: from gaussian mixture models to structured sparsity. IEEE Trans. Image Process. **21**(5), 2481–2499 (2011)
30. Zhang, K., Zuo, W., Chen, Y., Meng, D., Zhang, L.: Beyond a Gaussian denoiser: residual learning of deep CNN for image denoising. IEEE Trans. Image Process. **26**(7), 3142–3155 (2017)
31. Zhang, K., Zuo, W., Zhang, L.: FFDNet: toward a fast and flexible solution for CNN-based image denoising. IEEE Trans. Image Process. **27**(9), 4608–4622 (2018)
32. Zhang, Y., Tian, Y., Kong, Y., Zhong, B., Fu, Y.: Residual dense network for image super-resolution. In: Proceedings of the IEEE Conference on Computer Vision and Pattern Recognition, pp. 2472–2481 (2018)

Geometric Deformation on Objects: Unsupervised Image Manipulation via Conjugation

Changqing Fu[✉] and Laurent D. Cohen

CEREMADE, UMR CNRS 7534, Université Paris Dauphine, PSL,
Place du Marechal de Lattre de Tassigny, 75775 Paris cedex 16, France
{cfu,cohen}@ceremade.dauphine.fr
https://www.ceremade.dauphine.fr

Abstract. A novel two-stage approach is proposed for image manipulation and generation. User-interactive image deformation is performed through editing of contours. This is performed in the latent edge space with both color and gradient information. The output of editing is then fed into a multi-scale representation of the image to recover quality output. The model is flexible in terms of transferability and training efficiency.

Keywords: Machine learning · Image generation · Generative adversarial network · Image deformation · Contour editing

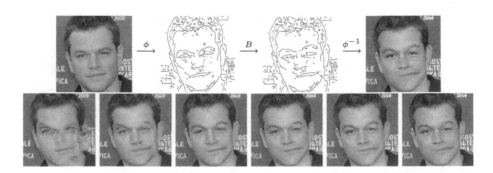

Fig. 1. Top row, from left to right: a) Input. b) Reconstruction from sparse contour. c) Reconstruction from deformed contour, moving the lip and eyebrow contours via user interaction. Bottom row, from left to right: Multi-scale Super-resolution of the deformed Reconstruction.

This work was funded in part by the French government under management of Agence Nationale de la Recherche as part of the "Investissements d'avenir" program, reference ANR-19-P3IA-0001 (PRAIRIE 3IA Institute).

A. Elmoataz et al. (Eds.): SSVM 2021, LNCS 12679, pp. 346–357, 2021.
https://doi.org/10.1007/978-3-030-75549-2_28

1   Introduction

Image manipulation task is among one of the fast-growing fields in Computer Vision. Many existing algorithms for image editing are supervised and not adaptive to new data, and often require fine-tuning and slow training to achieve desirable performance. Therefore, the motivation of this work is to propose an alternative unsupervised way to refine the reconstruction, providing flexibility for difficult training and limited dataset.

In our setting of image manipulation, we introduce two manifolds $\mathcal{M}$ and $\mathcal{N}$, where $\mathcal{M}$ denotes the space of natural images, and $\mathcal{N}$ is the space of contour representation of the images. Function A in $\mathcal{M}$ stands for the desired final editing effect, whereas B in $\mathcal{N}$ is user's editing in the latent space. Since A is often complicated to implement, we alternatively perform editing B in the hidden contour space $\mathcal{N}$ via a pull-back mapping.

Mathematically speaking, Conjugation, or Similarity Transformation, refers to a pair of transformations $A : \mathcal{M} \longrightarrow \mathcal{M}$ and $B : \mathcal{N} \longrightarrow \mathcal{N}$ that satisfies $A = \phi^{-1} \circ B \circ \phi$, where ϕ is a diffeomorphism between two manifolds $\mathcal{M}$ and $\mathcal{N}$. In our case, ϕ is a well-defined image contour detector. The invertibility of ϕ is necessary for the well-definedness of the conjugation, but it is an ill-posed inverse problem. With the recent advancement of CNN-based Image Translation models, the inversion map ϕ^{-1} can be learned in a data-driven manner.

One of the advantages of contour representation is its sparsity, since any moving or distorting operation will create no effect on flat or zero-valued areas, but will create discontinuity effect on a natural image. Moving or distorting operation can be easily manipulated by human interaction on a computer, and thus is practically meaningful. Furthermore, contour information is not enough to recover an image, and therefore color and gradient information are helpful to improve the contour's representation ability without adding to sparsity.

On top of the conjugation paradigm, a strategy similar to deep internal learning [23] is performed. Instead of storing all the information in a single model to produce a high-quality image, we make use of the input image itself to carve details on the output image. This method is especially good at producing textures similar to the input image, even if they are never seen in the training database.

The organization of this paper is as follows. After reviewing related work in Sect. 2, we will present the main ideas of our work in Sect. 3, including sparse contour and multi-scale representations, and then detail the algorithms involved in Sect. 4, illustrated by examples on various databases.

Our contribution is that a natural Downscaling-Reconstruction strategy is proposed as a post-processing approach to obtain high-fidelity output for image manipulation, and it requires very short training time, and at the same time provides better transferability comparing to the existing approach.

2   Related Works

There are lines of research that understand the latent space $\mathcal{M}$ as sparse representation [4], color representation [28], or both [19]. Recently there are works which deals with implicit hidden representation using inner features [7,8], providing the model with explainability. In this paper, we focus on geometric manipulation in the hidden space, and especially recovery in a new perspective under both supervised and unsupervised setting.

Sketch translation [3,14] aims to produce realistic images out of abstract sketches drawn by human. Natural images can be produced even with messy or cartoon-like input. Therefore, the precision of geometric constraints is compromised on as a systematic side effect of the trade-off. These methods are to tackle challenges like Sketchy [18], QuickDraw [10], or ShoeV2/ChairV2 datasets [27], whereas the contours in our method is similar to Edge2shoes [26] or Edges2-handbags [28], according to their realistic structure. Therefore, our focus is towards image synthesis, rather than image retrieval in the database or on the image manifold. It does not require any a priori edge information, since this is already incorporated inside the image itself, and could be computed using efficient context-aware edge detectors (Sect. 3.1).

On the other hand, the multi-scale image structure is motivated by SinGAN [21], a multiscale invariant of InGAN [22]. Other super-resolution techniques such as [6,25] exist but only tackle the standard super-resolution problem. The hierarchical structure of SinGAN enables more flexible input, and is able to recover not only inputs of low-resolution, but also color shapes, even those created from scratch by human.

3   Image Model

The intuition of our approach is twofold, consisting of two different ways of understanding for neural image perception.

3.1   Sparse Contour Representation

Fig. 2. Recovery from Sparse Representation. From left to right: Real image, Contour representation, Low frequency reconstruction, High frequency reconstruction.

Given image space $\mathcal{M}$ and contour space $\mathcal{N}$, in order to achieve a robust diffeomorphism $\phi : \mathcal{M} \longrightarrow \mathcal{N}$ between the edge representation space and the image space, as is mentioned, the invertibility of ϕ is the key to the recovery result. More precisely, ϕ^{-1} is naturally defined as $\phi^{-1} : \mathcal{N} \longrightarrow 2^{\mathcal{M}}$, $\phi^{-1}(y) = \{x \in \mathcal{M} | \phi(x) = y\}, \forall y \in \mathcal{N}$, and the pre-image of contour is not unique. Minimizing the GAN energy (Eq. 1) further finds the point in the pre-image which looks like the image database the most. This result in $\widehat{\phi^{-1}} : \mathcal{N} \longrightarrow \mathcal{M}$, a well-defined surrogate in a data-driven sense (Fig. 2).

In practice, we use an edge detector as ϕ, and apply a Generator network to fit the function ϕ^{-1}, together with its Discriminator counterpart. This is under the scope of Image Translation problem.

In general, image translation problem is described as follows. Let $\mathcal{X} = \mathcal{Y} = \mathbb{R}^{3 \times H \times W}$ be two image spaces of fixed size $H \times W$, with training examples $x_i \in \mathcal{X}, y_i \in \mathcal{Y}, i = 1, \ldots, N$. Supervised Learning approaches for image translation aims to learn a map $\widehat{\phi^{-1}}$ from $\mathcal{X}$ to $\mathcal{Y}$ by minimizing the loss function $\mathcal{L}(\phi(x_i), y_i)$. The fitted function is then used to generate images following the same distribution in $\mathcal{M}$. The restriction of this map on the sketch manifold $G := \widehat{\phi^{-1}_{|\mathcal{N}}} : \mathcal{N} \subseteq \mathcal{X} \longrightarrow \mathcal{M}$ is supposed to be an onto map to the image manifold $\mathcal{M} \subseteq \mathcal{Y}$, which is often called a Conditional Generator since it has a complex prior distribution on $\mathcal{X}$, and is obtained by minimizing the Adversarial Loss in Eq. (1).

Now we discuss some properties of G:

- Ideally, the onto property of G on $\mathcal{M}$ is often called generalization ability, whereas the lack of onto property on $\mathcal{M}$ and the fact that the image of G is restricted on $\{y_i\}_{i=1}^N \subseteq \mathcal{N}$ is called overfitting. Moreover, the lack of onto property of G on $\{y_i\}_{i=1}^N \subseteq \mathcal{N}$ is referred to as Mode Collapse.
- In practice, for the empirical $\widehat{\phi^{-1}}$, the pre-image of $\mathcal{M}$ is not necessarily equal to $\mathcal{N}$. In fact $\phi^{-1}(\mathcal{M}) \supset \mathcal{N}$. In other words, the map $\widehat{\phi^{-1}}^{-1}$ is not L^1-continuous, since the inverse of open ball in $\mathcal{M}$ is not open in $\mathcal{N}$ under L^1 topology. Illustrations are as follows in Fig. 3 We perform a differential attack or latent recovery [24] on a pretrained Pix2Pix [12] model, resulting in a non-sparse noisy pre-image. Instead of recovering the latent code as white noise in the original work, our input has meaningful structure. The fact that c) is not a clean sketch shows that for L^1 neighbours in $\mathcal{M}$, a) and d), their pre-image are not neighbours in $\mathcal{N}$, implying that $\widehat{\phi^{-1}}^{-1}$ is far from smooth. Therefore $\widehat{\phi^{-1}}$ is not an L^1-diffeomorphism itself. However, by restricting $\widehat{\phi^{-1}}$ on the contour manifold $\mathcal{N}$, the resulting $G = \widehat{\phi^{-1}_{|\mathcal{N}}}$ turns out to be a proper counterpart for ϕ to recover image from geometrical constraints.

(a)   (b)   (c)   (d)

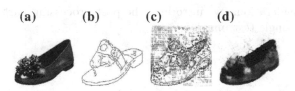

Fig. 3. From left to right: a) Original Image, b) Initialization for the Pre-Image Search, c) Pre-image Found, d) Reconstruction from c)

The edge-detector/translator pair ϕ and G between $\mathcal{M}$ and $\mathcal{N}$ are detailed as follows:

Edge Detector. A tree-based edge detection algorithm [5] is applied as edge extractor ϕ. The random forest is trained with samples from BSDS500 dataset with structured labels of edge and segmentation. Hard thresholding is then applied with a pre-defined threshold to produce an edge mask. We follow [4] to use an N-channel contour representation but with modifications. **a)** Different from their fixed sparsity rate, we used a fixed threshold since our image database naturally have contours of diverse sparsity. **b)** We extract both 3D-color and 6D-gradient information on the contour pixels without computing that of both side of the contour (proposed in the original paper). This solution reduces computation and at the same time avoids contour overlapping after user's editing to the greatest extend.

Image Generator. A U-Net[17] is applied as the baseline algorithm to approximate ϕ^{-1}. [4] refer to this model as the Low Frequency Network (LFN), which produces an intermediate output, and propose to apply another U-Net on top of the Low Frequency output and the contour input. The second network is called the High Frequency Network (HFN) since it produces finer details. It's worth noting that: **a)** During the optimization of the HFN, it's optional to update the weights of LFN in the meantime, since the computational graph of back-propagation can reach layers in the first network. This optional operation will distort the output of LFN, since the training process for the second network does not anymore aim to minimize the L^1 distance between the Low Frequency output and the training target. **b)** The second network adds complexity to the model, but it does not provide additional information to the model input. In fact, the concatenation operation in the U-Net plays a similar role, since it is nothing but concatenating inputs along with the intermediate outputs. HFN and concatenation operations are meaningful since they enforce the generator with intermediate layer information, and the secondary goal, low frequency network, is both meaningful and explainable. A recent work [15] is similar to this idea, and it inspires us that both networks could be trained simultaneously, encouraging LFN to be of even less frequency to be suitable for the input of the multi-scale model. In fact, flat and "quantized" input proves to be more effective in some cases [21]. **c)** Existing work does not tackle the cross-domain challenge, which motivates us to consider using the information outside the training database, namely the testing target itself.

With these considerations, we introduce the post-processing part of the model, based on every single test image.

3.2   Multi-scale Representation

Fig. 4. Optimal Reconstruction of an Image from coarse-grained scale.

We first introduce a sequence of RGB image space $V_n \subseteq \mathcal{X} = \mathbb{R}^{3 \times H_n \times W_n}$, $n \in \mathbb{Z}$, for images of height H_n and width W_n monotonically increasing with respect to n. The coarse-grained space V_0 is of some fixed size $H_0 \times W_0$, and $x_N \in V_N$ denotes an image with camera resolution $H_N \times W_N$. $V_N = \mathcal{M}$ is the output space, which is the previously defined image manifold. By definition, $V_{-\infty} := \lim_{n \to -\infty} V_n$ is a single RGB pixel, whereas $V_\infty := \lim_{n \to \infty} V_n$ is the perfect vision with infinite resolution. For each n, down-scaling $\pi : x_n \in V_n \longmapsto x_{n-1} \in V_{n-1}$ from $x_n \in V_n$ with scaling factor $r \in (0, 1)$, since the corresponding upsampling map π^{-1} is a bijection between V_{n-1} and a linear subspace of V_n (Fig. 4).

An image is represented by a Convolutional Neural Network. In other words, information in the image is memorized in weights of the neurons. More precisely, this is done by fitting a map G from randomly sampled noise to image data. An image generator is defined by $G : \mathcal{Z} \longrightarrow \mathcal{M} : z \longmapsto x_N$ and is trained to represent the image from the multi-scale code $z \in \mathcal{Z} = \prod_{n=1}^{N} \mathcal{Z}_n$, where $\mathcal{Z}_n = \mathbb{R}^d$ is the perturbation at each scale, for $n = 0, \ldots, N$. Mathematically, a map $G_\sharp$ between two probability distribution spaces is induced by the neural network G, where a smooth distribution in $\mathbb{R}^d$ is mapped to a probability distribution on the Image Manifold. This can be an empirical distribution, in the case of a database, or a Dirac distribution, in the case of a single image. A model that learns only to represent a single target is often referred to as Mode Collapse, which often causes low representation ability, signifying improper optimization. However, in our Super Resolution setting, our goal is to add texture details to the blurry input that are *unique* to the image. Therefore the flexibility over local perturbations on the input is the key to the model.

The unconditional GAN is defined as follows in an iterative form:

$$\begin{cases} \text{Refinement:} & x_n = \widetilde{x_{n-1}} + G_n(\widetilde{x_{n-1}} + z_n), \quad n = 0, \ldots, N \\ \text{Upsampling:} & \widetilde{x_n} = \pi(x_n) \\ \text{Initialization:} & x_0 = G_0(z_0) \end{cases}$$

where at scale n, x_n is the downsampled image at rate r^{N-n}, and z_n is white noise. As a result, the final Generator is given by

$$G(z) := G_N(z_0, \ldots, z_N)$$
$$= \underbrace{\underbrace{G_0(z_0)}_{\triangleq x_0} + G_1(G_0(z_0) + z_1)}_{\triangleq x_1} + \cdots$$

$$+ G_N(\underbrace{\underbrace{G_{N-1}(\cdots G_1(G_0(z_0) + z_1) \cdots + z_{N-1})}_{\triangleq x_{N-2}} + z_N)}_{\triangleq x_{N-1}}.$$

Upsampling is omitted here for simplicity, by assuming that every object lives in V_∞.

Note that $r = \frac{1}{2}$ is the special case. When $H_{n-1} = \frac{1}{2}H_n, W_{n-1} = \frac{1}{2}W_n$, suppose $x_N \in V_N$ is an image, and suppose the corresponding 2D Haar wavelet decomposition is $x_N = \sum_{n=0}^{N} a_n\varphi_n$, then $\{\varphi_n\}_{n=1}^N$ are *orthogonal* and $a_n\varphi_n \in V_n$,

However, in practice, $r = \frac{1}{2}$ does not produce high-fidelity recovery, and $r = 3/4$ is a good balance between model complexity and quality. Upsampling operation π is performed by spline interpolation. z_n is chosen to be $\mathcal{N}(0, \sigma_n)$ with $\sigma_n \propto \|\pi(x_{n-1}) - x_n\|$ in order to match the intensity of randomness at each scale. In reconstruction/super-resolution task, perturbation z_n can be set to zero. The input image can be fed into any scale by down-scaling/up-scaling to obtain output of size greater than the input image, which adds to the detail of the image and achieves super-resolutions.

4   Algorithm

The minimization formulation for the multi-scale reconstruction problem is adapted from [12]. For each scale $n = 1, \ldots, N$,

$$\min_G \max_D \mathcal{L} = \mathcal{L}_{adv}(G_n, D_n) + \alpha\mathcal{L}_{rec}(G_n) \tag{1}$$

The **Adversarial Loss** $\mathcal{L}_{adv}(G_n, D_n)$ is adapted from WGAN-GP [9]:

$$\mathcal{L}_{adv}(G_n, D_n) = \mathbb{E}_x[D_n(\mathbf{x_n})] + \mathbb{E}_z[-D_n(G_n(\mathbf{x_n} + \mathbf{z_n}))] \tag{2}$$

where $\mathbf{x_n} := (x_0, \ldots, x_n)$ and $\mathbf{z_n} := (z_0, \ldots, z_n)$ are sub-scale images and noise respectively.

In the sparse recovery case, the setting is similar and $\mathbf{x_n}$ is replaced with the real images y, and the perturbed down-scaled image $\mathbf{x_n} + \mathbf{z_n}$ is replaced with contour representation x.

$$\mathcal{L}_{adv}(G, D) = \mathbb{E}_y[D(y)] + \mathbb{E}_x[-D(G(x))] \tag{3}$$

This optimization objective is similar to the Binary Cross-Entropy Loss in the Vanilla GAN:

$$\mathcal{L}_{adv}(G_n, D_n) = \mathbb{E}_x[\log D_n(x)] + \mathbb{E}_z[\log(1 - D_n(G_n(z)))] \tag{4}$$

Note that in Vanilla GAN, the discriminator $D : \mathcal{N} \longrightarrow [0, 1]$ could be understood as the probability that the input image is fake, and thus the adversarial loss could be treated as the log likelihood function. In comparison, our discriminator aims to detect fake images as positive or negative otherwise, as opposed to the original case.

Reconstruction Loss is given by

$$L_{rec}(G_n) = \|G_n(\pi(\widetilde{x_{n-1}}) + 0) - x_n\|^2. \tag{5}$$

This deterministic term ensures that each layer performs proper refinement and adds high-frequency information to the image.

4.1 Cross-domain Transferability

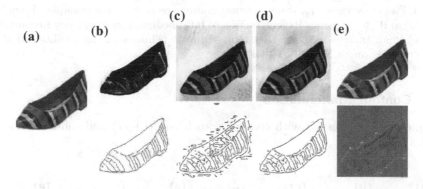

Fig. 5. Pretrained on VGG Face dataset. From left to right: a) Input. b) Reconstruction from supervised edge prior (Pix2Pix model trained on paired data). c) Reconstruction from detected edge (pretrained on Face Dataset). Original tensorflow implementation by [4]. d) Reconstruction from cleaned sparse edge. e) Our reconstruction result.

Figure 5 shows an example of cross-domain results of the model, using the original tensorflow implementation. The model was trained on the VGG Face Dataset [16], and tested on an image of a single object. Artefacts are produced with the cross-domain test sample, whereas our implementation trained on shoe dataset (right column) works well. However, this can be corrected by post-processing procedure (see Fig. 7 below). We also show the baseline result of Pix2Pix on the second left column as comparison (Fig. 7).

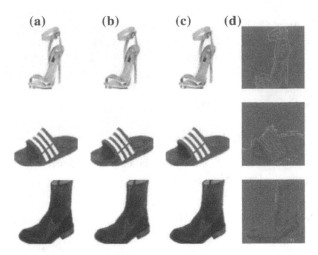

Fig. 6. Fast convergence for training –more contour reconstruction examples. From left to right: a) High Frequency Reconstruction b) Intermediate Low Frequency Reconstruction. c) Input Image. d) Sparse edge representation. This result can be obtained after *a few minutes* on an NVIDIA T4 GPU.

4.2   Contour Manipulation

We illustrate our results with contour translation (Fig. 7) and contour removal (Fig. 8) examples.

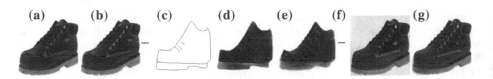

Fig. 7. From left to right: a) Input. b) Multi-scale Reconstruction. c,d) input and output of Pix2Pix model. e) Multi-scale Reconstruction of the Pix2Pix output. f) Reconstruction from contour representation, by manually moving the edge of the logo on the shoe. g) Multi-scale reconstruction of the deformation.

Figure 7 shows the robustness of the multi-scale post-processing in terms of transferability. The result is trained on face data and tested on shoe data. Even if the output of edge reconstruction has unexpected cool tone caused by cross-domain transfer learning (Second right in the figure), the information of the input image is still able to correct the details. The effectiveness of the post-processing for other tasks is presented as a bonus product (c, d, e in Fig. 7), still not perfect but showing significant improvement over the previous outcome.

(a)   (b)   (c)
                                    (d)   (e)

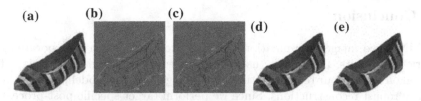

Fig. 8. From left to right: a) Input. b) Pre-image in parse edge space. c) Edge removal. d) Reconstruction of Low Frequency. d) Final Reconstruction.

Figure 8 shows the robustness of the sparse recovery method in terms of editing, quality and sparsity. Reconstruction is clean after a rough eraser edit. The figure is well recovered even from undertrained edge detector which produces noise on both sides of the contour. Recovery quality are not harmed by manually cleaning the noise on both sides.

Finally, Fig. 1 shows our final result where the supervised stage is trained independently on the VGG face dataset. The first row shows the validity of contour editing, and the second row presents the quality of post-processing. As can be seen, not only does the post-processing produce more natural skin color than the unprocessed reconstruction, it also adds to tiny randomness to the image so that the image is more diverse and privacy-protecting comparing to the input. The latter effect could be augmented by tuning α.

4.3   Implementation Details

Edge Detection. Edge detection is performed [5] by training on a few samples from the BSDS500 dataset [1]. We find that this version of edge detector, though under-trained, well preserves color information for image recovery.

Network Structure. For the Contour Reconstruction, we tested both U-net and ResNet Generators [13,29]. The U-net consists of Conv(3×3)-BatchNorm-LeakyReLU blocks of feature with concatenation operation. The ResNet [11] contains convolutions layers, several residual blocks, and then convolutions. For Multi-scale Reconstruction, we use ResNet of 5 convolution blocks of the form Conv(3×3)-BatchNorm-LeakyReLU [20]. The Discriminator are PatchGANs of fixed structure [12]. The number of patches depends on the input size. This is similar to a recent work [2] that propose to improve classification with Bag-of-words Patch features.

Training Strategy. For Contour Reconstruction, we use the Adam optimizer with learning rate 0.001 with Cosine Annealing learning rate for 50 epochs on a mini dataset of 200 samples. For Multi-scale Reconstruction, we train the hierarchical architecture of 5-layer ResNet by each scale, each with 2000 steps. Network and parameters are adapted from the original SinGAN paper. We use the Adam optimizer with learning rate 0.0005, $\beta_1 = 0.5, \beta_2 = 0.999$, and we apply Cosine Annealing to update the learning rate. To stabilize the training, we used WGAN-GP [9] to regularize the loss with gradient penalty.

5   Conclusion

A CNN-based image manipulation model is proposed, which incorporates geometric constraints. In practice, user performs editing through geometric deformation on the contour representation of the image, and the model produces high-quality robust reconstructions. Since we perform target-specific post-processing technique that does not require supervision, the model shows improvement in terms of transferability over existing work. Although our approach captures objects' textures automatically even if they are not a priori seen by the neural network in the training database, still more complex real-world image data (e.g. the BSDS database) are not within the scope. Future work includes adapting to more diverse dataset in the real life.

References

1. Arbelaez, P., Maire, M., Fowlkes, C., Malik, J.: Contour detection and hierarchical image segmentation. IEEE Trans. Pattern Anal. Mach. Intell. **33**(5), 898–916 (2011). https://doi.org/10.1109/TPAMI.2010.161
2. Brendel, W., Bethge, M.: Approximating CNNs with bag-of-local-features models works surprisingly well on ImageNet. In: International Conference on Learning Representations (2018)
3. Chen, W., Hays, J.: SketchyGAN: towards diverse and realistic sketch to image synthesis. In: Proceedings of the IEEE Conference on Computer Vision and Pattern Recognition, pp. 9416–9425 (2018)
4. Dekel, T., Gan, C., Krishnan, D., Liu, C., Freeman, W.T.: Sparse, smart contours to represent and edit images. In: Proceedings of the IEEE Conference on Computer Vision and Pattern Recognition, pp. 3511–3520 (2018)
5. Dollár, P., Zitnick, C.L.: Fast edge detection using structured forests. IEEE Trans. Pattern Anal. Mach. Intell. **37**(8), 1558–1570 (2014)
6. Dong, C., Loy, C.C., He, K., Tang, X.: Image super-resolution using deep convolutional networks. IEEE Trans. Pattern Anal. Mach. Intell. **38**(2), 295–307 (2015)
7. Ghorbani, A., Wexler, J., Zou, J., Kim, B.: Towards automatic concept-based explanations. In: Wallach, H.M., Larochelle, H., Beygelzimer, A., d'Alché-Buc, F., Fox, E.B., Garnett, R. (eds.) Advances in Neural Information Processing Systems 32: Annual Conference on Neural Information Processing Systems 2019, NeurIPS 2019, Vancouver, BC, Canada, 8–14 December 2019, pp. 9273–9282 (2019)
8. Guidotti, R., Monreale, A., Matwin, S., Pedreschi, D.: Black box explanation by learning image exemplars in the latent feature space. In: Brefeld, U., Fromont, E., Hotho, A., Knobbe, A., Maathuis, M., Robardet, C. (eds.) ECML PKDD 2019. LNCS (LNAI), vol. 11906, pp. 189–205. Springer, Cham (2020). https://doi.org/10.1007/978-3-030-46150-8_12
9. Gulrajani, I., Ahmed, F., Arjovsky, M., Dumoulin, V., Courville, A.C.: Improved training of Wasserstein GANs. In: Advances in Neural Information Processing Systems, pp. 5767–5777 (2017)
10. Ha, D., Eck, D.: A neural representation of sketch drawings. In: International Conference on Learning Representations (2018)
11. He, K., Zhang, X., Ren, S., Sun, J.: Deep residual learning for image recognition. In: Proceedings of the IEEE Conference on Computer Vision and Pattern Recognition, pp. 770–778 (2016)

12. Isola, P., Zhu, J.Y., Zhou, T., Efros, A.A.: Image-to-image translation with conditional adversarial networks. In: Proceedings of the IEEE Conference on Computer Vision and Pattern Recognition, pp. 1125–1134 (2017)
13. Johnson, J., Alahi, A., Fei-Fei, L.: Perceptual losses for real-time style transfer and super-resolution. In: Leibe, B., Matas, J., Sebe, N., Welling, M. (eds.) ECCV 2016. LNCS, vol. 9906, pp. 694–711. Springer, Cham (2016). https://doi.org/10.1007/978-3-319-46475-6_43
14. Liu, R., Yu, Q., Yu, S.X.: Unsupervised sketch to photo synthesis (2020)
15. Parekh, J., Mozharovskyi, P., d'Alche Buc, F.: A framework to learn with interpretation. arXiv preprint arXiv:2010.09345 (2020)
16. Parkhi, O.M., Vedaldi, A., Zisserman, A.: Deep face recognition. In: British Machine Vision Conference (2015)
17. Ronneberger, O., Fischer, P., Brox, T.: U-net: convolutional networks for biomedical image segmentation. In: Navab, N., Hornegger, J., Wells, W.M., Frangi, A.F. (eds.) MICCAI 2015. LNCS, vol. 9351, pp. 234–241. Springer, Cham (2015). https://doi.org/10.1007/978-3-319-24574-4_28
18. Sangkloy, P., Burnell, N., Ham, C., Hays, J.: The sketchy database: learning to retrieve badly drawn bunnies. ACM Trans. Graph. (Proceedings of SIGGRAPH) (2016)
19. Sangkloy, P., Lu, J., Fang, C., Yu, F., Hays, J.: Scribbler: controlling deep image synthesis with sketch and color. In: Proceedings of the IEEE Conference on Computer Vision and Pattern Recognition, pp. 5400–5409 (2017)
20. Santurkar, S., Tsipras, D., Ilyas, A., Madry, A.: How does batch normalization help optimization? In: Proceedings of the 32nd International Conference on Neural Information Processing Systems, pp. 2488–2498 (2018)
21. Shaham, T.R., Dekel, T., Michaeli, T.: SinGAN: learning a generative model from a single natural image. In: Proceedings of the IEEE International Conference on Computer Vision, pp. 4570–4580 (2019)
22. Shocher, A., Bagon, S., Isola, P., Irani, M.: InGAN: capturing and remapping the "DNA" of a natural image. arXiv preprint arXiv:1812.00231 (2018)
23. Shocher, A., Cohen, N., Irani, M.: "Zero-shot" super-resolution using deep internal learning. In: Proceedings of the IEEE Conference on Computer Vision and Pattern Recognition, pp. 3118–3126 (2018)
24. Webster, R., Rabin, J., Simon, L., Jurie, F.: Detecting overfitting of deep generative networks via latent recovery. In: Proceedings of the IEEE Conference on Computer Vision and Pattern Recognition, pp. 11273–11282 (2019)
25. Yang, F., Yang, H., Fu, J., Lu, H., Guo, B.: Learning texture transformer network for image super-resolution. In: Proceedings of the IEEE/CVF Conference on Computer Vision and Pattern Recognition, pp. 5791–5800 (2020)
26. Yu, A., Grauman, K.: Fine-grained visual comparisons with local learning. In: 2014 IEEE Conference on Computer Vision and Pattern Recognition (CVPR), June 2014
27. Yu, Q., Liu, F., SonG, Y.Z., Xiang, T., Hospedales, T., Loy, C.C.: Sketch me that shoe. In: 2016 IEEE Conference on Computer Vision and Pattern Recognition (2016)
28. Zhu, J.-Y., Krähenbühl, P., Shechtman, E., Efros, A.A.: Generative visual manipulation on the natural image manifold. In: Leibe, B., Matas, J., Sebe, N., Welling, M. (eds.) ECCV 2016. LNCS, vol. 9909, pp. 597–613. Springer, Cham (2016). https://doi.org/10.1007/978-3-319-46454-1_36
29. Zhu, J.Y., Park, T., Isola, P., Efros, A.A.: Unpaired image-to-image translation using cycle-consistent adversarial networks. In: Proceedings of the IEEE International Conference on Computer Vision, pp. 2223–2232 (2017)

Learning Local Regularization for Variational Image Restoration

Jean Prost[1(✉)], Antoine Houdard[1], Andrés Almansa[2], and Nicolas Papadakis[1]

[1] Univ. Bordeaux, Bordeaux INP, CNRS, IMB, UMR 5251, 33400 Talence, France
`jean.prost@math.u-bordeaux.fr`
[2] Université de Paris, MAP5, CNRS, 75006 Paris, France

Abstract. In this work, we propose a framework to learn a local regularization model for solving general image restoration problems. This regularizer is defined with a fully convolutional neural network that sees the image through a receptive field corresponding to small image patches. The regularizer is then learned as a *critic* between unpaired distributions of clean and degraded patches using a Wasserstein generative adversarial networks based energy. This yields a regularization function that can be incorporated in any image restoration problem. The efficiency of the framework is finally shown on denoising and deblurring applications.

1 Introduction

Inverse Problems and Convex Regularization. Many image restoration tasks require to solve an inverse problem. This can be addressed with a variational formulation involving a data-fidelity term and a regularization term encouraging the solution to satisfy given properties or to belong to a space of possible solutions. Some of the most famous regularization terms used for image restoration are convex non-smooth terms like the total variation [20], or ℓ^1 minimization of transform-domain coefficients such as Wavelet frames [4,6] or local Fourier or DCT representations [26]. However theses strategies tend to produce over-smoothed results, since they represent only a rough approximation of natural image statistics and geometry.

CNN-Based Non-convex Regularization. Later-on more accurate natural image priors emerged in the form of non-convex regularization terms, such as patch-based Gaussian mixture models (to be discussed below) or convolutional neural networks (CNN). Most common CNN-based regularizers are, however, trained in a way that the prior or regularizer itself is only partially and implicitly known via its gradient [2,18,19] or proximal operator [12,16,21,24,28]. Such implicit CNN regularizers, and the associated optimization algorithms, lack convergence guarantees or do so under overly restrictive conditions on the regularizer, the regularization parameter or the kind of inverse problems they can solve [18,21].

A. Elmoataz et al. (Eds.): SSVM 2021, LNCS 12679, pp. 358–370, 2021.
https://doi.org/10.1007/978-3-030-75549-2_29

To overcome these limitations a new breed of explicit CNN-based regularizers have been proposed, either in the form of the push-forward measure of a generative model [3], a variational autoencoder [7], or more directly as a discriminator network [15]. All these approaches are nevertheless limited to a particular class of image and do not generalize to images of arbitrary size.

Patch-Based Non-convex Regularization. Learning prior information has also been widely studied from the patch point-of-view. The main idea is to learn the prior knowledge from patches, that are local sub-images of small size, instead of learning a prior from whole images. This allows to avoid the high-dimensional issues faced when working with full-size image distributions. These approaches rely on parametric models of the patch distribution such as Gaussian mixture models [10,23,30]. However such simple models can not accurately represent the complexity of the patch space.

In this work, we introduce an explicit non-convex regularization function encoded with a fully convolutional neural network that acts as a local regularizer. This prior knowledge on the patch distribution can be applied to a whole image without size limitation. We propose (i) to learn the convolutional regularizer as a discriminator between patches using the Wasserstein GAN framework [1] as in [15], and (ii) to integrate this regularizer in patch-based models such as [30].

1.1   Setup of the Problem

The main goal of this paper is to perform image restoration by solving an inverse problem. That is, finding the underlying true image x^* from its perturbed observation y that we consider here to be of the form

$$y = Ax^* + \epsilon, \tag{1}$$

where $\epsilon \sim \mathcal{N}(0, \sigma^2)$ is a Gaussian white noise and A is a degradation operator that can typically be the identity (pure denoising), a mask (inpainting) or a blurring kernel (deconvolution). These inverse problems can be addressed with a variational formulation involving a regularization term. This amounts to find an estimate $\hat{x}$ of x^* of the form

$$\hat{x} \in \arg\min_x \frac{1}{2\sigma^2}\|Ax - y\|^2 + \lambda R(x), \tag{2}$$

where $\|Ax - y\|^2$ is the data-fidelity term ensuring that the recovered image $\hat{x}$ is close enough to the degraded observation y, $R(x)$ is the regularization term and $\lambda \geq 0$ monitors the influence of both terms. In the case where $R(x) = -\log(P_X(x)) + C$ is derived from a prior probability distribution P_X modeling the data x, then $\hat{x}$ from (2) corresponds to the *maximum a posteriori* estimator.

The choice of the regularization function R has a strong impact on the final result. We propose to learn R through a local regularization functional r acting on patches. Denoting as $\Omega_x = \{x_1, \cdots, x_n\}$ the set of all patches of size $p \times p$

from an image x, this function takes as input an image patch x_i and outputs a score $r(x_i)$ that indicates how likely the patch is to be a clean one. As in [30], we define the global regularization functional as the average value of the local scores on the set of all patches of image x:

$$R(x) = \frac{1}{|\Omega_x|} \sum_{x_i \in \Omega_x} r(x_i). \tag{3}$$

Working with patches yields three main advantages. It first makes the learning phase simpler, as a patch model contains far less parameters than a full image model. Next, the number of images required for training is reduced, as a single image already provides several thousands of patches. Finally, contrary to [15], the regularization function can be applied on images of any size.

In practice, we consider r as a CNN with perceptual size equal to the patch size $p \times p$ and taking values in $\mathbf{R}$. This representation is more general than Gaussian mixture models, and allows to encode complex distributions.

1.2 Contributions and Outline

We propose an image restoration method that relies on a regularization function learned on patches and applied to any image size. It gathers the advantages of previous CNN methods while avoiding the constraints of implicit plug & play priors (convergence guarantee) and of GAN or VAE priors (image size).

In addition, the regularization function is learned in an *unsupervised* manner, in the sense that it only relies on patch distributions of clean and degraded data and **it does not require paired data**. We can therefore deal with an unknown degradation model if a noisy dataset is available.

The organization of the paper is as follows. In Sect. 2, we propose an unsupervised framework for the learning of a compact convolutional neural network modeling the local patch regularity prior. We namely obtain the local regularization functional r as a critic trained to distinguish noisy patches from clean ones using the framework of Wasserstein generative adversarial models [1]. In Sect. 3, we provide implementation details to make the work fully reproducible. We show in Sect. 4 that the local functional r generalizes well to arbitrary levels of noise, i.e. noise level unseen during training. In Sect. 5, we demonstrate that the proposed framework is efficient for image denoising and deblurring.

2   Local Regularization for Image Inverse Problem

In this section we define our local image regularizer r_θ as a convolutional neural network and we describe how we use and train it.

Patch-based methods have shown to be efficient tools for solving inverse problems in imaging [30]. Hence we aim at defining a regularization function r_θ depending on parameters $\theta \in \Theta$ that encode prior knowledge at a patch level. In the patch-based literature, such regularizers rely on statistical modeling of the distribution of clean patches and the model parameters are usually inferred with a maximum likelihood estimation [10]. This leads to two main limitations. First, it requires to have access to the probability density function of the prior distribution and consequently it does not properly represent the intrinsic low dimensional manifold of clean patches. Second, maximizing the likelihood of a complex model leads to non-convex problems that are difficult to solve in practice.

In order to tackle these issues, we propose to take advantage of having two data sets of clean and degraded patches –not necessarily paired– and consider r_θ as a *critic* that tells us if a patch is more likely to be clean or degraded.

We first detail in Sect. 2.1 how the local regularization function is integrated as a global regularizer on images in order to solve the variational problem (2). In Sect. 2.2, we present the framework to learn the regularizer as a *critic* between two unpaired dataset of clean and degraded images.

2.1 Convolutional Regularizers for Variational Problems

We define, for the variational problem (2), a regularization term R that takes into account local prior knowledge of the images. To do so, we propose to consider a class of functions r_θ defined with a fully convolutional neural network with parameter $\theta \in \Theta$. We enforce the perceptual size of this network to be the patch size $p \times p$. That is, the successive convolutions operate on a window no larger than $p \times p$ pixels. Using this architecture permits to compute the global regularizer R from (3) by directly applying r_θ to the full image x and average the outputs. Once learned the local regularizer $r_\theta^\star$, the variational problem to solve becomes

$$\min_x \frac{1}{2\sigma^2}\|Ax - y\|^2 + \frac{\lambda}{|\Omega_x|}\sum_i r_\theta^\star(x_i). \tag{4}$$

We propose to find a local minimizer of (4) by performing an explicit gradient descent method. Let x^ℓ the image at iteration ℓ, a gradient step of step size η writes

$$x^{\ell+1} = x^\ell - \frac{\eta}{\sigma^2}A^*(Ax^\ell - y) - \frac{\eta\lambda}{|\Omega_x|}\sum_i \nabla r_\theta^\star(x_i^\ell), \tag{5}$$

where A^* is the adjoint operator of A. Contrary to plug & play methods that rely on implicit schemes [24], this explicit scheme converges for differentiable regularization functions and adequate time steps.

We now describe how the framework for learning the local regularization function.

2.2 Adversarial Local Regularizer (ALR)

In order to train r_θ as a critic between patch distributions, we consider the discriminator framework introduced for generative adversarial networks [8], without the generator network. Such approach nevertheless results in a critic r_θ approximating the hard clustering between clean and degraded patches. It therefore induces steep gradients ∇r_θ that may lead to numerical instabilities during the minimization of problem (2).

As a consequence, we rather rely on the Wasserstein GAN [1] formulation that amounts to approximate the optimal transport cost between the distribution of clean patches $\mathbb{P}_c$, and a distribution of degraded patches $\mathbb{P}_n$. Relying on the dual formulation of the optimal transport [22], an optimal critic $r_\theta^\star$ is seen as a Kantorovitch potential and shall satisfy

$$r_\theta^\star \in \arg\max_{\varphi \in \mathrm{Lip}_1} \mathbb{E}_{z \sim \mathbb{P}_n}\left[\varphi(z)\right] - \mathbb{E}_{z \sim \mathbb{P}_c}\left[\varphi(z)\right]. \tag{6}$$

Under the assumption that the support of the clean patches distribution $\mathcal{M}$ is compact [15], the solution of equation (6) corresponds to the distance function to the clean data manifold $\mathcal{M}$. Each iteration of the gradient descent on Eq. (2) thus brings our noisy data closer to the clean data.

In practice, imposing a neural network to be 1-Lipschitz is a difficult task and we therefore use the formulation proposed in [9] that encourages the gradient norm to be close to 1. This amounts to maximize the following quantity

$$D(\theta) = \mathbb{E}_{z \sim \mathbb{P}_n}\left[r_\theta(z)\right] - \mathbb{E}_{z \sim \mathbb{P}_c}\left[r_\theta(z)\right] - \mu \mathbb{E}_{z \sim \mathbb{P}_i}\left[(\|\nabla_z r_\theta(z)\|_2 - 1)^2\right] \tag{7}$$

where $\mathbb{P}_i$ is the distribution of all lines connecting samples in $\mathbb{P}_n$ and $\mathbb{P}_c$. In other words, the last term of (7) is a gradient penalty that makes the function 1-Lipschitz on the convex hull of the union of the support of $\mathbb{P}_c$ and $\mathbb{P}_n$. By enforcing the gradient ∇r_θ to be of norm close to 1, vanishing gradient issues are also avoided when solving problem (2) with gradient descent approaches.

We illustrate the properties of the regularization functional with a synthetic example in Fig. 1 containing random perturbations of clean data points located on a circle. The learned regularization function $r_\theta(z)$ therefore approximates the distance function to the circle. The gradient $\nabla r_\theta(z)$ thus indicates the direction to follow in order to transport z towards the a clean point within the circle.

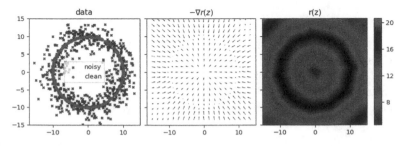

Fig. 1. Regularization functional $r(z)$ learned on a synthetic 2d denoising problem with clean data (blue dots) on the circle and noisy ones (red crosses). The gradient penalty ensures that the gradient ∇r is not flat close to the data manifold $\mathcal{M}$. (Color figure online)

3   Practical Considerations for Image Restoration

In this section, we provide implementation details to reproduce the proposed framework. After presenting the architecture of the regularization network r_θ, we explain the training strategy and describe how image restoration is performed.

3.1   Network Architecture

The local regularization functional r_θ is designed as a 6 layers convolutional network. Each layer is made of 3×3 convolution operations followed by ReLU activations [17]. This network has therefore a 15×15 receptive field. No padding is used. Hence, when a patch of the size of the network receptive field is fed to the network r_θ, the output is a scalar.

3.2   Training the Regularization Functional

The proposed regularization network is trained with patches matching the size of the receptive field of the network. We create the dataset $\mathcal{D}_c$ of clean patches by extracting all 15×15 patches from a 30000 image subset of the google landmarks dataset [25]. Similarly, we create the dataset $\mathcal{D}_n$ of noisy patches by extracting all 15×15 patches from another 30000 images subset of the landmarks dataset, to which we added an additive white Gaussian noise with standard deviation σ_{train}. Following [15], the local regularization network r_θ is trained to minimize the criterion (7) with Algorithm 1. We use the Adam optimizer [13] with hyperparameters $\beta_1 = 0.9$ and $\beta_2 = 0.999$, and an exponential learning rate decay, so that the learning rate α begins at a value of 10^{-3} for the first iteration, and ends up at 10^{-4} for the last iteration. We use a batch size of $m = 32$ and train the network for $K = 10^5$ iterations. The gradient-penalty parameter is set to $\mu = 5$.

Training samples of clean and noisy patches z, with their final regularizer value $r_\theta(z) \in \mathbb{R}$, are shown in Fig. 2. As can be observed from the functional values, there exists a slight ambiguity between texture patches ($r_\theta(z) = -0.23$ for the last patch of top row) and noisy homogeneous patches ($r_\theta(z) = -0.27$ for the first patches of bottom row). We nevertheless show in Fig. 3 that the distributions of clean and noisy patches are globally well separated.

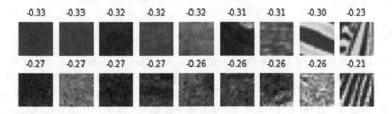

Fig. 2. Value of the local regularization functional r_θ trained with $\sigma_{train} = 0.1$ on clean (top row) and noisy (bottom row) patches ($\sigma = 0.1$).

Algorithm 1. Learning the local regularization r_θ

Input: Datasets $\mathcal{D}_c$ of clean patches and $\mathcal{D}_n$ of noisy patches; gradient penalty μ; batch size m, number of iterations K
Output: regularization function r_θ
for $k = 1$ **to** K **do**
    Sample minibatches of m clean patches $\{z_j^c\}_{j=1}^m$ from $\mathcal{D}_c$ and m noisy patches $\{z_j^n\}_{j=1}^m$ from $\mathcal{D}_n$ and a random number $\alpha \in [0;1]$
    Define interpolated patches $z_j^i = \alpha z_j^c + (1 - \alpha)z_j^n$
    for $j = 1$ **to** m **do**
        $D_j(\theta) = r_\theta(z_j^n) - r_\theta(z_j^c) - \mu(||\nabla_z r_\theta(z_j^i)||_2 - 1)^2$
    end for
    $\theta \leftarrow Adam(\nabla_\theta \sum_{j=1}^m D_j(\theta))$
end for

3.3 Solving the Variational Problem

Image restoration is realized by solving the variational problem (2). To do so, we search for the minimizing image x by performing 50 iterations of Adam [13], with the momentum parameter set to the default values $\beta_1 = 0.9$ and $\beta_2 = 0.999$, and an exponential learning rate decay, with an initial learning rate of 0.1 and a final learning rate of 0.01 at the last iteration. We implement the method with the pytorch deep learning framework, so that the gradient of the global regularization functional $R(x)$ can be easily computed using automatic differentiation.

4 Generalization to Unseen Noise Level

In this section, we study the robustness of the proposed regularization function to noise variations. The adversarial training of the regularization function, presented in the previous sections, requires to learn a different regularization function for every different noise level σ. We show how this limitation can be overcome.

We first analyze the behaviour of regularization functions trained on a single noise level σ_{train} and then used to denoise an image with a different noise level σ_{img}. Second, we propose to train the regularization functions with varying noise levels and demonstrate experimentally the superiority of this approach.

4.1 Robustness to Unseen Noise Level

To study the ability of the local regularization function to generalize to noise levels unseen during training, we train 4 regularization functions on 4 different noise levels $\sigma_{train} \in \{0.05, 0.1, 0.2, 0.4\}$. We then evaluate the quality of the regularization of those networks on denoising tasks, for 5 different noise levels $\sigma_{img} \in \{0.05, 0.1, 0.2, 0.3, 0.4\}$. The 4 networks share the same architecture and the same training procedure as described in Sect. 3.

While these regularizers have only been trained to distinguish between clean patches and noisy patches for a particular noise level, they generalize well to intermediate noise levels, in the sense that the regularizer value is an increasing function of the noise level of its input patch. Figure 3 illustrates this point for the noise level $\sigma_{train} = 0.1$. The overlap between the distribution for noise level $\sigma = 0$ (top) and $\sigma = 0.1$ (bottom) is small, showing the ability of the regularizer network to distinguish clean and noisy patches. Furthermore, the distribution for noise level $\sigma = 0.05$ is located in between the distributions $\sigma = 0$ and $\sigma = 0.1$, showing the ability of the network to generalize to intermediate noise levels but also to extrapolate to the noise level 0.15.

Next, we evaluate denoising quality by measuring the average PSNR on a validation set of 11 images. To that end, we solve problem (2) for $A = \text{Id}$. We denoise images with 5 noise levels $\sigma_{img} \in \{0.05, 0.1, 0.2, 0.3, 0.4\}$, and we respectively set the regularization parameter λ to $\{0.15, 0.35, 0.6, 0.8, 1\}$. The results displayed on Table 1 demonstrate that the trained regularization functions generalize well to unseen noise level, as for all 5 levels of noise σ_{img}, the 4 regularization functions yield average PSNR values that are contained in an interval of size smaller than 1 dB. Furthermore, regularization networks trained on small

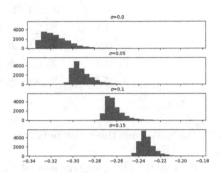

Fig. 3. Distribution of values $r_\theta(x)$ for a regularizer trained on noise level $\sigma_{train} = 0.1$. It generalizes well to patches x with an intermediate (0.05) or extrapolated (0.15) noise levels.

noise levels $\sigma_{train} \in \{0.05, 0.1\}$ generalize well to higher noise levels σ_{train} as they perform even better than networks trained on the specific noise level.

We suggest that this is due to the fact that, when trained on a small noise level, the regularization function is forced to learn a tight boundary between the clean and the noisy distribution which favors denoising performance. However, for the highest noise level $\sigma_{img} = 0.4$, the regularization function trained on a small noise level $\sigma_{train} = 0.05$ gives the worst results. As patches with very high noise levels are not seen during the training of the regularization function trained for $\sigma_{train} = 0.05$, we suggest that the gradient penalty is not enforced to 1 in this region of the patch space. Thus there is no guarantee that the gradient of the regularization function ∇r_θ is indeed directed towards the space of possible solutions. This prevents the optimization algorithm from finding a relevant local minimum of (2).

4.2   Robustness to Noise Variation During Training

We evaluate how robust the regularization function is to noise level variation during training. To do so, we train a regularization function on a distribution containing patches with noise level σ_{train} uniformly sampled in the interval

Table 1. Average PSNR on AWGN denoising, in function of the image noise level σ_{img}, and the noise level the regularization network was trained on σ_{train}. For each image noise level, best result is displayed in bold, and second best result is underlined. Regularization networks trained on small noise level σ_{train} generalizes well to higher noise levels σ_{img}. The regularization network trained on varying level of noise $\sigma_{train} \in [0.05; 0.3]$ performs better on high noise levels σ_{img}.

σ_{train} ╲ σ_{img}	0.05	0.1	0.2	0.3	0.4
0.05	**33.24**	**28.94**	<u>24.21</u>	<u>21.25</u>	18.91
0.1	<u>33.20</u>	28.82	24.17	<u>21.25</u>	<u>19.13</u>
0.2	32.42	28.23	23.80	21.03	18.96
0.4	33.01	28.23	23.84	21.03	18.96
[0.05; 0.3]	32.92	<u>28.90</u>	**24.91**	**22.58**	**20.84**

[0.05, 0.30]. We use the same network architecture and the same training procedure as in Sect. 3.

We evaluate the effectiveness of this regularization function by measuring the average PSNR when this function is used for denoising. We compare the performance with the prior trained on a single noise level in the last row of Table 1. Results show that the regularization function trained with a varying noise level has comparable performance with the regularization function trained on a single-noise level. Furthermore, for high noise level, the regularization function trained on a varying noise level significantly outperforms the regularization function trained on a single-noise level. This illustrates the fact that training the regularization function on varying noise level is actually beneficial.

We suggest that exposing the regularization function to various noise levels during training combines two advantages. It first learns a tight boundary around the clean patches distribution, as the networks trained on low noise levels. Second, the gradient-penalty is enforced even on highly noisy patches, as the networks trained on high noise levels.

5   Experiments

We evaluate the effectiveness of our learned regularization functional on two image restoration tasks, image denoising and image deblurring.

5.1   Denoising

We evaluate our method on additive white Gaussian noise denoising, which corresponds to solving (2) with $A = I$. We compare our method against two common patch-based denoising algorithms, BM3D [5] and EPLL [30], on 3 noise levels $\sigma_{img} \in \{0.1, 0.2, 0.4\}$. We use our model trained on varying noise level $\sigma_{train} \in [0.05, 0.3]$, with the regularization parameter λ respectively set to 0.15,

0.35 and 1. For BM3D, we use the implementation of [14] with default parameters, and for EPLL we use the implementation of [11] with default parameters and a prior GMM model learned on RGB patches.

The average PSNR and LPIPS [29] on the BSD68 dataset for the 3 methods are presented in Table 2, and examples of denoised images are shown on Fig. 4. For the 3 noise levels, the adversarial local regularization denoising outperform EPLL and BM3D in terms of PSNR, while having comparable perceptual quality. This illustrates the ability of convolutional neural networks to be used as local regularizers when trained the right way.

Table 2. Comparisons in terms of PSNR (left) and LPIPS (right) of the patch-based denoising algorithms ALR, EPLL and BM3D, for white Gaussian noise. Results are averaged on the 68 images of the BSD68 dataset.

σ	PSNR			LPIPS		
	ALR	EPLL	BM3D	ALR	EPLL	BM3D
0.1	**28.85**	<u>28.77</u>	28.26	<u>0.29</u>	**0.28**	0.30
0.2	<u>24.88</u>	**24.92**	24.69	0.44	**0.42**	<u>0.43</u>
0.4	**21.58**	19.75	<u>20.25</u>	**0.57**	0.61	<u>0.58</u>

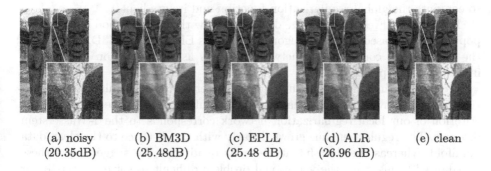

(a) noisy   (b) BM3D   (c) EPLL   (d) ALR   (e) clean
(20.35dB)   (25.48dB)   (25.48 dB)   (26.96 dB)

Fig. 4. Visual comparison of patch-based denoising methods for $\sigma = 0.1$.

5.2   Deblurring

To illustrate the adaptability of our local regularization function, we consider image deblurring. This corresponds to solving (2) with a linear degradation operator A taken as a convolution operation with a blur kernel k, that is $y = k * x + \varepsilon$. Figure 5 shows an example of image deblurring using our learned local regularization function. We provide visual comparison with the Plug-and-Play deblurring algorithm of [27].

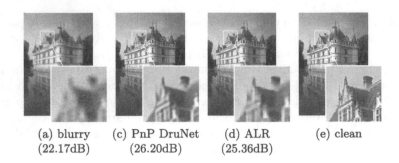

(a) blurry   (c) PnP DruNet   (d) ALR   (e) clean
(22.17dB)   (26.20dB)   (25.36dB)

Fig. 5. Illustration of deblurring using a 7×7 Gaussian kernel with standard deviation $\sigma_k = 3$.

6   Conclusion and Perspectives

We propose a new strategy to solve inverse problem in imaging using a convolutional neural network as a local regularization function. The local regularization network is trained to discriminate between clean and noisy patches, and the global regularization function is defined as the average value of the local function over the set of all image patches. Working with a local regularization function offers several advantages: it works with any image size, it requires less training data and has less parameters than a full size model. Furthermore, the fully convolutional architecture of the network makes it computationally efficient to compute the global regularization function and its gradient.

Experimental results on image denoising show that our method outperforms popular patch-based denoising algorithm such as EPLL and BM3D, illustrating the potential of convolutional networks to acts as regularization function for inverse imaging problems.

We believe that improving the training criterion of the regularization function could improves the performance of the regularization. Indeed, the training criterion of our local regularization network corresponds to the 1-Wasserstein distance. The regularizer thus grows linearly with the distance to the clean data manifold, whereas the data-fidelity term is quadratic. We suggest that these unbalanced terms make the variational problem difficult to solve, especially for high noise levels. We believe that learning a regularization term based on the 2-Wasserstein distance could help to overcome this limitation, as the learned regularization function would then grow with the square of the distance to the clean manifold.

Acknowledgements. This study has been carried out with financial support from the French Research Agency through the PostProdLEAP project (ANR-19-CE23-0027-01).

References

1. Arjovsky, M., Chintala, S., Bottou, L.: Wasserstein generative adversarial networks. In: International Conference on Machine Learning, pp. 214–223. PMLR (2017)

2. Bigdeli, S.A., Jin, M., Favaro, P., Zwicker, M.: Deep Mean-Shift Priors for Image Restoration. Adv. Neural Inf. Proces. Syst. **30**, 763–772 (2017)
3. Bora, A., Jalal, A., Price, E., Dimakis, A.G.: Compressed sensing using generative models. In: International Conference on Machine Learning, vol. 2, pp. 537–546 (2017)
4. Coifman, R.R., Donoho, D.L.: Translation-invariant de-noising. In: Antoniadis, A., Oppenheim, G. (eds,) Wavelets and Statistics. Lecture Notes in Statistics, vol. 103. Springer, New York (1995). https://doi.org/10.1007/978-1-4612-2544-7_9
5. Dabov, K., Foi, A., Katkovnik, V., Egiazarian, K.: Image denoising by sparse 3-D transform-domain collaborative filtering. IEEE Trans. Image Process. **16**, 2080–95 (2007)
6. Donoho, D.L., Johnstone, J.M.: Ideal spatial adaptation by wavelet shrinkage. Biometrika **81**(3), 425–455 (1994)
7. González, M., Almansa, A., Delbracio, M., Musé, P., Tan, P.: Solving Inverse Problems by Joint Posterior Maximization with a VAE Prior (2019)
8. Goodfellow, I.J., et al.: Generative adversarial networks. arXiv preprint arXiv:1406.2661 (2014)
9. Gulrajani, I., Ahmed, F., Arjovsky, M., Dumoulin, V., Courville, A.: Improved training of Wasserstein GANs. Adv. Neural. Inf. Process. Syst. **30**, 5767–5777 (2017)
10. Houdard, A., Bouveyron, C., Delon, J.: High-dimensional mixture models for unsupervised image denoising (HDMI). SIAM J. Imag. Sc. **11**(4), 2815–2846 (2018)
11. Hurault, S., Ehret, T., Arias, P.: EPLL: an image denoising method using a Gaussian mixture model learned on a large set of patches. Image Processing On Line **8**, 465–489 (2018)
12. Kamilov, U.S., Mansour, H., Wohlberg, B.: A plug-and-play priors approach for solving nonlinear imaging inverse problems. IEEE Signal Process. Lett. **24**(12), 1872–1876 (2017)
13. Kingma, D.P., Ba, J.: Adam: a method for stochastic optimization. In: International Conference on Learning Representations (2015)
14. Lebrun, M.: An analysis and implementation of the BM3D image denoising method. Image Processing On Line **2**, 175–213 (2012)
15. Lunz, S., Öktem, O., Schönlieb, C.B.: Adversarial regularizers in inverse problems. Adv. Neural. Inf. Process. Syst. **31**, 8507–8516 (2018)
16. Meinhardt, T., Moller, M., Hazirbas, C., Cremers, D.: Learning proximal operators: using denoising networks for regularizing inverse imaging problems. In: International Conference on Computer Vision, pp. 1781–1790 (2017)
17. Nair, V., Hinton, G.E.: Rectified linear units improve restricted Boltzmann machines. In: International Conference on Machine Learning (2010)
18. Reehorst, E.T., Schniter, P.: Regularization by Denoising: clarifications and New Interpretations. IEEE Trans. Comput. Imaging **5**(1), 52–67 (2019)
19. Romano, Y., Elad, M., Milanfar, P.: The little engine that could: regularization by denoising (RED). SIAM J. Imaging Sci. **10**(4), 1804–1844 (2017)
20. Rudin, L.I., Osher, S., Fatemi, E.: Nonlinear total variation based noise removal algorithms. Physica D **60**(1–4), 259–268 (1992)
21. Ryu, E.K., Liu, J., Wang, S., Chen, X., Wang, Z., Yin, W.: Plug-and-play methods provably converge with properly trained denoisers. In: International Conference on Machine Learning, pp. 5546–5557 (2019)
22. Santambrogio, F.: Optimal transport for applied mathematicians. Progress Nonlinear Diff. Eq. Appl. **87** (2015)

23. Teodoro, A.M., Bioucas-Dias, J.M., Figueiredo, M.A.T.: Scene-Adapted Plug-and-Play Algorithm with Guaranteed Convergence: Applications to Data Fusion in Imaging (2018)
24. Venkatakrishnan, S.V., Bouman, C.A., Wohlberg, B.: Plug-and-play priors for model based reconstruction. In: IEEE Global Conference on Signal and Information Processing, pp. 945–948 (2013)
25. Weyand, T., Araujo, A., Cao, B., Sim, J.: Google landmarks dataset v2 - a large-scale benchmark for instance-level recognition and retrieval (2020)
26. Yu, G., Sapiro, G.: DCT image denoising: a simple and effective image denoising algorithm. Image Processing On Line 1 (2011)
27. Zhang, K., Li, Y., Zuo, W., Zhang, L., Gool, L.V., Timofte, R.: Plug-and-play image restoration with deep denoiser prior (2020)
28. Zhang, K., Zuo, W., Gu, S., Zhang, L.: Learning deep CNN denoiser prior for image restoration. In: IEEE Conference on Computing Vision and Pattern Recognistion, pp. 2808–2817 (2017)
29. Zhang, R., Isola, P., Efros, A.A., Shechtman, E., Wang, O.: The unreasonable effectiveness of deep features as a perceptual metric (2018)
30. Zoran, D., Weiss, Y.: From learning models of natural image patches to whole image restoration. In: International Conference on Computer Vision, pp. 479–486 (2011)

Segmentation and Labelling

On the Correspondence Between Replicator Dynamics and Assignment Flows

Bastian Boll[1,2(✉)] [iD], Jonathan Schwarz[1,2] [iD], and Christoph Schnörr[1,2] [iD]

[1] Heidelberg Collaboratory for Image Processing, Heidelberg University,
Heidelberg, Germany
[2] Image and Pattern Analysis Group, Heidelberg University, Heidelberg, Germany
`bastian.boll@iwr.uni-heidelberg.de`

Abstract. Assignment flows are smooth dynamical systems for data labeling on graphs. Although they exhibit structural similarities with the well-studied class of replicator dynamics, it is nontrivial to apply existing tools to their analysis. We propose an embedding of the underlying assignment manifold into the interior of a single probability simplex. Under this embedding, a large class of assignment flows are pushed to much higher-dimensional replicator dynamics. We demonstrate the applicability of this result by transferring a spectral decomposition of replicator dynamics to assignment flows.

Keywords: Image labeling · Assignment flows · Replicator dynamics · Spectral decomposition

1 Introduction

Given a graph G with nodes $\mathcal{I}$ and a weighted adjacency matrix Ω, data labeling is the task of assigning a label from a discrete set $\mathcal{J}$ to each node in $\mathcal{I}$ such that both consistency with given data on $\mathcal{I}$ and spatial regularity with respect to Ω are simultaneously maximized. This constitutes a basic problem in image processing and formalizes e.g. image segmentation. In [1], a class of smooth geometric labeling systems is introduced which evolves high-entropy assignment states towards hard node–label decisions. *Assignment flows* are applicable to data in any metric space and regularize via geometric averaging according to Ω. If nodes are decoupled (i.e. $\Omega = \mathbb{I}$), assignment flows reduce to simple node-wise replicator equations which have been extensively studied as models of evolution in mathematical biology [6,9,10]. However, established theory is difficult to apply to the study of assignment flows, because it is unclear how to incorporate the coupling of nodes via geometric averaging which is at the core of their expressiveness. In the present work, we show how a large class of assignment flows can be seen as marginalization of replicator dynamics. We subsequently leverage this perspective to transfer a spectral decomposition result from the replicator setting to assignment flows.

© Springer Nature Switzerland AG 2021
A. Elmoataz et al. (Eds.): SSVM 2021, LNCS 12679, pp. 373–384, 2021.
https://doi.org/10.1007/978-3-030-75549-2_30

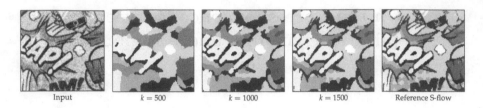

| Input | $k = 500$ | $k = 1000$ | $k = 1500$ | Reference S-flow |

Fig. 1. Spectral approximations to the (S)-assignment flow limit (right) by approximating (4.23) with $k \in \{500, 1000, 1500\}$ most dominant eigenvectors of Ω. Averaging weights are computed analogous to nonlocal means denoising [3].

1.1 Related Work

A nonlinear spectral framework has recently been developed for the treatment of regularizations induced by one-homogeneous functions [5,7] such as total variation (TV). Applicability of this framework was demonstrated for diverse data processing tasks such as image fusion [2]. However, because the underlying spectral theory is nonlinear, specialized methods are required to compute eigenfunctions [4]. The spectral decomposition we propose here can be seen as a tool to study *nonlinear* assignment flows which have been used in similar areas of application. However, unlike the above framework, the spectral analysis we propose is *linear*, making it amenable to standard methods.

2 Assignment Manifold and Assignment Flows

We briefly summarize assignment flows as introduced in [1] and refer to the recent survey [12] for more background, more details and a review of recent related work.

Let $(\mathcal{F}, d_\mathcal{F})$ be a metric space and $\mathcal{F}_n = \{f_i \in \mathcal{F} : i \in \mathcal{I}\}$, $|\mathcal{I}| = n$ be given data. Assume that a predefined set of $|\mathcal{J}| = c$ prototypes $\mathcal{F}_* = \{f_j^* \in \mathcal{F} : j \in \mathcal{J}\}$ is given. *Data labeling* denotes the task of finding assignments $j \to i$, $f_j^* \to f_i$ in a spatially regularized fashion. At each node $i \in \mathcal{I}$, a distribution $W_i = (W_{i1}, \ldots, W_{ic})^\top$ in the relative interior $\mathcal{S} = \mathrm{rint}\,\Delta_c$ of the probability simplex encodes assignment of $\mathcal{J}$ to i. We view $\mathcal{S}$ as a Riemannian manifold $(\mathcal{S}, g)$ endowed with the Fisher-Rao metric g from information geometry. The *assignment manifold* $(\mathcal{W}, g)$, $\mathcal{W} = \mathcal{S} \times \cdots \times \mathcal{S}$ ($n = |\mathcal{I}|$ factors) is the product manifold whose points encode label assignments at all nodes. By comparing given data to prototypes, the distance vector field $D_{\mathcal{F};i} = \left(d_\mathcal{F}(f_i, f_1^*), \ldots, d_\mathcal{F}(f_i, f_c^*)\right)^\top$, $i \in \mathcal{I}$ is a data representation which abstracts the specific feature space $\mathcal{F}$. It is lifted to the assignment manifold by the *likelihood map* and the *likelihood vectors*, respectively,

$$L_i \colon \mathcal{S} \to \mathcal{S}, \quad L_i(W_i) = \frac{W_i \odot e^{-\frac{1}{\rho} D_{\mathcal{F};i}}}{\langle W_i, e^{-\frac{1}{\rho} D_{\mathcal{F};i}} \rangle}, \quad i \in \mathcal{I}. \tag{2.1}$$

This map is based on the affine e-connection of information geometry and $\rho > 0$ is used to normalize the application-specific scale of distances $D_{\mathcal{F};i}$. Likelihood vectors are spatially regularized by the *similarity map* and *similarity vectors*, respectively,

$$S_i \colon \mathcal{W} \to \mathcal{S}, \quad S_i(W) = \mathrm{Exp}_{W_i}\Big(\sum_{k \in \mathcal{N}_i} w_{ik} \mathrm{Exp}_{W_i}^{-1}\big(L_k(W_k)\big)\Big), \quad i \in \mathcal{I}, \quad (2.2)$$

where $\mathrm{Exp}_p(v) = \frac{p e^{v/p}}{\langle p, e^{v/p}\rangle}$ is the exponential map corresponding to the e-connection. Positive weights w_{ik}, $k \in \mathcal{N}_i$, that sum up to 1 in every neighborhood $\mathcal{N}_i$, determine the regularization properties. We collect these weights into an adjacency matrix $\Omega \in \mathbb{R}^{n \times n}$. The (W-)*assignment flow* is induced on the assignment manifold $\mathcal{W}$ by the locally coupled system of nonlinear ODEs

$$\dot{W}_i = R_{W_i} S_i(W), \qquad W_i(0) = \mathbb{1}_{\mathcal{S}}, \quad i \in \mathcal{I}, \quad (2.3)$$

where the map $R_p = \mathrm{Diag}(p) - p p^\top$, $p \in \mathcal{S}$ turns the right-hand side into a tangent vector field and $\mathbb{1}_{\mathcal{W}} \in \mathcal{W}$ denotes the barycenter of the assignment manifold $\mathcal{W}$. The solution $W(t) \in \mathcal{W}$ is numerically computed by geometric integration [13] and determines a labeling $W(T)$ for sufficiently large T after a trivial rounding operation. Convergence and stability of the assignment flow have been studied by [14]. For symmetric weights $w_{ij} = w_{ji}$, it has been shown [11] that (2.3) can be parameterized entirely by similarity vectors

$$\dot{W}(t) = R_{W(t)}[S(t)], \qquad W(0) = \mathbb{1}_{\mathcal{W}} \quad (2.4)$$
$$\dot{S}(t) = R_{S(t)}[\Omega S(t)], \qquad S(0) = \exp_{\mathbb{1}_{\mathcal{W}}}(-\Omega D) . \quad (2.5)$$

We call the dynamics (2.5) *S-assignment flow* or *S-flow*. Here, the *lifting map* $\exp_W \colon T_0 \mathcal{W} \to \mathcal{W}$ is defined by

$$\exp_{W_i}(V_i)_i = \mathrm{Exp}_{W_i}(R_{W_i} V_i) = \frac{W_i \odot \exp(V_i)}{\langle W_i, \exp(V_i)\rangle}, \qquad i \in \mathcal{I} \quad (2.6)$$

on the flat tangent space $T_0 \mathcal{W} = \{V \in \mathbb{R}^{n \times c} \colon V \mathbb{1}_c = 0\}$ and R applies to each row of W resp. S separately. Note that $\exp_p(v + \alpha \mathbb{1}) = \exp_p(v)$ for any $\alpha \in \mathbb{R}$ as one easily checks.

Notation. In the following, we consider a graph G with n nodes and encode assignment of c classes to these nodes as assignment matrices $S \in \mathcal{W} \subseteq \mathbb{R}^{n \times c}$. For tensors $U \in \mathbb{R}^{n_1 \times \cdots \times n_r}$, $U^v = \mathrm{vec}\, U \in \mathbb{R}^{n_1 \cdots n_r}$ denotes vectorization. The action of linear operators T on U is written as $T(U)$ if U is regarded as a tensor and by juxtaposition if U is vectorized, i.e. $T(U)^v = T U^v$. We define the symbol Δ to refer to the relative interior of a single probability simplex with $N = c^n$ corners. Vectors $P \in \Delta$ are identified with tensors in $\mathbb{R}^{c \times \cdots \times c}$ such that their entries can be referred to by multi-indices $\alpha \in [c]^n$. The symbol $\mathbb{1}_{n \times c}$ denotes the matrix of size $n \times c$ filled with 1 and $\odot$ denotes componentwise multiplication.

3   S-Flow Embedding

Assignment matrices $S \in \mathcal{W}$ associate each node $i \in [n]$ with a distribution $S_i \in \mathcal{S}$. Up to a negligible set of pathological cases, assignment flows converge to integral assignments, i.e. $S_i(t) \to e_{k(i)} \in \mathbb{R}^c$ for $t \to \infty$ as shown in [14]. By enumerating all c^n possible assignments of c classes to n nodes, we may equivalently view the integral limit point as approached within Δ. In keeping with this perspective, the aim of this section is to find an embedding of $\mathcal{W}$ into Δ such that S-flows in $\mathcal{W}$ translate to replicator dynamics in Δ. To this end, define the maps

$$T \colon \mathcal{W} \to \Delta, \qquad T(S)_\alpha := \prod_{i \in [n]} S_{i,\alpha_i} \tag{3.1}$$

$$Q \colon \mathcal{W} \to \Delta, \qquad Q(S)_\gamma := \sum_{l \in [n]} S_{l,\gamma_l} . \tag{3.2}$$

T maps $S \in \mathcal{W}$ to a rank-1 tensor in Δ. The inverse process is marginalization for each node.

Lemma 1. *The map T defined by (3.1) is a diffeomorphism between $\mathcal{W}$ and a subset of Δ with inverse*

$$T^{-1}(P)_{i,j} = \sum_{\alpha_i = j} P_\alpha := \sum_{\alpha \in [c]^n} \delta_{\alpha_i = j} P_\alpha, \qquad (i,j) \in [n] \times [c] . \tag{3.3}$$

Proof. We check that the inverse of T has the form (3.3).

$$T^{-1}(T(S))_{i,j} = \sum_{\alpha_i = j} \prod_{r \in [n]} S_{r,\alpha_r} \tag{3.4a}$$

$$= \sum_{k_1 \in [c]} \cdots \sum_{k_{i-1} \in [c]} \sum_{k_{i+1} \in [c]} \cdots \sum_{k_n \in [c]} S_{1,k_1} \ldots S_{i-1,k_{i-1}} S_{i,j} S_{i+1,k_{i+1}} \ldots S_{n,k_n} \tag{3.4b}$$

$$= S_{i,j} \prod_{l \in [n] \setminus \{i\}} \langle S_l, \mathbb{1} \rangle = S_{i,j} . \tag{3.4c}$$

Clearly, both T and T^{-1} are smooth. $\square$

If individual nodes are decoupled, i.e. $\Omega = \mathbb{I}_n$, assignment flows reduce to simple replicator dynamics for each node. We will show that the case of coupled nodes via more general choices of Ω may likewise be seen as a single, much larger replicator equation. We start with a preparatory lemma.

Lemma 2 (Adjoint of Q). *T^{-1} and Q are adjoint linear operators.*

Proof. Let $P \in \mathbb{R}^N$ and $V \in \mathbb{R}^{n \times c}$, then

$$\langle P, Q(V) \rangle = \sum_\gamma P_\gamma Q(V)_\gamma = \sum_\gamma P_\gamma \sum_{l \in [n]} V_{l,\gamma_l} = \sum_{l \in [n]} \sum_{j \in [c]} \sum_{\gamma_l = j} P_\gamma V_{l,\gamma_l} \qquad (3.5a)$$

$$= \sum_{l \in [n]} \sum_{j \in [c]} V_{l,j} \sum_{\gamma_l = j} P_\gamma \stackrel{(3.3)}{=} \sum_{l \in [n]} \sum_{j \in [c]} V_{l,j} T^{-1}(P)_{l,j} \qquad (3.5b)$$

$$= \langle T^{-1}(P), V \rangle . \qquad (3.5c)$$

$\square$

Theorem 1 (Replicator dynamics induce S-Flow). *For any S-flow*

$$\dot{S}(t) = R_{S(t)}[\Omega S(t)] =: X(S(t)), \qquad S(0) = S_0 \qquad (3.6)$$

on $\mathcal{W}$ exists a matrix $\overline{\Omega} \in \mathbb{R}^{N \times N}$ such that (3.6) is induced by marginalization of the replicator dynamics

$$\dot{P}(t) = (T_\sharp X)(P) = R_{P(t)}[\overline{\Omega} P(t)], \qquad P(0) = T(S_0) \qquad (3.7)$$

on Δ. $\overline{\Omega}$ is symmetric exactly if Ω is symmetric.

Proof. We push forward the S-flow vector field $X(S) := R_S[\Omega S]$ via T. The components of the differential dT read

$$\frac{\partial T_\alpha}{\partial S_{l,m}} = \frac{\partial}{\partial S_{l,m}} \prod_{i \in [n]} S_{i,\alpha_i} = \prod_{i \in [n] \setminus \{l\}} S_{i,\alpha_i} \frac{\partial}{\partial S_{l,m}} S_{l,\alpha_l} = \delta_{\alpha_l = m} \prod_{i \in [n] \setminus \{l\}} S_{i,\alpha_i} .$$
$$(3.8)$$

By setting $P = T(S)$, we may rewrite this as

$$\frac{\partial T_\alpha}{\partial S_{l,m}} = \delta_{\alpha_l = m} \prod_{i \in [n] \setminus \{l\}} S_{i,\alpha_i} . \qquad (3.9)$$

In the following, every occurrence of S is meant as $S(P) = T^{-1}(P)$ (Lemma 1). The S-flow field X given by (3.6) has components

$$X(T^{-1}(P))_{l,m} = S_{l,m} \left(\sum_{j \in [n]} \omega_{lj} S_{j,m} - \langle S_l, (\Omega S)_l \rangle \right) \qquad (3.10)$$

and the pushforward via T consequently reads

$$(T_\sharp X)(P)_\gamma = \sum_{l \in [n]} \sum_{m \in [c]} \frac{\partial T_\gamma}{\partial S_{l,m}} X(T^{-1}(P))_{l,m} \qquad (3.11a)$$

$$= \sum_{l \in [n]} \left(\prod_{i \in [n] \setminus \{l\}} S_{i,\gamma_i} \right) S_{l,\gamma_l} \left[\sum_{j \in [n]} \omega_{lj} S_{j,\gamma_l} - \langle S_l, (\Omega S)_l \rangle \right] \qquad (3.11b)$$

$$= \left(\prod_{i \in [n]} S_{i,\gamma_i} \right) \sum_{l \in [n]} \left[(\Omega S)_{l,\gamma_l} - \langle S_l, (\Omega S)_l \rangle \right] \qquad (3.11c)$$

$$= P_\gamma \left(\sum_{l \in [n]} (\Omega S)_{l,\gamma_l} - \langle S, (\Omega S) \rangle \right) . \qquad (3.11d)$$

We define the linear operator $\overline{\Omega} = Q(\Omega \otimes \mathbb{I}_c)Q^\top$ on Δ with Q defined by (3.2). Clearly, $\overline{\Omega}$ is symmetric exactly if Ω is symmetric. Lemma 2 now implies

$$\langle P, \overline{\Omega}P \rangle = \langle TS^v, Q(\Omega \otimes \mathbb{I}_c)Q^\top TS^v \rangle = \langle S^v, (\Omega \otimes \mathbb{I}_c)S^v \rangle = \langle S, \Omega S \rangle \qquad (3.12)$$

because $S \in \mathcal{W}$, as well as

$$\sum_{l \in [n]} (\Omega S)_{l, \gamma_l} = Q(\Omega S)_\gamma = \big(Q(\Omega \otimes \mathbb{I})Q^\top TS^v \big)_\gamma = (\overline{\Omega}P)_\gamma . \qquad (3.13)$$

Returning to (3.11), this gives

$$(T_\sharp X)(P)_\gamma = P_\gamma \big((\overline{\Omega}P)_\gamma - \langle P, \overline{\Omega}P \rangle \big) \qquad (3.14)$$

which may be written more compactly as $(T_\sharp X)(P) = R_P[\overline{\Omega}P]$. Because the inverse map T^{-1} performs marginalization for each node, this shows the assertion. □

4 S-Flow Spectral Decomposition

Theorem 1 allows to view assignment flows as marginal dynamics of replicator systems. We aim to leverage this insight to transfer results from the study of replicator equations to the assignment flow setting. To this end, we first briefly describe a spectral decomposition result for replicator dynamics in Sect. 4.1 and subsequently apply it to the S-flow (2.5) in Sect. 4.2.

4.1 Selection Systems

Given a simplex $\Delta \subset \mathbb{R}^N$ and a symmetric matrix $\overline{\Omega} \in \mathbb{R}^{N \times N}$

$$\dot{P}(t) = R_{P(t)}[\overline{\Omega}P(t)], \qquad P(0) = P_0 \qquad (4.1)$$

is called replicator equation [6] for linear "fitness" $\overline{\Omega}$ and initial value P_0. Dynamics of this type have been studied in mathematical biology by [8]. Let $\overline{\Omega} = \sum_{k=1}^{\mathrm{rank}\,\overline{\Omega}} \lambda_k h_k h_k^\top$ denote the spectral decomposition of $\overline{\Omega}$. In mathematical models of evolution, P describes the relative frequencies of N traits in a given population. Let $l(t) \in \mathbb{R}^N$ model the absolute number of individuals exhibiting each trait such that $P(t) = \frac{l(t)}{\langle l(t), \mathbb{1}_N \rangle}$, $l(0) = l_0$. The selection system equivalent to (4.1) reads

$$l(t) = l_0 \odot K(t), \qquad K(t) = \exp \left(\sum_k s_k(t) h_k \right) \qquad (4.2)$$

with coefficients $s_k(t)$ following the so-called *escort system* dynamics [8]

$$\dot{s}_i(t) = \lambda_i \frac{\big\langle P_0, h_i \odot \exp \big(\sum_k s_k(t) h_k \big) \big\rangle}{\big\langle P_0, \exp \big(\sum_k s_k(t) h_k \big) \big\rangle} . \qquad (4.3)$$

Proposition 1. *The replicator dynamics (4.1) is equivalent to*

$$P(t) = \exp_{P_0} \left(\sum_k s_k(t) h_k \right) \tag{4.4}$$

$$\dot{s}_k(t) = \lambda_k \langle h_k, P(t) \rangle, \qquad s_k(0) = 0, \qquad k \in [\operatorname{rank} \overline{\Omega}] \tag{4.5}$$

Proof. The quantity l can be normalized to yield a corresponding assignment $P \in \Delta$. Let $l_0 = l(0)$ be such that $\langle l_0, \mathbb{1} \rangle = 1$. Then $P_0 = l_0$ and we find

$$P(t) = \frac{P_0 \odot \exp\left(\sum_k s_k(t) h_k \right)}{\left\langle \mathbb{1}, P_0 \odot \exp\left(\sum_k s_k(t) h_k \right) \right\rangle} \overset{(2.6)}{=} \exp_{P_0} \left(\sum_k s_k(t) h_k \right). \tag{4.6}$$

Escort system dynamics can be transformed to

$$\dot{s}_i(t) = \lambda_i \frac{\langle h_i, P_0 \odot \exp\left(\sum_k s_k(t) h_k \right) \rangle}{\langle P_0, \exp\left(\sum_k s_k(t) h_k \right) \rangle} \tag{4.7a}$$

$$= \lambda_i \left\langle h_i, \exp_{P_0} \left(\sum_k s_k(t) h_k \right) \right\rangle \tag{4.7b}$$

$$= \lambda_i \langle h_i, P(t) \rangle \tag{4.7c}$$

and the initial conditions (4.5) are consistent with (4.1). □

4.2 Spectral Decomposition

We aim to transfer the spectral decomposition of Proposition 1 to S-flows. To this end, the following lemmata describe further behavior of the maps T and Q introduced in (3.1) and (3.2).

Lemma 3. *For any matrix $V \in \mathbb{R}^{n \times c}$ it holds that $\langle Q(V), \mathbb{1}_N \rangle = c^{n-1} \langle V, \mathbb{1}_{n \times c} \rangle$.*

Proof. We directly compute

$$\sum_{\gamma} \sum_{l \in [n]} V_{l, \gamma_l} = \sum_{l \in [n]} \sum_{m \in [c]} \sum_{\gamma_l = m} V_{l, \gamma_l} = \sum_{l \in [n]} \sum_{m \in [c]} c^{n-1} V_{l, m} \tag{4.8a}$$

$$= c^{n-1} \langle V, \mathbb{1}_{n \times c} \rangle . \tag{4.8b}$$

□

Lemma 4. *For any $V \in T_0 \mathcal{W}$ it holds that $T^{-1}(Q(V)) = c^{n-1} V$.*

Proof. For arbitrary indices $(i, j) \in [n] \times [c]$, we find

$$T^{-1}(Q(V))_{i,j} \overset{(3.2),(3.3)}{=} \sum_{\gamma_i = j} \sum_{l \in [n]} V_{l, \gamma_l} = \sum_{\gamma_i = j} \left(V_{i, \gamma_i} + \sum_{l \in [n] \setminus \{i\}} V_{l, \gamma_l} \right) \tag{4.9a}$$

$$= c^{n-1} V_{i,j} + \sum_{\gamma_i = j} \sum_{l \in [n] \setminus \{i\}} V_{l, \gamma_l} . \tag{4.9b}$$

Let $\tilde{V}$ contain only the rows with indices $[n]\setminus\{i\}$ of V and let $\tilde{\gamma}$ denote multi-indices of its rows. Then the last sum in (4.9b) is proportional to

$$\sum_{l\in[n]\setminus\{i\}}\sum_{\tilde{\gamma}}\tilde{V}_{l,\tilde{\gamma}_l} = \sum_{l\in[n]\setminus\{i\}}\sum_{m\in[c]}\sum_{\tilde{\gamma}_l=m}\tilde{V}_{l,\tilde{\gamma}_l} \propto \sum_{l\in[n]\setminus\{i\}}\sum_{m\in[c]}\tilde{V}_{l,m} = 0 \qquad (4.10)$$

$\square$

Lemma 5. *If $V \in T_0 W$ and $\lambda \in \mathbb{R}$ satisfy $\Omega V = \lambda V$ then $\overline{V} = QV^v$ is an eigenvector of $\overline{\Omega} = Q(\Omega \otimes \mathbb{I}_c)Q^\top$ for eigenvalue $c^{n-1}\lambda$. Additionally, $\mathbb{1}_N$ is an eigenvalue of $\overline{\Omega}$ and $QU^v \propto \mathbb{1}_N$ if $U \propto \mathbb{1}_{n\times c}$.*

Proof. Let $V \in T_0 W$ and $\lambda \in \mathbb{R}$ satisfy $\Omega V = \lambda V$. Then by Lemma 4 it holds that

$$\overline{\Omega}QV^v = Q(\Omega \otimes \mathbb{I}_n)Q^\top QV^v = c^{n-1}Q(\Omega \otimes \mathbb{I}_n)V^v = c^{n-1}\lambda QV^v . \qquad (4.11)$$

Now let $U \propto \mathbb{1}_{n\times c}$. Then $QU^v \propto \mathbb{1}_N$ by Lemma 3 and we find

$$\overline{\Omega}\mathbb{1}_N = Q(\Omega \otimes \mathbb{I}_n)Q^\top \mathbb{1}_N \propto Q(\Omega \otimes \mathbb{I}_n)\mathbb{1}_{nc} = Q\mathbb{1}_{nc} \propto \mathbb{1}_N \qquad (4.12)$$

by using Lemma 2. $\square$

Lemma 6. *It holds $\ker Q = \{\mathrm{Diag}(d)\mathbb{1}_{n\times c}: d \in \mathbb{R}^n, \langle d, \mathbb{1}_n\rangle = 0\}$ as well as $\mathrm{rank}\, Q = nc - (n-1)$.*

Proof. Let $V \in \ker Q$ and let $\gamma, \tilde{\gamma}$ be two fixed multi-indices which differ exactly at position i but are otherwise arbitrary. We have $(QV)_\gamma = (QV)_{\tilde{\gamma}} = 0$ by assumption. Thus

$$(QV)_{\tilde{\gamma}} = V_{i,\tilde{\gamma}_i} + \sum_{l\in[n]\setminus\{i\}} V_{l,\tilde{\gamma}_l} = V_{i,\tilde{\gamma}_i} + \sum_{l\in[n]\setminus\{i\}} V_{l,\gamma_l} \qquad (4.13a)$$

$$= (QV)_\gamma = V_{i,\gamma_i} + \sum_{l\in[n]\setminus\{i\}} V_{l,\gamma_l} \qquad (4.13b)$$

which implies $V_{i,\tilde{\gamma}_i} = V_{i,\gamma_i}$, i.e. $V = \mathrm{Diag}(d)\mathbb{1}_{n\times c}$ for some $d \in \mathbb{R}^n$ since i was arbitrary. Let V have this form. Then

$$(QV)_\gamma = \sum_{l\in[n]} V_{l,\gamma_l} = \sum_{l\in[n]} d_l = \langle d, \mathbb{1}_n\rangle . \qquad (4.14)$$

so V is in the kernel of Q exactly if $\langle d, \mathbb{1}_n\rangle = 0$. There are $(n-1)$ linearly independent vectors $d \in \mathbb{R}^n$ with this property, therefore Q has the specified rank. $\square$

Proposition 2 (Eigenvectors of $\overline{\Omega}$). *Let $\Omega \in \mathbb{R}^{n\times n}$ be a symmetric matrix of full rank such that $\Omega\mathbb{1}_n = \mathbb{1}_n$. Then $\overline{\Omega}$ has rank $nc - (n-1)$ and there is an orthogonal matrix $V \in \mathbb{R}^{nc\times n(c-1)}$ such that the columns of V are eigenvectors of $\Omega \otimes \mathbb{I}_c$ and the columns of $\overline{V} = \sqrt{\frac{c}{N}}QV$ are pairwise orthonormal eigenvectors of $\overline{\Omega}$. The columns of $\overline{V}$ together with the vector $\frac{1}{\sqrt{N}}\mathbb{1}_N$ form an orthonormal basis of $\mathrm{img}\,\overline{\Omega}$.*

Proof. Let $G \in \mathbb{R}^{n \times n}$ be an orthogonal matrix of eigenvectors of Ω and let $\Lambda = \mathrm{Diag}(\lambda_1, \ldots, \lambda_n)$ have the respective eigenvalues as diagonal entries. Then $\Omega = G\Lambda G^\top$. Now let $H \in \mathbb{R}^{c \times c}$ be an orthogonal matrix with first column $h_1 = \frac{1}{\sqrt{c}} \mathbb{1}_c$ such that the remaining columns span $\{d \in \mathbb{R}^c \colon \langle d, \mathbb{1}_c \rangle = 0\}$. We find

$$(G \otimes H)(\Lambda \otimes \mathbb{I}_c)(G \otimes H)^\top = (G \otimes H)(\Lambda \otimes \mathbb{I}_c)(G^\top \otimes H^\top) \tag{4.15a}$$

$$= (G \otimes H)((\Lambda G^\top) \otimes H^\top) \tag{4.15b}$$

$$= (G\Lambda G^\top) \otimes (HH^\top) \tag{4.15c}$$

$$= \Omega \otimes \mathbb{I}_c \tag{4.15d}$$

as well as $(G \otimes H)(G \otimes H)^\top = (GG^\top) \otimes (HH^\top) = \mathbb{I}_{cn}$ so $(G \otimes H)$ is an orthogonal matrix whose columns are eigenvectors of $\Omega \otimes \mathbb{I}_c$. Viewing the eigenvectors V_i with indices $i \in I_1 = \{1 + c(k-1) \colon k \in [n]\}$ as matrices $V_i \in \mathbb{R}^{n \times c}$, we have $V_i^v = g_k \otimes h_1$ which gives $V_i = g_k h_1^\top = c^{-\frac{1}{2}} \mathrm{Diag}(g_k) \mathbb{1}_{n \times c}$ and thus $Q(V_i) = \frac{1}{\sqrt{c}} \langle g_k, \mathbb{1}_n \rangle \mathbb{1}_N \propto \mathbb{1}_N$. Therefore, the rank of $\overline{\Omega}$ is at most $nc - (n-1)$. For the remaining indices $j = l + c(k-1)$, $k \in [n]$, $l \in [c] \setminus \{1\}$ it holds that

$$V_j \mathbb{1}_c = g_k h_l^\top \mathbb{1}_c = g_k \langle h_l, \mathbb{1}_c \rangle = 0 \tag{4.16}$$

thus $V_j \in T_0 \mathcal{W}$. Denote the set of these indices by $I_2 = [nc] \setminus I_1$. By Lemma 4 it holds that

$$\langle \overline{V}_{j_1}, \overline{V}_{j_2} \rangle = \frac{c}{N} \langle Q V_{j_1}^v, Q V_{j_2}^v \rangle = \frac{c}{N} \langle V_{j_1}^v, Q^\top Q V_{j_2}^v \rangle = \langle V_{j_1}, V_{j_2} \rangle = 0 \tag{4.17}$$

for all $j_1, j_2 \in I_2$ and $\|\overline{V}_j\|_2^2 = \frac{c}{N} \langle Q V_j^v, Q V_j^v \rangle = \|V_j^v\|_2^2 = 1$ for $j \in I_2$. We additionally find

$$\langle \overline{V}_j, \frac{1}{\sqrt{N}} \mathbb{1}_N \rangle = \frac{\sqrt{c}}{N} \langle Q V_j^v, \mathbb{1}_N \rangle = \frac{1}{\sqrt{c}} \langle V_j, \mathbb{1}_{n \times c} \rangle = 0 \qquad j \in I_2 \tag{4.18}$$

by using Lemma 3. Because all columns of $\overline{V}$ are eigenvectors of $\overline{\Omega}$ by Lemma 5, this shows the assertion. $\qquad \square$

Lemma 7 (Lifting Map Lemma). *Let $S \in \mathcal{W}$ and $V \in \mathbb{R}^{n \times c}$. Then*

$$T(\exp_S(V)) = \exp_{T(S)}(Q(V)) . \tag{4.19}$$

Proof. We have $T(\exp(V)) = \exp(Q(V))$ because for any multi-index γ

$$\exp(Q(V))_\gamma = \exp\left(\sum_{l \in [n]} V_{l, \gamma_l} \right) = \prod_{l \in [n]} \exp\left(V_{l, \gamma_l} \right) = T(\exp(V))_\gamma . \tag{4.20}$$

Let $D \in \mathbb{R}^{n \times n}$ be a diagonal matrix with nonzero diagonal entries. Then $T(DR) \propto T(R)$ for any $R \in \mathbb{R}^{n \times c}$ because

$$T(DR)_\gamma = \prod_{l \in [n]} (DR)_{l, \gamma_l} = \left(\prod_{l \in [n]} D_{ll} \right) \left(\prod_{l \in [n]} R_{l, \gamma_l} \right) \propto T(R)_\gamma . \tag{4.21}$$

Due to $T(S \odot V) = T(S) \odot T(V)$ it follows

$$T(\exp_S(V)) \propto T(S \odot \exp(V)) = T(S) \odot \exp(Q(V)) \propto \exp_{T(S)}(Q(V)) . \quad (4.22)$$

Because both first and last term in (4.22) are clearly elements of Δ, this implies the assertion. □

Theorem 2 (Spectral Decomposition). *Let $\Omega \in \mathbb{R}^{n \times n}$ be a symmetric matrix of full rank such that $\Omega \mathbb{1}_n = \mathbb{1}_n$ and let $V \in \mathbb{R}^{nc \times nc}$ be given by Proposition 2. Then the S-flow (3.6) is equivalent to*

$$S(t) = \exp_{S_0} \left(\sum_{k \in I_2} s_k(t) V_k \right) \quad (4.23)$$

$$\dot{s}_k(t) = \lambda_k \langle V_k, S(t) \rangle, \qquad s_k(0) = 0, \qquad k \in I_2 \quad (4.24)$$

where $I_2 := [nc] \setminus \{1 + c(k-1) : k \in [n]\}$ and $V_k \in \mathbb{R}^{n \times c}$ denotes the unique matrix such that V_k^v is the k-th column of V.

Proof. By Propositions 1 and 2 it holds that

$$S(t) = T^{-1}(P(t)) = T^{-1} \left(\exp_{P_0} \left(\sum_{k \in I_1 \cup I_2} s_k(t) \overline{V}_k \right) \right) \quad (4.25a)$$

$$= T^{-1} \left(\exp_{P_0} \left(\sum_{k \in I_2} s_k(t) \overline{V}_k \right) \right) = \exp_{S_0} \left(\sum_{k \in I_2} s_k(t) V_k \right) \quad (4.25b)$$

where we used Lemma 7 as well as $\overline{V}_k \propto \mathbb{1}_N$ for $k \in I_1$ in (4.25b). Additionally, for $k \in I_2$ it holds

$$\dot{s}_k(t) = c^{n-1} \lambda_k \langle P(t), \overline{V}_k \rangle = \lambda_k \langle P(t), Q V_k \rangle \overset{\text{Lemma 2}}{=} \lambda_k \langle S(t), V_k \rangle . \quad (4.26)$$

□

5   Experiments

Because $\mathcal{W}$ is not a flat space, standard methods are not canonically suitable for numerical integration of assignment flows. A remedy is to construct a diffeomorphism $\varphi \colon W \to V$ and integrate the φ-related vector field in some flat space V. Theorem 2 implicitly achieves this by parameterizing S-flows through the coefficients $s_k(t)$, $k \in I_2$ which live in the unbounded flat space $\mathbb{R}$. Standard methods of numerical integration such as Runge-Kutta or linear multistep methods are therefore directly applicable to the dynamics (4.24). In our empirical examination, we consider a graph of 100×100 grid pixel nodes with adjacency Ω computed analogous to nonlocal means denoising [3], i.e. the averaging weight Ω_{i_1,i_2} is large if i_1 is close to i_2 in the image plane and if the patch of pixels around i_1 is similar to the patch of pixels around i_2. We added independent noise drawn from a normal distribution to each channel of the original cartoon image in Fig. 1 and aim to recover the original prototype colors in the RGB feature space. Euclidean distances for each pixel-prototype pair are collected in a distance matrix $D \in \mathbb{R}^{n \times c}$ (here, $n = 100 \cdot 100$, $c = 47$). The point $S_0 = \exp_{\mathbf{1}_\mathcal{W}}(-\frac{1}{\rho}D)$ is used to initialize a reference S-flow which we integrate numerically by the geometric Euler method with step-length $h = 0.1$ until a low-entropy assignment state is reached.

$\lambda_1 \approx 1.00$ $\lambda_{10} \approx 1.00$ $\lambda_{50} \approx 1.00$ $\lambda_{150} \approx 0.95$ $\lambda_{250} \approx 0.92$

Fig. 2. Eigenvectors of the adjacency matrix used in Fig. 1. More dominant parts of the spectrum correspond to low-frequency eigenvectors.

To visualize the spectral decomposition of Theorem 2, we first compute approximations to the dominant $k \in \{500, 1000, 1500\}$ eigenvectors of Ω by leveraging sparsity. A spectral approximation of the reference S-flow is obtained by dropping the remaining eigenvalues from (4.24), i.e. $\lambda_j \leftarrow 0$ for $j > k$. Numerical integration with the explicit Euler method (constant step-length $h = 0.1$) yields the integral label assignments shown in Fig. 1. As expected, the approximation becomes progressively more faithful to the reference S-flow if a larger fraction of the spectrum of Ω is considered. Note that Ω has full rank $n = 10^4$ so the approximations in Fig. 1 are obtained by considering at most 15% of its spectrum. Clearly, the described method can lead to improved computational efficiency, if the adjacency Ω can be well approximated by a low-rank matrix. The given example illustrates that informative regularizations may be achieved using relatively low-rank adjacency. We additionally observe that less dominant eigenvalues correspond to high-frequency components (see Fig. 2). This is consistent with the approximation behavior shown in Fig. 1; discarding the influence of less dominant eigenvalues leads to smoothing out high-frequency detail while retaining correct label assignment for largely uniform regions.

6   Conclusion

We have constructed an embedding of the assignment manifold $\mathcal{W}$ into a single probability simplex Δ. Under this embedding, S-assignment flows are pushed to replicator dynamics with linear fitness function. Because Δ has intractably large dimension, numerical integration can not be performed directly. However, we show that the embedding into Δ serves as a valuable tool to transfer structural results on replicator dynamics to the study of assignment flows. Conversely, Theorem 1 also identifies a class of replicator dynamics with linear fitness which can be decomposed into much lower-dimensional S-flows. A systematic study of applications is left for future work.

Acknowledgements. This work is supported by the Deutsche Forschungsgemeinschaft (DFG, German Research Foundation) under Germany's Excellence Strategy EXC 2181/1 - 390900948 (the Heidelberg STRUCTURES Excellence Cluster). The original artwork used in Fig. 1 was designed by dgim-studio/Freepik.

References

1. Åström, F., Petra, S., Schmitzer, B., Schnörr, C.: Image labeling by assignment. J. Math. Imaging Vis. **58**(2), 211–238 (2017). https://doi.org/10.1007/s10851-016-0702-4
2. Benning, M., et al.: Nonlinear spectral image fusion. In: Lauze, F., Dong, Y., Dahl, A.B. (eds.) SSVM 2017. LNCS, vol. 10302, pp. 41–53. Springer, Cham (2017). https://doi.org/10.1007/978-3-319-58771-4_4
3. Buades, A., Coll, B., Morel, J.M.: Image denoising methods. A new nonlocal principle. SIAM Rev. **52**(1), 113–147 (2010)
4. Bungert, L., Burger, M., Tenbrinck, D.: Computing nonlinear eigenfunctions via gradient flow extinction. In: Lellmann, J., Burger, M., Modersitzki, J. (eds.) SSVM 2019. LNCS, vol. 11603, pp. 291–302. Springer, Cham (2019). https://doi.org/10.1007/978-3-030-22368-7_23
5. Burger, M., Eckardt, L., Gilboa, G., Moeller, M.: Spectral representations of one-homogeneous functionals. In: Aujol, J.-F., Nikolova, M., Papadakis, N. (eds.) SSVM 2015. LNCS, vol. 9087, pp. 16–27. Springer, Cham (2015). https://doi.org/10.1007/978-3-319-18461-6_2
6. Fisher, R.A.: The Genetical Theory of Natural Selection. Dover, New York (1985)
7. Gilboa, G., Moeller, M., Burger, M.: Nonlinear spectral analysis via one-homogeneous functionals: overview and future prospects. J. Math. Imaging Vis. **56**(2), 300–319 (2016)
8. Karev, G.P.: On mathematical theory of selection: continuous time population dynamics. J. Math. Biol. **60**(1), 107–129 (2010)
9. Karev, G.P.: Replicator equations and the principle of minimal production of information. Bull. Math. Biol. **72**(5), 1124–1142 (2010)
10. Page, K.M., Nowak, M.A.: Unifying evolutionary dynamics. J. Theor. Biol. **219**(1), 93–98 (2002)
11. Savarino, F., Schnörr, C.: Continuous-domain assignment flows. Eur. J. Appl. Math. 1–28 (2020). https://doi.org/10.1017/S0956792520000273. https://www.cambridge.org/core/journals/european-journal-of-applied-mathematics/article/abs/continuousdomain-assignment-flows/4603DE52B18153DF85F9A35BA1BDED00
12. Schnörr, C.: Assignment flows. In: Grohs, P., Holler, M., Weinmann, A. (eds.) Handbook of Variational Methods for Nonlinear Geometric Data, pp. 235–260. Springer, Cham (2020). https://doi.org/10.1007/978-3-030-31351-7_8
13. Zeilmann, A., Savarino, F., Petra, S., Schnörr, C.: Geometric numerical integration of the assignment flow. Inverse Probl. **36**(3), 034003 (2020). (33pp)
14. Zern, A., Zeilmann, A., Schnörr, C.: Assignment flows for data labeling on graphs: convergence and stability. preprint: arXiv https://arxiv.org/abs/2002.11571 (2020)

Learning Linear Assignment Flows for Image Labeling via Exponential Integration

Alexander Zeilmann[1]([✉]) [ID], Stefania Petra[2] [ID], and Christoph Schnörr[1] [ID]

[1] Image and Pattern Analysis Group, Heidelberg University, Heidelberg, Germany
alexander.zeilmann@iwr.uni-heidelberg.de
[2] Mathematical Imaging Group, Heidelberg University, Heidelberg, Germany

Abstract. We introduce a novel algorithm for estimating optimal parameters of linear assignment flows for image labeling. This flow is determined by the solution of a linear ODE in terms of a high-dimensional integral. A formula of the gradient of the solution with respect to the flow parameters is derived and approximated using Krylov subspace techniques. Riemannian descent in the parameter space enables to determine optimal parameters for a 512×512 image in less than $10\,\mathrm{s}$, without the need to backpropagate errors or to solve an adjoint equation. Numerical experiments demonstrate a high generative model expressivity despite the linearity of the assignment flow parametrization.

Keywords: Image labeling · Assignment manifold · Linear assignment flows · Parameter learning · Exponential integration · Low-rank approximation

1 Introduction

Learning the parameters of large networks from training data constitutes a basic problem in imaging science, machine learning and other fields. The prevailing approach utilizes gradient descent or approximations thereof based on automatic differentiation software tools [4].

In this paper, we focus on a class of networks for image labeling, *linear assignment flows* introduced by [17], where parameter estimation can be based on an *exact* formula of the loss function gradient and its approximation, using established methods of large-scale numerical linear algebra. This enables easy implementations and better control of the approximation error. In particular, neither backpropagation, nor automatic differentiation or solving adjoint equations are required. Our parameter estimation is also time-efficient: for a 512×512 image, the runtime for estimating optimal parameters takes less than $10\,\mathrm{s}$. Besides demonstrating these properties, numerical experiments also reveal a high *expressivity* of linear assignment flows: multiscale image structure can be generated from *pure* noise by estimating corresponding parameters.

© Springer Nature Switzerland AG 2021
A. Elmoataz et al. (Eds.): SSVM 2021, LNCS 12679, pp. 385–397, 2021.
https://doi.org/10.1007/978-3-030-75549-2_31

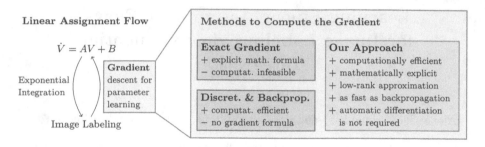

Fig. 1. Our approach & related work: pros and cons. Based on the linear assignment flow, our approach performs optimal parameter estimation using a low-rank approximation of the exact gradient. It is as efficient as applying automatic differentiation in frameworks like PyTorch. Unlike this latter methodology, however, our approach is mathematically explicit and hence supports further tasks like flow control.

Related Work. Linear assignment flows result from a *linear* tangent space parametrization of the nonlinear *assignment flows* introduced by [3]. While the linear assignment flow is still nonlinear, both the linearity of the corresponding tangent space ODE and of its parameter dependency facilitate optimal parameter estimation. This has been exploited in [8] by applying symplectic numerical integration for solving both the linear assignment flow and an adjoint ODE for determining the parameter sensitivities.

In this paper, we exploit the fact that the linear tangent space ODE admits a closed-form solution in terms of the high-dimensional integral (Duhamel's formula). By directly approximating an exact gradient formula using established numerical techniques [2,7], error backpropagation can be avoided altogether, and parameter estimation can be done by computing a Riemannian descent flow in the parameter space.

The standard approach to network parameter learning is using automatic differentiation [4]. Automatic differentiation is a general concept applicable to a variety of scenarios. However, the efficient handling of many scenarios requires a more targeted approach. In the current paper, we introduce such a specific algorithm.

The exact integration of the linear assignment flow requires the evaluation of the matrix exponential. The only machine learning frameworks which implement this function are PyTorch [13] and TensorFlow [1]. But even their implementations can not be applied directly to the large matrices required by imaging tasks. In contrast, our approach is able to learn the parameters for very large images or even 3D volumes as commonly found in medical image analysis.

Regarding the numerical techniques employed in our work, related work in scientific computing includes, e.g., [9,11] for computing the Fréchet derivative of the matrix exponential. However, these algorithms are only applicable if the gradient is computed using the *forward* mode of automatic differentiation (i.e., multiplying Jacobians from the right). For functions of the form $\mathbb{R}^n \to \mathbb{R}$ (with large n), the *reverse* mode of automatic differentiation (backpropagation, i.e., multiplying the Jacobians from the left) is significantly more efficient for computing gradients. This is in line with our approach, in which the Jacobians of the loss function and of the additive term of the linear assignment flow are multiplied from the left to the Jacobian of the matrix exponential. This is an essential component for reducing the dimensionality of the problem in our approach.

Contribution. We introduce a novel approach for *learning the model parameters* of the linear assignment flow [17]. These parameters control the regularization properties of image labelings determined by the flow. The problem of *learning* these parameters was raised in [3, Section 5 and Fig. 14]. Our approach has the following properties:

Linear complexity both in storage space and evaluation time with respect to image size and number of labels, which enables to handle large problem sizes.

No automatic differentiation is necessary. While automatic differentiation (and especially backpropagation) has been advancing significantly in recent years, there are not many comprehensive frameworks for it, and its implementation is nontrivial. Our algorithm does not require automatic differentiation and can be conveniently implemented in any numerical software framework.

Explicit parameter gradient formula. While automatic differentiation is a *procedure* computing the gradient, it does generally not lead to a succinct formula. By contrast, our approach is based on an explicit mathematical formula that can be analyzed theoretically and used in future tasks, e.g., for controlling the assignment flow.

Approximating the exact gradient instead of the exact gradient of an approximation. Approaches like exponential integration (see Sect. 2.3) or explicit Euler integration *approximate* the ordinary differential equation. Automatic differentiation, therefore, produces the exact gradient of this *approximation*. As a consequence, not the parameters of the linear assignment flow are learned but of its approximation. Our approach, on the other hand, directly approximates the exact gradient of the linear assignment flow for parameter estimation and hence enables a better control of the approximation error.

As efficient as backpropagation. In our experiments, our approach was as efficient as backpropagation of exponential integration and backpropagation of explicit Euler. In addition, it requires less storage space.

Low-rank approximation. An approximation is worked out in factorized form. This not only saves storage space but also enables to carry out subsequent computations more efficiently.

Applicability to any loss function and data. Besides being C^1, no further assumptions are made with respect to the loss function used for parameter estimation. Our approach shares with the general assignment flow the property that data in any metric space can be processed.

Organization. Section 2 summarizes the assignment flow, the linear assignment flow and exponential integration for computing it. Section 3 details the exact gradient of any loss function of the flow, with respect to the flow parameters. The application to the linear assignment flow is explained in Sect. 4 and experimental results are reported in Sect. 5. We conclude in Sect. 6.

2   Assignment Flow, Linear Assignment Flow

This section summarizes the *assignment flow* and its approximation, the *linear assignment flow*, introduced in [3] and [17], respectively. The linearized assignment flow provides the basis for our approach to parameter estimation developed in Sect. 3.

2.1   Assignment Flow

Let $G = (I, E)$ be a given undirected graph with vertices $i \in I$ indexing data $\mathcal{F}_I = \{f_i : i \in I\} \subset \mathcal{F}$ given in a metric space $(\mathcal{F}, d)$. The edge set E specifies neighborhoods $\mathcal{N}_i = \{k \in I : ik = ki \in E\} \cup \{i\}$ for every vertex $i \in I$ along with positive weight vectors $w_i \in \mathring{\Delta}_{|\mathcal{N}_i|}$, where $\mathring{\Delta}_n = \Delta_n \cap \mathbb{R}^n_{>0}$ denotes the relative interior of the probability simplex Δ_n.

Along with $\mathcal{F}_I$, *prototypical data (labels)* $\mathcal{L}_J = \{l_j \in \mathcal{F} : j \in J\}$ are given that represent classes $j = 1, \ldots, |J|$. *Supervised image labeling* denotes the task to assign precisely one prototype l_j to each datum f_i at every vertex i in a coherent way, depending on the label assignments in the neighborhoods $\mathcal{N}_i$. These assignments at i are represented by probability vectors

$$W_i \in \mathcal{S} := (\mathring{\Delta}_{|J|}, g_{FR}), \quad i \in I \tag{2.1}$$

on $\mathring{\Delta}_{|J|}$ that endowed with the Fisher-Rao metric g_{FR} becomes a Riemannian manifold denoted by $\mathcal{S}$. Collecting all assignment vectors as rows defines the strictly positive row-stochastic *assignment matrix*

$$W = (W_1, \ldots, W_{|I|})^\top \in \mathcal{W} = \mathcal{S} \times \cdots \times \mathcal{S} \subset \mathbb{R}^{|I| \times |J|}, \tag{2.2}$$

that we regard as point on the product *assignment manifold* $\mathcal{W}$. Image labeling is accomplished by geometrically integrating the *assignment flow* $W(t)$ solving

$$\dot{W} = R_W(S(W)), \qquad W(0) = \mathbb{1}_{\mathcal{W}} := \frac{1}{|J|} \mathbb{1}_{|I|} \mathbb{1}_{|J|}^\top \quad \text{(barycenter)}, \tag{2.3}$$

that provably converges towards a binary matrix [18], i.e., $\lim_{t \to \infty} W_i(t) = e_{j(i)}$, for every $i \in I$ and some $j(i) \in J$, which yields the label assignment $l_{j(i)} \to f_i$. In practice, geometric integration is terminated when $W(t)$ is ε-close to an integral point using the entropy criterion from [3], followed by trival safe rounding [18].

We specify the right-hand side of (2.3)—see (2.5) below—and refer to [3, 15] for more details and the background. With tangent space $T_0 = T_p \mathcal{S}$ independent of the base point $p \in \mathcal{S}$, we define

$$R_p \colon \mathbb{R}^{|J|} \to T_0, \qquad z \mapsto R_p(z) = \left(\mathrm{Diag}(p) - pp^\top\right)z, \qquad (2.4a)$$

$$\mathrm{Exp} \colon \mathcal{S} \times T_0 \to \mathcal{S}, \quad (p, v) \mapsto \mathrm{Exp}_p(v) = \frac{e^{\frac{v}{p}}}{\langle p, e^{\frac{v}{p}} \rangle} p, \qquad (2.4b)$$

$$\mathrm{Exp}^{-1} \colon \mathcal{S} \times \mathcal{S} \to T_0, \quad (p, q) \mapsto \mathrm{Exp}_p^{-1}(q) = R_p \log \frac{q}{p}, \qquad (2.4c)$$

where multiplication, division, exponentiation $e^{(\cdot)}$ and $\log(\cdot)$ apply *component-wise* to the vectors. Corresponding maps R_W and Exp_W in connection with the product manifold (2.2) are defined analogously, and likewise the tangent space $\mathcal{T}_0 = T_0 \times \cdots \times T_0$ to $\mathcal{W}$.

The vector field defining the assignment flow (2.3) arises through *coupling* flows for individual pixels through *geometric averaging* within the neighborhoods $\mathcal{N}_i$, $i \in I$, conforming to the underlying Fisher-Rao geometry

$$S(W) = \begin{pmatrix} \vdots \\ S_i(W)^\top \\ \vdots \end{pmatrix} = \mathcal{G}^\Omega\left(L(W)\right) \in \mathcal{W}, \qquad \begin{aligned} \Omega &= (\Omega_i)_{i \in I}, \\ \Omega_i &= (\omega_{ik})_{k \in \mathcal{N}_i}, \end{aligned} \qquad (2.5a)$$

$$S_i(W) = \mathcal{G}_i^\Omega\left(L(W)\right) = \mathrm{Exp}_{W_i}\left(\sum_{k \in \mathcal{N}_i} \omega_{ik} \mathrm{Exp}_{W_i}^{-1}\left(L_k(W_k)\right) \right), \quad i \in I. \qquad (2.5b)$$

The *similarity vectors* $S_i(W)$ are parametrized by *weight patches* $\Omega_i > 0$ that serve as *regularization parameters* and satisfy the constraints $\sum_{k \in \mathcal{N}_i} \omega_{ik} = 1$, $\forall i \in I$. Flattening these weight patches and complementing zero entries defines sparse row vectors of the matrix $\Omega \in \mathbb{R}_{\geq 0}^{|I| \times |I|}$. Estimating these parameters using the linear assignment flow is the subject of this paper.

2.2   Linear Assignment Flow

The *linear assignment flow*, introduced by [17], approximates (2.3) by

$$\dot{W} = R_W\left(S(W_0) + dS_{W_0} R_{W_0} \log \frac{W}{W_0}\right), \quad W(0) = W_0 \in \mathcal{W} \qquad (2.6)$$

around any point W_0. In what follows, we only consider the barycenter $W_0 = \mathbb{1}_\mathcal{W}$ which is the initial point of (2.3). The differential equation (2.6) is still *nonlinear* but can be parametrized by a *linear* ODE on the tangent space

$$W(t) = \mathrm{Exp}_{W_0}\left(V(t)\right), \qquad (2.7a)$$

$$\dot{V} = R_{W_0}\left(S(W_0) + dS_{W_0} V\right) =: B_{W_0} + A(\Omega)[V], \quad V(0) = 0, \qquad (2.7b)$$

where the linear operator $A(\Omega)$ linearly depends on the parameters Ω of (2.5) (see [17] for an explicit expression). The linear ODE (2.7b) admits a closed-form solution which in turn enables a different numerical approach (Sect. 2.3) and a novel approach to parameter learning (Sect. 3).

2.3 Exponential Integration

The solution to (2.7b) is given by a high-dimensional integral (Duhamel's formula) whose value in closed form is given by

$$V(t; \Omega) = t\varphi\big(tA(\Omega)\big)B_{W_0}, \qquad \varphi(x) = \frac{e^x - 1}{x}, \tag{2.8}$$

where φ is extended to matrix arguments in the standard way [6]. As the matrix A is already very large even for medium-sized images, however, it is not feasible in practice to compute $\varphi(tA)$. Exponential integration [7,12], therefore, was used in [17] for approximately evaluating (2.8).

Applying the row-stacking operator vec_r, that satisfies the general property $\mathrm{vec}_r(ABC) = (A \otimes C^\top)\mathrm{vec}_r(B)$, to both sides of (2.7b) and (2.8), respectively, yields with $v = \mathrm{vec}_r(V)$ and the Kronecker product $\otimes$ (cf. [16])

$$\dot{v} = b + A^J(\Omega)v, \qquad v(0) = 0, \qquad b = \mathrm{vec}_r(B_{W_0}), \tag{2.9a}$$

$$A^J = \big(A^J_{ik}\big)_{i,k\in I}, \qquad A^J_{ik} = \begin{cases} \omega_{ik}R_{S_i(W_0)}, & k \in \mathcal{N}_i, \\ 0, & k \notin \mathcal{N}_i. \end{cases} \tag{2.9b}$$

$$v(t; \Omega) = t\varphi\big(tA^J(\Omega)\big)b, \qquad n := \dim v(t) = |I||J|. \tag{2.9c}$$

Using the Arnoldi iteration [14] with initial vector $q_1 = b/\|b\|$, an orthonormal basis $Q_m = (q_1, \ldots, q_m) \in \mathbb{R}^{n\times m}$ of the Krylov space $\mathcal{K}_m(A^J, b)$ of dimension m is determined. As reported by [17], choosing $m \leq 10$ yields sufficiently accurate approximations of the actions of the matrix exponential expm and the φ operator on a vector, respectively, that using $H_m = Q_m^\top A^J(\Omega)Q_m$ are given by

$$\mathrm{expm}\big(tA^J(\Omega)\big)b \approx \|b\|Q_m\,\mathrm{expm}(tH_m)e_1, \tag{2.10a}$$

$$t\varphi\big(tA^J(\Omega)\big)b \approx t\|b\|Q_m\varphi(tH_m)e_1, \tag{2.10b}$$

where e_1 denotes the first unit vector. The expression $\varphi(tH_m)e_1$ results from computing the left-hand side of the relation [6, Section 10.7.4]

$$\mathrm{expm}\begin{pmatrix} tH_m & e_1 \\ 0 & 0 \end{pmatrix} = \begin{pmatrix} \mathrm{expm}(tH_m) & \varphi(tH_m)e_1 \\ 0 & 1 \end{pmatrix} \tag{2.11}$$

and extracting the upper-right vector. Since H_m is a small matrix, any standard method [10] can be used for computing the matrix exponential on the left-hand side.

3 Loss Function Gradient and Approximation

Fixing a point of time $t = T$, we set $v_T(\Omega) := v(T; \Omega)$ as given by (2.9c) and assume to be given a loss function

$$\mathcal{L}: W \longrightarrow \mathbb{R}, \qquad \mathcal{L}(\Omega) = f_\mathcal{L}\big(v_T(\Omega)\big) \tag{3.1}$$

that evaluates the solution (2.9c) to (2.7b) and hence also the corresponding labeling (2.7a), as a function of the weight parameters Ω; see Sect. 4 for a concrete example. In this section, various ways to approximate the gradient $\partial\mathcal{L}(\Omega)$ are discussed.

3.1 Loss Function Gradient

We compute differentials of matrix-valued functions $F(X)$ by using the row-stacking operator vec_r: if $\text{vec}_r(F(X)) = f(\text{vec}_r(X))$ defines the vector function f by F, then $\text{vec}_r(dF(X)Y) = df(\text{vec}_r(X))\text{vec}_r(Y)$, and we *define* the Jacobian of F by $dF(X) := df(\text{vec}_r(X))$. Thus, if $X \in \mathbb{R}^{m_1 \times m_2}$ and $F(X) \in \mathbb{R}^{n_1 \times n_2}$, then $dF(X) \in \mathbb{R}^{n_1 n_2 \times m_1 m_2}$ and $dF(X)Y \in \mathbb{R}^{n_1 \times n_2}$.

An example is the formula [6, Thm. 10.13] for the vectorized differential of the matrix exponential

$$\text{vec}_r(d\,\text{expm}(C)D) = (\text{expm}(C) \otimes I_n)\varphi(-C \oplus C^\top)\text{vec}_r(D), \qquad (3.2)$$

that we rearranged so as to conform to the row-stacking operator vec_r, rather than to the column-stacking operator used in [6]. $\oplus$ is the Kronecker sum $A \oplus B = A \otimes I_n + I_m \otimes B$ for $A \in \mathbb{R}^{m \times m}$, $B \in \mathbb{R}^{n \times n}$ with identity matrices $I_n \in \mathbb{R}^{n \times n}$.

Proposition 1. *Let $\mathcal{L}$ be a function of the form (3.1), where $v_T(\Omega) = v(T; \Omega)$ solves (2.9a) at $t = T$. Then the gradient of $\mathcal{L}$ is given by*

$$\partial \mathcal{L}(\Omega) = \text{vec}_r^{-1}(C(\Omega)^\top \partial f_{\mathcal{L}}(v_T(\Omega))), \qquad (3.3a)$$

$$C(\Omega) = ((e^{TA^J(\Omega)}, v_T(\Omega)) \otimes e_{n+1}^\top)\varphi(-B(\Omega) \oplus B(\Omega)^\top)df_B(\text{vec}_r(\Omega)) \qquad (3.3b)$$

where $n = |I||J|$, $e_{n+1}^\top = (0_n^\top, 1)$ and $A^J(\Omega)$ is given by (2.9b), and

$$B(\Omega) = \begin{pmatrix} TA^J(\Omega) & Tb \\ 0_n^\top & 0 \end{pmatrix}, \qquad f_B(\text{vec}_r(\Omega)) = \text{vec}_r(B(\Omega)). \qquad (3.3c)$$

Proof. By (3.1), we have

$$\langle \partial \mathcal{L}(\Omega), U \rangle = df_{\mathcal{L}}(v_T(\Omega))dv_T(\Omega)U. \qquad (3.4)$$

In order to determine the vector $dv_T(\Omega)U \in \mathbb{R}^n$, we compute analogous to (2.11),

$$\text{expm}(B(\Omega)) = \begin{pmatrix} \text{expm}(TA^J(\Omega)) & T\varphi(TA^J(\Omega))b \\ 0_n^\top & 1 \end{pmatrix}, \qquad (3.5)$$

and note that $v_T(\Omega) = v(T; \Omega)$ given by (2.9c) appears in the last column on the right-hand side. Thus, using $\text{vec}_r(dB(\Omega)U) = df_B(\text{vec}_r(\Omega))\text{vec}_r(U)$,

$$v_T(\Omega) \overset{(3.5)}{=} (I_n, 0_n)\,\text{expm}(B(\Omega))e_{n+1} \qquad (3.6a)$$

$$dv_T(\Omega)U = (I_n, 0_n)d\,\text{expm}(B(\Omega))dB(\Omega)Ue_{n+1} \qquad (3.6b)$$

$$\overset{(*)}{=} ((I_n, 0_n) \otimes e_{n+1}^\top)\text{vec}_r(d\,\text{expm}(B(\Omega))dB(\Omega)U) \qquad (3.6c)$$

$$\overset{(3.2)}{=} ((I_n, 0_n) \otimes e_{n+1}^\top)(\text{expm}(B(\Omega)) \otimes I_{n+1}) \qquad (3.6d)$$

$$\varphi(-B(\Omega) \oplus B(\Omega)^\top)df_B(\text{vec}_r(\Omega))\text{vec}_r(U) \qquad (3.6e)$$

$$\overset{(3.5)}{=} ((\text{expm}(TA^J(\Omega)), v_T(\Omega)) \otimes e_{n+1}^\top) \qquad (3.6f)$$

$$\varphi(-B(\Omega) \oplus B(\Omega)^\top)df_B(\text{vec}_r(\Omega))\text{vec}_r(U) \qquad (3.6g)$$

$$=: C(\Omega)\text{vec}_r(U) \qquad (3.6h)$$

where the right-hand side of equation (∗) is the vectorization of the left-hand side, i.e., the terms are rearranged, but the overall expression is unchanged. Since the left-hand side of (3.4) is a matrix inner product, we substitute the last equation into the right-hand side,

$$\langle \mathcal{L}(\Omega), U \rangle = df_{\mathcal{L}}(v_T(\Omega))C(\Omega)\mathrm{vec}_r(U) = \langle C(\Omega)^\top \partial f_{\mathcal{L}}(v_T(\Omega)), \mathrm{vec}_r(U) \rangle \quad (3.7)$$

and convert by vec_r^{-1} the linear form acting on $\mathrm{vec}_r(U)$.   □

Expression (3.3) is exact, but its evaluation is computationally infeasible for typical problem sizes. For example, $\varphi(-B(\Omega) \oplus B(\Omega)^\top)$ is a dense $(n+1)^2 \times (n+1)^2$ matrix, where $n = |I||J|$ is (number of pixels) × (number of labels). Therefore, approximations of the loss function gradient $\partial \mathcal{L}(\Omega)$ are studied next.

3.2   Approximation

Evaluating the gradient (3.3a) requires to multiply a large matrix by a vector,

$$C(\Omega)^\top \partial f_{\mathcal{L}}(v_T(\Omega)) = df_B(\mathrm{vec}_r(\Omega))^\top \varphi(-B(\Omega)^\top \oplus B(\Omega))\mathrm{vec}_r(F(\Omega)), \quad (3.8a)$$

$$F(\Omega) = \begin{pmatrix} \mathrm{expm}\left(TA^J(\Omega)^\top\right) \\ v_T(\Omega)^\top \end{pmatrix} \partial f_{\mathcal{L}}(v_T(\Omega))e_{n+1}^\top \quad (3.8b)$$

$$=: f_1 e_{n+1}^\top. \quad (3.8c)$$

The structure of this expression has the general form [5]

$$f(\mathcal{A})b = f(M_1 \oplus M_2)\mathrm{vec}(B). \quad (3.9)$$

Approximations for large problem sizes exploit the Kronecker sum structure of the matrix valued function f and the assumption that matrix B which generates the vector b has low rank. In our case (3.8a), function φ is applied to a Kronecker sum and the matrix $F(\Omega)$ (3.8b) has rank 1. Below, we *apply the approach* [5] *and refine it in terms of an additional approximation* that takes into account the structure of our problem (3.8).

The approach [5] amounts to determine two Krylov subspaces with orthonormal bases $P_m = (p_1, \ldots, p_m)$, $R_m = (r_1, \ldots, r_m)$ and the standard approximations [6, Section 13.2.1]

$$\mathcal{K}_m(-B(\Omega)^\top, f_1), \quad -B(\Omega)^\top P_m = P_m T_1 + T_{1;m+1,m}p_{m+1}e_m^\top, \quad (3.10a)$$

$$\mathcal{K}_m(B(\Omega), e_{n+1}), \quad B(\Omega)R_m = R_m T_2 + T_{2;m+1,m}r_{m+1}e_m^\top. \quad (3.10b)$$

Setting $U_m = P_m \otimes R_m$ yields

$$(-B(\Omega)^\top \oplus B(\Omega))U_m = -B(\Omega)^\top P_m \otimes R_m + P_m \otimes B(\Omega)R_m \qquad (3.11a)$$

$$\overset{(3.10)}{=} (P_m \otimes R_m)(T_1 \oplus T_2) + \text{low rank terms} \qquad (3.11b)$$

and with $T_m = T_1 \oplus T_2$ and $\mathcal{B}(\Omega) = -B(\Omega)^\top \oplus B(\Omega)$ the approximation of the matrix-vector product on the right-hand side of (3.8a)

$$\varphi(\mathcal{B}(\Omega))\mathrm{vec}_r(F(\Omega)) \approx U_m \varphi(T_m)U_m^\top \mathrm{vec}_r(F(\Omega)) \qquad (3.12a)$$

$$= U_m \varphi(T_m)\mathrm{vec}_r(P_m^\top F(\Omega)R_m) =: \mathrm{vec}_r(P_m Z R_m^\top), \qquad (3.12b)$$

where

$$Z = \mathrm{vec}_r^{-1}\big(\varphi(T_m)\mathrm{vec}_r(P_m^\top F(\Omega)R_m)\big) \qquad (3.13a)$$

$$\overset{(3.8c)}{=} \mathrm{vec}_r^{-1}\big(\varphi(T_m)\mathrm{vec}_r((P_m^\top f_1)(R_m^\top e_{n+1})^\top)\big). \qquad (3.13b)$$

The need to store $P_m Z R_m^\top \in \mathbb{R}^{n \times n}$ in the right-hand side of (3.12b) is problematic. Our **additional approximation**, therefore, exploits that fact that the matrix $F(\Omega)$ of (3.8) has rank 1 and that the singular values of $\varphi(T_m)$ typically decrease quickly. As a consequence, we directly approximate $Z \in \mathbb{R}^{m \times m}$ by a singular value decomposition (SVD),

$$Z \approx \sum_{i \in [r]} \sigma_i y_i \otimes z_i^\top, \qquad r \ll m, \qquad \text{to obtain} \qquad (3.14)$$

$$\varphi(\mathcal{B}(\Omega))\mathrm{vec}_r(F(\Omega)) \approx \mathrm{vec}_r\Big(P_m\Big(\sum_{i \in [r]} \sigma_i y_i \otimes z_i^\top\Big)R_m^\top\Big) \qquad (3.15a)$$

$$= \mathrm{vec}_r\Big(\sum_{i \in [r]} \sigma_i(P_m y_i) \otimes (R_m z_i)^\top\Big) = \sum_{i \in [r]} \sigma_i(P_m Y_i) \otimes (R_m Z_i). \qquad (3.15b)$$

The last expression shows that $U_m = P_m \otimes R_m$ does not have to be explicitly computed and stored. By leaving the Kronecker product unevaluated, the resulting vector (3.15b) can be conveniently multiplied with the Jacobian $df_B(\mathrm{vec}_r(\Omega))^\top$ of (3.8a). In addition, merely $\mathcal{O}(2rn)$ numbers have to be stored, which is feasible even for large problem sizes like 3D image labeling problems.

3.3   Computing the Gradient Using Automatic Differentiation

An entirely different approach for computing the gradient $\partial \mathcal{L}(\Omega)$ of the function (3.1) is using automatic differentiation. To this end, we implemented the explicit Euler integration of the linear assignment flow (2.7b)

$$V^{(k+1)} = V^{(k+1)} + h\left(A(\Omega)[V^{(k)}] + B_{W_0}\right), \qquad V^{(0)} = 0 \qquad (3.16)$$

in PyTorch and used PyTorch's automatic differentiation capabilities to compute the gradient. Similarly, we let PyTorch compute the gradient of the exponential integration (2.10b) of the linear assignment flow.

From a numerical point of view, we compare the approaches in Sect. 5.

4   Application to the Linear Assignment Flow

We provide further details on our implementation.

Loss Function. Because we can use trivial rounding to convert tangent vectors $V \in T_0$ and in turn assignment matrices $W \in \mathcal{W}$ to a labeling, only the direction of the tangent vectors is important, but not their length. Thus, we consider the angle between the solution of the linear assignment flow V and a ground truth direction $V^* = \mathrm{Exp}_{\mathbb{1}_\mathcal{W}}^{-1}(W^*)$, with ground truth labeling W^*, as loss function

$$f_{\mathcal{L}} : \mathbb{R}^n \longrightarrow \mathbb{R}, \qquad V \longmapsto 1 - \frac{\langle V^*, V \rangle}{\|V^*\| \|V\|}. \qquad (4.1)$$

We point out that our approach works with any C^1 loss function.

Riemannian Gradient Descent. As stated in Sect. 2.1, the parameters of the weight patches as nonzero entries of the row-stochastic matrix Ω of the linear assignment flow can be represented in the same way as the assignment vectors on the assignment manifold $\mathcal{W}$. Consequently, we convert the Euclidean gradient $\partial \mathcal{L}(\Omega)$ from Sect. 3 into the Riemannian gradient $\nabla \mathcal{L}(\Omega) = R_\Omega \partial \mathcal{L}(\Omega)$ using the mapping (2.4a) and perform Riemannian gradient descent

$$\Omega^{(k+1)} = \mathrm{exp}_{\Omega^{(k)}}(-h \nabla \mathcal{L}(\Omega^{(k)})), \quad k = 0, 1, 2, \ldots \qquad (4.2)$$

with step size $h > 0$. As initialization, we choose uniform weights $\Omega^{(0)} = \mathbb{1}_\mathcal{W}$.

Parameter Influence. Our implementation also takes into account the dependency of the vector b of (2.9a) and the block-matrices $R_{S_i(W_0)}$ of (2.9b) on $\Omega_i = (\omega_{ik})$. Due to lack of space, we did not include formulas of the corresponding gradient components in Sect. 3.

5   Experiments

Figures 2, 3 and 4 illustrate the applicability and performance of our approach. We refer to the captions for details.

A comparison of our algorithm and the two algorithms using backpropagation is depicted by Fig. 3. Generally speaking, regarding numerical computations (Fig. 3, left), the difference between the three algorithms is negligible compared to the influence of other hyperparameters like step size (learning rate) h for the

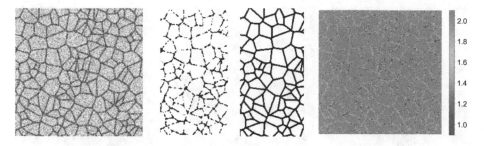

Fig. 2. Left: Noisy artificial vessel-like structures used as input data. **Center-left:** Left half of the labeling (labels: $J = \{\square, \blacksquare\}$) result using *uniform* weights. **Center-right:** The labeling result based on *learned* weights is very close to the ground truth ($<0.1\%$ wrong pixels). **Right:** Pseudo-color plot of entropies of learned weight patches for each center pixel. The algorithm learned where to denoise with uniform weights (high entropy) and where to enforce structure using non-uniform weights (low entropy).

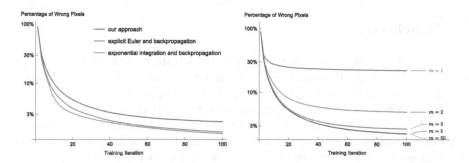

Fig. 3. Comparison of the loss for different algorithms and Krylov subspace dimensions. Left: We chose the Krylov subspace dimension resp. the discrete time step size such that all three algorithms required about the same computation time. Closeness of the curves demonstrates that all methods estimate the parameters accurately. The slightly worse accuracy of our algorithm is due to the fact that it uses an approximation of the gradient instead of the exact gradient of an approximated model. Our approach, however, is mathematically explicit and enjoys the pros displayed by Fig. 1. **Right:** For our approach to parameter learning based on the linear assignment flow, a Krylov subspace dimension of $m = 5$ turned out to be sufficient for most experiments, as illustrated by the plot.

gradient descent or neighborhood size $|\mathcal{N}_i|$, respectively. Our algorithm, however, has further advantageous properties as listed on page 3. In particular, it explicitly returns a subspace of low dimension m (Fig. 3, left) that will be useful for further tasks related to label prediction and flow control.

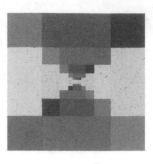

Fig. 4. Expressivity of the linear assignment flow. Left: Pure noise used as input data. **Center:** Labeling ($J = \{\blacksquare, \blacksquare, \blacksquare, \blacksquare, \blacksquare, \blacksquare, \blacksquare, \blacksquare\}$) with uniform weights returns more than 90% wrongly labeled pixels. **Right:** Our learning approach takes less than 10 s and produces a labeling at multiple spatial scales with well below 1% wrongly labeled pixels. These wrong label assignments can be corrected by choosing a larger neighborhood size or by iteratively applying the assignment flow again.

6   Conclusion

We presented a novel approach for learning the parameters of the linear assignment flow for image labeling. The algorithm approximates the computationally infeasible exact gradient of a linear dynamical system with respect to the regularizing weight parameters, using exponential integration, low-rank approximation and sparse matrix-vector multiplications. Regarding runtime, our research implementation is on par with highly tuned-machine learning toolboxes, and it additionally returns the essential information for image labeling in terms of a low-dimensional parameter subspace. Our future work will study these basic problems related to label prediction and flow control.

Acknowledgments. This work is supported by the Deutsche Forschungsgemeinschaft (DFG, German Research Foundation) under Germany's Excellence Strategy EXC 2181/1 - 390900948 (the Heidelberg STRUCTURES Excellence Cluster).

References

1. Abadi, M., et al.: TensorFlow: large-scale machine learning on heterogeneous systems. In: OSDI (2016)
2. Al-Mohy, A.H., Higham, N.J.: Computing the Fréchet derivative of the matrix exponential, with an application to condition number estimation. SIAM J. Matrix Anal. Appl. **30**(4), 1639–1657 (2009)
3. Åström, F., Petra, S., Schmitzer, B., Schnörr, C.: Image labeling by assignment. J. Math. Imaging Vis. **58**(2), 211–238 (2017). https://doi.org/10.1007/s10851-016-0702-4
4. Baydin, A., Pearlmutter, B., Radul, A., Siskind, J.: Automatic differentiation in machine learning: a survey. J. Mach. Learn. Res. **18**, 1–43 (2018)
5. Benzi, M., Simoncini, V.: Approximation of functions of large matrices with Kronecker structure. Numerische Mathematik **135**(1), 1–26 (2017)

6. Higham, N.J.: Functions of Matrices: Theory and Computation. SIAM, USA (2008)
7. Hochbruck, M., Lubich, C.: On Krylov subspace approximations to the matrix exponential operator. SIAM J. Numer. Anal. **34**(5), 1911–1925 (1997)
8. Hühnerbein, R., Savarino, F., Petra, S., Schnörr, C.: Learning adaptive regularization for image labeling using geometric assignment. J. Math. Imaging Vis. **63**, 186–215 (2021)
9. Kandolf, P., Koskela, A., Relton, S.D., Schweitzer, M.: Computing low-rank approximations of the Fréchet derivative of a matrix function using Krylov subspace methods. arXiv:2008.12926 (2020)
10. Moler, C., Loan, C.V.: Nineteen dubious ways to compute the exponential of a matrix, twenty-five years later. SIAM Rev. **45**(1), 3–49 (2003)
11. Najfeld, I., Havel, T.F.: Derivative of the matrix exponential and their computation. Adv. Appl. Math. **16**(3), 321–375 (1995)
12. Niesen, J., Wright, W.M.: Algorithm 919: a Krylov subspace algorithm for evaluating the φ-functions appearing in exponential integrators. ACM Trans. Math. Softw. **38**(3), 1–19 (2012)
13. Paszke, A., et al.: PyTorch: an imperative style, high-performance deep learning library. In: NIPS (2019)
14. Saad, Y.: Iterative Methods for Sparse Linear Systems. SIAM, USA (2003)
15. Schnörr, C.: Assignment flows. In: Grohs, P., Holler, M., Weinmann, A. (eds.) Handbook of Variational Methods for Nonlinear Geometric Data, pp. 235–260. Springer, Cham (2020). https://doi.org/10.1007/978-3-030-31351-7_8
16. Van Loan, C.F.: The ubiquitous Kronecker product. J. Comput. Appl. Math. **123**, 85–100 (2000)
17. Zeilmann, A., Savarino, F., Petra, S., Schnörr, C.: Geometric numerical integration of the assignment flow. Inverse Probl. **36**(3), 034003 (2020)
18. Zern, A., Zeilmann, A., Schnörr, C.: assignment flows for data labeling on graphs: convergence and stability. arXiv:2002.11571 (20 February 2020)

On the Geometric Mechanics of Assignment Flows for Metric Data Labeling

Fabrizio Savarino[1]([⊠])[iD], Peter Albers[2][iD], and Christoph Schnörr[1][iD]

[1] Institute of Applied Mathematics, Heidelberg University, Heidelberg, Germany
`fabrizio.savarino@iwr.uni-heidelberg.de`
[2] Mathematical Institute, Heidelberg University, Heidelberg, Germany

Abstract. Assignment flows are a general class of dynamical models for context dependent data classification on graphs. These flows evolve on the product manifold of probability simplices, called assignment manifold, and are governed by a system of coupled replicator equations. In this paper, we adopt the general viewpoint of Lagrangian mechanics on manifolds and show that assignment flows satisfy the Euler-Lagrange equations associated with an action functional. Besides providing a novel interpretation of assignment flows, our result rectifies the analogous statement of a recent paper devoted to uncoupled replicator equations evolving on a single simplex, and generalizes it to coupled replicator equations and assignment flows.

Keywords: Action functional · Assignment flows · Image labeling · Replicator equation · Evolutionary game dynamics

1 Introduction

Assignment flows, originally introduced by [4], are a general class of dynamical models evolving on a statistical manifold $\mathcal{W}$, called *assignment manifold*, for context dependent data classification on graphs. We refer to [13] for a recent survey on assignment flows and related work.

This approach is formulated for a general graph $\mathcal{G} = (\mathcal{V}, \mathcal{E})$ and can be summarized as follows. Assume for every node $i \in \mathcal{V}$ some data point f_i in a metric space $(\mathcal{F}, d_{\mathcal{F}})$ to be given, together with a set $\mathcal{F}_* = \{f_1^*, \dots, f_n^*\} \subset \mathcal{F}$ of predefined prototypes, also called *labels*. Context based *metric data labeling* refers to the task of assigning to each node $i \in \mathcal{V}$ a suitable label in $\mathcal{F}_*$ based on the metric distance to the given data f_i and the relation between data points encoded by the edge set $\mathcal{E}$.

In order to derive a geometric representation of this problem, the discrete label choice at each node $i \in \mathcal{V}$ is relaxed to a probability distribution over the label space $\mathcal{F}_*$ with full support, represented as a point on the manifold

$$\mathcal{S} := \{p \in \mathbb{R}^n : p > 0 \text{ and } \langle p, \mathbb{1}_n \rangle = 1\}. \tag{1.1}$$

A. Elmoataz et al. (Eds.): SSVM 2021, LNCS 12679, pp. 398–410, 2021.
https://doi.org/10.1007/978-3-030-75549-2_32

Accordingly, all probabilistic label choices on the graph are encoded as a single point $W \in \mathcal{W}$ on the assignment manifold

$$\mathcal{W} := \mathcal{S} \times \ldots \times \mathcal{S} \quad (m := |\mathcal{V}| \text{ factors}), \tag{1.2}$$

where the i-th component of $W = (W_k)_{k \in \mathcal{V}}$ represents the probability distribution of label assignments $W_i = (W_i^1, \ldots, W_i^n)^{\top} \in \mathcal{S}$ for the node $i \in \mathcal{V}$. Assignment flows are dynamical systems on $\mathcal{W}$ for inferring probabilistic label assignments that take the form of coupled replicator equations (see Sect. 4)

$$\dot{W}(t) = \mathcal{R}_{W(t)}[F(W(t))], \quad \text{with} \quad W(t) \in \mathcal{W}, \tag{1.3}$$

where the initial condition $W(0) \in \mathcal{W}$ contains information about the given data points $f_i \in \mathcal{F}$, $i \in \mathcal{V}$. These flows are derived by information geometric principles and usually consist of two interacting processes: non-local regularization of probabilistic label assignments and gradually enforcing unambiguous local decisions at every node $i \in \mathcal{V}$.

In [10, Thm. 2.1], the authors claim that all *uncoupled* replicator equations, i.e. $\dot{p} = R_p F(p)$, on a single simplex, $p(t) \in \mathcal{S}$, satisfy the Euler-Lagrange equation associated with the cost functional (again, see Sect. 4 for more details)

$$\mathcal{L}(p) := \int_{t_0}^{t_1} \frac{1}{2}\|\dot{p}(t)\|_g^2 + \frac{1}{2}\|R_{p(t)}F(p(t))\|_g^2 dt \quad \text{for curves } p: [t_0, t_1] \to \mathcal{S}. \tag{1.4}$$

In this paper, we (i) generalize this result to assignment flows and (ii) show that, in contrast to the claim of [10], the mentioned relation to extremal points of (1.4) holds if and only if condition (1.7) is fulfilled. Unlike the approach taken in [10], we derive this generalization from the more general viewpoint of Lagrangian mechanics on manifolds. This results in a better interpretable version of the Euler-Lagrange equation and leads to a characterization of critical points of the functional in terms of the function F governing the coupled replicator dynamics (1.3). Our main result is summarized in the following theorem.

Theorem 1. *Suppose $F: U \to \mathbb{R}^{m \times n}$ is a fitness function defined on an open set $U \subset \mathbb{R}^{m \times n}$ containing $\mathcal{W}$. If $W: I = [t_0, t_1] \to \mathcal{W}$ is a solution of the assignment flow (1.3), then $W(t)$ is a critical point of the action functional*

$$\mathcal{L}(W) = \int_{t_0}^{t_1} \frac{1}{2}\|\dot{W}(t)\|_g^2 + \frac{1}{2}\sum_{i \in \mathcal{V}} \text{Var}_{W_i(t)}\big(F_i(W(t))\big) dt, \tag{1.5}$$

that is, $W(t)$ fulfills the Euler-Lagrange equation

$$D_t^g \dot{W}(t) = \frac{1}{2}\sum_{i \in \mathcal{V}} \text{grad}^g \text{Var}_{W_i(t)}\big(F_i(W(t))\big) \quad \text{for } t \in I = [t_0, t_1], \tag{1.6}$$

if and only if the fitness function F fulfills the condition

$$0 = \mathcal{R}_{W(t)} \circ \big(dF|_{W(t)} - (dF|_{W(t)})^*\big) \circ \mathcal{R}_{W(t)}[F(W(t))], \quad \text{for } t \in I = [t_0, t_1], \tag{1.7}$$

where $(dF|_{W(t)})^*$ is the adjoint linear operator of $dF|_{W(t)}: \mathbb{R}^{m \times n} \to \mathbb{R}^{m \times n}$ with respect to the Frobenius inner product and $\mathcal{R}_{W(t)}$ is defined in (4.7).

The paper is organized as follows. In Sect. 2, we introduce our notation and list the necessary ingredients from differential geometry. Section 3 summarizes the required theory of Lagrangian systems on manifolds. Basic properties of assignment manifolds and flows are presented in Sect. 4, followed by the proof of Theorem 1 together with a counter example for the general claims of [10].

2 Preliminaries

Basic Notation. In accordance with the standard notation in differential geometry, coordinates of vectors have upper indices. For any $k \in \mathbb{N}$, we define $[k] := \{1, \ldots, k\} \subset \mathbb{N}$. The standard basis of $\mathbb{R}^d$ is denoted by $\{e_1, \ldots, e_d\}$ and we set $\mathbb{1}_d := (1, \ldots, 1)^\top \in \mathbb{R}^d$. The notation $\langle \cdot, \cdot \rangle$ is used for both, the standard and Frobenius inner product between vectors and matrices respectively. The identity matrix is denoted by $I_d \in \mathbb{R}^{d \times d}$ and the i-th row vector of any matrix A by A_i. The linear dependence of a function F on its argument x is indicated by square brackets $F[x]$. If x is a vector and F a matrix, then Fx is used instead of $F[x]$. For $a, b \in \mathbb{R}^d$, we denote componentwise multiplication (Hadamard product) by $a \diamond b := \mathrm{Diag}(a)b = (a^1 b^1, \ldots, a^d b^d)^\top$ and division, for $b > 0$, simply by $\frac{a}{b} = (\frac{a^1}{b^1}, \ldots, \frac{a^d}{b^d})^\top$. Similarly, inequalities between vectors or matrices are to be understood componentwise. We further set $a^{\diamond k} := a^{\diamond(k-1)} \diamond a$ with $a^{\diamond 0} := \mathbb{1}_d$. For later reference, we record the following statement here.

Lemma 1. *Assume for each $i \in [k]$ a matrix $Q^i \in \mathbb{R}^{d \times d}$ is given and let $\mathcal{Q} \colon \mathbb{R}^{k \times d} \to \mathbb{R}^{k \times d}$ be the linear map defined by $(\mathcal{Q}[X])_i := Q^i X_i$ for all rows $i \in [k]$. Then, the adjoint linear map $\mathcal{Q}^*$ with respect to the Frobenius inner product is given by $(\mathcal{Q}^*[Y])_i = Q^{i\top} Y_i$ for all $i \in [k]$.*

Proof. This is a direct consequence of $\langle X, \mathcal{Q}^*[Y] \rangle = \sum_{i \in [k]} \langle X_i, (\mathcal{Q}^*[Y])_i \rangle$ and $\langle X, \mathcal{Q}^*[Y] \rangle = \langle \mathcal{Q}[X], Y \rangle = \sum_{i \in [k]} \langle Q^i X_i, Y_i \rangle = \sum_{i \in [k]} \langle X_i, Q^{i\top} Y_i \rangle$ for arbitrary matrices $X, Y \in \mathbb{R}^{k \times d}$. □

Differential Geometry. We assume the reader is familiar with the basic concepts of Riemannian and symplectic manifolds as introduced in standard textbooks, e.g. [8,9] or [7]. The term "manifold" always means smooth manifold. The tangent and cotangent bundles of a d-dimensional manifold M are $TM = \cup_{x \in M} \{x\} \times T_x M$ and $T^* M = \cup_{x \in M} \{x\} \times T_x^* M$, together with their natural projections $\pi \colon TM \to M$ and $\pi^* \colon T^* M \to M$, sending $(x, v) \in TM$ and $(x, \alpha) \in T^* M$ to x. For local coordinates $(x^1, \ldots, x^d)$ on M, a tangent vector $v \in T_x M$ in these coordinates takes the form $v = \sum_{i \in [d]} v^i \frac{\partial}{\partial x^i}|_x$. The differential of a smooth map between manifolds $F \colon M \to N$ at $x \in M$ applied to a vector $v \in T_x M$ is denoted by $dF|_x[v]$. As usual (see e.g. [2, Sec. 3.5.7]), if $M \subset V$ is an embedded submanifold of a vector space V, such as $\mathbb{R}^d$ or $\mathbb{R}^{k \times d}$, then the tangent space at $x \in M$ is identified with the set of velocities of curves through x and, by abuse of notation, we again use $T_x M$ to denote this space

$$T_x M = \{\dot{\gamma}(0) \in V \colon \gamma \text{ curve in } M \text{ with } \gamma(0) = x\}. \tag{2.1}$$

If N is another submanifold of a vector space V', then the differential $dF|_x[v]$ of a map $F\colon M \to N$ at $x \in M$ can be calculated via a curve $\eta\colon (-\varepsilon, \varepsilon) \to M$, with $\eta(0) = x$ and $\dot{\eta}(0) = v \in T_x M$, by $dF|_x[v] = \frac{d}{dt} F(\eta(t))|_{t=0}$. Let $I \subset \mathbb{R}$ be an interval. If $\gamma\colon I \to TM$ is an integral curve of a vector field X on TM (or T^*M), i.e. $\dot{\gamma}(t) = X(\gamma(t))$, then $\pi \circ \gamma\colon I \to M$ (or $\pi^* \circ \gamma$) is called the *base integral curve*. For a Riemannian metric h (and similarly for a symplectic form ω) on M, there is a canonical isomorphisms $h^\flat\colon TM \to T^*M$, given by sending a tangent vector $v \in T_x M$ to the one-form $h_x(v, \cdot)\colon T_x^* M \to \mathbb{R}$. Its inverse $h^\sharp\colon T^*M \to TM$ sends a one-form $\alpha \in T_x^* M$ to a unique vector $v_\alpha \in T_x M$ such that $\alpha = h^\flat(v_\alpha) = h(v_\alpha, \cdot)$ holds. In particular, the Riemannian gradient of a function $f\colon M \to \mathbb{R}$ is defined as $\operatorname{grad}^h f := h^\sharp(df)$, where $df = \sum_i \frac{\partial f}{\partial x^i} dx^i$ is the differential of f. Thus, for all $x \in M$, $\operatorname{grad}^h f(x)$ is the unique vector with

$$df|_x[v] = h_x(\operatorname{grad}^h f(x), v), \quad \text{for all } v \in T_x M. \tag{2.2}$$

Furthermore, the Riemannian norm for $v \in T_x M$ is denoted by $\|v\|_h := \sqrt{h_x(v, v)}$ and the covariant derivative (with respect to the Riemannian metric h) of vector fields along curves by D_t^h.

3 Lagrangian Systems on Manifolds

In this section, we give a brief summary of Lagrangian systems on manifolds and mechanics on Riemannian manifolds from [1, Ch. 3].

Suppose M is a smooth manifold. Similar to Hamiltonian systems on momentum phase space T^*M, there is a related concept on the tangent bundle TM, interpreted as *velocity phase space*. In this context, a smooth function $L\colon TM \to \mathbb{R}$ is called *Lagrangian*. For a given point $x \in M$, denote the restriction of L to the fiber $T_x M$ by $L_x := L|_{T_x M}\colon T_x M \to \mathbb{R}$. The *fiber derivative* of L is defined as

$$\mathbb{F}L\colon TM \to T^*M, \quad (x, v) \mapsto \mathbb{F}L(x, v) := dL_x|_v, \tag{3.1}$$

where $dL_x|_v\colon T_x M \to \mathbb{R}$ is the differential of L_x at $v \in T_x M$. The function L is called a *regular Lagrangian* if $\mathbb{F}L$ is regular at all points (meaning that $\mathbb{F}L$ is a submersion), which is equivalent to $\mathbb{F}L\colon TM \to T^*M$ being a local diffeomorphism by [1, Prop. 3.5.9]. Furthermore, L is called *hyperregular Lagrangian* if $\mathbb{F}L\colon TM \to T^*M$ is a diffeomorphism. A class of hpyerregular Lagrangians, including the Lagrangian from Theorem 1, is given in (3.6) below.

The *Lagrange two-form* is defined as the pullback $\omega_L := (\mathbb{F}L)^* \omega^{\mathrm{can}}$ of the canonical symplectic form ω^{can} on the cotangent bundle T^*M under the fiber derivative $\mathbb{F}L$. According to [1, Prop. 3.5.9], ω_L is a symplectic form on T^*M if and only if L is a regular Lagrangian. In the following, we only consider regular Lagrangians. The *action* associated to the Lagrangian $L\colon TM \to \mathbb{R}$ is defined by

$$A\colon TM \to \mathbb{R}, \quad (x, v) \mapsto \mathbb{F}L(x, v)[v] = dL_x|_v[v], \tag{3.2}$$

and the *energy function* by $E := A - L$, having the form

$$E \colon TM \to \mathbb{R}, \quad (x, v) \mapsto \mathbb{F}L(x, v)[v] - L(x, v) = dL_x|_v[v] - L(x, v). \qquad (3.3)$$

The *Lagrangian vector field for* L is the unique vector field X_E on TM satisfying

$$dE|_{(x,v)}[u] = \omega_{L,(x,v)}(X_E, u) \quad \text{for all } (x, v) \in T_x M \text{ and } u \in T_{(x,v)}TM, \qquad (3.4)$$

that is $X_E = \omega_L^\sharp(dE)$. A curve $\gamma(t) = (x(t), v(t))$ on TM is an integral curve of X_E if $v(t) = \dot{x}(t)$ and the classical Euler-Lagrange equations in local coordinates are satisfied

$$\frac{d}{dt}\left(\frac{\partial L}{\partial v^i}(x(t), \dot{x}(t))\right) = \frac{\partial L}{\partial x^i}(x(t), \dot{x}(t)) \quad \text{for all } i \in [n]. \qquad (3.5)$$

Let $\gamma \colon I \to TM$ be any integral curve of X_E. Because of $\frac{d}{dt}E(\gamma) = dE|_\gamma[\dot{\gamma}] = dE|_\gamma[X_E(\gamma)] = \omega_{L,\gamma}(X_E(\gamma), X_E(\gamma)) = 0$, the energy E is constant along γ.

Now, assume (M, h) is a Riemannian manifold. Suppose a smooth function $G \colon M \to \mathbb{R}$, called *potential*, is given and consider the Lagrangian

$$L(x, v) := \frac{1}{2}\|v\|_h^2 - G(x), \quad \forall (x, v) \in TM. \qquad (3.6)$$

It then follows (see [1, Sec. 3.7] or by direct computation) that the fiber derivative of L is the canonical isomorphism $\mathbb{F}L = h^\flat \colon TM \to T^*M$. Hence, the Lagrangian L is hyperregular with action A and energy $E = A - L$ given by

$$A(x, v) = \|v\|_h^2 \quad \text{and} \quad E(x, v) = \frac{1}{2}\|v\|_h^2 + G(x) \quad \text{for all } (x, v) \in TM. \qquad (3.7)$$

Proposition 1. *([1, Prop. 3.7.4]). With L as defined in (3.6) on the Riemannian manifold (M, h), the curve $\gamma \colon I \to TM$ with $\gamma(t) = (x(t), v(t))$ is an integral curve of the Lagrangian vector field X_E, i.e. satisfies the Euler-Lagrange equation, if and only if the base integral curve $\pi \circ \gamma = x \colon I \to M$ satisfies*

$$D_t^h \dot{x}(t) = -\mathrm{grad}^h G(x(t)). \qquad (3.8)$$

4   Mechanics of Assignment Flows

We now return to the metric data labeling task on a graph $\mathcal{G} = (\mathcal{V}, \mathcal{E})$ from the beginning of this paper. In this section, we consistently use the notation $m = |\mathcal{V}|$ and $n = |\mathcal{F}_*|$ for the number of nodes and labels respectively.

We first give a brief summary of the most important properties of the statistical manifold $\mathcal{W}$, followed by a short description of assignment flows. A more detailed overview can be found in the original work [4] or the recent survey [13]. After this, we apply the general theory of Lagrangian Systems from Sect. 3 to prove our main result stated as Theorem 1.

4.1   Assignment Manifold and Flows

Assignment Manifold. In the following, we always identify the manifold $\mathcal{W}$ from (1.2) with its matrix embedding

$$\mathcal{W} = \{W \in \mathbb{R}^{m \times n} : W > 0 \text{ and } W \mathbb{1}_n = \mathbb{1}_m\}, \tag{4.1}$$

by sending the i-th component W_i of $W = (W_k)_{k \in \mathcal{V}} \in \mathcal{W}$ to the i-th row of a matrix in $\mathbb{R}^{m \times n}$. Therefore, points $W \in \mathcal{W}$ are row stochastic matrices with full support, called *assignment matrices*, with row vectors $W_i = (W_i^1, \ldots, W_i^n)^\top \in \mathcal{S}$ representing the relaxed label assignment for every $i \in [m]$. With the identification from (2.1), the tangent space of $\mathcal{S} \subset \mathbb{R}^n$ from (1.1) at any point $p \in \mathcal{S}$ is identified as

$$T_p\mathcal{S} = \{v \in \mathbb{R}^n : \langle v, \mathbb{1}_n \rangle = 0\} =: T. \tag{4.2}$$

Hence, $T_p\mathcal{S}$ is represented by the same vector space T for all $p \in \mathcal{S}$. In particular, the tangent bundle is trivial $T\mathcal{S} = \mathcal{S} \times T$. Viewing $\mathcal{W}$ as an embedded submanifold of $\mathbb{R}^{m \times n}$ by (4.1) and using the identification (2.1) for the tangent space, we identify

$$T_W\mathcal{W} = \{V \in \mathbb{R}^{m \times n} : V\mathbb{1}_n = 0\} =: \mathcal{T}, \quad \text{for all } W \in \mathcal{W} \subset \mathbb{R}^{m \times n}. \tag{4.3}$$

With this identification, the tangent bundle is also trivial $T\mathcal{W} = \mathcal{W} \times \mathcal{T}$.

From an information geometric viewpoint, e.g. [3] or [5], the Fisher-Rao (information) metric is a "canonical" Riemannian structure on $\mathcal{S}$, given by

$$g_p(u, v) := \langle u, \mathrm{Diag}(\tfrac{1}{p})v \rangle, \quad \text{for all } p \in \mathcal{S}, u, v \in T = T_p\mathcal{S}. \tag{4.4}$$

Next, we define two important matrices, the orthogonal projection of $\mathbb{R}^n$ onto T with respect to the Euclidean inner product

$$P_T := I_n - \tfrac{1}{n}\mathbb{1}_n\mathbb{1}_n \in \mathbb{R}^{n \times n} \quad \text{viewed as} \quad P_T : \mathbb{R}^n \to T \tag{4.5}$$

and for every $p \in \mathcal{S}$ the *replicator matrix*

$$R_p := \mathrm{Diag}(p) - pp^\top \in \mathbb{R}^{n \times n} \quad \text{viewed as} \quad R_p : \mathbb{R}^n \to T. \tag{4.6}$$

A simple calculation shows that $R_p = R_p P_T = P_T R_p$ as well as $\ker(R_p) = \mathbb{R}\mathbb{1}_n$ hold for all $p \in \mathcal{S}$. Furthermore, if R_p is restricted to the linear subspace $T \subset \mathbb{R}^n$, then $R_p|_T : T \to T$ is a linear isomorphism with inverse given by [12, Lem. 3.1]

$$(R_p|_T)^{-1}(u) = P_T \mathrm{Diag}(\tfrac{1}{p})u, \quad \text{for all } u \in T = T_p\mathcal{S}. \tag{4.7}$$

Now, suppose $J : \mathcal{S} \to \mathbb{R}$ is a smooth function defined on some open neighborhood U of $\mathcal{S}$, e.g. $U = \mathbb{R}^n_{>0}$. Then, according to [5, Prop. 2.2], the Riemannian gradient is given by $\mathrm{grad}^g J(p) = R_p \nabla J(p)$, for all $p \in \mathbb{R}^n$, where ∇J is the usual gradient of J on $U \subset \mathbb{R}^n$.

The product metric, again denoted by g, defined by

$$g_W(U, V) := \Sigma_{i \in [m]} g_{W_i}(U_i, V_i), \quad \text{for all } W \in \mathcal{W}, U, V \in \mathcal{T} = T_W\mathcal{W} \tag{4.8}$$

turns $\mathcal{W}$ into a Riemannian manifold. The orthogonal projection $\mathcal{P}_T \colon \mathbb{R}^{m \times n} \to \mathcal{T}$, $X \mapsto \mathcal{P}_T[X]$, with respect to the Frobenius inner product of matrices and, for each $W \in \mathcal{W}$, the replicator operator $\mathcal{R}_W \colon \mathbb{R}^{m \times n} \to \mathcal{T}$, $X \mapsto \mathcal{R}_W[X]$, are defined row-wise by

$$(\mathcal{P}_T[X])_i := P_T X_i \quad \text{and} \quad (\mathcal{R}_W[X])_i := R_{W_i} X_i \quad \text{for all } X \in \mathcal{T}, i \in [m]. \quad (4.9)$$

As a consequence, if a smooth function $J \colon \mathcal{W} \to \mathbb{R}$ is defined on some open neighborhood of $\mathcal{W}$, then the Riemannian gradient is given by

$$\operatorname{grad}^g J(W) = \mathcal{R}_W[\nabla J(W)] \in T_W \mathcal{W} = \mathcal{T}, \quad \text{for all } W \in \mathcal{W}, \quad (4.10)$$

where $\nabla J(W) \in \mathbb{R}^{m \times n}$ is the unique matrix fulfilling $dJ|_W[V] = \langle \nabla J(W), V \rangle$ for all $V \in \mathbb{R}^{m \times n}$. Therefore, $(\nabla J(W))_{ij} = \partial J / \partial W_i^j$, for all $i \in [m]$, $j \in [n]$.

Assignment Flows. The replicator equation is a well known differential equation for modeling various processes in fields such as biology, economy and evolutionary game dynamics, see [6] or [11]. In a typical game dynamics scenario, as described in [6], the labels correspond to different strategies of an agent playing a game and $p = (p^1, \ldots, p^n)^\top \in \mathcal{S}$ are the probabilities p^j of playing the j-th strategy, $j \in [n]$. The *fitness function* $F \colon \mathcal{S} \to \mathbb{R}^n$, also called *affinity measure*, represents the payoff $F^j(p)$ for each strategy j depending on the state $p \in \mathcal{S}$. The replicator equation is a consequence of the assumption that the growth rate $\dot{p}^j / p^j$ is given by the difference between the payoff $F^j(p)$ for strategy j and the average payoff $\sum_{k \in [n]} p^k F^k(p) = \langle F(p), p \rangle$, resulting in $\dot{p}^j = p^j(F^j(p) - \langle F(p), p \rangle)$. In vector notation, this can be written using the replicator matrix R_p from (4.6) as

$$\dot{p} = p \diamond F(p) - \langle F(p), p \rangle p = R_p F(p), \quad \text{for all } p \in \mathcal{S}.$$

The replicator dynamics therefore describes a selection process: over time, the agent selects successful strategies more often.

From this game dynamics perspective, assignment flows for data labeling can be seen as a game of interacting agents, where each node $i \in \mathcal{V}$ in the graph represents one agent and the strategies are the labels in $\mathcal{F}_*$. The fitness function (payoff) for node $i \in \mathcal{V}$ is a function $F_i \colon \mathcal{W} \to \mathbb{R}^n$ depending on the global label assignments $W \in \mathcal{W}$ and thereby coupling the label decisions between different nodes. Thus, for each $i \in \mathcal{V}$ the process of label selection on the corresponding simplex $\mathcal{S}$ is described by the replicator equation

$$\dot{W}_i = R_{W_i} F_i(W), \quad W_i(t) \in \mathcal{S},$$

coupled through the $F_i(W)$. In order to express this system of coupled replicator equations in a more compact way, we define the matrix valued fitness function $F \colon \mathcal{W} \to \mathbb{R}^{m \times n}$ with the i-th row given by $(F(W))_i := F_i(W)$. Together with the replicator operator $\mathcal{R}_W$ on $\mathcal{W}$ from (4.9), the coupled replicator equations are compactly expressed through (1.3). We again refer the reader to the survey [13] for applications of this framework to data labeling and related work.

4.2   Proof of Theorem 1

Let $I := [t_0, t_1]$ and suppose $F: U \to \mathbb{R}^{m \times n}$ is a fitness function defined on an open set $U \subset \mathbb{R}^{m \times n}$ containing $\mathcal{W}$. Since the squared Riemannian norm and the replicator operator are also defined on $\mathcal{W}$, the functional (1.4) from [10] can be easily extended to curves $W: I \to \mathcal{W}$ by simply replacing every occurrence of $p(t)$ with $W(t)$, resulting in

$$\mathcal{L}(W) := \int_{t_0}^{t_1} \frac{1}{2} \|\dot{W}(t)\|_g^2 + \frac{1}{2} \|\mathcal{R}_{W(t)}[F(W(t))]\|_g^2 dt. \qquad (4.11)$$

The term $\|\mathcal{R}_{W(t)}[F(W(t))]\|_g^2$ can be rewritten in a slightly more interpretable way. For this, we view the inner product between a vector $x \in \mathbb{R}^n$ and a point $p \in \mathcal{S}$ as the expected value $\langle x, p \rangle = \mathbb{E}_p[x]$ and similarly $\langle x^{\diamond 2}, p \rangle = \mathbb{E}_p[x^2]$. Thus, it is reasonable to talk about the variance of x with respect to p, given by

$$\mathrm{Var}_p(x) = \mathbb{E}_p[x^2] - (\mathbb{E}_p[x])^2 = \langle x^{\diamond 2}, p \rangle - \langle x, p \rangle^2. \qquad (4.12)$$

Lemma 2. *Let $p \in \mathcal{S}$ and $x \in \mathbb{R}^n$, then $\|R_p x\|_g^2 = \langle x, R_p x \rangle = \mathrm{Var}_p(x)$. Thus, for $W \in \mathcal{W}$ and $X \in \mathbb{R}^{m \times n}$, we have $\|\mathcal{R}_W[X]\|_g^2 = \sum_{i \in \mathcal{V}} \mathrm{Var}_{W_i}(X_i) = \langle X, \mathcal{R}_W[X] \rangle$.*

Proof. Since P_T is the orthogonal projection and $R_p x \in T$, the squared norm of the Fisher-Rao metric (4.4) is given by $\|R_p x\|_g^2 = \langle R_p x, P_T \mathrm{Diag}(1/p) R_p x \rangle$. As a result of $R_p = R_{p|T} P_T$ and the formula for the inverse of $R_{p|T}$ from (4.7), we have $P_T \mathrm{Diag}(\frac{1}{p}) R_p x = P_T \mathrm{Diag}(\frac{1}{p}) R_{p|T} P_T x = P_T x$. Therefore, $\|R_p\|_g^2 = \langle R_p x, P_T x \rangle = \langle R_p x, x \rangle$ follows. As a consequence of $R_p x = p \diamond x - \langle p, x \rangle p$ we also directly get $\langle x, R_p x \rangle = \langle x, p \diamond x - \langle p, x \rangle p \rangle = \langle x^{\diamond 2}, p \rangle - \langle x, p \rangle^2 = \mathrm{Var}_p(x)$. The statement for $\|\mathcal{R}_W[X]\|_g^2$ is a consequence of the product Riemannian metric (4.8) on $\mathcal{W}$ and the definition of $\mathcal{R}_W$ in (4.9) as a product map. $\qquad \square$

The result of the previous lemma explains the expression for $\mathcal{L}$ in Theorem 1. With this, we are in the regime of Lagrangian mechanics on Riemannian manifolds from Sect. 3 with $M = \mathcal{W}$, Riemannian metric $h = g$ and potential

$$G: \mathcal{W} \to \mathbb{R}, \quad G(W) := -\frac{1}{2} \|\mathcal{R}_W[F(W)]\|_g^2 = -\frac{1}{2} \sum_{k \in \mathcal{V}} \mathrm{Var}_{W_k}(F_k(W)). \qquad (4.13)$$

For $(W, V) \in T\mathcal{W} = \mathcal{W} \times T$, the corresponding Lagrangian (3.6) takes the form

$$L(W, V) = \frac{1}{2} \|V\|_g^2 - G(W) = \frac{1}{2} \|V\|_g^2 + \frac{1}{2} \sum_{k \in \mathcal{V}} \mathrm{Var}_{W_k}(F_k(W)).$$

Therefore, the Euler-Lagrange equation (1.6) in Theorem 1 is a direct consequence of Proposition 1. The corresponding energy function (3.7) takes the form $E(W(t), \dot{W}(t)) = \frac{1}{2} \|\dot{W}(t)\|_g^2 - \frac{1}{2} \|\mathcal{R}_{W(t)}[F(W(t))]\|_g^2$ and is constant along curves $W: I \to \mathcal{W}$ fulfilling the Euler-Lagrange equation (1.6). However, due to this specific form of the energy, it follows that $E(W(t), \dot{W}(t)) = 0$ holds for all assignment flows (1.3), irrespective of whether or not the Euler-Lagrange equation is satisfied. This fact was also reported in [10] for the uncoupled replicator dynamics on a single simplex.

In the remaining part, we derive the characterization (1.7) for which F the assignment flow fulfills the Euler-Lagrange equation (1.6). We start by considering $\mathcal{R}_W[F(W)]$ as a function of $W \in \mathcal{W}$, denoted by

$$\mathcal{R}[F]: \mathcal{W} \to \mathcal{T}, \quad W \mapsto \mathcal{R}[F](W) := \mathcal{R}_W[F(W)].$$

In order to calculate the differential of $\mathcal{R}[F]$, we define the $n \times n$-matrix

$$B(p, x) := \mathrm{Diag}(x) - \langle p, x \rangle I_n - px^\top, \quad \text{for } p \in \mathcal{S}, x \in \mathbb{R}^n \qquad (4.14)$$

and the linear map $\mathcal{B}(W, X): \mathbb{R}^{m \times n} \to \mathbb{R}^{m \times n}$ with i-th row

$$(\mathcal{B}(W, X)[V])_i := B(W_i, X_i)V_i, \quad \text{for } W \in \mathcal{W}, X \in \mathbb{R}^{m \times n} \qquad (4.15)$$

Lemma 3. *With the identifications $T_W\mathcal{W} = \mathcal{T}$ and $T_{\mathcal{R}_W[F(W)]}\mathcal{T} = \mathcal{T}$, the differential of $\mathcal{R}[F]$ is a linear map $d\mathcal{R}[F]|_W : \mathcal{T} \to \mathcal{T}$, given by*

$$d\mathcal{R}[F]\big|_W[V] = \mathcal{R}_W \circ dF|_W[V] + \mathcal{B}(W, F(W))[V], \quad \text{for } V \in \mathcal{T}.$$

Proof. A short calculation shows $\langle B(W_i, F_i(W))V_i, \mathbb{1}_n \rangle = 0$ for all $i \in \mathcal{V}$, proving that $\mathcal{B}(W, X)[V] \in \mathcal{T}$ holds. Let $\eta: (-\varepsilon, \varepsilon) \to \mathcal{W}$ be a curve with $\eta(0) = W$ and $\dot{\eta}(0) = V$. Keeping in mind $R_p = \mathrm{Diag}(p) - pp^\top$, we obtain for all rows $i \in \mathcal{V}$

$$\begin{aligned}
\left(d\mathcal{R}[F]|_W[V]\right)_i &= \tfrac{d}{dt}R_{\eta_i(t)}F_i(\eta(t))\big|_{t=0} = \tfrac{d}{dt}R_{\eta_i(t)}\big|_{t=0}F_i(W) + R_{W_i}\tfrac{d}{dt}F_i(\eta(t))\big|_{t=0} \\
&= \left(\mathrm{Diag}(V_i) - V_iW_i^\top - W_iV_i^\top\right)F_i(W) + \left(\mathcal{R}_W[\tfrac{d}{dt}F(\eta(t))\big|_{t=0}]\right)_i \\
&= \left(\mathcal{B}(W, F(W))[V]\right)_i + \left(\mathcal{R}_W \circ dF|_W[V]\right)_i,
\end{aligned}$$

where $\mathrm{Diag}(V_i)F_i(W) = \mathrm{Diag}(F_i(W))V_i$ and $V_i^\top F_i(W) = F_i(W)^\top V_i$ was used for the last equality. $\qquad\square$

Next, we consider the acceleration of curves on $\mathcal{S}$ and $\mathcal{W}$ with respect to the Riemannian metric g, that is the covariant derivative D_t^g of their velocities. Due to $T\mathcal{S} = \mathcal{S} \times T$, we can view the velocity of a curve $p: I \to \mathcal{S}$ as a map $\dot{p}: I \to T$. As T is a vector space, we can also consider its second derivative $\ddot{p}: I \to T$. Using the expression from [5, Eq. (2.60)] (with α set to 0), the acceleration $D_t^g\dot{p}$ of p is related to $\ddot{p}$ by

$$D_t^g\dot{p}(t) = \ddot{p}(t) - \frac{1}{2}\frac{(\dot{p}(t))^{\diamond 2}}{p(t)} + \frac{1}{2}\|\dot{p}(t)\|_g^2 p(t) = \ddot{p}(t) - \frac{1}{2}A(p(t), \dot{p}(t)),$$

with $A: \mathcal{S} \times T \to T$ defined as $A(p, v) := \frac{1}{p}v^{\diamond 2} - \|v\|_g^2 p$. Similarly, as a consequence of $T\mathcal{W} = \mathcal{W} \times \mathcal{T}$, the velocity of a curve $W: I \to \mathcal{W}$ can be viewed as a map $\dot{W}: I \to \mathcal{T}$, allowing for the second derivative $\ddot{W}$. Since the covariant derivative on a product manifold equipped with a product metric is the componentwise application of the individual covariant derivatives, the acceleration of $W(t)$ on $\mathcal{W}$ has the form

$$D_t^g\dot{W}(t) = \ddot{W}(t) - \frac{1}{2}\mathcal{A}(W(t), \dot{W}(t)), \qquad (4.16)$$

with i-th row of $\mathcal{A}: \mathcal{W} \times \mathcal{T} \to \mathcal{T}$ given by $(\mathcal{A}(W, X))_i := A(W_i, X_i)$ from above.

Lemma 4. *Suppose $W : I \to \mathcal{S}$ is a solution of the assignment flow (1.3). Then, the acceleration of $W(t)$, that is the covariant derivative of $\dot{W}(t)$, takes the form*
$$D_t^g \dot{W}(t) = \mathcal{R}_{W(t)} \circ dF|_{W(t)} \circ \mathcal{R}_{W(t)}[F(W(t))] + \tfrac{1}{2}\mathcal{A}(W(t), \mathcal{R}_{W(t)}[F(W(t))]).$$

Proof. Since $W(t)$ is a solution of $\dot{W}(t) = \mathcal{R}_{W(t)}[F(W(t))]$, the second derivative $\ddot{W} = \frac{d}{dt}\dot{W}(t)$ takes the form (to simplify notation we drop the dependence on t)

$$\ddot{W} = \frac{d}{dt}\mathcal{R}_W[F(W)] = d\mathcal{R}[F]|_W[\dot{W}] \overset{\text{Lem. 3}}{=} \mathcal{R}_W \circ dF|_W[\dot{W}] + \mathcal{B}(W, F(W))[\dot{W}]$$

The first term on the right-hand side equals $\mathcal{R}_W \circ dF|_W \circ \mathcal{R}_W[F(W)]$ and the second term $\mathcal{B}(W, F(W))[\mathcal{R}_W[F(W)]]$, where $\mathcal{B}$ is defined in terms of the matrix B from (4.14). Thus, consider $B(p, x)R_p x$, for $p \in \mathcal{S}$ and $x \in \mathbb{R}^n$. The relations $\langle x, R_p x \rangle = \|R_p x\|_g^2$ from Lemma 2 and $R_p x = p \diamond (x - \langle p, x \rangle \mathbb{1}_n)$ give $B(p, x)R_p x = (x - \langle p, x \rangle \mathbb{1}_n) \diamond R_p x - \langle x, R_p x \rangle p = \tfrac{1}{p}(R_p x)^{\diamond 2} - \|R_p x\|_g^2 p = A(p, R_p X)$. This implies $\mathcal{B}(W, F(W))[\mathcal{R}_W[F(W)]] = \mathcal{A}(W, \mathcal{R}_W[F(W)])$ and results in the identity $\ddot{W} = \mathcal{R}_W \circ dF|_W \circ \mathcal{R}_W[F(W)] + \mathcal{A}(W, \mathcal{R}_W[F(W)])$. Plugging this expression for $\ddot{W}$ into the one for $D_t^g \dot{W}$ in (4.16) finishes the proof.   □

In the final step, we calculate the Riemannian gradient for the potential G from (4.13). Since F is defined on an open set $U \subset \mathbb{R}^{m \times n}$, with $\mathcal{W} \subset U$, we identify $T_X U = \mathbb{R}^{m \times n}$ and $T_{F(X)}\mathbb{R}^{m \times n} = \mathbb{R}^{m \times n}$ for all $X \in U$. Accordingly, the differential of F at X is a linear map $dF|_X : \mathbb{R}^{m \times n} \to \mathbb{R}^{m \times n}$ and its adjoint with respect to the Frobenius inner product on $\mathbb{R}^{m \times n}$ are denoted by $(dF|_X)^*$.

Lemma 5. *The Riemannian gradient of the potential G from (4.13) is given by* $\mathrm{grad}^g G(W) = -\mathcal{R}_W \circ (dF|_W)^* \circ \mathcal{R}_W[F(W)] - \tfrac{1}{2}\mathcal{A}(W, \mathcal{R}_W[F(W)])$, *for $W \in \mathcal{W}$*

Proof. Let $W \in \mathcal{W}$. Since the i-th row of $\mathcal{R}_W$ is given by symmetric matrices $R_{W_i} = \mathrm{Diag}(W_i) - W_i W_i^{\top}$, Lemma 1 implies $\mathcal{R}_W^* = \mathcal{R}_W$. Next, we calculate an expression for $\nabla G(W)$. For this, assume $V \in \mathbb{R}^{m \times n}$ is arbitrary and let $\eta : (-\varepsilon, \varepsilon) \to \mathcal{W}$ be a curve with $\eta(0) = W$ and $\dot{\eta}(0) = V$. Then

$$dG|_W[V] = \frac{d}{dt}G(\eta(t))|_{t=0} \overset{\text{Lem. 2}}{=} -\frac{1}{2}\frac{d}{dt}\langle F(\eta(t)), \mathcal{R}_{\eta(t)}[F(\eta(t))]\rangle \Big|_{t=0}$$
$$= -\frac{1}{2}\langle dF|_W[V], \mathcal{R}_W[F(W)]\rangle - \frac{1}{2}\langle F(W), d\mathcal{R}[F]|_W[V]\rangle.$$

With the expression for $d\mathcal{R}[F]|_W$ from Lemma 3 together with $\mathcal{R}_W^* = \mathcal{R}_W$, the second inner product takes the form

$$\langle F(W), d\mathcal{R}[F]|_W[V]\rangle = \langle F(W), \mathcal{R}_W \circ dF|_W[V]\rangle + \langle F(W), \mathcal{B}(W, F(W))[V]\rangle$$
$$= \langle (dF|_W)^* \circ \mathcal{R}_W[F(W)], V\rangle + \langle \mathcal{B}^*(W, F(W))[F(W)], V\rangle.$$

Substituting this formula back into the above expression for $dG|_W$ together with $\langle dF|_W[V], \mathcal{R}_W[F(W)]\rangle = \langle V, (dF|_W)^* \circ \mathcal{R}_W[F(W)]\rangle$ for the first inner product, results in $dG|_W[V] = \langle -(dF|_W)^* \circ \mathcal{R}_W[F(W)] - \tfrac{1}{2}\mathcal{B}^*(W, F(W))[F(W)], V\rangle$.

Since V is arbitrary, $\nabla G(W) = -(dF|_W)^* \circ \mathcal{R}_W[F(W)] - \frac{1}{2}\mathcal{B}^*(W, F(W))[F(W)]$ follows. Due to (4.10), the Riemannian gradient is given by

$$\mathrm{grad}^g G(W) = -\mathcal{R}_W \circ (dF|_W)^* \circ \mathcal{R}_W[F(W)] - \frac{1}{2}\mathcal{R}_W[\mathcal{B}^*(W, F(W))[F(W)]].$$

Because $\mathcal{B}$ is defined in terms of the matrix B from (4.14), the adjoint $\mathcal{B}^*$ is determined by $B^\top$ through Lemma 1. For $p \in \mathcal{S}$ and $x \in \mathbb{R}^n$, we have

$$
\begin{aligned}
R_p B^\top(p, x)x &= R_p\big(\mathrm{Diag}(x) - \langle p, x\rangle I_n - xp^\top\big)x = R_p\big(x^{\diamond 2} - 2\langle p, x\rangle x\big) \\
&= p \diamond x^{\diamond 2} - \langle x^{\diamond 2}, p\rangle p - 2\langle p, x\rangle x \diamond p + 2\langle p, x\rangle^2 p \\
&= \big(p \diamond x^{\diamond 2} - 2\langle p, x\rangle x \diamond p + \langle p, x\rangle p\big) - \big(\langle x^{\diamond 2}, p\rangle - \langle p, x\rangle\big)p \\
&= \frac{1}{p}\big(p \diamond x - \langle p, x\rangle p\big)^{\diamond 2} - \|R_p x\|_g^2 p = A(p, R_p x),
\end{aligned}
$$

where the relation $\langle p, x^{\diamond 2}\rangle - \langle p, x\rangle^2 = \mathrm{Var}_p(x) = \|R_p x\|_g^2$ from (4.12) and Lemma 2 was used in the last line. Therefore, $\mathcal{R}_W[\mathcal{B}^*(W, F(W))[F(W)]] = A(W, \mathcal{R}_W[F(W)])$ holds which proves the statement. □

Proof (Theorem 1). Suppose $W(t)$ is a solution of the assignment flow (1.3). Due to Lemma 4 and 5, the expression for the acceleration of $W(t)$ and the Riemannian gradient of G at $W(t)$ both contain the term $\frac{1}{2}A(W(t), \mathcal{R}_{W(t)}[F(W(t))])$ which yields the relation

$$
\begin{aligned}
D_t^g \dot{W}(t) - \frac{1}{2}\sum_{k \in \mathcal{V}} \mathrm{grad}^g \mathrm{Var}_{W_k}(F_k(W)) &\overset{(4.13)}{=} D_t^g \dot{W}(t) + \mathrm{grad}^g G(W) \\
&= \mathcal{R}_{W(t)} \circ dF|_{W(t)} \circ \mathcal{R}_{W(t)}[F(W(t))] - \mathcal{R}_{W(t)} \circ (dF|_{W(t)})^* \circ \mathcal{R}_{W(t)}[F(W(t))] \\
&= \mathcal{R}_{W(t)} \circ \big(dF|_{W(t)} - (dF|_{W(t)})^*\big) \circ \mathcal{R}_{W(t)} F(W(t)).
\end{aligned}
$$

As a consequence, the characterization of F in (1.7) is equivalent to the Euler-Lagrange equation (1.6). □

Remark 1. As can be seen from the expression of $D_t^g \dot{W}(t)$ in (4.16), the Euler-Lagrange equation is a second-order differential equation. The reason why all second- and first-order terms disappear in the condition (1.7) for F is due to the fact that any solution of the assignment flow satisfies $\dot{W}(t) = \mathcal{R}_{W(t)}[F(W(t))]$, allowing to replace any occurrences of $\ddot{W}$ and $\dot{W}$ by alternative expressions in terms of the replicator operator. This basically is the statement of Lemma 4.

4.3   Counterexample

It can be shown that in the case of $n = 2$ labels any fitness function F fulfills condition (1.7) and therefore also the Euler-Lagrange equation. However, for $n > 2$ labels this is no longer true in general, as the example below demonstrates. Nevertheless, a large class of fitness functions always fulfilling condition (1.7) is given by those defined as the gradient $F = \nabla \beta$ of an objective function β. Since the corresponding derivative $dF|_x = \text{Hess}\,\beta(x)$ is self-adjoint, the condition is trivially fulfilled.

For the counterexample, assume $n > 2$. We first consider the case of $m = |\mathcal{V}| = 1$ nodes, that is an uncoupled replicator equation on a single simplex. Define the matrix $F := e_2 e_1^\top$, where e_i are the standard basis vectors of $\mathbb{R}^n$. Thus, the fitness is a linear map $p = (p^1, \ldots, p^n)^\top \mapsto Fp = p^1 e_2$, fulfilling $dF|_p = F$ and $(dF|_p)^* = F^\top$. After a short calculation, using the relation $R_p e_i = p^i(e_i - p)$ (Einstein summation convention is *not* used), the first coordinate of condition (1.7) takes the form

$$\left(R_p(F - F^\top)R_p Fp\right)^1 = -(p^1)^2 p^2(1 - p^1 - p^2) \neq 0, \quad \text{for all } p \in \mathcal{S}.$$

In the more general case $m > 1$, define the i-th row of the linear fitness $\mathcal{F}[W]$ by $(\mathcal{F}[W])_i := FW_i$. Since $(\mathcal{F}^*[W])_i = F^\top W_i$ by Lemma 1, the counterexample also extends to general coupled replicator equations on $\mathcal{W}$.

5   Conclusion

Starting from the viewpoint of Lagrangian mechanics on manifolds, we showed that assignment flows solve the Euler-Lagrange equations associated with an action functional. We further characterized those solutions in terms of the fitness function Γ, which allowed to rectify the result of [10] for uncoupled replicator equations on a single simplex.

Regarding future work, there is a relation to Hamiltonian mechanics via the Legendre transformation, which enables to analyze assignment flows as systems of interacting particles from a physics point of view. There also exists a connection to geodesic motion for a modified Riemannian metric on $\mathcal{W}$, the so called *Jacobi metric*, that provides yet another way of characterizing assignment flows.

Acknowledgments. This work is supported by Deutsche Forschungsgemeinschaft (DFG, German Research Foundation) under Germany's Excellence Strategy EXC-2181/1 - 390900948 (the Heidelberg STRUCTURES Cluster of Excellence). PA was also supported by the Transregional Collaborative Research Center CRC/TRR 191 (281071066).

References

1. Abraham, R., Marsden, J.: Foundations of Mechanics, 2nd edn. Addison-Wesley Publishing Company Inc., Redwood City (1987)
2. Absil, P.A., Mahony, R., Sepulchre, R.: Optimization Algorithms on Matrix Manifolds. Princeton University Press (2008)
3. Amari, S.I., Nagaoka, H.: Methods of Information Geometry. vol. 191. American Mathematical Soc. (2007)
4. Aström, F., Petra, S., Schmitzer, B., Schnörr, C.: Image labeling by assignment. J. Math. Imag. Vision **58**(2), 211–238 (2017)

5. Ay, N., Jost, J., Lê, H.V., Schwachhöfer, L.: Information Geometry. EMG-FASMSM, vol. 64. Springer, Cham (2017). https://doi.org/10.1007/978-3-319-56478-4

6. Hofbauer, J., Sigmund, K.: Evolutionary game dynamics. Bull. Am. Math. Soc. **40**(4), 479–519 (2003)

7. Jost, J.: Riemannian Geometry and Geometric Analysis. U. Springer, Cham (2017). https://doi.org/10.1007/978-3-319-61860-9

8. Lee, J.M.: Smooth manifolds. In: Introduction to Smooth Manifolds. Graduate Texts in Mathematics, vol 218. Springer, New York, NY (2013). https://doi.org/10.1007/978-1-4419-9982-5_1

9. Lee, J.M.: Introduction to Riemannian Manifolds. vol. 2. Springer International Publishing (2018)

10. Raju, V., Krishnaprasad, P.: A variational problem on the probability simplex. In: 2018 IEEE Conference on Decision and Control (CDC), pp. 3522–3528. IEEE (2018)

11. Sandholm, W.H.: Population Games and Evolutionary Dynamics. MIT press (2010)

12. Savarino, F., Schnörr, C.: Continuous-domain assignment flows. Europ. J. Appl. Math. 1–28 (2020)

13. Schnörr, C.: Assignment Flows. In: Grohs, P., Holler, M., Weinmann, A. (eds.) Handbook of Variational Methods for Nonlinear Geometric Data. GTM, pp. 235–260. Springer, Cham (2020). https://doi.org/10.1007/978-3-030-31351-7_8

A Deep Image Prior Learning Algorithm for Joint Selective Segmentation and Registration

Liam Burrows[1], Ke Chen[1(✉)], and Francesco Torella[2]

[1] Department of Mathematical Sciences and Centre for Mathematical Imaging Techniques, University of Liverpool, Liverpool, UK
k.chen@liv.ac.uk
[2] Liverpool Vascular and Endovascular Service, Liverpool University Hospitals, Liverpool L7 8XP, UK
https://www.liv.ac.uk/~cmchenke

Abstract. Effective variational models exist for either image segmentation or image registration for a given class of problems, though robustness is a longstanding issue. This paper proposes a new and effective variational model that aims to segment a pair of images through a joint model with registration, with the advantage of only requiring geometric prior information on one image (instead of two images) and obtaining selective segmentation on both images. Moreover we develop a deep image prior based learning algorithm to achieve the same segmentation and registration results by dropping the regularisation terms from the loss function. Numerical experiments show quality results obtained from the new approach.

Keywords: Image segmentation · Image registration · Deep image prior

1 Introduction

Image registration and segmentation are two fundamental tasks in image processing. Typically they are treated as two separate processes, however some efforts have been made to combine the two into a single formulation [11,14]. The aim of segmentation is to identify meaningful objects in an image, through their given intensity distributions. We are particularly interested in selective segmentation [1,5,9,16] where a set $\mathcal{M}$ of geometric markers is given by the user and the aim is to segment the object nearest to or containing $\mathcal{M}$. Hence to selectively segment two images, we need two sets of respective markers sets $\mathcal{M}_j$, $j = 1, 2$ for each image. The aim of registration is to find a transformation $\mathbf{y}(\mathbf{x}) : \mathbb{R}^2 \to \mathbb{R}^2$ which maps a template image T to a reference image R, with $T, R \in \Omega \subset \mathbb{R}^2$, such that: $T(\mathbf{y}(\mathbf{x})) \approx R(\mathbf{x})$. Typically the transformation is written as $\mathbf{y}(\mathbf{x}) = \mathbf{x} + \mathbf{u}(\mathbf{x})$, where $\mathbf{u}(\mathbf{x})$ is the displacement vector field. Here, to focus on modeling segmentation and registration jointly, this mono-modality registration setup is sufficient for our study, assuming that images T, R are acquired

© Springer Nature Switzerland AG 2021
A. Elmoataz et al. (Eds.): SSVM 2021, LNCS 12679, pp. 411–422, 2021.
https://doi.org/10.1007/978-3-030-75549-2_33

from the same source (e.g. both are digital or both are CTs). However, for multi-modal images where T, R sensing the same object come from different sources (e.g. T from infra-red and R from a digital camera) we simply employ suitably chosen transforms f_1, f_2 (e.g. gradient operations) such that $f_1(T)(\mathbf{y}(\mathbf{x})) \approx f_2(R)(\mathbf{x})$.

For a joint model, intuitively, the segmentation (of corresponding features) may be able to guide and improve the registration process while registration can transfer an accurate segmentation result of the reference image to the template.

Our study is in the variational framework. Take the image T as an example to define our notation. We are interested in finding its feature enclosed by boundary contour Γ. As Γ is unknown, a widely used idea is to embed it in a level set function $\phi(\mathbf{x})$ and to find this $\phi(\mathbf{x})$ first. Then segmentation of image T amounts to partitioning the image domain $\Omega = \Omega_1 \cup \Omega_2 \cup \Gamma$ as follows [4]:

$$\begin{cases} \Gamma = \{(\mathbf{x} \in \Omega : \phi(\mathbf{x}) = 0\} \\ \Omega_1 := \text{Inside}(\Gamma) = \{\mathbf{x} \in \Omega : \phi(\mathbf{x}) > 0\} \\ \Omega_2 := \text{Outside}(\Gamma) = \{\mathbf{x} \in \Omega : \phi(\mathbf{x}) < 0\} \end{cases}$$

with the region of interest being Ω_1.

Since the advantage of a level set functions ϕ enabling to model arbitrary changes in Ω_1's topology is well known, their use in modeling segmentation or other interface problems is common. However, variational models involving ϕ are almost always non-convex because integration over Ω_1 is replaced by integration over Ω involving $H(\phi)$, where H is the Heaviside function, which is non-convex; likewise integration over Ω_2 will be by over Ω involving $1 - H(\phi)$. Non-convex models suffer from the problems of non-unique solutions and strong dependence on initialisation. This apparent disadvantage may be overcome by the idea of relaxation [3] where one solves for the new variable $\theta = H(\phi)$ instead of ϕ, and then relaxes the constraint $\theta \in \{0, 1\}$ (binary set) to $\theta \in [0, 1]$ (continuous interval). Then once we found θ, a threshold of it will yield the required segmentation i.e. domain Ω_1. This is the approach that we shall adopt in this work.

Assume that two images T, R are given and a shape or feature of T is given in terms of a binary image (segmented), which can be converted to a level set function representation ϕ_0. To design a joint model of registration and segmentation, we have to understand how a given segmentation of image T defined in domain Ω_1 via $H(\phi_0(\mathbf{x}))$ interacts with the segmentation of image R of domain Ω_1 via $H(\phi(\mathbf{x}))$, and how registration $T(\mathbf{x} + \mathbf{u})$ comes into the formulation. It turns out that the key step is to evolve $H(\phi_0(\mathbf{x} + \mathbf{u}))$ to guide segmentation of R, and not to set up a new and unknown level set $\phi(\mathbf{x})$ [11,14,21]. Then it becomes natural to combine registration and segmentation.

Finally there is much freedom in choosing a regulariser, either by hand-crafting or learning from data, for both types of problems. While this facilitates optimised tuning for specific applications, it is generally hard to design a robust regulariser. For instance, Chan-Vese type models (or others based on Mumford-Shah framework) minimise the length of segmented boundaries via $\nabla H(\phi)$, while one may regularise the Euler elastica of the boundary [26]. There are even more choices of regularisers for image registration ranging from H_1 semi-norm to Euler elastica. Our work will utilise the implicit regularisation offered by the architecture of a neural network.

This paper presents an algorithm that segments two related images and find a mapping registering these two images, by using only a single marker set $\mathcal{M}$. Our work differs from previous models [11,14,21] in three aspects: i) we employ a relaxed formulation for segmentation part; ii) we incorporate geometric markers of T only to segment both T and R selectively; iii) we propose a deep image prior based approach for an accurate solution of our joint registration and segmentation model.

2   A New Variational Model

In this section, we outline the proposed joint segmentation-registration problem in the variational setting. We propose to utilise the selective segmentation framework of [18] and re-purpose it for joint segmentation and registration using the ideas from [14] and [11] models.

Most variational approaches to joint registration segmentation models usually assume that the region of interest on T is already segmented and given. The task then is focused on finding the displacement map using the given segmented region of T as additional information to aid the registration process. For our proposed model, we assume no prior knowledge and aim to selectively segment an object in an image, and simultaneously register the image to another. As this section will demonstrate, models of this class involve many terms, resulting in many parameters to tune for the user. In the next section, we propose implementing our model into a deep image prior framework, in which we can remove two regularisation terms (one in the segmentation problem and one in the registration problem). The removal of such terms reduces the complexity of tuning parameters and makes a joint framework with no prior knowledge more viable. This section however introduces the joint problem with explicit regularisation.

In variational setting, the model which we propose is given as follows:

$$\min_{\theta,\mathbf{u}} \mu \int_\Omega g|\nabla\theta(\mathbf{x})|d\Omega + \lambda_1 \int_\Omega \Phi(T,a_1)\theta(\mathbf{x})d\Omega + \lambda_2 \int_\Omega \Phi(R,c_1)\theta(\mathbf{x}+\mathbf{u})d\Omega$$

$$+ \xi \int_\Omega \mathcal{D}(\mathbf{x})\theta(\mathbf{x})d\Omega + \frac{1}{2}\int_\Omega |T_\mathbf{u} - R(\mathbf{x})|^2 d\Omega + \frac{\alpha}{2}\sum_\ell^2 \int_\Omega |\nabla u_\ell|^2 d\Omega, \quad (1)$$

where $\theta \in [0,1]$, $T_\mathbf{u} = T(\mathbf{x}+\mathbf{u})$, $\Phi(f,a) = \mu_1(f,a)-\mu_2(f,a)$, and $g = g(|\nabla T|) = \frac{1}{1+\iota|\nabla T|^2}$ is an edge detector. Here, as in [18], we define

$$\mu_1(f,c_1) = (f-c_1)^2, \quad \mu_2(f,c_1) = \begin{cases} 1 + \frac{f(x)-c_1}{\gamma_1}, & c_1 - \gamma_1 \le f(x) \le c_1, \\ 1 - \frac{f(x)-c_1}{\gamma_2}, & c_1 < f(x) \le c_1 + \gamma_2, \\ 0, & \text{else}, \end{cases} \quad (2)$$

where γ_1 and γ_2 are parameters fixed as detailed in [18], and we fix a_1 to be the average intensity of T inside the region defined by the user input $\mathcal{M}$. The parameter c_1 represents the average intensity of the region of interest on the image R, and as we only concern ourselves with mono-modal images we make the assumption that $c_1 \approx a_1$, and therefore fix $c_1 = a_1$.

In (1), the selective constraint $\mathcal{D}$ is the geodesic distance as proposed in [17] used to put a penalty on objects outside the region defined by the user in the marker set $\mathcal{M}$. It is given by $\mathcal{D} = \frac{\mathcal{D}_0}{||\mathcal{D}_0||_{L^\infty}}$, where

$$\begin{cases} |\nabla \mathcal{D}_0(x)| = \varepsilon_D + \beta_D |\nabla T(x)|^2 + \xi_D \mathcal{D}_E(x), & x \in \Omega \\ \mathcal{D}_0(x) = 0, & x \in \mathcal{M}, \end{cases} \tag{3}$$

where $\mathcal{D}_E$ is the Euclidean distance from $\mathcal{M}$ and the parameters are fixed as in [17] to be $\varepsilon_D = 10^{-3}, \beta_D = 1000$ and $\xi_D = 0.1$. The solution to the Eikonal equation (3) can be solved quickly using fast sweeping or fast marching methods.

Once the optimisation is performed and a minimiser θ^* is found, we define the foreground region representing the region of interest as:

$$\Sigma = \{x \in \Omega : \theta^*(x) > \gamma\},$$

where it is typical to select $\gamma = 0.5$.

Below we show some details of solving (1) in alternating minimization.

- Minimisation with respect to $\mathbf{u}$ yields this Euler-Lagrange equations:

$$0 = \lambda_2 \Phi(R, c_1) \nabla_{\mathbf{u}} \theta(\mathbf{x} + \mathbf{u}) + (T(\mathbf{x} + \mathbf{u}) - R(\mathbf{x})) \nabla_{\mathbf{u}} T(\mathbf{x} + \mathbf{u}) - \alpha \Delta \mathbf{u}.$$

Implementing a gradient descent and semi-implicit finite difference scheme, utilising additive operator splitting (AOS) [24, 25], yields the following iterative scheme.

$$\begin{cases} u_1^{(k+1)} = \frac{1}{2} \sum_{l=1}^{2} (I - 2\alpha\tau A_l)^{-1} (u_1^{(k)} - \tau g_1(u_1^{(k)}, u_2^{(k)}, \theta^{(k)})), \\ u_2^{(k+1)} = \frac{1}{2} \sum_{l=1}^{2} (I - 2\alpha\tau A_l)^{-1} (u_2^{(k)} - \tau g_2(u_1^{(k)}, u_2^{(k)}, \theta^{(k)})), \end{cases} \tag{4}$$

where $g_l(u_1, u_2, \theta) = \lambda_2 \Phi(R, c_1) \partial_{u_l} \theta(\mathbf{x} + \mathbf{u})(T(\mathbf{x} + \mathbf{u}) - R(\mathbf{x})) \partial_{u_l} T(\mathbf{x} + \mathbf{u})$, $l = 1, 2$, A_l is the discretisation of the Laplace operator Δ along the l−coordinate direction.

- The θ subproblem of (1), before deriving the Euler-Lagrange equation, is:

$$\min_{\theta \in [0,1]} \mu \int_\Omega g|\nabla\theta(\mathbf{x})|d\Omega + \lambda_1 \int_\Omega \Phi(T(\mathbf{x}), a_1)\theta(\mathbf{x})d\Omega + \lambda_2 \int_\Omega \Phi(R(\mathbf{x} - \mathbf{u}), c_1)\theta(\mathbf{x})d\Omega$$

$$+ \xi \int_\Omega \mathcal{D}(\mathbf{x})\theta(\mathbf{x})d\Omega, \qquad \Longrightarrow$$

$$0 = \lambda_1 \Phi(T(\mathbf{x}), a_1) + \lambda_2 \Phi(R(\mathbf{x} - \mathbf{u}), c_1) + \xi\mathcal{D}(\mathbf{x}) - \mu\nabla \cdot \left(g\frac{\nabla\theta}{|\nabla\theta|}\right).$$

Similar to solving the $\mathbf{u}$ subproblem, we implement a gradient descent with semi-implicit finite difference, yielding the following iterative scheme:

$$\theta^{(k+1)} = \frac{1}{2} \sum_{l=1}^{2} (I - 2\mu\tau A_l)^{-1} (\theta^{(k)} - \tau g_3(u_1^{(k)}, u_2^{(k)}, \theta^{(k)})), \tag{5}$$

where $g_3(u_1, u_2, \theta) = \lambda_1 \Phi(T(\mathbf{x}), a_1) + \lambda_2 \Phi(R(\mathbf{x} - \mathbf{u}), c_1) + \xi\mathcal{D}(\mathbf{x})$.

3   A Deep Image Prior Approach

The deep image prior (DIP) [20] approach can tackle inverse problems effectively. It is an unsupervised approach for training on a single data point, rather than having a large training set. The DIP method proposes to remove explicit regularisation from the traditional inverse problem and replace it with the implicit prior captured by the architecture of a neural network. This, combined with early stopping and varying the input to the network with noise has been shown to produce a regularisation effect on the solution, capturing generic regularity of a target image. It has been shown in [6,15,20,22] that the deep image prior approach for (inverse problems of) image restoration and superresolution achieve outstanding results. Here we extend the DIP idea to image segmentation and registration, especially the joint model of these two tasks.

In our case, removing explicit regularisation and exploiting the implicit regularisation offered by the architecture of a neural network allows us to remove two terms, simplifying parameter tuning dramatically. In addition, we find the deep image prior approach outperforms the same model implemented in the classical learning setting.

We utilise two deep image prior networks. As notation, let $\theta(\mathbf{x}) = \theta(\Theta_1)(\mathbf{x}) = \varphi_{\Theta_1}(z_1)$ be a network parametrised by weights Θ_1 with random noise z_1 as input dedicated to the segmentation task. To keep notation consistent with previous sections, we denote this as $\theta(\mathbf{x})$, as the first network performs the segmentation. Similarly, let $\mathbf{u} = \mathbf{u}(\Theta_2)(\mathbf{x}) = \psi_{\Theta_2}(z_2)$ be a network parametrised by weights Θ_2, with random noise input z_2, dedicated to the registration task. Our task is to minimise the energies in order to iteratively update the weights Θ_i:

$$\Theta_1^{(k+1)} = \arg\min_{\Theta_1} \mathcal{E}_1(\Theta_1, \Theta_2^{(k)}), \qquad \Theta_2^{(k+1)} = \arg\min_{\Theta_2} \mathcal{E}_2(\Theta_1^{(k)}, \Theta_2).$$

The energies $\mathcal{E}_1$ and $\mathcal{E}_2$, that serve as the loss functions for the two networks, are defined similarly as in the previous section, however without explicit regularisation:

$$L_1(\Theta_1) = \mathcal{E}_1(\Theta_1, \Theta_2^{(k)}) = \lambda_1 \int_\Omega \Phi(T(\mathbf{x}), a_1)\theta(\mathbf{x})d\Omega \qquad (6)$$

$$+ \lambda_2 \int_\Omega \Phi(R(\mathbf{x}), c_1)\theta(\mathbf{x} + \mathbf{u})d\Omega + \xi \int_\Omega \mathcal{D}(\mathbf{x})\theta(\mathbf{x})d\Omega,$$

$$L_2(\Theta_2) = \mathcal{E}_2(\Theta_1^{(k)}, \Theta_2) = \lambda_2 \int_\Omega \Phi(R(\mathbf{x}), c_1)\theta(\mathbf{x} + \mathbf{u})d\Omega \qquad (7)$$

$$+ \frac{1}{2}\int_\Omega (T(\mathbf{x} + \mathbf{u}) - R(\mathbf{x}))^2 d\Omega.$$

where we recall that $\theta(\mathbf{x})$ is the output image of the first network parametrised by weights Θ_1 and generated by input noise image z_1, and the displacement map $\mathbf{u} = (u_1, u_2)$ is the output (in 2 images) of the second network parametrised by weights Θ_2 and generated by input noise image z_2.

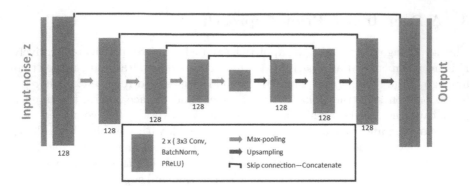

Fig. 1. An overview of the network architecture used.

The architecture used is a typical U-Net like [19] hourglass network featuring convolutions, batch normalisations and PReLU activation functions, see Fig. 1 for the full architecture. Both networks employ this architecture, only differing in the output layer. For the output layer, the segmentation network φ_{Θ_1} has a (1×1) convolution followed by a Sigmoid activation in the final layer, whereas the registration network just has a (1×1) convolution.

The initial inputs $z_i^{(0)}$ to the network are each drawn randomly from a uniform distribution between 0 and 0.1, and as in [20] has the same spatial size as T and R, i.e. $z_i^{(0)} \in \mathbb{R}^{N \times M \times C}$, where $N \times M$ is the image size and we typically fix $C = 32$. Likewise as in the original DIP approach, at each training iteration k we add additive noise $\hat{z}_i^{(k)}$ to the input $z_i^{(0)}$ drawn from a normal distribution, with mean 0 and standard deviations $\frac{1}{100}$ and $\frac{1}{30}$ for the segmentation and registration networks respectively i.e. $z_i^{(k)} = z_i^{(0)} + \hat{z}_i^{(k)}$.

Overall our workflow can be found in Fig. 2. Our implementation uses Keras with the Tensorflow backend, and each network uses ADAM optimiser with a learning rate of 0.001. The registration network was implemented with the aid of the voxelmorph package [2] to allow us to use a spatial transformer to interpolate the images using the output displacement map $\mathbf{u}(\mathbf{x})$, as shown in Fig. 2. Details of the DIP algorithm can be found in Algorithm 1.

We remark that multiple coupled DIP networks may provide a framework for a wide variety of applications [8]. Such applications include foreground/background segmentation, in which one network attempts to identify the foreground layer, another identifies the background layer, and a third identifies the binary mask. We note that this approach is similar to the approach we considered, although there is no discussion about using multiple DIP networks to tackle different inverse problems simultaneously. Moreover, Laves et al. [13] explored applying DIPs to medical image registration, exploiting the implicit regularisation that a randomly initialised CNN has to offer. Future work will look into multiple coupled DIP networks for our joint model.

Algorithm 1. Deep Image Prior Joint Selective Segmentation and Registration.

1: **Input**: Images T, R, user input $\mathcal{M}$, parameters $\lambda_1, \lambda_2, \xi$, learning rate (lr), max epochs.

2: **Initialise**: Θ_i randomly, $z_i \in \mathbb{R}^{N \times M \times 32} \sim U(0, \frac{1}{10})$, $i = 1, 2$.

3: **Calculate**: $\mathcal{D}$ from $\mathcal{M}$ on image T using (3).

4: **while** $Notconverged$ **do**

5:     Set $\hat{z}_1 \sim \mathcal{N}(0, \frac{1}{100})$ and $\hat{z}_2 \sim \mathcal{N}(0, \frac{1}{30})$.

6:     Update Θ_1^{k+1} by minimising (6) using back-propagation with network input $z_1 + \hat{z}_1$.

7:     Update Θ_2^{k+1} by minimising (7) using back-propagation with network input $z_2 + \hat{z}_2$.

8: **end while**

9: **Output**: Segmentation $\theta(\mathbf{x})$ from $\varphi_{\Theta_1}(z_1)$, and registration map from $\mathbf{u}(\mathbf{x})$ from $\psi_{\Theta_2}(z_2)$

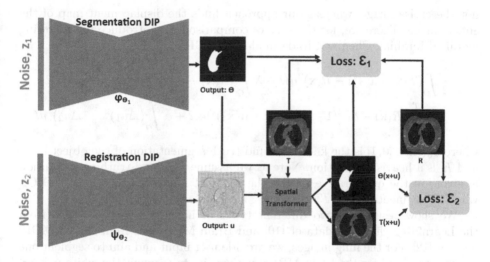

Fig. 2. An overview of the training of our deep image prior approach.

4   Numerical Experiments

In this section we conduct some experiments to show the two implementations of our model: the first as described in Sect. 2 with explicit regularisation implemented using gradient descent and time-marching, and the second as described in Sect. 3 implemented in a DIP framework. Additionally, as a further comparison we will use a model similar to the model by Ibrahim, Chen and Rada [11]. Their model makes use of a known segmentation of T given by a level set function ϕ_0. This is slightly different from our approach, as we do not assume a known segmentation of T, and we use an indicator function θ in the convex relaxed setting first introduced by Chan et al. [3]. In addition, the aim of [11] is to find the displacement map from the object in T to the object in R only, and

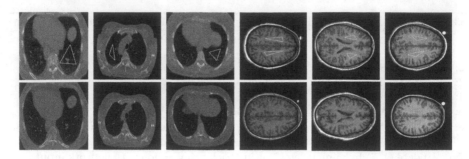

Fig. 3. Test images. Top row shows T with user input $\mathcal{M}$. Bottom row shows image R. From left to right, we refer to these pairs of images as: Lung 1 (L1), Lung 2 (L2), Lung 3 (L3), Brain 1 (B1), Brain 2 (B2), Brain 3 (B3).

not the entire image, whereas our approach finds the displacement map of the entire image. Therefore, for the sake of comparison, we introduce the following modified Ibrahim, Chen and Rada model (**M-ICR**):

$$\min_{\mathbf{u}} \frac{1}{2} \int_{\Omega} (T(\mathbf{x} + \mathbf{u}(\mathbf{x})) - R(\mathbf{x})^2 d\Omega + \lambda \int_{\Omega} |R(\mathbf{x}) - c_1|^2 H(\phi_0(\mathbf{x} + \mathbf{u}(\mathbf{x}))) d\Omega$$

$$+ \lambda \int_{\Omega} |R(\mathbf{x}) - c_2|^2 (1 - H(\phi_0(\mathbf{x} + \mathbf{u}(\mathbf{x})))) d\Omega + \alpha \int_{\Omega} (\Delta u_1)^2 + (\Delta u_2)^2 d\Omega,$$

where $\phi_0(\mathbf{x}) \in [0, 1]$ is the known ground truth segmentation of the object in T, and H is a heaviside function. Note as we assume a ground truth segmentation is given, in the quantitative results the **M-ICR** model will not have a DICE value for segmentation of T.

We show results on two different types of images: Lung CT scans from the Learn2Reg challenge dataset [10] and Brain MRI scans from the CUMC12 dataset [12]. For the lung images, we provide user input and aim to segment one of the lungs, and for the brain MRI scans we aim to segment the white matter, while in both cases registering the image to another. We show the 6 test images in Fig. 3, in the top row is T with the user input $\mathcal{M}$ to indicate the region of interest, the bottom row displays the reference image, R.

In Figs. 4, 5 and 6 we display some results on the lung images. The image pairs are from the same patient at roughly the same slice, the difference is that T and R are from expiration and inspiration scans respectively. In all three examples, we find that the approach with explicit regularisation struggles to smooth out some nodules in the lung segmentation, which the DIP result does effectively. We note that, while increasing the parameter μ in the model (1) solves this, doing so makes the contour at the edges of the object not as sharp as currently displayed.

Moreover, in Figs. 7, 8 and 9 we display the results on the brain images. In these images, we aim to segment the white matter in the brain. Interestingly the model with explicit regularisation is competitive with the DIP approach for the segmentation task for this set of images, however for the registration task both visually and quantitatively (see Table 1), the DIP approach performs better.

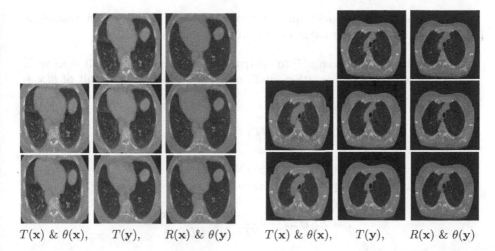

$T(\mathbf{x})$ & $\theta(\mathbf{x})$, $T(\mathbf{y})$, $R(\mathbf{x})$ & $\theta(\mathbf{y})$ $T(\mathbf{x})$ & $\theta(\mathbf{x})$, $T(\mathbf{y})$, $R(\mathbf{x})$ & $\theta(\mathbf{y})$

Fig. 4. Lung 1 Results: Top row - **M-ICR** model. Middle row - Explicit Regularisation. Bottom row - DIP framework.

Fig. 5. Lung 2 Results: Top row - **M-ICR** model. Middle row - Explicit Regularisation. Bottom row - DIP framework.

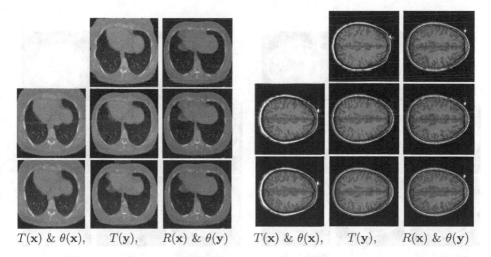

$T(\mathbf{x})$ & $\theta(\mathbf{x})$, $T(\mathbf{y})$, $R(\mathbf{x})$ & $\theta(\mathbf{y})$ $T(\mathbf{x})$ & $\theta(\mathbf{x})$, $T(\mathbf{y})$, $R(\mathbf{x})$ & $\theta(\mathbf{y})$

Fig. 6. Lung 3 Results: Top row - **M-ICR** model. Middle row - Explicit Regularisation. Bottom row - DIP framework.

Fig. 7. Brain 1 Results: Top row - **M-ICR** model. Middle row - Explicit Regularisation. Bottom row - DIP framework.

For all examples we provide quantitative results in Table 1. To determine the performance of the three models we use three evaluation metrics:

1. DICE similarity coefficient [7] to compare the segmentation result $\theta(\mathbf{x})$ with the ground truth segmentation on T, and the segmentation result of $\theta(\mathbf{x} + \mathbf{u}(\mathbf{x}))$ on R, which is defined by:

$$DICE(\Sigma, GT) = \frac{2|\Sigma \cap GT|}{|\Sigma| + |GT|}.$$

2. The structure similarity measure (SSIM) [23].
3. The relative sum of squared differences (Rel_SSD), defined by:

$$Rel\_SSD = \frac{||T(\mathbf{x} + \mathbf{u}(\mathbf{x})) - R(\mathbf{x})||^2}{||T(\mathbf{x}) - R(\mathbf{x})||^2}.$$

It is clear to see from Table 1, the DIP approach outperforms both the approach with explicit regularisation optimised using time-marching, and the **M-ICR** model. In general, the **M-ICR** displays competitive segmentation results, however this is largely due to the fact that it is supplied with the ground truth segmentation of T, whereas the other two approaches aren't.

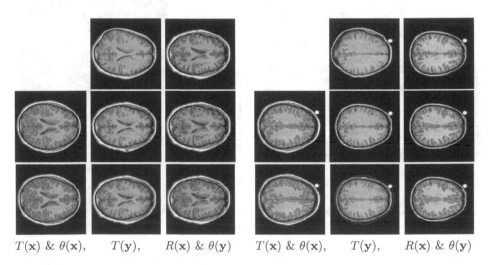

$T(\mathbf{x})$ & $\theta(\mathbf{x})$, $T(\mathbf{y})$, $R(\mathbf{x})$ & $\theta(\mathbf{y})$ $T(\mathbf{x})$ & $\theta(\mathbf{x})$, $T(\mathbf{y})$, $R(\mathbf{x})$ & $\theta(\mathbf{y})$

Fig. 8. Brain 2 Results: Top row - **M-ICR** model. Middle row - Explicit Regularisation. Bottom row - DIP framework.

Fig. 9. Brain 3 Results: Top row - **M-ICR** model. Middle row - Explicit Regularisation. Bottom row - DIP framework.

Table 1. Comparison of M-ICR (M), Explicit Reg (ER) and the DIP approach for Lung (Lj) and Brain (Bj) images.

	Dice of T		Dice of R			SSIM			rel_ssd		
	ER	DIP	M	ER	DIP	M	ER	DIP	MR	ER	DIP
L1	0.95	0.95	0.97	0.97	**0.98**	0.46	0.57	**0.84**	0.86	0.32	**0.09**
L2	0.95	**0.96**	0.97	0.96	**0.99**	0.48	0.54	**0.73**	0.30	0.15	**0.08**
L3	0.96	0.96	0.97	0.97	**0.98**	0.45	0.55	**0.77**	0.63	0.18	**0.07**
B1	**0.87**	0.86	0.79	**0.84**	0.82	0.62	0.72	**0.81**	0.45	0.17	**0.07**
B2	0.84	**0.90**	0.77	**0.83**	0.82	0.59	0.71	**0.79**	0.40	0.12	**0.04**
B3	**0.84**	0.80	0.77	**0.88**	0.87	0.50	0.67	**0.71**	0.43	**0.06**	0.08

5 Conclusion

In this work, we proposed a new model for joint segmentation and registration. Our model assumes no prior knowledge, only user supplied input in the form of a marker set $\mathcal{M}$ to indicate the region of interest. We then selectively segment the object of interest and simultaneously register the image T to R. We first introduce our model in the variational setting and then insert it into a deep image prior (DIP) framework. Numerical experiments show that the DIP approach generally outperforms the explicit regularisation approach, and an existing similar model (**M-ICR**).

References

1. Badshah, N., Chen, K.: Image selective segmentation under geometrical constraints using an active contour approach. Commun. Comput. Phys. **7**, 759–778 (2010)
2. Balakrishnan, G., Zhao, A., Sabuncu, M., Guttag, J., Dalca, A.V.: Voxelmorph: a learning framework for deformable medical image registration. IEEE TMI: Trans. Med. Imaging **38**, 1788–1800 (2019)
3. Chan, T.F., Esedoglu, S., Nikolova, M.: Algorithms for finding global minimizers of image segmentation and denoising models. SIAM J. Appl. Math. **66**(5), 1632–1648 (2006)
4. Chan, T.F., Vese, L.A.: Active contours without edges. IEEE Trans. Image Process. **10**(2), 266–277 (2001)
5. Chen, K., Debroux, N., Guyader, C.L.: A survey of topology and geometry-constrained segmentation methods in weakly supervised settings. In: Chen, K., Schönlieb, C.B., Tai, X.C., Younes, L. (eds.) Mathematical Models and Algorithms in Computer Vision and Imaging. Springer (2022 to appear). https://meteor.springer.com/project/dashboard.jsf?id=839&tab=About
6. Cheng, Z., Gadelha, M., Maji, S., Sheldon, D.: A Bayesian perspective on the deep image prior. In: Proceedings of the IEEE Conference on Computer Vision and Pattern Recognition, pp. 5443–5451 (2019)
7. Dice, L.R.: Measures of the amount of ecologic association between species. Ecology **26**(3), 297–302 (1945)

8. Gandelsman, Y., Shocher, A., Irani, M.: "Double-dip": unsupervised image decomposition via coupled deep-image-priors. In: IEEE Conference on Computer Vision and Pattern Recognition (CVPR), vol. 6, p. 2 (2019)
9. Gout, C., Le Guyader, C., Vese, L.A.: Segmentation under geometrical conditions with geodesic active contour and interpolation using level set methods. Numer. Algorithms **39**(1), 155–173 (2005)
10. Hering, A., Murphy, K., van Ginneken, B.: Learn2reg challenge: Ct lung registration - training data (2020)
11. Ibrahim, M., Chen, K., Rada, L.: An improved model for joint segmentation and registration based on linear curvature smoother. J. Algorithms Comput. Technol. **10**(4), 314–324 (2016)
12. Klein, A., et al.: Evaluation of 14 nonlinear deformation algorithms applied to human brain MRI registration. Neuroimage **46**(3), 786–802 (2009)
13. Laves, M.-H., Ihler, S., Ortmaier, T.: Deformable medical image registration using a randomly-initialized CNN as regularization prior. arXiv preprint arXiv:1908.00788 (2019)
14. Le Guyader, C., Vese, L.A.: A combined segmentation and registration framework with a nonlinear elasticity smoother. Comput. Vis. Image Underst. **115**(12), 1689–1709 (2011)
15. Mataev, G., Milanfar, P., Elad, M.: Deepred: deep image prior powered by red. In: Proceedings of the IEEE International Conference on Computer Vision Workshops (2019)
16. Rada, L., Chen, K.: A new variational model with dual level set functions for selective segmentation. Commun. Comput. Phys. **12**(1), 261–283 (2012)
17. Roberts, M., Chen, K., Irion, K.L.: A convex geodesic selective model for image segmentation. J. Math. Imaging Vis. **61**(4), 482–503 (2019)
18. Roberts, M., Spencer, J.: Chan-vese reformulation for selective image segmentation. J. Math. Imaging Vis. **61**(8), 1173–1196 (2019)
19. Ronneberger, O., Fischer, P., Brox, T.: U-Net: convolutional networks for biomedical image segmentation. In: Navab, N., Hornegger, J., Wells, W.M., Frangi, A.F. (eds.) MICCAI 2015. LNCS, vol. 9351, pp. 234–241. Springer, Cham (2015). https://doi.org/10.1007/978-3-319-24574-4_28
20. Ulyanov, D., Vedaldi, A., Lempitsky, V.: Deep image prior. In: Proceedings of the IEEE Conference on Computer Vision and Pattern Recognition, pp. 9446–9454 (2018)
21. Unal, G., Slabaugh, G.: Coupled PDEs for non-rigid registration and segmentation. In: CVPR (2005)
22. Van Veen, D., Jalal, A., Soltanolkotabi, M., Price, E., Vishwanath, S., Dimakis, A.G.: Compressed sensing with deep image prior and learned regularization. arXiv preprint arXiv:1806.06438 (2018)
23. Wang, Z., Bovik, A.C., Sheikh, H.R., Simoncelli, E.P.: Image quality assessment: from error visibility to structural similarity. IEEE Trans. Image Process. **13**(4), 600–612 (2004)
24. Weickert, J.: Applications of nonlinear diffusion in image processing and computer vision. Acta Mathematica Universitatis Comenianae **LXX**, 33–50 (2001)
25. Weickert, J., Romeny, B.T.H., Viergever, M.A.: Efficient and reliable schemes for nonlinear diffusion filtering. IEEE Trans. Image Process. **7**(3), 398–410 (1998)
26. Zhu, W., Tai, X.-C., Chan, T.: Image segmentation using Euler's elastica as the regularization. J. Sci. Comput. **57**, 414–438 (2013)

Restoration, Reconstruction
and Interpolation

Inpainting-Based Video Compression in FullHD

Sarah Andris$^{(\boxtimes)}$, Pascal Peter$^{(\boxtimes)}$, Rahul Mohideen Kaja Mohideen$^{(\boxtimes)}$, Joachim Weickert$^{(\boxtimes)}$, and Sebastian Hoffmann$^{(\boxtimes)}$

Mathematical Image Analysis Group, Faculty of Mathematics and Computer Science, Saarland University, Campus E1.7, 66041 Saarbrücken, Germany
{andris,peter,rakaja,weickert,hoffmann}@mia.uni-saarland.de

Abstract. Compression methods based on inpainting are an evolving alternative to classical transform-based codecs for still images. Attempts to apply these ideas to video compression are rare, since reaching real-time performance is very challenging. Therefore, current approaches focus on simplified frame-by-frame reconstructions that ignore temporal redundancies. As a remedy, we propose a highly efficient, real-time capable prediction and correction approach that fully relies on partial differential equations (PDEs) in all steps of the codec: Dense variational optic flow fields yield accurate motion-compensated predictions, while homogeneous diffusion inpainting is applied for intra prediction. To compress residuals, we introduce a new highly efficient block-based variant of pseudodifferential inpainting. Our novel architecture outperforms other inpainting-based video codecs in terms of both quality and speed. For the first time in inpainting-based video compression, we can decompress FullHD (1080p) videos in real-time with a fully CPU-based implementation, outperforming previous approaches by roughly one order of magnitude.

Keywords: Inpainting-based compression · Hybrid video coding · Optical flow · Homogeneous diffusion inpainting · Pseudo-differential inpainting

1 Introduction

In today's world, videos are a vital part of our communication, be it personal or professional. Video sharing is constantly rising, making up a large portion of total IP traffic. It is therefore an important task to constantly continue research on improving codecs for video compression.

Currently, the most well-known and widely used codecs belong to the MPEG family [15]. They are based on hybrid video coding which combines a prediction

This work has received funding from the European Research Council (ERC) under the European Union's Horizon 2020 research and innovation programme (grant agreement no. 741215, ERC Advanced Grant INCOVID). We thank Matthias Augustin for sharing his in-depth knowledge in pseudodifferential inpainting.

A. Elmoataz et al. (Eds.): SSVM 2021, LNCS 12679, pp. 425–436, 2021.
https://doi.org/10.1007/978-3-030-75549-2_34

and a correction step where prediction depends on the frame type. Intra frames rely solely on information from within themselves. Propagating values from preceding or subsequent frames along motion vectors approximates inter frames. The correction step is identical for both types: The residual contains the difference to the original frame and can be efficiently compressed, if the initial prediction was of high quality. Bull [7] gives an approachable introduction into all important ideas leading up to current standards.

Traditional video codecs use transform-based image compression for residual storage. For still images, however, codecs of this type have been successfully challenged by inpainting-based techniques. They keep values only at a few carefully selected points (the so-called *inpainting mask*) and reconstruct the missing image parts via inpainting. The state of the art for colour images is a PDE-based method by Peter et al. [23] that extends work by Galić et al. [13] and Schmaltz et al. [28] and performs on par with JPEG2000 [32] for images with a small to medium amount of texture. As Jost et al. [18] have shown, these methods can even outperform HEVC/intra on piecewise smooth images.

Classical inpainting problems only reconstruct small amounts of missing image content, but inpainting-based compression relies on sparse known data. Therefore, it tends to be fairly slow, and meeting the real-time requirements of video compression is very challenging. Most existing approaches only focus on specific parts of the video coding pipeline without presenting a full codec [5,12,21,31,34]. Fully inpainting-based codecs almost exclusively ignore the temporal dimension. They only work on a frame-by-frame basis [19,24] and mainly established speed-up strategies. However, inpainting-based video compression methods will only reach their full potential if they consequently exploit temporal redundancies. So far, only Andris et al. [1] have proposed a modular framework that combines the successful idea of prediction and correction with inpainting-based methods. This framework provides a useful basis for codec design, but the accompanying codec in [1] does not use it to its full capacity: It can only reconstruct colour videos with up to a size of 854×364 in real-time and is limited to small compression ratios due to its simplistic residual storage.

1.1 Our Contribution

Our goal is to design a codec with significantly improved performance both in quality and speed compared to [1] and other inpainting-based codecs. To this end, we use the framework from [1] as a basis and consequently implement inpainting-based methods for all prediction and compression submodules. Our hybrid inpainting-based video codec (HIVC) employs the methods of Brox et al. [6] for optic flow field computation, homogeneous diffusion inpainting [8] for intra prediction, and finite state entropy (FSE) coding [10]. For representing the residuals, we design a block-based variant of pseudodifferential inpainting [3] which we can fully describe in fast cosine transforms on the decoder side. Through efficient algorithms, we are able to push real-time reconstruction of colour videos to FullHD 1080p resolution without resorting to parallelisation on the GPU. Furthermore, we can reach compression ratios which are over fifty times larger than in [1] while even slightly increasing quality.

1.2 Related Work

Inpainting-based techniques have been applied for several parts of the coding pipeline or even for complete codecs. Köstler et al. [19] were the first to achieve real-time performance with a codec based on homogeneous diffusion inpainting on a *Playstation 3*. The state of the art in PDE-based image compression, the R-EED codec by Schmaltz et al. [28], has been adapted by Peter et al. [24] for video compression. However, these codecs work on a strict frame-by-frame basis. First attempts to incorporate motion information were made by Schmaltz and Weickert [29] who combine pose tracking with static background compression by anisotropic diffusion inpainting. Breuß et al. [5] still perform frame-based inpainting, but they acquire masks by shifting one optimised mask along motion vectors. They do not present a full video compression pipeline. Another stand-alone codec is proposed by Wu et al. [33] who employ deep learning to interpolate inter frames between two intra frames. Liu et al. [21], Doshkov et al. [12], and Zhang and Lin [34] incorporate inpainting ideas directly into the intra prediction of a hybrid video coder. Their methods combine homogeneous diffusion inpainting with edge information, template matching, and adaptive boundary values, respectively. For the same task, Tan et al. [31] perform inpainting via template matching. Jost et al. [18] focus on the compression of dense optic flow fields and design a well-performing method employing edge-aware homogeneous diffusion inpainting. However, their method is not real-time capable.

Since high quality motion prediction is essential for efficient video compression, dense optic flow fields have been investigated in multiple works. Li et al. [20] employ the classical Horn and Schunck [17] approach for bidirectional prediction. More complex techniques such as Bayesian methods in Han and Podilchuk [14] and velocity field modeling in Chen and Mied [9] yield optic flow fields with higher prediction quality. Both methods take also the compressibility of the acquired motion into account. Going further into this direction, Ottaviano and Kohli [22] represent optic flow in a wavelet basis and obtain the corresponding coefficients by minimising the residual after inter prediction. All of these works consider flow fields independent of compression or as an augmentation for transform-based video codecs. In contrast, we embed them into a fully inpainting-based approach that addresses all relevant coding steps.

1.3 Paper Structure

We present our video codec in Sect. 2 and discuss corresponding experiments in Sect. 3. We conclude our paper in Sect. 4.

2 A Fully Inpainting-Based Video Codec

We generalise the framework by Andris et al. [1], but keep the same central concept of prediction and correction. In a first step, we compute and subsequently compress dense *backwards optic flow fields (BOFFs)* between frames. We represent one scene by a *group of pictures (GOP)*, which includes one intra frame at the beginning and subsequent inter frames. The size of these GOPs may vary according to video content. In a second step, we predict intra frames with some inpainting technique and inter frames with motion compensation based on the previously computed BOFFs. For each frame, a corresponding correction (residual) contains the difference to the originals. These residuals, again, have to be compressed. Finally, we store and encode all data needed for decoding.

In the following, we present a concrete codec implementation for this general framework. The most important ingredients are the intra prediction and residual compression techniques, which we investigate in Sect. 2.1 and 2.2, respectively. We then show our final codec design with a description of all relevant components in Sect. 2.3.

2.1   Global Homogeneous Diffusion Inpainting

Let f be a 1D representation of a 2D discrete image of size $n_x \times n_y$, i.e. we sort the image pixels row-wise into the vector $f \in \mathbb{R}^N$ with $N = n_x n_y$. We store values only at a few selected locations represented by the binary inpainting mask $m \in \mathbb{R}^N$ and discard all other values. Then, we can acquire an approximation $u \in \mathbb{R}^N$ of the original image by solving the general discrete inpainting problem

$$M(u - f) - (I - M)Au = 0. \tag{1}$$

The matrix $M \in \mathbb{R}^{N \times N}$ contains the entries of the inpainting mask m on its diagonal, and is zero everywhere else. I is the identity matrix, and A represents a discrete inpainting operator with reflecting boundary conditions. The first term ensures that u adopts the original values from f at mask positions, the second term realises inpainting steered by the operator A in regions inbetween. The choice of A influences the reconstruction quality immensely and simultaneously affects the complexity of the algorithms solving the inpainting problem. Choosing $A = -L$ with discrete Laplacian L results in homogeneous diffusion inpainting [8], which presents a good balance between quality and simplicity.

We use a coarse-to-fine algorithm similar to the cascadic conjugate gradient method by Deuflhard [11] to solve the arising system of equations. The basic idea is to build an image pyramid by subsampling, solving the system on a coarse level, and use this solution as input for the next finer level. This yields a much faster convergence (both on the individual levels and in total) compared to classically solving the system only on the finest level.

2.2   Block-Based Pseudodifferential Inpainting

Pseudodifferential inpainting has been established by Augustin et al. [3] as a connecting concept between inpainting with rotationally invariant PDEs and radial basis function (RBF) interpolation. Their paper extends results of Hoffmann et al. [16] on harmonic and biharmonic inpainting with Green's functions. We use their work to build a highly efficient algorithm for inpainting based on the discrete cosine transform, which yields major improvements for the final codec compared to [1]. For images, mask, and operator, we stick to the notation used in Sect. 2.1.

Introducing the theoretical background of Green's functions is beyond the scope of this paper. We refer the interested reader to [16]. Instead, we provide an intuitive interpretation: For a discrete symmetric inpainting operator A, the corresponding discrete Green's function g_k characterises the influence of an impulse at pixel position k. Instead of solving a system of equations for the general inpainting problem, we can then directly obtain a solution via a weighted sum of the operator's Green's functions at mask positions:

$$u_k = \sum_{i=1}^{n_x n_y} (m_i \cdot c_i \cdot (g_i)_k) + a \quad \text{for } k = 1, ..., N. \tag{2}$$

Recall that $m = (m_i)_{i=1,...,N}$ is the inpainting mask and $u = (u_i)_{i=1,...,N}$ the corresponding inpainting solution in vector notation. The coefficient vector c and the constant a can be acquired by solving a linear system of equations of size $(K+1) \times (K+1)$, where K is the number of mask points; see [16] for more details. In contrast to the sparse but large matrix used for solving the general inpainting problem directly, the system matrix is fairly small and densely populated.

It is well-known that the discrete Green's functions build up the pseudo-inverse of the corresponding operator. Therefore, they are symmetric, i.e. $(g_i)_k = (g_k)_i$ for all i, k, since we assumed A to be symmetric. We define G as the matrix containing the Green's functions g_k as its columns. Then we can reformulate Eq. (2) as

$$u - GMc \mid a \tag{3}$$

where the vector a has the constant a in every entry. Since our inpainting operator A is a finite difference matrix and the Green's functions are its pseudo-inverse, G is also a difference matrix. Hence, we can apply results by Strang and MacNamara [30], proving that G is Toeplitz-plus-Hankel. Sanchez et al. showed in [27] that matrices of this type are diagonalisable by the even discrete cosine transform (DCT) of type II. Denoting the corresponding transform matrix by $\mathcal{C}$, we can rewrite Eq. (3) as

$$u = \mathcal{C}^{-1} \operatorname{diag}(\lambda) \mathcal{C}(Mc) + a \tag{4}$$

where the vector λ contains the eigenvalues of the Green's functions. Thus, if we have the coefficients c, we can acquire the final inpainting solution by multiplying the coefficients with the eigenvalues in the transform domain, computing the backtransform, and adding the constant to all pixels.

All these considerations hold for arbitrary image sizes. If we now subdivide the image domain into 8×8 blocks and perform inpainting on each block independently, we can employ a dedicated fast DCT algorithm. We opt for the method by Arai et al. [2] which is also used in JPEG. We treat each block as an independent image, i.e. impose mirrored boundary conditions. Note that if we store the coefficients and the constant on the encoder side, the decoder only has to perform two fast DCTs, $n_x n_y$ multiplications, and $n_x n_y$ additions to acquire

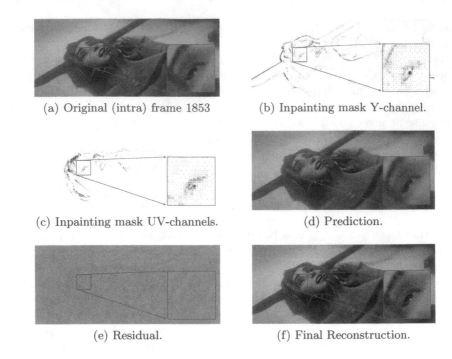

(a) Original (intra) frame 1853

(b) Inpainting mask Y-channel.

(c) Inpainting mask UV-channels.

(d) Prediction.

(e) Residual.

(f) Final Reconstruction.

Fig. 1. Intermediate results for intra coding at a compression rate of roughly 100:1. Global homogeneous diffusion inpainting is especially suited for smooth regions. Thus, the mask concentrates at edges. With the residual we can correct remaining errors.

the final inpainting result. This brings our method close to the main concept of JPEG. However, in contrast to JPEG, our coefficients do not correspond to frequencies, but give a connection to local structures. Moreover, the inpainting operator is now completely defined by the eigenvalues of the Green's functions and can be easily replaced without changing the algorithm or influencing the speed on the decoder side.

2.3   The Final Codec

Our final model combines methods explained in the previous sections to obtain a fully inpainting-based video codec. The codec by Andris et al. [1] includes homogeneous diffusion inpainting for intra prediction, the method of Brox et al. [6] for inter prediction, quantisation for residual compression, and arithmetic coding for entropy coding. It is able to outperform other inpainting-based codecs, however, it has a limited range of possible compression ratios due to the simple residual compression technique, and real-time decoding has only been shown for a resolution of 854×364. We aim at extending the compression range, improving quality, and increase decoding speed by consequently filling all submodules of the compression pipeline with suitable methods.

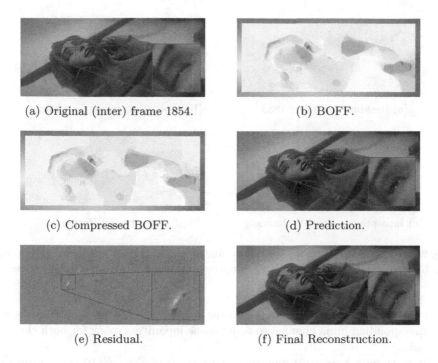

(a) Original (inter) frame 1854.

(b) BOFF.

(c) Compressed BOFF.

(d) Prediction.

(e) Residual.

(f) Final Reconstruction.

Fig. 2. Intermediate results for inter coding at a compression rate of roughly 100:1. The colour coding of the BOFF is adapted from [4]. The flow field yields a prediction of high accuracy and only struggles at occlusions and disocclusions (e.g. the closing eyes).

In a first step, we compute backwards optic flow fields between all frames. As in [1], we employ the method of Brox et al. [6]. It produces piecewise smooth motion fields of high quality while still being computationally manageable. Applying their algorithm back to front for all frames in a GOP, we obtain BOFFs for inter prediction that are highly compressible due to their piecewise smoothness but can still yield accurate predictions at motion boundaries. We also tested the much simpler method by Horn and Schunck [17], but got consistently worse results both in terms of reconstruction error and final compression ratio of the video. Thus, it is worthwhile to invest into more advanced methods to acquire accurate flow fields.

The human visual system combines structural with colour information and is less sensitive to errors in the colour domain. Thus, most established codecs transform the input to a colour space with a luma and two chroma channels and compress the chroma channels more strongly. According to this principle, we convert the frame sequence to YUV space with the reversible colour transform (RCT) employed in JPEG2000 and carry out all computations in this domain. In order to realise a stronger compression in the chroma channels, we take half the amount of mask points there compared to the luma channel for all our inpainting

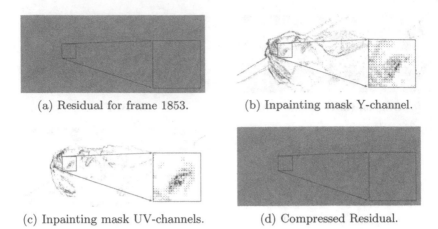

(a) Residual for frame 1853.   (b) Inpainting mask Y-channel.

(c) Inpainting mask UV-channels.   (d) Compressed Residual.

Fig. 3. Intermediate results for residual coding at a compression rate of roughly 100:1. The block-based inpainting is able to compensate small-scale errors at image edges.

methods. More precisely, we use the same mask optimisation technique as for the corresponding luma channel to acquire one inpainting mask for both chroma channels.

Since we only have to reconstruct one intra frame per GOP, we choose the global homogeneous diffusion inpainting method (Sect. 2.1) for prediction. This results in a lower prediction error at the cost of an increased runtime. We acquire the corresponding inpainting mask with a rectangular subdivision scheme introduced by Schmaltz et al. [28], which allows to concentrate mask points in regions with large reconstruction errors. To further reduce storage, we quantise colour values with a simple uniform quantisation.

For flow fields and residuals, we have more severe restrictions regarding runtime, since we have to recover one of each for every reconstructed frame. Experiments showed that typical BOFFs consist of very large regions with barely changing values. Thus, we use the rectangular subdivision scheme from [28] and assign each region its average value. Residuals tend to contain more structure, since both intra and inter prediction work best on smooth regions. Therefore, we opt for block-based pseudodifferential inpainting with the harmonic operator as described in Sect. 2.2. For every 8×8 block we again acquire mask positions with rectangular subdivision. The resulting coefficients of the Green's functions in general lie in a larger range than the standard colour values and also attain negative values. Therefore, we map the coefficients to $[-127, 127]$ and apply uniform quantisation centered around zero, also called dead-zone quantisation (see e.g. Chapter 3 in [32]). This way, we can ensure that zero coefficients are reconstructed as zero again and thus avoid flickering.

After dead-zone quantisation, we can expect to produce data that have a high probability of being zero or close to zero. Therefore, we adopt an idea from JPEG and define categories which describe larger ranges for exceeding distance

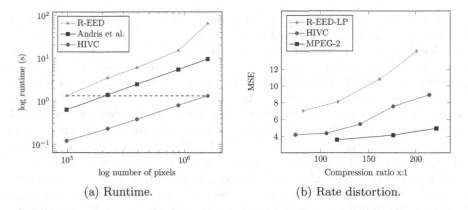

(a) Runtime.   (b) Rate distortion.

Fig. 4. Comparison with other codecs in terms of runtime and rate distortion. **Left:** Runtime depending on the number of pixels in one frame. The dashed line marks real-time decoding with 24 frames per second. The proposed HIVC approach outperforms the other codecs by roughly one order of magnitude. **Right:** Quality comparisons over several compression ratios. HIVC outperforms R-EED-LP consistently.

of its values from zero. We represent the categories with Huffman codes and store additional bits for the exact value. Finally, we remove remaining redundancies with an entropy coder. The well-known concept of arithmetic coding (AC) by Rissanen [25] and variants thereof are still successfully used in various codecs. However, its problem of requiring costly arithmetic operations has only partly been solved. Collet's Finite State Entropy (FSE) coder [10] presents an interesting alternative. It performs on par with arithmetic coding, but is consistenly faster. Our implementation is a lightweight version of FSE specifically designed to fit our video codec structure.

3   Experiments

We perform experiments on a sequence of 32 frames of the *Sintel* video by Rosendaal [26] with an *Intel Xeon CPU W3565@3.20* GHz.

In order to provide an insight into how our codec's different modules operate, we show intermediate results of the individual coding steps for an intra frame in Fig. 1. Inter frame processing (Fig. 2) looks similar, with the difference that we use motion compensation for prediction instead of inpainting.

For decoder run-time comparison, we use the 1080p version of *Sintel*. Figure 4a shows results over resolutions ranging from 480×205 to 1920×818 for our proposed codec HIVC, the codec by Andris et al. [1], and R-EED [24]. For the R-EED video codec we have to perform the runtime tests on greyscale videos, but still use colour videos for HIVC and Andris et al., giving R-EED a significant advantage. Our new codec outperforms the other methods by roughly one order of magnitude and is the only codec capable of real-time decoding beyond a resolution of 720×307.

In Fig. 4b, we compare the reconstruction quality of our approach with the best possible inpainting-based video compression method that provides an encoder as well as a decoder. Since the R-EED video codec only works on greyvalue videos, we supplement it by the state-of-the-art colour image compression codec R-EED-LP [23]. This is straight-forward since the original codec is completely frame-based, however, the resulting codec is significantly slower. Although it is not yet our goal to beat the established codecs of the MPEG family, we provide results for H.262/MPEG-2 [15] as a point of reference. *Sintel* at a resolution of 960×409 serves as test video. We do not include the codec by Andris et al., since it is only able to compress at small compression ratios up to 35:1. In contrast, we are able to reach much larger ratios since our codec consequently integrates inpainting ideas in every prediction and compression step. Especially for BOFF and residual storage, our methods are much more efficient compared to [1]. Furthermore, [1] can only compete with R-EED-LP on highly textured videos. Here, we pick a sequence with a low amount of texture that is very well suited for R-EED-LP compression, but HIVC is still able to outperform it consistently over several compression ratios.

4   Conclusion and Outlook

Real-time decoding presents a significant challenge for codecs based on inpainting ideas that so far could not even be reached in Full-HD with parallelisation on GPUs. Our approach is not only the first fully inpainting-based codec to achieve this, but also relies only on a standard CPU. This is possible due to a new pseuodifferential inpainting technique that allows highly efficient reconstruction through fast cosine transforms and carefully selected components for coding and prediction. This allows us to outperform previous inpainting-based video codecs [1] by one order of magnitude in speed, while simultaneously providing better quality than state-of-the-art inpainting [23].

Our significant advances show the high potential of inpainting-based codecs. In future work, we aim at real-time decoding of 4K video by pushing efficient engineering even further and exploiting parallelisation on GPUs. Furthermore, our pseudodifferential inpainting may give an insight into connections between transform-based and inpainting-based methods and inherently incorporates a variety of linear operators.

References

1. Andris, S., Peter, P., Weickert, W.: A proof-of-concept framework for PDE-based video compression. In: Proceedings of the 2016 Picture Coding Symposium. IEEE Computer Society Press, Nürnberg, Germany (2016)
2. Arai, Y., Agui, T., Nakajima, M.: A fast DCT-SQ scheme for images. IEICE Trans. **71**(11), 1095–1097 (1988)
3. Augustin, M., Weickert, J., Andris, S.: Pseudodifferential inpainting: the missing link between PDE- and RBF-based interpolation. In: Lellmann, J., Burger, M., Modersitzki, J. (eds.) SSVM 2019. LNCS, vol. 11603, pp. 67–78. Springer, Cham (2019). https://doi.org/10.1007/978-3-030-22368-7_6

4. Baker, S., Scharstein, D., Lewis, J., Roth, S., Black, M., Szeliski, R.: A database and evaluation methodology for optical flow. Int. J. Comput. Vision **92**(1), 1–31 (2011)
5. Breuß, M., Hoeltgen, L., Radow, G.: Towards PDE-based video compression with optimal masks prolongated by optic flow. J. Math. Imag. Vision **62**, 1–13 (2020)
6. Brox, T., Bruhn, A., Papenberg, N., Weickert, J.: High accuracy optical flow estimation based on a theory for warping. In: Pajdla, T., Matas, J. (eds.) ECCV 2004. LNCS, vol. 3024, pp. 25–36. Springer, Heidelberg (2004). https://doi.org/10.1007/978-3-540-24673-2_3
7. Bull, D.: Communicating Pictures: A Course in Image and Video Coding. Academic Press, Cambridge, MA (2014)
8. Carlsson, S.: Sketch based coding of grey level images. Signal Process. **15**(1), 57–83 (1988)
9. Chen, W., Mied, R.: Optical flow estimation for motion-compensated compression. Image Vision Comput. **31**(3), 275–289 (2013)
10. Collet, Y.: Finite state entropy (2013). https://github.com/Cyan4973/FiniteStateEntropy
11. Deuflhard, P.: Cascadic conjugate gradient methods for elliptic partial differential equations: algorithm and numerical results. In: Keyes, D.E., Xu, J. (eds.) Contemporary Mathematics, vol. 180, pp. 29–29. American Mathematical Society, Providence, RI (1994)
12. Doshkov, D., Ndjiki-Nya, P., Lakshman, H., Koppel, M., Wiegand, T.: Towards efficient intra prediction based on image inpainting methods. In: Proceedings of the 27th Picture Coding Symposium, pp. 470–473. IEEE Computer Society Press, Nagoya, Japan (2010)
13. Galić, I., Weickert, J., Welk, M., Bruhn, A., Belyaev, A., Seidel, H.P.: Image compression with anisotropic diffusion. J. Math. Imag. Vision **31**(2–3), 255–269 (2008)
14. Han, S.C., Podilchuk, C.: Video compression with dense motion fields. IEEE Trans. Image Process. **10**(11), 1605–1612 (2001)
15. Haskell, B.G., Puri, A., Netravali, A.N.: Digital Video: An Introduction to MPEG-2. Springer, Berlin (1996)
16. Hoffmann, S., Plonka, G., Weickert, J.: Discrete Green's functions for harmonic and biharmonic inpainting with sparse atoms. In: Tai, X.-C., Bae, E., Chan, T.F., Lysaker, M. (eds.) EMMCVPR 2015. LNCS, vol. 8932, pp. 169–182. Springer, Cham (2015). https://doi.org/10.1007/978-3-319-14612-6_13
17. Horn, B., Schunck, B.: Determining optical flow. Artif. Intell. **17**, 185–203 (1981)
18. Jost, F., Peter, P., Weickert, J.: Compressing flow fields with edge-aware homogeneous diffusion inpainting. In: Proceedings of the 45th International Conference on Acoustics, Speech, and Signal Processing (ICASSP), pp. 2198–2202. IEEE Computer Society Press, Barcelona, Spain (2020)
19. Köstler, H., Stürmer, M., Freundl, C., Rüde, U.: PDE based video compression in real time. Tech. Rep. 07–11, Lehrstuhl für Informatik 10, University Erlangen-Nürnberg, Germany (2007)
20. Li, B., Han, J., Xu, Y.: Co-located reference frame interpolation using optical flow estimation for video compression. In: Proceedings of the 2018 Data Compression Conference, pp. 13–22. IEEE Computer Society Press, Snowbird, UT (2018)
21. Liu, D., Sun, X., Wu, F., Zhang, Y.Q.: Edge-oriented uniform intra prediction. IEEE Trans. Image Process. **17**(10), 1827–1836 (2008)
22. Ottaviano, G., Kohli, P.: Compressible motion fields. In: Proceedings of the 2013 IEEE Conference on Computer Vision and Pattern Recognition, pp. 2251–2258. IEEE Computer Society Press, Oregon, OH (2013)

23. Peter, P., Kaufhold, L., Weickert, J.: Turning diffusion-based image colorization into efficient color compression. IEEE Trans. Image Process. **26**(2), 860–869 (2016)
24. Peter, P., Schmaltz, C., Mach, N., Mainberger, M., Weickert, J.: Beyond pure quality: progressive modes, region of interest coding, and real time video decoding for PDE-based image compression. J. Vis. Commun. Image Represent. **31**(4), 253–265 (2015)
25. Rissanen, J.J.: Generalized Kraft inequality and arithmetic coding. IBM J. Res. Dev. **20**(3), 198–203 (1976)
26. Roosendaal, T.: Sintel. In: ACM SIGGRAPH 2011 Computer Animation Festival, p. 71. New York, NY, USA (2011)
27. Sanchez, V., Garcia, P., Peinado, A.M., Segura, J.C., Rubio, A.J.: Diagonalizing properties of the discrete cosine transforms. IEEE Trans. Signal Process. **43**(11), 2631–2641 (1995)
28. Schmaltz, C., Peter, P., Mainberger, M., Ebel, F., Weickert, J., Bruhn, A.: Understanding, optimising, and extending data compression with anisotropic diffusion. Int. J. Comput. Vision **108**(3), 222–240 (2014)
29. Schmaltz, C., Weickert, J.: Video compression with 3-D pose tracking, PDE-based image coding, and electrostatic halftoning. In: Pinz, A., Pock, T., Bischof, H., Leberl, F. (eds.) DAGM/OAGM 2012. LNCS, vol. 7476, pp. 438–447. Springer, Heidelberg (2012). https://doi.org/10.1007/978-3-642-32717-9_44
30. Strang, G., MacNamara, S.: Functions of difference matrices are Toeplitz plus Hankel. SIAM Rev. **56**(3), 525–546 (2014)
31. Tan, T.K., Boon, C.S., Suzuki, Y.: Intra prediction by template matching. In: Proceedings of the 2006 IEEE International Conference on Image Processing, pp. 1693–1696. IEEE Computer Society Press, Atlanta, GA, USA (2006)
32. Taubman, D.S., Marcellin, M.W. (eds.): JPEG 2000: Image Compression Fundamentals. Standards and Practice. Kluwer, Boston (2002)
33. Wu, C.-Y., Singhal, N., Krähenbühl, P.: Video compression through image interpolation. In: Ferrari, V., Hebert, M., Sminchisescu, C., Weiss, Y. (eds.) ECCV 2018. LNCS, vol. 11212, pp. 425–440. Springer, Cham (2018). https://doi.org/10.1007/978-3-030-01237-3_26
34. Zhang, Y., Lin, Y.: Improving HEVC intra prediction with PDE-based inpainting. In: Asia-Pacific Signal and Information Processing Association Annual Summit and Conference (APSIPA), IEEE Computer Society Press, Chiang Mai, Thailand (2014)

Sparsity-Aided Variational Mesh Restoration

Martin Huska, Serena Morigi$^{(\boxtimes)}$, and Giuseppe Antonio Recupero

Department of Mathematics, University of Bologna, Bologna, BO 40126, Italy
{martin.huska,serena.morigi}@unibo.it, giuseppe.recupero@studio.unibo.it

Abstract. We propose a variational method for recovering discrete surfaces from noisy observations which promotes sparsity in the normal variation more accurately than ℓ_1 norm (total variation) and ℓ_0 pseudo-norm regularization methods by incorporating a parameterized non-convex penalty function. This results in denoised surfaces with enhanced flat regions and maximally preserved sharp features, including edges and corners. Unlike the classical two-steps mesh denoising approaches, we propose a unique, effective optimization model which is efficiently solved by an instance of Alternating Direction Method of Multipliers. Experiments are presented which strongly indicate that using the sparsity-aided formulation holds the potential for accurate restorations even in the presence of high noise.

Keywords: Non-convex optimization · Surface denoising · Sparse variational formulation

1 Introduction

The goal of a surface denoising algorithm is to remove undesirable noise or spurious information on a 3D mesh, while preserving original features, including edges, creases and corners. The restored surface is a 3D mesh that represents as faithfully as possible a piecewise smooth surface, where edges appear as discontinuities in the normals.

Through time, three main numerical approaches have been developed to solve the mesh denoising problem. Initially, linear/nonlinear diffusion equations were proposed in which the evolution of vertices is guided by Partial Differential Equations, see [7,12]. In general, the isotropic/anisotropic diffusion flows are known to have a strong regularization effect, failing, therefore, in the accurate recovery of sharp mesh features, despite the expedients proposed to preserve the local curvature of the mesh, [8]. More recently, also thanks to the considerable impact in the image processing field, two major challenges have emerged for mesh denoising: data driven and optimization-based methods. Approaches belonging to the former class aim to learn the relationship between noisy geometry and the

Research is supported in part by INDaM-GNCS research project 2020.

A. Elmoataz et al. (Eds.): SSVM 2021, LNCS 12679, pp. 437–449, 2021.
https://doi.org/10.1007/978-3-030-75549-2_35

ground-truth geometry from a training dataset, [13]. Optimization-based mesh denoising methods formulate the mesh restoration as a minimization problem and seek for a denoised mesh that can best fit to the input mesh while satisfying a prior knowledge of the ground-truth geometry and the noise distribution. These approches grew in popularity more and more also thanks to the last sparsity-inducing extraordinary results. This work belongs to this latter class of optimization-based methods.

Assuming an observed noisy triangulated surface with vertex set V^0 be corrupted by additive white Gaussian noise, an estimate $V^* \in \mathbb{R}^{n_V \times 3}$ of the noisy-free vertex set V can be obtained as a solution of the following variational model

$$V^* \in \arg\min_{V \in \mathbb{R}^{n_V \times 3}} \mathcal{J}(V), \qquad \mathcal{J}(V) := \frac{\lambda}{2} \|V - V^0\|_2^2 + \mathcal{R}(V), \qquad (1)$$

where $\|v\|_2$ denotes the Frobenius norm of matrix v. $\mathcal{J}(V)$ is the sum of a regularization term $\mathcal{R}(V)$ and a convex smooth (quadratic) fidelity term, accompanied by the classical regularization parameter λ that controls the trade-off between fidelity to the observation and regularity in the solutions V^* of (1). The regularizer $\mathcal{R}(V)$ encodes a priori knowledge on the solution. In particular, to promote solutions that have piecewise constant normals with sharp discontinuities in the normal map the regularizer can be designed in a way to penalize a measure of the "roughness" or bumpiness (curvature) of a mesh, or, equivalently, to promote sparsity on this measure. A natural bumpiness measure for a surface is the normal deviations, represented by $y_i := \|(\nabla N)_i\|_2$, with $\nabla \in \mathbb{R}^{n_E \times n_T}$ a gradient (linear) operator and $(\nabla N)_i$ the normal variation between two adjacent triangles sharing the i-th edge. The ideal regularizer to induce sparsity on the vector y is the ℓ_0 pseudo-norm, but its combinatorial nature makes the minimization of (1) an NP-hard problem. Nevertheless, in [2] ℓ_0 optimization is directly applied to denoise mesh vertices and in [11] a similar strategy is applied to smooth point clouds. However, even under small amounts of noise and/or with non-uniformly shaped triangles, the strong effect of ℓ_0 can produce spurious overshoots and fold-backs, and the method becomes extremely computationally inefficient. The alternative ℓ_1 norm is the convex relaxation of the ℓ_0 pseudo-norm, and plays a fundamental role in sparse image/signal processing. In [1], ℓ_1-sparsity has been adopted to denoise point sets in a two-phase minimization strategy. However, the ℓ_1 norm tends to underestimate high-amplitude values, thus struggling in the recovery under high-level noise and presents undesired staircase and shrinkage effects. A substantial amount of recent works has argued for classes of sparsity-promoting parametrized nonconvex regularizers in favor of their superior theoretical properties and excellent practical performances [5,9]. In this direction, the Minimax Concave (MC) penalty $\phi(\cdot; a) : \mathbb{R} \to \mathbb{R}$ see [4], provides a recognized alternative to the ℓ_1 norm and this motivated us to use it in the construction of our regularizer $\mathcal{R}(V)$. The parameter a allows to tune the degree of non-convexity, such that $\phi(\cdot; a)$ tends to ℓ_0 pseudo-norm for $a \to \infty$. The proposed regularizer $\mathcal{R}(V)$ controls sparsity of the normal deviation magnitudes more accurately than the ℓ_1 norm, while mitigating the strong effect and the numerical difficulties of ℓ_0

pseudo-norm. It can handle higher level noise than [2], produce better shaped triangles, while faithful recovering straight or smoothly curved edges.

Most of the variational mesh denoising approaches, split the process into two optimization phases - the normal smoothing phase followed by a vertex position update. The second phase suffers from foldovers problems and normal inconsistency, [15]. We propose a one-phase, effective, sparse variational model to directly smooth vertices, while preserving sharp features, keeping the normal consistency, and reducing foldovers problems. An efficient algorithm for minimizing the (non-convex) formulation is proposed which is based on the Alternating Direction Method of Multipliers (ADMM). Numerical experiments show the effectiveness of the proposed method for the solution of several mesh denoising examples.

2   Sparsity-Aided Variational Model

Let us assume a surface embedded in $\mathbb{R}^3$, which is approximated by a triangulated mesh (V, T, E), where $V \in \mathbb{R}^{n_V \times 3}$, $V = \{v_i\}_{i=1}^{n_V}$ represents the set of vertices, $T \in \mathbb{R}^{n_T \times 3}$, $T = \{\tau_m\}_{m=1}^{n_T}$ is the set of triangles and $E \in \mathbb{R}^{n_E \times 2}$, $E = \{e_j\}_{j=1}^{n_E}$ is the set of edges. We denote the first disk, i.e. triangle neighbors of a vertex v_i, by $\mathcal{D}(v_i) = \{\tau_m \mid v_i \in \tau_m\}$. Let $\mathcal{N}(V) : \mathbb{R}^{n_V \times 3} \to \mathbb{R}^{n_T \times 3}$ be the mapping that computes the piecewise-constant normal field over the mesh, where the m-th element being the outward unit normal at face $\tau_m = (v_i, v_j, v_k)$, is defined as

$$\mathcal{N}_m(V) := \left(\frac{(v_j - v_i) \times (v_k - v_i)}{\|(v_j - v_i) \times (v_k - v_i)\|_2} \right)^T \in \mathbb{R}^3 , \qquad m = 1, \ldots, n_T . \quad (2)$$

Focusing on the recovery of surfaces characterized by piecewise constant normals with sharp discontinuities in the normal map, we propose the following sparsity-inducing variational model to determine solutions V^* which are close to the noisy data V^0 according to the observation model and, at the same time, for which the vector of components $y_i^* = \|(\nabla N^*)_i\|_2$, $i = 1, \cdots, n_E$, is sparse

$$V^* \in \arg \min_{V \in \mathbb{R}^{n_V \times 3}} \mathcal{J}(V; \lambda, a) \quad (3)$$

$$\mathcal{J}(V; \lambda, a) := \frac{\lambda}{2} \|V - V^0\|_2^2 + \sum_{j=1}^{n_E} \phi \left(\left\| (\nabla \mathcal{N}(V))_j \right\|_2 ; a \right) .$$

At the aim to construct a parameterized sparsity-promoting regularizers characterized by tunable degree of non-convexity $a \in \mathbb{R}_+$, the function $\phi(t; a)$ is chosen among the wide class of parameterized, scalar, non-convex penalty functions, which mimic the asymptotically constant behaviour of the ℓ_0 pseudo-norm. In particular we consider one of the most effective representatives, the so-called minimax concave (MC) penalty function, $\phi(t; a) : \mathbb{R} \to \mathbb{R}$, defined as

$$\phi(t; a) = \begin{cases} -\dfrac{a}{2} t^2 + \sqrt{2a}\, t & \text{for } t \in \left[0, \sqrt{2/a} \right), \\ 1 & \text{for } t \in \left[\sqrt{2/a}, +\infty \right) \end{cases} \quad (4)$$

which, for any value of the parameter a, satisfies the following properties:

- $\phi(t;a) \in \mathcal{C}^0(\mathbb{R}) \cap \mathcal{C}^2(\mathbb{R} \setminus \{0\})$
- $\phi'(t;a) \geq 0$, $\phi''(t;a) \leq 0$, $\forall t \in [0,\infty) \setminus \{\sqrt{2/a}\}$
- $\phi(0;a) = 0$, $\inf\limits_{t}\phi''(t;a) = -a$.

The proposed non-convex penalty plays a key role in controlling, by the parameter $a > 0$, the normal variation more accurately than total variation (TV) and ℓ_0 regularizations, and induces sparsity more effectively.

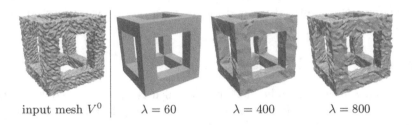

input mesh V^0 | $\lambda = 60$ $\lambda = 400$ $\lambda = 800$

Fig. 1. Example 1: influence of the fidelity parameter λ for fixed $a = 0.8$.

Finally, we introduce the discretization of the gradient operator on the 3D mesh. Since the normal field is piecewise-constant over the mesh triangles, the gradient operator vanishes to zero everywhere but the mesh edges along which it is constant. Therefore, the gradient operator discretization is represented by a global sparse matrix $D \in \mathbb{R}^{n_E \times n_T}$ defined as

$$
D_{ij} = \begin{cases} \sqrt{l_i} & \text{if } \tau_j \bigcap \tau_k = e_i \, , \ k > j, \\ -\sqrt{l_i} & \text{if } \tau_j \bigcap \tau_k = e_i \, , \ k < j, \\ 0 & \text{otherwise} \, , \end{cases} \tag{5}
$$

where $l_i = |e_i|$, $i = 1, \ldots, n_E$ represents the length of i-th edge.

In Sect. 3 we describe an iterative optimization algorithm to solve (3).

As the mesh topology does not change during the iterations, the matrix D can be decomposed as $D = L\bar{D}$, with $L = diag\{\sqrt{l_1}, \sqrt{l_2}, \ldots, \sqrt{l_{n_E}}\}$ being the diagonal matrix of edge lengths, updated during the iterations, and $\bar{D} \in \mathbb{R}^{n_E \times n_T}$ an edge-length independent sparse matrix.

3   Numerical Solution of the Sparse Variational Model

In this section we provide details of the Alternating Direction Method of Multipliers (ADMM)- based numerical method for the solution of the nonconvex optimization problem (3). Introducing a matrix variable $N \in \mathbb{R}^{n_T \times 3}$ with row components defined in (2), and utilizing the variable splitting technique for

$t \in \mathbb{R}^{n_E \times 3}$, $t = DN$, where D is discretized as in (5), the optimization problem (3) is reformulated as

$$\{V^*, N^*, t^*\} \in \arg \min_{V, N, t} \left\{ \frac{\lambda}{2} \|V - V^0\|_2^2 + \sum_{j=1}^{n_E} \phi\left(\|t_j\|_2; a\right) \right\}, \text{ s.t. } \begin{array}{c} t = DN, \\ N = \mathcal{N}(V). \end{array}$$

(6)

We define the augmented Lagrangian for (6) as

$$\mathcal{L}(V, N, t, \rho_1, \rho_2; \lambda, \beta_1, \beta_2, a) = \frac{\lambda}{2} \|V - V^0\|_2^2$$

$$+ \sum_{j=1}^{n_E} \left[\phi\left(\|t_j\|_2; a\right) - \langle \rho_{1_j}, t_j - (DN)_j \rangle + \frac{\beta_1}{2} \|t_j - (DN)_j\|_2^2 \right]$$

$$+ \sum_{\substack{m=1 \\ \tau_m = (v_i, v_j, v_k)}}^{n_T} \left[-\langle \rho_{2_m}, N_m - \mathcal{N}_m(V) \rangle + \frac{\beta_2}{2} \|N_m - \mathcal{N}_m(V)\|_2^2 \right], \quad (7)$$

where $\beta_1, \beta_2 > 0$ are scalar penalty parameters, and $\rho_1 \in \mathbb{R}^{n_E \times 3}$, $\rho_2 \in \mathbb{R}^{n_T \times 3}$ represent the matrices of Lagrange multipliers associated with the constraints. We then consider the following saddle-point problem:

Find $\quad (V^*, N^*, t^*, \rho_1^*, \rho_2^*) \in \mathbb{R}^{n_V \times 3} \times \mathbb{R}^{n_T \times 3} \times \mathbb{R}^{n_E \times 3} \times \mathbb{R}^{n_E \times 3} \times \mathbb{R}^{n_T \times 3}$

s.t. $\quad \mathcal{L}(V^*, N^*, t^*, \rho_1, \rho_2) < \mathcal{L}(V^*, N^*, t^*, \rho_1^*, \rho_2^*) \leq \mathcal{L}(V, N, t, \rho_1^*, \rho_2^*),$

$\quad \forall (V, N, t, \rho_1, \rho_2) \in \mathbb{R}^{n_V \times 3} \times \mathbb{R}^{n_T \times 3} \times \mathbb{R}^{n_E \times 3} \times \mathbb{R}^{n_E \times 3} \times \mathbb{R}^{n_T \times 3}.$ (8)

An ADMM-based iterative scheme is applied to approximate the solution of the saddle-point problem (7)–(8). Initializing to zeros both the dual variables $\rho_1^{(0)}$, $\rho_2^{(0)}$ and $N_m^{(0)} = \mathcal{N}_m(V^{(0)})$, $m = 1, \ldots, n_T$, the k-th iteration of the proposed alternating iterative scheme reads as follows:

$$t^{(k+1)} = \arg \min_{t \in \mathbb{R}^{n_E \times 3}} \mathcal{L}(V^{(k)}, N^{(k)}, t; \rho_1^{(k)}, \rho_2^{(k)}), \qquad (9)$$

$$N^{(k+1)} = \arg \min_{N \in \mathbb{R}^{n_T \times 3}} \mathcal{L}(V^{(k)}, N, t^{(k+1)}; \rho_1^{(k)}, \rho_2^{(k)}), \qquad (10)$$

$$V^{(k+1)} = \arg \min_{V \in \mathbb{R}^{n_V \times 3}} \mathcal{L}(V, N^{(k+1)}, t^{(k+1)}; \rho_1^{(k)}, \rho_2^{(k)}), \qquad (11)$$

$$\rho_1^{(k+1)} = \rho_1^{(k)} - \beta_1 \left(t^{(k+1)} - DN^{(k+1)} \right), \qquad (12)$$

$$\rho_2^{(k+1)} = \rho_2^{(k)} - \beta_2 \left(N^{(k+1)} - \mathcal{N}\left(V^{(k+1)}\right) \right). \qquad (13)$$

The updates of Lagrangian multipliers ρ_1 and ρ_2 have closed form, while the solutions to the remaining subproblems will be described in detail in the following sections.

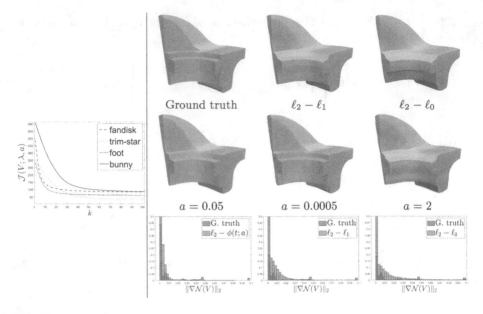

Fig. 2. Example 1: (left) empirical convergence of ADMM algorithm for some reconstructed meshes; (right) sensitivity to the penalty.

Subproblem for t. Omitting the constant terms in (7), we can rewrite the subproblem (9) as

$$t^{(k+1)} = \arg\min_{t \in \mathbb{R}^{n_E \times 3}} \sum_{j=1}^{n_E} \left[\phi\left(\|t_j\|_2 \, ; a\right) - \langle \rho_{1_j}, t_j - (DN)_j \rangle + \frac{\beta_1}{2} \|t_j - (DN)_j\|_2^2 \right]. \tag{14}$$

Due to the separability property of $\phi(\cdot\,; a)$, problem (14) is equivalent to n_E three-dimensional problems for each t_j, $j = 1, \ldots, n_E$ in form

$$t_j^{(k+1)} = \arg\min_{t_j \in \mathbb{R}^3} \left\{ \frac{1}{\beta_1} \phi\left(\|t_j\|_2 \, ; a\right) + \frac{1}{2} \|t_j - r_j^{(k+1)}\|_2^2 \right\}, \tag{15}$$

where $r_j^{(k+1)} := \left(DN^{(k)}\right)_j + \frac{1}{\beta_1} \left(\rho_1^{(k)}\right)_j$.

Necessary and sufficient conditions for strong convexity of the cost functions in (15) are demonstrated in [3]. In particular, the problems in (15) are strongly convex if and only if the following condition holds:

$$\beta_1 > a \implies \beta_1 = \varepsilon a, \quad \text{for} \quad \varepsilon > 1. \tag{16}$$

Under the assumption (16), the unique minimizers of (15) can be obtained in closed form as

$$t_j^{(k+1)} = \min(\max(\nu - \zeta/\|r_j\|_2, 0), 1)\, r_j\,,$$

where $\nu = \dfrac{\beta_1}{\beta_1 - a}$ and $\zeta = \dfrac{\sqrt{2a}}{\beta_1 - a}$. We remark that the condition on β_1 in (7) only ensures the convexity conditions (16) of t-subproblem (15), but does not guarantee convergence of the overall ADMM scheme.

Subproblem for N. Gathering the non-constant terms w.r.t. N in (7), we can reformulate (10) as

$$N^{(k+1)} = \arg\min_{N \in \mathbb{R}^{n_T \times 3}} \left\{ \frac{\beta_1}{2}\|t^{(k+1)} - DN\|_2^2 + \langle \rho_1^{(k)}, DN \rangle - \langle \rho_2^{(k)}, N \rangle \right.$$
$$\left. + \frac{\beta_2}{2}\left\| N - \mathcal{N}\left(V^{(k)}\right)\right\|_2^2 \right\},$$

for which the first optimality conditions lead to the following three linear systems, one for each spatial coordinate of $N \in \mathbb{R}^{n_T \times 3}$

$$\left(D^T D + \frac{\beta_2}{\beta_1} I \right) N = \frac{\beta_2}{\beta_1} \mathcal{N}\left(V^{(k)}\right) + \frac{\rho_2^{(k)}}{\beta_1} + D^T \left(t^{(k+1)} - \frac{1}{\beta_1}\rho_1^{(k)} \right). \tag{17}$$

Since $\beta_1, \beta_2 > 0$, the linear system coefficient matrix is sparse, symmetric, positive definite and identical for all three coordinate vectors, therefore, the system can be solved by applying a unique Cholesky decomposition. At each iteration, the edge lengths diagonal matrix L in $D = L\bar{D}$, needs to be updated as the vertices V move to their new position. For large meshes, an iterative solver warm-started with the solution of the last ADMM iteration, is rather preferred. A normalization is finally applied as N represents the normal field.

Subproblem for V. Omitting the constant terms in (7), the subproblem for V reads as

$$V^{(k+1)} = \arg\min_{V \in \mathbb{R}^{n_V \times 3}} \{\mathcal{J}_V(V)\}$$

$$\mathcal{J}_V(V) = \frac{\lambda}{2}\|V - V^0\|_2^2 + \sum_{m=1}^{n_T}\left[\langle \rho_{2_m}^{(k)}, \mathcal{N}_m(V) \rangle + \frac{\beta_2}{2}\left\| N_m^{(k+1)} - \mathcal{N}_m(V)\right\|_2^2 \right]. \tag{18}$$

The functional $\mathcal{J}_V(V)$ is proper, smooth, non-convex and bounded from below by zero. A minimum can be obtained applying the gradient descent algorithm with backtracking satisfying the Armijo condition or using the BFGS method. For the experimental section we used the gradient descent algorithm for efficiency.

The partial derivative of (18) w.r.t vertex $v_i \in V$, $i = 1, \ldots, n_V$ reduces the sum in (18) to the sum over the first disk $\mathcal{D}(v_i)$ and is given as

$$\nabla_{v_i} \mathcal{J}_V(V) = \lambda(v_i - v_i^0) + \sum_{\substack{\tau_m \in \mathcal{D}(v_i) \\ \tau_m = (v_i, v_j, v_k)}} \left[\frac{\left(\rho_{2_m}^{(k)} - \beta_2 N_m^{(k+1)}\right) \times (v_k - v_j)}{\|(v_j - v_i) \times (v_k - v_i)\|_2} \right.$$
$$\left. - \frac{\langle \rho_{2_m}^{(k)} - \beta_2 N_m^{(k+1)}, (v_j - v_i) \times (v_k - v_i) \rangle \left[(v_j - v_i) \times (v_k - v_i) \times (v_k - v_j) \right]}{\|(v_j - v_i) \times (v_k - v_i)\|_2^3} \right],$$

$$(19)$$

which simplifies as follows, for all triangles $m = 1, \ldots, n_T$,

$$\nabla_{v_i} \mathcal{J}_V(V) = \lambda(v_i - v_i^0) +$$
$$\sum_{\tau_m \in \mathcal{D}(v_i)} \frac{\left[\left(\rho_{2_m}^{(k)} - \beta_2 N_m^{(k+1)} \right) - \langle \rho_{2_m}^{(k)} - \beta_2 N_m^{(k+1)}, \mathcal{N}_m(V) \rangle \mathcal{N}_m(V) \right]}{2 s_{\tau_m}} \times (v_k - v_j),$$

with $s_{\tau_m} := \|(v_j - v_i) \times (v_k - v_i)\|_2 / 2$ the area of triangle τ_m with updated vertices in V, and $\mathcal{N}_m(V) = ((v_j - v_i) \times (v_k - v_i))/(2 s_{\tau_m})$.

The convergence of our proposed three block ADMM scheme is not easy to derive relying on the results presented so far, see [14]. However, we will provide some evidence of the numerical convergence in the experimental section.

Many two-phase mesh denoising algorithms present the normal orientation ambiguity problem in the vertex updating phase, which provokes ambiguous shifts of the vertex position due to direction inconsistency of the normal vectors [10,16]. In [6] this issue is solved by an orientation aware vertex updating scheme which considers the parallelism of the normal determined by triangle vertices to a given normal vector.

Proposition 1. *The reconstructed normal map N^* obtained by solving (6) via the proposed ADMM, satisfies the orientation consistency.*

Proof. The vertex update is computed by solving the ADMM sub-problem (18). The second term in function (18) induces for each triangle τ_m the orthogonality between the triangle normal $\mathcal{N}_m$ and ρ_2, that is, it minimizes the volume of the parallelepiped defined by the edges ρ_2, $(v_j - v_i)$ and $(v_k - v_i)$ thus imposing ρ_2 to lie in the triangle plane generated by $(v_j - v_i)$ and $(v_k - v_i)$. The third term in the objective function (18) penalizes the discrepancy between the restored normal $(N_m^{(k+1)})$ and the triangle face normal obtained by its updated vertices, both in orientation and in magnitude.

4   Numerical Examples

We validate the proposed method both qualitatively and quantitatively on a variety of benchmark triangulated surfaces characterized by different sharpness and smoothness features. The noisy meshes have been synthetically corrupted following the degradation model

$$v_i^0 = v_i + c_i d_i , \qquad i = 1, \ldots, n_V , \tag{20}$$

where $c_i \in \mathbb{R}$ is Gaussian noise distributed with zero mean and standard deviation $\sigma = \gamma \bar{l}$ where $\bar{l}$ is the average edge length and $\gamma \geq 0$ represents the noise level. The vectors d_i, $i = 1, \ldots, n_V$ can be random directions or the vertex normal itself. All the meshes are rendered in flat-shading model.

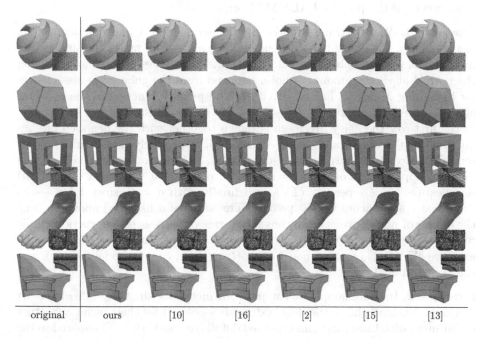

| original | ours | [10] | [16] | [2] | [15] | [13] |

Fig. 3. Example 2: Denoising results from input meshes corrupted by noise levels $\gamma = \{0.15, 0.3, 0.3, 0.2, 0.2\}$, from top to bottom.

Example 1 presents various aspects of our algorithm, and illustrates the sparsity-promoting benefits introduced by the penalty $\phi(\cdot; a)$ with respect to ℓ_1 (TV) and ℓ_0 penalty terms. In Example 2 the performance of the proposed method is compared with some other variational methods for mesh denoising, namely [2,10,15,16], which have been kindly provided by authors of [15] at `https://github.com/bldeng/GuidedDenoising`, and a learning-based approach [13]. For each method, we show their best results we achieved by tuning the corresponding set of parameters.

The quantitative evaluation regards the following error metrics, which measure the discrepancy of the computed V^*, N^* to the noisy-free mesh V_{GT}, N_{GT}:

- **Mean squared angular error (MSAE)** $MSAE = \mathbb{E}[\angle(N_{GT}, N^*)^2]$,
- L_2**vertex to vertex error** (E_V) $E_V = \frac{\|V^* - V_{GT}\|_F}{n_V}$.

For all the tests, the iterations k of the ADMM algorithm are stopped as soon as either of the two following conditions is fulfilled:

$$k > \mathrm{TH}_1 = 200\,, \qquad \left\|V^{(k+1)} - V^{(k)}\right\|_2 / \left\|V^{(k)}\right\|_2 < \mathrm{TH}_2 = 10^{-6}. \qquad (21)$$

Figure 2(left) shows the energy evolution curve in terms of the number of iterations for some of the meshes reported in this section, which returns the empirical convergence of the proposed ADMM algorithm.

Example 1. We illustrate how the model parameters λ and a influence the result quality. The value for λ depends on the amount of noise: the smaller the noise the bigger has to be λ. In Fig. 1, for a fixed noise level $\gamma = 0.3$ and d_i being the normal at v_i, the amount of noise removed from the cube_hole mesh is less for increasing λ values. The sensitivity to the penalty function in the recovery of a corrupted fandisk mesh (noise level $\gamma = 0.2$) is illustrated in Fig. 2(right). In the first row the noisy free mesh is shown together with the denoised meshes obtained by the $\ell_2 - \ell_1$ and the $\ell_2 - \ell_0$ models, respectively. The results present remarkable losses of sharp features and creases. In the second row the $\ell_2 - \phi(t; a)$ model is applied with optimal $a = 5 \times 10^{-2}$, $a = 5 \times 10^{-4}$ as for $a \to 0$, $\phi(t; a)$ behaves like the ℓ_1- penalty (TV), and finally with $a = 2$, since for $a \to \infty$ the penalty approaches the ℓ_0- penalty. The value of a in $\phi(\cdot; a)$ allows to tune the degree of non-convexity and thus the degree of flatness reconstruction, while preserving features. Figure 2 (third row) shows the histograms of the sparsified measure $\|\nabla\mathcal{N}(V)\|_2$ for our model $\ell_2 - \phi(t; a)$ (left), $\ell_2 - \ell_1$ model (center), and the $\ell_2 - \ell_0$ model (right).

Example 2. In this example we compare our method with other state of-the-art methods. Figure 3 shows the denoised meshes colored by their mean curvature scalar map, with fixed range, and zoomed details on mesh edges. Compared to the visually better denoising results obtained by our method, remarkable overlaps appear in the other results and severe triangle perturbations are introduced in the reconstructed meshes. To further demonstrate the robustness to noise, in Fig. 4 we test our sparsity-inducing variational framework on increasing levels of noise $\gamma = \{0.2, 0.3, 0.4, 0.5, 0.6\}$ from top to bottom, and in the last two rows also for arbitrary noise directions (d_i) in (20) and a real 3D scanned data, respectively. Below each synthetic result, we report the quantitative evaluations according to the error metrics ($MSAE \times 10^2$, $E_V \times 10^6$). Both quantitatively and qualitatively the results confirm the effectiveness of our sparse variational proposal. Finally, we can comment on the efficiency of our algorithm which computational time is, on average, one order less than the $\ell_2 - \ell_0$ denoising method, and comparable to the other tested methods.

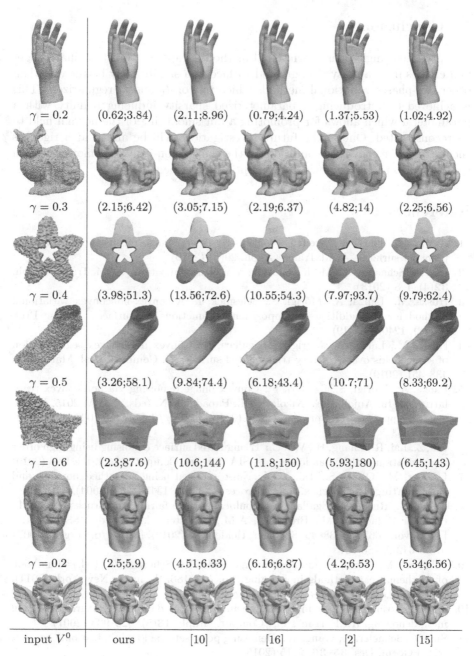

	$\gamma = 0.2$	(0.62;3.84)	(2.11;8.96)	(0.79;4.24)	(1.37;5.53)	(1.02;4.92)
	$\gamma = 0.3$	(2.15;6.42)	(3.05;7.15)	(2.19;6.37)	(4.82;14)	(2.25;6.56)
	$\gamma = 0.4$	(3.98;51.3)	(13.56;72.6)	(10.55;54.3)	(7.97;93.7)	(9.79;62.4)
	$\gamma = 0.5$	(3.26;58.1)	(9.84;74.4)	(6.18;43.4)	(10.7;71)	(8.33;69.2)
	$\gamma = 0.6$	(2.3;87.6)	(10.6;144)	(11.8;150)	(5.93;180)	(6.45;143)
	$\gamma = 0.2$	(2.5;5.9)	(4.51;6.33)	(6.16;6.87)	(4.2;6.53)	(5.34;6.56)
input V^0	ours	[10]	[16]	[2]	[15]

Fig. 4. Example 2: Comparison of our method with related works.

5   Conclusion

The proposed single-phase variational method is capable to restore sharp edges and creases from a noisy triangulated surface in a significantly better way than other two-phases variational methods which rely on ℓ_0 and ℓ_1 regularizers. This is achieved by introducing a parameterized sparsity inducing penalty with a parameter a which allows for promoting a fair smoothing of the normal field to be reconstructed. One of the future investigations will be aimed at a rigorous theory to derive convexity conditions in the convex non-convex framework which would lead to the well-known convex optimization benefits.

References

1. Avron, H., Sharf, A., Greif, C., Cohen-Or, D.: ℓ_1-sparse reconstruction of sharp point set surfaces. ACM Trans. Graph. 29(5) (2010)
2. He, L., Schaefer, S.: Mesh denoising via l0 minimization. ACM Trans. Graph. **32**(4), 1–8 (2013)
3. Huska, M., Lanza, A., Morigi, S., Selesnick, I.: A convex-nonconvex variational method for the additive decomposition of functions on surfaces. Inverse Prob **35**(12), 124008 (2019)
4. Huska, M., Lanza, A., Morigi, S., Sgallari, F.: Convex non-convex segmentation of scalar fields over arbitrary triangulated surfaces. J. Comput. Appl. Math. **349**, 438–451 (2019)
5. Lanza, A., Morigi, S., Sgallari, F.: Convex image denoising via non-convex regularization. In: Aujol, J.-F., Nikolova, M., Papadakis, N. (eds.) SSVM 2015. LNCS, vol. 9087, pp. 666–677. Springer, Cham (2015). https://doi.org/10.1007/978-3-319-18461-6_53
6. Liu, Z., Lai, R., Zhang, H., Wu, C.: Triangulated surface denoising using high order regularization with dynamic weights. SIAM J. Sci. Comput. **41**(1), B1–B26 (2019)
7. Lysaker, M., Osher, S., Tai, X.-C.: Noise removal using smoothed normals and surface fitting. IEEE Trans. Image Process. **13**(10), 1345–1357 (2004)
8. Morigi, S., Rucci, M., Sgallari, F.: Nonlocal surface fairing. In: Bruckstein, A.M., ter Haar Romeny, B.M., Bronstein, A.M., Bronstein, M.M. (eds.) SSVM 2011. LNCS, vol. 6667, pp. 38–49. Springer, Heidelberg (2012). https://doi.org/10.1007/978-3-642-24785-9_4
9. Nikolova, M.: Energy minimization methods. In: Scherzer, O. (ed.) Handbook of Mathematical Methods in Imaging, pp. 138–186. Springer, New York (2011). https://doi.org/10.1007/978-0-387-92920-0_5
10. Sun, X., Rosin, P.L., Martin, R., Langbein, F.: Fast and effective feature-preserving mesh denoising. IEEE Trans. Vis. Comput. Graph. **13**(5), 925–938 (2007)
11. Sun, Y., Schaefer, S., Wang, W.: Denoising point sets via l0 minimization. Comput. Aided Geom. Des. **35–36**, 2–15 (2015)
12. Tasdizen, T., Whitaker, R., Burchard, P., Osher, S.: Geometric surface processing via normal maps. ACM Trans. Graph. **22**(4), 1012–1033 (2003)
13. Wang, P.S., Liu, Y., Tong, X.: Mesh denoising via cascaded normal regression. ACM Trans. Graph. **35**(6), 232–1 (2016)

14. Wang, Y., Yin, W., Zeng, J.: Global convergence of admm in nonconvex nonsmooth optimization. J. Sci. Comput. **78**(12), 29–63 (2019)
15. Zhang, W., Deng, B., Zhang, J., Bouaziz, S., Liu, L.: Guided mesh normal filtering. Comput. Graph. Forum **34**(7), 23–34 (2015)
16. Zheng, Y., Fu, H., Au, O.K., Tai, C.: Bilateral normal filtering for mesh denoising. IEEE Trans. Vis. Comput. Graphics **17**(10), 1521–1530 (2011)

Lossless PDE-based Compression of 3D Medical Images

Ikram Jumakulyyev[ID] and Thomas Schultz[(✉)][ID]

University of Bonn, Bonn, Germany
{ijumakulyyev,schultz}@cs.uni-bonn.de

Abstract. Inpainting with Partial Differential Equations (PDEs) has previously been used as a basis for lossy image compression. For medical images, lossless compression is often considered to be safer, given that even subtle details could be diagnostically relevant. In this work, we introduce a PDE-based codec that achieves competitive compression rates for lossless image compression. It is based on coding the differences between the original image and its PDE-based reconstruction. These differences often have lower entropy than the original image, and can therefore be coded more efficiently. We optimize this idea via an iterative reconstruction scheme, and a separate coding of empty space, which takes up a considerable fraction of the field of view in many 3D medical images. We demonstrate that our PDE-based codec compares favorably to previously established lossless codecs. We also investigate the individual benefit from each ingredient of our codec on multiple examples, explore the effect of using homogeneous, edge enhancing, and fourth-order anisotropic diffusion, and discuss the choice of contrast parameters.

1 Introduction

The overall size of neuroimaging data that is acquired each year has been reported to grow exponentially [5], due to the proliferation of medical imaging devices, their increased resolution, and the increasing use of multiple contrasts or channels. This makes the development of compression schemes for the storage of 3D medical images an important and timely research goal.

The use of diffusion-based inpainting has been explored for the lossy compression of images [6,23,24], videos [1,11,22], and audio [21]. This paradigm is based on storing information only for a sparse subset of the original samples, and interpolating it to approximate the remaining parts of the original signal. Interpolation is often done via Partial Differential Equations (PDEs) that are inspired by the well-known heat transfer equation, in analogy to how radiators that are sparsely distributed in a room would heat up the space in between them.

Supported by the German Academic Exchange Service (DAAD). The brain MR images were kindly provided by Tobias Schmidt-Wilcke, University of Düsseldorf. The foot CT dataset is courtesy of Philips Research.

A. Elmoataz et al. (Eds.): SSVM 2021, LNCS 12679, pp. 450–462, 2021.
https://doi.org/10.1007/978-3-030-75549-2_36

Almost all previous works on PDE-based compression have focused on 2D natural images or videos. Only a single example has considered a 3D extension [20]. Even more importantly, all above-mentioned codecs are for lossy compression, and their benefit relative to established transform-based codecs like JPEG [18] and JPEG2000 [25] tends to be most pronounced at high compression rates [24]. However, compression schemes that lead to visually noticeable changes are less suitable for medical images, since potentially subtle, but diagnostically relevant details might be perturbed. Therefore, lossless compression is often preferred and is sometimes even legally required [9,10,16], since it guarantees not to interfere with interpretation or quantification of the image contents.

Our work is the first to explore the potential of PDE-based methods for the lossless compression of 3D medical images. In Sect. 3, we present a lossless PDE-based codec that stores the residuals between the PDE-based reconstruction and the original values. Its success rests on three key ideas: First, we use a simple regular grid as the initial inpainting mask, so that the locations of the mask voxels do not have to be stored explicitly. Second, we encode and decode the image iteratively, alternating between PDE-based reconstructions and a dilation of the inpainting mask. Compared to a single reconstruction, this further reduces the entropy of the residuals that have to be stored. Third, we optionally code regions of empty space separately, since they take up a substantial fraction of the field of view in many medical images.

In Sect. 4, we demonstrate that our codec achieves a higher compression rate than several established codecs on three Magnetic Resonance Images with different characteristics, as well as a Computed Tomography image. Moreover, we study the effect of several variations of our codec, using different PDEs, iteration modes, and contrast parameters.

2   Related Work

Several lossless compression standards are widely used in medical imaging. The Digital Imaging and Communications in Medicine (DICOM) standard defines a unified image file format for different devices, manufacturers, and modalities [12]. DICOM accounts for lossless compression with JPEG-LS, as well as lossy and lossless JPEG and JPEG2000. Consequently, these are most often used as a reference to which new codecs are compared: Lossy JPEG and JPEG2000 in case of lossy and hybrid or near-lossless medical image compression schemes [20,28], the lossless JPEG family for lossless compression [9,10].

The Neuroimaging Informatics Technology Initiative (NIfTI) defines an alternative file format that has been widely adopted for brain imaging. It consists of a header, followed by a binary representation of voxel intensities. NIfTI files are commonly compressed by simply applying GZIP [4] to them. GZIP is based on the Deflate algorithm [3], which is in turn based on the LZ77 and Huffman compression schemes, which have occasionally been used as an additional reference for lossless image compression [9]. In Sect. 4.1, we will compare our own codec to JPEG-LS, lossless JPEG and JPEG2000, as well as GZIP.

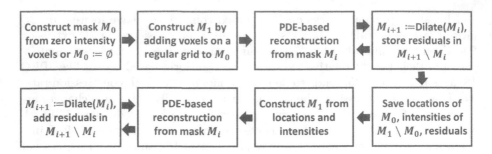

Fig. 1. An overview of the individual steps taken to encode (blue) or decode (red) a 3D image. At the core of our codec is an iteration that alternates between PDE-based reconstruction from an inpainting mask, and a dilation of that mask. (Color figure online)

Prior to our work, the only PDE-based image compression codec for 3D medical images was C-EED [20]. It is based on edge-enhancing diffusion (EED) [26] and a cuboid subdivision scheme that extends the rectangular subdivision in the previously proposed R-EED codec [24]. Since it aims for lossy compression, C-EED is based on very different design decisions than our codec. In particular, it applies brightness optimizations and quantization to the mask voxel values, which makes it more efficient to store them but, in our context, would require storing residuals even for the voxels in the inpainting mask.

Recent work has demonstrated the potential of deep learning for lossless compression of natural images [15]. Adapting such an approach to 3D medical images will have to account for data privacy, which makes it difficult to obtain large-scale training data and raises concerns about inference attacks [17]. We consider this to be a separate line of research which is outside of our scope.

3   Our Proposed Lossless Codec

Figure 1 shows a high-level overview of our PDE-based lossless codec. Section 3.1 will provide details on the first two steps, in which the encoder (blue) constructs an initial inpainting mask. The next two steps are the core of our approach. They alternate between a PDE-based reconstruction and a dilation of the mask, and will be discussed in Sect. 3.2. Finally, the initial mask and residuals are stored in compressed form (Sect. 3.4). The decoder (red) mirrors the encoder in that it again alternates between reconstruction and mask dilation.

3.1   Constructing the Initial Mask

In most lossy PDE-based image compression schemes, a substantial effort goes into selecting a suitable small subset of pixels as an inpainting mask from which the original image can be approximately reconstructed. To increase image quality, semantically important image features such as edges and corners are typically included in the mask [13,14], and optimal inpainting masks have been approximated by sophisticated mask selection methods [6,7,23,24].

Our lossless PDE-based codec restores the original image exactly by also coding the residual with respect to the PDE-based reconstruction. This strategy yields an advantage in terms of compression rate since the residuals are more compressible than the original intensities. However, our strategy only achieves a net benefit as long as the cost of coding the initial mask does not exceed the gain from increased compressibility of the residuals.

Therefore, we simply use a regular grid as our initial mask, which has the advantage of not having to store any voxel locations. In particular, for a 3D input image of size $n_x \times n_y \times n_z$, our initial mask is the hexahedral grid consisting of voxels $(4i, 4j, 4k)$, where $i \in \{0, 1, \ldots, \lfloor (n_x - 1)/4 \rfloor\}$, and j, k are defined accordingly. This amounts to storing the intensities of approximately 1.6% of all voxels. We attempted to use more sophisticated masks that exploit edge information, but found that, even though it yielded even more compressible residuals, the cost of coding the masks grew disproportionally.

Our codec exploits the fact that many medical images contain a substantial amount of empty space, which typically yields the lowest possible intensity, and can be coded efficiently as a run length encoded binary mask. In the following, we assume that the minimum intensity will be zero. In practice, our encoder deals with negative intensities, as they arise in computed tomography (CT), by subtracting the minimum from the original input and storing it, so that the decoder can add it again to its output. In some cases, the gain from including voxels with zero intensity in the preliminary inpainting mask M_0 is substantial. In others, its cost outweighs its benefit, because intensities within empty space are perturbed by strong measurement noise, or the image contains little or no empty space. In this case, our encoder simply sets $M_0 := \emptyset$. The initial inpainting mask M_1 arises as the union of M_0 and the voxels on the above-described grid. We only store the intensities of grid voxels outside of M_0.

3.2   Iterative Reconstruction and Residual Coding

A straightforward lossless PDE-based codec would reconstruct the image from the inpainting mask M_1, and it would code the residuals in all non-mask voxels. However, we found the initial reconstruction to be so coarse that this does not yet yield a competitive compression rate. This reflects the fact that our initial mask does not adequately sample all semantically relevant image structures. We compensate for this by an iterative reconstruction and coding of residuals.

In each iteration, we first reconstruct the image from the current inpainting mask M_i. We then store the residuals in the immediate vicinity of the current mask. Those residuals are typically the most compressible, since the uncertainty in the inpainting result tends to increase with distance away from the known part of the image. Voxels whose residuals are stored are added to the inpainting mask M_{i+1} that will be used in the next iteration. The decoder mirrors this iterative reconstruction, again starting with the initial mask M_1, then adding the stored residuals from the immediate neighbors to the reconstruction results.

This yields the original intensities in a subdomain of the image that grows with each iteration, until all voxels have become part of the mask.

We grow the mask by applying a morphological dilation to it. We experimented with two different structuring elements. The first is a cube, which amounts to a $3 \times 3 \times 3$ neighborhood. We call this Mode 0. The second is a cross, which amounts to the six face-connected neighbors. We call this Mode 1. Compression in Mode 0 requires two or three iterations, while Mode 1 takes six or seven iterations, depending on boundary effects. Section 4.3 will investigate the effect of the two different modes on the final compression rates. The computational cost of later iterations decreases with the number of remaining unknown voxels, and because we initialize them with the inpainting result from the previous iteration.

Residuals could be positive or negative. We avoid having to store them as signed integers by performing subtractions (in the encoder) and additions (in the decoder) in modular arithmetic, with the maximum value as the modulus. As mentioned above, the minimum intensity at this point will always be zero.

3.3   Choice of PDE and Its Parameters

Our compression strategy bears a certain conceptual resemblance to some established lossless codecs, such as JPEG-LS, which predict the values that have not yet been coded from the ones that are already known, and only code the residuals. Whether we can beat their compression rate should partly depend on whether PDE-based predictions are more successful at decreasing residual entropy compared to the simpler predictor used in JPEG-LS.

We experimentally determined the suitability of three different PDEs for lossless compression: Linear homogeneous diffusion as a simple baseline, edge-enhancing diffusion (EED), which is a popular choice for PDE-based lossy compression [24], and a recently introduced fourth-order generalization of EED [8].

Second-order diffusion can be stated as

$$\partial_t u = \mathrm{div}(\mathbf{D} \cdot \nabla u), \tag{1}$$

where u denotes the image intensity as a function of location within the image domain, and of diffusion time t. Diffusion-based inpainting uses the intensities in the mask voxels as Dirichlet boundary conditions, and obtains the inpainted result as the steady state that is attained as $t \to \infty$ [7].

In linear homogeneous diffusion, the diffusion tensor $\mathbf{D}$ is the identity. For EED, it is a symmetric matrix field that encodes directional dependence, so that diffusion across images edges is decreased depending on the gradient magnitude, while diffusion along the edge is free. In fourth-order EED, the first-order divergence and gradient operators in Eq. (1) are replaced with second-order counterparts, and the second-order diffusion tensor $\mathbf{D}$ is replaced with a fourth-order tensor that acts on the Hessian. For the sake of brevity, we refer the reader to [8,26] for the full mathematical details and the numerical implementation of these PDEs.

The definitions of diffusion tensors for second- and fourth-order EED involve a diffusivity function, which determines the diffusivity across the edge as a function of gradient magnitude. As it is customary in PDE-based inpainting, we select the Charbonnier diffusivity function,

$$g(s^2) = \frac{1}{\sqrt{1 + \frac{s^2}{\lambda^2}}} , \qquad (2)$$

where $s = \|\nabla_\sigma u\|$ is the gradient magnitude, computed with a certain amount of pre-smoothing. In our experiments, we fixed it at $\sigma = 1$. We also tried other values of σ, but found that this had only a very minor effect on the compression rate. This agrees with experimental findings in lossy image compression [23].

A second parameter in Eq. (2) is the contrast parameter λ, which corresponds to the scale of $\|\nabla_\sigma u\|$ at which g switches from high to low diffusivity. This parameter affects the quality of the PDE-based reconstructions, and therefore, the compression rate. Which contrast parameter value is optimal depends on the image contents, inpainting mask, and PDE. Some lossy PDE-based codecs have optimized λ by trying out different candidate values [23,24].

Empirically optimizing λ causes a noticeable computational expense and, as will be reported in more detail in Sect. 4.4, we found its benefit in the context of lossless compression to be relatively minor. Therefore, we rely on a heuristic choice of λ. It is based on one suggested by Perona and Malik [19], who proposed to set λ to the 90th percentile of the gradient magnitudes in the input image. We adapt this in two ways: First, we only consider gradient magnitudes outside of the initial mask M_1, in order to exclude the potentially large flat regions of empty space. Second, we need to account for the fact that inpainting from a sparse mask results in an image that is much smoother than the original one. For this reason, we divide the value at the 90th percentile by the empirical divisor 25. All reported results are based on this simple heuristic.

3.4 Compressed File Format

Our compressed files consist of a header, the locations of zero intensity voxels (if separating them yielded a benefit), the values at the initial mask voxels, as well as the residuals at non-mask voxels. We assume that intensity values are 16 bit integers, as it is common in medical imaging. If zero intensity voxels are coded separately, this is done as a binary mask, which is compressed using run length encoding, followed by the Deflate algorithm. The mask intensities and residuals are compressed using the Deflate algorithm or pure Huffman coding, depending on which choice resulted in the smaller size.

To ensure that comparisons to compressed NIfTI files are fair, we add the full NIfTI header (348 bytes), which includes the image dimensions among other information. In addition, we have to store the original minimum and maximum values (4 bytes), sizes of the compressed data streams for zero voxel binary mask and mask intensities (8 bytes), the contrast parameter (4 bytes), as well as single byte that encodes the type of PDE (2 bits), the dilation mode (1 bit), and the types of encoding for mask intensities and residuals (2 bits).

Table 1. A comparison of different variants of our PDE-based codec to established lossless standards. Positive percentages indicate a relative benefit from our codec.

Image	PDE Codec	GZIP	JPEG	JPEG2000	JPEG-LS
B0	R-ILH-1	+26.489%	+29.747%	+17.238%	+2.980%
B0	R-IEED-1	+28.036%	+31.225%	+18.979%	+5.022%
B0	R-IFOEED-1	+28.784%	+31.940%	+19.821%	+6.009%
B700	R-ILH-1	+23.778%	+6.922%	−4.461%	+7.123%
B700	R-IEED-1	+27.167%	+11.061%	+0.184%	+11.253%
B700	R-IFOEED-1	+27.552%	+11.530%	+0.711%	+11.721%
T1	R-ILH-1	+32.294%	+31.912%	−5.142%	−1.650%
T1	R-IEED-1	+35.615%	+35.252%	+0.015%	+3.336%
T1	R-IFOEED-1	+37.925%	+37.575%	+3.602%	+6.804%
CT	R-ILH-1	+16.954%	+37.111%	+11.198%	+3.527%
CT	R-IEED-1	+19.886%	+39.332%	+14.334%	+6.934%
CT	R-IFOEED-1	+20.658%	+39.916%	+15.158%	+7.830%

4 Results

For our experiments, we chose four 3D medical images which are illustrated in Fig. 2. Even though three of them are from brain imaging, they have been chosen to represent diverse contrasts and properties.

1. A scan from diffusion MRI [2] with diffusion weight $b = 0$ and $136 \times 136 \times 84$ voxels. A brain extraction algorithm has zeroed out 73% of the voxels. This results in a test case that is analogous to hybrid compression, in which only a clinically relevant region of interest (ROI) is losslessly compressed [27,28].
2. A diffusion MRI scan with $b = 700$ and $104 \times 104 \times 72$ voxels. This time, no brain extraction has been performed, and there is substantial measurement noise in the background, yielding less than 10% voxels with exactly zero intensity. This should provide a challenging test case for our codec, since the noisy background region should be difficult to compress losslessly.
3. A T1 weighted MR image with $256 \times 256 \times 220$ voxels. No brain extraction has been performed, but there is much less noise, leading to more than 65% zero voxels. Due to the higher spatial resolution, we expected a larger degree of spatial dependencies which could be exploited by a PDE-based inpainting.
4. A foot CT image with $256 \times 256 \times 256$ voxels, which we expected to be challenging due to the noisy appearance within the foreground region.

4.1 Comparison to Other Codecs

Table 1 compares the compression rate of our proposed codec to four established alternatives, by specifying the relative differences in final file sizes. Positive values

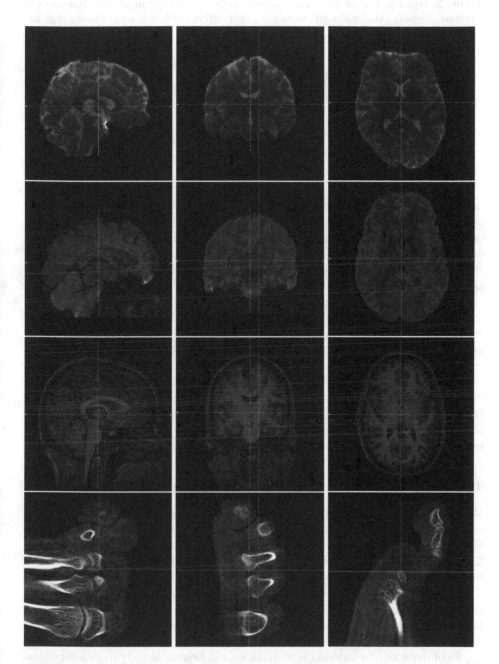

Fig. 2. The four 3D medical images used in our experiments. Top to bottom: Three brain MR images with *B0*, *B700*, and *T1* weighting, and a foot *CT* image. Right to left: The middle slices on the sagittal, coronal, and axial planes.

Table 2. Delta coding of intensities already reduced entropy in all test images. We separately report this for the whole images, and within their non empty space regions.

Image	Image entropy	Delta coded image entropy	Nonzero region entropy	Delta coded nonzero region entropy
B0	3.63997	3.50099	10.33056	9.63958
B700	5.34271	5.01050	5.39311	5.11973
T1	3.14553	2.73150	6.46612	5.54785
CT	2.81012	2.66688	6.71209	6.15826

indicate a benefit of our codec. Details on the four lossless codecs included in our comparison are given in Sect. 2. Results consider different variants of our codec, using linear homogeneous (LH), second-order edge enhancing (EED), or fourth-order edge enhancing diffusion (FOEED). All cases use iterative reconstruction from a regular grid (R-I) in Mode 1 (see Sect. 3.2). The effect of using fewer or no iterations will be studied separately in Sect. 4.3.

In all four examples, we observe a clear improvement when moving from basic linear homogeneous diffusion to anisotropic diffusion. Highest compression rates were achieved with the recently introduced fourth-order EED. It allowed us to achieve higher lossless compression rates than any established codec. In many cases, the margin was considerable.

However, FOEED also had the highest computational cost. For the B700 image, iterative 3D reconstruction on a single 3.3 GHz CPU core took 27 s with LH diffusion, 478 s with EED, and 6185 s with FOEED. We expect that these times could be shortened significantly by a parallel implementation. This was not the focus of our current work.

4.2   Non-PDE Baseline for Further Comparisons

We require a suitable baseline to pinpoint the effect that the iterative reconstruction and tuning of the contrast parameter have on the final compression rate. Since we use Huffman coding or the Deflate algorithm for the final encoding, file sizes achieved with GZIP are a natural reference. In addition, we designed a baseline codec that is "in between" GZIP and our PDE-based codec.

First, we found that, when 3D images contain a substantial amount of empty space, coding it separately can increase compression rates. Therefore, our baseline codec performs the same run length encoding as our PDE-based codec if it decreases the overall file size. This was the case in all examples except B700.

Second, we were wondering how much we can benefit already from a very simple, non PDE-based prediction of voxel intensities. To this end, we performed a delta coding, i.e., we fed differences between subsequent voxel intensities instead of the intensities themselves into the final compression. Table 2 shows that, in all cases, this decreased the entropy. It also slightly increased compression rates.

Table 3. Our non-PDE baseline that makes use of delta coding and optional empty space coding already results in a clear improvement over GZIP.

Image	GZIP (bytes)	Non-PDE Baseline	Zero Density	Zero Mask (bytes)
B0	692.372	+21.431%	72.92%	9.928
B700	610.968	+18.015%	9.07%	43.681
T1	5.207.535	+26.955%	65.70%	290.969
CT	4.515.257	+8.347%	71.10%	315.173

Table 4. Compared to the non-PDE baseline, direct coding of residuals after a single reconstruction with second-order EED does not yet result in a clear benefit. However, an iterative reconstruction as described in Sect. 3.2 does.

Image	Direct residual	Iteration in Mode 0	Iteration in Mode 1
B0	+0.095%	+4.367%	+8.406%
B700	+1.784%	+6.847%	+11.164%
T1	−0.698%	+6.715%	+11.845%
CT	−6.476%	+7.286%	+12.590%

Finally, as in our PDE-based codec, our baseline codec uses either Deflate or pure Huffman coding, depending on what results in a smaller file. We used the same implementations from zlib and dippykit, respectively. Table 3 shows that this baseline already improves considerably over GZIP.

4.3 Effect of Iterative Construction of Residuals

Table 4 shows how the iterative alternation between reconstruction and residual coding that is described in Sect. 3.2 affects the overall file sizes achieved with our codec. Differences are relative to the non-PDE baseline from the previous section. In this experiment, second-order EED has been used for reconstruction. Positive values indicate a benefit of PDE-based predictions over delta coding.

Results indicate that a single reconstruction from a sparse regular grid is not sufficient to obtain a benefit from second-order anisotropic diffusion. On the other hand, the proposed iterative reconstruction achieves a clear additional reduction in compressed file size. It is most pronounced in Mode 1, which dilates the inpainting mask with a cross-shaped structuring element and consequently requires more iterations than Mode 0, which dilates with a box.

4.4 Effect of Contrast Parameter

Even though it can be seen from Table 1 that moving from isotropic to anisotropic diffusion noticeably improved compression rates, we found that fine-tuning the contrast parameter in the diffusivity function is less important. Table 5 explores the effect of varying the ad-hoc threshold value of 90% that was used in Sect. 3.3

Table 5. Results from varying the threshold that our heuristic uses to set the contrast parameter λ. Despite a noticeable effect on λ itself, the corresponding differences in compression rates are rather small. Improvements are relative to the non-PDE baseline.

Image	30% Threshold		60% Threshold		90% Threshold	
	λ	Improvement	λ	Improvement	λ	Improvement
B0	2.04741	+4.272%	4.42198	+4.362%	11.83980	+4.367%
B700	0.01379	+6.468%	0.10905	+6.794%	0.56127	+6.847%
T1	0.12347	+7.108%	0.30572	+7.017%	0.58445	+6.715%
CT	0.12464	+7.060%	0.22596	+7.225%	0.65364	+7.286%

to two other values, 60% and 30%. For each image and threshold, the table reports the corresponding values of contrast parameter λ, as well as the resulting improvement over the non-PDE baseline.

All differences due to the contrast parameter are below 0.5%. This supports our decision to rely on a heuristic setting of the contrast parameter for lossless compression, rather than spending computational resources on trying to optimize it. Results in Table 5 used second-order EED with iteration mode 0, because fine-tuning the contrast parameter would be even more costly in mode 1.

5   Conclusion

PDE-based inpainting has previously been shown to have a strong potential for lossy image compression, especially at high compression rates. We demonstrated that this approach also holds promise for lossless compression of 3D medical images. In particular, we propose a codec that beats state-of-the-art alternatives by combining a simple yet efficient to code initial inpainting mask with iterative reconstruction and coding of residuals, as well as a separate coding of empty space. In the future, we are planning to extend our work to exploit redundancies along the fourth axis that arises in diffusion MRI, i.e., orientation of the diffusion gradient [2]. This will require operating on the space of positions and orientation.

References

1. Andris, S., Peter, P., Weickert, J.: A proof-of-concept framework for PDE-based video compression. In: Picture Coding Symposium (PCS), pp. 1–5. IEEE (2016)
2. Basser, P.J., Mattiello, J., Le Bihan, D.: Estimation of the effective self-diffusion tensor from the NMR spin echo. J. Magnet. Resonan. **B**(103), 247–254 (1994)
3. Deutsch, P.: RFC1951: DEFLATE compressed data format specification version 1.3 (1996)
4. Deutsch, P.: RFC1952: GZIP file format specification version 4.3 (1996)
5. Dinov, I.D.: Volume and value of big healthcare data. J. Med. Stat. Inform. **4**, 3 (2016)

6. Galić, I., Weickert, J., Welk, M., Bruhn, A., Belyaev, A., Seidel, H.-P.: Towards PDE-based image compression. In: Paragios, N., Faugeras, O., Chan, T., Schnörr, C. (eds.) VLSM 2005. LNCS, vol. 3752, pp. 37–48. Springer, Heidelberg (2005). https://doi.org/10.1007/11567646_4
7. Galić, I., Weickert, J., Welk, M., Bruhn, A., Belyaev, A., Seidel, H.P.: Image compression with anisotropic diffusion. J. Math. Imaging Vis. **31**(2–3), 255–269 (2008)
8. Jumakulyyev, I., Schultz, T.: Fourth-order anisotropic diffusion for inpainting and image compression. In: Anisotropy Across Fields and Scales. Mathematics and Visualization, pp. 99–123. Springer (2021)
9. Kil, S.K., Lee, J.S., Shen, D., Ryu, J., Lee, E., Min, H., Hong, S.: Lossless medical image compression using redundancy analysis. Int. J. Comput. Sci. Network Secur. **6**(1), 50–56 (2006)
10. Kim, Y.S., Pearlman, W.A.: Lossless volumetric medical image compression. In: Applications of Digital Image Processing XXII. vol. 3808, pp. 305–312. International Society for Optics and Photonics (1999)
11. Köstler, H., Stürmer, M., Freundl, C., Rüde, U.: PDE based video compression in real time. In: Technical report, 07–11, Lehrstuhl für Informatik 10. University Erlangen-Nürnberg (2007)
12. Larobina, M., Murino, L.: Medical image file formats. J. Digit. Imaging **27**(2), 200–206 (2014)
13. Liu, D., Sun, X., Wu, F., Li, S., Zhang, Y.Q.: Image compression with edge-based inpainting. IEEE Trans. Circuits Syst. Video Technol. **17**(10), 1273–1287 (2007)
14. Mainberger, M., Weickert, J.: Edge-based image compression with homogeneous diffusion. In: International Conference on Computer Analysis of Images and Patterns, pp. 476–483 (2009)
15. Mentzer, F., Agustsson, E., Tschannen, M., Timofte, R., Gool, L.V.: Practical full resolution learned lossless image compression. In: Proceedings of the IEEE Conference on Computer Vision and Pattern Recognition (CVPR), pp. 10629–10638 (2019)
16. Miaou, S.G., Ke, F.S., Chen, S.C.: A lossless compression method for medical image sequences using JPEG-LS and interframe coding. IEEE Trans. Inf Technol. Biomed. **13**(5), 818–821 (2009)
17. Nasr, M., Shokri, R., Houmansadr, A.: Comprehensive privacy analysis of deep learning: passive and active white-box inference attacks against centralized and federated learning. In: Proceedings of the IEEE Symposium on Security and Privacy, pp. 739–753 (2019)
18. Pennebaker, W.B., Mitchell, J.L.: JPEG: Still Image Data Compression Standard. Springer, Heidelberg (1993)
19. Perona, P., Malik, J.: Scale-space and edge detection using anisotropic diffusion. IEEE Trans. Pattern Anal. Mach. Intell. **12**(7), 629–639 (1990)
20. Peter, P.: Three-dimensional data compression with anisotropic diffusion. In: Weickert, J., Hein, M., Schiele, B. (eds.) GCPR 2013. LNCS, vol. 8142, pp. 231–236. Springer, Heidelberg (2013). https://doi.org/10.1007/978-3-642-40602-7_24
21. Peter, P., Contelly, J., Weickert, J.: Compressing audio signals with inpainting-based sparsification. In: Lellmann, J., Burger, M., Modersitzki, J. (eds.) SSVM 2019. LNCS, vol. 11603, pp. 92–103. Springer, Cham (2019). https://doi.org/10.1007/978-3-030-22368-7_8
22. Peter, P., Schmaltz, C., Mach, N., Mainberger, M., Weickert, J.: Beyond pure quality: progressive modes, region of interest coding, and real time video decoding for PDE-based image compression. J. Vis. Commun. Image Represent. **31**, 253–265 (2015)

23. Schmaltz, C., Peter, P., Mainberger, M., Ebel, F., Weickert, J., Bruhn, A.: Understanding, optimising, and extending data compression with anisotropic diffusion. Int. J. Comput. Vis. **108**(3), 222–240 (2014)
24. Schmaltz, C., Weickert, J., Bruhn, A.: Beating the quality of JPEG 2000 with anisotropic diffusion. In: Denzler, J., Notni, G., Süße, H. (eds.) DAGM 2009. LNCS, vol. 5748, pp. 452–461. Springer, Heidelberg (2009). https://doi.org/10.1007/978-3-642-03798-6_46
25. Taubman, D., Marcellin, M.: JPEG2000: Image Compression Fundamentals, Standards and Practice. Springer (2002). https://doi.org/10.1007/978-1-4615-0799-4
26. Weickert, J.: Anisotropic diffusion in image processing. Teubner Stuttgart (1998)
27. Yee, D., Soltaninejad, S., Hazarika, D., Mbuyi, G., Barnwal, R., Basu, A.: Medical image compression based on region of interest using better portable graphics (BPG). In: IEEE International Conference on Systems, Man, and Cybernetics, pp. 216–221 (2017)
28. Zukoski, M.J., Boult, T., Iyriboz, T.: A novel approach to medical image compression. Int. J. Bioinform. Res. Appl. **2**(1), 89–103 (2006)

Splines for Image Metamorphosis

Jorge Justiniano, Marko Rajković[(✉)], and Martin Rumpf

Institute for Numerical Simulation, University of Bonn, Bonn, Germany
marko.rajkovic@ins.uni-bonn.de

Abstract. Cubic splines are a classical tool for higher order interpolation of points in Euclidean space known to minimize the integral of the squared acceleration along the interpolation path. This paper transfers this method to the smooth interpolation of key frames in the space of images. To this end the metamorphosis model based on a simultaneous transport of image intensities and a modulation of intensities along motion trajectories is generalized. The proposed spline energy combines quadratic functionals of the Eulerian motion acceleration and of the second material derivative representing an acceleration in the change of intensities along motion paths. A variational time discretization of this spline model is proposed and the convergence to a suitably relaxed time continuous model is discussed using the tool of Γ-convergence. In particular, this also allows to establish the existence of an optimal spline path interpolating given key frame images. The spatial discretization is based on a finite difference and a stable spline interpolation. A variety of numerical examples demonstrates the robustness and versatility of the proposed method for real images using a variant of the iPALM algorithm for the minimization of the fully discrete energy functional.

Keywords: Image metamorphosis · Image morphing · Spline interpolation · Γ-convergence · iPALM algorithm

1 Introduction

Image metamorphosis is a flexible model for image morphing generalizing the flow of diffeomorphism approach (cf. the textbook by Younes [24]) and investigated extensively by Trouvé, Younes and coworkers [22,23]. The approach is based on the minimization of a path energy functional over all regular paths connecting a pair of input images. Minimizers can be understood as geodesic paths in the space of images considered as a Riemannian manifold. The underlying metric associates a cost both to the transport of image intensities via a viscous flow and to image intensity variations along motion paths.

In this paper, we introduce a spline energy as the second order extension of the first order path energy. Given a set of key frames at disjoint times a spline path is given as a minimizer of the spline energy subject to the key frame interpolation constraint. In Euclidean space cubic splines $t \mapsto u(t)$ are known to be minimizers of the integral of the squared acceleration $\int_0^1 |\ddot{u}(t)|^2 \, dt$ due to the

© Springer Nature Switzerland AG 2021
A. Elmoataz et al. (Eds.): SSVM 2021, LNCS 12679, pp. 463–475, 2021.
https://doi.org/10.1007/978-3-030-75549-2_37

famous result by de Boor [5]. In a Riemannian context, Noakes et al. [17] introduced Riemannian cubic splines as stationary paths of the integrated squared covariant derivative of the velocity.

Today, there is a variety of spline approaches in non-linear spaces and with applications to shape spaces. Trouvé and Vialard [21] studied a second-order shape functional in landmark space based on an optimal control approach. Singh et al. [19] introduced an optimal control method involving a functional measuring the motion acceleration in a flow of diffeomorphisms ansatz for image regression. Benamou et al. [3] and Chen et al. [6] independently discuss spline interpolation in the shape space of probability measures endowed with the Wasserstein metric. Thereby, energy splines are defined as minimizers of the action functional on Wasserstein space which involves the acceleration of measure-valued paths, sharing similarities with the spline functional in the space of images introduced in this paper. To cure computational intractability of such approaches, both aforementioned articles share the idea of relaxation by using multi-marginal optimal transport and entropic regularization. This approach, however, is not only computationally still relatively expensive, but also the transport problem it aims at solving might not have a Monge solution. These limitations are remedied by a new method introduced by Chewi et al. [7] to construct measure-valued splines, dubbed transport splines, which enjoys substantial computational advantages over the preceding approaches. Tahraoui and Vialard [20] consider a second-order variational model involving the Eulerian acceleration in the context of diffeomorphic flow. They propose a relaxed model leading to a Fisher-Rao functional, as a convex functional on the space of measures.

Here, we will introduce a spline energy as a generalization of the path energy in the metamorphosis model. This path energy consists of a first term measuring the dissipation caused by the Eulerian motion velocity field and a second term measuring the material derivative of the image intensities along motion lines. Our spline model introduces second order variants of both terms, i.e. involving the Eulerian motion acceleration and the second order material derivative of image intensities. Furthermore, we will study a corresponding time discrete variational model, which generalizes the time discrete metamorphosis model proposed in [4, 10] and we show the convergence of this time discrete model to the original metamorphosis model in the sense of Mosco [16]. As a consequence, one obtains existence of spline paths and further discretizing the model in space we derive a numerical scheme to solve for spline paths in the space of images. Let us remark that the proposed spline model is not Riemannian in the sense that splines are minimizers of the squared covariant derivative of the path velocity as in [17,20] or in [12], where a related time discretization is proposed for Riemannian splines.

This paper is organized as follows. In Sect. 2 the time continuous spline energy is derived and the proper interplay between Lagrangian and Eulerian perspective is discussed. Then, in Sect. 3 a variational time discretization of the continuous spline energy is introduced and the convergence to the time continuous model is presented. Section 4 explains the fully discrete scheme and Sect. 5 shows how to set up a suitable iPALM algorithm to numerically solve for a spline interpolation given a set of key frames. Finally, Sect. 6 experimentally demonstrates properties of the spline approach for image metamorphosis and shows applications of the proposed method.

2 Time Continuous Metamorphosis Splines

The image metamorphosis model is a refinement of the large deformation diffeo-morphic metric mapping (LDDMM) method [2,9,14,15] based on the diffeomor-phic flow paradigm by Arnold [1]. Thereby, the temporal evolution of images $u : \Omega \to \mathbb{R}^c$ on an image domain $\Omega \subset \mathbb{R}^d$ with $d \geq 1$ into a c-dimensional color or feature space is governed by a flow defined via a family of diffeomor-phisms $(\psi_t)_{t\in[0,1]} : \overline{\Omega} \to \overline{\Omega}$. Under the brightness constancy assumption, such a flow induces a family of images $(u_t)_{t\in[0,1]}$ with $u_t = u_0 \circ \psi_t^{-1}$. Image interpo-lation is based on the minimization of a path energy which is quadratic in the Eulerian motion velocity $v_t = \dot{\psi}_t \circ \psi_t^{-1}$ and measures the flow induced dissipa-tion. Now, in the metamorphosis model intensities might vary along transport paths and the path energy functional for a sufficiently regular family of images $(u_t)_{t\in[0,1]} : \Omega \to \mathbb{R}^c$ and a constant $\delta > 0$ is given by

$$\mathcal{E}[u] := \inf_{(v,z)[u]} \int_0^1 \int_\Omega L^m[v,v] + \frac{1}{\delta}z^2 \, dx \, dt,$$

where $(v, z)[u]$ is the set of Eulerian velocity fields and material derivatives of the image map satisfying $z = \dot{u} + \nabla u \cdot v$. Here, in the simplest case $L^m[v,v] = |\varepsilon[v]|^2 + \gamma|D^m v|^2$ with $m > 2 + \frac{d}{2}$, $\gamma > 0$ and $\varepsilon[v] = \frac{1}{2}(Dv + Dv^T)$. Morphing two images u_A, u_B amounts to minimizing this path energy in the class of regular curves with $u_0 = u_A$ and $u_1 = u_B$ as studied analytically by Trouvé, Younes [22].

On this background let us now derive a physically sound model for a spline interpolation $(u_t)_{t\in[0,1]}$ given a set of J key frame constraints

$$u_{t_j} = u_j^I \tag{1}$$

for $j = 1, \ldots J$ and disjoint times $t_j \in [0, 1]$. To this end, we recall that cubic splines in Euclidean space minimize the integral over the squared motion acceler-ation subject to position constraints [5], whereas linear interpolation is associated with the minimization of the integral over the squared motion velocity. As image morphing via path energy minimization corresponds to this linear interpolation and the derivation of a spline model requires to minimize integrals over quadratic acceleration quantities, one is naturally led to the following still formal spline energy:

$$\mathcal{F}[u] := \inf_{(a,w)[u]} \int_0^1 \int_\Omega L^{m-1}[a,a] + \frac{1}{\delta}w^2 \, dx \, dt,$$

where $(a, w)[u]$ is the set consisting of Eulerian flow acceleration $a_t(x) = \ddot{\psi}_t \circ \psi_t^{-1}(x)$ and the second order material derivative $w = \frac{D^2 u}{dt^2} = \ddot{u} + 2v \cdot D\dot{u} + Du \cdot$

$a + v \cdot D^2 u \cdot v$. As proposed in [12], we introduce a combination of spline energy and path energy

$$\mathcal{F}^\sigma := \mathcal{F} + \sigma \mathcal{E} \,,$$

with $\sigma > 0$ to ensure existence of minimizers.

For image paths $u \in L^2([0,1], \mathcal{I} := L^2(\Omega, \mathbb{R}^c))$ first and second order material derivatives are in general not defined. Furthermore, the notion of regular image paths requires particular care (cf. [22]).

Thus, in what follows we give a rigorous formulation of the spline interpolation problem. We consider Eulerian motion fields $v \in L^2((0,1), \mathcal{V})$ with $\mathcal{V} := H_0^1(\Omega, \mathbb{R}^d) \cap H^m(\Omega, \mathbb{R}^d)$, where H^m is the usual Sobolev space of order m. Integration of such motion fields via solving $\dot{\psi}_t = v_t \circ \psi_t$ results in a family of diffeomorphisms $\psi \in H^1((0,1), C^1(\overline{\Omega}, \overline{\Omega}))$. For the Eulerian motion acceleration $a_t \circ \psi_t = \ddot{\psi}_t$ we take into account the ansatz space $\mathcal{A} := H^{m-1}(\Omega, \mathbb{R}^d)$.

A key point in the definition of path and spline energy is that one switches from the Eulerian formulation to the Lagrangian formulation to obtain a notion of the material derivative z and the material acceleration w. In fact, one obtains (cf. [22, Theorem 2, Lemma 6])

$$\int_s^t z_r \circ \psi_r \, \mathrm{d}r = u_t \circ \psi_t - u_s \circ \psi_s \,, \tag{2}$$

$$\int_0^\tau \int_s^t w_{r+l} \circ \psi_{r+l} \, \mathrm{d}r \, \mathrm{d}l = u_{t+\tau} \circ \psi_{t+\tau} - u_t \circ \psi_t - u_{s+\tau} \circ \psi_{s+\tau} + u_s \circ \psi_s \,, \tag{3}$$

for all $s, t, r, l, \tau \in (0,1)$. Because there is no differentiation involved in these definitions, they work for general image paths.

In a further relaxation, we replace the equalities in (2) and (3) by inequalities for a scalar absolute first and second material derivative $\bar{z}$ and $\bar{w}$ respectively, i.e.

$$\int_s^t \bar{z}_r \circ \psi_r \, \mathrm{d}r \geq |u_t \circ \psi_t - u_s \circ \psi_s| \,, \tag{4}$$

$$\int_0^\tau \int_s^t \bar{w}_{r+l} \circ \psi_{r+l} \, \mathrm{d}r \, \mathrm{d}l \geq |u_{t+\tau} \circ \psi_{t+\tau} - u_t \circ \psi_t - u_{s+\tau} \circ \psi_{s+\tau} + u_s \circ \psi_s| \,, \tag{5}$$

and finally obtain

$$\mathcal{F}^\sigma[u] := \inf_{(v, a, \bar{z}, \bar{w}) \in \mathcal{C}[u]} \int_0^1 \int_\Omega L^{m-1}[a, a] + \frac{1}{\delta} \bar{w}^2 + \sigma \left(L^m[v, v] + \frac{1}{\delta} \bar{z}^2 \right) \, \mathrm{d}x \, \mathrm{d}t \,, \tag{6}$$

where $\bar{z}$ and $\bar{w}$ are supposed to obey (4) and (5), respectively. Here, the admissible set of tuples $(v, a, \bar{z}, \bar{w})$ for given image path u is

$$\mathcal{C}[u] \subset L^2((0,1), \mathcal{V}) \times L^2((0,1), \mathcal{A}) \times L^2((0,1) \times \Omega) \times L^2((0,1) \times \Omega) \,.$$

Let us remark that with $\mathcal{C}[u]$ being non-empty $u \in C^0([0,1], \mathcal{I})$ (cf. [10, Remark 1]) which justifies the postulation of the key frame constraint (1) and that for minimizers of $\mathcal{F}^\sigma$ one gets that $\bar{z} = |z|$ and $\bar{w} = |w|$.

Furthermore, the existence of a minimizer of $\mathcal{F}^\sigma$ is a corollary of the convergence result for the time discrete spline interpolation stated in the next section.

3   Time Discrete Metamorphosis Splines

Next, we derive a variational time discretization of the time continuous spline energy $\mathcal{F}^\sigma$ picking up the approach in [4,10]. To this end, we consider a discrete image curve $\mathbf{u} = (u_0, \ldots, u_K)$ with $u_k \in \mathcal{I}$ and define a set of admissible deformations $\mathcal{D} := \{\phi \in H^m(\Omega, \Omega), \ \det(D\phi) \geq \epsilon, \ \phi = \mathbb{1} \text{ on } \partial\Omega\}$ for a fixed small $\epsilon > 0$ which consists of $C^1(\Omega, \Omega)$–diffemorphisms [8, Theorem 5.5-2]. For $\mathbf{u} \in \mathcal{I}^{K+1}$ and $\mathbf{\Phi} = (\phi_1, \ldots, \phi_K) \in \mathcal{D}^K$ we consider the discrete path energy

$$\mathbf{E}^{K,D}[\mathbf{u}, \mathbf{\Phi}] := K \sum_{k=1}^{K} \int_\Omega W_D(D\phi_k) + \gamma |D^m \phi_k|^2 + \frac{1}{\delta} |u_k \circ \phi_k - u_{k-1}|^2 \, dx,$$

with a simple elastic energy density $W_D(B) := |B^{sym} - \mathbb{1}|^2$, where we use $\mathbb{1}$ for both the identity map and the identity matrix. In this image matching functional, $z_k := K(u_k \circ \phi_k - u_{k-1})$ is a discrete material derivative and $K(\phi_k - \mathbb{1})$ a discrete Eulerian velocity. The discrete counterpart of the spline energy is defined as

$$\mathbf{F}^{K,D}[\mathbf{u}, \mathbf{\Phi}] := \sum_{k=1}^{K-1} \int_\Omega \frac{1}{K} \left(W_A(Da_k) + \gamma |D^{m-1} a_k|^2 \right) + \frac{K}{\delta} |z_{k+1} \circ \phi_k - z_k|^2 \, dx,$$

with the energy density $W_A(B) := |B^{sym}|^2$, a discrete motion acceleration $a_k := K^2((\phi_{k+1} - \mathbb{1}) \circ \phi_k - (\phi_k - \mathbb{1}))$ which is in $H^m(\Omega, \mathbb{R}^d)$ by [13], and a discrete second material derivative $w_k := K(z_{k+1} \circ \phi_k - z_k)$. Finally, the combined spline and path energy is given by

$$\mathbf{F}^{\sigma, K, D}[\mathbf{u}, \mathbf{\Phi}] := \mathbf{F}^{K,D}[\mathbf{u}, \mathbf{\Phi}] + \sigma \mathbf{E}^{K,D}[\mathbf{u}, \mathbf{\Phi}]. \tag{7}$$

Generalizing the arguments from the calculus of variations in [4] one can show that for $t_j = k_j/K$ with integers $0 \leq k_j \leq K$ there exists a discrete spline as a minimizer of

$$\mathbf{F}^{\sigma, K}[\mathbf{u}] := \inf_{\mathbf{\Phi} \in \mathcal{D}^K} \mathbf{F}^{\sigma, K, D}[\mathbf{u}, \mathbf{\Phi}],$$

where the set $\mathcal{I}^K_{adm} := \{\mathbf{u} \in \mathcal{I}^{K+1}, \ u_{k_j} = u^I_j, \ j = 1, \ldots J\}$ ensures the interpolation constraints.

Figure 1 shows a first test case. As key frames we consider three images showing two dimensional Gaussian distribution with small variance at different positions and of different mass. Spline interpolation is compared with piecewise

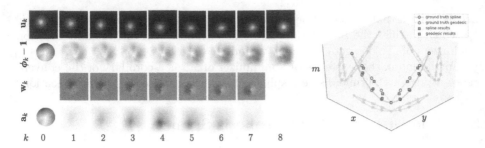

Fig. 1. Left: Time discrete spline with framed key frame images (first row), color-coded displacement field (second row), discrete second order material derivative (third row) and color-coded discrete acceleration field (fourth row) for the Gaussian example and values of the parameters $\delta = 5 \cdot 10^{-3}$, $\sigma = 1$, $\theta = 5 \cdot 10^{-5}$, $N = 64$ (cf. Sect. 4). Right: Euclidean splines in (x, y, m) coordinates for the input parameters versus splines for metamorphosis in (x, y, m) extracted from the numerical results in post-processing. (Color figure online)

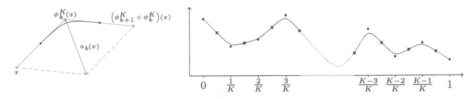

Fig. 2. Left: Schematic drawing of the Hermite interpolation (blue) on the time interval $[(k - \frac{1}{2})/K, (k + \frac{1}{2})/K]$ together with the discrete acceleration $a_k^K(x)$. Right: Image extension $\mathcal{U}^K[\mathbf{u}^K, \mathbf{\Phi}^K](\cdot, x)$ along the extension paths from the left plotted against time. Dots represent the values u_k^K, and crosses the "half-way" values $\frac{1}{2}(u_k^K + u_{k-1}^K)$. (Color figure online)

geodesic interpolation. Furthermore, it is depicted that for the metamorphosis spline, the curve in (x, y, m)-space (position, mass) corresponds almost perfectly to the cubic spline interpolation of the parameters of the Gaussian distribution on the key frames.

To study convergence of discrete spline paths, we have to introduce a sufficiently smooth time continuous extension. Following [12] we define the discrete transport path as a cubic Hermite interpolation on intervals $[(k - \frac{1}{2})/K, (k + \frac{1}{2})/K]$ and affine linear interpolation on $[0, \frac{1}{2K}]$ and $[1 - \frac{1}{2K}, 1]$ as depicted in Fig. 2. We use the analogous Hermite interpolation $\mathcal{U}^K[\mathbf{u}^K, \mathbf{\Phi}^K] \in L^2([0, 1], \mathcal{I})$ for image intensities along the resulting discrete transport path. Based on this, the discrete spline functional can be extended. For $u \in L^2([0, 1], \mathcal{I})$ we define $\mathcal{F}^{\sigma, K}[u] := \inf_{\mathbf{\Phi}^K \in \mathcal{D}^K} \mathbf{F}^{\sigma, K, D}[\mathbf{u}^K, \mathbf{\Phi}^K]$ in case $u = \mathcal{U}[\mathbf{u}^K, \mathbf{\Phi}^K]$ and ∞ otherwise. For this extension we then obtain the following theorem

Theorem 1. *For every $\sigma > 0$, the energy functional $\mathcal{F}^{\sigma, K}$ converges to $\mathcal{F}^\sigma$ in the sense of Mosco as $K \to \infty$ in the topology $L^2([0, 1], \mathcal{I})$. Furthermore, the subsequence of minimizers $\{u^K\}_{K \in \mathbb{N}}$ converges weakly in $L^2([0, 1], \mathcal{I})$ to a min-*

imizer of $\mathcal{F}^\sigma$ subject to the interpolation constraint and the associated sequence of $\mathcal{F}^{\sigma,K}$ converges to the minimal continuous spline energy.

Let us remark that Mosco-convergence can be regarded as a stronger convergence notion than Γ-convergence, since only weak convergence in the liminf inequality is required.

The proof picks up ideas from [4] and [11]. We give a brief outline by listing some of the most crucial steps[1]

- *liminf inequality:* At first, a priori estimates for the discrete material derivatives, the discrete image sequence, and the discrete Eulerian velocity and acceleration fields have to be verified. Based on these, a Taylor expansion arguments enable to relate the discrete energies and the continuous energies. Next, we have to prove the lower semi-continuity of both the weak material acceleration and the acceleration flow dissipation. To this end, one has to show that the discrete scalar second material derivative converges to a weak, scalar material acceleration, and the discrete flow acceleration converges as well. Finally, we verify that these limits belong to $\mathcal{C}[u]$.
- *limsup inequality:* First, one proves the existence of deformations ϕ_k associated to an image path u if the functional $\mathcal{F}^\sigma[u]$ is finite. From this, one deduces the actual recovery sequence. Making use of Taylor arguments one is then able to find upper bounds of the sequence of functionals $\mathcal{F}^{\sigma,K}[u^K]$, which readily implies the limsup inequality. Thereby, the core argument is an application of Jensen's inequality to bound the discrete dissipation and the discrete acceleration derived from the extracted discrete deformations by the continuous counterpart in the limit. This justifies why discrete deformations have to be recovered first. Finally, it remains to verify that the recovery sequence actually converges to u in $L^2((0,1),\mathcal{I})$. In that respect, it turns out to be decisive to switch to a Lagrangian formulation using the definitions (4) and (5) for the first and second material derivative.

The convergence of the discrete minimizers follows from usual Γ-convergence arguments and the equicoercivity of the set of discrete minimizers.

4   Fully Discrete Metamorphosis Splines

To numerically implement splines for image metamorphosis we have to further discretize in space. We restrict ourselves here to scalar images ($c = 1$) and $\Omega := [0,1]^2$ ($d = 2$). A generalization to multi-channel images is straightforward. At first, we increase the robustness by explicitly introducing a variable $\hat{z} \in L^2((0,1), L^2(\Omega, \mathbb{R}))$ and obtain a relaxation of (6):

$$\mathcal{F}^\sigma[u, \hat{z}] = \inf_{(v,a,z,w)} \int_0^1 \int_\Omega L^{m-1}[a,a] + \frac{1}{\delta}w^2 + \sigma(L^m[v,v] + \frac{1}{\delta}\hat{z}^2) + \frac{1}{\theta}(\hat{z}-z)^2 \, \mathrm{d}x \, \mathrm{d}t$$

---

[1] The comprehensive proof is available for review on request and will appear elsewhere.

with a penalty on the misfit of the new variable $\hat{z}$ and the actual material derivative z given by (2), while w is the first material derivative of $\hat{z}$ in the sense of (2). To adapt the time discrete counterpart $\mathbf{F}^{\sigma,K}[\mathbf{u},\hat{\mathbf{z}}]$ with $\hat{\mathbf{z}} = (\hat{\mathbf{z}}_1,\dots,\hat{\mathbf{z}}_K)$ accordingly, in (7) we replace $K|\mathbf{u}_k \circ \boldsymbol{\phi}_k - \mathbf{u}_{k-1}|^2$ by $\frac{1}{K}|\hat{\mathbf{z}}_k|^2$ and $\frac{K}{\delta}|\mathbf{z}_{k+1} \circ \boldsymbol{\phi}_k - \mathbf{z}_k|^2$ by $\frac{K}{\delta}|\hat{\mathbf{z}}_{k+1} \circ \boldsymbol{\phi}_k - \hat{\mathbf{z}}_k|^2$ to penalize second material derivative. Finally, as discretization of the last term in $\mathcal{F}^\sigma[\mathbf{u},\hat{z}]$ we add terms $\frac{1}{\theta K}|K(\mathbf{u}_k \circ \boldsymbol{\phi}_k - \mathbf{u}_{k-1}) - \hat{\mathbf{z}}_k|^2$.

The $N \times N$ computational grid is defined as $\Omega_N := \left\{ \frac{0}{N-1}, \frac{1}{N-1}, \dots, \frac{N-1}{N-1} \right\}^2$ ($N \geq 3$) with discrete boundary $\partial\Omega_N = \{x \subset \Omega_N : x \in \partial([0,1]^2)\}$ and the L^p-norm of a function $\mathbf{u}$ on Ω_N is given by $\|\mathbf{u}\|_{L_N^p}^p := \frac{1}{N^2}\sum_{(i,j)\in\Omega_N}|\mathbf{u}(i,j)|^p$. The discrete image space is $\mathcal{I}_N := \{\mathbf{u} : \Omega_N \to \mathbb{R}\}$ and the set of admissible deformations $\mathcal{D}_N := \{\boldsymbol{\phi} : \Omega_N \to [0,1]^2 : \boldsymbol{\phi} = \mathbb{1} \text{ on } \partial\Omega_N\}$. The discrete Jacobian operator of $\boldsymbol{\phi}$ at $(i,j) \in \Omega_N$ is defined as the forward finite difference operator with discrete Neumann boundary conditions. To stabilize the computation, the Jacobian operator applied to the images is approximated using a Sobel filter. A spatial warping operator $\mathbf{T}[\mathbf{u},\boldsymbol{\phi}](x,y) := \sum_{(i,j)\in\Omega_N} s(\boldsymbol{\phi}_1(x,y) - i)\,s(\boldsymbol{\phi}_2(x,y) - j)\,\mathbf{u}(i,j)$ approximates the pull-back of an image $\mathbf{u} \circ \boldsymbol{\phi}$ at a point $(x,y) \in \Omega_N$, where s is the third order B-spline interpolation kernel. This form of warping is also used for derivatives and for composition of deformations.

In summary, the fully discrete spline energy in the metamorphosis model for a $(K+1)$-tuple $(\mathbf{u}_k)_{k=0}^K$ of discrete images and a K-tuple $(\hat{\mathbf{z}}_k)_{k=1}^K$ of discrete derivatives reads as

$$\mathbf{F}_N^{\sigma,K}[(\mathbf{u}_k)_{k=0}^K,(\hat{\mathbf{z}}_k)_{k=1}^K] = \inf_{\boldsymbol{\Phi}\in\mathcal{D}_N^K} \mathbf{F}_N^{\sigma,K,D}[(\mathbf{u}_k)_{k=0}^K,(\hat{\mathbf{z}}_k)_{k=1}^K,(\boldsymbol{\phi}_k)_{k=1}^K] =$$

$$\inf_{\boldsymbol{\Phi}\in\mathcal{D}_N^K} \sum_{k=1}^{K-1} \frac{1}{K}\|W_A(D\mathbf{a}_k)\|_{L_N^1} + \frac{K}{\delta}\mathbf{D}_N^s[\hat{\mathbf{z}}_k,\hat{\mathbf{z}}_{k+1},\boldsymbol{\phi}_k]$$

$$+ \sum_{k=1}^K \sigma\left(K\|W_D(D\boldsymbol{\phi}_k)\|_{L_N^1} + \frac{1}{\delta K}\|\hat{\mathbf{z}}_k\|_{L_N^2}^2\right) + \frac{1}{\theta K}\mathbf{D}_N^g[\mathbf{u}_{k-1},\mathbf{u}_k,\hat{\mathbf{z}}_k,\boldsymbol{\phi}_k],$$

where $\mathbf{D}_N^s[\mathbf{z},\tilde{\mathbf{z}},\boldsymbol{\phi}] := \frac{1}{2}\|\mathbf{T}[\tilde{\mathbf{z}},\boldsymbol{\phi}] - \mathbf{z}\|_{L_N^2}^2$, $\mathbf{D}_N^g[\mathbf{u},\tilde{\mathbf{u}},\mathbf{z},\boldsymbol{\phi}] := \frac{1}{2}\|K(\mathbf{T}[\tilde{\mathbf{u}},\boldsymbol{\phi}] - \mathbf{u}) - \mathbf{z}\|_{L_N^2}^2$ and $\mathbf{a}_k := K^2(\mathbf{T}[\boldsymbol{\phi}_{k+1} - \mathbb{1},\boldsymbol{\phi}_k] - (\boldsymbol{\phi}_k - \mathbb{1}))$. For simplicity, we neglect the higher order Sobolev norm terms in this fully discrete model with grid dependent regularity ensured by the use of cubic B-splines. To improve the numerical computation of minimizing deformations, we use a multi-level scheme where the initial deformations and images are obtained via a bilinear interpolation of the preceding coarser scale solutions.

5   Numerical Optimization Using the iPALM Algorithm

In this section, we discuss the numerical solution of the above fully discrete variational problem based on the application of a variant of the inertial proximal alternating linearized minimization algorithm (iPALM, [18]). Following [10], to enhance the stability the warping operation is linearized w.r.t the deformation at $\phi^{[\beta]} \in \mathcal{D}_N$ from the previous iteration leading to the modified energies

$$\tilde{\mathbf{D}}_N^s[\mathbf{z}, \tilde{\mathbf{z}}, \phi, \phi^{[\beta]}] := \tfrac{1}{2}\|\mathbf{T}[\tilde{\mathbf{z}}, \tilde{\phi}] + \langle \Lambda(\mathbf{z}, \tilde{\mathbf{z}}, \phi^{[\beta]}), \phi - \phi^{[\beta]} \rangle - \mathbf{z}\|_{L_N^2}^2,$$

$$\tilde{\mathbf{D}}_N^g[\mathbf{u}, \tilde{\mathbf{u}}, \mathbf{z}, \phi, \phi^{[\beta]}] := \tfrac{1}{2}\|KT[\tilde{\mathbf{u}}, \phi^{[\beta]}] + \langle \Lambda(K\mathbf{u} + \mathbf{z}, K\tilde{\mathbf{u}}, \phi^{[\beta]}), \phi - \phi^{[\beta]} \rangle$$
$$- (K\mathbf{u} + \mathbf{z})\|_{L_N^2}^2.$$

based on the gradient $\Lambda(\mathbf{u}, \tilde{\mathbf{u}}, \phi^{[\beta]}) := \frac{1}{2}(\nabla_N \mathbf{T}[\tilde{\mathbf{u}}, \phi^{[\beta]}] + \nabla_N \mathbf{u})$. We consider the proximal mapping $\text{prox}_\tau^f[x] := \text{argmin}_{y \in \mathcal{D}_N}(\frac{\tau}{2}\|x - y\|_{L_N^2}^2 + f(y))$ of a functional $f : \mathcal{D}_N \to (-\infty, \infty]$. We are interested in the functional $f = \frac{K}{\delta}\tilde{\mathbf{D}}_N^s + \frac{1}{K\theta}\tilde{\mathbf{D}}_N^g$ for which the operator $\text{prox}_\tau^f[\phi]$ is given by

$$\left(\mathbb{1} + \tfrac{K}{\tau\delta}|\Lambda^s|^2 + \tfrac{1}{\tau\theta K}|\Lambda^g|^2\right)\left(\phi - \tfrac{K\Lambda^s}{\tau\delta}\left(\mathbf{T}[\hat{\mathbf{z}}_{k+1}, \phi_k^{[\beta]}] - (\Lambda^s)^T\phi_k^{[\beta]} - \hat{\mathbf{z}}_k\right)\right.$$
$$\left. - \tfrac{\Lambda^g}{\tau\theta K}\left(\mathbf{T}[K\mathbf{u}_k, \phi_k^{[\beta]}] - (\Lambda^g)^T\phi_k^{[\beta]} - K\mathbf{u}_{k-1} - \hat{\mathbf{z}}_k\right)\right),$$

where $\Lambda^s := \Lambda(\hat{\mathbf{z}}_k, \hat{\mathbf{z}}_{k+1}, \phi_k^{[\beta]})$ and $\Lambda^g := \Lambda(K\mathbf{u}_{k-1} + \hat{\mathbf{z}}_k, K\mathbf{u}_k, \phi_k^{[\beta]})$. The first terms in both brackets are activated only for $k < K$.

In the jth iteration of the algorithm for the minimization of the spline energy $\mathbf{F}_N^{\sigma, K}$ the kth path elements are updated in the following order and manner

$$\phi_k^{j,t} = \phi_k^{[\beta,j]} - \tfrac{1}{L[\phi_k^{[j]}]}\nabla_{\phi_k}\left(\sigma K\|W_D(D\phi_k^{[\beta,j]})\|_{L_N^1}\right.$$
$$\left. + \tfrac{1}{K}\|W_A(D\mathbf{a}_k^{[\beta,j]}) + W_A(D\mathbf{a}_{k-1}^{[\beta,j]})\|_{L_N^1}\right),$$

$$\phi_k^{[j+1]} = \text{prox}_{L[\phi_k^{[j]}]}^{\frac{K}{\delta}\tilde{\mathbf{D}}_N^s + \frac{1}{K\theta}\tilde{\mathbf{D}}_N^g}[\phi_k^{j,t}],$$

$$\hat{\mathbf{z}}_k^{[j+1]} = \hat{\mathbf{z}}_k^{[\beta,j]} - \tfrac{1}{L[\hat{\mathbf{z}}_k^{[j]}]}\nabla_{\hat{\mathbf{z}}_k}\mathbf{F}_N^{\sigma,K,D}[\mathbf{u}^{[k,j]}, \hat{\mathbf{z}}^{[\beta,k,j]}, \phi^{[k+1,j]}],$$

$$\mathbf{u}_k^{[j+1]} = \mathbf{u}_k^{[\beta,j]} - \tfrac{1}{L[\mathbf{u}_k^{[j]}]}\nabla_{\mathbf{u}_k}\mathbf{F}_N^{\sigma,K,D}[\mathbf{u}^{[\beta,k,j]}, \hat{\mathbf{z}}^{[k+1,j]}, \phi^{[k+1,j]})].$$

Here, the extrapolation with $\beta > 0$ of the kth path element in the jth iteration step is given by $h_k^{[\beta,j]} = h_k^{[j]} + \beta(h_k^{[j]} - h_k^{[j-1]})$ and we use the notation $h^{[k,j]} = (h_{1,...,k-1}^{[j+1]}, h_{k,...,K}^{[j]})$, $h^{[\beta,k,j]} = (h_1^{[j+1]}, \ldots, h_{k-1}^{[j+1]}, h_k^{[\beta,j]}, h_{k+1}^{[j]}, \ldots, h_K^{[j]})$, while the acceleration $\mathbf{a}^{[\beta,j]}$ is computed with correspondingly updated $\phi^{[\beta,j]}$ values and we do not change values of the fixed images. Furthermore, we denote by $L[h]$ the Lipschitz constant of the gradient of function h, determined by backtracking.

6   Applications

In what follows, we investigate and discuss qualitative properties of the spline interpolation in the space of images, being aware that the superior temporal smoothness of this interpolation is difficult to show with series of still images. For all the examples we use 5 levels in the multi-levels approach and $\beta = \frac{1}{\sqrt{2}}$. For the first example $N = 64$ and $K = 8$, while for the other two $N = 128$ and $K = 16$. For plotting of images we crop the values to $[0, 1]$ and for the displacements plots hue refers to the direction and the intensity is proportional to its norm as indicated by the color wheel.

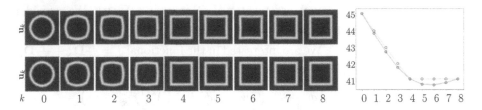

Fig. 3. Left: Time discrete spline (top row) and piecewise geodesic (bottom row) interpolation with framed key frames. The influence of the circle's curvature on the spline segment between the two identical squares is still visible via concave 'edges', while for the piecewise geodesic interpolation any memory of the circle is lost between the squares. Right: Width of the interpolated shape measured at the horizontal axis of symmetry (in number of pixels) for a spline interpolation (orange) and piecewise geodesic interpolation (green) showing the concavities ($K = 8$, $\delta = 5 \cdot 10^{-3}$, $\sigma = 1$, $\theta = 5 \cdot 10^{-4}$). (Color figure online)

At first we conceptually compare splines for metamorphosis and piecewise geodesic paths in Fig. 3.

Next, we consider spline interpolation between three human portraits on Fig. 4. The plots of second material derivative and acceleration show a strong concentration around the key frames, where the spline is expected to be smooth and the piecewise geodesic path just Lipschitz. In Fig. 5 we see analogous results for spline interpolation between the letters P, A, and Q.

A comparison of selected frames of splines and piecewise geodesic paths is shown in Fig. 6. One particularly observes that for the faces the shown spline image is thicker and for the letters the spline image shows more round contours than the ones for the piecewise geodesic counterpart. This is again the non-local impact of key frames beyond those bounding the current interpolation.

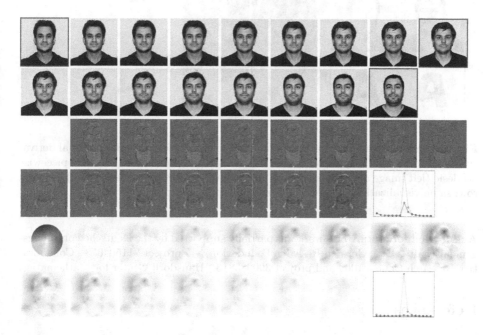

Fig. 4. Time discrete spline with framed fixed images (first and second row), second order material derivative with energies comparison (third and fourth row) and color-coded acceleration field with energies comparison (fifth and sixth row) for values of the parameters $\delta = 2 \cdot 10^{-2}$, $\sigma = 2$, $\theta = 8 \cdot 10^{-4}$. The graphics on the right in row four and six show for the spline (orange) time plots of the L^2-norm of the second order material derivative and the dissipation energy density reflecting motion acceleration, respectively. This is compared to the corresponding piecewise geodesic interpolation (green) (not visualized here, cf. Fig. 6). (Color figure online)

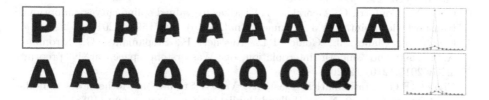

Fig. 5. Left: Time discrete spline with framed fixed images (top and bottom row). Right: Energy density function of acceleration flow $W_A(D\mathbf{a}_k)$ (top), and L^2-norm of second order material derivative $\mathbf{w}_k$ (bottom). Parameter values: $\delta = 10^{-3}, \sigma = 2, \theta = 2 \cdot 10^{-5}$.

Fig. 6. Top row: Image u_{11} for the human face example, second order material derivative and acceleration displacement field for $k = 8$ (right) for the time discrete piecewise geodesic (left image of each panel pair) and spline (right image of each panel). Bottom row: same visualization of image u_4 from the letter example.

Acknowledgements. This work was partially supported by the Deutsche Forschungsgemeinschaft (DFG, German Research Foundation) via project 211504053 - Collaborative Research Center 1060 and project 390685813 - Hausdorff Center for Mathematics.

References

1. Arnold, V.: Sur la géométrie différentielle des groupes de Lie de dimension infinie et ses applications à l'hydrodynamique des fluides parfaits. Ann. Inst. Fourier (Grenoble) 16(fasc., fasc. 1), 319–361 (1966)
2. Beg, M.F., Miller, M.I., Trouvé, A., Younes, L.: Computing large deformation metric mappings via geodesic flows of diffeomorphisms. Int. J. Comput. Vis. **61**(2), 139–157 (2005)
3. Benamou, J.D., Gallouët, T.O., Vialard, F.X.: Second-order models for optimal transport and cubic splines on the Wasserstein space. Found. Comput. Math. **19**(5), 1113–1143 (2019)
4. Berkels, B., Effland, A., Rumpf, M.: Time discrete geodesic paths in the space of images. SIAM J. Imaging Sci. **8**(3), 1457–1488 (2015)
5. de Boor, C.: Best approximation properties of spline functions of odd degree. J. Math. Mech. **12**, 747–749 (1963)
6. Chen, Y., Conforti, G., Georgiou, T.T.: Measure-valued spline curves: an optimal transport viewpoint. SIAM J. Numer. Anal. **50**(6), 5947–5968 (2018)
7. Chewi, S., Clancy, J., Gouic, T.L., Rigollet, P., Stepaniants, G., Stromme, A.J.: Fast and smooth interpolation on Wasserstein space. arXiv preprint arXiv:2010.12101 (2020)
8. Ciarlet, P.G.: Mathematical Elasticity. Vol. I, Studies in Mathematics and its Applications, vol. 20. North-Holland Publishing Co., Amsterdam (1988)
9. Dupuis, P., Grenander, U., Miller, M.I.: Variational problems on flows of diffeomorphisms for image matching. Quart. Appl. Math. **56**(3), 587–600 (1998)
10. Effland, A., Kobler, E., Pock, T., Rajković, M., Rumpf, M.: Image morphing in deep feature spaces: theory and applications. J. Math. Imaging Vis. **63**(2), 309–327 (2021)

11. Effland, A., Neumayer, S., Rumpf, M.: Convergence of the time discrete meta-morphosis model on Hadamard manifolds. SIAM J. Imaging Sci. **13**(2), 557–588 (2020)
12. Heeren, B., Rumpf, M., Wirth, B.: Variational time discretization of Riemannian splines. IMA J. Numer. Anal. **39**(1), 61–104 (2018)
13. Inci, H., Kappeler, T., Topalov, P.: On the regularity of the composition of diffeo-morphisms. Mem. Am. Math. Soc. **226**(1062), vi+60 (2013)
14. Joshi, S.C., Miller, M.I.: Landmark matching via large deformation diffeomor-phisms. IEEE Trans. Image Process. **9**(8), 1357–1370 (2000)
15. Miller, M.I., Trouvé, A., Younes, L.: On the metrics and Euler-Lagrange equations of computational anatomy. Annu. Rev. Biomed. Eng. **4**(1), 375–405 (2002)
16. Mosco, U.: Convergence of convex sets and of solutions of variational inequalities. Adv. Math. **3**, 510–585 (1969)
17. Noakes, L., Heinzinger, G., Paden, B.: Cubic splines on curved spaces. IMA J. Math. Control Inform. **6**(4), 465–473 (1989)
18. Pock, T., Sabach, S.: Inertial proximal alternating linearized minimization (iPALM) for nonconvex and nonsmooth problems. SIAM J. Imaging Sci. **9**(4), 1756–1787 (2016)
19. Singh, N., Vialard, F.X., Niethammer, M.: Splines for diffeomorphisms. Med. Image Anal. **25**(1), 56–71 (2015)
20. Tahraoui, R., Vialard, F.X.: Minimizing acceleration on the group of diffeomor-phisms and its relaxation. ESAIM Control Optim. Calc. Var. **25** (2019)
21. Trouvé, A., Vialard, F.X.: Shape splines and stochastic shape evolutions: a second order point of view. Quart. Appl. Math. **70**(2), 219–251 (2012)
22. Trouvé, A., Younes, L.: Local geometry of deformable templates. SIAM J. Math. Anal. **37**(1), 17–59 (2005)
23. Trouvé, A., Younes, L.: Metamorphoses through Lie group action. Found. Comput. Math. **5**(2), 173–198 (2005)
24. Younes, L.: Shapes and Diffeomorphisms. Applied Mathematical Sciences, vol. 171. Springer, Heidelberg (2010). https://doi.org/10.1007/978-3-642-12055-8

Residual Whiteness Principle for Automatic Parameter Selection in ℓ_2-ℓ_2 Image Super-Resolution Problems

Monica Pragliola[2(✉)], Luca Calatroni[1], Alessandro Lanza[2], and Fiorella Sgallari[2]

[1] CNRS, UCA, INRIA, Morpheme, I3S, Sophia-Antipolis, France
calatroni@i3s.unice.fr
[2] Department of Mathematics, University of Bologna, Bologna, Italy
{alessandro.lanza2,monica.pragliola2,fiorella.sgallari}@unibo.it

Abstract. We propose an automatic parameter selection strategy for variational image super-resolution of blurred and down-sampled images corrupted by additive white Gaussian noise (AWGN) with unknown standard deviation. By exploiting particular properties of the operators describing the problem in the frequency domain, our strategy selects the optimal parameter as the one optimising a suitable residual whiteness measure. Numerical tests show the effectiveness of the proposed strategy for generalised ℓ_2-ℓ_2 Tikhonov problems.

Keywords: Single image super-resolution · Residual whiteness principle · Adaptive parameter selection

1 Introduction

The problem of single-image Super-Resolution (SR) consists in finding a high-resolution (HR) image starting from low-resolution (LR) blurred and noisy data. The huge number of applications which benefits from the recovery of HR information, ranging from remote sensing to biomedical imaging, motivates the large amount of research still ongoing in this field.

Mathematically, the problem can be described as follows. Let $\mathbf{X} \in \mathbb{R}^{N_r \times N_c}$ denote the original HR image, with $\mathbf{x} = \mathrm{vec}(\mathbf{X}) \in \mathbb{R}^N$, $N = N_r N_c$, being its vectorisation. The process describing the mapping from HR to LR data can be described by the following linear observation model

$$\mathbf{b} = \mathbf{SKx} + \mathbf{e}, \quad \text{with } \mathbf{e} \text{ realisation of } \mathbf{E} \sim \mathcal{N}(0, \sigma^2 \mathbf{I}_n), \tag{1}$$

LC acknowledges the support received by the EU H2020 RISE NoMADS, GA 777826. Research of AL, MP and FS was supported by ex60 project by the University of Bologna. All the authors acknowledge the "National Group for Scientific Computation (GNCS-INDAM)".

A. Elmoataz et al. (Eds.): SSVM 2021, LNCS 12679, pp. 476–488, 2021.
https://doi.org/10.1007/978-3-030-75549-2_38

where $\mathbf{b}, \mathbf{e} \in \mathbb{R}^n$, $n = n_r n_c$, are the vectorised observed and noise image, respectively, both consisting of $n_r \times n_c$ pixels, $\mathbf{S} \in \mathbb{R}^{n \times N}$ is the down-sampling operator inducing a pixel decimation with factor d_r and d_c along the rows and the columns of $\mathbf{X}$, respectively - i.e., $N_r = n_r d_r$, $N_c = n_c d_c$ - $\mathbf{K} \in \mathbb{R}^{N \times N}$ represents a space-invariant blurring operator, $\mathbf{I}_n \in \mathbb{R}^{n \times n}$ denotes the n-dimensional identity matrix and $\mathbf{E}$ is an n-variate Gaussian-distributed random vector with zero mean and scalar covariance matrix, with σ indicating the (unknown) noise standard deviation. We set $d = d_r d_c$, so that $N = nd$.

To overcome the ill-posedness of problem (1), one can seek an estimate $\mathbf{x}^*$ of $\mathbf{x}$ by minimising a suitable cost function $\mathcal{J} : \mathbb{R}^N \to \mathbb{R}^+$. In this work, we consider in particular a generalised ℓ_2-ℓ_2 Tikhonov-regularised model of the form

$$\mathbf{x}^*(\mu) = \arg \min_{\mathbf{x} \in \mathbb{R}^N} \left\{ \mathcal{J}(\mathbf{x}; \mu) := \frac{\mu}{2} \|\mathbf{SKx} - \mathbf{b}\|_2^2 + \frac{1}{2} \|\mathbf{Lx} - \mathbf{v}\|_2^2 \right\}, \qquad (2)$$

where the operator $\mathbf{L} \in \mathbb{R}^{M \times N}$ and the vector $\mathbf{v} \in \mathbb{R}^M$ are known. The data term $\|\mathbf{SKx} - \mathbf{b}\|_2^2$ encodes the AWGN assumption on $\mathbf{e}$, while the regularisation term $\|\mathbf{Lx} - \mathbf{v}\|_2^2$ encodes prior information on the unknown target. Finally, the *regularisation parameter* $\mu \in \mathbb{R}_*^+$, with $\mathbb{R}_*^+ = \mathbb{R}^+ \setminus \{0\}$, (2) balances the action of the fidelity against regularisation; its choice is of crucial importance for high quality reconstructions.

When $\mathbf{S} = \mathbf{I}_N$, under general assumptions - see (A3)-(A4) in Sect. 1.1 - the problem in (2) can be solved very efficiently. However, the presence of a nontrivial $\mathbf{S}$ makes the computation of the least-squares solution very costly. In [12], upon a specific choice of $\mathbf{S}$, the authors proposed an efficient strategy for the solution of (2), for which Generalised Cross Validation [3] is used to select the optimal μ. This is known to be impractical for large-scale problems [2].

A popular strategy which aims at overcoming the downsides of empirical parameter selection rules while exploiting the information available on the noise corruption is the celebrated discrepancy principle (DP) (see [1,4] for general problems and [9] for applications to super-resolution problems), which can be formulated as follows:

$$\text{select } \mu = \mu^* \text{ such that } \|\mathbf{r}^*(\mu^*)\|_2 = \|\mathbf{SKx}^*(\mu) - \mathbf{b}\|_2 = \tau \sqrt{n} \sigma, \qquad (3)$$

with $\mathbf{x}^*(\mu)$ being the solution of (2) and τ denoting the discrepancy coefficient. When σ is known, τ is set equal to 1, otherwise a value slightly greater than 1 is typically chosen to avoid noise under-estimation. Clearly, in real world applications an accurate estimate of σ is not available, which often limits the applicability of DP strategies.

Recently, in the context of image restoration problems, a number of works has focused on the design of variational models explicitly exploiting in their formulations the assumed whiteness of the corrupting noise - see, e.g., [5,7]. Based on these promising results, in [6], the authors propose a strategy named *residual whiteness principle* (RWP), that relies on the whiteness property of the noise to properly set the regularisation parameter μ. The RWP automatically selects a value for μ that maximises the whiteness of the residual image

$\mathbf{r}^*(\mu) = \mathbf{SKx}^*(\mu) - \mathbf{b}$, or equivalently minimises the squared Euclidean norm of the normalised auto-correlation of $\mathbf{r}^*(\mu)$. The RWP has there been applied to the automatic selection of μ in Tikhonov-regularised least squares problems which are frequently encountered in iterative *alternating direction method of multipliers* (ADMM) optimisation frameworks when used to larger classes of non-smooth regularisation models.

In this paper, we extend the results obtained in [6] to SR problems of the form (2). As in [6], the proposed strategy can be easily extended to models more general than the one in (2).

1.1 Notations, Preliminaries and Assumptions

In the following, for $c \in \mathbb{C}$ we use $\bar{c}, |c|$ to indicate the conjugate and the modulus of c, respectively. We denote by $\mathbf{F}, \mathbf{F}^H$ the 2D Fourier transform and its inverse, respectively. For any $\mathbf{v} \in \mathbb{R}^N$ and any $\mathbf{A} \in \mathbb{R}^{N \times N}$, we use the notations $\tilde{\mathbf{v}} = \mathbf{Fv}$ and $\tilde{\mathbf{A}} = \mathbf{FAF}^H$ to denote the action of the 2D Fourier transform operator $\mathbf{F}$ on vectors and matrices, respectively. Given a permutation matrix $\mathbf{P} \in \mathbb{R}^{N \times N}$, we denote by $\hat{\mathbf{v}} = \mathbf{P}\tilde{\mathbf{v}}$ and by $\hat{\mathbf{A}} = \mathbf{P}\tilde{\mathbf{A}}\mathbf{P}^T$ the action of $\mathbf{P}$ on the Fourier-transformed vector $\tilde{\mathbf{v}}$ and matrix $\tilde{\mathbf{A}}$, respectively. Finally, by $\check{\mathbf{A}}$ we denote the product $\check{\mathbf{A}} = \mathbf{P}\tilde{\mathbf{A}}^H\mathbf{P}^T$, i.e. the action of $\mathbf{P}$ on $\tilde{\mathbf{A}}^H$.

We recall some results that will be useful in the following discussion and a well-known property of the Kronecker product '$\otimes$'.

Lemma 1 ([10]). *Let* $\mathbf{J}_d \in \mathbb{R}^{d \times d}$ *denote a matrix of ones. We have:*

$$\widetilde{\mathbf{S}^H \mathbf{S}} = \frac{1}{d}(\mathbf{J}_{d_r} \otimes \mathbf{I}_{n_r}) \otimes (\mathbf{J}_{d_c} \otimes \mathbf{I}_{n_c}). \tag{4}$$

Lemma 2. *Let* $\mathbf{A}, \mathbf{B}, \mathbf{C}, \mathbf{D}$ *be given matrices such that* $\mathbf{AC}, \mathbf{BD}$ *exist. We have:*

$$(\mathbf{A} \otimes \mathbf{B})(\mathbf{C} \otimes \mathbf{D}) = (\mathbf{AC} \otimes \mathbf{BD}). \tag{5}$$

Lemma 3 (Woodbury formula). *Let* $\mathbf{A}_1, \mathbf{A}_2, \mathbf{A}_3, \mathbf{A}_4$ *matrices and let* $\mathbf{A}_1$ *and* $\mathbf{A}_3$ *be invertible. Then, the following inversion formula holds:*

$$(\mathbf{A}_1 + \mathbf{A}_2\mathbf{A}_3\mathbf{A}_4)^{-1} = \mathbf{A}_1^{-1} + \mathbf{A}_1^{-1}\mathbf{A}_2(\mathbf{A}_3^{-1} + \mathbf{A}_4\mathbf{A}_1^{-1}\mathbf{A}_2)^{-1}\mathbf{A}_4\mathbf{A}_1^{-1}. \tag{6}$$

The results recalled and proposed in this paper rely on the following assumptions on the image formation model and on the linear operators $\mathbf{S}, \mathbf{K}, \mathbf{L}$.

(A1) The original image $\mathbf{X}$ is assumed to be square, i.e. $N_r = N_c$, and $d_c = d_r$.
(A2) The conjugate transpose $\mathbf{S}^H \in \mathbb{R}^{N \times n}$ of the down-sampling operator interpolates the decimated image with zeros, and $\mathbf{SS}^H = \mathbf{I}_n$.
(A3) The matrices $\mathbf{S}, \mathbf{K}$ and $\mathbf{L}$ in (2) are such that $\text{null}(\mathbf{SK}) \cap \text{null}(\mathbf{L}) = \mathbf{0}_N$, with $\mathbf{0}_N$ denoting the N-dimensional null vector.
(A4) As a consequence of the space-invariance of the blur, the matrix $\mathbf{K}$ represents a 2D discrete convolution operator. Also the regularisation matrix $\mathbf{L}$

is required to represent a 2D convolutional operator, so that $\mathbf{K}$ and $\mathbf{L}$ can be diagonalised by the 2D discrete Fourier transform. In formula:

$$\mathbf{K} = \mathbf{F}^H \mathbf{\Lambda} \mathbf{F} \quad \text{and} \quad \mathbf{L} = \mathbf{F}^H \mathbf{\Gamma} \mathbf{F}, \quad \text{with} \quad \mathbf{F}^H \mathbf{F} = \mathbf{F} \mathbf{F}^H = \mathbf{I}_N, \quad (7)$$

where $\mathbf{\Lambda}, \mathbf{\Gamma} \in \mathbb{C}^{N \times N}$ are diagonal matrices defined by

$$\mathbf{\Lambda} = \text{diag}(\tilde{\lambda}_1, \dots, \tilde{\lambda}_N), \quad \mathbf{\Gamma} = \text{diag}(\tilde{\gamma}_1, \dots, \tilde{\gamma}_N).$$

Notice that assumption (A3) guarantees the existence of global minimisers for the cost function $\mathcal{J}(\cdot; \mu) : \mathbb{R}^N \to \mathbb{R}^+$ in (2).

2 Residual Whiteness Principle

Let us consider the noise realisation $\mathbf{e}$ in (1) in its original $n_r \times n_c$ matrix form:

$$\mathbf{e} = \{e_{i,j}\}_{(i,j) \in \Omega}, \quad \Omega := \{0, \dots, n_r - 1\} \times \{0, \dots, n_c - 1\}.$$

The *sample auto-correlation* $a : \mathbb{R}^{n_r \times n_c} \to \mathbb{R}^{(2n_r - 1) \times (2n_c - 1)}$ of realisation $\mathbf{e}$ is

$$a(\mathbf{e}) = \{a_{l,m}(\mathbf{e})\}_{(l,m) \in \Theta}, \quad \Theta := \{-(n_r - 1), \dots, n_r - 1\} \times \{-(n_c - 1), \dots, n_c - 1\},$$

with each scalar component $a_{l,m}(\mathbf{e}) : \mathbb{R}^{n_r \times n_c} \to \mathbb{R}$ given by

$$a_{l,m}(\mathbf{e}) = \frac{1}{n} \left(\mathbf{e} \star \mathbf{e} \right)_{l,m} = \frac{1}{n} \left(\mathbf{e} * \mathbf{e}' \right)_{l,m} = \frac{1}{n} \sum_{(i,j) \in \Omega} e_{i,j}\, e_{i+l,j+m}, \; (l,m) \in \Theta, \quad (8)$$

where index pairs (l, m) are commonly called *lags*, $\star$ and $*$ denote the 2-D discrete correlation and convolution operators, respectively, and where $\mathbf{e}'(i, j) = \mathbf{e}(-i, -j)$. The noise realisation $\mathbf{e}$ is padded with at least $n_r - 1$ samples in the vertical direction and $n_c - 1$ samples in the horizontal direction by assuming periodic boundary conditions, such that $\star$ and $*$ in (8) denote 2-D circular correlation and convolution, respectively. This allows to consider only lags

$$(l, m) \in \overline{\Theta} := \{0, \dots, n_r - 1\} \times \{0, \dots, n_c - 1\}.$$

If the corruption $\mathbf{e}$ in (1) is the realisation of a white Gaussian noise process - as in our case - it is well known that as $n \to +\infty$, the sample auto-correlation $a_{l,m}(\mathbf{e})$ vanishes for all $(l, m) \neq (0, 0)$, while $a_{0,0}(\mathbf{e}) = \sigma^2$ - see, e.g., [5].

The DP exploits only the information at lag $(0, 0)$. In fact, the standard deviation recovered by the residual image is required to be equal to σ. Imposing whiteness of the restoration residual by constraining the residual auto-correlation at non-zero lags to be small is a much stronger requirement.

In [6], the authors introduce the following non-negative scalar measure of whiteness $\mathcal{W} : \mathbb{R}^{n_r \times n_c} \to \mathbb{R}^+$ of noise realisation $\mathbf{e}$:

$$\mathcal{W}(\mathbf{e}) := \|\mathbf{e} \star \mathbf{e}\|_2^2 / \|\mathbf{e}\|_2^4 = \widetilde{\mathcal{W}}(\tilde{\mathbf{e}}), \quad (9)$$

where $\| \cdot \|_2$ denotes the Frobenius norm, while the second equality comes from Proposition 1 below, with $\widetilde{\mathcal{W}} : \mathbb{C}^{n_r \times n_c} \to \mathbb{R}^+$ the function defined in (10). Notice that the presence of the denominator in the function in (9) makes the whiteness principle completely independent of the noise level.

Proposition 1. *Let* $\mathbf{e} \in \mathbb{R}^{n_r \times n_c}$ *and* $\tilde{\mathbf{e}} \in \mathbb{C}^{n_r \times n_c}$. *Then, under the assumption of periodic boundary conditions for* $\mathbf{e}$, *the function* $\mathcal{W}$ *defined in (9) satisfies:*

$$\mathcal{W}(\mathbf{e}) = \widetilde{\mathcal{W}}(\tilde{\mathbf{e}}) := \sum_{(l,m)\in\overline{\Theta}} |\tilde{e}_{l,m}|^4 \Big/ \Big(\sum_{(l,m)\in\overline{\Theta}} |\tilde{e}_{l,m}|^2 \Big)^2 . \tag{10}$$

3 RWP for Super-Resolution

By now looking at (2), we observe that the nearer the super-resolved image $\mathbf{x}^*(\mu)$ is to the original image $\mathbf{x}$, the closer the associated residual image $\mathbf{r}^*(\mu) = \mathbf{SKx}^*(\mu) - \mathbf{b}$ is to the white noise realisation $\mathbf{e}$ in (1) and, hence, the whiter is the residual image according to the scalar measure in (9).

This motivates the choice of the RWP for automatically selecting the regularisation parameter μ in variational models of the form (2), which reads:

$$\text{Select } \mu = \mu^* \text{ s.t. } \mu^* \in \arg\min_{\mu\in\mathbb{R}_*^+} W(\mu) := \mathcal{W}(\mathbf{r}^*(\mu)) , \tag{11}$$

where the scalar non-negative cost function $W : \mathbb{R}_*^+ \to \mathbb{R}^+$ in (11), from now on referred to as the *residual whiteness function*, takes the following form:

$$W(\mu) = \| \mathbf{r}^*(\mu) \star \mathbf{r}^*(\mu) \|_2^2 / \| \mathbf{r}^*(\mu) \|_2^4 . \tag{12}$$

Let us now give a closer look to the function in (12). First, we observe

$$\mathbf{r}^*(\mu) = \mathbf{SKx}^*(\mu) - \mathbf{b} = \mathbf{SKx}^*(\mu) - \mathbf{SS}^H \mathbf{b} = \mathbf{Sr}_H^*(\mu) ,$$

where $\mathbf{r}_H^*(\mu) = \mathbf{Kx}^*(\mu) - \mathbf{b}_H$ is the high-resolution residual, while $\mathbf{b}_H = \mathbf{S}^H \mathbf{b}$. The denominator in (12) can be thus expressed as follows

$$\|\mathbf{r}^*(\mu)\|_2^4 = \|\mathbf{Sr}_H^*(\mu)\|_2^4 = \|\mathbf{S}^H \mathbf{Sr}_H^*(\mu)\|_2^4 = \|\mathbf{F}^H(\mathbf{FS}^H\mathbf{SF}^H)\mathbf{Fr}_H^*(\mu)\|_2^4, \tag{13}$$

where the second equality comes from recalling that $\mathbf{S}^H$ interpolates $\mathbf{Sr}_H^*(\mu)$ with zeros giving null contribution when computing the norm. From Lemma 1 and by applying the Parseval's theorem, we get the following chain of equalities:

$$\|\mathbf{r}^*(\mu)\|_2^4 = \big\|(1/d)\mathbf{F}^H(\mathbf{J}_{d_r} \otimes \mathbf{I}_{n_r}) \otimes (\mathbf{J}_{d_c} \otimes \mathbf{I}_{n_c})\tilde{\mathbf{r}}_H^*(\mu)\big\|_2^4$$
$$= \big\|(1/d)(\mathbf{J}_{d_r} \otimes \mathbf{I}_{n_r}) \otimes (\mathbf{J}_{d_c} \otimes \mathbf{I}_{n_c})\tilde{\mathbf{r}}_H^*(\mu)\big\|_2^4 . \tag{14}$$

The non-zero entries of the matrix introduced in Lemma 1, which are all equal to 1, are arranged along replicated patterns; this particular structure can be exploited by considering a permutation matrix $\mathbf{P} \in \mathbb{R}^{N\times N}$ such that:

$$\mathbf{P}\left[(\mathbf{J}_{d_r} \otimes \mathbf{I}_{n_r}) \otimes (\mathbf{J}_{d_c} \otimes \mathbf{I}_{n_c})\right]\mathbf{P}^T = (\mathbf{I}_n \otimes \mathbf{J}_d) . \tag{15}$$

The designed permutation acts on the matrix of interest by gathering together the replicated rows and columns. In Fig. 1, we show the structure of the matrix in (4) and of the permuted matrix in (15) for $n_r = n_c = 3$ and $d_r = d_c = 2$.

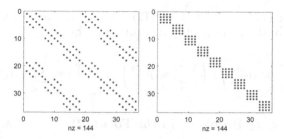

Fig. 1. Structure of the matrix in (4) (left) and of the permutation induced by **P** (right) for $n_r = n_c = 3$, $d_r = d_c = 2$.

Hence, the expression in (14) can be rewritten as

$$\|\mathbf{r}^*(\mu)\|_2^4 = \left\|\frac{1}{d}\mathbf{P}\left[(\mathbf{J}_{d_r} \otimes \mathbf{I}_{n_r}) \otimes (\mathbf{J}_{d_c} \otimes \mathbf{I}_{n_c})\right]\mathbf{P}^T\mathbf{P}\tilde{\mathbf{r}}_H^*(\mu)\right\|_2^4 = \left\|\frac{1}{d}(\mathbf{I}_n \otimes \mathbf{J}_d)\hat{\mathbf{r}}_H^*(\mu)\right\|_2^4,$$

where

$$((\mathbf{I}_n \otimes \mathbf{J}_d)\hat{\mathbf{r}}_H^*(\mu))_i = \sum_{j=0}^{d-1} (\hat{\mathbf{r}}_H^*(\mu))_{\iota+j}, \quad \text{with } \iota := 1 + \left\lfloor\frac{i-1}{d}\right\rfloor d,$$

for every $i = 1, \ldots N$. The denominator in (12) can be thus expressed as

$$\|\mathbf{r}^*(\mu)\|_2^4 = \frac{1}{d^4}\left(\sum_{i=1}^{N}\left|\sum_{j=0}^{d-1}(\hat{\mathbf{r}}_H^*(\mu))_{\iota+j}\right|^2\right)^2.$$

Let us now consider the numerator of the function $W(\mu)$ in (12), which, based on the definitions of auto-correlation given in (8) and of $\mathbf{S}^H$, reads

$$\|\mathbf{r}^*(\mu) \star \mathbf{r}^*(\mu)\|_2^2 = \|\mathbf{Sr}_H^*(\mu) \star \mathbf{Sr}_H^*(\mu)\|_2^2 = \|\mathbf{S}^H\mathbf{Sr}_H^*(\mu) \star \mathbf{S}^H\mathbf{Sr}_H^*(\mu)\|_2^2.$$

By applying again the Parseval's theorem and the convolution theorem, we get

$$\begin{aligned}
\|\mathbf{r}^*(\mu) \star \mathbf{r}^*(\mu)\|_2^2 &= \|\mathbf{F}\left((\mathbf{S}^H\mathbf{Sr}_H^*(\mu)) \star (\mathbf{S}^H\mathbf{Sr}_H^*(\mu))\right)\|_2^2 \\
&= \|\mathbf{F}(\mathbf{S}^H\mathbf{Sr}_H^*(\mu)) \odot \overline{\mathbf{F}(\mathbf{S}^H\mathbf{Sr}_H^*(\mu))}\|_2^2 \\
&= \|\mathbf{F}(\mathbf{S}^H\mathbf{S})\mathbf{F}^H\mathbf{Fr}_H^*(\mu) \odot \overline{\mathbf{F}(\mathbf{S}^H\mathbf{S})\mathbf{F}^H\mathbf{Fr}_H^*(\mu)}\|_2^2,
\end{aligned} \tag{16}$$

where $\odot$ denotes the Hadamard matrix product operator. The expression in (16) is manipulated by applying Lemma 1 and the permutation in (15), so as to give

$$\|\mathbf{r}^*(\mu) \star \mathbf{r}^*(\mu)\|_2^2 = \frac{1}{d^4}\sum_{i=1}^{N}\left|\sum_{j=0}^{d-1}(\hat{\mathbf{r}}_H^*(\mu))_{\iota+j}\right|^4. \tag{17}$$

Finally, plugging (17) and (13) into (12), we get the following form for the whiteness measure $W(\mu)$ for a super-resolution problem

$$W(\mu) = \left(\sum_{i=1}^{N} |w_i(\mu)|^4\right) \Big/ \left(\sum_{i=1}^{N} |w_i(\mu)|^2\right)^2, \quad w_i(\mu) = \sum_{j=0}^{d-1} (\hat{\mathbf{r}}_H(\mu))_{\iota+j}. \quad (18)$$

3.1   RWP for ℓ_2-ℓ_2 Problems in the Form (2)

Here, we derive the analytical expression of the whiteness function $W(\mu)$ defined in (18) when addressing Tikhonov-regularised least squares problems as the one in (2). We start following [12] to deduce an explicit and easily-computable expression of $\mathbf{x}^*(\mu)$. By optimality, we get:

$$\mathbf{x}^*(\mu) = (\mu(\mathbf{SK})^H(\mathbf{SK}) + \mathbf{L}^H\mathbf{L})^{-1}(\mu(\mathbf{SK})^H\mathbf{b} + \mathbf{L}^H\mathbf{v}),$$

which can be manipulated in terms of $\mathbf{F}$ and $\mathbf{F}^H$ to deduce

$$\mathbf{x}^*(\mu) = (\mu\mathbf{F}^H\mathbf{F}\mathbf{K}^H\mathbf{F}^H\mathbf{F}\mathbf{S}^H\mathbf{S}\mathbf{F}^H\mathbf{F}\mathbf{K}\mathbf{F}^H\mathbf{F} + \mathbf{F}^H\mathbf{F}\mathbf{L}^T\mathbf{F}^H\mathbf{F}\mathbf{L}\mathbf{F}^H\mathbf{F})^{-1}(\mu\mathbf{K}^H\mathbf{S}^H\mathbf{b} + \mathbf{L}^H\mathbf{v})$$

$$= (\mu\mathbf{F}^H\mathbf{\Lambda}^H(\mathbf{F}\mathbf{S}^H\mathbf{S}\mathbf{F}^H)\mathbf{\Lambda}\mathbf{F} + \mathbf{F}^H\mathbf{\Gamma}^H\mathbf{\Gamma}\mathbf{F})^{-1}(\mu\mathbf{K}^H\mathbf{S}^H\mathbf{b} + \mathbf{L}^H\mathbf{v}), \quad (19)$$

where $\mathbf{\Lambda}, \mathbf{\Gamma}$ are defined in (7). Lemma 1 provides a useful expression for the product $(\mathbf{F}\mathbf{S}^H\mathbf{S}\mathbf{F}^H)$, by which (19) becomes:

$$\mathbf{x}^*(\mu) = \left(\frac{\mu}{d}\mathbf{F}^H\mathbf{\Lambda}^H\mathbf{P}^T(\mathbf{I}_n \otimes \mathbf{J}_d)\mathbf{P}\mathbf{\Lambda}\mathbf{F} + \mathbf{F}^H\mathbf{\Gamma}^H\mathbf{\Gamma}\mathbf{F}\right)^{-1}(\mu\mathbf{K}^H\mathbf{S}^H\mathbf{b} + \mathbf{L}^H\mathbf{v})$$

$$= \mathbf{F}^H\left(\frac{\mu}{d}\mathbf{\Lambda}^H\mathbf{P}^T(\mathbf{I}_n \otimes \mathbf{J}_d)\mathbf{P}\mathbf{\Lambda} + \mathbf{\Gamma}^H\mathbf{\Gamma}\right)^{-1}\mathbf{F}(\mu\mathbf{K}^H\mathbf{F}^H\mathbf{F}\mathbf{S}^H\mathbf{b} + \mathbf{L}^H\mathbf{F}^H\mathbf{F}\mathbf{v})$$

$$= \mathbf{F}^H\left(\frac{\mu}{d}\mathbf{\Lambda}^H\mathbf{P}^T(\mathbf{I}_n \otimes \mathbf{J}_d)\mathbf{P}\mathbf{\Lambda} + \mathbf{\Gamma}^H\mathbf{\Gamma}\right)^{-1}(\mu\mathbf{\Lambda}^H\tilde{\mathbf{b}}_H + \mathbf{\Gamma}^H\tilde{\mathbf{v}}), \quad (20)$$

where $\tilde{\mathbf{b}}_H = \mathbf{F}\mathbf{b}_H = \mathbf{F}\mathbf{S}^H\mathbf{b}$ contains d replication of $\tilde{\mathbf{b}}$ - see, e.g., [8]. We now introduce the following operators

$$\underline{\mathbf{\Lambda}} := (\mathbf{I}_n \otimes \mathbf{1}_d^T)\mathbf{P}\mathbf{\Lambda} \qquad \underline{\mathbf{\Lambda}}^H := \mathbf{\Lambda}^H\mathbf{P}^T(\mathbf{I}_n \otimes \mathbf{1}_d) \quad (21)$$

where $\mathbf{1}_d \in \mathbb{R}^d$ is a vector of ones. In compact form, equation (20) reads:

$$\mathbf{x}^*(\mu) = \mathbf{F}^H\left(\frac{\mu}{d}\underline{\mathbf{\Lambda}}^H\underline{\mathbf{\Lambda}} + \mathbf{\Gamma}^H\mathbf{\Gamma}\right)^{-1}(\mu\mathbf{\Lambda}^H\tilde{\mathbf{b}}_H + \mathbf{\Gamma}^H\tilde{\mathbf{v}}). \quad (22)$$

Proceeding as in [12], we can now apply the Woodbury formula (6) and perform few manipulations, so as to obtain that the expression in (22) becomes:

$$\mathbf{x}^*(\mu) = \mathbf{F}^H\left[\mathbf{\Psi} - \mu\mathbf{\Psi}\underline{\mathbf{\Lambda}}^H\left(d\mathbf{I} + \mu\underline{\mathbf{\Lambda}}\mathbf{\Psi}\underline{\mathbf{\Lambda}}^H\right)^{-1}\underline{\mathbf{\Lambda}}\mathbf{\Psi}\right](\mu\mathbf{\Lambda}^H\tilde{\mathbf{b}}_H + \mathbf{\Gamma}^H\tilde{\mathbf{v}}), \quad (23)$$

whence the Fourier transform of the high resolution residual $\mathbf{r}_H^*(\mu) = \mathbf{K}\mathbf{x}^*(\mu) - \mathbf{b}$, with $\mathbf{x}^*(\mu)$ given in (23), can be written as

$$\tilde{\mathbf{r}}_H^*(\mu) = \mathbf{\Lambda}\left[\mathbf{\Psi} - \mu\mathbf{\Psi}\underline{\mathbf{\Lambda}}^H\left(d\mathbf{I} + \mu\underline{\mathbf{\Lambda}}\mathbf{\Psi}\underline{\mathbf{\Lambda}}^H\right)^{-1}\underline{\mathbf{\Lambda}}\mathbf{\Psi}\right](\mu\mathbf{\Lambda}^H\tilde{\mathbf{b}}_H + \mathbf{\Gamma}^H\tilde{\mathbf{v}}) - \tilde{\mathbf{b}}_H, \quad (24)$$

where $\boldsymbol{\Psi} = (\boldsymbol{\Gamma}^H\boldsymbol{\Gamma} + \epsilon)^{-1}$ and the parameter $0 < \epsilon \ll 1$ guarantees the inversion of $\boldsymbol{\Gamma}^H\boldsymbol{\Gamma}$. Recalling Lemma 2 and the property (15), we prove the following result.

Proposition 2. *Let* $\boldsymbol{\Phi} \in \mathbb{R}^{n \times n}$ *be a diagonal matrix and consider the matrix* $\underline{\boldsymbol{\Lambda}}$ *defined in* (21). *Then, the following equality holds:*

$$\underline{\boldsymbol{\Lambda}}^H\boldsymbol{\Phi}\underline{\boldsymbol{\Lambda}} = \mathbf{P}^T(\boldsymbol{\Phi} \otimes \mathbf{I}_d)\mathbf{P}\underline{\boldsymbol{\Lambda}}^H\underline{\boldsymbol{\Lambda}}.$$

Proof. Recalling property (5) in Lemma 2, we get the following chain of equalities

$$\begin{aligned}
\underline{\boldsymbol{\Lambda}}^H\boldsymbol{\Phi}\underline{\boldsymbol{\Lambda}} &= \boldsymbol{\Lambda}^H\mathbf{P}^T(\mathbf{I}_n \otimes \mathbf{1}_d)\boldsymbol{\Phi}(\mathbf{I}_n \otimes \mathbf{1}_d^T)\mathbf{P}\boldsymbol{\Lambda} = \boldsymbol{\Lambda}^H\mathbf{P}^T(\mathbf{I}_n \otimes \mathbf{1}_d)(\boldsymbol{\Phi} \otimes \mathbf{1}_d^T)\mathbf{P}\boldsymbol{\Lambda} \\
&= \boldsymbol{\Lambda}^H\mathbf{P}^T(\mathbf{I}_n\boldsymbol{\Phi} \otimes \mathbf{1}_d\mathbf{1}_d^T)\mathbf{P}\boldsymbol{\Lambda} = \boldsymbol{\Lambda}^H\mathbf{P}^T(\boldsymbol{\Phi}\mathbf{I}_n \otimes \mathbf{J}_d)\mathbf{P}\boldsymbol{\Lambda} \\
&= \boldsymbol{\Lambda}^H\mathbf{P}^T(\boldsymbol{\Phi}\mathbf{I}_n \otimes \mathbf{I}_d\mathbf{J}_d)\mathbf{P}\boldsymbol{\Lambda} = \boldsymbol{\Lambda}^H\mathbf{P}^T(\boldsymbol{\Phi} \otimes \mathbf{I}_d)(\mathbf{I}_n \otimes \mathbf{J}_d)\mathbf{P}\boldsymbol{\Lambda} \\
&= \boldsymbol{\Lambda}^H\mathbf{P}^T(\boldsymbol{\Phi} \otimes \mathbf{I}_d)\mathbf{P}\mathbf{P}^T(\mathbf{I}_n \otimes \mathbf{J}_d)\mathbf{P}\boldsymbol{\Lambda},
\end{aligned}$$

where the sparse block-diagonal matrix $\mathbf{P}^T(\boldsymbol{\Phi} \otimes \mathbf{I}_d)\mathbf{P} \in \mathbb{R}^{N \times N}$ commutes with $\boldsymbol{\Lambda}^H$, so that $\boldsymbol{\Lambda}^H\mathbf{P}^T(\boldsymbol{\Phi} \otimes \mathbf{I}_d)\mathbf{P} = \mathbf{P}^T(\boldsymbol{\Phi} \otimes \mathbf{I}_d)\mathbf{P}\boldsymbol{\Lambda}^H$. Recalling (21), this yields:

$$\underline{\boldsymbol{\Lambda}}^H\boldsymbol{\Phi}\underline{\boldsymbol{\Lambda}} = \mathbf{P}^T(\boldsymbol{\Phi} \otimes \mathbf{I}_d)\mathbf{P}\underline{\boldsymbol{\Lambda}}^H\underline{\boldsymbol{\Lambda}},$$

which completes the proof.   □

Corollary 1. *Let* $\boldsymbol{\Phi} = \left(d\mathbf{I} + \mu\underline{\boldsymbol{\Lambda}}\boldsymbol{\Psi}\underline{\boldsymbol{\Lambda}}^H\right)^{-1}$. *Then, the expression in* (24) *turns into*

$$\tilde{\mathbf{r}}_H^*(\mu) = \underline{\boldsymbol{\Lambda}}\left[\boldsymbol{\Psi} - \mu\boldsymbol{\Psi}\mathbf{P}^T\left((d\mathbf{I} + \mu\underline{\boldsymbol{\Lambda}}\boldsymbol{\Psi}\underline{\boldsymbol{\Lambda}}^H)^{-1} \otimes \mathbf{I}_d\right)\mathbf{P}\underline{\boldsymbol{\Lambda}}^H\underline{\boldsymbol{\Lambda}}\boldsymbol{\Psi}\right](\mu\underline{\boldsymbol{\Lambda}}^H\check{\mathbf{b}}_H + \boldsymbol{\Gamma}^H\tilde{\mathbf{v}}) - \check{\mathbf{b}}_H.$$

Proof. We first notice that

$$\underline{\boldsymbol{\Lambda}}\boldsymbol{\Psi}\underline{\boldsymbol{\Lambda}}^H = (\mathbf{I}_n \otimes \mathbf{1}_d^T)\widehat{\boldsymbol{\Lambda}\boldsymbol{\Psi}\boldsymbol{\Lambda}^H}(\mathbf{I}_n \otimes \mathbf{1}_d), \tag{25}$$

is diagonal as $\widehat{\boldsymbol{\Lambda}\boldsymbol{\Psi}\boldsymbol{\Lambda}^H} = \mathbf{P}\boldsymbol{\Lambda}\boldsymbol{\Psi}\boldsymbol{\Lambda}^H\mathbf{P}^T$ is. The matrix in (25) can thus be written as

$$\underline{\boldsymbol{\Lambda}}\boldsymbol{\Psi}\underline{\boldsymbol{\Lambda}}^H = \mathrm{diag}(\omega_1, \ldots, \omega_n), \qquad \omega_i = \sum_{j=0}^{d-1}\frac{|\hat{\lambda}_{\iota+j}|^2}{|\hat{\gamma}_{\iota+j}|^2 + \epsilon}.$$

Hence, since $\boldsymbol{\Phi}$ is the inverse of the sum of two diagonal matrices, it is diagonal so we can apply Proposition 2 and deduce the thesis.   □

Recalling now the action of the permutation matrix $\mathbf{P}$ on vectors, we have that the product $\hat{\mathbf{r}}_H^*(\mu) = \mathbf{P}\tilde{\mathbf{r}}_H^*(\mu)$ reads

$$\hat{\mathbf{r}}_H^*(\mu) = \left[\widehat{\boldsymbol{\Lambda}\boldsymbol{\Psi}} - \mu\widehat{\boldsymbol{\Lambda}\boldsymbol{\Psi}}\left((d\mathbf{I} + \mu\underline{\boldsymbol{\Lambda}}\boldsymbol{\Psi}\underline{\boldsymbol{\Lambda}}^H)^{-1} \otimes \mathbf{I}_d\right)\widehat{\underline{\boldsymbol{\Lambda}}^H\underline{\boldsymbol{\Lambda}}\boldsymbol{\Psi}}\right](\mu\check{\boldsymbol{\Lambda}}\hat{\mathbf{b}}_H + \check{\boldsymbol{\Gamma}}\hat{\mathbf{v}}) - \hat{\mathbf{b}}_H, \tag{26}$$

where the matrix $\widehat{\underline{\boldsymbol{\Lambda}}^H\underline{\boldsymbol{\Lambda}}\boldsymbol{\Psi}} = \mathbf{P}\boldsymbol{\Lambda}^H\mathbf{P}^T(\mathbf{I}_n \otimes \mathbf{J}_d)\mathbf{P}\boldsymbol{\Lambda}\boldsymbol{\Psi}\mathbf{P}^T$ acts on $\mathbf{g} \in \mathbb{R}^N$ as

$$(\widehat{\underline{\boldsymbol{\Lambda}}^H\underline{\boldsymbol{\Lambda}}\boldsymbol{\Psi}}\mathbf{g})_i = \bar{\hat{\lambda}}_i\sum_{j=0}^{d-1}\frac{\hat{\lambda}_{\iota+j}}{|\hat{\gamma}_{\iota+j}|^2 + \epsilon}g_{\iota+j}.$$

Combining altogether, we finally deduce:

$$\hat{\mathbf{r}}_H^*(\mu) = \mu\widehat{\underline{\mathbf{\Lambda}}\mathbf{\Psi}}\check{\mathbf{\Lambda}}\hat{\mathbf{b}}_H + \widehat{\underline{\mathbf{\Lambda}}\mathbf{\Psi}\check{\mathbf{\Gamma}}\hat{\mathbf{v}}} - \mu^2\widehat{\underline{\mathbf{\Lambda}}\mathbf{\Psi}}\left[(d\mathbf{I} + \mu\underline{\mathbf{\Lambda}}\mathbf{\Psi}\underline{\mathbf{\Lambda}}^H)^{-1} \otimes \mathbf{I}_d\right]\widehat{\underline{\mathbf{\Lambda}}^H\underline{\mathbf{\Lambda}}\mathbf{\Psi}}\check{\mathbf{\Lambda}}\hat{\mathbf{b}}_H$$

$$-\mu\widehat{\underline{\mathbf{\Lambda}}\mathbf{\Psi}}\left[(d\mathbf{I} + \mu\underline{\mathbf{\Lambda}}\mathbf{\Psi}\underline{\mathbf{\Lambda}}^H)^{-1} \otimes \mathbf{I}_d\right]\widehat{\underline{\mathbf{\Lambda}}^H\underline{\mathbf{\Lambda}}\mathbf{\Psi}\check{\mathbf{\Gamma}}\hat{\mathbf{v}}} - \hat{\mathbf{b}}_H\,,$$

whence we can explicitly compute the expression for each component $i = 1, \ldots, n$:

$$(\hat{\mathbf{r}}_H^*(\mu))_i = \mu\left[\frac{|\hat{\lambda}_i|^2}{|\hat{\gamma}_i|^2 + \epsilon}\,\hat{b}_{H,i}\right] + \frac{\hat{\lambda}_i\bar{\hat{\gamma}}_i\hat{v}_i}{|\hat{\gamma}_i|^2 + \epsilon} - \left[\mu^2\sum_{j=0}^{d-1}\frac{|\hat{\lambda}_{\iota+j}|^2\hat{b}_{H,\iota+n}}{|\hat{\gamma}_{\iota+j}|^2 + \epsilon}\right.$$

$$\left.+ \mu\sum_{j=0}^{d-1}\frac{\hat{\lambda}_{\iota+j}\bar{\hat{\gamma}}_{\iota+j}\hat{v}_{\iota+j}}{|\hat{\gamma}_{\iota+j}|^2 + \epsilon}\right]\frac{|\hat{\lambda}_i|^2}{|\hat{\gamma}_i|^2 + \epsilon}\left(d + \mu\sum_{j=0}^{d-1}\frac{|\hat{\lambda}_{\iota+j}|^2}{|\hat{\gamma}_{\iota+j}|^2 + \epsilon}\right)^{-1} - \hat{b}_{H,i}\,.$$

We can thus deduce the following expression of the terms in formula (18)

$$\sum_{j=0}^{d-1}(\hat{\mathbf{r}}_H^*(\mu))_{\iota+j} = \frac{1}{d + \mu\displaystyle\sum_{j=0}^{d-1}\frac{|\hat{\lambda}_{\iota+j}|^2}{|\hat{\gamma}_{\iota+j}|^2 + \epsilon}}\left[\mu\left(d\sum_{j=0}^{d-1}\frac{|\hat{\lambda}_{\iota+j}|^2}{|\hat{\gamma}_{\iota+j}|^2 + \epsilon}\,\hat{b}_{H,\iota+j}\right.\right.$$

$$\left.\left.- \sum_{j=0}^{d-1}\hat{b}_{H,\iota+j}\sum_{j=0}^{d-1}\frac{|\hat{\lambda}_{\iota+j}|^2}{|\hat{\gamma}_{\iota+j}|^2 + \epsilon}\right) + d\left(\sum_{j=0}^{d-1}\frac{\hat{\lambda}_{\iota+j}\bar{\hat{\gamma}}_{\iota+j}\hat{v}_{\iota+j}}{|\hat{\gamma}_{\iota+j}|^2 + \epsilon} - \sum_{j=0}^{d-1}\hat{b}_{H,\iota+j}\right)\right]. \quad (27)$$

In light of its replicating structure, we observe that the action of the permutation $\mathbf{P}$ on $\tilde{\mathbf{b}}_H$ will cluster the identical entries, so that the $\hat{b}_{H,\iota+j}$ can be written as the mean of the set of d values $\{\hat{b}_{H,\iota}, \ldots, \hat{b}_{H,\iota+d-1}\}$. This allows to simplify formula (27) as the difference in the first bracket vanishes. By now setting

$$\eta_i := \frac{1}{d}\sum_{j=0}^{d-1}\frac{|\hat{\lambda}_{\iota+j}|^2}{|\hat{\gamma}_{\iota+j}|^2 + \epsilon}, \quad \varrho_i := \sum_{j=0}^{d-1}\hat{b}_{H,\iota+j}, \quad \nu_i := \sum_{j=0}^{d-1}\frac{\hat{\lambda}_{\iota+j}\bar{\hat{\gamma}}_{\iota+j}\hat{v}_{\iota+j}}{|\hat{\gamma}_{\iota+j}|^2 + \epsilon}\,,$$

which can all be computed beforehand, and plugging (27) into (18) we finally get

$$W(\mu) = \left(\sum_{i=1}^{N}\left|\frac{\nu_i - \varrho_i}{1 + \eta_i\mu}\right|^4\right) \Big/ \left(\sum_{i=1}^{N}\left|\frac{\nu_i - \varrho_i}{1 + \eta_i\mu}\right|^2\right)^2. \quad (28)$$

Note that when $d = 1$, i.e. when no decimation is considered, this formula corresponds exactly to the one considered in [6] in the context of image deblurring.

According to the RWP, the optimal μ^* is selected as the one minimising the whiteness measure function in (28). We remark that the action of the permutation matrix $\mathbf{P}$ can be efficiently replicated without deriving its explicit expression; as a result, the overall computational cost for the evaluation of $W(\mu)$ amounts to $O(N \log N)$, namely the cost of the 2D fast Fourier transform and of its inverse, and the value μ^* can be efficiently detected via grid-search. Finally, the optimal μ^* is used for the computation of the reconstruction $\mathbf{x}^*(\mu^*)$ based on (23).

The main steps of the proposed procedure are summarised in Algorithm 1.

Algorithm 1: SR for (2) with automatic parameter selection via RWP

---

inputs: observed image $\mathbf{b} \in \mathbb{R}^n$, forward model operator $\mathbf{K} \in \mathbb{R}^{nd \times nd}$,
down-sampling operator $\mathbf{S} \in \mathbb{R}^{n \times nd}$

- **Compute Fourier diagonalisations:** $\boldsymbol{\Lambda} = \mathbf{F}\mathbf{K}\mathbf{F}^H$, $\boldsymbol{\Gamma} = \mathbf{F}\mathbf{L}\mathbf{F}^H$
- **Compute matrices:** $\underline{\boldsymbol{\Lambda}} = (\mathbf{I}_n \otimes \mathbf{1}_d^T)\mathbf{P}\boldsymbol{\Lambda}$, $\boldsymbol{\Psi} = (\boldsymbol{\Gamma}^H\boldsymbol{\Gamma} + \epsilon)^{-1}$
- **Residual whiteness principle for the selection of μ^* :**
  - · Compute $W(\mu)$ in (28) for different values of μ, based on Corollary 1 and (26)
  - · Select $\mu^* \in \arg\min W(\mu)$
- **Compute the reconstruction:** $\mathbf{x}^*(\mu^*)$ by (23)

---

4   Numerical Results

We evaluate the proposed RWP-based automatic procedure for selecting the regularisation parameter μ in variational models of the form (2) when $\mathbf{v} = \mathbf{0}_N$ and $\mathbf{L} = \mathbf{D} := \left(\mathbf{D}_h^T, \mathbf{D}_v^T\right)^T \in \mathbb{R}^{2N \times N}$, with $\mathbf{D}_h, \mathbf{D}_v \in \mathbb{R}^{N \times N}$ representing the finite difference operators discretising the first-order horizontal and vertical partial derivatives, respectively. Note that $\mathbf{D}$ verifies assumptions (A3)-(A4) in Sect. 1.1.

Our goal is to highlight that the proposed RWP selects a regularisation parameter value μ^* yielding high quality restorations. The RWP is compared with the DP, defined in (3) when $\tau = 1$. There is a one-to-one relationship between the μ-value and the norm of the associated residual image. Hence, in all the presented results we will substitute the μ-values with the corresponding τ-values, with τ defined according to (3) by $\tau^*(\mu) := \|\mathbf{S}\mathbf{H}\mathbf{x}^*(\mu) - \mathbf{b}\|_2/(\sqrt{n}\sigma)$.

The quality of the restorations $\mathbf{x}^*$, for different values of τ^*, with respect to the original undecimated image $\mathbf{x}$, will be assessed by means of three scalar measures, namely the Structural Similarity Index (SSIM) [11], the Peak-Signal-to-Noise-Ratio (PSNR) and the Improved-Signal-to-Noise Ratio (ISNR), defined by $\text{PSNR} = 20\log_{10}(\sqrt{N}\max(\mathbf{x}, \mathbf{x}^*)/\|\mathbf{x} - \mathbf{x}^*\|_2)$ and $\text{ISNR} = 10\log_{10}(\|\mathbf{x} - \bar{\mathbf{b}}\|_2/\|\mathbf{x} - \mathbf{x}^*\|_2)$, respectively, with $\max(\mathbf{x}, \mathbf{x}^*)$ representing the largest value of $\mathbf{x}$ and $\mathbf{x}^*$, while $\bar{\mathbf{b}}$ denotes the bicubic interpolation of $\mathbf{b}$.

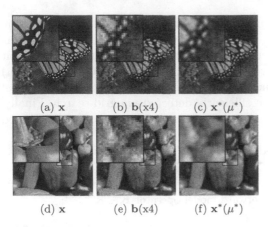

(a) x   (b) b(x4)   (c) x*(μ*)

(d) x   (e) b(x4)   (f) x*(μ*)

Fig. 2. From left to right: original **x**, observed **b**, our reconstruction **x***(μ*) for **monarch** (top) and **peppers** (bottom).

Table 1. Achieved PSNR, ISNR, SSIM values for **monarch** and **peppers** for the two degradation settings.

monarch

	PSNR	ISNR	SSIM	
x*	21.4050	1.3452	0.6736	Test 1
b̄	20.0598	–	0.5435	
x*	18.9277	1.1561	0.6297	Test 2
b̄	17.7716	–	0.3105	

peppers

	PSNR	ISNR	SSIM	
x*	23.5674	1.9147	0.6757	Test 1
b̄	21.6526	–	0.5187	
x*	21.3034	2.6078	0.6240	Test 2
b̄	18.6956	–	0.3032	

We consider two test images of size 512×512 with pixel values normalised in $[0, 1]$, namely **monarch** and **peppers**, shown in Figs. 2a–2d, respectively. The decimation factors along the rows and the columns of the original images are set as $d_c = d_r = 4$. As a first example, the original test images are corrupted by Gaussian blur, generated by the Matlab routine fspecial with input parameters band $= 9$ and sigma $= 2$. The band parameter represents the side length (in pixels) of the square support of the kernel, whereas sigma is the standard deviation (in pixels) of the isotropic bivariate Gaussian distribution defining the kernel in the continuous setting. Finally, the decimated and blurred images are corrupted by an AWGN with standard deviation $\sigma = 0.05$. The observed data for the test images **monarch** and **peppers** are displayed in Fig. 2b–2e, respectively.

In Figs. 3a, we report the behavior of the whiteness measure $W(\mu)$ as a function of $\tau^*(\mu)$ for the test images **monarch** (solid blue line) and **peppers** (solid black line), respectively. The plotted values have been obtained by solving the model (2) for a fine grid of different μ values, and then computing for each μ the associated $\tau^*(\mu)$ and $W(\mu)$. The optimal τ^*s corresponding to μ^*s are indicated by the vertical dashed magenta and green lines for **monarch** $(\tau^*(\mu^*) = 0.9398)$ and **peppers** $(\tau^*(\mu^*) = 0.9633)$, respectively, while $\tau = 1$, representing the DP, is depicted by the vertical dotted black line. Notice that the whiteness curves computed *a posteriori* admit a minimiser over the considered domain which coincides with the τ^* selected by the RWP. Moreover, the proximity of the optimal τ^*s to 1 indicates that the noise level estimated starting from $\mathbf{r}^*(\mu)$ is close to the true one.

In Figs. 3b–3c, we graphic the achieved ISNR and SSIM for the two test images. Note that the RWP tends to automatically select a μ-value returning the best trade-off between the two quality measures. The reconstructed $\mathbf{x}^*(\mu^*)$ for the two test images are shown in Figs. 2c–2f. Finally, the PSNR, ISNR and

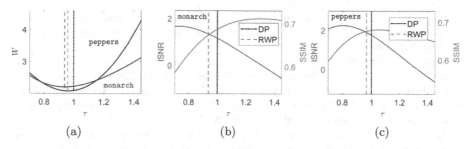

Fig. 3. Whiteness measure functions (left column) and ISNR/SSIM values for different τs (second column) for the monarch and peppers test images.

SSIM values achieved by the proposed strategy are reported in the top part of Table 1 (Test 1), together with the PSNR and SSIM of the bicubic interpolation.

As a second example, we perform the same reconstructions with different degradation levels. More specifically, we consider a Gaussian blur with parameters band = 13, sigma = 3, and AWGN with standard deviation $\sigma = 0.1$. The achieved quality measures are reported in the bottom part of Table 1 (Test 2). In this case, the RWP automatically selects an optimal τ^* corresponding to a very accurate estimate of the original noise standard deviation, namely $\tau^*(\mu^*) = 0.9938$ for monarch and $\tau^*(\mu^*) = 0.9915$ for peppers.

5 Conclusions and Outlook

We extended the residual whiteness principle introduced in [6] for automatic parameter selection with unknown noise level in image deblurring to image super-resolution problems solved by generalised Tikhonov regularisation models in the form (2) whose solution can be efficiently computed by means of the approach outlined in [12]. By exploiting carefully technical properties of the operators involved in the model in the frequency domain, a compact formula for the whiteness measure can be found. Its minimisation provides an accurate estimate of the unknown noise level. As a future work, we plan to explicitly formalise the extension of the RWP to non-smooth super-resolution models as well as to explicitly tackle the minimisation of the whiteness measure with more sophisticated techniques.

References

1. Chen, A.Z., Huo, B.X., Wen, C.Y.: Adaptive regularization for color image restoration using discrepancy principle. ICSPCC **2013**, 1–6 (2013)
2. Clason, C.: Regularization of Inverse Problems. Kluwer, Dordrecht (1996)
3. Craven, P., Wahba, G.: Smoothing noisy data with spline functions. Numer. Math. **31**, 377–403 (1978)
4. Hansen, P.: Rank-deficient and discrete ill-posed problems: numerical aspects of linear inversion (1987)

5. Lanza, A., Morigi, S., Sciacchitano, F., Sgallari, F.: Whiteness constraints in a unified variational framework for image restoration. J. Math. Imaging Vis. **60**, 1503–1526 (2018)
6. Lanza, A., Pragliola, M., Sgallari, F.: Residual whiteness principle for parameter-free image restoration. Electron. Trans. Numer. Anal. **53**, 329–351 (2020)
7. Paul Riot. Blancheur du résidu pour le débruitage d'image. Ph.D. thesis (2018)
8. Robinson, M.D., Farsiu, S., Lo, J.Y., Milanfar, P., Toth, C.: Efficient registration of aliased x-ray images. ACSSC 215–219 (2007)
9. Toma, A., Sixou, B., Peyrin, F.: Iterative choice of the optimal regularization parameter in TV image restoration. Inverse Probl. Imaging **9**, 1171 (2015)
10. Tuador, N.K., Pham, D., Michetti, J., Basarab, A., Kouamé, D.: A novel fast 3D single image super-resolution algorithm. arXiv:abs/2010.15491 (2020)
11. Wang, Z., Bovik, A., Sheikh, H.R., Simoncelli, E.P.: Image quality assessment: from error visibility to structural similarity. IEEE Trans. Image Process. **13**, 600–612 (2004)
12. Zhao, N., Wei, Q., Basarab, A., Dobigeon, N., Kouamé, D., Tourneret, J.: Fast single image super-resolution using a new analytical solution for ℓ_2-ℓ_2 problems. IEEE Trans. Image Process. **25**, 3683–3697 (2016)

Inverse Problems in Imaging

Total Deep Variation for Noisy Exit Wave Reconstruction in Transmission Electron Microscopy

Thomas Pinetz[1,2]([✉]) [iD], Erich Kobler[1] [iD], Christian Doberstein[3] [iD],
Benjamin Berkels[3] [iD], and Alexander Effland[1,4,5] [iD]

[1] Institute of Computer Graphics and Vision, Graz University of Technology,
Graz, Austria
thomas.pinetz@icg.tugraz.at
[2] Austrian Institute of Technology, Vienna, Austria
[3] AICES Graduate School, RWTH Aachen University, Aachen, Germany
[4] Silicon Austria Labs (TU Graz SAL DES Lab), Graz, Austria
[5] Institute for Applied Mathematics, University of Bonn, Bonn, Germany

Abstract. Transmission electron microscopes (TEMs) are ubiquitous devices for high-resolution imaging on an atomic level. A key problem related to TEMs is the reconstruction of the exit wave, which is the electron signal at the exit plane of the examined specimen. Frequently, this reconstruction is cast as an ill-posed nonlinear inverse problem. In this work, we integrate the data-driven total deep variation regularizer to reconstruct the exit wave in this inverse problem. In several numerical experiments, the applicability of the proposed method is demonstrated for different materials.

Keywords: Total deep variation · Transmission electron microscopy · Deep learning · Exit wave reconstruction · Nonlinear inverse problem

1 Introduction

Nowadays, transmission electron microscopes (TEMs) are standard devices for imaging on an atomic level. In a TEM, beams of electrons (emitted by a cathode) are diffracted by a specimen and subsequently recorded at the image plane after traversing electromagnetic lenses. A common problem in this context is the reconstruction of the so-called *exit wave*, i.e. the electron wave straight after passing through the specimen, which can be cast as a highly ill-posed and nonlinear inverse problem. Since interpreting TEM images is in general challenging, the exit wave is used to gain further insight into the material structure. Due to the beam sensitivity of many materials, images acquired by TEMs are often strongly deteriorated by noise. Thus, we focus on noisy exit wave reconstruction in this work.

Two classical methods for exit wave reconstruction are the *multiple input maximum a posteriori (MIMAP)* algorithm [15] and the *maximum-likelihood*

© Springer Nature Switzerland AG 2021
A. Elmoataz et al. (Eds.): SSVM 2021, LNCS 12679, pp. 491–502, 2021.
https://doi.org/10.1007/978-3-030-75549-2_39

(MAL) algorithm [4]. More specifically, let $(X_{Z_j}^{\exp})_{j=1}^N$ be a family of experimental images, where $Z_j \in \mathbb{R}$ is the focus of the objective lens used when $X_{Z_j}^{\exp}$ was acquired. In both approaches, the sought exit wave Ψ then defines (using a forward model) a synthesized image X_{Ψ,Z_j} depending on the focus Z_j. In MIMAP, an *a priori* estimate Ψ^{estimate} for the reconstruction of the exit wave Ψ is additionally required. Overall, both algorithms rely on solving

$$\min_{\Psi} \frac{1}{N} \sum_{j=1}^N \|X_{\Psi,Z_j} - X_{Z_j}^{\exp}\|_2^2 + \alpha\|\Psi - \Psi^{\text{estimate}}\|_2^2 \tag{1}$$

with $\alpha > 0$ in MIMAP and $\alpha = 0$ in MAL. Recently, the case $\alpha > 0$ with $\Psi^{\text{estimate}} = 0$ has been analyzed in [5].

We remark that the last term in (1) is a regularizer of Tikhonov type. Commonly in image processing, inverse problems have been solved by variational methods incorporating hand-crafted regularizers including the total variation [24] or the total generalized variation [1]. Recently, learned regularizers have improved the performance drastically [20,23].

In this work, we propose a variational approach for the exit wave reconstruction, in which the energy functional is composed of a model-based nonlinear data fidelity term and a data-driven regularizer. Here, the data fidelity term [3,5,9] represents the underlying physical model and is detailed in Sect. 2. Throughout this paper, we use the total deep variation (TDV) [17,18] as learned regularizer, which is a deep multiscale convolutional neural network embedded in an optimal control formulation [8]. So far, TDV has successfully been applied to *linear* inverse problems in imaging including denoising, deblurring, and demosaicing [21]. The transfer to *nonlinear* inverse problems such as exit wave reconstruction is indeed novel. The overall approach is elaborated on in Sect. 3. Finally, we demonstrate the applicability of our method to noise-free and noisy exit wave reconstruction in Sect. 4.

Notation. We denote the absolute value of a complex scalar by $|\cdot|$. The complex conjugate of a vector $v \in \mathbb{C}^n$ is denoted by v^*, and the Fourier transform by $\mathcal{F}$.

2   Transmission Electron Microscopy

In this section, we recall the basic operating principle of TEMs with a particular focus on the derivation of the forward model of the associated nonlinear inverse problem. The subsequent presentation roughly follows [5]. For further details, we refer the reader to [2,3,9,16].

A simplified prototypic structure of a TEM is depicted in Fig. 1 along with selected adjustable hyperparameters in the form of microscope settings, which are relevant for this work. Initially, an electron beam with a wavelength λ is emitted by an electron source with a half angle of beam convergence α, which is subsequently diffracted when traversing through a specimen. Then, the generated unknown exit wave Ψ is focused by an objective lens with focus value Z and spherical aberration coefficient C_s. Afterwards, the exit wave passes through the aperture allowing for a maximum semiangle $\alpha_{\max}$. Finally, the output image is generated by counting the electrons hitting the sensor in the image plane in a time period. Thus, there the upper bound for the intensity values of a given pixel depends on the physical configuration.

The forward model of TEM assumes that the image $X_{\Psi,Z} \in L^2(\mathbb{R}^2, \mathbb{R}_0^+)$ is a function of the exit wave $\Psi \in L^2(\mathbb{R}^2, \mathbb{C})$, the transmission cross-coefficient (TCC) $T_Z \in L^\infty(\mathbb{R}^2 \times \mathbb{R}^2, \mathbb{C})$ and the focus value $Z \in \mathbb{R}$ and is given by

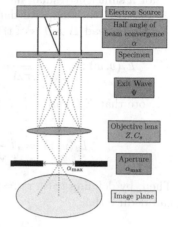

$$X_{\Psi,Z} = \mathcal{F}^{-1}(\Psi \star_{T_Z} \Psi). \tag{2}$$

Recall that the weighted cross-correlation for $\omega \in \mathbb{R}^2$ is defined as

$$(\Psi \star_{T_Z} \Psi)(\omega) := \int_{\mathbb{R}^2} \Psi^*(y)\Psi(\omega+y)T_Z(\omega+y, y)\, dy. \tag{3}$$

We note that the exit wave is not unique due to a rotational invariance, i.e. $\widetilde{\Psi}_q = e^{iq}\Psi$ for $q \in \mathbb{R}$ are equivalent exit waves for an optimal exit wave Ψ.

Fig. 1. Prototypic TEM.

Throughout this paper, we exclusively use an adaption of the prominent Ishizuka approximation [13] of the TCC advocated in [4], which reads as

$$T_Z(v, w) = p(v)p^*(w)a(v)a^*(w)E_{s,Z}(v)E_{s,Z}^*(w)E_t(v, w).$$

Note that this adaption is only allowed for very small values of α (commonly $\alpha < 2 \cdot 10^{-5}$, see [4]), which is valid for modern TEMs [3].

The components defining the Ishizuka approximation are given as follows:

- The phase transfer function $p \in L^\infty(\mathbb{R}^2, \mathbb{C})$ quantifies the aberrations of the lens, i.e.

$$p_Z(v) := \exp(-2\pi i \chi_Z(v))$$

with $\chi_Z \colon \mathbb{R}^2 \to \mathbb{R}$ being the isotropic aberration function approximated as

$$\chi_Z(v) := \tfrac{1}{2}Z\lambda\|v\|_2^2 + \tfrac{1}{4}C_s\lambda^3\|v\|_2^4.$$

Here, χ_Z depends on the focus Z, the electron wavelength $\lambda > 0$ and on the coefficient of the spherical aberration C_s. Note that the wavelength is proportional to the energy of the beam, which is in our case $200\,\mathrm{kV}$.

- The aperture function $a \in L^\infty(\mathbb{R}^2, \mathbb{R})$ is assumed to be of the form

$$a(v) := \begin{cases} 1 & \lambda\|v\|_2 < \alpha_{\max}, \\ 0 & \text{else}, \end{cases}$$

with $\alpha_{\max} > 0$ denoting the maximum semiangle allowed by the aperture.

- The spatial damping envelope $E_{s,Z} \in L^\infty(\mathbb{R}^2, \mathbb{R})$ is given by

$$E_{s,Z}(v) := \exp\left(-\left(\frac{\pi\alpha}{\lambda}\right)^2 \|\nabla\chi_Z(v)\|_2^2\right)$$

for a sufficiently small half angle of beam convergence $\alpha > 0$.

- Likewise, the temporal damping envelope $E_t \in L^\infty(\mathbb{R}^2 \times \mathbb{R}^2, \mathbb{R})$ for a fixed focus spread $\Delta > 0$ is of the form

$$E_t(v, w) := \int_{\mathbb{R}} \frac{1}{\sqrt{2\pi}\Delta} \exp\left(-\frac{x^2}{2\Delta^2}\right) \exp\left(-\pi i\lambda x(\|v\|_2^2 - \|w\|_2^2)\right) \mathrm{d}x. \quad (4)$$

Note that $E_t(v, w) = \int_{\mathbb{R}} \widetilde{E}_t(v, x)\widetilde{E}_t^*(w, x)\,\mathrm{d}x$ with

$$\widetilde{E}_t(v, x) := \sqrt{\frac{1}{\sqrt{2\pi}\Delta} \exp\left(-\frac{x^2}{2\Delta^2}\right)} \exp\left(-\pi i\lambda x\|v\|_2^2\right).$$

Then, by defining $t_Z(v, x) := p(v)a(v)E_{s,Z}(v)\widetilde{E}_t(v, x)$ for $v \in \mathbb{R}^2$ and $x \in \mathbb{R}$ we can deduce that

$$T_Z(v, w) = \int_{\mathbb{R}} t_Z(v, x)t_Z^*(w, x)\,\mathrm{d}x.$$

Thus, the weighted cross-correlation (3) can be recast by exploiting axial symmetry as follows

$$(\Psi \star_{T_Z} \Psi)(\omega) = \mathcal{F}\left(\int_{\mathbb{R}} |\mathcal{F}^{-1}(\Psi(\omega)t_Z(\omega, x))|^2\,\mathrm{d}x\right). \quad (5)$$

The values of the physical parameters used in all numerical experiments are listed in Table 1.

Table 1. Physical parameters of the TEM used throughout all numerical experiments.

Parameter	Symbol	Value	Unit
Wavelength	λ	$2.5 \cdot 10^{-3}$	nm
Focus spread	Δ	4	nm
Half angle of beam convergence	α	10^{-5}	mrad
Focus value	Z	$\{-10, -8, -6, \ldots, 8, 10\} \cdot 10^4$	nm
Spherical aberration	C_s	-70	nm
Objective aperture semiangle	α_{max}	125	mrad

From a physical point of view, the Ishizuka transmission cross-coefficient relies on the subsequent assumptions:

1. The diameter of the aperture is sufficiently large compared to the highest frequency of interest, making the aperture function independent of the integration variable.

2. The specimen is weakly scattering (typically, the thickness of the specimen is negligible).
3. Both the intensity distribution of the illumination and the normalized focus spread can roughly be identified as Gaussian probability density functions.

3   Exit Wave Reconstruction Using Total Deep Variation

As sketched earlier, we minimize an energy functional composed of a data fidelity term and a learned regularizer to reconstruct the exit wave in the forward model (2). Here, the associated minimizer of the energy defines the reconstructed exit wave. The data fidelity term quantifies the mismatch between the observed experimental images $X_{Z_j}^{\exp}$ and the result of the forward model $\mathcal{F}^{-1}(\Psi \star_{T_{Z_j}} \Psi)$ in terms of the squared L^2-norm, i.e.

$$\frac{\xi}{N} \sum_{j=1}^{N} \|\mathcal{F}^{-1}(\Psi \star_{T_{Z_j}} \Psi) - X_{Z_j}^{\exp}\|_{L^2(\mathbb{R}^2)}^2$$

for a weight parameter $\xi > 0$. Note that we work with properly aligned images and thus no translation, rotation or registration is required. We approximate the data fidelity term $\mathcal{D}$ in a discrete domain $\Omega_h = [-\frac{\alpha_{\max}}{\lambda}, \frac{\alpha_{\max}}{\lambda}]^2 \cap h\mathbb{Z}^2$ with d lattice cells and fixed corresponding grid size $h > 0$ as follows

$$\mathcal{D}(\Psi, \{\mathbf{X}_{Z_j}^{\exp}\}_{j=1}^{N}) = \frac{\xi h^2}{N} \sum_{j=1}^{N} \|\mathbf{F}^{-1}(\Psi \star_{T_{Z_j}} \Psi) - \mathbf{X}_{Z_j}^{\exp}\|_2^2.$$

Here, $\mathbf{F} \in \mathbb{R}^{d \times d}$ denotes the matrix representation of the discrete Fourier transform in 2D, $\Psi \in \mathbb{C}^d$ is the discrete exit wave, and $\mathbf{X}_{Z_j}^{\exp} = \mathcal{I}_{\Omega_h}(X_{Z_j}^{\exp}) \in \mathbb{R}^d$ denote the discrete experimental images with $\mathcal{I}_{\Omega_h} \colon L^2(\mathbb{R}^2, \mathbb{R}) \to \mathbb{R}^d$ being the local mean projection on the lattice cells. The discrete weighted cross-correlation reads as

$$\mathbf{F}^{-1}(\Psi \star_{T_Z} \Psi)(\omega) = \sum_{x \in \mathcal{C}_{R,M}} h_{R,M} \left(|\mathbf{F}^{-1}(\Psi(\omega)t_Z(\omega, x))|^2\right)$$

for $\omega \in \Omega_h$, where we tacitly identified two-dimensional matrices with their corresponding one-dimensional flattened vectors for notational simplicity. Here, we denote by $\mathcal{C}_{R,M}$ a set of M equidistant control points in the range $[-R, R]$ for an odd $M \in \mathbb{N}$ and $R > 0$ depending on the focus spread Δ, and $h_{R,M}$ is the associated grid size. In all experiments, we set $M = 7$ and $R = 10.5$ implying $h_{R,M} = 3.5$, which is already sufficient following [5].

Throughout this work, we use the *total deep variation (TDV)* [17,18] as a regularizer in image space for this inverse problem, which is a deep convolutional neural network with a sophisticated structure operating on several scales as depicted in Fig. 2. The learned parameters $\Theta \in \Theta$ for a compact *regularizer parameter space* $\Theta \subset \mathbb{R}^{n_\Theta}$ are later computed in a sampled optimal control problem following [6,8].

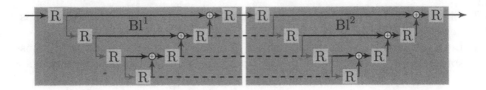

Fig. 2. Network architecture used throughout this work. (Color figure online)

In detail, the TDV regularizer $\mathcal{R} \in C_c^3(\mathbb{C}^d \times \Theta, \mathbb{R})$ is the sum of the pixel-wise deep variation $r : \mathbb{C}^d \times \Theta \to \mathbb{R}^d$, i.e. $\mathcal{R}(x, \theta) = \sum_{i=1}^d r(x, \theta)_i$, where r has the form $r(x, \theta) = w\mathcal{N}(Kx)$. The components of the pixel-wise deep variation are as follows:

- $K \in \mathbb{R}^{md \times 2d}$ is the matrix representation of a learned 3×3 convolution kernel k with m feature channels and zero-mean constraint (i.e. $\sum_{i,j=1}^3 k_{i,j} = 0$) to enforce mean-invariance using the identification $\mathbb{C}^d \cong \mathbb{R}^{2d}$,
- $\mathcal{N} : \mathbb{R}^{md} \to \mathbb{R}^{md}$ is a sufficiently smooth multiscale convolutional neural network with compact support,
- $w \in \mathbb{R}^{d \times md}$ is the matrix representation of a learned 1×1 convolution kernel.

Thus, each Θ encodes the learnable parameters of K, $\mathcal{N}$ and w.

In this subsection, we present the convolutional neural network $\mathcal{N}$ as a vital part of $\mathcal{R}$, which is adapted from [17, 18]. On the largest scale, the total deep variation is composed of 2 blocks Bl^i, $i = 1, 2$, where each block is designed as a U-Net [22] with 7 residual blocks $\mathrm{R}_1^i, \ldots, \mathrm{R}_7^i$. Residual blocks on the same scale are linked with skip connections (solid vertical arrows). To enhance the expressiveness of $\mathcal{N}$, residual connections (dashed vertical lines) linking residual blocks to consecutive blocks on the same scale are added. Each residual block R_j^i for $j = 1, \ldots, 7$ is modeled as

$$\mathrm{R}_j^i(x) = x + K_{j,2}^i \Phi(K_{j,1}^i x)$$

for 3×3 convolution operators with $m = 32$ feature channels and no bias, which are represented by matrices $K_{j,1}^i, K_{j,2}^i$. Here, we use the log-Student-t-distribution $\phi(x) = \frac{1}{2} \log(1 + x^2)$ as the componentwise activation function of $\Phi = (\phi, \ldots, \phi)$, which is inspired by the work of Huang and Mumford [12] on the statistics of natural images. To avoid aliasing, we follow the recent approach by Zhang [25], who advocated the use of 3×3 convolutions and transposed convolutions with stride 2 and a blur kernel to realize downsampling (red arrows) and upsampling (blue arrows). In total, we use $|\Theta| \approx 4 \cdot 10^5$ learnable parameters.

In summary, the energy can be computed as

$$\mathcal{E}(\mathbf{\Psi}, \{\mathbf{X}_{Z_j}^{\exp}\}_{j=1}^N, \Theta) = \mathcal{D}(\mathbf{\Psi}, \{\mathbf{X}_{Z_j}^{\exp}\}_{j=1}^N) + \mathcal{R}(\mathbf{F}^{-1}\mathbf{\Psi}, \Theta). \tag{6}$$

To approximate the minimizer, we consider the Landweber iteration [19], which can be interpreted as a time-discretized gradient flow [8] given by

$$\mathbf{\Psi}_{s+1} = \mathbf{\Psi}_s - \nabla_1 \mathcal{D}(\mathbf{\Psi}_s, \{\mathbf{X}_{Z_j}^{\exp}\}_{j=1}^N) - \mathbf{F}(\nabla_1 \mathcal{R}(\mathbf{F}^{-1}\mathbf{\Psi}_s, \Theta)) \tag{7}$$

for $s = 0, \ldots, S - 1$ with S being the total number of iterations. For simplicity, we set the step size parameter of this iteration to 1 since both $\mathcal{D}$ and $\mathcal{R}$ have learned scaling parameters (ξ and w respectively). Finally, we denote by $\boldsymbol{\Psi}_S(\{\mathbf{X}_{Z_j}^{\exp}\}_{j=1}^N, \boldsymbol{\Theta})$ the terminal state of the Landweber iteration.

Next, we learn the parameters $\boldsymbol{\Theta} \in \Theta$ and $\xi > 0$ using a supervised optimal control problem. To this end, let $\{\boldsymbol{\Psi}_l, \{\mathbf{X}_{Z_j,l}^{\exp}\}_{j=1}^N\}_{l=1}^L$ be the entirety of training data comprised of L pairs of ground truth exit waves $\boldsymbol{\Psi}_l$ and corresponding observations $\{\mathbf{X}_{Z_j}^{\exp}\}_{j=1}^N$. Following [6,18], we cast the training problem as the subsequent sampled optimal control problem

$$\overline{\boldsymbol{\Theta}} \in \inf_{\boldsymbol{\Theta} \in \Theta} \sum_{l=1}^L \|\boldsymbol{\Psi}_S(\{\mathbf{X}_{Z_j,l}^{\exp}\}_{j=1}^N, \boldsymbol{\Theta}) - \boldsymbol{\Psi}_l\|_2^2. \tag{8}$$

The existence of an optimal parameter $\overline{\boldsymbol{\Theta}}$ is implied by the compactness of Θ and the regularity of the problem. For further details of existence and consistency proofs we refer the reader to [17,21].

4   Numerical Results

In this section, we present numerical results for (noisy) exit wave reconstruction.

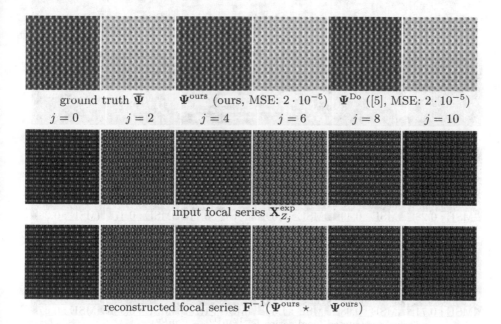

Fig. 3. Results for exit wave reconstruction in the noise-free case.

For training purposes, we use the first 100 exit waves of the dataset [7], which was originally synthesized to predict the amplitude from a phase image. For each exit wave, we generate a focal series containing 11 images with varying focus values Z_j for the reconstruction (see Table 1). The dataset itself has not yet been used for this purpose, but competing methods have so far only been evaluated on a few hand-crafted exit waves [4,5]. This dataset also includes 963 seperate exit waves with different materials for validation purposes. Due to the non-uniqueness of the exit wave (see Sect. 2) we have to rotate the ground truth exit waves of the dataset in a pre-processing step to ensure a proper alignment of the exit waves in (8). As an initialization, we use the root mean intensity value of the focal series, as advocated in [5].

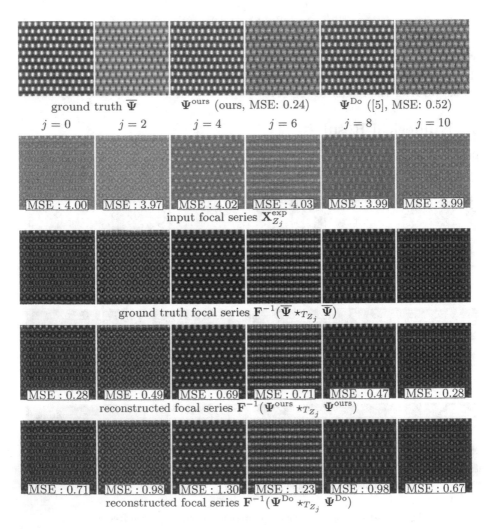

Fig. 4. Results for exit wave reconstruction deteriorated by additive white Gaussian noise.

The baseline model for the benchmark utilizes the same data term with a Tikhonov regularization with $\alpha = 10^{-5}$ for Do [5] and $\alpha = 0$ for MAL [4]. Here, a gradient descent on the energy functional for 200 steps is performed while performing a grid search on the step size for each exit wave separately.

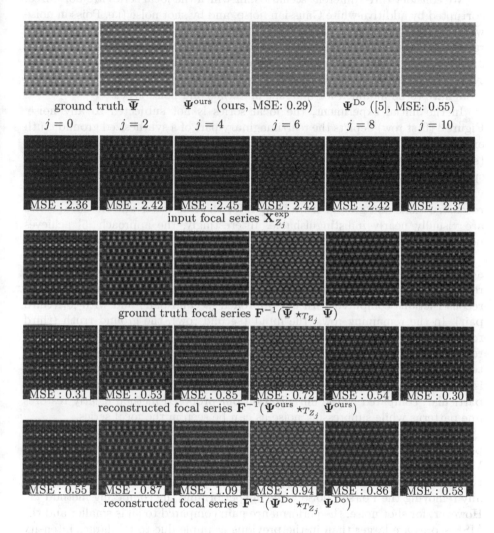

ground truth $\overline{\Psi}$ Ψ^{ours} (ours, MSE: 0.29) Ψ^{Do} ([5], MSE: 0.55)

$j = 0$ $j = 2$ $j = 4$ $j = 6$ $j = 8$ $j = 10$

MSE : 2.36   MSE : 2.42   MSE : 2.45   MSE : 2.42   MSE : 2.42   MSE : 2.37

input focal series $\mathbf{X}_{Z_j}^{\text{exp}}$

ground truth focal series $\mathbf{F}^{-1}(\overline{\Psi} \star_{T_{Z_j}} \overline{\Psi})$

MSE : 0.31   MSE : 0.53   MSE : 0.85   MSE : 0.72   MSE : 0.54   MSE : 0.30

reconstructed focal series $\mathbf{F}^{-1}(\Psi^{\text{ours}} \star_{T_{Z_j}} \Psi^{\text{ours}})$

MSE : 0.55   MSE : 0.87   MSE : 1.09   MSE : 0.94   MSE : 0.86   MSE : 0.58

reconstructed focal series $\mathbf{F}^{-1}(\Psi^{\text{Do}} \star_{T_{Z_j}} \Psi^{\text{Do}})$

Fig. 5. Results for exit wave reconstruction in case of shot noise.

For the optimization of (8), we use $S = 40$ discretization steps of the gradient flow. This large number of iteration steps is necessitated by the nonlinearity of the data fidelity term and clearly exceeds the common choice of S, which is below 20 in [17,18]. The ADAM optimizer [14] with a learning rate of 10^{-4} for ξ and 10^{-6} for the remaining optimization variables as well as the momentum variables $(\beta_1, \beta_2) = (0.9, 0.999)$ is used in all experiments. Furthermore, we train for 20.000 iterations with a batch size of 1 due to memory limitations. The weights of the convolutional layers of the TDV regularizer are initialized with a Gaussian distribution according to [11], while ξ is initialized with 0.1.

We consider three different scenarios, in which the focal series are noise-free, corrupted by additive white Gaussian noise, and by shot noise (i.e. Poisson noise with intensity-dependent parameters). Hence, the forward model (2) incorporates additional noise. We emphasize that all learned parameters of our model are identical for different materials and are only recomputed for different noise types. Additionally, we stress that our approach is generic, i.e. no hyperparameters have to be modified for different materials and noise types.

In the initial experiment, the focal series is not subjected to any noise. Figure 3 (first row) depicts the real/imaginary part of a synthesized ground truth exit wave (left pair), a corresponding reconstructed exit wave with our approach (second pair) as well as the reconstruction following [5] (third pair). The corresponding input focal series $\mathbf{X}_{Z_j}^{\exp}$ and reconstructed focal series $\mathbf{F}^{-1}(\mathbf{\Psi} \star_{T_{Z_j}} \mathbf{\Psi})$ for varying Z_j are shown in the second and third row, respectively. As a result, the reconstructions and ground truth images for the exit waves and the focal series are visually nearly indistinguishable. Interestingly, the approach [5] achieves nearly identical results for this noise-free case.

Next, we present numerical experiments if the focal series is deteriorated by additive white Gaussian noise with standard deviation $\sigma = 2$. Figure 4 (first row) contains the synthesized ground truth exit wave (first pair) as well as reconstructions computed with our approach (second pair) and with [5] (third pair). In the remaining rows, the input (second row), the ground truth (third row) as well as the reconstructed focal series computed with our approach (fourth row) and [5] (fifth row) are displayed for varying values of Z_j. Compared to [5] our technique significantly reduces the MSE. Moreover, the noise in the focal series is visually clearly reduced proving the effectiveness of our proposed method. In particular, compared to the initialization our proposed method clearly restores fine patterns while suppressing noise as can be seen in the focal series.

Figure 5 depicts results if the focal series is corrupted by shot noise with peak 0.5, where the images are arranged in exactly the same way as in Fig. 4. We consider Poisson noise since this statistics properly models photon arrival and is therefore prevalent in microscopy [10]. Again, our method is capable of visually removing the Poisson noise while outperforming the competing method [5]. However, for shot noise, the performance gain compared to [5] is smaller and the MSE scores are larger than in the previous example due to the larger intensity spread of the shot noise.

Finally, we show quantitative results of our method for all exit waves in the validation dataset. Table 2 summarizes the MSE and the runtimes for the entire dataset and all noise instances. The initialization and ground truth exit waves are not affected by corrupting the focal series with zero mean noise and therefore the distance from the initialization to the ground truth is identical. We remark that the runtime is dominated by the calculation of $\nabla_1 \mathcal{D}$.

Table 2. Performance (mean squared error, MSE) for all considered noise types and the corresponding runtime in seconds per exit wave.

	Noise-free	Additive Gaussian noise	Shot noise	Runtime
Initialization	1.27	1.27	1.27	–
MAL [4]	0.01	0.83	0.66	3.37
Do [5]	0.01	0.82	0.65	3.42
Ours	0.01	0.42	0.46	2.69

5   Conclusions

In this work, we have shown that the total deep variation can be successfully applied to the nonlinear inverse problems of exit wave reconstruction for transmission electron microscopy. Compared to competing methods, our approach clearly yields superior results while reducing the required runtime, and no hyperparameters have to be adjusted for each material. Finally, we empirically validated that our approach is stable in the presence of noise and is therefore better suited for more realistic applications.

In future work, we expect to gain additional performance by incorporating the non-uniqueness of the exit wave directly in the loss functional.

References

1. Bredies, K., Kunisch, K., Pock, T.: Total generalized variation. SIAM J. Imaging Sci. **3**(3), 492–526 (2010). https://doi.org/10.1137/090769521
2. Buseck, P., Cowley, J., Eyring, L.: High-Resolution Transmission Electron Microscopy And Associated Techniques. Oxford University Press, Oxford (1989)
3. Carter, C.B., Williams, D.B. (eds.): Transmission Electron Microscopy. Springer, Cham (2016). https://doi.org/10.1007/978-3-319-26651-0
4. Coene, W., Thust, A., Op de Beeck, M., Van Dyck, D.: Maximum-likelihood method for focus-variation image reconstruction in high resolution transmission electron microscopy. Ultramicroscopy **64**(14), 109–135 (1996)
5. Doberstein, C., Berkels, B.: A least-squares functional for joint exit wave reconstruction and image registration. Inverse Problems **35**(5), 054004, 31 (2019). https://doi.org/10.1088/1361-6420/ab0b04
6. E, W., Han, J., Li, Q.: A mean-field optimal control formulation of deep learning. Res. Math. Sci. **6**(1), 1–41 (2018). https://doi.org/10.1007/s40687-018-0172-y
7. Ede, J.M., Peters, J.J., Sloan, J., Beanland, R.: Exit wavefunction reconstruction from single transmission electron micrographs with deep learning. arXiv (2020)
8. Effland, A., Kobler, E., Kunisch, K., Pock, T.: Variational networks: an optimal control approach to early stopping variational methods for image restoration. J. Math. Imaging Vis. **62**(3), 396–416 (2020). https://doi.org/10.1007/s10851-019-00926-8
9. Egerton, R.F.: Physical Principles of Electron Microscopy. Springer, Boston (2005). https://doi.org/10.1007/b136495

10. Haider, S.A., Cameron, A., Siva, P., Lui, D., Shafiee, M.J., Boroomand, A., Haider, N., Wong, A.: Fluorescence microscopy image noise reduction using a stochastically-connected random field model. Sci. Rep. (2016). https://doi.org/10.1038/srep20640

11. He, K., Zhang, X., Ren, S., Su, J.: Delving deep into rectifiers:surpassing human-level performance on ImageNet classification. In: ICCV (2015)

12. Huang, J., Mumford, D.: Statistics of natural images and models. In: CVPR, vol. 1, pp. 541–547 (1999). https://doi.org/10.1109/CVPR.1999.786990

13. Ishizuka, K.: Contrast transfer of crystal images in tem. Ultramicroscopy 5(1), 55–65 (1980). https://doi.org/10.1016/0304-3991(80)90011-X

14. Kingma, D.P., Ba, J.L.: ADAM: a method for stochastic optimization. In: International Conference on Learning Representations (2015)

15. Kirkland, E.J.: Improved high resolution image processing of bright field electron micrographs: I. theory. Ultramicroscopy 15(3), 151–172 (1984). https://doi.org/10.1016/0304-3991(84)90037-8, http://www.sciencedirect.com/science/article/pii/0304399184900378

16. Kirkland, E.J.: Advanced Computing in Electron Microscopy (2010). https://doi.org/10.1007/978-1-4419-6533-2

17. Kobler, E., Effland, A., Kunisch, K., Pock, T.: Total deep variation: a stable regularizer for inverse problems. arXiv (2020)

18. Kobler, E., Effland, A., Kunisch, K., Pock, T.: Total deep variation for linear inverse problems. In: CVPR (2020)

19. Landweber, L.: An iteration formula for fredholm integral equations of the first kind. Am. J. Math. 73(3), 615–624 (1951). https://doi.org/10.2307/2372313

20. Li, H., Schwab, J., Antholzer, S., Haltmeier, M.: NETT: solving inverse problems with deep neural networks. Inverse Problems 36(6), 065005, 23 (2020). https://doi.org/10.1088/1361-6420/ab6d57

21. Pinetz, T., Kobler, E., Pock, T., Effland, A.: Shared prior learning of energy-based models for image reconstruction. arXiv preprint arXiv:2011.06539 (2020)

22. Ronneberger, O., Fischer, P., Brox, T.: U-Net: convolutional networks for biomedical image segmentation. In: Navab, N., Hornegger, J., Wells, W.M., Frangi, A.F. (eds.) MICCAI 2015. LNCS, vol. 9351, pp. 234–241. Springer, Cham (2015). https://doi.org/10.1007/978-3-319-24574-4_28

23. Roth, S., Black, M.J.: Fields of experts. Int. J. Comput. Vis. 82(2), 205–229 (2009). https://doi.org/10.1007/s11263-008-0197-6

24. Rudin, L.I., Osher, S., Fatemi, E.: Nonlinear total variation based noise removal algorithms. Phys. D 60(1–4), 259–268 (1992). https://doi.org/10.1016/0167-2789(92)90242-F

25. Zhang, R.: Making convolutional networks shift-invariant again. ICML. 97, 7324–7334 (2019)

GMM Based Simultaneous Reconstruction and Segmentation in X-Ray CT Application

Shi Yan[1,2]([✉])[ID] and Yiqiu Dong[2][ID]

[1] School of Mathematical Science, Nankai University, Tianjin 300071, China
ysicesword@mail.bnu.edu.cn
[2] Department of Applied Mathematics and Computer Science, Technical University of Denmark, 2800 Kgs., Lyngby, Denmark
yido@dtu.dk

Abstract. In this paper, we propose a new simultaneous reconstruction and segmentation (SRS) model in X-ray computed tomography (CT). The new SRS model is based on the Gaussian mixture model (GMM). In order to transform non-separable log-sum term in GMM into a form that can be easy solved, we introduce an auxiliary variable, which in fact plays a segmentation role. The new SRS model is much simpler comparing with the models derived from the hidden Markov measure field model (HMMFM). Numerical results show that the proposed model achieves improved results than other methods, and the CPU time is greatly reduced.

Keywords: Simultaneous reconstruction and segmentation · Inverse problem · X-ray CT · Alternating minimization method · Gaussian mixture model · Fast algorithm

1 Introduction

X-ray computed tomography (CT) is a widely used technique to obtain the inside of an object without damaging the object. When a beam of X-ray photons go through the object, the photons interact with the objects and are absorbed or deviated from the original direction. Attenuation coefficients are used to measure the absorption or deviation, which is different for different materials. Thus, by measuring the dump of X-rays from multiple angles through the object, one could model the X-ray CT as a classical inverse problem, and solve it using reconstruction methods, such as the filtered back projection algorithm [9], and algebraic reconstruction techniques [7].

Prior information plays an important role in image reconstruction methods. In this paper, we consider a special set of prior: the object could be segmented

The work was supported by Villum Investigator grant 25893 from the Villum Foundation.

into several meaningful classes with some known statistic parameters, e.g., in the medical applications objects may consist of soft tissue and bone, while in the industrial applications objects may include plastics and metals. In this work, we assume that the object is piecewise smooth, and the percentage of each class inside the object as well as the mean and variation of each class are known. Note that the segmentation of the classes is unknown here.

The assumption on the distributions of the classes in the object is also applied in simultaneous reconstruction and segmentation (SRS) methods, e.g., [4,8,10,16,17,20]. The advantage of the SRS methods is the improved accuracy of reconstruction and segmentation results due to the simultaneous structure and capability of avoiding error propagations. Recent years, several SRS methods were proposed. The first SRS method for CT was proposed in [16], where the Mumford-Shah model [14] was applied on the CT reconstruction problem. Later, hidden Markov measure field model (HMMFM) [13] is used instead of Mumford-Shah model for segmentation in [19]. As another type of multi-class segmentation method, Potts model [15] is used in SRS in [20], which does not need prior knowledge on the gray levels nor the number of segments. In [10], an SRS method for CT with limited field of view or shadowed data was proposed. To improve the accuracy of reconstruction and segmentation, the means and variances of the segmentation classes are used as prior in [17]. Then in [8], the SRS method in [17] was improved by updating the variances of the segmentation classes. In addition, a lot of other techniques are used in the SRS methods, such as dictionary learning [4], the Bregman distance [2], and the graph cut method [12]. We briefly review two closely related methods as follows.

In [17], by using Bayes' rule and the maximum a posteriori (MAP) estimate the authors proposed the following model

$$\min_{x,\delta \in \mathcal{A}} \left\{ E_0(x, \delta) = \lambda_n \|Ax - b\|_2^2 + \lambda_c \sum_{k=1}^{K} R(\delta_k) - \sum_{j=1}^{N} \ln p_H(x_j|\delta, \mu, \sigma) \right\}, \quad (1)$$

where $x \in \mathbb{R}^N$ is the reconstructed image with N pixels, $A \in \mathbb{R}^{M \times N}$ is the system matrix in the CT reconstruction problem, and $b \in \mathbb{R}^M$ is the measured data. Furthermore, $\delta = (\delta_1, \cdots, \delta_K) = \{\delta_{jk}\} \in \mathbb{R}^{N \times K}$ is the probability map from HMMFM with the number of classes K. Each element δ_{jk} represents the probability of the jth pixel in x belonging to the kth class. The constraint on δ is defined as

$$\mathcal{A} = \left\{ \delta \in \mathbb{R}^{N \times K} \; \middle| \; \sum_{k=1}^{K} \delta_{jk} = 1, \text{ for all } j \text{ and } \delta_{jk} \in [0,1], \text{ for all } j, k \right\}. \quad (2)$$

One assume that in each class the pixels in x follow a Gaussian distribution $\mathcal{N}(\mu_k, \sigma_k^2)$ with mean μ_k and standard deviation σ_k, then the last term in (1) is defined as

$$p_H(x_j|\delta, \mu, \sigma) = \sum_{k=1}^{K} \delta_{jk} p(x_j|\mu_k, \sigma_k) \quad (3)$$

with

$$p(x_j|\mu_k, \sigma_k) = \frac{1}{\sqrt{2\pi}\sigma_k} \exp\left(-\frac{(x_j - \mu_k)^2}{2\sigma_k^2}\right). \tag{4}$$

The regularization parameters λ_n and λ_c in (1) control the balance among the three terms. The regularization on $\boldsymbol{\delta}$, $R(\cdot)$, can be the total variation (TV) regularization [18] or Tikhonov regularization [22]. Numerical results shown in [17] illustrate good performance of the method by solving (1). But due to a non-separable logarithmic-summation (log-sum) operator in (1), the method is very time-consuming.

In [5], by introducing a transformation to the log-sum term, the model (1) is transformed into

$$\min_{x,\boldsymbol{\delta}\in\mathcal{B},\boldsymbol{\phi}\in\mathcal{B}}\left\{\begin{array}{l} E_1(\boldsymbol{x},\boldsymbol{\delta},\boldsymbol{\phi}) = \lambda_n\|\boldsymbol{A}\boldsymbol{x} - \boldsymbol{b}\|_2^2 + \lambda_c \sum_{k=1}^{K} R(\boldsymbol{\delta}_k) \\ \\ + \sum_{j=1}^{N}\left[-\sum_{k=1}^{K}\phi_{jk}\ln[\delta_{jk}p(x_j|\mu_k,\sigma_k)] + \sum_{k=1}^{K}\phi_{jk}\ln\phi_{jk}\right]\end{array}\right\} \tag{5}$$

with

$$\mathcal{B} = \left\{\boldsymbol{\phi}\in\mathbb{R}^{N\times K}\ \middle|\ \sum_{k=1}^{K}\phi_{jk} = 1,\text{ for all } j,\text{ and }\phi_{jk}\in(0,1),\text{ for all } j,k\right\}. \tag{6}$$

Although one more variable, $\boldsymbol{\phi}$, was introduced in (5), the minimization problem in (5) is much easier to solve, which leads to more efficient method.

In the model (5), we find that in fact $\boldsymbol{\phi}$ and $\boldsymbol{\delta}$ play very similar roles, and both represent the segmentation. Therefore, if we can remove one of them, the efficiency of the method will be further improved. In this paper, we derive a new SRS model based on Gaussian mixture model (GMM), where only one variable indicates the segmentation result. GMM for image segmentation have been studied for a long time [6]. According to GMM, the distribution of x_j can be represented as

$$p_G(x_j|\boldsymbol{\mu},\boldsymbol{\sigma}) = \sum_{k=1}^{K}\alpha_k p(x_j|\mu_k,\sigma_k), \tag{7}$$

where α_k denotes the percentage of k-th class and satisfies $\sum_{k=1}^{K}\alpha_k = 1$, and $p(x_j|\mu_k,\sigma_k)$ is the same as in (4). The HMMFM model (3) and GMM model (7) are very similar. One major difference is on $\boldsymbol{\delta}$ and $\boldsymbol{\alpha}$. In (3) δ_{jk} represents the probability of the jth pixel belonging to the kth class, i.e., $\boldsymbol{\delta}$ is the probability map for each point. But in (7) $\boldsymbol{\alpha}$ represents the percentage of each class, which is not related to a single point j. In order to incorporate segmentation into GMM, we introduce an auxiliary variable to transform the log-sum term from GMM and show that this variable in fact represents the segmentation result. The new proposed model is much simpler comparing with the ones in (1) and (5). We apply the alternating minimization method to solve the problem in the new model and show that the proposed method can provide comparable results as the methods in [17] and [5] with much less CPU time.

The rest of the paper is organized as follows. In Sect. 2, we introduce the new SRS model based on GMM and the corresponding numerical algorithm. Numerical results are shown in Sect. 3. In the last section, we conclude the paper.

2   Proposed Method

In this paper, we use the same assumption as in [17] that the parameter set $\Theta = \{\alpha, \mu, \sigma\}$ is given as priori. In addition, we assume that the measurements b are contaminated by additive white Gaussian noise $\mathcal{N}(0, \sigma_n^2)$ and the pixel values in x are independent to each other. According to the definition in (7), the negative logarithm of GMM can be written as

$$- \log \left(\prod_{j=1}^{N} \sum_{k=1}^{K} \alpha^k p(x_j | \mu_k, \sigma_k) \right) = - \sum_{j=1}^{N} \log \sum_{k=1}^{K} \alpha_k p(x_j | \mu_k, \sigma_k). \qquad (8)$$

After combining with the least-squares data fidelity term, we obtain the GMM-based CT reconstruction model

$$\min_x F_1(x) = \lambda_n \|Ax - b\|_2^2 - \sum_{j=1}^{N} \log \sum_{k=1}^{K} \alpha_k p(x_j; \mu_k, \sigma_k), \qquad (9)$$

where λ_n is a positive regularization parameter. Note that the second term in (9) consists of a non-separable log-sum term, which is hard to deal with. Based on the work in [11,21], the log-sum term can be transformed into a sum-log term by introducing an auxiliary variable. We include this result in the following lemma.

Lemma 1 *(Commutativity of log-sum operations [11,21]). Given $f_k > 0$, we have*

$$- \log \sum_{k=1}^{K} f_k = \min_{\psi \in \mathcal{C}} - \sum_{k=1}^{K} \psi_k \log f_k + \sum_{k=1}^{K} \psi_k \log \psi_k,$$

where $\psi = (\psi_1, \psi_2, \cdots, \psi_K)$ is a vector, and the feasible set $\mathcal{C}$ is defined as

$$\mathcal{C} = \left\{ \psi \in \mathbb{R}^{1 \times K} \,\middle|\, \sum_{k=1}^{K} \psi_k = 1, \psi_k \in (0,1) \right\}. \qquad (10)$$

Remark 1. Note that $\mathcal{B}$ and $\mathcal{C}$ are slightly different from $\mathcal{A}$ defined in (2) to ensure that the logarithm makes sense.

Applying Lemma 1 with $f_k = \alpha_k p(x_j; \mu_k, \sigma_k)$ for each j, we get

$$\min_{x, \phi \in \mathcal{B}} F(x, \phi) = \lambda_n \|Ax - b\|_2^2 - \sum_{j=1}^{N} \sum_{k=1}^{K} \phi_{jk} \ln[\alpha_k p(x_j; \mu_k, \sigma_k)] + \sum_{j=1}^{N} \sum_{k=1}^{K} \phi_{jk} \ln \phi_{jk}.$$
$$(11)$$

where $\mathcal{B}$ is defined in (6). In [11], it was shown that ϕ in fact represents the classification.

Remark 2. The model (9) is a CT reconstruction model, but the model (11) can be considered as an SRS model, where $\boldsymbol{x}$ is the reconstruction and $\boldsymbol{\phi}$ represents the segmentation.

In order to reduce the influence of the measurement noise, we enforce smoothness on $\boldsymbol{x}$ and $\boldsymbol{\phi}$ by introducing additional regularization terms in (11). We assume that the segmentation is piecewise constant and the reconstruction is piecewise smooth. Then, the regularized model is

$$\min_{\boldsymbol{x},\boldsymbol{\phi}\in\mathcal{B}} F(\boldsymbol{x},\boldsymbol{\phi}) = \lambda_n\|A\boldsymbol{x}-\boldsymbol{b}\|_2^2 + \lambda_{tik}\|\nabla\boldsymbol{x}\|_2^2 - \sum_{j=1}^{N}\sum_{k=1}^{K}\phi_{jk}\ln[\alpha_k p(x_j;\mu_k,\sigma_k)]$$
$$+ \sum_{j=1}^{N}\sum_{k=1}^{K}\phi_{jk}\ln\phi_{jk} + \lambda_{TV}\sum_{k=1}^{K}R_{TV}(\boldsymbol{\phi}_k),$$

$$(12)$$

where according to the assumptions we add the TV regularization [18] on the segmentation $\boldsymbol{\phi}$ and the Tikhonov regularization on $\nabla\boldsymbol{x}$ with ∇ as the discrete gradient operator, which is computed via a forward finite difference scheme with reflexive boundary conditions.

The minimization problem in (12) is non-convex, but both the sub-problems with respect to $\boldsymbol{x}$ and $\boldsymbol{\phi}$ are convex. We apply the alternating minimization method introduced in [3] to solve the minimization problem in (12), and the algorithm is described in Algorithm 1.

Algorithm 1. Algorithm for solving the minimization problem (12)

1, Set λ_n, λ_{tik} and λ_{TV}, initialize $\boldsymbol{x}^0 = \boldsymbol{0}$, and $\boldsymbol{\phi}^0$ with all $\phi_{j,k}^0 = \frac{1}{K}$.

2, Update $\boldsymbol{x}^{n+1}$, by

$$\boldsymbol{x}^{n+1} = \arg\min_{\boldsymbol{x}}\left\{\lambda_n\|A\boldsymbol{x}-\boldsymbol{b}\|_2^2 + \lambda_{tik}\|\nabla\boldsymbol{x}\|_2^2 - \sum_{j=1}^{N}\sum_{k=1}^{K}\phi_{jk}^n\ln[\alpha_k p(x_j;\mu_k,\sigma_k)]\right\}.$$

$$(13)$$

3, Update $\boldsymbol{\phi}^{n+1}$, by

$$\phi_{jk}^{n+1} = \arg\min_{\boldsymbol{\phi}\in\mathcal{B}}\left\{-\sum_{j=1}^{N}\sum_{k=1}^{K}\phi_{jk}\ln[\alpha_k p(x_j^{n+1};\mu_k,\sigma_k)] + \sum_{j=1}^{N}\sum_{k=1}^{K}\phi_{jk}\ln\phi_{jk} + \lambda_{TV}\sum_{k=1}^{K}R_{TV}(\boldsymbol{\phi}_k)\right\}.$$

$$(14)$$

4, If $\frac{\|\boldsymbol{x}^{n+1}-\boldsymbol{x}^n\|_2}{\|\boldsymbol{x}^n\|_2} < 10^{-4}$, then, stop. Otherwise, let $n = n+1$, and go to 2.

For the x-sub-problem in (13), we substitute $p(x_j; \mu_k, \sigma_k)$ defined in (4) into (13), and obtain

$$x^{n+1} = \arg\min_x \left\{ \lambda_n \|Ax - b\|_2^2 + \lambda_{tik} \|\nabla x\|_2^2 + \sum_{j=1}^{N} \sum_{k=1}^{K} \frac{\phi_{jk}^n}{2\sigma_k^2} (x_j - \mu_k)^2 \right\}. \quad (15)$$

Note that the objective function in (15) is quadratic. Thus, it can be easily solved by applying the CGLS method [1].

The sub-problem with respect to ϕ, i.e., (14), is exactly the same as the ϕ-sub-problem in [24]. We apply the method proposed in [24], and ϕ can be obtained from the following expression

$$\phi_{jk}^{n+1} = \frac{\alpha_k p(x_j^{n+1} \mid \mu_k, \sigma_k) \exp(\lambda_{TV} [\nabla^T \eta_k^*]_j)}{\sum_{i=1}^{K} \alpha_i p(x_j^{n+1} \mid \mu_i, \sigma_i) \exp(\lambda_{TV} [\nabla^T \eta_i^*]_j)}, \quad (16)$$

where $[\cdot]_j$ denotes the jth element in the vector $j \in \{1, 2, \cdots, N\}$, and $\eta^* = (\eta_1^*, \cdots, \eta_K^*) = \{\eta_{j,k}^*\} \in \mathbb{R}^{2N \times K}$ is a fixed point of the following iteration

$$\eta_{jk}^{m+1} = \text{Proj}_{\mathcal{D}} \left(\eta_{jk}^m - \Delta t \nabla \left(\frac{\left(\alpha_k p(x_j^{n+1} \mid \mu_k, \sigma_k) \right)^{\frac{1}{\lambda_p}} \exp\left(\frac{\lambda_{TV}}{\lambda_p} [\nabla^T \eta_k^m]_j \right)}{\sum_{i=1}^{K} (\alpha_i p(x_j^{n+1} \mid \mu_i, \sigma_i))^{\frac{1}{\lambda_p}} \exp\left(\frac{\lambda_{TV}}{\lambda_p} [\nabla^T \eta_i^m]_j \right)} \right) \right). \quad (17)$$

In (17), λ_p is a positive parameter and Δt is a chosen step size. The set $\mathcal{D}$ is given by

$$\left\{ \eta \in \mathbb{R}^{2N \times K} \mid \max_{j,k} \sqrt{(\eta_{j,k})^2 + (\eta_{j+N,k})^2} \leq 1 \text{ for all } j \in \{1, 2, \cdots, N\}, k \in \{1, 2, \cdots, K\} \right\},$$

and the projection on $\mathcal{D}$ can be obtained by

$$\text{Proj}_{\mathcal{D}}(\eta_{j,k}) = \begin{cases} \eta_{j,k}, & \text{if } \sqrt{(\eta_{j,k})^2 + (\eta_{j+N,k})^2} \leq 1, \\ \frac{\eta_{j,k}}{\sqrt{(\eta_{j,k})^2 + (\eta_{j+N,k})^2}}, & \text{otherwise}, \end{cases} \quad \text{for } j \leq N,$$

and

$$\text{Proj}_{\mathcal{D}}(\eta_{j,k}) = \begin{cases} \eta_{j,k}, & \text{if } \sqrt{(\eta_{j-N,k})^2 + (\eta_{j,k})^2} \leq 1, \\ \frac{\eta_{j,k}}{\sqrt{(\eta_{j-N,k})^2 + (\eta_{j,k})^2}}, & \text{otherwise}, \end{cases} \quad \text{for } j > N.$$

For the convergence analysis of the proposed algorithm, we have the following proposition.

Proposition 1. *For the sequence $\{(x^n, \phi^n)\}$ generated by the Algorithm 1, every cluster point is a coordinatewise minimum point of $F(x, \phi)$.*

The proof of this proposition is similar with the proof in [5]. Please see [5, 23] for more details.

3   Experimental Results

In this section, numerical experimental results of the proposed method is demonstrated with the comparison to the methods in [17] and [5]. All numerical tests are performed using MATLAB R2018a, on a linux server with CPU 2.30 Hz. In CGLS method for solving (15), we set the stopping rule as

$$\frac{\|x^{m+1} - x^m\|_2}{\|x^m\|_2} \leq 10^{-4},$$

and the maximum iteration number is 100. In the η sub-problem (17), the step size Δt is chosen as 0.01, and the stopping rule is

$$\frac{\|\eta^{m+1} - \eta^m\|_2}{\|\eta^m\|_2} < 10^{-3}.$$

3.1   Experimental Settings

In the experiments, we use the AIR Tools package [7] to generate the phantoms and the system matrix with the command **phantomgallery** and **paralleltomo**, respectively. The size of the phantom is 64-by-64 except the last experiment in Sect. 3.4. For the detector, the number of pixels is set to 91 and the number of projections is 30 where the angles are from $6°$ to $180°$ with equal space $6°$. According to the detector setting, the underdetermined rate of our inverse problem $Ax = b$ is 0.667. σ_n is chosen as $\varepsilon \|A\bar{x}\|_2$, where ε is a pre-defined parameter and differs from experiments, and $\bar{x}$ is the ground truth attenuation coefficient for the object.

To compare, we use the following defined reconstruction error and segmentation error

$$rec_{err} = \frac{\|x - \bar{x}\|_2}{\|x\|_2}, \quad seg_{err} = \frac{1}{N} \sum_{j \in \Omega} I(l_j - l_j^*), \tag{18}$$

where,

- l_j denotes the segmentation label of the jth point, given by $l_j = \arg\max_k \delta_{jk}$,
  where l_j^* denotes the true segmentation label for pixel j.
- The function $I(\cdot)$ is given by

$$I(l_j - l_j^*) = \begin{cases} 0 & \text{if } l_j - l_j^* = 0, \\ 1 & \text{otherwise.} \end{cases}$$

The parameters $\lambda_n, \lambda_{tik}, \lambda_{TV}$, and λ_p are chosen to minimize $rec_{err} + seg_{err}$, which is the same rule as in the methods [17] and [5].

3.2   Comparison on an 8-Class Piecewise Constant Phantom

In this example, an 8-class piecewise constant phantom is generated using the parameter **grain** in the command **phantomgallery**. The noise level ε is set to 0.05. The mean and standard deviation for the kth class is given as $\mu_k = \frac{k-1}{7}$ and $\sigma_k = 0.1$, for $k = 1, \cdots, 8$ respectively. In the proposed method, we set the parameters as $\lambda_n = 0.04, \lambda_{tik} = 3.3, \lambda_{TV} = 1.1$ and $\lambda_p = 0.05$.

Firstly, we compare the reconstruction error, the segmentation error and the CPU time for all three methods. We generate noise using different seeds for 50 times to get the average errors and time. The results are listed in Table 1. From Table 1, we can see that the proposed method achieves the smallest reconstruction error while the method in [5] achieves the smallest segmentation error, and the total error is comparable among all three methods. For the CPU time, the proposed method consumes much less than the other two methods, about $\frac{1}{15}$ of the method in [5] and $\frac{1}{90}$ of the method in [17].

Table 1. Comparison on an 8-class piecewise constant phantom

Average through 50 tests	Method in [17]	Method in [5]	Proposed method
Reconstruction error	0.086	0.088	**0.071**
Segmentation error	0.033	**0.026**	0.033
Total error	0.119	0.114	**0.104**
CPU Time (in seconds)	469.7	93.9	**5.9**

Secondly, we show the best and worst reconstruction result and the segmentation result with smallest $(0.066 + 0.026)$ and largest $(0.074 + 0.042)$ total error among the 50 runs in Fig. 1. For the smallest total error case, it is obvious that the reconstruction result by the proposed method is much smoother than the other two methods. The segmentation result is comparable among all three methods, and no obvious difference could be seen. For the largest total error case, there are some isolated points in the reconstruction result of the proposed method and method [5]. Furthermore, the reconstruction by using the proposed method is slightly over-smoothed, which is due to the use of Tikhonov regularization. Note that for such a piecewise constant test image, the TV regularization would be more suitable for x.

3.3   Comparison on a 3-Class Smooth Phantom

In this example, we use a 3-class smooth phantom to test three methods. The phantom is generated by using the parameter **threephasessmooth** in command **phantomgallery**. The noise level ε is set to 0.01, the mean and standard deviation for each class is $\mu = (0.08, 0.27, 0.61)$ and $\sigma_k = 0.22$ for $k = 1, 2, 3$. The parameters in the proposed method are set as $\lambda_n = 7.5, \lambda_{tik} = 0.44, \lambda_{TV} = 4$ and $\lambda_p = 0.05$.

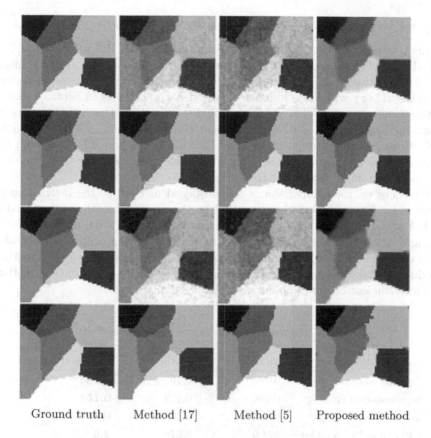

Ground truth   Method [17]   Method [5]   Proposed method

Fig. 1. Reconstruction and segmentation results on an 8-class phase smooth phantom with the best and worst cases. The first row: reconstruction results of the best case, the second row: segmentation result of the best case. The third row: reconstruction result of the worst case, the fourth row: segmentation result of the worst case.

The experiment is also repeated on 50 different noise realizations. In Table 2, the average reconstruction error, segmentation error and CPU time for all three methods are listed. From Table 2, it can be seen that the proposed method achieves both the smallest reconstruction error and the segmentation error. In the mean time, the CPU time consumed by the proposed method is much less than the other two methods, about over $\frac{1}{100}$ of the method in [17] and $\frac{1}{10}$ of the method in [5].

In Fig. 2, we demonstrate the results of the smallest $(0.174 + 0.139)$ and largest $(0.192 + 0.155)$ total error in the 50 runs. For the visual comparison, there are no obvious difference.

3.4   CPU Time Comparison on Different Resolution

In this test, phantoms with different resolution (64×64, 128×128, 256×256, and 512×512) are generated to compare the CPU time among all three methods. The phantoms are generated using the parameter **grain** in the command **phantomgallery** with class number 8, which is the same as in Sect. 3.2 expect for the image resolution. The projection angles are adjusted such that the underdetermined rate is kept as 0.667. We repeat the experiment on 3 different noise realizations and the experiment for the 512×512 in method [17] is missing, due to the runtime and memory limitation. The image resolution, the CPU time and the projection angle setting are listed in Table 3.

We focus on the comparison of the proposed method to the method introduced in [5]. From Table 3, the time used by the proposed method is reduced to $\frac{1}{15}, \frac{1}{4}, \frac{1}{3}$ and $\frac{1}{2}$ of the method from [5] for the resolution $64 \times 64, 128 \times 128, 256 \times 256$, and 512×512, respectively. This indicates that the proposed method is faster than the other two methods. But the improvement of the proposed method is reducing when the resolution is increasing. This means the proposed method are more efficient on small phantoms than large phantoms.

Table 2. Comparison on a 3-class smooth phantom

Average through 50 tests	Method in [17]	Method in [5]	Proposed method
Reconstruction error	0.203	0.195	**0.184**
Segmentation error	0.166	0.172	**0.151**
Total error	0.369	0.367	**0.335**
CPU time (in seconds)	374.9	33.7	**2.5**

Table 3. Comparison of CPU time (second) on an 8-class piecewise image

Resolution	Projection angle	CPU time (second)		
		Method [17]	Method [5]	Proposed method
64×64	$6° : 6° : 180°$	469.7	93.9	**5.9**
128×128	$3° : 3° : 180°$	1981.4	161.2	**39.3**
256×256	$1.5° : 1.5° : 180°$	9447.3	783.5	**222.6**
512×512	$0.75° : 0.75° : 180°$	–	3224.2	**1753.0**

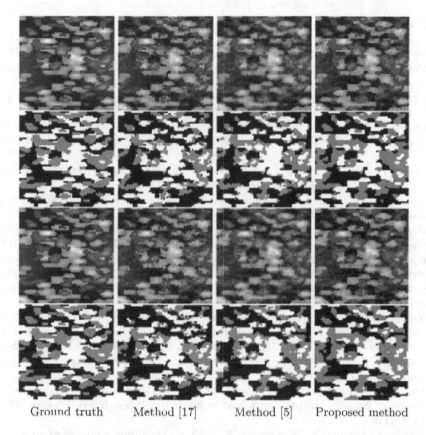

Ground truth   Method [17]   Method [5]   Proposed method

Fig. 2. Reconstruction and segmentation results on a 3-class smooth phantom with the best and worst cases. The first row: reconstruction results of the best case, the second row: segmentation result of the best case. The third row: reconstruction result of the worst case, the fourth row: segmentation result of the worst case.

4   Conclusion and Future Work

In this paper, a new simultaneous reconstruction and segmentation method for X-ray CT is proposed. The new SRS model is based on the Gaussian mixture model and inspired by commutativity of log-sum operation. The optimization problem in the proposed model is solved by the alternating minimization method. Numerical experiments demonstrate that the proposed method achieves comparable results but uses much less CPU time than the other compared methods. One future work is to study the sensitivity and selection of the parameters.

References

1. Bjorck, A.: Numerical Methods for Least Squares Problems, vol. 51. SIAM (1996)
2. Corona, V., et al.: Enhancing joint reconstruction and segmentation with non-convex Bregman iteration. Inverse Prob. **35**(5), 055001 (2019)
3. Csiszár, I., Tusnády, G.: Information geometry and alternating minimization procedures. Stat. Decisions **1**, 205–237 (1984)
4. Dong, Y., Hansen, P.C., Kjer, H.M.: Joint CT reconstruction and segmentation with discriminative dictionary learning. IEEE Trans. Comput. Imaging **4**(4), 528–536 (2018)
5. Dong, Y., Wu, C., Yan, S.: A fast method for simultaneous reconstruction and segmentation in X-ray CT application. arXiv:2102.00250
6. Gupta, L., Sortrakul, T.: A gaussian-mixture-based image segmentation algorithm. Pattern Recogn. **31**(3), 315–325 (1998)
7. Hansen, P., Saxild-Hansen, M.: AIR tools–a MATLAB package of algebraic iterative reconstruction methods. J. Comput. Appl. Math. **236**(8), 2167–2178 (2012)
8. Kjer, H.M., Dong, Y., Hansen, P.C.: User-friendly simultaneous tomographic reconstruction and segmentation with class priors. In: Lauze, F., Dong, Y., Dahl, A.B. (eds.) SSVM 2017. LNCS, vol. 10302, pp. 260–270. Springer, Cham (2017). https://doi.org/10.1007/978-3-319-58771-4_21
9. Kuchment, P.: The Radon Transform and Medical Imaging, vol. 85. SIAM (2014)
10. Lauze, F., Quéau, Y., Plenge, E.: Simultaneous reconstruction and segmentation of CT scans with shadowed data. In: Lauze, F., Dong, Y., Dahl, A.B. (eds.) SSVM 2017. LNCS, vol. 10302, pp. 308–319. Springer, Cham (2017). https://doi.org/10.1007/978-3-319-58771-4_25
11. Liu, J., Tai, X., Huang, H., Huan, Z.: A weighted dictionary learning model for denoising images corrupted by mixed noise. IEEE Trans. Image Process. **22**(3), 1108–1120 (2013)
12. Maeda, S., Fukuda, W., Kanemura, A., Ishii, S.: Maximum a posteriori x-ray computed tomography using graph cuts. J. Phys: Conf. Ser. **233**, 012023 (2010)
13. Marroquin, J., Santana, E., Botello, S.: Hidden Markov measure field models for image segmentation. IEEE Trans. Pattern Anal. Mach. Intell. **25**(11), 1380–1387 (2003)
14. Mumford, D., Shah, J.: Optimal approximations by piecewise smooth functions and associated variational problems. Commun. Pure Appl. Math. **42**(5), 577–685 (1989)
15. Potts, R.: Some generalized order-disorder transformations. In: Mathematical Proceedings of the Cambridge Philosophical Society, vol. 48, pp. 106–109. Cambridge University Press (1952)
16. Ramlau, R., Ring, W.: A Mumford-Shah level-set approach for the inversion and segmentation of X-ray tomography data. J. Comput. Phys. **221**(2), 539–557 (2007)
17. Romanov, M., Dahl, A.B., Dong, Y., Hansen, P.C.: Simultaneous tomographic reconstruction and segmentation with class priors. Inverse Prob. Sci. Eng. **24**(8), 1432–1453 (2016)
18. Rudin, L., Osher, S., Fatemi, E.: Nonlinear total variation based noise removal algorithms. Physica D **60**(1), 259–268 (1992)
19. de Sompel, D.V., Brady, M.: Simultaneous reconstruction and segmentation algorithm for positron emission tomography and transmission tomography. In: 2008 5th IEEE International Symposium on Biomedical Imaging: From Nano to Macro, ISBI 2008, pp. 1035–1038. IEEE (2008)

20. Storath, M., Weinmann, A., Frikel, J., Unser, M.: Joint image reconstruction and segmentation using the Potts model. Inverse Prob. **31**(2), 025003 (2015)
21. Teboulle, M.: A unified continuous optimization framework for center-based clustering methods. J. Mach. Learn. Res. **8**, 65–102 (2007)
22. Tikhonov, A.N.: On the stability of inverse problems. Dokl. Akad. Nauk SSSR **39**, 195–198 (1943)
23. Tseng, P.: Convergence of a block coordinate descent method for nondifferentiable minimization. J. Optim. Theor. Appl. **109**(3), 475–494 (2001)
24. Yan, S., Liu, J., Huang, H., Tai, X.: A dual EM algorithm for TV regularized Gaussian mixture model in image segmentation. Inverse Prob. Imaging **13**(3), 653–677 (2019)

Phase Retrieval via Polarization in Dynamical Sampling

Robert Beinert$^{(\boxtimes)}$ (iD) and Marzieh Hasannasab (iD)

Technische Universität Berlin, Institut für Mathematik, Straße des 17. Juni 136,
10623 Berlin, Germany
{beinert,hasannas}@math.tu-berlin.de

Abstract. In this paper we consider the nonlinear inverse problem of phase retrivial in the context of dynamical sampling. Where phase retrieval deals with the recovery of signals & images from phaseless measurements, dynamical sampling was introduced by Aldroubi et al. in 2015 as a tool to recover diffusion fields from spatiotemporal samples. Considering finite-dimensional signals evolving in time under the action of a known matrix, our aim is to recover the signal up to global phase in a stable way from the absolute value of certain space-time measurements. First, we state necessary conditions for the dynamical system of sampling vectors to make the recovery of the unknown signal possible. The conditions deal with the spectrum of the given matrix and the initial sampling vector. Then, assuming that we have access to a specific set of further measurements related to aligned sampling vectors, we provide a feasible procedure to recover almost every signal up to global phase using polarization techniques. Moreover, we show that by adding extra conditions like full spark, the recovery of all signals is possible without exceptions.

Keywords: Phase retrieval · Dynamical frames · Vandermonde matrix · Polarization identity · Dynamical sampling

1 Introduction

The purpose of the current paper is to consider dynamical sampling [3,4,6,8] in the context of phase retrieval. The research field of dynamical sampling originates back to [23,28], where the trade-off between the spatial and temporal sampling densities of a sensor network is studied in order to recover a linear diffusion fields stably. In optics, dynamical sampling with phaseless measurements occurs if the diffraction at an object is measured in several distances [21]. In the following, we consider an unknown vector $x \in \mathbb{C}^d$ that evolves under the action of a matrix $A \in \mathbb{C}^{d \times d}$ meaning that at time $\ell \in \mathbb{N}$ the signal becomes $x_\ell = (A^*)^\ell x$. Our aim is to recover x up to global phase from phaseless measurements. More precisely, we want to recover x from

$$|\langle (A^*)^\ell x, \phi \rangle| = |\langle x, A^\ell \phi \rangle|, \tag{1}$$

© Springer Nature Switzerland AG 2021
A. Elmoataz et al. (Eds.): SSVM 2021, LNCS 12679, pp. 516–527, 2021.
https://doi.org/10.1007/978-3-030-75549-2_41

where $\ell = 0, \ldots, L - 1$ with $L \geq d$, and where $\phi \in \mathbb{C}^d$ is some sampling vector. Phase retrieval in dynamical sampling has been already considered for real Hilbert spaces in [5,7], where the authors provided conditions to ensure that $\{A^\ell \phi\}_{\ell=0}^{L-1}$ has the complementary property meaning that each subset or its complement spans the entire space. The results have then be generalized to several sampling vectors. The complementary property is here equivalent to the uniqueness (up to global phase) of phase retrieval from (1). However, their techniques cannot be immediately generalized to the complex setting since here the complementary property is not sufficient [11,12].

To ensure that we can do phase retrieval in $\mathbb{C}^d$, we will assume that $\{A^\ell \phi\}_{\ell=0}^{L-1}$ is a frame, and we will align ϕ with specifically chosen additional sampling vectors to exploit polarization techniques. This idea is inspired by interferometry used in [9,13]. Using the extra information, we first recover the frame coefficients $\langle x, A^\ell \phi \rangle$ up to global phase and then recover x in a stable way via the dual frame of $\{A^\ell \phi\}_{\ell=0}^{L-1}$.

The paper is organized as follows. In Sect. 2 we set the stage by providing the necessary background information about polarization identities, frames and Vandermonde matrices. In Sect. 3 we find conditions on the spectrum of A and the vector ϕ such that the iterated set $\{A^\ell \phi\}_{\ell=0}^{L-1}$ is a frame. Moreover, in Sect. 4, we provide conditions under which this frame has full spark. In Sect. 5 we prove that the aligned sampling vectors allow phase retrieval for almost all $x \in \mathbb{C}^d$; moreover, if the underlying dynamical frame has full spark, the recovery of all $x \in \mathbb{C}^d$ is possible.

2 Preliminaries

2.1 Polarization and Relative Phases

Our main results are based on the following polarization technique, which allow the recovery of lost phases from certain phaseless information.

Theorem 1 (Polarization, [13]). *Let* $\alpha_1, \alpha_2 \in \mathbb{R}$ *satisfy* $\alpha_1 - \alpha_2 \notin \pi\mathbb{Z}$. *Then, for every* $z_1, z_2 \in \mathbb{C} \setminus \{0\}$, *the product* $\bar{z}_1 z_2$ *is uniquely determined by*

$$|z_1|, \quad |z_2|, \quad |z_1 + e^{i\alpha_1} z_2|, \quad |z_1 + e^{i\alpha_2} z_2|.$$

Remark 1 (Real polarization). For every $z_1, z_2 \in \mathbb{R} \setminus \{0\}$, the product $z_1 z_2$ is uniquely determined by $|z_1|$, $|z_2|$, and $|z_1 + \alpha z_2|$ with $\alpha \in \{-1, 1\}$ because $|z_1 + \alpha z_2|^2 = z_1^2 + z_2^2 + 2\alpha z_1 z_2$.

Remark 2 (Polarization identities [9,13]). For certain α_1 and α_2 as in Theorem 1, the phase of z_1, z_2 can be computed without solving a linear equation system. More generally, for every integer $K > 2$, if ζ_K is chosen to be the Kth root of unity, $\zeta_K = e^{2\pi i/K}$, then we have

$$\bar{z}_1 z_2 = \frac{1}{K} \sum_{k=0}^{K-1} \zeta_K^k |z_1 + \zeta_K^{-k} z_2|^2.$$

2.2 Frames

Given a matrix $A \in \mathbb{C}^{d \times d}$ and a vector $\phi \in \mathbb{C}^d$, the set $\{A^\ell \phi\}_{\ell=0}^{L-1}$ is called a *dynamical frame* if it spans $\mathbb{C}^d$. Dynamical frames were first introduced in [3] in order to recover a signal evolving in time from certain time-space measurements, where also the infinite dimensional problem is addressed. The topic was further developed in [2–4, 6, 8, 15, 17–19, 27]. An arbitrary vector $x \in \mathbb{C}^d$ can be recovered from the set $\{\langle A^\ell x, \phi \rangle\}_{\ell=0}^{L-1}$ in an stable way if there exists $\alpha, \beta > 0$ such that

$$\alpha \|y\|^2 \leq \|\{\langle A^\ell y, \phi \rangle\}_{\ell=0}^{L-1}\|^2 \leq \beta \|y\|^2, \quad \text{for all } y \in \mathbb{C}^d,$$

i.e., when the set $\{(A^T)^\ell \phi\}_{\ell=0}^{L-1}$ is a frame for $\mathbb{C}^d$.

For any dynamical frame $\{A^\ell \phi\}_{\ell=0}^{L-1}$ there exists a set of vectors in the form $\{B^\ell \tilde{\phi}\}_{\ell=0}^{L-1}$ such that every $x \in \mathbb{C}^d$ can be written as

$$x = \sum_{\ell=0}^{L-1} \langle x, A^\ell \phi \rangle B^\ell \tilde{\phi}. \tag{2}$$

Indeed for a given frame $\{A^\ell \phi\}_{\ell=0}^{L-1}$, the frame matrix $T := \sum_{\ell=0}^{L-1} A^\ell \phi (A^\ell \phi)^*$ is symmetric and positive definite and the canonical dual frame $\{T^{-1} A^\ell \phi\}_{\ell=0}^{L-1}$ can be written in the form $\{B^\ell \tilde{\phi}\}_{\ell=0}^{L-1}$ where $B := T^{-1} A T$ and $\tilde{\phi} := T^{-1} \phi$. For more information about frames, we refer to [16].

Example 1. Let $d = 2$, and consider the rotation matrix $A = \left(\begin{smallmatrix} \cos \theta & -\sin \theta \\ \sin \theta & \cos \theta \end{smallmatrix} \right)$ with $\theta \neq k\pi$ for $k \in \mathbb{Z}$. For every nonzero vector $\phi \in \mathbb{C}^2$ and $L \geq 2$ the set $\{A^\ell \phi\}_{\ell=0}^{L-1}$ is a dynamical frame for $\mathbb{C}^2$.

2.3 Vandermonde Matrices

As we will see in Sect. 3, the frame property of the set $\{A^\ell \phi\}_{\ell=0}^{L-1}$ is highly related to the Vandermonde matrix generated by the vector λ whose coordinates consists of the eigenvalues of the matrix A. There are different types of Vandermonde matrices in the literature. We will need the following kinds.

The Classical Vandermode Matrix. For $\lambda := (\lambda_0, \ldots, \lambda_{d-1})^{\mathrm{T}} \in \mathbb{C}^d$, the Vandermonde matrix $V_\lambda \in \mathbb{C}^{d \times L}$ generated by λ is defined as

$$V_\lambda := (\lambda_k^\ell)_{k,\ell=0}^{d-1, L-1}.$$

The determinant of a square Vandermonde matrix $V_\lambda \in \mathbb{C}^{d \times d}$ equals to

$$\det V_\lambda = \prod_{0 \leq k < j \leq d-1} (\lambda_j - \lambda_k).$$

Generalization of the First Kind. A generalized Vandermonde matrix of the first kind is a matrix consisting of selective columns of V_λ. More precisely, for a vector $\lambda := (\lambda_0, \ldots, \lambda_{d-1})^{\mathrm{T}} \in \mathbb{C}^d$ and $m := (m_0, \ldots, m_{L-1})^{\mathrm{T}} \in \mathbb{N}_0^L$, the Vandermonde matrix $V_{\lambda,m} \in \mathbb{C}^{d \times L}$ is defined as

$$V_{\lambda,m} := \left(\lambda_k^{m_\ell}\right)_{k,\ell=0}^{d-1,L-1},$$

The Vandermonde determinant of the first kind may be factorized by

$$\det V_{\lambda,m} = \left(\prod_{k>j}(\lambda_k - \lambda_j)\right) S(\lambda). \tag{3}$$

where S is a symmetric polynomial in λ with non-negative, integer coefficients [20]. The occurring polynomials S are better known as Schur functions [24,25].

Generalization of the Second Kind. The second kind generalized Vandermonde matrix $\widetilde{V}_{\lambda,m} \in \mathbb{C}^{d \times L}$ is defined as

$$\widetilde{V}_{\lambda,m} := \begin{bmatrix} R_0 \\ \vdots \\ R_{M-1} \end{bmatrix} \quad \text{with} \quad R_j := \left(\binom{\ell}{k} \lambda_j^{\ell-k}\right)_{k,\ell=0}^{m_j-1,L-1},$$

where $M \in \mathbb{N}$, $\lambda \in \mathbb{C}^M$ and $m \in \mathbb{N}^M$ such that $|m| := \sum_{j=0}^{M-1} |m_j| = d$. Clearly if m is the unite vector, i.e., $m = (1, \ldots, 1)^{\mathrm{T}} \in \mathbb{N}^M$, then $\widetilde{V}_{\lambda,m}$ equals the Vandermonde matrix V_λ. The determinant is given by

$$\det(\widetilde{V}_{\lambda,m}) = \prod_{0 \le k < j \le M-1} (\lambda_j - \lambda_k)^{m_k m_j}, \tag{4}$$

see [1,22]. Obviously, a square Vandermonde matrix $\widetilde{V}_{\lambda,m}$ is invertible precisely when λ has distinct elements.

Example 2. For $\lambda := (\lambda_0, \lambda_1, \lambda_2)^{\mathrm{T}}$, $m := (3, 1, 2)^{\mathrm{T}}$ and $L = d = 6$, we have

$$\widetilde{V}_{\lambda,m} := \begin{bmatrix} 1 & \lambda_0 & \lambda_0^2 & \lambda_0^3 & \lambda_0^4 & \lambda_0^5 \\ 0 & 1 & 2\lambda_0 & 3\lambda_0^2 & 4\lambda_0^3 & 5\lambda_0^4 \\ 0 & 0 & 1 & 3\lambda_0 & 6\lambda_0^2 & 10\lambda_0^3 \\ 1 & \lambda_1 & \lambda_1^2 & \lambda_1^3 & \lambda_1^4 & \lambda_1^5 \\ 1 & \lambda_2 & \lambda_2^2 & \lambda_2^3 & \lambda_2^4 & \lambda_2^5 \\ 0 & 1 & 2\lambda_2 & 3\lambda_2^2 & 4\lambda_2^3 & 5\lambda_2^4 \end{bmatrix}.$$

3 Dynamical Frames

To recover a signal from (1), we first study conditions on the matrix $A \in \mathbb{C}^{d \times d}$ and the vector $\phi \in \mathbb{C}^d$ such that $\{A^\ell \phi\}_{\ell=0}^{L-1}$ is a frame for $\mathbb{C}^d$. The cornerstone is here the Jordan canonical form of A. More precisely, every matrix $A \in \mathbb{C}^{d \times d}$

is similar to a so-called Jordan matrix meaning that there exists an invertible matrix $S \in \mathbb{C}^{d \times d}$ such that $A = SJS^{-1}$ and $J \in \mathbb{C}^{d \times d}$ is a blocked diagonal matrix of the form

$$J = \operatorname{diag}(J_0, \ldots, J_{M-1}) \qquad \text{with} \qquad J_j = \begin{pmatrix} \lambda_j & 1 & & \\ & \lambda_j & \ddots & \\ & & \ddots & 1 \\ & & & \lambda_j \end{pmatrix} \in \mathbb{C}^{m_j \times m_j},$$

where λ_j is the jth eigenvalue and m_j the corresponding algebraic multiplicity, and where the columns of $S = [S_0 | \ldots | S_{M-1}]$ with blocks $S_j = [s_{j,0} | \ldots | s_{j,m_j-1}]$ span the generalized eigenspaces of A. Further, we have $(A - \lambda_j I)^{k+1} s_{j,k} = 0$ but $(A - \lambda_j I)^k s_{j,k} \neq 0$ for $k = 0, \ldots, m_j - 1$. The Jordan chain S_j related to λ_j is generated by s_{j,m_j-1} via $s_{j,k} = (A - \lambda I)^{m_j-k-1} s_{j,m_j-1}$. We say that ϕ depends on the Jordan generator or leading generalized eigenvector s_{j,m_j-1} if $(S^{-1}\phi)_{k-1} \neq 0$ where $k = \sum_{i=0}^{j-1} m_i$. For pairwise distinct eigenvalues λ_j as usually assumed in the following, the generators are unique up to scaling. In this case, S is unique up to scaling and permutation of the blocks S_j. Finally we notice that the ℓth power of a Jordan matrix and the corresponding Jordan blocks are given by

$$J^\ell = \operatorname{diag}(J_0^\ell, \ldots, J_{M-1}^\ell) \qquad \text{with} \qquad J_j^\ell = \left(\binom{\ell}{n-k} \lambda_j^{\ell-n+k} \right)_{k,n=0}^{m_j-1}.$$

The following two theorems are special cases of [3], where the construction of a frame by iterated actions of A on a finite set of sampling vectors $\{\phi_j\} \subset \mathbb{C}^d$ is studied. In difference to [3], we provide brief, direct proofs based on the Vandermonde determinant for the case that A acts on a single generator ϕ.

Theorem 2 (Dynamical basis). *Let $A \in \mathbb{C}^{d \times d}$ be arbitrary. Then $\{A^\ell \phi\}_{\ell=0}^{d-1}$ is a basis if and only if the eigenvalues of the Jordan blocks of A are pairwise distinct and ϕ depends on all Jordan generators.*

Proof. Assume that A has the Jordan decomposition $A = SJS^{-1}$. We represent the vector ϕ with respect to the column-wise basis in S according to the size of the Jordan blocks in J. More precisely, we denote by ψ_j the coordinates corresponding to the basis vectors in S_j. The coefficients are thus given by

$$\psi := \begin{pmatrix} \psi_0 \\ \vdots \\ \psi_{M-1} \end{pmatrix} = S^{-1}\phi.$$

Next, we consider the generated vectors $\phi_\ell := A^\ell \phi$ with $\ell = 0, \ldots, d-1$. On the basis of the Jordan canonical form, they are given by $\phi_\ell = SJ^\ell \psi$. Considering only the jth Jordan block, we notice

$$J_j^\ell \psi_j = H(\psi_j) \left(\binom{\ell}{k} \lambda_j^{\ell-k} \right)_{k=0}^{m_j-1}, \tag{5}$$

where

$$H(\psi_j) = \begin{bmatrix} (\psi_j)_0 & (\psi_j)_1 & \cdots & (\psi_j)_{m_j-1} \\ \vdots & \vdots & \cdots & \vdots \\ (\psi_j)_{m_j-2} & (\psi_j)_{m_j-1} & & \\ (\psi_j)_{m_j-1} & 0 & \cdots & 0 \end{bmatrix},$$

is an upper-left Hankel matrix in $\mathbb{C}^{m_j \times m_j}$. The vector on the right-hand side of (5) is here the ℓth column of R_j within the definition of generalized Vandermonde matrix $\widetilde{V}_{\lambda,m}$. The matrix of the generated vectors may hence be written as

$$[\phi_0 | \ldots | \phi_{d-1}] = S \begin{bmatrix} H(\psi_0) & \cdots & 0 \\ \vdots & \ddots & \vdots \\ 0 & \cdots & H(\psi_{m-1}) \end{bmatrix} \widetilde{V}_{\lambda,m}.$$

This matrix is invertible if and only if the generalized Vandermonde matrix $\widetilde{V}_{\lambda,m}$ is invertible, i.e. if the eigenvalues are pairwise distinct, see (4), and if the Hankel matrices $H(\psi_j)$ are regular, i.e. if the coefficients $(\psi_j)_{m_j-1}$ of the highest-order generalized eigenvectors do not vanish. □

Theorem 3 (Dynamical frame). *Let $L \geq d$, and let $A \in \mathbb{C}^{d \times d}$ be arbitrary. Then $\{A^\ell \phi\}_{\ell=0}^{L-1}$ is a frame if and only if the eigenvalues of the Jordan blocks of A are pairwise distinct and ϕ depends on all Jordan generators.*

Proof. If the vector ϕ is independent of one Jordan generator, then the images $A^\ell \phi$ are also independent of this generator; so $\{A^\ell \phi\}_{\ell=0}^{L-1}$ can not be a frame for $\mathbb{C}^d$. Now assume that some eigenvalues of A coincide, i.e. the Jordan block to this eigenvalue decompose into several smaller Jordan blocks. Assume that the eigenvalues λ_{j_0} and λ_{j_1} coincide, and that the corresponding Jordan blocks have dimension $m_{j_0} \times m_{j_0}$ and $m_{j_1} \times m_{j_1}$. Using the notation in the proof of Theorem 2, the coordinates of ϕ in $E := \operatorname{span}\{s_{j_0,m_{j_0}-1}, s_{j_1,m_{j_1}-1}\}$ are $(\psi_{j_0})_{m_{j_0}-1}$ and $(\psi_{j_1})_{m_{j_1}-1}$. Applying A^ℓ to ϕ, we get the coordinates $\lambda_{j_0}^\ell (\psi_{j_0})_{m_{j_0}-1}$ and $\lambda_{j_1}^\ell (\psi_{j_1})_{m_{j_1}-1}$ with $\lambda_{j_0} = \lambda_{j_1}$ regarding the subspace E. Thus the projections $\operatorname{proj}_E(\{A^\ell \phi\}_{\ell=0}^{L-1})$ only span a one-dimensional subspace. As a consequence $\{A^\ell \phi\}_{\ell=0}^{L-1}$ cannot span $\mathbb{C}^d$, and we cannot obtain a frame. The opposite direction has already be proven with Theorem 2. □

Since generic matrices $A \in \mathbb{C}^{d \times d}$ are diagonalizable with pairwise distinct eigenvalues, for almost all matrices holds the following special case.

Corollary 1 (Dynamical frame). *Let $L \geq d$, and let $A \in \mathbb{C}^{d \times d}$ be diagonalizable. Then $\{A^\ell \phi\}_{\ell=0}^{L-1}$ is a frame if and only if the eigenvalues of A are pairwise distinct and ϕ depends on all eigenvectors.*

Proof. Since the Jordan blocks here reduces to size 1×1, the matrix of the generated vectors in the proof of Theorem 2 simplifies to

$$[\phi | A\phi | \ldots | A^{d-1}\phi] = S \operatorname{diag}(\psi) V_\lambda.$$

This matrix is invertible if and only if the classical Vandermonde matrix $V_\lambda \in \mathbb{C}^{d \times d}$ is invertible and none of the coordinates of ψ vanishes.   □

For $a \in \mathbb{C}^d$, let $\text{circ}(a)$ denote the circulant matrix whose first column is given by the vector a. All circulant matrices are diagonalizable with respect to the discrete Fourier transform, i.e.,

$$\text{circ}(a) = \tfrac{1}{d} F \operatorname{diag}(\hat{a}) F^{-1},$$

where $\hat{a} = Fa$ is given via the Fourier matrix $F = (e^{-\frac{2\pi ijk}{d}})_{j,k=0}^{d-1}$.

Corollary 2 (Repeated convolution). *Let $L \geq d$, and let $\phi, a \in \mathbb{C}^d$ be arbitrary. Then the family*

$$\{\underbrace{a * \cdots * a}_{\ell \text{ times}} * \phi\}_{\ell=0}^{L-1}$$

is a frame for $\mathbb{C}^d$ if and only if the coordinates of $\hat{\phi}$ do not vanish and the coordinates of $\hat{a}$ are pairwise distinct.

Proof. Note that $a * \phi = \text{circ}(a)\phi$ and $A := \text{circ}(a)$ is a diagonalizable matrix that by hypothesis has pairwise distinct eigenvalues $\{\hat{a}_k\}_{k=0}^{d-1}$. The result follows now from Corollary 1.   □

4   Full-Spark Dynamical Frames

A frame $\{f_k\}_{k=0}^{L-1}$ has full spark if every subset embracing d elements spans $\mathbb{C}^d$. This property makes full-spark frames attractive in phase retrieval and more generally in signal processing [9–11,26]. In the following, we study conditions ensuring that frames generated via diagonalizable matrices have full spark. We show that a dynamical frame has full spark precisely when the Vandermonde matrix V_λ related to the eigenvalues of A has full spark.

Theorem 4. *Let $A \in \mathbb{C}^{d \times d}$ be diagonalizable with eigenvalues λ. For every $L \geq d$, the set $\{A^\ell \phi\}_{\ell=0}^{L-1}$ is a full spark frame if and only if ϕ depends on all eigenvectors and $V_\lambda \in \mathbb{C}^{d \times L}$ has full spark.*

Proof. Assume that A has the eigenvalue decomposition $A = SJS^{-1}$, where J is a diagonal matrix, and denote the coordinates of ϕ with respect to S by $\psi := S^{-1}\phi$. Consider an arbitrary subset $\{A^{m_\ell}\phi\}_{\ell=0}^{d-1}$ of $\{A^\ell \phi\}_{\ell=0}^{L-1}$ with $m = (m_0, \ldots, m_{d-1})^T$. Then the matrix

$$[A^{m_0}\phi | A^{m_1}\phi | \ldots | A^{m_{d-1}}\phi] = S \operatorname{diag}(\psi) V_{\lambda,m}.$$

is invertible if and only if all elements of ψ are non-zero and if $V_{\lambda,m}$ is invertible, which means that V_λ has full spark.   □

The following result specializes Theorem 4 for A with eigenvalues $\lambda_k = \lambda^k$.

Corollary 3. *Let $A \in \mathbb{C}^{d \times d}$ be diagonalizable with eigenvalues $\boldsymbol{\lambda} = (\lambda^k)_{k=0}^{d-1}$ with $\lambda^k \neq 1$ for some $\lambda \in \mathbb{C}$. For every $L \geq d$, the set $\{A^\ell \phi\}_{\ell=0}^{L-1}$ is a full spark frame if and only if ϕ depends on all eigenvectors.*

Proof. For the chosen $\boldsymbol{\lambda}$, every $d \times d$ sub-matrix of $V_{\boldsymbol{\lambda}}$ is an invertible Vandermonde matrix. □

Example 3. Let $L \geq d$ and $\lambda = e^{2\pi i/L}$ be the Lth unit root. Consider $A = \operatorname{diag}(\lambda^0, \ldots, \lambda^{d-1})$ and $\phi = 1$. Then the set $\{A^\ell \phi\}_{\ell=0}^{L-1}$ is a frame for $\mathbb{C}^d$ and has full spark by Corollary 1 and Corollary 3. This frame is called harmonic and is related to a submatrix of the discrete Fourier matrix. Indeed if L is prime, Chebotarëv's theorem implies that every submatrix consists of d rows of the discrete Fourier transform matrix forms a full-spark frame. For more information we refer to [10, 29].

Example 4. The dynamical frame $\{A^\ell \phi\}_{\ell=0}^{L-1}$ in Example 1 generated by iterated action of a rotation matrix $A \in \mathbb{C}^2$ is a full-spark frame if $\theta \in (-\pi, \pi]$ is chosen such that $\ell \theta - \theta \bmod \pi \neq 0$, for every $\ell \in \{1, \ldots, L-1\}$.

Theorem 5. *Let $L \geq d$, and let $A \in \mathbb{C}^{d \times d}$ be diagonalizable with distinct real and non-negative eigenvalues. Then $\{A^\ell \phi\}_{\ell=0}^{L-1}$ is a full-spark frame if ϕ depends on all eigenspaces.*

Proof. Due to Theorem 4, the set $\{A^\ell \phi\}_{\ell=0}^{L-1}$ has full spark if and only if the generalized Vandermonde matrices $V_{\boldsymbol{\lambda},m}$ are invertible for every $m \in \mathbb{N}_0^d$ with distinct coordinates. Since the Schur functions in (3) have only non-negative coefficients, the generalized Vandermonde determinant is here positive for all $\boldsymbol{\lambda}$ with non-negative, distinct coordinates, which establishes the assertion. □

5   Phase Retrieval in Dynamical Sampling

As mentioned in the introduction, the complementary property can be exploited to ensure phases retrieval for real signals [5,7] Since this approach fails in the complex setting, we align ϕ with further sampling vectors allowing polarization. This allow us to recover the frame coefficient $\langle x, A^\ell \phi \rangle$ up to global phase and then using the frame property we can reconstruct x.

Theorem 6. *Let $\{A^\ell \phi\}_{\ell=0}^{L-1}$ be a frame for $\mathbb{C}^d$, and let $\alpha_1, \alpha_2 \in \mathbb{R}$ be real numbers with $\alpha_1 - \alpha_2 \notin \pi\mathbb{Z}$. Then almost all $x \in \mathbb{C}^d$ can be recovered from*

$$\{|\langle x, A^\ell \phi \rangle|\}_{\ell=0}^{L-1} \cup \{|\langle x, A^\ell (\phi + e^{i\alpha_k} A\phi) \rangle|\}_{\ell=0,k=1}^{L-2,2}$$

up to global phase.

Proof. We consider the dense set of $x \in \mathbb{C}^d$ for which $\langle x, A^\ell \phi \rangle \neq 0$ for $\ell = 0, \ldots, L-1$. Using the polarization in Theorem 1, we determine the products

$$\overline{\langle x, A^\ell \phi \rangle} \langle x, A^{\ell+1} \phi \rangle \qquad (\ell = 0, \ldots, L-2).$$

Considering the phase of the above identity, we calculate the relative phases

$$\arg \langle \boldsymbol{x}, \boldsymbol{A}^{\ell+1}\boldsymbol{\phi} \rangle - \arg \langle \boldsymbol{x}, \boldsymbol{A}^{\ell}\boldsymbol{\phi} \rangle \quad \bmod 2\pi \qquad (\ell = 0, \dots, L-2).$$

Choosing the phase of $\langle \boldsymbol{x}, \boldsymbol{\phi} \rangle$ arbitrary, we may thus recover the frame coefficients $\langle \boldsymbol{x}, \boldsymbol{A}^{\ell}\boldsymbol{\phi} \rangle$ up to global phase and thus $\boldsymbol{x}$. □

Corollary 4. *If* $\{\boldsymbol{A}^{\ell}\boldsymbol{\phi}\}_{\ell=0}^{L-1}$ *is a frame for* $\mathbb{R}^d$, *and* $\alpha \in \{-1, 1\}$, *then almost every* $\boldsymbol{x} \in \mathbb{R}^d$ *can be recovered from*

$$\{|\langle \boldsymbol{x}, \boldsymbol{A}^{\ell}\boldsymbol{\phi} \rangle|\}_{\ell=0}^{L-1} \cup \{|\langle \boldsymbol{x}, \boldsymbol{A}^{\ell}(\boldsymbol{\phi} + \alpha \boldsymbol{A}\boldsymbol{\phi}) \rangle|\}_{\ell=0}^{L-2}$$

up to sign.

Although the extended measurement set allows the extraction of relative phases, the proposed procedure may fail in rare cases, where some of the coefficients $|\langle \boldsymbol{x}, \boldsymbol{A}^{\ell}\boldsymbol{\phi} \rangle|$ are zero for some ℓ which means that we are not able to recover $\boldsymbol{x}$ if it lies in the union of finitely many hyperplanes. For instance, if $\{\boldsymbol{A}^{\ell}\boldsymbol{\phi}\}_{\ell=0}^{d-1}$ is an orthonormal basis for $\mathbb{C}^d$ and $|\langle \boldsymbol{x}, \boldsymbol{A}^{j}\boldsymbol{\phi} \rangle|$ is the only zero coefficient for some $j \in \{1, \dots, d-2\}$, then we can recover the coefficients in $\{\langle \boldsymbol{x}, \boldsymbol{A}^{\ell}\boldsymbol{\phi} \rangle\}_{\ell=0}^{j-1}$ up to a joint phase and in $\{\langle \boldsymbol{x}, \boldsymbol{A}^{\ell}\boldsymbol{\phi} \rangle\}_{\ell=j+1}^{d-1}$ up to a joint phase. Due to the zero coefficient, we can however not propagate the phase from one set to the other; so every signal of the form

$$\boldsymbol{y} = \sum_{\ell=0}^{j-1} e^{i\alpha_1} \langle \boldsymbol{x}, \boldsymbol{A}^{\ell}\boldsymbol{\phi} \rangle \boldsymbol{A}^{\ell}\boldsymbol{\phi} + \sum_{\ell=j+1}^{d-1} e^{i\alpha_2} \langle \boldsymbol{x}, \boldsymbol{A}^{\ell}\boldsymbol{\phi} \rangle \boldsymbol{A}^{\ell}\boldsymbol{\phi}$$

with $\alpha_1, \alpha_2 \in (-\pi, \pi]$ is a solution.

On the contrary, if the generated frame has full-spark, one do not need all of the coefficients to recover the wanted signal.

Theorem 7. *Let* $\{\boldsymbol{A}^{\ell}\boldsymbol{\phi}\}_{\ell=0}^{L-1}$ *be a full-spark frame for* $\mathbb{C}^d$, *and let* $\alpha_1, \alpha_2 \in \mathbb{R}$ *be real numbers with* $\alpha_1 - \alpha_2 \notin \pi\mathbb{Z}$. *If* $L \geq {}^{d^2}\!/4 + {}^d\!/2$, *then every* $\boldsymbol{x} \in \mathbb{C}^d$ *can be recovered from the samples*

$$\{|\langle \boldsymbol{x}, \boldsymbol{A}^{\ell}\boldsymbol{\phi} \rangle|\}_{\ell=0}^{L-1} \cup \{|\langle \boldsymbol{x}, \boldsymbol{A}^{\ell}(\boldsymbol{\phi} + e^{i\alpha_k} \boldsymbol{A}\boldsymbol{\phi}) \rangle|\}_{\ell=0, k=1}^{L-2, 2}$$

up to global phase.

Proof. Since $\{\boldsymbol{A}^{\ell}\boldsymbol{\phi}\}_{\ell=0}^{L-1}$ is a full-spark frame, we only need to know the phase of d coefficients $\langle \boldsymbol{x}, \boldsymbol{A}^{\ell}\boldsymbol{\phi} \rangle$ to recover $\boldsymbol{x}$. Obviously, if at least d coefficients are zero, then the unknown signal is zero everywhere. Now assume that $m < d$ measurements $|\langle \boldsymbol{x}, \boldsymbol{A}^{\ell}\boldsymbol{\phi} \rangle|$ are zero. As soon as we find $d - m$ consecutive non-zero measurements, we can transfer the relative phases to enough frame elements to recover $\boldsymbol{x}$ using the extended measurement set. In the worst case, we measure $d - m - 1$ consecutive non-zeros followed by a zero. After this pattern has been repeated m times, the remaining measurements have to be non-zero, otherwise we will get more than m zero measurements which is not possible by the hypothesis. If we thus have at least $L \geq (d-m)m + (d-m) = (m+1)(d-m)$ measurements, the existence of at least $d-m$ non-zero consecutive measurements is guaranteed. Notice that the maximum of $(m+1)(d-m)$ over m is attained at the integer $m := (d-1)/2$ for odd d and $m := d/2$ for even d. Plugging in $m := d/2$ for even d, we obtain immediately the assertion. For odd d and $m := (d-1)/2$, we obtain

$$\left(\tfrac{d-1}{2}+1\right)\left(d - \tfrac{d-1}{2}\right) = \left(\tfrac{d+1}{2}\right)^2 = \tfrac{d^2}{4} + \tfrac{d}{2} + \tfrac{1}{4}.$$

Noticing that this is the smallest integer greater than $d^2/4 + d/2$ finishes the proof.   □

Theorem 8. *Let $\{A^\ell \phi\}_{\ell=0}^{L-1}$ be a full-spark frame for $\mathbb{C}^d$, let $\alpha_1, \alpha_2 \in \mathbb{R}$ be real numbers with $\alpha_1 - \alpha_2 \notin \pi\mathbb{Z}$, and let $J \in \{0, \ldots, d-2\}$. If $L \geq d^2/4(J+1)+d$, then every $x \in \mathbb{C}^d$ can be recovered from the samples*

$$\{|\langle x, A^\ell \phi\rangle|\}_{\ell=0}^{L-1} \cup \{|\langle x, A^\ell(\phi + e^{i\alpha_k} A^j\phi))|\}_{\ell=0, k=1, j=1}^{L-2, 2, J+1}$$

up to global phase.

Proof. The difference to the proof of Theorem 7 is that we may here jump over J consecutive zeros while calculating the relative phases. Thus, if m measurements $|\langle x, A^\ell\phi\rangle|$ are zero, the worst case scenario is that $d-m-J-1$ consecutive non-zero measurements are followed by $J+1$ zeros. Repeating this pattern $\lfloor m/(J+1)\rfloor$ times, and placing the remaining $m \bmod (J+1) \leq m$ zeros and $d-m$ non-zeros at the end – so at most d elements, we require at most

$$\left\lfloor \tfrac{m}{J+1} \right\rfloor (d-m) + (d-m) + m \bmod (J+1) \leq \tfrac{m}{J+1}(d-m) + d$$

measurements to transfer the relative phases far enough to recover x. The maximum on the right-hand side with respect to m is attained at $m := (d+1)/2$ for odd d and $m := d/2$ for even d. Inserting $m := d/2$ for even d, we obtain the bound on L in the assertion. For odd d and $m := (d+1)/2$, the estimate becomes

$$\tfrac{1}{J+1}\tfrac{d+1}{2}\left(d - \tfrac{d+1}{2}\right) + d = \tfrac{1}{J+1}\tfrac{d+1}{2}\tfrac{d-1}{2} + d = \tfrac{1}{J+1}\tfrac{d^2}{4} + d - \tfrac{1}{4(J+1)};$$

so if $L \geq d^2/4(J+1) + d$, than phase propagation and retrieval is possible.   □

6   Conclusion

Phase retrieval from dynamical samples is a severely ill-posed nonlinear inverse problem. In dynamical sampling the trade-off between the number of spatial and temporal samples has been always of great importance, as usually the spacial measurements corresponds to the sampling devices while the temporal measurements corresponds to the sampling over time. We start by minimum number of spatial samples, namely one sampling vector ϕ. First we provide conditions on the system which is represented by a matrix A and also the number of temporal measurements that ensure unique solution of the underlying sampling problem, namely the conditions that guarantee that the sequence $\{A\phi\}_{\ell,i=0}^{L-1}$ is a frame. As the second step, we showed that by adding only few more specially designed sampling vectors, one can recover almost all signals from the phaseless dynamical samples uniquely and stably. The extra sampling vectors that is suggested, depend on the structure of the system. Employing full-spark dynamical frames, we show that the recovery for all signals is possible at the price of increasing the number of temporal samples. Finally a relation between the required number of measurements in time and space is presented. Surely, we can consider phase retrieval from dynamical samples with more spatial sampling vectors, i.e. $\{|\langle x, A\phi_i\rangle|\}_{\ell,i=0}^{L-1,J-1}$. Especially, for diffraction measurements in optics, specific measurement vectors may be physically constructed using additional masks in the optical system. In this context, the dynamical sampling and phase retrieval problem based on several diffraction measurements with respect to a series of distances to the unknown object as considered in [21] may be solved by polarization techniques. So far we suppose that the matrix A in known. An interesting future work would be to combine phase retrieval of the signal with the system identification, in case that only partial information about the matrix A is available. The result will appear in a forthcoming paper [14].

References

1. Aitken, A.C.: Determinants and Matrices, 3rd edn. University Mathematical Texts, Oliver and Boyd, Edinburgh (1944)
2. Aldroubi, A., Cabrelli, C., Cakmak, A.F., Molter, U., Petrosyan, A.: Iterative actions of normal operators. J. Funct. Anal. **272**(3), 1121–1146 (2017)
3. Aldroubi, A., Cabrelli, C., Molter, U., Tang, S.: Dynamical sampling. Appl. Comput. Harmon. Anal. **42**(3), 378–401 (2017)
4. Aldroubi, A., Huang, L., Petrosyan, A.: Frames induced by the action of continuous powers of an operator. J. Math. Anal. Appl. **478**(2), 1059–1084 (2019)
5. Aldroubi, A., Krishtal, I., Tang, S.: Phaseless reconstruction from space-time samples. Appl. Comput. Harmon. Anal. **48**(1), 395–414 (2020)
6. Aldroubi, A., Krishtal, I.: Krylov subspace methods in dynamical sampling. Sampl. Theor. Sig. Image Process **15**, 9–20 (2016)
7. Aldroubi, A., Krishtal, I., Tang, S.: Phase retrieval of evolving signals from space-time samples. In: 2017 Proceedings of the SampTA, pp. 46–49 (2017)
8. Aldroubi, A., Petrosyan, A.: Dynamical sampling and systems from iterative actions of operators, chap. 2. In: Frames and Other Bases in Abstract and Function Spaces, pp. 15–26. Birkhäuser, Cham (2017)
9. Alexeev, B., Bandeira, A.S., Fickus, M., Mixon, D.G.: Phase retrieval with polarization. SIAM J. Imaging Sci. **7**(1), 35–66 (2014)
10. Alexeev, B., Cahill, J., Mixon, D.G.: Full spark frames. J. Fourier Anal. Appl. **18**(6), 1167–1194 (2012). https://doi.org/10.1007/s00041-012-9235-4
11. Balan, R., Casazza, P., Edidin, D.: On signal reconstruction without phase. Appl. Comput. Harmon. Anal. **20**(3), 345–356 (2006)
12. Bandeira, A.S., Cahill, J., Mixon, D.G., Nelson, A.A.: Saving phase: injectivity and stability for phase retrieval. Appl. Comput. Harmon. Anal. **37**(1), 106–125 (2014)
13. Beinert, R.: One-dimensional phase retrieval with additional interference measurements. Results Math. **72**(1), 1–24 (2017)

14. Beinert, R., Hasannasab, M.: Phase retrieval and system identification in dynamical sampling via prony's method (2021). arXiv:2103.10086
15. Cabrelli, C., Molter, U., Paternostro, V., Philipp, F.: Dynamical sampling on finite index sets. J. d'Analyse Mathématique **140**(2), 637–667 (2020). https://doi.org/10.1007/s11854-020-0099-2
16. Christensen, O.: An Introduction to Frames and Riesz Bases. Springer, New York (2016). https://doi.org/10.1007/978-0-8176-8224-8
17. Christensen, O., Hasannasab, M.: Operator representations of frames: boundedness, duality, and stability. Integr. Equ. Oper. Theor. **88**(4), 483–499 (2017)
18. Christensen, O., Hasannasab, M., Philipp, F.: Frame properties of operator orbits. Math. Nachr. **293**(1), 52–66 (2020)
19. Christensen, O., Hasannasab, M., Rashidi, E.: Dynamical sampling and frame representations with bounded operators. J. Math. Anal. Appl. **463**(2), 634–644 (2018)
20. Delvaux, S., Van Barel, M.: Rank-deficient submatrices of Fourier matrices. Linear Algebra Appl. **429**(7), 1587–1605 (2008)
21. Jaming, P.: Uniqueness results in an extension of Pauli's phase retrieval problem. Appl. Comput. Harmon Anal. **37**(3), 413–441 (2014)
22. Kalman, D.: The generalized Vandermonde matrix. Math. Mag. **57**(1), 15–21 (1984)
23. Lu, Y.M., Vetterli, M.: Spatial super-resolution of a diffusion field by temporal oversampling in sensor networks. In: 2009 Proceedings of the ICASSP, pp. 2249–2252 (2009)
24. Macdonald, I.G.: Schur functions: theme and variations. Séminaire Lotharingien de Combinatoire (Saint-Nabor. 1992), Publ Inst Rech Math Av, vol. 498, pp. 5–39. Univ Louis Pasteur, Strasbourg (1992)
25. Macdonald, I.G.: Symmetric Functions and Hall Polynomials. Oxford Mathematical Monographs, 2nd edn. Oxford University Press, Oxford (1995)
26. Malikiosis, R.D., Oussa, V.: Full spark frames in the orbit of a representation. Appl. Comput. Harmon. Anal. **49**(3), 791–814 (2020)
27. Philipp, F.: Bessel orbits of normal operators. J. Math. Anal. Appl. **448**(2), 767–785 (2017)
28. Ranieri, J., Chebira, A., Lu, Y.M., Vetterli, M.: Sampling and reconstructing diffusion fields with localized sources. In: 2011 Proceedings of the ICASSP 2011, pp. 4016–4019 (2011)
29. Stevenhagen, P., Lenstra, H.W.: Chebotarëv and his density theorem. Math. Intell. **18**(2), 26–37 (1996). https://doi.org/10.1007/BF03027290

Invertible Neural Networks Versus MCMC for Posterior Reconstruction in Grazing Incidence X-Ray Fluorescence

Anna Andrle[1], Nando Farchmin[1], Paul Hagemann[1,2](✉),
Sebastian Heidenreich[1], Victor Soltwisch[1], and Gabriele Steidl[2]

[1] Department of Mathematical Modelling and Data Analysis and Department
of Radiometry with Synchrotron Radiation, Physikalisch-Technische Bundesanstalt
Braunschweig und Berlin, Abbestrasse 2-12, 10587 Berlin, Germany
[2] Institute of Mathematics, TU Berlin, Straße des 17. Juni 136,
10623 Berlin, Germany
p.hagemann@campus.tu-berlin.de

Abstract. Grazing incidence X-ray fluorescence is a non-destructive technique for analyzing the geometry and compositional parameters of nanostructures appearing e.g. in computer chips. In this paper, we propose to reconstruct the posterior parameter distribution given a noisy measurement generated by the forward model by an appropriately learned invertible neural network. This network resembles the transport map from a reference distribution to the posterior. We demonstrate by numerical comparisons that our method can compete with established Markov Chain Monte Carlo approaches, while being more efficient and flexible in applications.

Keywords: GIXRF · Inverse problem · Invertible neural networks · MCMC · Transport maps · Bayesian inversion

1 Introduction

Computational progress is deeply tied with making the structure of computer chips smaller and smaller. Hence there is a need for efficient methods that investigate the critical dimensions of a microchip. Optical scattering techniques are frequently used for the characterization of periodic nanostructures on surfaces in the semiconductor industry [7,11]. As a non-destructive technique, grazing incidence X-ray fluorescence (GIXRF) is of particular interest for many industrial applications. Mathematically, the reconstruction of nanostructures, i.e., of their geometrical parameters, can be rephrased in an inverse problem. Given grazing incidence X-ray fluorescence measurements y, we want to recover the distribution of the parameters x of a grating. To account for measurement errors, it appears to be crucial to take a Bayesian perspective. The main cause of uncertainty is due to inexact measurements y, which are assumed to be corrupted by additive Gaussian noise with different variance in each component.

© Springer Nature Switzerland AG 2021
A. Elmoataz et al. (Eds.): SSVM 2021, LNCS 12679, pp. 528–539, 2021.
https://doi.org/10.1007/978-3-030-75549-2_42

The standard approach to recover the distribution of the parameters are Markov Chain Monte Carlo (MCMC) based algorithms [1]. Instead, in this paper, we make use of invertible neural networks (INNs) [4,10] within the general concept of transport maps [12]. This means that we sample from a reference distribution and seek a diffeomorphic transport map, or more precisely its approximation by an INN, which maps this reference distribution to the problem posterior. This approach has some advantages over standard MCMC–based methods: i) Given a transport map, which is computed in an offline step, the generation of independent posterior samples essentially reduces to sampling the freely chosen reference distribution. Additionally, observations indicate that learning the transport map requires less time than generating a sufficient amount of independent samples via MCMC. ii) Although the transport map is conditioned on a specific measurement, it can serve as a good initial guess for the transport related to similar measurements or as a prior in related inversion problems. Hence the effort to find a transport for different runs within the same experiment reduces drastically. An even more sophisticated way of using a pretrained diffeomorphism has been recently suggested in [16].

Having trained the INN, we compare its ability to recover the posterior distribution with the established MCMC method for fluorescence experiments. Although, a similar INN approach with a slightly simpler noise model was recently also reported for reservoir characterization in [14], we are not aware of any comparison of this kind in the literature.

The outline of the paper is as follows: We start with introducing INNs with an appropriate loss function to sample posterior distributions of inverse problems in Sect. 2. Here we follow the lines of an earlier version of [10]. In particular, the likelihood function has to be adapted to our noise model with different variances in each component of the measurement for the application at hand. Then, in Sect. 3, we describe the forward model in GIXRF in its experimental, numerical and surrogate setting. The comparison of INN with MCMC posterior sampling in done in Sect. 4. Finally, conclusions are drawn and topics of further research are addressed in Sect. 5.

2   Posterior Reconstruction by INNs

In this section, we explain, based on [10], how the posterior of an inverse problems can be analyzed using INNs. In the following, products, quotients and exponentials of vectors are meant componentwise. Denote by p_x the density function of a distribution P_X of a random variable $X \colon \Omega \to \mathbb{R}^d$. Further, let $p_{x|y}(\cdot|y)$ be the density function of the conditional distribution $P_{X|Y=y}$ of X given the value of a random variable $Y \colon \Omega \to \mathbb{R}^n$ at $Y = y \in \mathbb{R}^n$. We suppose that we have a differentiable forward model $f \colon \mathbb{R}^d \to \mathbb{R}^n$. In our applications the forward model will be given by the GIXRF method. We assume that the measurements y are corrupted by additive Gaussian noise $\mathcal{N}(0, b^2 \operatorname{diag}(w^2))$, where the (positive) weight vector $w \in \mathbb{R}^n$ accounts for the different scales of the measurement components. The factor $b^2 > 0$ models the intensity of the noise. In other words,

$$y = f(x) + \eta,$$

where η is a realization of a $\mathcal{N}(0, b^2 \operatorname{diag}(w^2))$ distributed random variable. Then the sampling density reads as

$$p_{y|x}(y|x) = \frac{1}{(2\pi)^{\frac{n}{2}} b^n (\prod_{i=1}^{n} w_i)} \exp\left(-\frac{1}{2} \left\| \frac{y - f(x)}{bw} \right\|^2\right). \tag{1}$$

Given a measurement y from the forward model, we are interested in the inverse problem posterior distribution $P_{X|Y=y}$. By Bayes' formula the inverse problem posterior density can be rewritten as

$$p_{x|y} = \frac{p_{y|x} p_x}{\int_{\mathbb{R}^d} p_{y|x}(y|x) p_x(x) \, dx} \propto p_{y|x} \, p_x.$$

Let p_ξ be the density function of an easy to sample distribution P_Ξ of a random variable $\Xi \colon \Omega \to \mathbb{R}^d$. Following for example [12], we want to find a differentiable and invertible map $T \colon \mathbb{R}^d \to \mathbb{R}^d$, such that T pushes P_Ξ to $P_{X|Y=y}$, i.e.,

$$P_{X|Y=y} = T_\# P_\Xi := P_\Xi \circ T^{-1}.$$

Recall that $T_\# p_\xi = p_\xi \circ T^{-1} |\det \nabla T^{-1}|$ for the corresponding density functions, where ∇T^{-1} denotes the Jacobian of T^{-1}.

Once T is learned for some measurement y, sampling from the posterior $P_{X|Y=y}$ can be approximately done by evaluating T at samples from the reference distribution P_Ξ (see also [12]). Since it is in general hard or even impossible to find the analytical map T, we aim to approximate T by an invertible neural network $T = T(\cdot; \theta) \colon \mathbb{R}^d \to \mathbb{R}^d$ with network parameters θ. In this paper, we use a variation of the INN proposed in [4], see [3]. More precisely, T is a composition

$$T = T_L \circ P_L \circ \cdots \circ T_1 \circ P_1, \tag{2}$$

where P_ℓ are permutation matrices and T_ℓ are invertible mappings of the form

$$T_\ell(\xi_1, \xi_2) = (x_1, x_2) := \left(\xi_1 e^{s_{\ell,2}(\xi_2)} + t_{\ell,2}(\xi_2), \, \xi_2 e^{s_{\ell,1}(x_1)} + t_{\ell,1}(x_1) \right)$$

for some splitting $(\xi_1, \xi_2) \in \mathbb{R}^d$ with $\xi_i \in \mathbb{R}^{d_i}$, $i = 1, 2$. Here $s_{\ell,2}, t_{\ell,2} \colon \mathbb{R}^{d_2} \to \mathbb{R}^{d_1}$ and $s_{\ell,1}, t_{\ell,1} \colon \mathbb{R}^{d_1} \to \mathbb{R}^{d_2}$ are ordinary feed-forward neural networks. The parameters θ of $T(\cdot; \theta)$ are specified by the parameters of these subnetworks. The inverse of the layers T_ℓ is analytically given by

$$T_\ell^{-1}(x_1, x_2) = (\xi_1, \xi_2) := \left((x_1 - t_{\ell,2}(\xi_2)) e^{-s_{\ell,2}(\xi_2)}, \, (x_2 - t_{\ell,1}(x_1)) e^{-s_{\ell,1}(x_1)} \right)$$

and does not require an inversion of the feed-forward subnetworks. Hence the whole map T is invertible and allows for a fast evaluation of both forward and inverse map.

In order to learn the INN, we utilize the Kullback-Leibler divergence as a measure of distance between two distribution as loss function

$$L(\theta) := \mathrm{KL}(T_{\#}p_\xi, p_{x|y}) = \int_{\mathbb{R}^d} T_{\#}p_\xi \log\left(\frac{T_{\#}p_\xi}{p_{x|y}}\right) \mathrm{d}x. \qquad (3)$$

Minimizing the loss function L by e.g. a standard stochastic gradient descent algorithm requires the computation of the gradient of L. To ensure that this is feasible, we rewrite the loss L in the following way.

Proposition 1 *Let $T = T(\cdot\,;\theta)\colon \mathbb{R}^d \to \mathbb{R}^d$ be a Lebesgue measurable diffeomorphism parameterized by θ. Then, up to an additive constant, (3) can be written as*

$$L(\theta) = -\mathbb{E}_\xi\Big[\log p_{y|x}\big(y|T(\xi)\big) + \log p_x\big(T(\xi)\big) + \log|\det \nabla T(\xi)|\Big]. \qquad (4)$$

For the Gaussian likelihood (1), this simplifies to

$$L(\theta) = \mathbb{E}_\xi\Big[\frac{1}{2b^2}\Big\|\frac{y - (f \circ T)(\xi)}{w}\Big\|^2 - \log p_x\big(T(\xi)\big) - \log|\det \nabla T(\xi)|\Big]. \qquad (5)$$

Proof. By definition of the push-forward density and the transformation formula [15, Theorem 7.26], we may rearrange

$$\mathrm{KL}(T_{\#}p_\xi, p_{x|y})$$
$$= \int_{\mathbb{R}^d} (p_\xi \circ T^{-1})(x)\,|\det \nabla T^{-1}(x)|\,\log\left(\frac{(p_\xi \circ T^{-1})(x)|\det \nabla T^{-1}(x)|}{p_{x|y}(x|y)}\right) \mathrm{d}x$$
$$= \int_{\mathbb{R}^d} p_\xi(\zeta)\,\log\left(\frac{p_\xi(\zeta)}{p_{x|y}(T(\xi)|y)\,|\det \nabla T(\xi)|}\right) \mathrm{d}\xi$$
$$= \mathrm{KL}(p_\xi, T_{\#}^{-1}p_{x|y})$$
$$= -\mathbb{E}_\xi\big[\log p_{x|y}(T(\xi)|y) + \log|\det \nabla T(\xi)|\big] + \mathbb{E}_\xi[\log p_\xi].$$

Now Bayes' formula yields

$$\log p_{x|y}(T(\xi)|y) = \log\Big(p_{y|x}(y|T(\xi))\,p_x(T(\xi))\Big) - \log p_y(y).$$

Ignoring the constant terms since they are irrelevant for the minimization of L, we obtain (4). The rest of the assertion follows by (1). □

The different terms on the right-hand side of (4), resp. (5) are interpretable: The first term forces the samples pushed through T to have the correct forward mapping, the second assures that the samples are pushed to the support of the prior distribution and the last term employs a counteracting force to the first. To see this note that the first term is minimized if T pushes p_ξ to a delta distribution, whereas the log determinant term would be unbounded in that case. Hence there is an equilibrium between those terms directly influenced by the error parameter b, i.e. as b tends to zero, the push-forward tends to a delta distribution.

The computation of the gradient of the empirical loss function L corresponding to (5) requires besides standard differentiations of elementary functions and of the network T, the differentiation of i) the forward model f within the chain rule of $f \circ T$, ii) of p_x, and iii) of $|\det \nabla T|$. This can be done by the following observations:

i) In the next section, we describe how a feed-forward neural network can be learned to approximate the forward mapping in GIXRF. Then this network will serve as forward operator and its gradient can be computed by standard backpropagation.

ii) The prior density p_x has to be known. In our application, we can assume that the geometric parameters x are uniformly distributed in a compact set which is specified for each component of $x \in \mathbb{R}^d$ in the numerical section. This has the consequence that the term $\log p_x$ is constant within the support of p_x and is not defined outside. Therefore, we impose an additional boundary loss that penalizes samples out of the support of the prior, more precisely, if the prior is supported in $[s_1, t_1] \times \ldots \times [s_d, t_d]$, then we use

$$L_{\mathrm{bd}}(x) = \lambda_{\mathrm{bd}} \sum_{i=1}^{d} \left(\mathrm{ReLU}(x_i - t_i) + \mathrm{ReLU}(s_i - x_i) \right), \qquad \lambda_{\mathrm{bd}} > 0.$$

Note that the non-differentiable ReLU function at zero can be replaced by various smoothed variants.

iii) For general networks, $\log |\det \nabla T|$ in the loss function is hard to compute and is moreover either not differentiable or has a huge Lipschitz constant. However, it becomes simple for INNs due to their special structure. Since $T_\ell = T_{2,\ell} \circ T_{1,\ell}$ with $T_{1,\ell}(\xi_1, \xi_2) = (x_1, \xi_2) := \left(\xi_1 e^{s_{\ell,2}(\xi_2)} + t_{\ell,2}(\xi_2), \xi_2 \right)$, and $T_{2,\ell}(x_1, \xi_2) = (x_1, x_2) := \left(x_1, \xi_2 e^{s_{\ell,1}(x_1)} + t_{\ell,1}(x_1) \right)$ we have

$$\nabla T_{1,\ell}(\xi_1, \xi_2) = \begin{pmatrix} \mathrm{diag}\left(e^{s_{\ell,2}(\xi_2)} \right) & \mathrm{diag}\left(\nabla_{\xi_2} \left(\xi_1 e^{s_{\ell,2}(\xi_2)} + t_{\ell,2}(\xi_2) \right) \right) \\ 0 & I_{d_2} \end{pmatrix}$$

so that $\det \nabla T_{1,\ell}(\xi_1, \xi_2) = \prod_{k=1}^{d_1} e^{(s_{\ell,2}(\xi_2))_k}$ and similarly for $\nabla T_{2,\ell}$. Applying the chain rule in (2), noting that the Jacobian of P_ℓ is just P_ℓ^T with $\det P_\ell^\mathsf{T} = 1$, and that $\det(AB) = \det(A)\det(B)$, we conclude

$$\log(|\det(\nabla T(\xi))|) = \sum_{\ell=1}^{L} \left(\mathrm{sum}\left(s_{\ell,2}\left((P_\ell \xi^\ell)_2 \right) \right) + \mathrm{sum}\left(s_{\ell,1}\left((T_{1,\ell} P_\ell \xi^\ell)_1 \right) \right) \right),$$

where sum denotes the sum of the components of the respective vector, $\xi^1 := \xi$ and $\xi^\ell = T_{\ell-1} P_{\ell-1} \xi^{\ell-1}$, $\ell = 2, \ldots, L$.

3   Forward Model from GIXRF

In this section, we consider a silicon nitride (Si_3N_4) lamellar grating on a silicon substrate. The grating oxidized in a natural fashion resulting in a thin SiO_2

layer. A cross-section of the lamellar grating is shown in Fig. 1, left. It can be characterized by seven parameters $x \in \mathbb{R}^d$, $d = 7$, namely the height (h) and middle-width (cd) of the line, the sidewall angle (swa), the thickness of the covering oxide layer (t_t), the thickness of the etch offset of the covering oxide layer beside the lamella (t_g) and additional layers on the substrate (t_s, t_b).

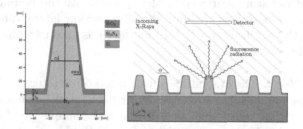

Fig. 1. Left: Cross-section of one grating line with characterizing parameters. Right: Cross-section of the grating with incoming X-rays (typically in z_3 direction) and emitted fluorescence radiation.

To determine the parameters, we want to apply the GIXRF technique recently established to find the geometry parameters of nanostructures and its atomic composition [2,8].

Experimental Setting. In GIXRF, the angles α_i, $i = 1, \dots, n$ between an incident monochromatic X-ray beam and the sample surface is varied around a critical angle for total external reflection. Depending on the local field intensity of the X-ray radiation, atoms are ionized and are emitting a fluorescence radiation. The resulting detected fluorescence radiation $(F(\alpha_i; x))_{i=1}^n$ is characteristic for the atom type.

Besides direct experimental measurements of the fluorescence F, its mathematical modeling at each angle α can be done in two steps, namely by computing the intensity of the local electromagnetic field E arising from the incident wave (X-ray) with angle α and then to use its modulus to obtain the fluorescence value $F(\alpha, x)$. The propagation of E is in general described by Maxwell's equations and simplifies for our specific geometry to the partial differential equation

$$\nabla_z \cdot \left(\mu(z; x)^{-1} \nabla_z E(z) \right) - \omega^2 \varepsilon(z; x) \, E(z) = 0, \quad z \in \mathbb{R}^2. \tag{6}$$

Here ω is the frequency of the incident plane wave, ε and μ are the permittivity and permeability depending on the grating parameters x, resp. From the 2D distribution of E in (6), more precisely from the resulting field intensities $|E|$, the fluorescence radiation $F(\alpha; x)$ at the detector can be calculated by an extension of the Sherman equation [17]. This computation requires just the appropriately scaled summation of the values of $|E|^2$ on the FEM mesh of the Maxwell solver used. The really time consuming part is the numerical computation of E for each angle α_i, $i = 1, \dots, n$ by solving the PDE (6).

Numerical Treatment. In order to compute E given by (6) with appropriate boundary conditions, we employ the finite element method (FEM) implemented in the JCMsuite software package to discretize and solve the corresponding scattering problem on a bounded computational unit cell in the weak formulation as described in [13]. This formulation yields a splitting of the complete $\mathbb{R}^2$ into an interior domain hosting the total field and an exterior domain, where only the purely outward radiating scattered field is present. At the boundaries, Bloch-periodic boundary conditions are applied in the periodic lateral direction and an adaptive perfectly matched layer (PML) method was used to realize transparent boundary conditions.

Surrogate NN Model. The evaluation of the fluorescence intensity F for a single realization of the parameters x involves solving (6) for each angle α_i, $i = 1, \ldots, n$. Since this is very time consuming, we learn instead a simple feed-forward neural network with one hidden layer with 256 nodes and ReLU activation as surrogate $f \colon \mathbb{R}^d \to \mathbb{R}^n$ of F such that the L^2 error between $(f_i(x))_{i=1}^n$ and $(F(\alpha_i; x))_{i=1}^n$ becomes minimal. The network was trained on roughly 10^4 sample pairs (F, x) which were numerically generated as described above in a time consuming procedure. The L^2-error of the surrogate, evaluated on a separate test set containing about 10^3 sample pairs was smaller than $2 \cdot 10^{-3}$. This is sufficient for the application, since we have measurement noise on the data of at least one order of magnitude larger for both the experimental data and the synthetic study. Hence we neglect the approximation error of the FE model and the surrogate further on.

4   Numerical Results

In this section, we solve the statistical inverse problem of GIXRF using the Bayesian approach with an INN and the MCMC method based on our surrogate NN forward model for virtual and experimental data. Note that obtaining training data for the forward NN took multiple days of computation on a compute server with 120 CPUs. We compare the resulting posterior distributions and the computational performance. To the best of our knowledge, this was not done in the literature so far for any forward model.

The fluorescence intensities for the silicon nitride layer of the lamellar grating depicted in Fig. 1 were measured for $n = 178$ different incidence angles α_i ranging from 0.8° to 14.75°. The seven parameters of the grating were considered to be uniformly distributed according to the domains listed in Table 1.

Table 1. Domains of the different parameters. Units are given in [nm] for all parameters except the sidewall angle (swa), which is given in [°].

Parameter:	h	cd	swa	t_t	t_b	t_g	t_s
Domain:	[85, 100]	[45, 55]	[76, 88]	[2, 4]	[0, 5]	[2, 10]	[0.1, 3]

To gain maximal performance of the MCMC method, we utilize an affine invariant ensemble sampler for the Markov-Chain Monte Carlo algorithm [5]. This allows parallel computations with multiple Markov chains and reduces the number of method specific free parameters for the MCMC steps. The error parameter b is usually a priori unknown and is thus subject to expert knowledge. However, MCMC algorithms can introduce those as additional posterior hyperparameters for reconstruction by a slight modification of the prior. Define $\tilde{x} := (x, b) \sim P_{\tilde{x}}$, where $P_{\tilde{x}}$ is given by the density $p_{(x,b)} = p_x \, p_b$ for uniform p_b. Using this error model in the likelihood in (1), we obtain $p_{y|(x,b)}$. Then the MCMC algorithm applied to $p_{(x,b)|y}$ yields the distribution of the parameter b as well.

To approximate the INN that pushes the example density forward to the posterior one, we learn an INN with $L = 10$ layers. Each subnetwork of each layer is chosen as a two hidden layer ReLU feedforward network with 256 nodes in each hidden layer. The network is trained on the empirical counterpart of the loss (5) by sampling from a standard Gaussian distribution using an adaptive moment estimation optimization (Adam) algorithm, see [9]. We trained for 80 epochs, an epoch consists of 40 parameter updates with a batch-size of 200. The learning rate is lowered by a factor of 0.1 every 20 epochs. The INN model is built and trained with the freely available FrEIA software package[1].

The time that is required to obtain $2 \cdot 10^4$ posterior samples via the MCMC algorithm varies between 1.5 and 3.5 h on a standard Laptop. In comparison the training of the INN takes less than 20 min, and $2 \cdot 10^4$ independent posterior samples are generated in less than one second.

4.1   Synthetic Data

As a first application, we perform a virtual experiment for the GIXRF to obtain a problem with known true values x_{true}. To approach that, we pick a pair $(x_{\text{true}}, f(x_{\text{true}}) = y_{\text{true}})$, where parameter x_{true} was chosen as the means of the respective domains and vary $b \in \{10^{-2}, 3 \cdot 10^{-2}, 10^{-1}\}$. Using these values of b we obtain a synthetic noisy measurement according to $y_{\text{meas}} = y_{\text{true}} + \varepsilon$ for various realizations of ε of $\mathcal{N}(0, b^2 \text{diag}(y_{\text{true}}^2))$. For the computation of the INN, we set b according to the true value, whereas MCMC is able to estimate a distribution of the parameter b for given y_{meas}. For the application of both MCMC and INN, we set $y = y_{\text{meas}}$ and $w = y_{\text{meas}}$ (note that we regard $w = y_{\text{meas}}$ as a constant). Figure 2 displays the one dimensional marginals of the posterior for both the MCMC and the INN approach alongside the ground truth for three different values of b.

First one sees that the width of the marginals decreases as b gets smaller. Comparing the INN and MCMC marginals shows an almost identical shape and support for most of the parameters, where the uncertainties obtained by the INN tend to be a bit smaller. The posterior means of the MCMC and INN approach are in proximity of each other relative to the domain size, in particular in the

---

[1] https://github.com/VLL-HD/FrEIA.

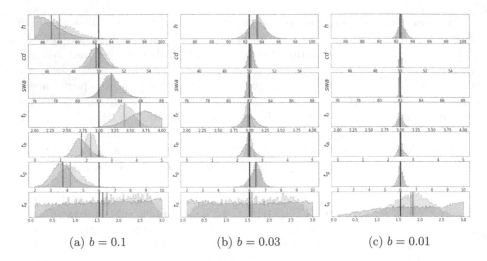

(a) $b = 0.1$ (b) $b = 0.03$ (c) $b = 0.01$

Fig. 2. One dimensional marginals of posterior samples for different values of b generated by MCMC (blue) and INN (orange) together with their means as solid lines. The values of the ground truth x_i, $i = 1, \ldots, 7$ are displayed as solid black lines. (Color figure online)

cases, where b is smaller. It is remarkable that although the reconstruction in $b = 0.1$ is far-off from the ground truth, both methods agree in their estimate. This can be explained by the large magnitude of noise. The noise realization for $b = 0.1$ changes the measurements drastically such that a different parameter configuration becomes more likely. Interestingly, the INNs seems to transport less mass to the boundary, which is probably due to the boundary loss preventing mass to be transported outside of the boundary. Furthermore, both methods identify the last parameter to be the least sensitive.

4.2   Experimental Data

In the next experiment, we use fluorescence measurements $y_{\text{meas}} = (F(\alpha_i; x))_{i=1}^{n}$ obtained from an GIXRF experiment. Here neither exact values of the parameters x nor of the noise level b are available. In the noise model (1) we use $w = y_{\text{meas}}$. To train the INN we set the value b to the mean 0.02 of the reconstructed b obtained by the MCMC algorithm. Figure 3 depicts the one and two dimensional marginals of the posterior for both the MCMC and the INN calculations on a smaller subset of the parameter domain. The means and supports of the posteriors agree well. Figure 3 also shows that the posterior is mostly a sharp Gaussian. The Si_3N_4 height etch offset t_b is known to be close to zero for the real grating, which explains the non-Gaussian accumulation at the boundary of the interval. Similar holds true for the height of the lower SiO_2 layer.

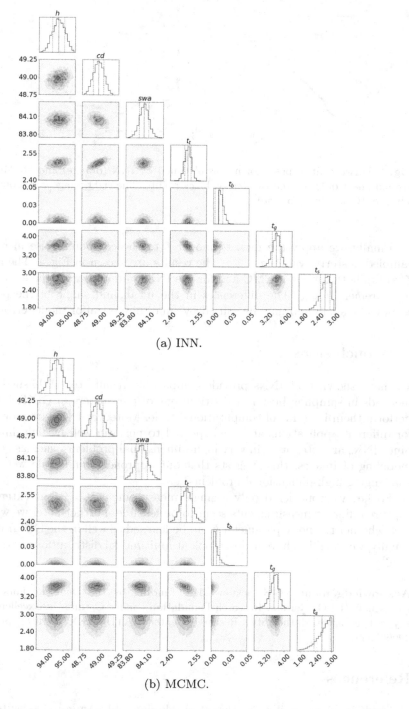

(a) INN.

(b) MCMC.

Fig. 3. Comparison of 2D densities of the posterior for y_{meas} calculated via the different methods. The straight vertical line depicts the mean, the dashed ones the standard deviation.

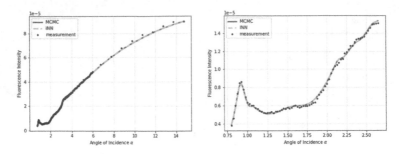

Fig. 4. Forward fit of posterior means obtained by INN (orange) and MCMC (blue). Measurement data are represented by black points. The right image is a zoom into the left one. (Color figure online)

Finally, we apply the forward model f to the componentwise mean of the sampled posterior values $x \in \mathbb{R}^7$. The results are shown in Fig. 4. The resulting $f(x)$ fits both for the MCMC and for the INN quite well with the experimental measurement y_{meas}. The differences in the mean and shape of the posteriors seem to be small and may be caused by small numerical or model errors.

5   Conclusions

We have shown that INNs provide comparable results to established MCMC methods in sampling posterior distributions of parameters in GIXRF, but outperform them in terms of computational time. Moreover INNs are more flexible for different applications and are expected to perform well in high dimensions. Since INNs are often used in very high-dimensional problems, such as generative modeling of images, this suggests that our approach could scale well to more challenging high-dimensional problems.

So far, we considered only a single measurement y_{meas}, but extensions to capture different measurements seem to be feasible. Furthermore, we will figure out, whether the noise parameter b can be learned as well in the INN framework. Finally, we intend to have a closer look at multimodal distributions as done, e.g., in [6].

Acknowledgements. G.S. acknowledges funding by the Deutsche Forschungsgemeinschaft (DFG, German Research Foundation) under Germany's Excellence Strategy – The Berlin Mathematics Research Center MATH+ (EXC-2046/1, project ID: 390685689).

References

1. Andrieu, C., de Freitas, N., Doucet, A., Jordan, M.I.: An introduction to MCMC for machine learning. Mach. Learn. **50**, 5–43 (2003)

2. Andrle, A., et al.: Grazing incidence x-ray fluorescence based characterization of nanostructures for element sensitive profile reconstruction. In: Modeling Aspects in Optical Metrology VII, vol. 11057, p. 110570M. International Society for Optics and Photonics (2019)

3. Ardizzone, L., Kruse, J., Rother, C., Köthe, U.: Analyzing inverse problems with invertible neural networks. In: 7th International Conference on Learning Representations, ICLR 2019, New Orleans, LA, USA, (2019)

4. Dinh, L., Sohl-Dickstein, J., Bengio, S.: Density estimation using real NVP. In: 5th International Conference on Learning Representations, ICLR 2017, Toulon, France, April 24–26, 2017, Conference Track Proceedings (2017)

5. Foreman-Mackey, D., Hogg, D.W., Lang, D., Goodman, J.: EMCEE: the MCMC hammer. Publ. Astron. Soc. Pacific 125(925), 306–312 (2013)

6. Hagemann, P.L., Neumayer, S.: Stabilizing invertible neural networks using mixture models. Inverse Problems (2021)

7. Henn, M.-A., Gross, H., Heidenreich, S., Scholze, F., Elster, C., Bär, M.: Improved reconstruction of critical dimensions in extreme ultraviolet scatterometry by modeling systematic errors. Meas. Sci. Technol. 25(4), 044003/1–9 (2014)

8. Hönicke, P., et al.: Grazing incidence-x-ray fluorescence for a dimensional and compositional characterization of well-ordered 2D and 3D nanostructures. Nanotechnology 31(50), 505709/1–8 (2020)

9. Kingma, D.P., Ba, J.: Adam: a method for stochastic optimization. In: Bengio, Y., LeCun, Y. (eds.) 3rd International Conference on Learning Representations, ICLR 2015, San Diego, CA, USA, May 7–9, 2015, Conference Track Proceedings (2015)

10. Kruse, J., Detommaso, G., Scheichl, R., Köthe, U.: HINT: hierarchical invertible neural transport for density estimation and Bayesian inference. arXiv preprint arXiv:1905.10687 (2020)

11. Mack, C.: Fundamental principles of optical lithography: the science of microfabrication. John Wiley & Sons (2008)

12. Marzouk, Y., Moselhy, T., Parno, M., Spantini, A.: Sampling via measure transport: an introduction. Handbook of Uncertainty Quantification, pp. 1–41 (2016)

13. Pomplun, J., Burger, S., Zschiedrich, L., Schmidt, F.: Adaptive finite element method for simulation of optical nano structures. Phys. Status Solidi (B) 244(10), 3419–3434 (2007)

14. Rizzuti, G., Siahkoohi, A., Witte, P., Herrmann, F.: Parameterizing uncertainty by deep invertible networks: an application to reservoir characterization. SEG Tech. Program Expanded Abs. 2020, 1541–1545 (2020)

15. Rudin., W.: Real Analysis. McGraw-Hill, 3rd edition (1987)

16. Siahkoohi, A., Rizzuti, G., Louboutin, M., Witte, P., Herrmann, F.J.: Preconditioned training of normalizing flows for variational inference in inverse problems. In: 3rd Symposium on Advances in Approximate Bayesian Inference (2021)

17. Soltwisch, V., et al.: Element sensitive reconstruction of nanostructured surfaces with finite elements and grazing incidence soft x-ray fluorescence. Nanoscale 10(13), 6177–6185 (2018)

Adversarially Learned Iterative Reconstruction for Imaging Inverse Problems

Subhadip Mukherjee[1(✉)], Ozan Öktem[2], and Carola-Bibiane Schönlieb[1]

[1] Department of Applied Mathematics and Theoretical Physics,
University of Cambridge, Cambridge, UK
{sm2467,cbs31}@cam.ac.uk
[2] Department of Mathematics, KTH – Royal Institute of Technology,
Stockholm, Sweden
ozan@kth.se

Abstract. In numerous practical applications, especially in medical image reconstruction, it is often infeasible to obtain a large ensemble of ground-truth/measurement pairs for supervised learning. Therefore, it is imperative to develop unsupervised learning protocols that are competitive with supervised approaches in performance. Motivated by the maximum-likelihood principle, we propose an unsupervised learning framework for solving ill-posed inverse problems. Instead of seeking pixel-wise proximity between the reconstructed and the ground-truth images, the proposed approach learns an iterative reconstruction network whose output matches the ground-truth in distribution. Considering tomographic reconstruction as an application, we demonstrate that the proposed unsupervised approach not only performs on par with its supervised variant in terms of objective quality measures, but also successfully circumvents the issue of over-smoothing that supervised approaches tend to suffer from. The improvement in reconstruction quality comes at the expense of higher training complexity, but, once trained, the reconstruction time remains the same as its supervised counterpart.

Keywords: Generative adversarial networks (GANs) · Iterative reconstruction · Unsupervised learning · Inverse problems

1 Introduction

Inverse problems arise in a wide range of applications, especially in medical imaging. The aim is to recover an unknown model parameter $x^* \in \mathbb{X}$ containing critical information about the structural details of an underlying subject from data $y = \mathcal{A}(x^*) + e \in \mathbb{Y}$, representing noisy, indirect, and potentially incomplete set of measurements. The forward operator $\mathcal{A} : \mathbb{X} \to \mathbb{Y}$ and the distribution of the measurement noise $e \in \mathbb{Y}$ are typically known and they jointly form a simulator for the data acquisition process. Inverse problems are typically ill-posed

© Springer Nature Switzerland AG 2021
A. Elmoataz et al. (Eds.): SSVM 2021, LNCS 12679, pp. 540–552, 2021.
https://doi.org/10.1007/978-3-030-75549-2_43

in the absence of any further information apart from the measurements alone, meaning that different model parameters can give rise to the same measurement. Variational reconstruction [17, Part II] is a generic, yet adaptable framework for solving inverse problems, wherein ill-posedness is tackled by solving

$$\min_{x \in \mathbb{X}} \|y - \mathcal{A}(x)\|_2^2 + \lambda \mathcal{R}(x). \tag{1}$$

The goal here is to alleviate the aforementioned inherent indeterminacy by encoding some prior knowledge about the model parameter using a hand-crafted regularizer $\mathcal{R}$ in addition to seeking data-consistency.

While variational methods enjoy rigorous theoretical guarantees for stability and convergence, and have remained the state-of-the-art for several decades, they are limited in their ability to adapt to a particular application at hand. With the emergence of deep learning, modern approaches for solving inverse problems have increasingly shifted towards data-driven reconstruction [5], which generally offers significantly superior reconstruction quality as compared to the traditional variational methods. Data-adaptive reconstruction methods can broadly be classified into two categories: (i) end-to-end trained over-parametrized models that either attempt to map the measured data to the true model parameter (such as AUTOMAP proposed in [18]), or remove artifacts from the output of an analytical reconstruction method [10], and (ii) learning the image prior using a neural network based on training data of images and then using such a learned regularizer in a variational model for reconstruction [11,12,14,16]. The first approach relies on learning the reconstruction method from a large training dataset that consists of many ordered pairs of model-parameter and corresponding noisy data. Since obtaining a vast amount of paired examples is difficult in medical imaging applications, over-parametrized models trained end-to-end in a supervised manner might run into the danger of overfitting and generalize poorly on unseen data. The second category still requires one to solve a high-dimensional variational problem where the objective involves a trained neural network, a task that is typically computationally demanding.

One promising way to circumvent the limited data problem is to build network architectures by incorporating the physics of the acquisition process [2,3]. The learned primal-dual (LPD) approach proposed in [3] is data-efficient as compared to fully data-driven approaches and can generalize well when trained on a moderate amount of examples. However, an unrolled LPD network trained by minimizing the squared-ℓ_2 error between the network output and the target essentially returns an approximation to the conditional-mean estimator, which is the statistical expectation of the target image conditioned on the measurement. Owing to this implicit averaging, the resulting estimate tends to suffer from blurring artifacts with the loss of important details in the reconstruction. The proposed adversarially trained LPD method, referred to as ALPD, circumvents this problem by seeking proximity in the space of distribution instead of aiming to minimize the squared-ℓ_2 distortion in the image space.

2   Main Contributions: Training Objective and Protocol

For supervised learning, one needs paired training examples of the form $\{x_i, y_i\}_{i=1}^{n}$ sampled i.i.d. from the joint probability distribution $\pi_X(x)\pi_{\text{data}}(y|x)$ of the image and the measurement. In contrast, the proposed training protocol is unsupervised, i.e., it assumes availability of i.i.d. samples $\{x_i\}_{i=1}^{n_1}$ and $\{y_i\}_{i=1}^{n_2}$ from the marginal distributions π_X and π_Y of the ground-truth image and measurement data, respectively. The image and the data samples are unpaired, i.e., y_i does not necessarily correspond to the noisy measurement of x_i. In the context of CT, x_i's could be the high-/normal-dose reconstructions obtained using the classical filtered back-projection algorithm, whereas y_i's correspond to low-dose projection. We begin with a description of the proposed unsupervised approach and how it differs from and relates to supervised and classical variational methods. Subsequently, we motivate the training loss using the maximum-likelihood (ML) principle and explain the reconstruction network parametrization, which follows the same philosophy proposed in [3].

2.1   Proposed Training Protocol for ALPD

Similar to supervised training, one key component of the proposed unsupervised approach is to first build a parametric reconstruction network $\mathcal{G}_\theta : \mathbb{Y} \to \mathbb{X}$ (see Sect. 2.3 for details) that takes the measurement as input and produces a reconstructed image as the output. However, unlike supervised training, it is not possible to train $\mathcal{G}_\theta$ by minimizing a chosen distortion measure between $\mathcal{G}_\theta(y_i)$ and x_i, since x_i is not the ground-truth image corresponding to y_i. Our training framework essentially seeks to achieve the following three objectives:

1. The reconstructions produced by $\mathcal{G}_\theta$ should be close to the ground-truth images in the training dataset in terms of distribution (measured with respect to the Wasserstein distance);
2. $\mathcal{G}_\theta$ should be encouraged to be the right-inverse of $\mathcal{A}$, so that the forward operator applied on the output of $\mathcal{G}_\theta$ is close to the measured data; and
3. $\mathcal{G}_\theta$ should approximately be a left-inverse of $\mathcal{A}$, i.e., $\mathcal{G}_\theta$ must recover the ground-truth from noise-free measurement.

More concretely, we propose to learn $\mathcal{G}_\theta : \mathbb{Y} \to \mathbb{X}$ by minimizing the training loss

$$J(\theta) = \mathcal{W}\left((\mathcal{G}_\theta)_{\#}\pi_Y, \pi_X\right) + \lambda_{\mathbb{Y}}\, \mathbb{E}_{\pi_Y}\left[\|\mathcal{A}(\mathcal{G}_\theta(Y)) - Y\|_2^2\right]$$
$$+ \lambda_{\mathbb{X}}\, \mathbb{E}_{\pi_X}\left[\|\mathcal{G}_\theta\left(\mathcal{A}(X)\right) - X\|_2^2\right]. \quad (2)$$

The penalty parameters $\lambda_{\mathbb{X}}$ and $\lambda_{\mathbb{Y}}$ control the relative weighting of the three objectives. Notably, in the absence of noise in the measurement, the first objective becomes superfluous, i.e., any $\mathcal{G}_\theta$ that satisfies the third objective automatically satisfies the first one too. For noisy measurements, the combination of the first and the third objectives helps compute a stable estimate which does not overfit to noise, whereas the second objective ensures that the reconstruction explains the data well. Similar to [4], we make use of the Kantorovich-Rubinstein (KR) duality for approximating the Wasserstein distance term in (2). This requires training a critic network $\mathcal{D}_\alpha : \mathbb{X} \to \mathbb{R}$ that scores an image on the real line based on how closely it resembles the ground-truth images in the dataset. More precisely, the KR duality helps estimate the Wasserstein distance by solving

$$\mathcal{W}\left((\mathcal{G}_\theta)_\# \pi_Y, \pi_X\right) = \sup_\alpha \mathbb{E}_{\pi_X}\left[\mathcal{D}_\alpha\left(X\right)\right] - \mathbb{E}_{(\mathcal{G}_\theta)_\# \pi_Y}\left[\mathcal{D}_\alpha\left(X\right)\right] \text{ where } \mathcal{D}_\alpha \in \mathbb{L}_1. \quad (3)$$

Here, $\mathbb{L}_1$ denotes the space of 1-Lipschitz functions. In practice, both $\mathcal{G}_\theta$ and $\mathcal{D}_\alpha$ are updated in an alternating manner instead of fully solving (3) for each $\mathcal{G}_\theta$ update. The 1-Lipschitz condition is enforced by penalizing the gradient of the critic with respect to the input [9]. Estimating the Wasserstein distance in (3) and the training loss for $\mathcal{G}_\theta$ in (2) requires samples from the marginals, thereby rendering the training framework unsupervised. The detailed steps involved in training the networks are listed in Algorithm 1.

At this point, it is instructive to interpret the training objective (2) through the lens of the variational framework, by recasting the variational problem as a minimization over the parameter θ of $\mathcal{G}_\theta$ instead of x. Given the data distribution π_Y, it is natural to estimate θ that minimizes the expected variational loss:

$$\min_\theta J_1(\theta) := \mathbb{E}_{\pi_Y}\left[\|\mathcal{A}(\mathcal{G}_\theta(Y)) - Y\|_2^2 + \lambda \mathcal{R}\left(\mathcal{G}_\theta(Y)\right)\right]. \quad (4)$$

Now, suppose the existence of an *ideal* regularizer in (4), which returns a small score when the input is drawn from π_X and a large score when the distribution of the input differs from π_X. For such a regularizer, the difference

$$\mathcal{L}(\mathcal{R}) = \mathbb{E}_{\pi_X}\left[\mathcal{R}(X)\right] - \mathbb{E}_{\pi_Y}\left[\mathcal{R}(\mathcal{G}_\theta(Y))\right], \quad (5)$$

should be small. As a matter of fact, a consequence of the KR duality is that

$$\inf_{\mathcal{R}} \mathcal{L}(\mathcal{R}) = -\mathcal{W}\left((\mathcal{G}_\theta)_\# \pi_Y, \pi_X\right),$$

provided that $\mathcal{R}$ is constrained to be 1-Lipschitz. Substituting this in (5) and ignoring terms independent of θ reduces (4) to minimizing $\theta \mapsto \hat{J}_1(\theta)$ where

$$\hat{J}_1(\theta) := \mathbb{E}_{\pi_Y}\left[\|\mathcal{A}(\mathcal{G}_\theta(Y)) - Y\|_2^2\right] + \lambda \mathcal{W}\left((\mathcal{G}_\theta)_\# \pi_Y, \pi_X\right). \quad (6)$$

If $\lambda := 1/\lambda_\mathbb{Y}$, then we obtain the inequality $J(\theta) \geq \lambda_\mathbb{Y} \hat{J}_1(\theta)$ for the training loss J. This indicates that our training loss majorizes (up to a scaling) the objective $\hat{J}_1$ in (6), which emerges naturally from the variational loss under the assumption of a 1-Lipschitz ideal regularizer. The penalty terms in the $\mathbb{X}$- and $\mathbb{Y}$-domains in (2) are unmistakably reminiscent of the cycle-consistency losses in cycle-GANs [19] that are widely used learning paradigms for unpaired image-to-image translation problems. Similar unsupervised approaches involving GANs were also proposed in [13] for conditional image-to-image synthesis tasks. The proposed approach can indeed be thought of as a simpler variant of cycle-GAN with the generator learned in only one direction ($\mathbb{Y} \to \mathbb{X}$) instead of two. Notably, it was recently shown in [6] that the cycle-GAN training loss can be derived as the optimal transport loss corresponding to the case where the transport cost is equal to the variational loss with $\mathcal{R}(x) = \|x - \mathcal{G}_\theta(y)\|_2$.

2.2 A Maximum-Likelihood (ML) Perspective

The ML principle seeks to solve

$$\max_\theta \left[\frac{1}{n_1} \sum_{i=1}^{n_1} \log \pi_X^{(\theta)}(x_i) + \frac{1}{n_2} \sum_{i=1}^{n_2} \log \pi_Y^{(\theta)}(y_i) \right], \tag{7}$$

where $\pi_X^{(\theta)}$ and $\pi_Y^{(\theta)}$ are the distributions induced by appropriately postulated probabilistic models on X and Y, respectively. In the following, we explain the statistical models and use them to derive a tractable lower-bound (generally referred to as the evidence lower-bound (ELBO)) on the ML objective in (7). Our analysis reveals that the resulting ELBO is equivalent to the ALPD training loss in (2) in spirit, except for the measure of distance for comparing the distributions of the reconstruction and the ground-truth. The KL-divergence-based distance measure is replaced with the Wasserstein-1 distance since it lends itself to continuous differentiability with respect to the parameters of the reconstruction network, thereby facilitating a stable gradient-based parameter update.

Bound on the Data Likelihood. The ELBO for Y is derived by treating Y as the observed variable and X as the unobserved/latent variable. We then derive two expressions for the conditional distribution of X given $Y = y$, one from the measurement process and the other from the reconstruction process. The lower-bound is tight when these are close to each other.

Measurement process: Model parameters are generated by $X \sim \pi_X$ whereas the measured data are generated by the $\mathbb{Y}$-valued random variable $(Y|X = x) \sim \mathcal{N}\left(\mathcal{A}(x), \sigma_e^2 I\right)$ for given $x \in \mathbb{X}$. Let $\pi_{X,Y}^{(m)}$ denote the induced joint distribution of (X, Y) with $\pi_X^{(m)}(x) := \pi_X(x)$ and $\pi_Y^{(m)}(y)$ denoting its marginals. Also, let $\pi_{X|Y}^{(m)}(x|y)$ and $\pi_{Y|X}^{(m)}(y|x)$ denote the corresponding conditional distributions.

Reconstruction process: Data are generated by $Y \sim \pi_Y$ and reconstructed model parameters are generated by the $\mathbb{X}$-valued random variable $(X|Y = y) \sim \mathcal{N}\left(\mathcal{G}_\theta(y), \sigma_1^2 I\right)$ for given $y \in \mathbb{Y}$. Denote the associated joint distribution by $\pi_{X,Y}^{(r)}(x, y)$ and its marginals and corresponding conditionals are denoted similarly as for the measurement process.

The log-likelihood of Y is given by $\mathcal{L}_{\mathrm{ML}}^{(y)}(\theta) = \frac{1}{n_2} \sum_{i=1}^{n_2} \log \pi_Y^{(m)}(y_i)$, which is the empirical average of the natural logarithm of the model-induced probability density computed over samples of the true distribution of Y. Using the statistical model above, $\mathcal{L}_{\mathrm{ML}}^{(y)}(\theta)$ can be expressed as

$$\log \pi_Y^{(m)}(\boldsymbol{y}) = \log \left(\int_{\mathsf{X}} \pi_{X,Y}^{(m)}(\boldsymbol{x}, \boldsymbol{y}) \, d\boldsymbol{x} \right) = \log \left(\mathbb{E}_{\pi_{X|Y}^{(r)}} \left[\frac{\pi_{X,Y}^{(m)}(X, \boldsymbol{y})}{\pi_{X|Y}^{(r)}(X|\boldsymbol{y})} \right] \right).$$

Since log is a concave function, applying Jensen's inequality leads to

$$\log \pi_Y^{(m)}(\boldsymbol{y}) \geq \mathbb{E}_{\pi_{X|Y}^{(r)}} \left[\log \left(\frac{\pi_X(X)\pi_{Y|X}^{(m)}(\boldsymbol{y}|X)}{\pi_{X|Y}^{(r)}(X|\boldsymbol{y})} \right) \right].$$

The above can further be simplified as

$$\log \pi_Y^{(m)}(\boldsymbol{y}) \geq \mathbb{E}_{\pi_{X|Y}^{(r)}} \left[\log \left(\pi_{Y|X}^{(m)}(\boldsymbol{y}|X) \right) \right] + \mathbb{E}_{\pi_{X|Y}^{(r)}} \left[\log \left(\frac{\pi_X(X)}{\pi_{X|Y}^{(r)}(X|\boldsymbol{y})} \right) \right]$$

$$= \mathbb{E}_{\pi_{X|Y}^{(r)}} \left[\log \left(\pi_{Y|X}^{(m)}(\boldsymbol{y}|X) \right) \right] - \mathrm{KL} \left(\pi_{X|Y=\boldsymbol{y}}^{(r)}, \pi_X \right). \tag{8}$$

Under the postulated statistical model, we have that

$$\pi_{Y|X}^{(m)}(\boldsymbol{y}|\boldsymbol{x}) := \mathcal{N} \left(\mathcal{A}(\boldsymbol{x}), \sigma_e^2 \, I \right) \text{ and } \pi_{X|Y}^{(r)}(\boldsymbol{x}|\boldsymbol{y}) := \mathcal{N} \left(\mathcal{G}_\theta(\boldsymbol{y}), \sigma_1^2 \, I \right). \tag{9}$$

If the forward operator $\mathcal{A}$ is linear (which reduces to a matrix in the finite-dimensional case), then by (9) one can simplify the bound in (8):

$$- \log \pi_Y^{(m)}(\boldsymbol{y}) \leq \mathrm{KL} \left(\pi_{X|Y=\boldsymbol{y}}^{(r)}, \pi_X \right) + \frac{1}{2\sigma_e^2} \left\| \boldsymbol{y} - \mathcal{A}(\mathcal{G}_\theta(\boldsymbol{y})) \right\|_2^2 + c_1. \tag{10}$$

Here, c_1 is a constant independent of θ (see Proposition 1).

Bound on the Image Likelihood. The ELBO corresponding to X can be derived by treating X as the observed variable and the clean (synthetic) data $U = \mathcal{A}(X)$ as the latent variable.

Backward process: Here $U \sim \pi_U(\boldsymbol{u})$ and $(X|U = \boldsymbol{u}) \sim \mathcal{N} \left(\mathcal{G}_\theta(\boldsymbol{u}), \sigma_2^2 \, I \right)$ for given $\boldsymbol{u}$, with possibly $\sigma_2 \ll \sigma_1$.

Forward process: $X \sim \pi_X$ and $(U|X = \boldsymbol{x}) \sim \delta(\boldsymbol{u} - \mathcal{A}(\boldsymbol{x}))$ (Dirac measure concentrated at $\boldsymbol{u} = \mathcal{A}(\boldsymbol{x})$).

Proceeding similarly to the analysis used for deriving a bound on the data likelihood, we can show that (with the superscripts (f) and (b) indicating the forward and backward processes, respectively)

$$\log \pi_X^{(b)}(\boldsymbol{x}) \geq \mathbb{E}_{U \sim \pi_{U|X=\boldsymbol{x}}^{(f)}} \left[\log \left(\pi_{X|U}^{(b)}(\boldsymbol{x}|U) \right) \right] - \underbrace{\mathrm{KL} \left(\pi_{U|X=\boldsymbol{x}}^{(f)}, \pi_U \right)}_{\text{does not depend } \theta}. \tag{11}$$

Using the postulated distributions to simplify the first term in (11) leads to

$$- \log \pi_X^{(b)}(\boldsymbol{x}) \leq \frac{1}{2\sigma_2^2} \left\| \boldsymbol{x} - \mathcal{G}_\theta(\mathcal{A}(\boldsymbol{x})) \right\|_2^2 + \mathrm{KL} \left(\pi_{U|X=\boldsymbol{x}}^{(f)}, \pi_U \right). \tag{12}$$

Algorithm 1. Adversarial training of an iterative reconstruction network.

---

1. Input: Gradient penalty λ_{gp}, initial reconstruction network parameter θ and critic parameter α, batch-size n_b, Adam optimizer parameters (η, β_1, β_2), the number of $\mathcal{D}$ updates per $\mathcal{G}$ update (denoted as K), penalty parameters λ_{X} and λ_{Y}.

2. for mini-batches $m = 1, 2, \cdots$, **do (until convergence):**

  – Sample $x_j \sim \pi_X$, $y_j \sim \pi_Y$, and $\epsilon_j \sim$ uniform $[0, 1]$; for $1 \le j \le n_b$. Compute $x_j^{(\epsilon)} = \epsilon_j x_j + (1 - \epsilon_j) \mathcal{G}_\theta(y_j)$.

  – Critic loss: $\mathcal{L}_{\mathcal{D}} = \frac{1}{n_b} \sum_{j=1}^{n_b} \left[\mathcal{D}_\alpha(\mathcal{G}_\theta(y_j)) - \mathcal{D}_\alpha(x_j) + \lambda_{\text{gp}} \left(\left\| \nabla \mathcal{D}_\alpha \left(x_j^{(\epsilon)} \right) \right\|_2 - 1 \right)_+^2 \right]$.

  – **for** $k = 1, \cdots, K$, **do:** update critic as $\alpha \leftarrow \text{Adam}_{\eta, \beta_1, \beta_2} (\alpha, \nabla_\alpha \mathcal{L}_{\mathcal{D}})$.

  – Compute the loss for the reconstruction network for the current mini-batch:

$$\mathcal{L}_{\mathcal{G}} = \frac{1}{n_b} \sum_{j=1}^{n_b} \left[-\mathcal{D}_\alpha(\mathcal{G}_\theta(x_j)) + \lambda_{\text{X}} \|\mathcal{G}_\theta(\mathcal{A}(x_j)) - x_j\|_2^2 + \lambda_{\text{Y}} \|\mathcal{A}(\mathcal{G}_\theta(y_j)) - y_j\|_2^2 \right].$$

  – Update reconstruction network parameters: $\theta \leftarrow \text{Adam}_{\eta, \beta_1, \beta_2} (\theta, \nabla_\theta \mathcal{L}_{\mathcal{G}})$.

3. Output: The trained iterative reconstruction network $\mathcal{G}_\theta$.

Evidence Bound. The idea is now to combine (10) and (12). Then, observe that minimizing the so-called (negative) evidence bound on the overall negative log-likelihood in (7) can be phrased as follows:

$$\min_\theta \text{KL} \left(\pi_{X|Y=y}^{(r)}, \pi_X \right) + \frac{1}{2\,\sigma_e^2} \left\| y - \mathcal{A}\big(\mathcal{G}_\theta(y)\big) \right\|_2^2 + \frac{1}{2\sigma_2^2} \left\| x - \mathcal{G}_\theta(\mathcal{A}(x)) \right\|_2^2. \quad (13)$$

This is identical to minimizing the ALPD training loss in (2) but using the KL divergence instead of the Wasserstein-1 to quantify similarity in distribution.

Proposition 1. *Let $X \in \mathbb{R}^{d_x}$, $Y \in \mathbb{R}^{d_y}$, and let $\mathcal{A} \in \mathbb{R}^{d_y \times d_x}$ be a $d_y \times d_x$ matrix. Then, $\mathbb{E}_{X \sim \mathcal{N}(\mu, K)} \left[\|Y - \mathcal{A}X\|_2^2 \right] = \|Y - \mathcal{A}\mu\|_2^2 + \text{trace}\,(\mathcal{A}^\top \mathcal{A}K)$.*

Proof: Expanding the squared ℓ_2-norm, we have that

$$\mathbb{E}\left[\|Y - \mathcal{A}X\|_2^2 \right] = \mathbb{E}\left[Y^\top Y - 2Y^\top \mathcal{A}X + X^\top \mathcal{A}^\top \mathcal{A}X \right]. \quad (14)$$

Since the expectation is a linear operation, the expected value of the second term in (14) is $\left(-2Y^\top \mathcal{A}\mu \right)$. The expected value of the third term can be evaluated as

$$\mathbb{E}\left[X^\top \mathcal{A}^\top \mathcal{A}X \right] = \mathbb{E}\left[\text{trace}\,(\mathcal{A}^\top \mathcal{A}XX^\top) \right] = \text{trace}\,(\mathcal{A}^\top \mathcal{A}(K + \mu\mu^\top))$$
$$= \text{trace}\,(\mathcal{A}^\top \mathcal{A}K) + \mu^\top \mathcal{A}^\top \mathcal{A}\mu. \quad (15)$$

Substituting (15) in (14) leads to the desired result.   ∎

2.3   Parametrizing the Reconstruction and the Critic Networks

For parametrizing the reconstruction network $\mathcal{G}_\theta$, we adopt the same strategy as in [3], which is briefly explained here to make the exposition self-contained. The architecture of $\mathcal{G}_\theta$ is built upon the idea of *iterative unrolling*, the origin of which can be traced back to the seminal work by Gregor and LeCun [8] on learned sparse approximation. Specifically, our reconstruction network $\mathcal{G}_\theta$ is parametrized by unrolling the Chambolle-Pock (CP) algorithm [7] for non-smooth convex optimization. The CP algorithm is an iterative primal-dual scheme aimed at minimizing objectives of the form $f(\mathcal{K}x) + g(x)$, where $\mathcal{K}$ is a bounded linear operator, and g and f^* (the convex conjugate of f) are proper, convex, and lower semi-continuous. For convex $\mathcal{R}(x)$ in (1), a wide range of problems are solvable by the CP algorithm, the update rules of which are given by

$$h^{(\ell+1)} = \text{prox}_{\sigma f^*}(h^{(\ell)} + \sigma \mathcal{K}(\bar{x}^{(\ell)})), x^{(\ell+1)} = \text{prox}_{\tau g}(x^{(\ell)} - \tau \mathcal{K}^*(h^{(\ell+1)})),$$
$$\bar{x}^{(\ell+1)} = x^{(\ell+1)} + \gamma(x^{(\ell+1)} - x^{(\ell)}), \text{for} 0 \leq \ell \leq L - 1, \qquad (16)$$

starting from a suitable initial point $x^{(0)} = \bar{x}^{(0)}$. In order to construct an architecture for $\mathcal{G}_\theta$, we essentially replace the proximal operators in (16) by trainable convolutional neural networks (CNNs). More specifically, the output of $\mathcal{G}_\theta$ is computed by applying the following two steps repeatedly L times:

$$h^{(\ell+1)} = \Gamma_{\theta_{\text{d}}^{(\ell)}}(h^{(\ell)}, \sigma^{(\ell)} \mathcal{A}(x^\ell), y), \text{and} x^{(\ell+1)} = \Lambda_{\theta_{\text{p}}^{(\ell)}}(x^{(\ell)}, \tau^{(\ell)} \mathcal{A}^*(h^{\ell+1})),$$

The learnable parameters $\left\{ \theta_{\text{p}}^{(\ell)}, \theta_{\text{d}}^{(\ell)}, \sigma^{(\ell)}, \tau^{(\ell)} \right\}_{\ell-0}^{L-1}$ are denoted using the shorthand notation θ. The CNNs Γ and Λ are composed of a cascade of convolutional layers followed by a parametric ReLU activation. For the CT reconstruction experiment conducted in Sect. 3, we set $h^{(0)} = 0$, and take the initial estimate $x^{(0)}$ as the filtered back-projection (FBP) reconstruction. The number of layers is selected as $L = 15$ and the filters in Γ and Λ are taken to be of size 5×5 to increase the overall receptive field of the model to make it suitable for sparse-view CT. The critic $\mathcal{D}_\alpha$ (consisting of ~ 2.76 million parameters) is a simple feed-forward CNN with four cascaded modules; each consisting of a convolutional layer, an instance-normalization layer, and a leaky-ReLU activation with negative-slope 0.2; followed by a global average-pooling layer in the end.

3   Numerical Results

For numerical evaluation of the proposed approach, we consider the classical inverse problem of sparse-view CT reconstruction. First, we demonstrate a proof-of-concept using phantoms containing random ellipses of different intensities for training the networks. Subsequently, we present a comparative study of the proposed ALPD approach with state-of-the-art model- and data-driven reconstruction methods. Parallel-beam projection data along 200 uniformly spaced angular directions, with 400 lines/angle, are simulated using the ODL library [1] with a GPU-accelerated *astra* back-end. Subsequently, white Gaussian noise with a standard-deviation of $\sigma = 2.0$ is added to the projection data to simulate noisy measurements. For supervised training, the phantoms and their corresponding noisy parallel-beam projections are aligned, whereas they are shuffled for unsupervised learning to eliminate the pairing information.

The penalty parameters in (2) are selected as $\lambda_{\mathbb{X}} = \lambda_{\mathbb{Y}} = 10.0$, and the gradient penalty in Algorithm 1 is also taken as $\lambda_{\mathrm{gp}} = 10.0$. The parameters in the Adam optimizer for updating both $\mathcal{G}_\theta$ and $\mathcal{D}_\alpha$ are chosen as $(\eta, \beta_1, \beta_2) = (5 \times 10^{-5}, 0.50, 0.99)$. The same set of hyper-parameters are used for training on both phantoms and real CT images. The critic $\mathcal{D}_\alpha$ is updated once per $\mathcal{G}_\theta$ update and the batch-size is taken as one (i.e., $K = 1$ and $n_b = 1$ in Algorithm 1).

3.1   Training on Ellipse Phantoms

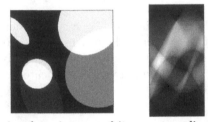

Example of training data, image and its corresponding projection data

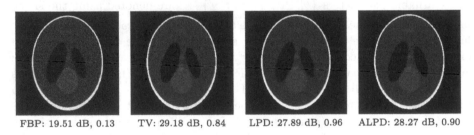

FBP: 19.51 dB, 0.13   TV: 29.18 dB, 0.84   LPD: 27.89 dB, 0.96   ALPD: 28.27 dB, 0.90

Fig. 1. Comparison of supervised and unsupervised training on the Shepp-Logan phantom. The PSNR (dB) and SSIM are indicated below the images. ALPD does a better job of alleviating over-smoothing, unlike its supervised variant (LPD).

In this experiment, we generate a set of 2000 2D phantoms, each of size 512×512 and containing 5 ellipses of random eccentricities at random locations and orientations, for training the networks. Each ellipse has an intensity value chosen uniformly at random in the range $[0.1, 1]$. The intensity of the background is taken as 0.0 and the intensities of the ellipses add up in the regions where they overlap. A representative phantom and its corresponding noisy sparse-view parallel-beam projection are shown in Figs. 1(a) and 1(b), respectively.

The main objective of this experiment is to study the differences between supervised and unsupervised learning in terms of their ability to reproduce images containing homogeneous regions separated by sharp edges. For performance evaluation, we consider reconstruction of the Shepp-Logan phantom which essentially consists of elliptical homogeneous regions delineated by sharp boundaries. Since the total-variation (TV) regularizer, which seeks sparsity in the gradient image, is tailor-made for such phantoms, we consider the reconstructed image produced by TV as the 'gold-standard' in this case and compare the proposed ALPD approach with its supervised counterpart vis-à-vis the TV reconstruction. To compute the TV solution, we use the ADMM-based solver in the ODL library with the penalty parameter $\lambda = 10.0$, which leads to the best reconstruction in our setting.

A visual comparison of the reconstructed images using LPD and ALPD (in Fig. 1) indicates that ALPD does a better job of recovering the three small tumors on the top region of the Shepp-Logan phantom. The ALPD reconstruction, although slightly inferior to TV in terms of PSNR, looks almost identical, while the supervised LPD reconstruction looks significantly blurry, making it difficult to discern the small tumors.

Table 1. Average performance of reconstruction methods. LPD has the best overall performance (in terms of PSNR and SSIM), but this reconstruction method needs to be trained against supervised data, i.e., pairs of high quality images (ground-truth) and corresponding noisy data. ALPD has slightly worse PSNR and SSIM values, but it can be trained against unsupervised data, which is vastly easier to get hold of as compared to supervised data. Note also that ALPD has significantly fewer parameters than AR, indicating that it can be trained against smaller datasets.

Method	PSNR (dB)	SSIM	# param.	Time (ms)
FBP	21.2866	0.2043	1	14.0
TV	30.3476	0.8110	1	21 315.0
Trained against supervised data				
FBP + U-Net	31.8008	0.7585	7 215 233	18.6
LPD	35.1561	0.9048	854 040	184.4
Trained against unsupervised data				
AR	33.6207	0.8750	19 347 890	41 058.3
ALPD	33.7386	0.8559	854 040	183.9

3.2 Sparse-View CT on Mayo-Clinic Data

We perform a comparison of the proposed ALPD method with competing model- and data-driven reconstruction techniques on human abdominal CT scans released by the Mayo Clinic for the low-dose CT grand challenge [15].

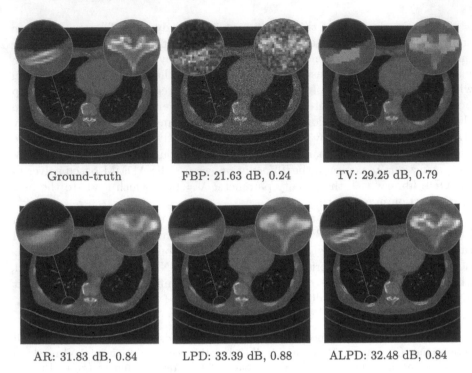

Ground-truth   FBP: 21.63 dB, 0.24   TV: 29.25 dB, 0.79

AR: 31.83 dB, 0.84   LPD: 33.39 dB, 0.88   ALPD: 32.48 dB, 0.84

Fig. 2. Comparison of ALPD with state-of-the-art model- and data-driven reconstruction methods on Mayo clinic data. The corresponding PSNR (dB) and SSIM are indicated below the images and the key differences in the reconstructed images are highlighted. The ALPD reconstruction is visibly sharper as compared to LPD, enabling easier identification of clinically important features.

The dataset consists of CT scans corresponding to 10 patients, from which we extract 2D slices of size 512 × 512 for our experiment. A total of 2250 slices extracted from 3D scans for 9 patients are used to train the networks in the data-driven methods, while 128 slices extracted from the scan for the remaining one patient are used for performance validation and comparison. The acquisition geometry and measurement noise distribution are kept the same as stated in Sect. 3.1. For the sake of bench-marking the performance, we consider two model-based techniques, namely the classical FBP and TV reconstruction. As two representative state-of-the-art data-driven methods, we consider adversarial regularization (AR) [14], and the LPD method [3] trained on paired data. The performance of a U-Net-based learned post-processing applied on FBP is reported in Table 1 along with the aforementioned techniques as a baseline for fully data-driven methods.

Similar to what we noted for the Shepp-Logan phantom, the ALPD reconstruction outperforms LPD in terms of recovering sharp boundaries in the images, thus facilitating better delineation of clinically important features (see Fig. 2).

In terms of PSNR and SSIM, ALPD performs slightly worse than LPD, but it outperforms other competing techniques both qualitatively and quantitatively, as seen from the average PSNR and SSIM values reported in Table 1. Notably, ALPD has the same reconstruction time as LPD, which is a couple of orders of magnitude lower than variational methods such as TV and AR that require computing iterative solutions to a high-dimensional optimization problem.

4   Conclusions

We proposed an unsupervised training protocol that learns a parametric reconstruction operator for solving imaging inverse problems from samples of the marginal distributions of the image and the measurement. The reconstruction operator is parametrized by an unrolled iterative scheme, namely the Chambolle-Pock method, originally developed for solving non-smooth convex optimization in [7] and subsequently adopted for network parametrization in the supervised learning framework in [3]. The proposed learning strategy, nevertheless, is not limited to the specific parametrization of the reconstruction operator chosen in this work and extends, in principle, to other iterative reconstruction schemes. Experimental evidence suggests that the proposed method does not suffer from the curse of over-smoothing as it minimizes a distortion measure in the distribution space instead of seeking pixel-wise proximity. Minimizing the Wasserstein-1 distance requires the introduction of a critic network, leading to a more resource-intensive training, which pays off in terms of superior performance and a more flexible training framework that it offers.

References

1. Adler, J., Kohr, H., Öktem, O.: Operator discretization library (ODL) (2017). Software https://github.com/odlgroup/odl
2. Adler, J., Öktem, O.: Solving ill-posed inverse problems using iterative deep neural networks. Inverse Probl. **33**(12), 1–24 (2017)
3. Adler, J., Öktem, O.: Learned primal-dual reconstruction. IEEE Trans. Med. Imaging **37**(6), 1322–1332 (2018)
4. Arjovsky, M., Chintala, S., Bottou, L.: Wasserstein GAN. arXiv:1701.07875v3, December 2017
5. Arridge, S., Maass, P., Öktem, O., Schönlieb, C.B.: Solving inverse problems using data-driven models. Acta Numerica **28**, 1–174 (2019)
6. Byeongsu, S., Gyutaek, O., Jeongsol, K., Chanyong, J., Ye, J.C.: Optimal transport driven CycleGAN for unsupervised learning in inverse problems. arXiv:1909.12116v4, August 2020
7. Chambolle, A., Pock, T.: A first-order primal-dual algorithm for convex problems with applications to imaging. J. Math. Imaging Vis. **40**(1), 120–145 (2010)
8. Gregor, K., LeCun, Y.: Learning fast approximations of sparse coding. In: International Conference on Machine Learning (2010)
9. Gulrajani1, I., Ahmed, F., Arjovsky, M., Dumoulin, V., Courville, A.: Improved training of Wasserstein GANs. arXiv:1704.00028v3, December 2017

10. Jin, K.H., McCann, M.T., Froustey, E., Unser, M.: Deep convolutional neural network for inverse problems in imaging. IEEE Trans. Image Process. **26**(9), 4509–4522 (2017)

11. Kobler, E., Effland, A., Kunisch, K., Pock, T.: Total deep variation for linear inverse problems. In: Proceedings of the IEEE Conference on Computer Vision and Pattern Recognition, pp. 7549–7558 (2020)

12. Li, H., Schwab, J., Antholzer, S., Haltmeier, M.: NETT: solving inverse problems with deep neural networks. arXiv:1803.00092v3, December 2019

13. Lin, J., Xia, Y., Qin, T., Chen, Z., Liu, T.: Conditional image-to-image translation. In: IEEE/CVF Conference on Computer Vision and Pattern Recognition, pp. 5524–5532 (2018)

14. Lunz, S., Öktem, O., Schönlieb, C.B.: Adversarial regularizers in inverse problems. In: Advances in Neural Information Processing Systems, pp. 8507–8516 (2018)

15. McCollough, C.: TFG-207a-04: overview of the low dose CT grand challenge. Med. Phys. **43**(6), 3759–3760 (2014)

16. Meinhardt, T., Moller, M., Hazirbas, C., Cremers, D.: Learning proximal operators: using denoising networks for regularizing inverse imaging problems. In: Proceedings of the IEEE International Conference on Computer Vision pp. 1781–1790 (2017)

17. Scherzer, O., Grasmair, M., Grossauer, H., Haltmeier, M., Lenzen, F.: Variational Methods in Imaging. AMS, vol. 167. Springer, Cham (2009). https://doi.org/10.1007/978-0-387-69277-7

18. Zhu, B., Liu, J.Z., Cauley, S.F., Rosen, B.R., Rosen, M.S.: Image reconstruction by domain-transform manifold learning. Nature **555**, 487–492 (2018)

19. Zhu, J.Y., Park, T., Isola, P., Efros, A.A.: Unpaired image-to-image translation using cycle-consistent adversarial networks. arXiv:1703.10593v7, August 2020

Towards Off-the-grid Algorithms for Total Variation Regularized Inverse Problems

Yohann De Castro[1], Vincent Duval[2,3], and Romain Petit[2,3(✉)]

[1] Institut Camille Jordan, CNRS UMR 5208, École Centrale de Lyon,
69134 Écully, France
[2] CEREMADE, CNRS, UMR 7534, Université Paris-Dauphine,
PSL University, 75016 Paris, France
`romain.petit@inria.fr`
[3] INRIA-Paris, MOKAPLAN, 75012 Paris, France

Abstract. We introduce an algorithm to solve linear inverse problems regularized with the total (gradient) variation in a gridless manner. Contrary to most existing methods, that produce an approximate solution which is piecewise constant on a fixed mesh, our approach exploits the structure of the solutions and consists in iteratively constructing a linear combination of indicator functions of simple polygons.

Keywords: Off-the-grid imaging · Inverse problems · Total variation

1 Introduction

By promoting solutions with a certain specific structure, the regularization of a variational inverse problem is a way to encode some prior knowledge on the signals to recover. Theoretically, it is now well understood which regularizers tend to promote signals or images which are sparse, low rank or piecewise constant. Yet, paradoxically enough, most numerical solvers are not designed with that goal in mind, and the targeted structural property (sparsity, low rank or piecewise constancy) only appears "in the limit", at the convergence of the algorithm.

Several recent works have focused on incorporating structural properties in optimization algorithms. In the context of ℓ^1-based sparse spikes recovery, it was proposed to switch from, e.g. standard proximal methods (which require the introduction of an approximation grid) to algorithms which operate directly in a continuous domain: interior point methods solving a sum-of-squares reformulation of the problem [7] or a Frank-Wolfe/conditional gradient algorithm [6] which approximates a solution in a greedy way. More generally, the conditional gradient algorithm has drawn a lot of interest from the data science community, for it provides iterates which are a sum of a small number of atoms which are promoted by the regularizer (see the review paper [11]).

In the present work, we explore the extension of these fruitful approaches to the total (gradient) variation regularized inverse problem

$$\min_{u \in L^2(\mathbb{R}^2)} T_\lambda(u) \stackrel{\text{def.}}{=} \frac{1}{2} \|\Phi u - y\|^2 + \lambda \, |Du| \, (\mathbb{R}^2) \tag{1}$$

© Springer Nature Switzerland AG 2021
A. Elmoataz et al. (Eds.): SSVM 2021, LNCS 12679, pp. 553–564, 2021.
https://doi.org/10.1007/978-3-030-75549-2_44

where $\Phi\colon L^2(\mathbb{R}^2) \to \mathbb{R}^m$ is a continuous linear map such that

$$\forall u \in L^2(\mathbb{R}^2), \ \Phi u = \int_{\mathbb{R}^2} u(x)\,\varphi(x)\,\mathrm{d}x \tag{2}$$

with $\varphi \in \left[L^2(\mathbb{R}^2)\right]^m \cap C^0(\mathbb{R}^2, \mathbb{R}^m)$ and $|Du|(\mathbb{R}^2)$ denotes the total variation of (the gradient of) u. Such variational problems have been widely used in image processing for the last decades, following the pioneering works of Rudin, Osher and Fatemi [15].

Many algorithms have been proposed to solve (1). With the notable exception of [17], most of them rely on the introduction of a fixed discrete grid, which yields reconstruction artifacts such as anisotropy or blur (see the experiments in [16]). On the other hand, it is known that some solutions of (1) are sums of a finite number of indicator functions of simply connected sets [4,5], yielding piecewise constant images. Our goal is to design an algorithm which does not suffer from some grid bias while providing solutions built from the above-mentioned atoms.

2   A Modified Frank-Wolfe Algorithm

In the spirit of [3,6,9] which introduced variants of the conditional gradient algorithm for sparse spikes recovery in a continuous domain, we derive a modified Frank-Wolfe algorithm allowing to iteratively solve (1) in a gridless manner.

2.1   Description

The Frank-Wolfe algorithm allows to minimize a convex differentiable function f over a weakly compact convex subset C of a Banach space. It is possible to recast problem (1) into that framework, e.g. by using an epigraphical lift (see [9, Sec. 4.1] for the case of the sparse spikes problem), or using a direct analysis (see [6]).

Each step of the algorithm consists in minimizing a linearization of f on C, and building the next iterate as a convex combination of the obtained point and the current iterate. In our case, the minimization of the linearized function is of the form

$$\inf_{u \in L^2(\mathbb{R}^2)} \int_{\mathbb{R}^2} \eta\,u \quad \text{s.t.} \quad |Du|\,(\mathbb{R}^2) \leq 1 \tag{3}$$

for an iteration-dependent function $\eta \in L^2(\mathbb{R}^2)$. Since the objective is linear and the total variation unit ball is convex and compact (in the weak $L^2(\mathbb{R}^2)$ topology), at least one of its extreme points is optimal. A result due to Fleming [10] (see also [2]) states that those extreme points are exactly the functions of the form $\pm 1_E/P(E)$ where $E \subseteq \mathbb{R}^2$ is a simple set (the measure-theoretic analog of simply connected sets) with $0 < |E| < +\infty$ and $P(E) < \infty$, and $P(E)$ denotes the perimeter of E. This means the linear minimization step can be carried out by finding a simple set solving the following geometric variational problem:

$$J(E) \stackrel{\text{def}}{=} \sup_{E \subseteq \mathbb{R}^2} \frac{\left|\int_E \eta\right|}{P(E)} \quad \text{s.t.} \quad 0 < P(E) < \infty. \tag{4}$$

As a result, the iterates produced by the algorithm may be constructed as linear combinations of indicators of simple sets. Since Problem (4) is reminiscent of the well-known Cheeger problem [13], which, given a domain $\Omega \subseteq \mathbb{R}^2$, consists in finding the subsets E of Ω minimizing the ratio $P(E)/|E|$, we refer to it as the "Cheeger problem" in the rest of the paper.

An important feature of the Frank-Wolfe algorithm is that, when building the next iterate, one can pick any admissible point that decreases the objective more than the standard update, without breaking convergence guarantees. Exploiting this, [3,6,9] have introduced "sliding steps" which help identify the sparse structure of the sought-after signal. Following the above-mentioned works, we derive an update step which provably decreases the objective more than the standard update, and derive Algorithm 1. As already mentioned, adding the sliding step

---

Algorithm 1: Modified Frank-Wolfe algorithm

Data: measurement operator Φ, obs. y, max. num. of iter. n, reg. param. λ

Result: function u^*

1   $u^{[0]} \leftarrow 0$;

2 **for** k *from* 0 *to* $n-1$ **do**

3     $\eta^{[k]} \leftarrow \frac{1}{\lambda} \Phi^* \left(y - \Phi u^{[k]} \right)$;

4     $E_*^{[k]} \leftarrow \underset{E}{\operatorname{argmax}} \ \frac{1}{P(E)} \left| \int_E \eta^{[k]} \right|$;

5   **if** $\left| \int_{E_*^{[k]}} \eta^{[k]} \right| \leq P(E_*^{[k]})$ **then**

6   output $u^* \leftarrow u^{[k]}$, which is optimal;

7   **else**

8        $E^{[k+1]} \leftarrow (E_1^{[k]}, ..., E_{N^{[k]}}^{[k]}, E_*^{[k]})$;

9        $a^{[k+1]} \leftarrow \underset{a \in \mathbb{R}^{N^{[k]}+1}}{\operatorname{argmin}} \ T_\lambda \left(\sum_{i=1}^{N^{[k]}+1} a_i \mathbf{1}_{E_i^{[k+1]}} \right)$;

10   remove atoms with zero amplitude, $N^{[k+1]} \leftarrow$ num. of atoms in $E^{[k+1]}$;

11   update $(E^{[k+1]}, a^{[k+1]})$ by performing a local descent on

         $(E, a) \mapsto T_\lambda \left(\sum_{i=1}^{N^{[k+1]}} a_i \mathbf{1}_{E_i} \right)$ initialized with $(E^{[k+1]}, a^{[k+1]})$;

12        $u^{[k+1]} \leftarrow \sum_{i=1}^{N^{[k+1]}} a_i^{[k+1]} \mathbf{1}_{E_i^{[k+1]}}$;

13   **end**

14 **end**

---

to the Frank-Wolfe algorithm does not break its convergence properties. In particular the following property holds (see [11]):

Proposition 1. *Let* $(u^{[k]})_{k \geq 0}$ *be a sequence produced by Algorithm 1. Then there exists* $C > 0$ *such that for any solution* u_* *of Problem* (1),

$$\forall k \in \mathbb{N}^*, \ T_\lambda(u^{[k]}) - T_\lambda(u_*) \leq \frac{C}{k}. \tag{5}$$

Moreover, $(u^{[k]})_{k \geq 0}$ has an accumulation point in the weak $L^2(\mathbb{R}^2)$ topology and that point is a solution of Problem (1).

The introduction of the sliding step (Line 11 of Algorithm 1), first proposed in [6], allows to considerably improve the convergence speed in practice (the variant proposed in [9], that we use here, even provides finite time convergence in the setting of the sparse spikes problem). It also produces sparser solutions: if the solution is expected to be a linear combination of a few indicator functions, removing the sliding step will typically produce iterates made of a much larger number of indicator functions, the majority of them correcting the crude approximations of the support of the solution made over the first iterations. The "local descent" mentioned at Line 11 is discussed in the next section.

2.2 Implementation

The implementation of Algorithm 1[1] requires two oracles to carry out the operations described on Lines 4 and 11: a first one that, given a weight function η, returns a solution of (4), and a second one that, given a set of initial amplitudes and atoms, returns another such set which is obtained by performing a local descent on the objective. Our approach for designing these oracles relies on polygonal approximations: we look for a maximizer of J among polygons, and perform the sliding step by finding a sequence of amplitudes and polygons so as to iteratively decrease the objective. This choice is mainly motivated by our goal to solve (1) "off-the-grid", which naturally leads us to consider purely Lagrangian methods which do not rely on the introduction of a pre-defined discrete grid.

Given an integer $n \geq 3$, we denote by $\mathcal{X}_n$ the set of all $x \in (\mathbb{R}^2)^n$ such that the polygon E_x defined by the list of vertices x is simple and counterclockwise oriented. We also define J_n by setting, for all $x \in \mathcal{X}_n$, $J_n(x) \stackrel{\text{def.}}{=} J(E_x)$. The area and the perimeter functionals, and hence J_n, are continuously differentiable on the open set $\mathcal{X}_n$.

Polygonal Approximation of Cheeger Sets. Our method for carrying out Line 4 in Algorithm 1 consists of several steps. First, we solve a discrete version of (3), where the minimization is performed over the set of piecewise constant functions on a given mesh. Then, we extract a level set of the solution, and obtain a simple polygon whose edges are located on the edges of the mesh. Finally, we use a first order method initialized with the previously obtained polygon to locally maximize J_n. Given a simple polygon $E_{x^t} = (x_1^t, ..., x_n^t)$ and a step size α^t, the next iterate is defined as the polygon $E_{x^{t+1}} = (x_1^{t+1}, ..., x_n^{t+1})$ such that[2]:

$$x_j^{t+1} \stackrel{\text{def.}}{=} x_j^t - \frac{\alpha^t}{P(E_{x^t})^2}\left(P(E_{x^t})(w_j^{t-}\nu_{j-1}^t + w_j^{t+}\nu_j^t) + \left[\int_{E_{x^t}} \eta\right](\tau_j^t - \tau_{j-1}^t)\right) \quad (6)$$

---

[1] A repository containing a complete implementation of Algorithm 1 can be found online at https://github.com/rpetit/tvsfw.

[2] Details on the first variations of perimeter and area can be found in [12, Sect. 17.3].

where, for all j, τ_j^t and ν_j^t are respectively the unit tangent and outer normal vectors on $[x_j^t, x_{j+1}^t]$ and

$$w_j^{t+} \stackrel{\text{def.}}{=} \int_{[x_j^t, x_{j+1}^t]} \eta(x) \frac{||x - x_{j+1}^t||}{||x_j^t - x_{j+1}^t||} d\mathcal{H}^1(x),$$

$$w_j^{t-} \stackrel{\text{def.}}{=} \int_{[x_j^t, x_{j-1}^t]} \eta(x) \frac{||x - x_{j-1}^t||}{||x_j^t - x_{j-1}^t||} d\mathcal{H}^1(x).$$

We compute the weights w_j^{t+}, w_j^{t-} and the integral of η over E_{x^t} using numerical integration methods. We found that integrating the weight function η on each triangle of a sufficiently fine triangulation of E_{x^t} yields good numerical results (this triangulation must be updated at each iteration, and sometimes re-computed from scratch to avoid the presence of ill-shaped triangles).

Comments. Two potential concerns regarding the above procedure are whether iterates remain simple polygons and whether it converges to a global maximizer of J_n. As the mesh used to solve the discrete version of (3) gets finer, the level sets of any of the solutions can be guaranteed to be arbitrarily close (in terms of the Lebesgue measure of the symmetric difference) to a solution of (4) (see [8]).

We could not prove that the iterates remain simple polygons along the process, but since the initial polygon can be taken arbitrarily close to a simple set solving (4), we do not expect nor observe any change of topology during the optimization.

Moreover, even if the non-concavity of J_n makes its maximization difficult, the above initialization allows us to start our local descent with a polygon that hopefully lies in the basin of attraction of a global maximizer. However, let us stress that at Line 4 of Algorithm 1, we only need to find a set with near optimal energy J in (4) (see [11, Algorithm 2]).

An interesting problem is to quantify the distance (e.g. in the Haussdorff sense) of a maximizer of J_n to a maximizer of J. We discuss in Sect. 4 the simpler case of radial measurements. In the general case, if the sequence of polygons defined above converges to a simple polygon E_x, then E_x is such that

$$w_j^+ = w_j^- = \frac{\int_{E_x} \eta}{P(E_x)} \tan\left(\frac{\theta_j}{2}\right) \tag{7}$$

for all j, where θ_j is the angle between $x_j - x_{j-1}$ and $x_{j+1} - x_j$ (the j-th exterior angle of the polygon). This can be seen as a discrete version of the following first order optimality condition for solutions of (4):

$$\eta = \frac{\int_E \eta}{P(E)} H_E \text{ on } \partial^* E \tag{8}$$

where H_E denotes the distributional curvature of E. Note that (8) is similar to the optimality condition for the classical Cheeger problem (i.e. with $\eta = 1$ and the additional constraint $E \subseteq \Omega$), namely $H_E = P(E)/|E|$ in the free boundary of E (see [1] or [13, Prop. 2.4]).

Sliding Step. The implementation of the local descent (Line 11 in Algorithm 1) is similar to what is described above for refining crude approximations of Cheeger sets. We use a first order method to minimize the function that maps a list of amplitudes and a list of polygons (each seen as a list of vertices) to the objective value of the linear combination of indicator functions they define. Given a step size α^t, N polygons $E_{x_1^t}, ..., E_{x_N^t}$, and a vector of amplitudes $a^t \in \mathbb{R}^N$, we set $u^t \overset{\text{def.}}{=} \sum_{i=1}^N a_i^t \mathbf{1}_{E_{x_i^t}}$, and perform the following update:

$$a_i^{t+1} \overset{\text{def.}}{=} a_i^t - \alpha^t \left[\left\langle \Phi \mathbf{1}_{E_{x_i^t}}, \Phi u^t - y \right\rangle + \lambda P \left(E_{x_i^t} \right) \operatorname{sign}\left(a_i^t \right) \right]$$

$$x_{i,j}^{t+1} \overset{\text{def.}}{=} x_{i,j}^t - \alpha^t a_i^t \left[\quad \left\langle \Phi u^t - y, w_{i,j}^{t-} \right\rangle \nu_{i,j-1}^t \right.$$
$$\left. + \left\langle \Phi u^t - y, w_{i,j}^{t+} \right\rangle \nu_{i,j}^t - \lambda \operatorname{sign}(a_i^t) \left(\tau_{i,j}^t - \tau_{i,j-1}^t \right) \right]$$

where $\tau_{i,j}^t$, $\nu_{i,j}^t$ are respectively the unit tangent and outer normal vectors on the edge $[x_{i,j}^t, x_{i,j+1}^t]$ and

$$w_{i,j}^{t+} \overset{\text{def.}}{=} \int_{[x_{i,j}^t, x_{i,j+1}^t]} \varphi(x) \frac{\|x - x_{i,j+1}^t\|}{\|x_{i,j}^t - x_{i,j+1}^t\|} d\mathcal{H}^1(x),$$

$$w_{i,j}^{t-} \overset{\text{def.}}{=} \int_{[x_{i,j}^t, x_{i,j-1}^t]} \varphi(x) \frac{\|x - x_{i,j-1}^t\|}{\|x_{i,j}^t - x_{i,j-1}^t\|} d\mathcal{H}^1(x).$$

Comments. Unlike the local descent we perform to approximate Cheeger sets, the sliding step may tend to induce topology changes (see Sect. 3.2 for an example). Typically, a simple set may tend to split in two simple sets over the course of the descent. This is a major difference (and challenge) compared to the sliding steps used in sparse spikes recovery (where the optimization is carried out over the space of Radon measures) [3,6,9]. This phenomenon is closely linked to topological properties of the faces of the total gradient variation unit ball: its extreme points do not form a closed set for any reasonable topology (e.g. the weak $L^2(\mathbb{R}^2)$ topology), nor do its faces of dimension $d \le k$ for any $k \in \mathbb{N}$. As a result, when moving continuously on the set of faces of dimension $d = k$, it is possible to "stumble upon" a point which only belongs to a face of dimension $d > k$.

Let us stress that these dimension changes have a clear interpretation in terms of the topology of the sets which evolve through the descent: they typically correspond to a splitting. Our current implementation does not allow to handle them in a consistent way, and finding a way to deal with them "off-the-grid" is the subject of an ongoing work.

It is important to note that not allowing topological changes during the descent is not an issue, since all convergence guarantees of Algorithm 1 are preserved as soon as the output of the sliding step decreases the energy more than the standard Frank-Wolfe update. One can stop the local descent at any point before any change of topology occurs, which avoids having to treat them. Still, in order to yield iterates that are as sparse as possible (and probably to decrease the objective as quickly as possible), it seems preferable to handle topological changes.

3 Numerical Experiments

3.1 Recovery Examples

Here, we investigate the practical performance of Algorithm 1. We focus on the case where Φ is a sampled Gaussian convolution operator i.e.:

$$\forall x \in \mathbb{R}^2, \ \varphi(x) = \left(\exp\left(-\frac{\|x - x_i\|^2}{2\sigma^2} \right) \right)_{i \in \{1,\dots,m\}}$$

for a given $\sigma > 0$ and a sampling grid $(x_i)_{i \in \{1,\dots,m\}}$. The noise is drawn from a multivariate Gaussian with zero mean and isotropic covariance matrix $\tau^2 I_m$. We take the regularization parameter λ of the order of $\sqrt{2 \log(m) \tau^2}$.

Numerically certifying that a given function is an approximate solution of (1) is difficult. However, as the sampling grid becomes finer, Φ tends to the convolution with the Gaussian kernel, which is injective. Relying on a Γ-convergence argument, one may expect that if u_0 is a piecewise constant image and w is some small additive noise, the solutions of (1) with $y = \Phi u_0 + w$ are close to u_0, modulo the regularization effects of the total variation.

Our first experiment consists in recovering a function u_0 that is a linear combination of three indicator functions (see Fig. 1). During each of the three iterations required to obtain a good approximation of u_0, a new atom is added to its support. One can see the sliding step (Line 11) is crucial: the large atom on the left, added during the second iteration, is significantly refined during the sliding step of the third iteration, when enough atoms have been introduced.

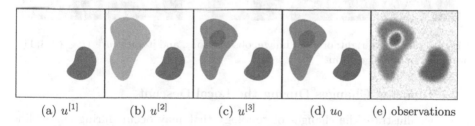

(a) $u^{[1]}$ (b) $u^{[2]}$ (c) $u^{[3]}$ (d) u_0 (e) observations

Fig. 1. First experiment ($u^{[k]}$ is the k-th iterate produced by Algorithm 1).

The second experiment (see Fig. 2) consists in recovering the indicator function of a set with a hole (which can also be seen as the sum of two indicator functions of simple sets). The support of u_0 and its gradient are accurately estimated. Still, the typical effects of total (gradient) variation regularization are noticeable: corners are slightly rounded, and there is a "loss of contrast" in the eye of the pacman.

The third experiment (Fig. 3) also showcases the rounding of corners, and highlights the influence of the regularization parameter: as λ decreases, the curvature of the edge set increases.

Finally, we provide in Fig. 4 the results of an experiment on a more challenging task, which consists in reconstructing a natural grayscale image.

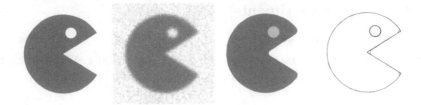

Fig. 2. From left to right: unknown function, observations, final estimate, gradients support (red: estimate, black: unknown) (Color figure online)

Fig. 3. Left: unknown function, middle: observations, right: output of the algorithm for different values of λ

Fig. 4. From left to right: original image, observations, and iterates $u^{[k]}$ ($k \in \{2, 4, 11\}$) produced by the algorithm

3.2 Topology Changes During the Local Descent

Here, we illustrate the changes of topology that may occur during the sliding step (Line 11 of Algorithm 1). All relevant plots are given in Fig. 5. The unknown function (see (a)) is the sum of two indicator functions:

$$u_0 = \mathbf{1}_{B((-1,0),0.6)} + \mathbf{1}_{B((1,0),0.6)}$$

and observations are shown in (b). The Cheeger set computed at Line 4 of the first iteration covers the two disks (see (c)).

In this setting, our implementation of the local descent (Line 11) converges to a function similar to (f)[3], and we obtain a valid update that decreases the objective more than the standard Frank-Wolfe update. The next iteration of the algorithm will then consist in adding a new atom to the approximation, with negative amplitude, so as to compensate for the presence of the small bottleneck.

---

[3] This only occurs when λ is small enough. For higher values of λ, the output is similar to (d) or (e).

However, it seems natural that the support of (f) should split into two disjoint simple sets, which is not possible with our current implementation. To investigate what would happen in this case, we manually split the two sets (see (g)) and let them evolve independently. The support of the approximation converges to the union of the two disks, which produces an update that decreases the objective even more than (f).

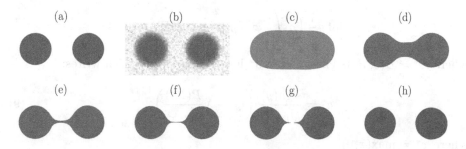

Fig. 5. Topology change experiment. (a): unknown signal, (b): observations, (c): weighted Cheeger set, (d, e, f, g): sliding step iterations (with splitting), (h): final function.

4   The Case of a Single Radial Measurement

In this section, we study a particular setting, where the number of observations m is equal to 1, and the unique sensing function is radial. To be more precise we assume that the measurement operator is given by (2) with $\varphi : \mathbb{R}^2 \to \mathbb{R}_+^*$ a positive radial function with $\varphi(x) = \tilde{\varphi}(||x||)$. We may invoke the following assumption:

Assumption 1. The function $f : r \mapsto r\tilde{\varphi}(r)$ is continuously differentiable on $\mathbb{R}_+^*$, there exists $\rho_0 > 0$ s.t. $f'(r) > 0$ on $]0, \rho_0[$, $f'(r) < 0$ on $]\rho_0, +\infty[$, and $rf(r) \to 0$ when $r \to +\infty$.[4]

We first explain what each step of Algorithm 1 should theoretically return, without worrying about approximations made for implementation matters. Then, we compare those with the output at each step of the practical algorithm. The proofs of all results, which are omitted here, will be given in a longer version of this work in preparation.

4.1   Theoretical Behavior of the Algorithm

The first step of Algorithm 1 consists in solving the Cheeger problem (4) associated to $\eta \overset{\text{def}}{=} \frac{1}{\lambda}\Phi^* y = \frac{y}{\lambda}\varphi$ (or equivalently to φ). Using the Steiner symmetrization principle, and denoting by $B(0, R)$ the disk of radius R centered at the origin, we may state:

---

[4] Assumption 1 is for example satisfied by $\varphi : x \mapsto \exp\left(-||x||^2/(2\sigma^2)\right)$ for any $\sigma > 0$.

Proposition 2. *All the solutions of the Cheeger problem* (4) *associated to* $\eta \overset{def}{=} \varphi$ *are disks centered at the origin. Under Assumption 1 the unique solution is the disk* $B(0, R^*)$ *with* R^* *the unique maximizer of* $R \mapsto \left[\int_0^R r\,\tilde{\varphi}(r)\,dr\right]/R.$

The second step (Line 9) of the algorithm then consists in solving

$$\inf_{a\in\mathbb{R}} \frac{1}{2}\left(a\int_{E^*}\varphi - y\right)^2 + \lambda\,P(E^*)\,|a|$$

where $E^* = B(0, R^*)$. The solution a^* has a closed form which writes:

$$a^* = \frac{\text{sign}(y)}{\int_{E^*}\varphi}\left(|y| - \lambda\frac{P(E^*)}{\int_{E^*}\varphi}\right)^+ \tag{9}$$

where $x^+ = \max(x, 0)$.

The next step should be the sliding one (Line 11). However, in this specific setting, one can show that the constructed function is already optimal, as stated by the following proposition:

Proposition 3. *Under Assumption 1, Problem* (1) *has a unique solution, which is of the form* $a^*\mathbf{1}_{E^*}$ *with* $E^* = B(0, R^*)$ *the solution of the Cheeger problem given by Proposition 2, and* a^* *given by* (9).

To summarize, with a single observation and a radial sensing function, a solution is found in a single iteration, and its support is directly identified by solving the Cheeger problem.

4.2   Study of Implementation Approximations

In practice, instead of solving (4), we look for an n-gon minimizing J, for some given integer n. Investigating the proximity of this n-gon with $B(0, R^*)$ is hence natural. Solving classical geometric variational problems restricted to the set of n-gons is more involved, as the Steiner symmetrization procedure does not preserve n-gons in general [14, Sect. 7.4]. However, using a trick from Pólya and Szegö, one may prove:

Proposition 4. *Let* $n \in \{3, 4\}$. *Then all the maximizers of* J_n *are regular n-gons inscribed in a circle centered at the origin.*

Our proof does not extend to $n \geq 5$, but the following conjecture is natural:

Conjecture 1. The result of Prop. 4 holds for all $n \geq 3$.

For $n \in \{3, 4\}$ or if Conjecture 1 holds, it remains to compare the optimal n-gon with $B(0, R^*)$. If we define $\mathcal{J}(R) \overset{\text{def.}}{=} J(B(0, R))$ and $\mathcal{J}_n(R)$ the value of J at any regular n-gon inscribed in a circle of radius R centered at the origin, then we can state the following:

Proposition 5. *Under Assumption 1, we have that* $||\mathcal{J}_n - \mathcal{J}||_\infty = O\left(\frac{1}{n^2}\right)$. *Moreover, if f is of class C^2 and $f''(\rho_0) < 0$, then for n large enough, there exists a unique maximizer R_n^* of $\mathcal{J}_n$, and $|R_n^* - R^*| = O\left(\frac{1}{n}\right)$. If φ is the function defined by $\varphi : x \mapsto exp\left(-||x||^2/(2\sigma^2)\right)$, then this last result holds for all $n \geq 3$.*

Now, the output of our method for approximating Cheeger sets, described in Sect. 2.2, is a polygon that is obtained by locally maximizing J_n using a first order method. Even if we carefully initialize this first order method, the possible existence of non-optimal critical points makes its analysis challenging. However, in our setting (a radial weight function), the critical points of J_n coincide with the global maximizers (at least for small n).

Proposition 6. *Let $n \in \{3,4\}$. Under Assumption 1, if f is of class C^2 and $f''(\rho_0) < 0$, the critical points of J_n are the regular n-gons inscribed in the circle of radius R_n^* centered at the origin.*

We make the following conjecture:

Conjecture 2. The result of Proposition 6 holds for all $n \geq 3$.

If $n \in \{3, 4\}$, or if Conjecture 2 holds, we may therefore expect our polygonal approximation to be at Hausdorff distance of order $O\left(\frac{1}{n}\right)$ to $B(0, R^*)$.

5   Conclusion

As shown in the present exploratory work, solving total variation regularized inverse problems in a gridless manner is highly beneficial, as it allows to preserve structural properties of their solutions, which cannot be achieved by traditional numerical solvers. The price to pay for going "off-the-grid" is an increased complexity of the analysis and the implementation of the algorithms. Furthering their theoretical study and improving their practical efficiency and reliability is an interesting avenue for future research.

Acknowledgments. This work was supported by a grant from Région Ile-De-France and by the ANR CIPRESSI project, grant ANR-19-CE48-0017-01 of the French Agence Nationale de la Recherche. RP warmly thanks the owners of Villa Margely for their hospitality during the writing of this paper.

References

1. Alter, F., Caselles, V., Chambolle, A.: Evolution of characteristic functions of convex sets in the plane by the minimizing total variation flow. Interfaces Free Boundaries **7**(1), 29–53 (2005). https://doi.org/10.4171/IFB/112
2. Ambrosio, L., Caselles, V., Masnou, S., Morel, J.M.: Connected components of sets of finite perimeter and applications to image processing. J. Eur. Math. Soc. **3**(1), 39–92 (2001). https://doi.org/10.1007/PL00011302
3. Boyd, N., Schiebinger, G., Recht, B.: The alternating descent conditional gradient method for sparse inverse problems. SIAM J. Optim. **27**(2), 616–639 (2017). https://doi.org/10.1137/15M1035793

4. Boyer, C., Chambolle, A., De Castro, Y., Duval, V., de Gournay, F., Weiss, P.: On representer theorems and convex regularization. SIAM J. Optim. **29**(2), 1260–1281 (2019). https://doi.org/10.1137/18M1200750

5. Bredies, K., Carioni, M.: Sparsity of solutions for variational inverse problems with finite-dimensional data. Calc. Var. Partial. Differ. Equ. **59**(1), 1–26 (2019). https://doi.org/10.1007/s00526-019-1658-1

6. Bredies, K., Pikkarainen, H.K.: Inverse problems in spaces of measures. ESAIM Control Optim. Calc. Var. **19**(1), 190–218 (2013). https://doi.org/10.1051/cocv/2011205

7. Candès, E.J., Fernandez-Granda, C.: Towards a mathematical theory of super-resolution. Commun. Pure Appl. Math. **67**(6), 906–956 (2014). https://doi.org/10.1002/cpa.21455

8. Carlier, G., Comte, M., Peyré, G.: Approximation of maximal Cheeger sets by projection. ESAIM Math. Model. Numer. Anal. **43**(1), 139–150 (2009). https://doi.org/10.1051/m2an/2008040

9. Denoyelle, Q., Duval, V., Peyre, G., Soubies, E.: The sliding Frank-Wolfe algorithm and its application to super-resolution microscopy. Inverse Prob. (2019). https://doi.org/10.1088/1361-6420/ab2a29

10. Fleming, W.H.: Functions with generalized gradient and generalized surfaces. Annali di Matematica **44**(1), 93–103 (1978). https://doi.org/10.1007/BF02415193

11. Jaggi, M.: Revisiting Frank-Wolfe: projection-free sparse convex optimization. In: International Conference on Machine Learning, pp. 427–435. PMLR, February 2013

12. Maggi, F.: Sets of Finite Perimeter and Geometric Variational Problems: An Introduction to Geometric Measure Theory. Cambridge Studies in Advanced Mathematics, Cambridge University Press, Cambridge (2012). https://doi.org/10.1017/CBO9781139108133

13. Parini, E.: An introduction to the Cheeger problem. Surv. Math. Appl. **6**, 9–21 (2011)

14. Pólya, G., Szegö, G.: Isoperimetric Inequalities in Mathematical Physics. No. 27 in Annals of Mathematics Studies, Princeton University Press (1951)

15. Rudin, L.I., Osher, S., Fatemi, E.: Nonlinear total variation based noise removal algorithms. Physica D **60**(1), 259–268 (1992). https://doi.org/10.1016/0167-2789(92)90242-F

16. Tabti, S., Rabin, J., Elmoata, A.: Symmetric upwind scheme for discrete weighted total variation. In: 2018 IEEE International Conference on Acoustics, Speech and Signal Processing (ICASSP), pp. 1827–1831, April 2018. https://doi.org/10.1109/ICASSP.2018.8461736

17. Viola, F., Fitzgibbon, A., Cipolla, R.: A unifying resolution-independent formulation for early vision. In: 2012 IEEE Conference on Computer Vision and Pattern Recognition, pp. 494–501, June 2012. https://doi.org/10.1109/CVPR.2012.6247713

Multi-frame Super-Resolution
from Noisy Data

Kireeti Bodduna[✉], Joachim Weickert, and Marcelo Cárdenas

Mathematical Image Analysis Group, Faculty of Mathematics and Computer Science,
Saarland University, 66041 Saarbrücken, Germany
{bodduna,weickert,cardenas}@mia.uni-saarland.de

Abstract. Obtaining high resolution images from low resolution data with clipped noise is algorithmically challenging due to the ill-posed nature of the problem. So far such problems have hardly been tackled, and the few existing approaches use simplistic regularisers. We show the usefulness of two adaptive regularisers based on anisotropic diffusion ideas: Apart from evaluating the classical edge-enhancing anisotropic diffusion regulariser, we introduce a novel non-local one with one-sided differences and superior performance. It is termed sector diffusion. We combine it with all six variants of the classical super-resolution observational model that arise from permutations of its three operators for warping, blurring, and downsampling. Surprisingly, the evaluation in a practically relevant noisy scenario produces a different ranking than the one in the noise-free setting in our previous work (SSVM 2017).

Keywords: Super-resolution · Denoising · Anisotropic diffusion ·
Non-local methods · One-sided derivatives

1 Introduction

Super-resolution (SR) is an image processing technique designed to overcome the resolution limits of cameras. Generating a high resolution (HR) image from one single low resolution (LR) image is referred to as single-frame super-resolution [7,15,24]. In this work, we concentrate on multi-frame super-resolution, where information from multiple LR images is fused into a single HR image [4,6,8–10,12–18,25].

In bio-medical and bio-physical applications we encounter images that possess a significant amount of noise. Multi-frame super-resolution in the presence of noise is thus practically relevant and also a very challenging research field. Algorithms that are designed to solve this problem compute derivative information on noisy data which showcases the ill-posed nature of the problem. In view of these algorithmic challenges, it is not surprising that very little efforts have been put into obtaining high resolution images from noisy low resolution data.

Deep learning-based methods are less suitable for bio-physical applications like electron microscopic imaging due to three main reasons: Firstly, there is

© Springer Nature Switzerland AG 2021
A. Elmoataz et al. (Eds.): SSVM 2021, LNCS 12679, pp. 565–577, 2021.
https://doi.org/10.1007/978-3-030-75549-2_45

very little ground truth data available. Secondly, the raw noise type is not well understood, unlike normal cameras. Finally, this imaging pipeline employs a huge amount of steps to obtain the final structure of the specimen under observation. After each one of these steps, the noise type changes. This makes deep-learning models that are specifically trained for a particular kind of noise, sub-optimal.

In the present paper we specifically aim at reconstructing a HR image from its LR versions that have been corrupted by additive white Gaussian noise (AWGN).

Formalisation of the Problem. For multi-frame super-resolution, we want to find a HR image u of resolution $N_H = H_1 \times H_2$ from N low resolution images $\{f_i\}_{i=1}^{N}$ of resolution $N_L = L_1 \times L_2$. The low resolution images are assumed to be degraded versions of the real world HR scene. The standard formulation of the relation between the high resolution scene and its LR realisations is [5]

$$f_i = DBW_iu + e_i. \tag{1}$$

In this observational model, we express the motion of the objects in the image using the warping operator W_i (size: $N_H \times N_H$). The operator B (size: $N_H \times N_H$) denotes the blur due to the point spread function of the camera. We represent the downsampling of the HR scene by the camera detector system using D (size: $N_L \times N_H$). The vector e_i depicts the noise (error) acquired due to the imaging system. The operators B and D do not have an index i as we assume the same camera conditions for all images.

The standard model (1), however, has a disadvantage: The operator W_i acts on the high-resolution scene u. Hence, the model assumes that we have motion information at the high-resolution scale. In practice, we just have the downsampled and blurred images f_i at our disposal. Motion at high-resolution must be approximated by upsampling the one computed on a lower resolution. Thus, the following question arises: Can one improve the practical performance of the SR approach by permuting the order of the operators? The seminal work of Wang and Qi [19] and the paper by Bodduna and Weickert [1] made progress in this direction. In [1], the authors tried to evaluate the SR observational model in a noise-free scenario by studying the six different alternatives that arise from permutations of its three operators D, B and W_i. This has led to improvements in terms of both quality and speed, the former of which was also observed in [19]. However, such an evaluation is missing for the practically relevant scenario with noisy images. Our paper will address this problem.

Moreover, there is a second problem: For super-resolution of noisy data, the ideal observational model in (1) should be stabilised with a regulariser. In most cases, this is done by embedding it into the following quadratic energy minimisation framework:

$$E(u) = \frac{1}{2} \sum_{i=1}^{N} |DBW_iu - f_i|^2 + \frac{1}{2}\alpha|Au|^2. \tag{2}$$

Here, A is a discrete approximation of the continuous gradient operator, α is the regularisation constant, and $|\cdot|$ denotes the Euclidean norm. The first term is the data term that encapsulates the observational model. The second one serves as smoothness term which eliminates noise. Minimising (2) by setting its gradient to zero gives

$$\sum_{i=1}^{N} \boldsymbol{W}_i^{\top} \boldsymbol{B}^{\top} \boldsymbol{D}^{\top} (\boldsymbol{DBW}_i \boldsymbol{u} - \boldsymbol{f}_i) - \alpha \boldsymbol{A}_{\mathrm{HD}} \boldsymbol{u} = \boldsymbol{0}, \tag{3}$$

where $\boldsymbol{A}_{\mathrm{HD}} = \boldsymbol{A}^{\top} \boldsymbol{A}$ is the discrete approximation of the continuous Laplacian operator. In this paper, we use a Gaussian blur kernel, such that $\boldsymbol{B}^{\top}$ equals $\boldsymbol{B}$. We denote the upsampling and downsampling matrices by $\boldsymbol{D}^{\top}$ and $\boldsymbol{D}$, respectively. The operator $\boldsymbol{W}_i$ represents forward warping, while $\boldsymbol{W}_i^{\top}$ encodes backward registration. The explicit gradient descent scheme with parameters τ (the time step size) and k_{max} (the number of iterations) to solve Eq. (3) is given by

$$\boldsymbol{u}^{k+1} = \boldsymbol{u}^k + \tau \Big(\alpha \boldsymbol{A}_{\mathrm{HD}} \boldsymbol{u}^k - \sum_{i=1}^{N} \boldsymbol{W}_i^{\top} \boldsymbol{B}^{\top} \boldsymbol{D}^{\top} (\boldsymbol{DBW}_i \boldsymbol{u}^k - \boldsymbol{f}_i) \Big). \tag{4}$$

In this evolution equation, $\boldsymbol{A}_{\mathrm{HD}}$ acts as the denoiser. However, such a noise elimination scheme uses a simple homogeneous diffusion process that also blurs important structures. As far as the usage of diffusion-based regularisers for super-resolution is concerned, only a few papers with simplistic models are available. Thus, it is highly desirable to introduce more advanced structure preserving regularisers. This is our second challenge.

Our Contribution. To address the first problem, we investigate the performance of all permutations of the standard observational model in an AWGN setting that is clipped to the dynamic range $[0, 255]$. This practically relevant noise model covers over- and under-exposed image acquisition conditions.

To incorporate structure-preserving regularisers, we start with replacing the homogeneous diffusion operator by the classical model of edge-enhancing anisotropic diffusion (EED) [21]. Although this model is around since a long time, its performance for super-resolution has not been examined so far. Moreover, we also introduce a more sophisticated non-local anisotropic model that offers better structure preservation and superior noise elimination than EED. We call it sector diffusion (SD). It differs from all other diffusion models by the fact that it is fully based on one-sided derivatives.

We first compare the denoising performance of EED and SD for real-world images with clipped-AWGN, before we embed them as regularisers in the SR framework. We deliberately do not evaluate popular denoising methods such as 3D block matching [3] and non-local Bayes [11]: Most of these techniques rely heavily on a correct noise model, which renders them inferior for clipped noise, in particular with large amplitudes.

Paper Structure. Our paper is organised as follows: We introduce our novel sector diffusion model in Sect. 2. Here, we also review various super-resolution

Table 1. Left: The seven SR observational models. **Right:** Parameter settings for optical flow calculation. We have two model parameters: α_{OF} (smoothness parameter) and σ_{OF} (Gaussian pre-smoothing). Numerical parameters are chosen as $\eta = 0.95$ (downsampling factor), $\eta_1 = 10$ (inner fixed point iterations), $\eta_2 = 10$ (outer fixed point iterations) and $\omega = 1.95$ (successive over-relaxation parameter).

Model	Equation	Dataset	σ_{OF}	α_{OF}
M1	$DBW_iu + e_i = f_i$	Text1	2.6	13.3
M2	$DW_iBu + e_i = f_i$	Text2	1.0	15.6
M3	$BDW_iu + e_i = f_i$	Text3	2.3	6.3
M4	$W_iDBu + e_i = f_i$	House1	3.8	13.5
M5	$BW_iDu + e_i = f_i$	House2	1.2	17.0
M6	$W_iBDu + e_i = f_i$	House3	2.7	16.5
M2.1	$Bu + e_i = D^\top W_i^\top f_i$			

observational models and the EED-based image evolution equation. In Sect. 3, we present several denoising and SR reconstruction experiments along with some discussions. Finally, in Sect. 4 we conclude with a summary about robust multi-frame SR reconstruction as well as an outlook on future work.

2   Modeling and Theory

In this section, we first review the various possible permutations of the super-resolution observational model in (1). Afterwards, we introduce the different regularisation schemes utilised for both denoising and SR reconstruction purposes.

2.1   Super-Resolution Observational Models

Table 1 shows the various permutations of the original observational model M1 [1]. While models M2-M6 depict the five other possible permutations, M2.1 represents a technique that is derived from M2. The motivation behind the modelling of M1-M6 is quality reasons. M2.1, on the other hand, is designed to exploit the precomputable nature of the term on the right hand side of the corresponding equation. Such a design is faster than any of the other models.

2.2   Edge-Enhancing Diffusion

Edge-enhancing diffusion was proposed by Weickert [21] with the goal to enhance smoothing along edges while inhibiting it across them. To achieve this, one designs a diffusion tensor D with eigenvectors v_1 and v_2 that are parallel and perpendicular to a Gaussian smoothed image gradient. This is followed by setting the eigenvalue corresponding to the eigenvector perpendicular to the gradient to one, indicating full flow. The eigenvalue corresponding to the eigenvector parallel to the gradient is determined by a diffusivity function. Using this idea, one can inhibit smoothing across edges. The following is the continuous mathematical formulation of the evolution of image u under EED:

$$\partial_t u = \text{div}(D(\nabla u_\sigma)\nabla u), \tag{5}$$

$$D(\nabla u_\sigma) = g(|\nabla u_\sigma|^2) \cdot v_1 v_1^T + 1 \cdot v_2 v_2^T, \tag{6}$$

$$v_1 \parallel \nabla u_\sigma, \quad |v_1| = 1 \quad \text{and} \quad v_2 \perp \nabla u_\sigma, \quad |v_2| = 1. \tag{7}$$

Here, div is the 2D divergence operator and ∇u the spatial gradient. The Gaussian-smoothed image is u_σ. Computing the gradient on u_σ makes the diffusion process robust under the presence of noise. Both EED and SD evolution equations are initialised with the noisy image f. Finally, the diffusivity function $g(x)$ is chosen as [22]

$$g(x) = 1 - \exp\left(\frac{-3.31488}{\left(\frac{x}{\lambda}\right)^8}\right). \tag{8}$$

Thus, by replacing the Laplacian A_{HD} in (4) with the space discrete version A_{EED} of the EED operator in (5), we arrive at the EED-based scheme for reconstructing the high resolution scene:

$$u^{k+1} = u^k + \tau\left(\alpha(A_{EED}(u^k)) - \sum_{i=1}^{N} W_i^T B^T D^T (DBW_i u^k - f_L^i)\right). \tag{9}$$

The details regarding the discretisation of the EED operator can be found in [23].

2.3   Sector Diffusion

Continuous Model. Our goal is to design a diffusion method with a higher adaptation to image structures than previous anisotropic models such as EED. To this end, we start with Weickert's integration model from [20]:

$$\partial_t u(x, t) = \frac{1}{\pi} \int_0^\pi \partial_\theta \left(g\left(\partial_\theta u_\sigma\right) \partial_\theta u\right) d\theta. \tag{10}$$

Here, ∂_θ stands for the directional derivative in the direction represented by angle θ, g is the diffusivity function and u_σ denotes a convolution of u with a Gaussian of standard deviation σ. This model considers each orientation separately and is thus capable of diffusing along edges, but not across them.

In order to improve its structure adaptation even further, we replace the directional derivatives by one-sided directional derivatives and integrate over $[0, 2\pi]$ instead of $[0, \pi]$:

$$\partial_t u(\boldsymbol{x}, t) = \frac{1}{2\pi} \int_0^{2\pi} \partial_\theta^+ \left(g \left(\partial_\theta^+ u_\sigma^\theta \right) \partial_\theta^+ u \right) d\theta. \tag{11}$$

Here, u_σ^θ represents a one-sided smoothing of u in the orientation given by the angle θ, and ∂_θ^+ denotes a one-sided derivative in the same orientation. In contrast to the usual Gaussian smoothing applied in (10), this one-sided smoothing allows the filter to distinguish two different derivatives for a given direction: One in the orientation of θ, and the other in the orientation of $\theta + \pi$. A formal definition of these concepts can be realised by considering the restriction of u to the corresponding ray starting at $\boldsymbol{x}$, in the orientation of each θ. Namely, for fixed $\boldsymbol{x}, t, \theta$, we consider $u(h; \boldsymbol{x}, t) := u(\boldsymbol{x} + h(\cos(\theta), \sin(\theta))^T, t)$, for $h \in [0, \infty]$. Then, the one-sided directional derivative ∂_θ^+ is formally defined as

$$\partial_\theta^+ u := \lim_{h \to 0^+} \frac{u(h; \boldsymbol{x}, t) - u(\boldsymbol{x}, t)}{h}. \tag{12}$$

To our knowledge, diffusion filters that are explicitly based on one-sided directional derivatives have not been described in the literature so far.

In order to introduce a second alteration of model (10), we incorporate the concept of non-locality. This leads to

$$\partial_t u(\boldsymbol{x}, t) = \int_{B_{x,\rho}} J(|\boldsymbol{y} - \boldsymbol{x}|) \, g \left(\frac{u_\sigma(\boldsymbol{y}; \boldsymbol{y} - \boldsymbol{x}) - u_\sigma(\boldsymbol{x}; \boldsymbol{y} - \boldsymbol{x})}{|\boldsymbol{y} - \boldsymbol{x}|} \right) (u(\boldsymbol{y}) - u(\boldsymbol{x})) \, d\boldsymbol{y}.$$

Here, $B_{x,\rho}$ denotes the disc with center x and radius ρ. The diffusivity g has already been defined in (8). Also, the function $J(s)$ is a slightly Gaussian-smoothed version of $F(s) := \frac{1}{s^2}$. Moreover, we assume that it decreases fast but smoothly to zero such that $J(s) = 0$ for $|s| \geq \rho$. The slight Gaussian smoothing of F is required for avoiding the singularity of J as $s \to 0$. The value $u_\sigma(z; y - x)$ corresponds to a one-dimensional Gaussian smoothing of u inside the segment $\lambda_{xy}(s) := \left\{ x + s \frac{(y-x)}{|y-x|} : s \in [0, \rho] \right\}$ evaluated at z. This idea of making the diffusivity dependent on values inside an orientation dependent segment determines the structure preservation capabilities of the model. In the next paragraph we will see how to translate this non-local filter into a space-discrete version by dividing the disc $B_{x,\rho}$ into sectors. This explains the name *sector diffusion*.

Discrete Model. In order to properly adapt our filter to the local image structure, we first divide a disc shaped neighborhood $B_{i,\rho}$ of radius ρ centered around pixel i, into M sectors. With the objective of reducing interactions between regions of dissimilar grey values we employ robust smoothing within these sectors. This mirrors the continuous modelling idea of smoothing within the segments λ_{xy}. The final design objective is that we employ one-sided finite differences instead of central differences for discretisation purposes. The latter have a

Table 2. MSE values of denoised images including parameters used. L40 stands for Lena image with $\sigma_{\text{noise}} = 40$. B, H, P denote Bridge, House and Peppers respectively.

Image	EED				SD			
	σ	λ	$k_{\max}$	MSE	σ	λ	$k_{\max}$	MSE
L40	1.2	7.5	34	98.67	0.6	3.1	7	**92.99**
L60	1.8	5.0	63	156.24	0.6	3.3	11	**138.48**
L80	2.0	4.6	87	230.28	0.6	2.9	18	**180.66**
B40	0.9	14.4	12	294.32	0.5	3.3	4	**261.62**
B60	1.1	13.4	20	418.71	0.5	4.1	6	**360.87**
B80	1.4	10.4	28	514.23	0.6	4.0	9	**436.60**
H40	0.9	11.1	34	**96.62**	0.7	2.6	9	104.31
H60	1.1	12.1	33	167.72	0.7	2.7	14	**152.24**
H80	1.8	5.8	72	247.09	0.6	2.7	19	**207.65**
P40	1.2	8.1	28	102.97	0.6	2.1	10	**86.57**
P60	1.7	5.6	51	200.31	0.6	1.8	19	**133.19**
P80	1.9	5.1	68	353.61	0.6	1.7	30	**188.86**

property of smoothing over the central pixel, thus destroying image structures. This idea is again a direct consequence of considering orientations rather that directions, in the continuous model. With these motivations in mind, we define the space-discrete formulation of the sector diffusion model as

$$\frac{du_i}{dt} = A_{\text{SD}}(u) = \sum_{\ell=1}^{M} \sum_{j \in S_\ell} g_{i,j} \cdot \frac{u_j - u_i}{|x_j - x_i|^2}. \tag{13}$$

Here, $g_{i,j} = g\left(\frac{u_{\sigma j\ell} - u_{\sigma i\ell}}{|x_j - x_i|}\right)$, S_ℓ is the set of pixels within a particular sector ℓ, and x_i and x_j denote the position of the pixels i and j in the image grid. The sector-restricted smoothing is defined as

$$u_{\sigma j\ell} = \frac{1}{c} \sum_{k \in S_\ell} h(k, j, \sigma) u_k. \tag{14}$$

Here, c is a normalisation constant and

$$h(k, j, \sigma) = \exp\left(\frac{-|x_k - x_j|^2}{2\sigma^2}\right). \tag{15}$$

Similar to EED, we can now define the SD-based SR framework as

$$u^{k+1} = u^k + \tau\left(\alpha(A_{\text{SD}}(u^k)) - \sum_{i=1}^{N} W_i^\top B^\top D^\top (DBW_i u^k - f_L^i)\right). \tag{16}$$

3   Experiments and Discussion

In this section, we first evaluate EED and SD in terms of their denoising as well as SR regularisation capability. Then we choose the best of the two as regulariser for evaluating the operator orders in the SR observational model.

3.1   Denoising Experiments

Datasets. The test images for denoising experiments Lena, House, Peppers and Bridge[1] were corrupted with clipped-AWGN ($\sigma_{\text{noise}} = 40$, 60 and 80).

Table 3. MSE values of SR reconstructed images including parameters used. T2 stands for Text2 dataset with ground truth optical flow, while T2-S was computed using sub-optimal calculated flow. Ground truth image size for Text: 512×512. T1-T3 represent images downsized by factors 1, 2 and 3, respectively. Image size for House: 256×256. H1-H3 represent images downsized by factors 1, 1.5 and 2, respectively. Every dataset has 30 images each, with the last of them being the reference frame for registration.

EED							SD					
Dataset	σ	σ_B	λ	α	$k_{\max}$	MSE	σ	σ_B	λ	α	$k_{\max}$	MSE
H1	0.6	0.8	11.0	118.0	37	110.45	0.9	1.0	2.0	1.6	17	**83.12**
H2	0.7	0.5	12.0	120.0	9	162.64	0.8	0.7	1.7	5.3	17	**133.35**
H2-S	0.7	0.5	13.0	115.0	9	172.94	0.9	0.8	1.8	4.5	17	**141.62**
H3	0.6	0.4	14.0	127.0	48	201.91	0.6	0.8	2.3	2.9	49	**161.26**
T1	1.0	1.1	9.0	14.0	136	164.72	0.6	1.1	3.0	0.3	48	**158.50**
T2	1.3	0.9	7.0	18.0	11	397.09	0.6	1.0	2.7	0.6	34	**378.72**
T2-S	1.3	1.0	7.0	18.0	14	510.80	0.6	1.1	2.9	0.5	32	**499.60**
T3	1.2	0.4	7.0	14.0	13	674.65	0.6	0.6	2.3	0.6	49	**657.82**

Table 4. Data term evaluation. **Left:** Text2 with ground truth flow. **Right:** Text2 with sub-optimal flow.

Model	σ	σ_B	λ	α	$k_{\max}$	MSE	Model	σ	σ_B	λ	α	$k_{\max}$	MSE
1	0.6	1.0	2.7	0.6	34	**378.72**	1	0.6	1.1	2.9	0.5	32	**499.60**
2	0.6	1.0	2.7	0.6	34	382.63	2	0.6	1.1	3.4	0.4	33	502.78
3	0.6	0.6	2.9	1.2	20	381.75	3	0.3	0.8	3.8	1.0	21	500.50
4	0.6	0.8	2.8	0.6	35	392.66	4	0.6	0.9	3.0	0.5	32	511.09
5	0.6	0.5	2.7	0.7	33	391.91	5	0.4	0.6	4.6	0.3	43	513.29
6	0.3	0.5	4.1	0.4	60	403.50	6	0.4	0.6	4.6	0.3	48	518.04
2.1	0.6	1.5	3.3	0.2	55	394.85	2.1	0.6	1.6	3.5	0.2	56	523.34

---

[1] http://sipi.usc.edu/database/.

Original   Noisy

EED, MSE=510.80   SD, MSE=499.60

Fig. 1. Zoom into SR reconstructions for the Text2 dataset with sub-optimal flow.

Original   Noisy   EED, MSE=172.94   SD, MSE=141.62

Fig. 2. Zoom into SR reconstructions for the House2 dataset with sub-optimal flow.

Parameter Selection. Our experience indicates that 36 sectors gives a reasonable directional resolution. Also, we have chosen the radius of the disc-shaped neighborhood to be 7 pixels. The time step size of the explicit scheme for SD was chosen such that the maximum–minimum principle is not violated. Our grid size is set to 1. For EED, we choose the time step size $\tau = 0.2$. All other parameters (Gaussian smoothing σ, diffusivity parameter λ, and number of iterations k_{max}) have been optimised with respect to the mean squared error (MSE).

Denoising Performance. In the first experiment, using Eq. (5) and (13), we evaluate the denoising performance of EED and SD, respectively. It is clear from MSE values in Table 2 that SD produces better results. The superior performance of SD compared to EED can be attributed to its higher adaptivity towards image structures like edges.

3.2   Super-Resolution Reconstruction Experiments

Datasets. For the SR reconstruction experiments, we have considered two high-resolution scenes in the form of 'Text'[2] and 'House' images. The ground truth HR images have been warped (randomly generated deformation motion), blurred (Gaussian blur with standard deviation 1.0), downsampled (with bilinear interpolation), and degraded by noise (clipped-AWGN with $\sigma_{noise} = 40$).

Parameter Selection. To account for a large spectrum of optical flow qualities, we have used both the ground truth flow as well as a simplified approach of Brox et al. [2] without gradient constancy assumption. The parameters for different datasets are shown in Table 1. We optimise these parameters just once, but not after every super-resolution iteration. For SR reconstruction, we additionally optimise the parameters α (smoothness) and σ_B (Gaussian blur operator) apart from the already mentioned denoising parameters. Again the grid size is 1. As time step we choose $\tau = 0.05$ for EED and $\tau = 0.012$ for SD, giving experimental stability and convergence to a plausible reconstruction. We initialise u with a bilinearly upsampled image.

Smoothness Term Evaluation. The SR reconstruction quality of the two regularisers is evaluated using Eq. (9) and (16). From Table 3 and Figs. 1,2 we observe that SD outperforms EED consistently. This holds both for ground truth and suboptimal optical flow, over all downsampling factors.

Data Term Evaluation. Since we have observed a superior performance of SD for regularisation purposes, we also use it in the smoothness term while evaluating the data term with the observational model. Table 4 shows the MSE values of the reconstructed high resolution scene with all observational models from Table 1.

For ground truth flow, the observational model M1 performs best. This is in accordance with [1,19]. For suboptimal flow, M1 also outperforms M2. Interestingly, this is in contrast to the findings in [1,19], where M2 gave superior results for SR problems without noise. We explain this by the fact that we first warp the HR scene in M1. This introduces an error by applying a motion field computed from blurred LR images to sharp HR images. On the other hand, such an error does not occur for M2, as we first blur the HR scene. However, swapping blur and warp operators induces errors since matrix multiplication is not commutative. The error magnitude depends on the images and their noise. In our case, we conjecture that the latter error is higher than the former. Therefore, M1 outperforms M2.

In [1], model M2.1 was much faster than M2 with only little loss in reconstruction quality. However, this model becomes irrelevant in the noisy scenario, as M1 outperforms M2, and we also encounter a further quality loss when replacing M2 by M2.1.

---

[2] https://pixabay.com/en/knowledge-book-library-glasses-1052014/.

4   Conclusion and Outlook

Our paper belongs to the scarce amount of literature that ventures to investigate super-resolution models in the practically relevant scenario of substantial amounts of clipped noise. In contrast to classical least squares approaches with homogeneous diffusion regularisation we have paid specific attention to structure preserving regularisers such as edge-enhancing anisotropic diffusion (EED). Interestingly, EED has not been used for super-resolution before, in spite of the fact that alternatives such as BM3D and NLB are less suited for super-resolution from data with clipped noise. More importantly, we have also proposed a novel anisotropic diffusion model called sector diffusion. It is the first diffusion method that consequently uses only one-sided directional derivatives. In its local formulation, this is a model that offers also structural novelties from a mathematical perspective, since it cannot be described in terms of a partial differential equation. From a practical perspective, the non-local sector diffusion possesses a higher structural adaptivity and a better denoising performance than simpler diffusion models. Thus, it appears promising to study its usefulness also in applications beyond super-resolution. This is part of our ongoing work.

Acknowledgements. J.W. has received funding from the European Research Council (ERC) under the European Union's Horizon 2020 research and innovation programme (grant no. 741215, ERC Advanced Grant INCOVID). We thank our colleagues Dr. Matthias Augustin and Dr. Pascal Peter for useful comments on a draft version of the paper.

References

1. Bodduna, K., Weickert, J.: Evaluating Data Terms for Variational Multi-frame Super-Resolution. In: Lauze, F., Dong, Y., Dahl, A.B. (eds.) SSVM 2017. LNCS, vol. 10302, pp. 590–601. Springer, Cham (2017). https://doi.org/10.1007/978-3-319-58771-4_47
2. Brox, T., Bruhn, A., Papenberg, N., Weickert, J.: High Accuracy Optical Flow Estimation Based on a Theory for Warping. In: Pajdla, T., Matas, J. (eds.) ECCV 2004. LNCS, vol. 3024, pp. 25–36. Springer, Heidelberg (2004). https://doi.org/10.1007/978-3-540-24673-2_3
3. Dabov, K., Foi, A., Katkovnik, V., Egiazarian, K.: Image denoising by sparse 3D transform-domain collaborative filtering. IEEE Transactions on Image Processing **16**(8), 2080–2095 (2007)
4. Doulamis, A., Doulamis, N., Ioannidis, C., Chrysouli, C., Grammalidis, N., Dimitropoulos, K., Potsiou, C., Stathopoulou, E., Ioannides, M.: 5d modelling: An efficient approach for creating spatiotemporal predictive 3d maps of large-scale cultural resources. ISPRS Annals of the Photogrammetry, Remote Sensing and Spatial Information Sciences **2**(5), 61–68 (2015)
5. Elad, M., Feuer, A.: Restoration of a single superresolution image from several blurred, noisy, and undersampled measured images. IEEE Transactions on Image Processing **6**(12), 1646–1658 (1997)

6. Farsiu, S., Robinson, M.D., Elad, M., Milanfar, P.: Fast and robust multi-frame super resolution. IEEE Transactions on Image Processing **13**(10), 1327–1364 (2004)
7. Glasner, D., Shai, B., Irani, M.: Super-resolution from a single image. In: Proc. IEEE International Conference on Computer Vision (ICCV). pp. 349–356. Kyoto, Japan (Sep 2009)
8. Knoll, F., Bredies, K., Pock, T., Stollberger, R.: Second-order total generalized variation (TGV) for MRI. Magnetic Resonance in Medicine **65**(2), 480–491 (2011)
9. Kosmopoulos, D., Doulamis, N., Voulodimos, A.: Bayesian filter based behavior recognition in workflows allowing for user feedback. Computer Vision and Image Understanding **116**, 422–434 (2012)
10. Laghrib, A., Hakim, A., Raghay, S.: A combined total variation and bilateral filter approach for image robust super resolution. EURASIP Journal on Image and Video Processing **2015**(1), 1–10 (2015). https://doi.org/10.1186/s13640-015-0075-4
11. Lebrun, M., Buades, A., Morel, J.: A nonlocal Bayesian image denoising algorithm. SIAM Journal on Imaging Sciences **6**(3), 1665–1688 (2013)
12. Li, X., Mooney, P., Zheng, S., Booth, C., Braunfeld, M., Gubbens, S., Agard, D., Cheng, Y.: Electron counting and beam-induced motion correction enable near-atomic-resolution single-particle cryo-EM. Nature Methods **10**(6), 584–590 (2013)
13. Lin, F., Fookes, C., Chandran, V., Sridharan, S.: Investigation into optical flow super-resolution for surveillance applications. In: Lovell, B.C., Maeder, A.J. (eds.) APRS Workshop on Digital Image Computing: Pattern Recognition and Imaging for Medical Applications. pp. 73–78. Brisbane (Feb 2005)
14. Makantasis, K., Karantzalos, K., Doulamis, A., Doulamis, N.: Deep supervised learning for hyperspectral data classification through convolutional neural networks. In: Proc. IEEE International Geoscience and Remote Sensing Symposium (IGARSS). pp. 4959–4962 (Jul 2015)
15. Marquina, A., Osher, S.: Image super-resolution by TV-regularization and Bregman iteration. Journal of Scientific Comuputing **37**(3), 367–382 (2008)
16. Pham, T., Vliet, L., Schutte, K.: Robust fusion of irregularly sampled data using adaptive normalized convolution. EURASIP Journal on Advances in Signal Processing **2006**(083268), (Dec 2006)
17. Tatem, A., Lewis, H., Atkinson, P., Nixon, M.: Super-resolution target identification from remotely sensed images using a Hopfield neural network. IEEE Transactions on Geoscience and Remote Sensing **39**(4), 781–796 (2001)
18. Tatem, A., Lewis, H., Atkinson, P., Nixon, M.: Super-resolution land cover pattern prediction using a Hopfield neural network. Remote Sensing of Environment **79**(1), 1–14 (2002)
19. Wang, Z., Qi, F.: On ambiguities in super-resolution modeling. IEEE Signal Processing Letters **11**(8), 678–681 (2004)
20. Weickert, J.: Anisotropic diffusion filters for image processing based quality control. In: Fasano, A., Primicerio, M. (eds.) Proc. Seventh European Conference on Mathematics in Industry, pp. 355–362. Teubner, Stuttgart (1994)
21. Weickert, J.: Theoretical foundations of anisotropic diffusion in image processing. In: Kropatsch, W., Klette, R., Solina, F., Albrecht, R. (eds.) Theoretical Foundations of Computer Vision, Computing Supplement, vol. 11, pp. 221–236. Springer, Vienna (1996)
22. Weickert, J.: Anisotropic Diffusion in Image Processing. Teubner, Stuttgart (1998)
23. Weickert, J., Welk, M., Wickert, M.: L2-stable nonstandard finite differences for anisotropic diffusion. In: Kuijper, A., Bredies, K., Pock, T., Bischof, H. (eds.) Scale Space and Variational Methods in Computer Vision, pp. 380–391. Lecture Notes in Computer Science, Springer, Berlin (Jun (2013)

24. Yang, J., Wright, J., Huang, T., Ma, Y.: Image super-resolution via sparse representation. IEEE Transactions on Image Processing **19**(11), 2681–2873 (2010)
25. Yuan, Q., Zhang, L., Shen, H.: Multiframe super-resolution employing a spatially weighted total variation model. IEEE Transactions on Circuits, Systems and Video Technology **22**(3), 379–392 (2012)

Author Index

Achddou, Raphaël 333
Aggrawal, Hari Om 216
Albers, Peter 398
Almansa, Andrés 358
Alt, Tobias 294
Andris, Sarah 425
Andrle, Anna 528
Arslan, Mazlum Ferhat 91
Aujol, Jean-François 281
Azad, Amitoz 320

Barth, Andrea 140
Bednarski, Danielle 229
Beinert, Robert 516
Bekkers, Erik 27
Berkels, Benjamin 491
Berkov, Tom 52
Beschle, Cedric 140
Bodduna, Kireeti 565
Boll, Bastian 373
Breuß, Michael 65, 165
Brokman, Jonathan 40
Bruhn, Andrés 140
Bungert, Leon 307
Burrows, Liam 411

Calatroni, Luca 242, 476
Cárdenas, Marcelo 565
Chen, Ke 411
Cohen, Ido 52
Cohen, Laurent D. 346
Cremers, Daniel 204

Darkner, Sune 177
De Castro, Yohann 553
Debroux, Noémie 115
Delplancke, Claire 254
Diop, El Hadji S. 100
Doberstein, Christian 491
Dong, Yiqiu 503
Drexler, Wolfgang 128
Duits, Remco 27
Duval, Vincent 553

Effland, Alexander 491
Ehrhardt, Matthias J. 254
Elmoataz, Abderrahim 153, 320
Estatico, Claudio 242

Farchmin, Nando 528
Fu, Changqing 346

Gilboa, Guy 40, 52
Gousseau, Yann 333
Gutiérrez, Eric B. 254

Hagemann, Paul 528
Hasannasab, Marzieh 516
Heidenreich, Sebastian 528
Hoffmann, Sebastian 425
Houdard, Antoine 269, 358
Hubmer, Simon 128
Huska, Martin 437

Jumakulyyev, Ikram 450
Justiniano, Jorge 463

Kahra, Marvin 65
Kobler, Erich 491
Kočvara, Michal 191
Köhler, Alexander 165
Krainz, Lisa 128

Ladjal, Saïd 333
Lanza, Alessandro 476
Laude, Emanuel 204
Lauze, François 177
Lazzaretti, Marta 242
Le Guyader, Carole 115
Leclaire, Arthur 269
Lellmann, Jan 229
Lindeberg, Tony 3

Manga, Bakary 100
Mbengue, Alioune 100
Mehl, Lukas 140
Modersitzki, Jan 216
Mohideen Kaja Mohideen, Rahul 425
Morigi, Serena 437

Mukherjee, Subhadip 540
Mukkamala, Mahesh Chandra 204

Ochs, Peter 204
Öktem, Ozan 540

Papadakis, Nicolas 269, 358
Peter, Pascal 15, 294, 425
Petit, Romain 553
Petra, Stefania 191, 385
Pinetz, Thomas 491
Pizenberg, Matthieu 153
Plier, Jan 191
Portegies, Jim 27
Pragliola, Monica 476
Prost, Jean 358

Quéau, Yvain 153

Raab, René 307
Rabin, Julien 269, 320
Rajković, Marko 463
Rebegoldi, Simone 242
Recupero, Giuseppe Antonio 437
Roith, Tim 307
Rumpf, Martin 463

Savarino, Fabrizio 191, 398
Scherzer, Otmar 128
Schnörr, Christoph 373, 385, 398

Schönlieb, Carola-Bibiane 540
Schrader, Karl 294
Schultz, Thomas 450
Schwarz, Jonathan 373
Schwinn, Leo 307
Seck, Diaraf 100
Sgallari, Fiorella 476
Sherina, Ekaterina 128
Shi, Hui 281
Smets, Bart 27
Soltwisch, Victor 528
Sridhar, Vivek 65
Steidl, Gabriele 528

Tari, Sibel 91
Tenbrinck, Daniel 307
Torella, Francesco 411
Traonmilin, Yann 281

Vese, Luminita A. 115
Vidarte, José D. T. 177

Weickert, Joachim 294, 425, 565
Welk, Martin 78
Westerkamp, Felix 204

Yan, Shi 503

Zeilmann, Alexander 385